Geochemical Kinetics

Geochemical Kinetics

Youxue Zhang

PRINCETON UNIVERSITY PRESS · PRINCETON AND OXFORD

Copyright © 2008 by Princeton University Press

Published by Princeton University Press, 41 William Street, Princeton, New Jersey 08540

In the United Kingdom: Princeton University Press, 6 Oxford Street, Woodstock, Oxfordshire OX20 1TW

All Rights Reserved

Library of Congress Cataloging-in-Publication Data

Zhang, Youxue, 1957–

Geochemical kinetics / Youxue Zhang.

  p. cm.

Includes bibliographical references and index.

ISBN-13: 978-0-691-12432-2 (hardcover: alk. paper)

ISBN-10: 0-691-12432-9 (alk. paper)

1. Chemical kinetics—Textbooks.   2. Geochemistry—Textbooks.   I. Title.

QE515.5.K55 Z43 2008

551.9—dc22   2008062105

British Library Cataloging-in-Publication Data is available

This book has been composed in ITC Stone Serif and ITC Stone Sans

Printed on acid-free paper. ∞

press.princeton.edu

Printed in the United States of America

10 9 8 7 6 5 4 3 2 1

Contents

3 Mass Transfer: Diffusion and Flow 173

Figures

Tables

Preface

Geochemical Kinetics is a textbook for graduate students and advanced undergraduate students. This book is based on my courses on geochemical kinetics at the University of Michigan, Ann Arbor. Its aim is to provide a comprehensive introduction to the principles and theories of geochemical kinetics. It is hoped that students and scientists in geochemical kinetics will use this book as a standard reference.

Geochemical kinetics is a spin-off from chemical kinetics, and may be viewed as the application of chemical kinetics to geology, as it has been by many previous authors. Just as geochemistry has distinguished itself from chemistry, in the last 40 years geochemical kinetics has begun to distinguish it from chemical kinetics in at least three aspects. First, whereas chemical kineticists are only interested in forward problems, geochemical kineticists are also interested in inverse problems, and have developed theories of geochronology, thermochronology, and geospeedometry to infer age and thermal history of rocks. Secondly, while chemical kineticists work almost exclusively on isothermal reaction kinetics, geochemical kineticists have advanced methods to treat nonisothermal kinetics, such as reaction and diffusion during cooling. Thirdly, while chemical kineticists focus on homogeneous reactions, geochemical kineticists mostly investigate heterogeneous reactions.

The need to apply geochemical kinetics led to numerous papers, monographs, and books. The Carnegie Institution of Washington sponsored a conference on geochemical transport and kinetics (Hofmann et al., 1974). The Mineralogical Society of America organized a short course on the kinetics of geochemical processes (Lasaga and Kirkpatrick, 1981). However, these earlier books on geochemical kinetics did not cover the subject of geochemical kinetics in a systematic way. Lasaga (1998) published a systematic treatise on *Kinetic Theory in the Earth Sciences*. This book differs from that of Lasaga (1998) in that I emphasize the "geo" aspect of geochemical kinetics, and de-emphasize some chemical aspects. Geochemical inverse problems are elucidated in detail, including geochronology, thermochronology, and geospeedometry. For example, geospeedometry based on homogeneous reaction kinetics (including order–disorder reactions) is elaborated in this book, and thermochronology, an increasingly important tool in geochemistry, is treated in a more thorough manner. On the other hand, transition-state theory is covered only briefly in this book. There are numerous other differences in terms of coverage and organizations (e.g., I provide homework problems at the end of each chapter). Furthermore, since the book of Lasaga (1998), progress has been made and is included in this book when appropriate.

This book aims to cover all basic theories in geochemical kinetics. The in-depth elaborations are mostly on high-temperature geochemical kinetic problems, although some astrophysical and room-temperature examples are also included. This bias is because my own research is mainly on high-temperature geochemical kinetics.

This book is organized as follows: overview of geochemical kinetics, homogeneous reactions, mass transfer, heterogeneous reactions, and inverse problems. Homogeneous reactions are relatively simple in terms of both concepts and mathematical requirement (ordinary differential equations). Mass transfer through diffusion and fluid flow is more complicated, and requires the handling of partial differential equations. Heterogeneous reactions are the most complicated, and involve component processes such as interface reaction and mass transfer. Hence, the general flow of the book goes from homogeneous reactions to mass transfer to heterogeneous reactions. Because this book is on *geo*chemical kinetics, the most important geological applications of kinetics (inverse problems including geochronology, thermochronology, and geospeedometry) are emphasized in a separate chapter (Chapter 5) of the book.

After much consideration, the first chapter is more than a traditional first chapter with brief introduction of the subject and historical developments. Rather, it provides a lengthy introduction of the whole field but at a lower level. One may consider the first chapter as a coverage of geochemical kinetics at the undergraduate level. Therefore, in this book, readers will learn geochemical kinetics twice: first at the basic level (Chapter 1), and then at an advanced level (Chapters 2 to 5). The function of the first chapter is threefold. First, it provides the big picture of the whole field of geochemical kinetics, which should help students to get an overview in a short time. Secondly, it may be used as a stand-alone chapter to teach geochemical kinetics to undergraduate students. Thirdly, although kinetics can be classified as homogeneous reactions, mass transfer and heterogeneous reactions, and the complexity generally increases in that order, the theories nevertheless do not flow linearly and cross references are necessary.

For example, diffusion may play a role in some homogeneous reactions. A brief introduction to diffusion in the first chapter is hence useful in dealing with the diffusion aspect in homogeneous reaction kinetics. For convenience and to make reading easier, each section was designed to be roughly independent, which led to some repetition (rather than repeatedly referring to other sections for derivation).

Thermodynamics is a prerequisite for understanding kinetics; it is assumed that readers have a basic knowledge of thermodynamics, e.g., at the level of an undergraduate physical chemistry course. The thermodynamic concepts needed to understand this book include chemical equilibrium, thermodynamic functions such as enthalpy, entropy, and free energy, and the relation between Gibbs free energy and equilibrium constant. When a thermodynamic topic is critical to the development of kinetic concepts, it is reviewed briefly, but not thoroughly. In terms of mathematical background, the readers are assumed to know ordinary differential equations and some linear algebra. Knowledge of partial differential equations is a great plus, but not required (key partial differential equation problems are introduced in this book).

In this book, boxes are used for specific derivations and may be viewed as "appendixes" placed in the text. Examples are given to illustrate how to apply the concepts and equations. Homework problems are provided at the end of every chapter. Appendixes offer additional information related to the presentation of the text. A lengthy reference list is at the end of the book.

This book is a major undertaking that took over two years, including my sabbatical at Caltech in 2005 and my Jiangzuo Professor appointment at Peking University in 2005–2007, and would be impossible without the help of my family, friends, and colleagues. I thank Ed Stolper at Caltech, Mao Pan, Lifei Zhang, Ping Guan, and Haifei Zheng at Peking University for hosting my visits; Jibamitra Ganguly (University of Arizona), Chih-An Huh (Academia Sinica), and Xiaomei Xu (University of California, Irvine) for providing data and examples; Chuck Cowley (the University of Michigan) for helping me understand the reaction kinetics of nuclear hydrogen burning; Jim Walker (the University of Michigan) for help with the ozone hole information; Jim Kubicki (Pennsylvania State University), Eric Essene (the University of Michigan) and an anonymous reviewer for constructive and insightful comments of the book; Dale Austin (the University of Michigan) for preparation of some of the figures; Charles W. Carrigan (Olivet Nazarene University) for providing BSE images of zircon zonation; and my students Huaiwei Ni, Haoyue Wang, Yang Chen, and Hejiu Hui for comments and help. (Errors are of course my own responsibility, and I would greatly appreciate comments and corrections; please send them to youxue@umich.edu.) I also thank my editor at Princeton University Press, Ingrid Gnerlich. Last but not least, I thank my wife, Zhengjiu Xu, and my sons, Dan and Ray, for assisting me with the book and for putting up with me during the years it took to write it.

Youxue Zhang, Ann Arbor, 31 December 2006

Notation

A	Chemical species A
A:	The pre-exponential factor; absorbance; radioactivity; surface area [A], that is, concentration of species A; the mass number; arbitrary constant
a:	Radius: some constant
a, b, c:	Crystallographic directions
B:	Arbitrary constant
b:	Parameter such as $(w_0 - w_\infty)/(w^{cryst} - w_0)$
C:	Concentration; the unit is M (mol/L) unless otherwise specified
C_p:	Heat capacity
D:	Diffusion coefficient (often interdiffusivity or chemical diffusivity); daughter nuclide
D_0:	Diffusion coefficient at the initial temperature T_0 or concentration C_0
$\mathcal{D}$:	Self-diffusivity; tracer diffusivity; intrinsic diffusivity
d:	Thickness
E:	Energy (often activation energy); electric potential
e:	Value of 2.7182818 . . .
e:	Charge of a proton
F:	Faraday constant (96,485 C/mol); flux (such as diffusion flux); degree of partial melting; fraction (such as fractional mass loss, isotopic fraction)
f:	Fugacity; frictional coefficient

G:	Gibbs free energy; shape factor; modulus
H:	Enthalpy
h:	Planck constant; depth
$\mathbf{J}$:	Flux vector
J:	Magnitude of the flux vector
K:	Equilibrium constant; partition coefficient
K_{ae}:	Apparent equilibrium constant
k:	Reaction rate coefficient; permeability; heat conductivity
k_B:	Boltzmann constant
k_f, k_b:	Reaction rate coefficient for the forward and backward reactions
L:	Dissolution or growth distance; half-thickness
l:	Thickness (temporary parameter)
M:	Mass
m:	Mass
N:	Number of components
N_{av}:	Avogadro's number (6.0221×10^{23})
n:	Often $n = N - 1$
P:	Pressure; parent nuclides
p:	Production rate; temporary parameter
q:	Cooling rate; temporary parameter
R:	Universal gas constant ($R = 8.31447$ J mol^{-1} K^{-1}); if R is given in J kg^{-1} K^{-1}, then the gas constant depends on the gas species
r:	Radial coordinate; reaction rate
S:	Entropy
T:	Temperature; nondimensional time
T_{ae}:	Apparent equilibrium temperature
T_c:	Closure temperature
T_g:	Glass transition temperature
t:	Time
U:	Flow velocity
$\mathbf{u}$:	Fluid flow velocity vector
u:	Growth or dissolution rate; boundary motion velocity; flow velocity
V:	Volume; growth or dissolution rate
W:	Molar mass; mass of dry rhyolite per mole of oxygen (32.49 g/mol)
w:	Mass fraction or percent; degree of saturation; temporary variable such as $w = rC$.
X:	Mole fraction
x:	Distance coordinate (especially for the one-dimensional case)
Z:	Atomic number (or number of protons)
z:	Charge of an ion; depth

α: Isotopic fractionation factor; $\int k dt$; also used as temporary parameters of various sorts

β: Used as temporary coefficients of various sorts

γ: Coefficients (such as activity coefficients)

δ: Boundary layer thickness used in δ-notation of stable isotope ratios; also used for δ functions

ε: Molar absorptivity; temporary constant

η: Viscosity; dummy variable

θ: Angle, such as contact angle

κ: Heat diffusivity; transmission coefficient in the transition-state theory

Λ: Total molar conductivity of electrolyte

λ: Ionic molar conductivity

μ: Chemical potential

ν: Fundamental frequency ($k_B T/h$); coefficients in chemical reactions

ξ: Reaction progress parameter

ρ: Density; common units are kg/m^3, mol/L, kg/L, and mol/m^3

σ: Surface energy; collision cross section; standard deviation; entropy production rate

τ: Timescale; time constant

τ_c: Cooling timescale in $T = T_0/(1 + t/\tau_c)$

τ_r: Reaction timescale

ϕ: Porosity; mobility

The units are SI units unless otherwise specified with the following common exceptions: the unit of volume may be L (liter); the unit of concentration and density may be mol/L; the unit of pressure and fugacity may be bar or atmosphere. Unfortunately, different units are a fact of life, and it is difficult to avoid them.

Physical Constants

Speed of light in vacuum	c	2.99792458×10^8 m/s
Avogadro's number	N_{av}	6.022142×10^{23}
Planck constant	h	6.62607×10^{-34} J·s
Gravitational constant	G	6.673×10 mol^{-1} N m^2 kg^{-2}
Stefan-Boltzmann constant	σ	5.6704×10^{-8} W m^{-2} K^{-4}
Change of proton	e	1.602176×10^{-19} C
Faraday constant	$F = N_{av}e$	96485.3 C/mol
Boltzmann constant	k_B	1.38065×10^{-23} J/K
Universal gas constant	$R = N_{av}k_B$	8.31447 J K^{-1} mol^{-1}
Vacuum permittivity	ε_0	$8.8541878 \times 10^{-12}$ C^2 N^{-1} m^{-2}
Atomic mass unit	$u = 10^{-3}$ kg/mol	1.66054×10^{-27} kg

Geochemical Kinetics

1 Introduction and Overview

Geochemists study chemical processes on and in the Earth as well as meteorites and samples from the other planetary bodies. In geochemical kinetics, chemical kinetic principles are applied to Earth sciences. Many theories in geochemical kinetics are from chemical kinetics, but the unique nature of Earth sciences, especially the inference of geological history, requires development of theories that are specific for geochemical kinetics.

Although classical *thermodynamics* provides a powerful tool for understanding the equilibrium state (end point) of a chemical process, it is *kinetics* that elucidates the timescale, steps, and paths to approach the equilibrium state. For example, thermodynamics tells us that diamond is not stable at room temperature and pressure, but kinetics and experience tell us that diamond persists at room temperature and pressure for billions of years. Transition from diamond to graphite or oxidation of diamond to carbon dioxide is extremely slow at room temperature and pressure. Another example is the existence of light elements. According to thermodynamics, if the universe were to reach equilibrium, there would be no light elements such as H, C, and O (and hence no life) because they should react to form Fe. The fact is (fortunately) that an equilibrium state would never be reached. Hence, one may say that thermodynamics determines the direction and equilibrium state of a reaction or process, but only at the mercy of kinetics. Thermodynamics is sometimes a good approximation, but kinetics rules in many cases. Therefore, it is critical to understand kinetics.

Chemists have been studying kinetics for a long time, but early geochemists mostly applied thermodynamics to terrestrial chemical processes because long geologic times (and high temperatures in many cases) presumably would allow many reactions to reach equilibrium. However, it became evident that many processes could not be understood in terms of equilibrium thermodynamics alone. The need to apply kinetics in geochemistry led to numerous papers, monographs, and books. The Carnegie Institution of Washington sponsored a conference on geochemical transport and kinetics (Hofmann et al., 1974). The Mineralogical Society of America organized a short course on the kinetics of geochemical processes (Lasaga and Kirkpatrick, 1981). The series *Advances in Physical Geochemistry* covers many aspects of kinetics, especially in Volumes 2, 3, 4, and 8 (Saxena, 1982, 1983; Thompson and Rubie, 1985; Ganguly, 1991). The early diagenesis books by Berner (1980) and Boudreau (1997) included much kinetics. Lasaga (1998) published a tome on *Kinetic Theory in the Earth Sciences.* Many chemical kinetics textbooks are also available.

Chemical reactions may be classified by the number of phases involved in the reaction. If the reaction takes place inside one single phase, it is said to be a *homogeneous reaction*. Otherwise, it is a *heterogeneous reaction*. For homogeneous reactions, there are no surface effects and *mass transfer* usually does not play a role. Heterogeneous reactions, on the other hand, often involve surface effects, formation of new phases (*nucleation*), and mass transfer (*diffusion* and *convection*). Hence, the theories for the kinetics of homogeneous and heterogeneous reactions are different and are treated in different sections.

All geochemical methods and tools (such as the isochron method in geochronology) inferring time and rate are based on kinetics. Applications of geochemical kinetics to geology may be classified into two categories. One category may be referred to as *forward problems*, in which one starts with the initial conditions and tries to understand the subsequent reaction progress. This is an important goal to geochemists who aim to understand the kinetics of geological processes, such as reaction kinetics in aqueous solutions, the kinetics of magma crystallization, bubble growth during volcanic eruptions, weathering rate and mechanisms, and metamorphic reaction rate and mechanisms. The second category of applications may be called *inverse problems*, in which one starts from the end products (rocks) and tries to infer the past. This second category is unique in geology, and is particularly important to geochemists who aim to infer the age, thermal history, and initial conditions from the rock assemblage (that is, treating a rock as a history book).

One specific application in the first category is to estimate the time required for a reaction to reach equilibrium in nature. If equilibrium is assumed in modeling a geochemical process, it is important to know the limitations (e.g., the timescale for the assumption to be valid). For example, in acid–base reactions, the reaction is rapid and the timescale to reach equilibrium is much less than one second. Hence, pH measurement of natural waters is usually meaningful and can be used to estimate species concentrations of various pH-related reactions. However, in redox reactions, the reaction is often slow and it may take days or years to reach equilibrium. Therefore, pe (or Eh) measurement of natural waters may not mean much, and each half-reaction may result in a distinct pe value (negative of base-10 logarithm of electron activity).

Another application in the first category is for experimentalists investigating equilibrium processes (such as the determination of equilibrium constants) to evaluate whether equilibrium is reached. The experimental duration must be long enough to reach equilibrium. To estimate the required experimental duration to insure that equilibrium is reached, one needs to have a rough idea of the kinetics of the reaction to be studied. Or experiments of various durations can be conducted to evaluate the attainment of equilibrium.

A specific example of applications in the second category is the dating of rocks. Age determination is an inverse problem of radioactive decay, which is a first-order reaction (described later). Because radioactive decay follows a specific law relating concentration and time, and the decay rate is independent of temperature and pressure, the extent of decay is a measure of time passed since the radioactive element is entrapped in a crystal, hence its age. In addition to the age, the initial conditions (such as initial isotopic ratios) may also be inferred, which is another example of inverse problems.

A second example of applications in the second category is to estimate cooling history of a rock given the mineral assemblage with abundances and compositions. For example, the presence of glass in a rock or the retention of a high-temperature polymorph such as sanidine in a rock means that it cooled down rapidly. With quantitative understanding of the rate of chemical reactions or diffusion, it is possible to quantify the cooling rate, as well as the rate for a subducted slab to return to the surface, by studying the mineral assemblage, such as (i) the core and rim composition as well as the compositional gradient of each mineral, and (ii) the intracrystalline elemental distribution.

This chapter provides a general discussion of kinetics versus thermodynamics, chemical kinetics versus geochemical kinetics, and an overview of the basics of various kinetic processes and applications. Subsequent chapters will provide in-depth development of theories and applications of specific subjects. The purpose of the overview in this chapter is to provide the big picture of the whole field before in-depth exploration of the topics. Furthermore, this chapter is a stand-alone chapter that may be used in a general geochemistry course to introduce kinetics to students.

1.1 Thermodynamics versus Kinetics

Thermodynamics is a powerful tool. It states that at constant temperature and pressure, the system always moves to a state of lower *Gibbs free energy*. Equilibrium is achieved when the lowest Gibbs free energy of the system is attained. Given an initial state, thermodynamics can predict the direction of a chemical reaction, and the maximum extent of the reaction. Macroscopically, reactions opposite to the predicted direction cannot happen spontaneously. Hence, thermodynamics is widely applied to predict yields in chemical industry and to understand reactions in nature. For example, at 25°C, if the pH of an aqueous solution is 5 (meaning that H^+ activity is 10^{-5} M), we know that the OH^- activity of the solution must be 10^{-9} M.

However, thermodynamics is not enough. It cannot predict the time to reach equilibrium, or even whether the equilibrium state will ever be reached. Some equilibria may never be reached (and we also hope so). For example, if the universe reached equilibrium, there would be no light elements such as hydrogen, helium, lithium, beryllium, and boron, because they would react to form Fe. It is the high activation energy for these reactions that prevents them from happening. Some equilibria take such a long time that practically it can be said that the reaction is not happening, such as homogenization of a zoned crystal at room temperature. Other equilibria take place slowly, such as weathering of rocks under surface conditions. Some equilibria are rapidly reached, such as acid–base reactions in water.

Consider, for another example, a diamond ring. Thermodynamically the diamond crystal is unstable, and should convert to graphite, or react with oxygen in air to become carbon dioxide. Graphite in itself is also unstable in air and should burn in air to become carbon dioxide. Nonetheless, kinetically the reaction is very slow because of the strong C–C bonds in diamond and graphite. Breaking these bonds requires high *activation energy* (this concept is explored in detail later) and does not happen at room temperature, except in the presence of a strong oxidant. Or one could also say that the reaction is extremely slow at room temperatures, and, for practical purposes, it can be regarded that "a diamond is forever."

A beauty of thermodynamics is that it is not concerned with the detailed processes, and its predictions are independent of such details. Thermodynamics predicts the extent of a reaction when equilibrium is reached, but it does not address or care about reaction mechanism, i.e., how the reaction proceeds. For example, thermodynamics predicts that falling tree leaves would decompose and, in the presence of air, eventually end up as mostly CO_2 and H_2O. The decomposition could proceed under dry conditions, or under wet conditions, or in the presence of bacteria, or in a pile of tree leaves that might lead to fire. The reaction paths and kinetics would be very different under these various conditions. Because thermodynamics does not deal with the processes of reactions, it cannot provide insight on reaction mechanisms.

In a similar manner, in thermodynamics, often it is not necessary to know the detailed or actual species of a component. For example, in thermodynamic treatment, dissolved CO_2 in water is often treated as $H_2CO_3(aq)$, although most of the dissolved $CO_2(aq)$ is in the form of molecular $CO_2(aq)$ and only about 0.2–0.3% of dissolved $CO_2(aq)$ is in the form of $H_2CO_3(aq)$. Another example is for dissolved SiO_2 in water. In thermodynamic treatment, $SiO_2(aq)$ is commonly used for H_4SiO_4(aq) or other species of dissolved SiO_2(aq). As long as consistency is maintained, assuming the wrong species would not cause error in thermodynamic treatment. However, in kinetics, knowing speciation is crucial.

An equilibrium Earth would be extremely boring (e.g., there would be no life, no oxygen in the air, no plate tectonics, etc). Disequilibrium is what makes the world so diverse and interesting. Hence, kinetics may also be regarded, especially by kineticists, as our friends. Without kinetic barriers, there would be no geochemists to study kinetics or science because all human beings, and in fact all life forms, "should burst into flames!" Some geochemists have a more positive attitude and understand that "geochemists never die, but merely reach equilibrium" (Lasaga, 1998).

The goals of geochemical kinetics are to understand (i) the reaction rate and how long it would take to reach equilibrium for a specific reaction or system, (ii) atoministic mechanisms for a reaction to proceed, and (iii) the history (such as age and cooling rate) of rocks based on reaction extents.

The kinetics of a reaction is inherently much more difficult to investigate than the equilibrium state of the reaction. The first step in studying the kinetics of a reaction is to know and stoichiometrically balance the reaction, and to understand the thermodynamics. If the reaction cannot even be written down and balanced, then it's premature to study the kinetics (this may sound trivial but there are authors who try to model the kinetics of undefined reactions). Equilibrium and kinetics can be studied together. For example, one may carry out a series of experiments at different durations, and examine how the reaction reaches equilibrium. This time series would provide information on both equilibrium and kinetics of the reaction. In addition to the macroscopic (or thermodynamic) understanding, reaction kinetics also requires an understanding at the molecular or atomic level. A reaction may be accomplished by several steps or through several paths. It may involve intermediate species that are neither reactants nor products. Catalysts can change reaction path and, hence, reaction rates. The equilibrium state is independent of these steps, paths, intermediate species, and/or catalysts, but the reaction rate may depend on all these. Hence, a seemingly simple reaction may have complicated reaction rate laws.

Understanding the kinetics of reactions can be rewarding. First, knowing reaction rates allows prediction of how quickly reactions reach equilibrium. To a thermodynamicist, this is the most important application. For example, when a reaction is used as a *geothermometer*, it is important to master the kinetics of the reaction so that the limitations and the meaning of the inferred temperature in these applications are understood. Second, since the rate of a reaction depends on the detailed path or mechanism of the reaction, insight into the reaction at the molecular level can be gained. Third, quantification of reaction rates and their dependence or lack of dependence on temperature allows geochemists to infer the age, thermal history, and initial conditions of the system. This class of applications is probably the most important to Earth scientists.

1.2 Chemical Kinetics versus Geochemical Kinetics

The scope of kinetics includes (i) the rates and mechanisms of homogeneous chemical reactions (reactions that occur in one single phase, such as ionic and molecular reactions in aqueous solutions, radioactive decay, many reactions in silicate melts, and cation distribution reactions in minerals), (ii) diffusion (owing to random motion of particles) and convection (both are parts of mass transport; diffusion is often referred to as kinetics and convection and other motions are often referred to as dynamics), and (iii) the kinetics of phase transformations and heterogeneous reactions (including nucleation, crystal growth, crystal dissolution, and bubble growth).

Geochemical kinetics can be viewed as applications of chemical kinetics to Earth sciences. Geochemists have borrowed many theories and concepts from chemists. Although fundamentally similar to chemical kinetics, geochemical kinetics distinguishes itself from chemical kinetics in at least the following ways:

(1) Chemists mostly try to understand the processes that would happen under a given set of conditions (such as temperature, pressure, and initial conditions), which may be termed *forward problems*. Geochemists are interested in the forward problems, but also *inverse problems* from the product (usually a rock) to infer the initial conditions and history, including the age, the peak temperature and pressure, the temperature–pressure history, and the initial isotopic ratio or mineral composition. If the extent of a reaction depends on time but not on temperature and pressure (such as radioactive decay and growth), then the reaction can be used to infer the age (geochronology). If the extent of a reaction depends on time and temperature (such as a chemical reaction, or the diffusive loss of a radiogenic daughter), then the reaction may be used as a geothermometer and cooling rate indicator (geospeedometer). If the extent of a reaction depends on pressure, then that reaction may be used as a geobarometer. (Because chemical reaction rate usually does not depend strongly on pressure, few decompression rate indicators are developed.) In other words, the inverse problems in geochemical kinetics include geochronology based on radioactive decay and radiogenic growth, thermochronology based on radiogenic growth and diffusive loss, and geospeedometry based on temperature-dependent reaction rates.

(2) Chemists mostly deal with kinetics under isothermal conditions. However, due to the nature of many geological problems, geochemists often must deal with kinetics of reactions and diffusion during cooling. Furthermore, the inverse problems (thermochronology and geospeedometry) also require an understanding of kinetic problems during cooling. The investigation of reactions and diffusion during cooling led to kinetic concepts such as apparent equilibrium temperature, closure temperature, and apparent age, which are unique to geochemistry. Dealing with kinetics under cooling also requires more complicated mathematics and numerical simulations.

(3) The goal of chemical kinetics is to understand principles using laboratory and theoretical tools. Hence, chemists often use the simplest reactions for experimental and theoretical work to elucidate the principles. Geochemists investigate natural kinetic processes in the atmosphere, rivers, oceans, weathering surfaces, magma, and rocks, as well as processes crossing the boundaries of various systems, and, hence, must deal with complicated reactions and processes. Although experimental studies are often necessary, experimental work is motivated by and designed to address a geological problem. Because geological problems are complicated, such applications often involve approximations and assumptions so that a simple model of the complicated system is developed. One has to understand kinetics as well as the geological problem to make the right approximations and assumptions.

(4) Chemical kinetic textbooks mainly deal with kinetics of homogeneous reactions to elucidate the principles of kinetics. Some chemical kinetics texts are entirely on homogeneous reactions. Because most geochemical reactions are heterogeneous reactions and because geochemists need to treat realistic reactions in nature, geochemical kinetic textbooks must treat heterogeneous reactions more thoroughly.

In short, geochemical kineticists do not have the luxury of chemical kineticists and must deal with real-world and more complicated systems. Geochemists developed the theories and concepts to deal with inverse kinetic problems, reaction kinetics during cooling, and other geologically relevant questions. These new scopes, especially the inverse theories, reflect the special need of Earth sciences, and make geochemical kinetics much more than merely chemical kinetic theories applied to Earth sciences.

1.3 Kinetics of Homogeneous Reactions

A homogeneous reaction is a reaction inside a single phase, that is, all reactants and products as well as intermediate species involved in the reaction are part of a single phase. The phase itself may be homogeneous, but does not have to be so. For example, there may be concentration gradients in the phase. Homogeneous reactions are defined relative to heterogeneous reactions, meaning reactions involving two or more phases. The following are some examples of homogeneous reactions, and how to distinguish homogeneous versus heterogeneous reactions.

(1) *Radioactive decay.* Two examples are

$$^{87}Rb \rightarrow ^{87}Sr, \tag{1-1}$$

$$^{147}Sm \rightarrow ^{143}Nd + ^4He. \tag{1-2}$$

Even though the above reactions are at the level of nuclei, in the notation adopted in this book, each nuclide is treated as a neutral atomic species including both the nucleus and the full number of electrons. Hence, electrons that eventually would become part of the atom do not appear separately in the products or reactants. Furthermore, emission of γ-rays or other forms of energy does not appear because it is a form of energy, and it is understood that a reaction will always be accompanied by energy changes.

In the above radioactive decays, a parent nuclide shakes itself to become another nuclide or two nuclides. A unidirectional arrow indicates that there is no reverse reaction; or if there is any reverse reaction, it is not considered. He produced by the homogeneous reaction (radioactive decay) may subsequently escape into another phase, which would be another kinetic process.

(2) *Other nuclear reactions.* The Sun is powered by *nuclear hydrogen burning* in the Sun's core:

$$4\,{}^{1}\mathrm{H} \rightarrow {}^{4}\mathrm{He}. \tag{1-3}$$

This might be said to be the most important reaction in the solar system because energy from this reaction powers the Sun, lights up the planets, warms the Earth's surface, and nourishes life on the Earth. This is a complicated reaction, with several pathways to accomplish it, and each pathway involving several steps.

(3) *Chemical reactions in the gas phase.* One reaction is the *chemical hydrogen burning*:

$$2\mathrm{H}_2(\mathrm{gas}) + \mathrm{O}_2(\mathrm{gas}) \rightarrow 2\mathrm{H}_2\mathrm{O}(\mathrm{gas}). \tag{1-4}$$

Another reaction is ozone decomposition reaction:

$$2\mathrm{O}_3(\mathrm{gas}) \rightarrow 3\mathrm{O}_2(\mathrm{gas}). \tag{1-5}$$

This reaction and the ozone production reaction determine the ozone level in the stratosphere. Note that in geochemistry, for accurate notation, a reaction species is in general followed by the phase the species is in. The advantage of this notation will be clear later when multiple phases are involved.

Another example of gas-phase reaction is the oxidation of the toxic gas CO (released when there is incomplete burning of natural gas or coal) by oxygen in air:

$$2\mathrm{CO}(\mathrm{gas}) + \mathrm{O}_2(\mathrm{gas}) \rightleftharpoons 2\mathrm{CO}_2(\mathrm{gas}). \tag{1-6}$$

The two-directional arrow $\rightleftharpoons$ indicates that there is reverse reaction. The final product will satisfy the equilibrium constant $K_6 = f_{\mathrm{CO}_2}^2 / [f_{\mathrm{CO}}^2 f_{\mathrm{O}_2}] = \exp(-20.72 + 67{,}997/T)$ (obtained using data in Robie and Hemingway, 1995), where T is temperature in kelvins and the subscript 6 means that it is for Reaction 1-6. At room temperature and pressure, the equilibrium constant is large (about 10^{45}), and, hence, the reaction goes all the way to CO_2. At higher temperatures, the equilibrium constant decreases. At higher pressures, the equilibrium constant increases. This reaction is important for experimental geochemists because they vary the ratio of CO gas (often the minor gas) to CO_2 gas (often the major gas) in a gas-mixing furnace to generate the desired f_{O_2} at high temperatures, with $f_{O_2} = (CO_2/CO)^2/K_6$. Reversible reactions such as Reaction 1-6 may be viewed as two reactions moving in opposite directions. The forward reaction goes from the left-hand side to the right-hand side, and will be referred to as Reaction 1-6f. The backward reaction goes from the right-hand side to the left-hand side, and will be referred to as Reaction 1-6b. That is, Reactions 1-6f and 1-6b are as follows:

$$2CO(gas) + O_2(gas) \rightarrow 2CO_2(gas). \tag{1-6a}$$

$$2CO_2(gas) \rightarrow 2CO(gas) + O_2(gas). \tag{1-6b}$$

(4) *Chemical reactions in an aqueous solution.* One example is

$$CO_2(aq) + H_2O(aq) \rightleftharpoons H_2CO_3(aq). \tag{1-7}$$

$CO_2(aq)$ means dissolved CO_2 in the aqueous solution. The following reaction

$$CO_2(gas) + H_2O(aq) \rightleftharpoons H_2CO_3(aq), \tag{1-8}$$

is different and is not a homogeneous reaction because CO_2 is in the gas phase. By comparing Reactions 1-7 and 1-8, the importance of denoting the phases is clear.

An aqueous solution contains many ionic species and one can write numerous reactions in it. A fundamental chemical reaction in all aqueous solutions is the ionization of water:

$$2H_2O(aq) \rightleftharpoons H_3O^+(aq) + OH^-(aq). \tag{1-9}$$

The above reaction can also be written as $H_2O(aq) \rightleftharpoons H^+(aq) + OH^-(aq)$, depending on how one views the proton species in water. More aqueous reactions can be found in Table 1-1a.

(5) *Chemical reactions in silicate melts.* One example is

$$H_2O(melt) + O(melt) \rightleftharpoons 2OH(melt). \tag{1-10}$$

In the above reaction, H_2O(melt) is molecular H_2O dissolved in the melt, O(melt) is a bridging oxygen in the melt, and OH(melt) is a hydroxyl group in the melt. The charges are ignored but the oxidation state for each species is understood in the context. The above reaction may also be written as

$$H_2O(melt) + O^{2-}(melt) \rightleftharpoons 2OH^-(melt). \tag{1-10a}$$

Chemists may prefer the notation of Reaction 1-10a, and cry over the notation of Reaction 1-10 because the charges are not indicated. However, in geochemistry, often O^{2-}(melt) is used to indicate a free oxygen ion (i.e., oxygen ion not bonded to Si^{4+} ion) in melt, O^-(melt) is used to indicate nonbridging oxygen (oxygen ion bonded to only one Si^{4+} ion), and O(melt) is used to indicate bridging oxygen

(oxygen ion in between two Si^{4+} ions, as Si–O–Si). Reaction 1-10 may also be written more specifically as

$$H_2O + SiOSi \rightleftharpoons 2SiOH. \tag{1-10c}$$

Another homogeneous reaction in silicate melt is the silicon speciation reaction:

$$Q_4(\text{melt}) + Q_2(\text{melt}) \rightleftharpoons 2Q_3(\text{melt}), \tag{1-11}$$

where Q_n means that a SiO_4^{4-} tetrahedral unit in which n oxygen anions are bridging oxygens.

(6) *Chemical reactions in a mineral.* One example is the Mg–Fe order–disorder reaction in an orthopyroxene (opx) crystal:

$$Fe^{2+}_{M2}(\text{opx}) + Mg^{2+}_{M1}(\text{opx}) \rightleftharpoons Fe^{2+}_{M1}(\text{opx}) + Mg^{2+}_{M2}(\text{opx}), \tag{1-12}$$

where M1 and M2 are two octahedral crystalline sites with slightly different size and symmetry, and $Fe^{2+}_{M2}(\text{opx})$ means Fe^{2+} in M2 site of opx. The Mg^{2+} prefers the M1 site and Fe^{2+} prefers the M2 site. Hence, the forward reaction is the disordering reaction, and the backward reaction is the ordering reaction. Another way to write the above chemical reaction is

$$FeMgSi_2O_6(\text{opx}) \rightleftharpoons MgFeSi_2O_6(\text{opx}), \tag{1-12a}$$

where $FeMgSi_2O_6(\text{opx})$ means that Fe is in M2 site (that is, the first element in the formula is in M2 site) and Mg is in M1 site.

There are many other order–disorder reactions in minerals, for example,

$$Mg^{2+}_{M1}(\text{opx}) + Mn^{2+}_{M2}(\text{opx}) \rightleftharpoons Mn^{2+}_{M1}(\text{opx}) + Mg^{2+}_{M2}(\text{opx}).$$
$$Mg^{2+}_{M1}(\text{oliv}) + Fe^{2+}_{M2}(\text{oliv}) \rightleftharpoons Fe^{2+}_{M1}(\text{oliv}) + Mg^{2+}_{M2}(\text{oliv}).$$
$$^{18}O_{OH}(\text{alunite}) + {}^{16}O_{SO_4}(\text{alunite}) \rightleftharpoons {}^{16}O_{OH}(\text{alunite}) + {}^{18}O_{SO_4}(\text{alunite}).$$

(7) *Heterogeneous reactions.* Many reactions encountered in geology are not homogeneous reactions, but are heterogeneous reactions. For example, phase transition from diamond to graphite is not a homogeneous reaction but a heterogeneous reaction:

$$C(\text{diamond}) \rightarrow C(\text{graphite}). \tag{1-13}$$

Oxidation of magnetite to hematite is also a heterogeneous reaction:

$$4Fe_3O_4(\text{magnetite}) + O_2(\text{gas}) \rightarrow 6Fe_2O_3(\text{hematite}). \tag{1-14}$$

Fe–Mg exchange between two minerals is another heterogeneous reaction:

$$Mg^{2+}(\text{oliv}) + Fe^{2+}(\text{garnet}) \rightleftharpoons Fe^{2+}(\text{oliv}) + Mg^{2+}(\text{garnet}). \tag{1-15}$$

There are many other exchange reactions between two minerals, for example,

$Mg^{2+}(opx) + Fe^{2+}(garnet) \rightleftharpoons Fe^{2+}(opx) + Mg^{2+}(garnet).$

$Mg^{2+}(biotite) + Fe^{2+}(garnet) \rightleftharpoons Fe^{2+}(biotite) + Mg^{2+}(garnet).$

$Mn^{2+}(biotite) + Fe^{2+}(garnet) \rightleftharpoons Fe^{2+}(biotite) + Mn^{2+}(garnet).$

$^{18}O(magnetite) + {}^{16}O(quartz) \rightleftharpoons {}^{16}O(magnetite) + {}^{18}O(quartz).$

Two more heterogeneous reactions are as follows:

$$MgAl_2SiO_6(opx) + Mg_2Si_2O_6(opx) \rightleftharpoons Mg_3Al_2Si_3O_{12}(garnet). \tag{1-16}$$

$$TiO_2(rutile) + MgSiO_3(opx) \rightleftharpoons SiO_2(quartz) + MgTiO_3(ilmenite). \tag{1-17}$$

The dissolution of a mineral in water or in a silicate melt is also a heterogeneous reaction. Heterogeneous reactions will be discussed separately from homogeneous reactions.

For the kinetics of a reaction, it is critical to know the rough time to reach equilibrium. Often the term "mean reaction time," or "reaction timescale," or "relaxation timescale" is used. These terms all mean the same, the time it takes for the reactant concentration to change from the initial value to $1/e$ toward the final (equilibrium) value. For unidirectional reactions, half-life is often used to characterize the time to reach the final state, and it means the time for the reactant concentration to decrease to half of the initial value. For some reactions or processes, these times are short, meaning that the equilibrium state is easy to reach. Examples of rapid reactions include $H_2O \rightleftharpoons H^+ + OH^-$ (timescale $\leq 67\ \mu s$ at 298 K), or the decay of ^6He (half-life 0.8 s) to ^6Li. For some reactions or processes, the equilibrium state takes a very long time to reach. For example, ^{26}Al decays to ^{26}Mg with a half-life of 730,000 years, and ^{144}Nd decays to ^{140}Ce with a half-life of 2100 trillion years. Converting ^1H to ^4He does not happen at all at room temperature, but can occur at extreme temperatures in the core of the Sun.

1.3.1 Reaction progress parameter ξ

Consider the forward reaction of Reaction 1-10f, $H_2O(melt) + O(melt) \rightarrow 2OH(melt)$. To describe the reaction rate, one can use the concentration of any of the species involved in the reaction, such as $d[H_2O]/dt$, $d[O]/dt$, and $d[OH]/dt$, where brackets mean concentration in the melt (e.g., mol/L). Because in this case the reaction is going to the right-hand side, $d[OH]/dt$ is positive, and $d[H_2O]/dt$ and $d[O]/dt$ are negative. Furthermore, the absolute value of $d[OH]/dt$ and that of $d[H_2O]/dt$ differ by a factor of 2, because one mole of H_2O reacts with one mole of network O to form two moles of OH in the melt. In general, we can write that $d[OH]/dt = -2d[H_2O]/dt = -2d[O]/dt$. In other words, $(\frac{1}{2})\ d[OH]/dt = -d[H_2O]/dt = -d[O]/dt$.

Without a standard definition of reaction progress, one would have to be specific about which species is used in describing the reaction rate. To standardize the description and to avoid confusion, a standard reaction progress parameter ξ is defined as

$$\frac{d\xi}{dt} \equiv \frac{d[OH]}{2dt} \equiv \frac{-d[H_2O]}{dt} \equiv \frac{-d[O]}{dt}, \tag{1-18}$$

$$\xi|_{t=0} = 0, \tag{1-19}$$

where the stoichiometric coefficients 2 and 1 are in the denominator, and a negative sign accompanies the reaction rate of the reactants. By this definition, ξ is positive if the reaction goes to the right-hand side. If the reaction goes to the left-hand side, the above treatment also works, but ξ would be negative. That is, if ξ is found to be negative, or $d\xi/dt$ is found to be negative, then the reaction goes to the left-hand side. The species concentrations are related to ξ as follows:

$$[H_2O] = [H_2O]_0 - \xi, \tag{1-20}$$

$$[O] = [O]_0 - \xi, \tag{1-21}$$

$$[OH] = [OH]_0 + 2\xi, \tag{1-22}$$

where the subscript "0" means at the initial time. Hence, after solving for ξ, the concentration evolution of all species can be obtained.

1.3.2 Elementary versus overall reactions

If a reaction is a one-step reaction, that is, if it occurs on the molecular level as it is written, then the reaction is called an *elementary reaction*. In an elementary reaction, either the particles collide to produce the product, or a single particle shakes itself to become something different. For example, Reactions 1-1 and 1-2 occur at the atomic scale as they are written. That is, a parent nuclide shakes itself to become a more stable daughter nuclide (or two daughter nuclides).

If a reaction is not an elementary reaction, i.e., if the reaction does not occur at the molecular level as it is written, then it is called an *overall reaction*. An overall reaction may be accomplished by two or more steps or paths and/or with participation of intermediate species. For example, nuclear hydrogen burning Reaction 1-3, $4\,{}^1H \rightarrow {}^4He$, is an overall reaction, not an elementary reaction. There are several paths to accomplish the reaction, with every path still an overall reaction accomplished by three or more steps. One path is called a *PP I chain* and involves the following steps:

$$2\,{}^1H \rightarrow {}^2H, \quad (1.442, MeV), \tag{1-23}$$

$$^1H + {}^2H \rightarrow {}^3He, \quad (5.493, MeV), \tag{1-24}$$

$$2\,{}^3He \rightarrow {}^4He + 2\,{}^1H, \quad (12.86, MeV). \tag{1-25}$$

Each of the above three reactions is an elementary reaction. During the first step, two ${}^1H$ nuclides collide to form one ${}^2H$ (in the process, one proton plus one electron become a neutron). In the second step, one ${}^2H$ collides with ${}^1H$ to form one ^{3}He. In the third step, two ^{3}He collide to form one ^{4}He and two ^{1}H. ^{2}H and ^{3}He are intermediate species, which are produced and consumed. The net result is 4^1H $\rightarrow$ ^{4}He (which can be obtained by 2 times the first step, plus 2 times the second step, plus the third step), releasing 26.73 MeV energy. In the presence of carbon, a second path to accomplish nuclear hydrogen burning is through the *CNO cycle* (carbon–nitrogen–oxygen cycle). This cycle involves the following steps:

$$^{12}C + {}^1H \rightarrow {}^{13}N, \quad (1.943, MeV), \tag{1-26}$$

$$^{13}N \rightarrow {}^{13}C, \quad (\beta - decay, 2.221, MeV), \tag{1-27}$$

$$^{13}C + {}^1H \rightarrow {}^{14}N, \quad (7.551, MeV), \tag{1-28}$$

$$^{14}N + {}^1H \rightarrow {}^{15}O, \quad (7.297, MeV), \tag{1-29}$$

$$^{15}O \rightarrow {}^{15}N, \quad (\beta - decay, 2.754, MeV), \tag{1-30}$$

$$^{15}N + {}^1H \rightarrow {}^{12}C + {}^4He, \quad (4.966, MeV). \tag{1-31}$$

The net result by adding up all the above reactions is four ^{1}H reacting to form a ^{4}He. The ^{12}C is first used for reaction and then returned unchanged. Hence, ^{12}C acts as a *catalyst*, a substance that helps a reaction to take place without itself being consumed.

Note that in the above notation, nuclear reactions are written in the same format as chemical reactions. Physicists would include extra information in writing these reactions such as energy released or required, e.g., as γ-particles or neutrinos. If needed, energy information is given separately as shown above. (Sometimes, the energy required is explicitly included to highlight that the reaction would not be possible without energy input, such as photochemical reactions.) Furthermore, because physicists treat ^{1}H as the nucleus of a hydrogen atom (i.e., without the electron), they also include the electron or positron released or required. In this book, ^{1}H means a hydrogen atom (i.e., including the electron). Hence, electrons (which are part of an atom) and positrons (which would annihilate electrons) are not needed.

Reaction 1-5, $2O_3$(gas) $\rightarrow$ $3O_2$(gas), is an overall reaction. Both Reactions 1-6f and 1-6b, $2CO$(gas) $+ O_2$(gas) $\rightleftharpoons$ $2CO_2$(gas), are also overall reactions. Both Reactions 1-9f and 1-9b are elementary reactions. Whether a reaction is an elementary reaction or an overall reaction can only be determined experimentally, and cannot be determined by simply looking at the reaction. Many simple gas-phase reactions in the atmosphere involve intermediate radicals and, hence, are complicated overall reactions.

1.3.3 Molecularity of a reaction

The *molecularity* of an elementary reaction refers to the number of particles in the reactants (left-hand side). If the molecularity is 1, the elementary reaction is said to be unimolecular. If the molecularity is 2, the elementary reaction is said to be bimolecular. If the molecularity is 3, the elementary reaction is said to be tri-molecular (or termolecular). No example is known for higher molecularities because it is basically impossible for 4 particles to collide simultaneously.

For example, the molecularity is 1 for radioactive decay reactions (1-1) and (1-2). The molecularity of the forward reaction does not have to be the same as that of the backward reaction.

Although elementary reactions and overall reactions can only be distinguished in the laboratory, a few simple guidelines can be used to guess. If the number of particles of the reaction is 4 or more, it is an overall reaction. If the number of particles is 3, then most likely the reaction is an overall reaction because there are only a limited number of trimolecular reactions. Almost all elementary reactions have molecularities of 1 or 2. However, the reverse is not true. For example, Reaction 1-5, $2O_3(gas) \rightarrow 3O_2(gas)$, has a "molecularity" of 2 but is not an ele-mentary reaction.

In thermodynamics, a reaction can be multiplied by a constant factor without changing the meaning of the reaction. However, in kinetics, an elementary reaction is written according to how the reaction proceeds, and cannot be multiplied by a constant. For example, if Reaction 1-7, $CO_2(aq) + H_2O(aq) \rightleftharpoons H_2CO_3(aq)$, is multi-plied by 2, thermodynamic treatment stays the same, but kinetically the forward reaction would have a molecularity of 4, and is different from Reaction 1-7f.

1.3.4 Reaction rate law, rate constant, and order of a reaction

The reaction rate law is an empirical relation on how the reaction rate depends on the various species concentrations. For example, for the following reaction,

$$H_2(gas) + I_2(gas) \rightarrow 2HI(gas), \tag{1-32}$$

the experimentally determined reaction rate law is

$$d\xi/dt = k_{32}[H_2][I_2], \tag{1-33}$$

where k_{32} is a constant called the reaction rate constant or reaction rate coeffi-cient. It depends on temperature as $k_{32} = \exp(25.99 - 20{,}620/T)$ $L\,mol^{-1}\,s^{-1}$ in the temperature range of 400–800 K (Baulch et al., 1981, p. 521; Kerr and Drew, 1987, p. 209).

For another reaction,

$$H_2(gas) + Br_2(gas) \rightarrow 2HBr(gas), \tag{1-34}$$

although it looks simple and similar to Reaction 1-32, the experimentally deter-mined reaction rate law is very different and contains two constants (k_{34} and k'_{34}):

$$\frac{d\xi}{dt} = \frac{k_{34}[H_2][Br_2]^{1/2}}{1 + k'_{34}\frac{[HBr]}{[Br_2]}}, \tag{1-35}$$

with $k_{34} = \exp(30.24 - 20{,}883/T)$ and $k'_{34} = (275/T)\exp(990/T)$ (Baulch et al., 1981, pp. 348–423).

Another simple reaction with a complicated reaction rate law is Reaction 1-5, $2O_3(\text{gas}) \rightarrow 3O_2(\text{gas})$, which may be accomplished thermally or by photochemical means. The reaction rate law for the thermal decomposition of ozone is $d\xi/dt = k_5[O_3]^2/[O_2]$ when $[O_2]$ is very high, and is $d\xi/dt = k'_5[O_3]$ when $[O_2]$ is low.

For an unknown reaction, the reaction law cannot be written down simply by looking at the reaction equation. Instead, experimental study must be carried out on how the reaction rate depends on the concentration of each species. For elementary reactions, the reaction rate follows the law of mass action and can be written by looking at the reaction. If the following reaction is an elementary reaction

$$\alpha A + \beta B \rightarrow \text{product,} \tag{1-36}$$

then the reaction rate law (i.e., the law of mass action) is

$$d\xi/dt = k[A]^\alpha[B]^\beta, \tag{1-37}$$

where k is the reaction rate constant. The value of k depends on the specific reaction and on temperature. The overall order of the reaction is $\alpha + \beta$. The order of the reaction with respect to species A is α. The order of the reaction with respect to species B is β. If the concentration of one species does not vary at all (e.g., concentration of H_2O in a dilute aqueous solution), the concentration raised to some power becomes part of the reaction rate constant. A reaction does not have to have an order. For example, Reaction 1-34 does not have an order.

In summary, when a reaction is said to be an elementary reaction, the reaction rate law has been experimentally investigated and found to follow the above rate law. One special case is single-step radioactive decay reactions, which are elementary reactions and do not require further experimental confirmation of the reaction rate law. For other reactions, no matter how simple the reaction may be, without experimental confirmation, one cannot say a priori that it is an elementary reaction and cannot write down the reaction rate law, as shown by the complicated reaction rate law of Reaction 1-34. On the other hand, if the reaction rate law of Reaction 1-36 is found to be Equation 1-37, Reaction 1-36 may or may not be an elementary reaction. For example, Reaction 1-32 is not an elementary reaction even though the simple reaction law is consistent with an elementary reaction (Bamford and Tipper, 1972, p. 206).

The rate law for the radioactive decay of ^{87}Rb (Reaction 1-1), $^{87}\text{Rb} \rightarrow {}^{87}\text{Sr}$, is

$$d\xi/dt = k_1[^{87}\text{Rb}], \tag{1-38}$$

which is equivalent to the familiar expression of

$$d[^{87}Rb]/dt = -\lambda_{87}[^{87}Rb],$$ (1-39)

with $k_1 = \lambda_{87}$. Rate constants for radioactive decay are special in that they do not vary with temperature or pressure or chemical environment (an exception to this rule is found for decay by electron capture). The rate law for other radioactive decay systems can be written down similarly.

The rate law for Reaction 1-7f, $CO_2(aq) + H_2O(aq) \rightarrow H_2CO_3(aq)$, follows that of an elementary reaction (but Lewis and Glaser (2003) presented a quantum mechanical study that suggests the reaction is not elementary):

$$d\xi/dt = k_{7f}[CO_2],$$ (1-40)

where $[CO_2]$ is the concentration of dissolved CO_2 in the aqueous solution, and k_{7f} is the forward reaction rate constant (k_{7b} will denote the backward reaction rate constant). The concentration of H_2O does not appear because it is a constant and is absorbed into k_{7f}. That is, Reaction 1-7f has a molecularity of 2 but an order of 1. Hence, even for elementary reactions, the molecularity of a reaction does not have to be the same as the order of the reaction. When the molecularity is not the same as the order of a reaction because the concentration of one or two species is kept constant either due to the concentration of the species is high or because the concentration of the species is buffered, the reaction order is also referred to as pseudo-order. Therefore, one may also say that Reaction 1-7f is a pseudo-first-order reaction. The rate law for the backward reaction (Reaction 1-7b) is

$$d\xi/dt = k_{7b}[H_2CO_3].$$ (1-41)

Elementary reaction $2H_2O(aq) \rightarrow H_3O^+(aq) + OH^-(aq)$ (Reaction 1-9f) is a zeroth-order reaction (or pseudo-zeroth-order reaction):

$$d\xi/dt = k_{9f},$$ (1-42)

because $[H_2O]^2$ is a constant absorbed into k_{9f}. This is another example in which the molecularity (2) is not the same as the order (0) of a reaction. Another pseudo-zeroth-order reaction is the decomposition of PH_3 on hot tungsten at high pressures (which is a heterogeneous reaction but has a simple order); PH_3 decomposes at a constant rate until its disappearance.

The backward reaction (Reaction 1-9b) is, on the other hand, a second-order reaction with the following rate law:

$$d\xi/dt = k_{9b}[H_3O^+][OH^-].$$ (1-43)

The units of the reaction rate constant depend on the order of reaction. The units can be determined by knowing that the left-hand side must have a unit

Table 1-1a Reaction rate coefficients for some chemical reactions in aqueous solutions

Reaction	T (K)	Order	k_f	k_b	K	Ref.
$H_2O \rightleftharpoons H^+ + OH^-$	298	0; 2	$10^{-2.85}$	$10^{11.15}$	$10^{-14.00}$	1
$D_2O \rightleftharpoons D^+ + OD^-$	298	0; 2	$\sim 10^{-3.79}$	$10^{10.92}$	$\sim 10^{-14.71}$	2
$H_3O^+ + NH_3 \rightarrow NH_4^+ + H_2O$	293	2; 1	$10^{10.63}$	$10^{1.37}$	$10^{9.26}$	2
$CO_2 + H_2O \rightleftharpoons H_2CO_3$	298	1; 1	0.043	15	$10^{-2.54}$	3
$CO_2 + H_2O \rightleftharpoons H_2CO_3$	273	1; 1	0.002			4
$H_2CO_3 \rightleftharpoons H^+ + HCO_3^-$	298	1; 2	$10^{6.9}$	$10^{10.67}$	$10^{-3.77}$	3, 5
$HCO_3^- \rightleftharpoons CO_2 + OH^-$	298	1; 2	$10^{-4.00}$	$10^{3.65}$	$10^{-7.65}$	3, 5
$HCO_3^- \rightleftharpoons H^+ + CO_3^{2-}$	298	1; 2			$10^{-10.33}$	5
$HCO_3^- + OH^- \rightleftharpoons CO_3^{2-} + H_2O$	293	2; 1	$\sim 10^{9.8}$	$\sim 10^{6.1}$	$10^{3.67}$	3
$H_2CO_3 + OH^- \rightleftharpoons HCO_3^- + H_2O$	298	2; 1			$10^{10.23}$	
$^{56}Fe^{2+} + {}^{55}Fe^{3+} \rightarrow$ $^{56}Fe^{3+} + {}^{55}Fe^{2+}$		2; 2	0.87	0.87	$1.000x$	4
$^{56}Fe^{2+} + {}^{55}FeCl^{2+} \rightarrow$ $^{56}FeCl^{2+} + {}^{55}Fe^{2+}$		2; 2	5.4	5.4	$1.000x$	4

Note. Units of k and K are customary with concentrations in M and time in s. In the "Order" column the first number indicates the reaction order of the forward reaction, and the second number for the backward reaction.

References. 1, Laidler (1987, p. 39); 2, Pilling and Seakins (1995, p. 169); 3, Bamford and Tipper (1972, p. 284); 4, Lasaga and Kirkpatrick (1981, p. 23, p. 12); 5, Drever (1997, p. 42).

of concentration (in M) per unit time (in s), or $M s^{-1}$. Hence, the units of k are $M s^{-1}$ for zeroth-order reactions, s^{-1} for first-order reactions, $M^{-1} s^{-1}$ (or $L mol^{-1} s^{-1}$) for second-order reactions, etc. For reactions in silicate melt or mineral, the concentration may be given by mole fractions that are dimensionless; then the unit of k would always be s^{-1}. Table 1-1 lists the values of k for some reactions.

For overall reactions, the reaction rate law cannot be written down by simply looking at the reaction, but has to be determined from experimental studies. (Whether a reaction is elementary must be determined experimentally, which means that reaction rate laws for all chemical reactions must be experimentally determined.) The reaction rate law may take complicated forms, which might mean that the order of the reaction is not defined.

Table 1-1b Reaction rate coefficients for some gas-phase reactions

Reaction	T (K)	Order	k_{f} (L mol^{-1} s^{-1})	k_{b} (L mol^{-1} s^{-1})	Ref.
$H_2 + I_2 \rightleftharpoons 2HI$	400–800	2; 2	$\exp(26.00 - 20{,}620/T)$	$\exp(23.97 - 22{,}020/T)$	1
$NO + O_3 \rightarrow$ $NO_2 + O_2$	283–442	2	$\exp(21.67 - 1598/T)$		1
$NO_2 + O_3 \rightarrow$ $NO_3 + O_2$	230–360	2	$\exp(18.10 - 2450/T)$		1

Note. Notation as in Table 1-1a.
References. 1, Kerr and Drew (1987, pp. 209–212).

Table 1-1c Reaction rate coefficients for some solid-phase reactions

Reaction	T (°C)	Order	k_{f} (s^{-1})	k_{b} (s^{-1})	K	Ref.
$H_2O(ice) \rightleftharpoons H^+(ice) + OH^-(ice)$	263	0; 2			$10^{12.93}$	1
$Fe^{(M2)}(opx) + Mg^{(M1)}(opx) \rightleftharpoons$ $Fe^{(M1)}(opx) + Mg^{(M2)}(opx)$	800	2, 2	$10^{-3.1}$	$10^{-2.4}$	0.189	2

Note. Notation as in Table 1-1a. Unit of concentration is mole fraction.
References. 1, Pilling and Seakins (1995, p. 169); 2, Data are for an opx with Fe/(Fe + Mg) = 0.011 (Wang et al., 2005).

Strictly speaking, the concepts of elementary versus overall reactions, reaction rate law, and orders of a reaction apply only to homogeneous reactions. For heterogeneous reactions, the reaction rate is often discussed in terms of inter-face reaction and mass transfer. Hence, the order of a heterogeneous reaction, such as Reaction 1-8, $CO_2(gas) + H_2O(aq) \rightarrow H_2CO_3(aq)$, or Reaction 1-14, $4Fe_3O_4$ (magnetite) $+ O_2(gas) \rightarrow 6Fe_2O_3$(hematite), or the dissolution of a mineral in water, may be meaningless. (For part of the heterogeneous reaction process, the interface reaction, it may be possible to define the order.) There are other ways to describe the overall rates of heterogeneous reactions. For example, if a mineral dissolves at a constant rate (which could be due to convection for a falling mineral in water or in a well-stirred solution, or due to slow interface reaction rate), it may be called a linear dissolution law, and should not be called a zeroth-order nor pseudo-zeroth-order reaction. If the dissolution distance is proportional

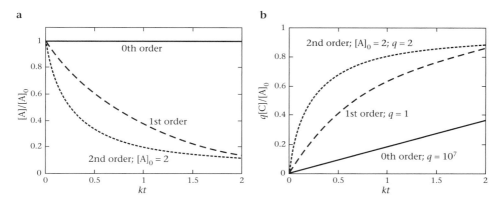

Figure 1-1 Comparison of (a) reactant and (b) product concentration evolution for zeroth-order, first-order, and the first type of second-order reactions. The horizontal axis is kt, and the vertical axis is normalized reactant concentration in (a) and normalized product concentration multiplied by a parameter q so that the comparison can be more clearly seen in (b).

to square root of time, then it may be called a parabolic reaction law (not a $-\frac{1}{2}$-order reaction), which usually implies diffusion control.

1.3.5 Concentration evolution for reactions of different orders

Only unidirectional elementary reactions are considered in this overview chapter because these reactions are relatively simple to treat. More complicated homogeneous reactions are discussed in Chapter 2.

1.3.5.1 Zeroth-order reactions

An example of a zeroth-order reaction is Reaction 1-9f, $2H_2O(aq) \rightarrow H_3O^+(aq) + OH^-(aq)$. For zeroth-order reactions, the concentrations of the reactants do not vary (which is why they are zeroth-order reactions). Use the reaction rate progress parameter ξ. Then

$$d\xi/dt = k_{9f}. \tag{1-44}$$

The unit of k_{9f} is $M\,s^{-1}$. Integration of the above leads to $\xi = k_{9f}t$. That is, $[H_3O^+] = [H_3O^+]_0 + k_{9f}t$, and $[OH^-] = [OH^-]_0 + k_{9f}t$, where subscript 0 means the initial concentration. The concentration of the reactant stays the same, $[H_2O] = [H_2O]_0 - 2k_{9f}t \approx [H_2O]_0$ (Figure 1-1). The concentration of each of the products increases linearly with time (Figure 1-1). Because of the backward reaction (Reaction 1-9b), $H_3O^+(aq) + OH^-(aq) \rightarrow 2H_2O(aq)$, the linear concentration increase would not continue for long. The concentration evolution for this reversible reaction is discussed in Chapter 2.

1.3.5.2 First-order reactions

There are many examples of first-order reactions. The most often encountered in geochemistry is the radioactive decay of an unstable nuclide. For example, the rate law for the decay of ^{147}Sm (Reaction 1-2) can be written as

$$d\xi/dt = k_2[^{147}\text{Sm}] = k_2([^{147}\text{Sm}]_0 - \xi). \tag{1-45}$$

That is, $d\xi/[^{147}\text{Sm}]_0 - \xi) = k_2 dt$. The unit of k_2 (i.e., λ_{147}) is s^{-1}. Remember that $\xi|_{t=0} = 0$ by the definition of ξ. Integration leads to

$$\ln([^{147}\text{Sm}]_0 - \xi) - \ln[^{147}\text{Sm}]_0 = -k_2.$$

That is, $([^{147}\text{Sm}]_0 - \xi)/[^{147}\text{Sm}]_0 = \exp(-k_2 t)$. Hence,

$$\xi = [^{147}\text{Sm}]_0\{1 - \exp(-k_2 t)\}. \tag{1-46}$$

Written in terms of species concentrations,

$$[^{147}\text{Sm}] = [^{147}\text{Sm}]_0 - \xi = [^{147}\text{Sm}]_0 \exp(-k_2 t), \tag{1-47a}$$
$$[^{143}\text{Nd}] = [^{143}\text{Nd}]_0 + [^{147}\text{Sm}]_0\{1 - \exp(-k_2 t)\}, \tag{1-47b}$$
$$[^{143}\text{Nd}] = [^{143}\text{Nd}]_0 + [^{147}\text{Sm}]\{\exp(k_2 t) - 1\}, \tag{1-47c}$$
$$[^{4}\text{He}] = [^{4}\text{He}]_0 + \xi = [^{4}\text{He}]_0 + [^{147}\text{Sm}]\{\exp(k_2 t) - 1\}. \tag{1-47d}$$

The concentration of the radioactive nuclide (reactant, such as ^{147}Sm) decreases exponentially, which is referred to as *radioactive decay*. The concentration of the daughter nuclides (products, including ^{143}Nd and ^{4}He) grows, which is referred to as *radiogenic growth*. Note the difference between Equations 1-47b and 1-47c. In the former equation, the concentration of ^{143}Nd at time t is expressed as a function of the initial ^{147}Sm concentration. Hence, from the initial state, one can calculate how the ^{143}Nd concentration would evolve. In the latter equation, the concentration of ^{143}Nd at time t is expressed as a function of the ^{147}Sm concentration also at time t. Let's now define time t as the present time. Then $[^{143}\text{Nd}]$ is related to the present amount of ^{147}Sm, the age (time since ^{147}Sm and ^{143}Nd were fractionated), and the initial amount of ^{143}Nd. Therefore, Equation 1-47b represents forward calculation, and Equation 1-47c represents an *inverse problem* to obtain either the age, or the initial concentration, or both. Equation 1-47d assumes that there are no other α-decay nuclides. However, U and Th are usually present in a rock or mineral, and their contribution to ^{4}He usually dominates and must be added to Equation 1-47d.

Similarly, for Reaction 1-1, the concentration evolution with time can be written as

$$[^{87}\text{Rb}] = [^{87}\text{Rb}]_0 \exp(-k_1 t), \tag{1-48a}$$

$$[^{87}Sr] = [^{87}Sr]_0 + [^{87}Rb]_0\{1 - \exp(-k_1 t)\}, \tag{1-48b}$$

$$[^{87}Sr] = [^{87}Sr]_0 + [^{87}Rb]\{\exp(k_1 t) - 1\}, \tag{1-48c}$$

where $k_1 = \lambda_{87}$.

The most important geologic applications of radioactive decay and radiogenic growth are to determine the age of materials and events, in a branch of geochemistry called *geochronology*. Unlike the *forward problems* of calculating the concentration evolution with time given the initial conditions, in geochronology, the age and the initial conditions are inferred from what can be observed today. These inverse problems are especially important in geology. Equations of type Equation 1-47a to 1-47d are the basic equations for dating. For example, in ^{14}C dating, an equation of type Equation 1-47a is used. For $^{40}K-^{40}Ar$ dating, it is often assumed that $[^{40}Ar]_0$ is known (often assumed to be zero) and hence age can be determined.

To use Equation 1-47c for dating, one has to overcome the difficulty that there are two unknowns, the initial amount of ^{143}Nd and the age. With this in mind, the most powerful method in dating, the *isochron method*, is derived. To obtain the *isochron* equation, one divides Equation 1–47c by the stable isotope of the product (such as ^{144}Nd):

$$\left(\frac{^{143}Nd}{^{144}Nd}\right) = \left(\frac{^{143}Nd}{^{144}Nd}\right)_0 + \left(\frac{^{147}Sm}{^{144}Nd}\right)(e^{k_2 t} - 1), \tag{1-49}$$

where $(^{143}Nd/^{144}Nd)$, $(^{147}m/^{144}Nd)$, $(^4He/^3He)$, and $(^{147}Sm/^3He)$ are present-day ratios that can be measured. Equation 1-49 is referred to as an *isochron equation*, which is the most important equation in isotope geochronology. Its application is as follows. A rock usually contains several minerals. If they formed at the same time (hence isochron, where *iso* means same and *chron* means time), which excludes inherited minerals in a sedimentary or metamorphic rock, and if they have the same initial isotopic ratio $(^{143}Nd/^{144}Nd)_0$, then a plot of $y = (^{143}Nd/^{144}Nd)$ versus $x = (^{147}Sm/^{144}Nd)$ would yield a straight line. The slope of the straight line is $(e^{k_2 t} - 1)$ and the intercept is $(^{143}Nd/^{144}Nd)_0$. From the slope, the age t can be calculated. From the intercept, the initial isotopic ratio is inferred. Comparison of Equations 1-47c and 1-49 reveals the importance of dividing by ^{144}Nd: different minerals formed from a common source (such as a melt) would rarely have the same $[^{143}Nd]_0$ concentration, but they would have the same isotopic ratio $(^{143}Nd/^{144}Nd)_0$. Hence, Equation 1-47c would not yield a straight line (because the "intercept" is not a constant), but Equation 1-49 would yield a straight line. The use of radioactive decay and radiogenic growth in geochronology and thermochronology is covered more extensively in Chapter 5.

A good example of a first-order (pseudo-first-order) chemical reaction is the hydration of CO_2 to form carbonic acid, Reaction 1-7f, $CO_2(aq) + H_2O(aq) \rightarrow H_2CO_3(aq)$. Because this is a reversible reaction, the concentration evolution is considered in Chapter 2.

1.3.5.3 Second-order reactions

Most elementary reactions are second-order reactions. There are two types of second-order reactions: $2A \to C$ and $A + B \to C$. The first type (special case) of second-order reactions is

$$2A \to C. \tag{1-50}$$

The reaction rate law is

$$d\xi/dt = k[A]^2 = k([A]_0 - 2\xi)^2. \tag{1-51}$$

The solution can be found as follows:

$$d\xi/([A]_0 - 2\xi)^2 = k\,dt. \tag{1-51a}$$

Then

$$-d([A]_0 - 2\xi)/([A]_0 - 2\xi)^2 = 2k\,dt. \tag{1-51b}$$

Then

$$1/([A]_0 - 2\xi) - 1/[A]_0 = 2kt. \tag{1-51c}$$

That is

$$1/[A] - 1/[A]_0 = 2kt. \tag{1-52}$$

Or

$$[A] = [A]_0/(1 + 2k[A]_0 t). \tag{1-53}$$

The concentration of the reactant varies with time hyperbolically.

The second type (general case) of second-order reactions is

$$A + B \to C. \tag{1-54}$$

The reaction rate law is

$$d\xi/dt = k[A][B] = k([A]_0 - \xi)([B]_0 - \xi). \tag{1-55}$$

If $[A]_0 = [B]_0$, The solution is the same as Equation 1-53. For $[A]_0 \neq [B]_0$, the solution can be found as follows:

$$d\xi/\{([A]_0 - \xi)([B]_0 - \xi)\} = k\,dt. \tag{1-56}$$

$$u\,d\xi/([A]_0 - \xi) - u\,d\xi/([B]_0 - \xi) = k\,dt, \quad \text{where } u = 1/([B]_0 - [A]_0).$$

$$u\ln\{([A]_0 - \xi)/[A]_0\} - u\ln\{([B]_0 - \xi)/[B]_0\} = -kt.$$

$$\ln\{([A]_0 - \xi)/[A]_0\} - \ln\{([B]_0 - \xi)/[B]_0\} = -k([B]_0 - [A]_0)t$$

$$\xi = [A]_0[B]_0(q - 1)/(q[A]_0 - [B]_0), \text{ where } q = \exp\{-k([B]_0 - [A]_0)t\}. \tag{1-57}$$

Figure 1-1 compares the concentration evolution with time for zeroth-, first-, and the first type of second-order reactions. Table 1-2 lists the solutions for concentration evolution of most elementary reactions.

1.3.5.4 Half-lives and mean reaction times

A simple way to characterize the rate of a reaction is the time it takes for the concentration to change from the initial value to halfway between the initial and final (equilibrium). This time is called the *half-life* of the reaction. The half-life is often denoted as $t_{1/2}$. The longer the half-life, the slower the reaction. The half-life is best applied to a first-order reaction (especially radioactive decay), for which the half-life is independent of the initial concentration. For example, using the decay of ^{147}Sm as an example, $[^{147}\text{Sm}] = [^{147}\text{Sm}]_0 \exp(-kt)$ (derived above). Now, by definition,

$$[^{147}\text{Sm}] = [^{147}\text{Sm}]_0/2 \text{ at } t = t_{1/2}.$$

That is,

$$[^{147}\text{Sm}]_0/2 = [^{147}\text{Sm}]_0 \exp(-kt_{1/2}).$$

Solving $t_{1/2}$, we obtain

$$t_{1/2} = (\ln 2)/k. \tag{1-58}$$

For reactions with a different order, the half-life depends on the initial concentrations. For example, for a second-order reaction, $2A \rightarrow$ product, with $d[A]/dt = -2k[A]$, then

$$t_{1/2} = 1/\{2k[A]_0\}. \tag{1-59}$$

That is, the higher the initial concentration, the shorter the half-life! This counterintuitive result is due to the reaction rate being proportional to the square of the concentration, meaning that the rate increases more rapidly than the concentration itself. Nonetheless, for [A] to reach 0.01 M, it takes a longer time starting from 0.2 M than starting from 0.1 M by the extra time for [A] to attain from 0.2 to 0.1 M. The half-lives of various reactions are listed in Table 1-2.

The *mean reaction time* or *reaction timescale* (also called *relaxation timescale*; relaxation denotes the return of a system to equilibrium) is another characteristic time for a reaction. Roughly, the mean reaction time is the time it takes for the concentration to change from the initial value to 1/e toward the final (equilibrium) value. The mean reaction time is often denoted as τ (or τ_r where subscript "r" stands for reaction). The rigorous definition of τ is through the following equation (Scherer, 1986; Zhang, 1994):

$$\frac{d\xi}{dt} = \frac{\xi_\infty - \xi}{\tau}, \tag{1-60}$$

Table 1-2 Concentration evolution and half-life for elementary reactions

Order	Type of reaction	Reaction rate law	Expression of $\xi(t)$	Linear plot	Half-life ($t_{1/2}$) and mean reaction time (τ)
0	$A \rightarrow C$	$d\xi/dt = k$	$\xi = kt$ $A = A_0 - kt \approx A_0$ $C = C_0 + kt$	$C = C_0 + kt$	$t_{1/2} = \dfrac{A_0}{2k}; \tau = A/k$
1	$A \rightarrow C$	$d\xi/dt = kA$	$\xi = A_0(1 - e^{-kt})$; $A = A_0 e^{-kt}$ $C = C_0 + A_0(1 - e^{-kt})$	$\ln A = \ln A_0 - kt$	$t_{1/2} = \dfrac{\ln 2}{k}; \tau = 1/k$
2	$2A \rightarrow C$	$d\xi/dt = kA^2$ $-dA/dt = 2kA^2$	$A = \dfrac{A_0}{1 + 2kA_0 t}$	$1/A = 1/A_0 + 2kt$	$t_{1/2} = \dfrac{1}{2kA_0}$; $\tau = 1/(2kA)$

2	$A + B \rightarrow C$	$d\xi/dt = kAB$	If $A_0 \neq B_0$, then $\xi = \dfrac{A_0 B_0 (q-1)}{(qA_0 - B_0)}$
			
			where $q = e^{-k(B0-A0)t}$.
			$A_0 = A_0 - \xi;\ B = B_0 - \xi$
			If $A_0 = B_0$, then $A = B = \dfrac{B_0}{1 + kB_0 t}$
				If $A_0 = B_0$, then
				$\ln \dfrac{A}{B} = \ln \dfrac{A_0}{B_0} + (A_0 - B_0)kt$
				
				If $A_0 = B_0$, then
				$1/A = 1/B = 1/B_0 + kt$
					If $A_0 = B_0$, then
					$t_{1/2} = 1/(kA_0)$;
					if $A_0 \ll B_0$, then;
					$t_{1/2} = 1/(kB_0)$ for A;
					if $A_0 \gg B_0$, then
					$t_{1/2} = 1/(kA_0)$ for B

3	$3A \rightarrow C$	$d\xi/dt = kA^3$ $-dA/dt = 3kA^3$	$\dfrac{1}{A^2} = \dfrac{1}{A_0^2} + 6kt$	$\dfrac{1}{A^2} = \dfrac{1}{A_0^2} + 6kt$	$t_{1/2} = \dfrac{1}{2kA_0^2}; \tau = 1/(3kA^2)$
3	$2A + B \rightarrow C$	$d\xi/dt = kA^2 B$ $-dA/dt = 2kA^2 B$	If $A_0 \neq 2B_0$, then $kt(A_0 - 2B_0) = \left(\dfrac{1}{A_0} - \dfrac{1}{A} \right) + \ln \dfrac{AB_0}{A_0 B}$		
3	$A + B + C \rightarrow$	$d\xi/dt = kABC$			
n	$nA \rightarrow C$	$d\xi/dt = kA^n$ $-dA/dt = nkA^n$	$\dfrac{1}{A^{n-1}} = \dfrac{1}{A_0^{n-1}} + n(n-1)kt$ $n \neq 0,\ 1$	$\dfrac{1}{A^{n-1}} = \dfrac{1}{A_0^{n-1}} + n(n-1)kt$ $n \neq 0,\ 1$	$t_{1/2} = \dfrac{2^{n-1} - 1}{n(n-1)kA_0^{n-1}}$ $n \neq 0,\ 1$

A, B, and C denote the species, and A, B, and C (italicized) denote the concentration of the species (i.e., $A = [A]$). Subscript "0" denotes the initial concentration. ξ is the reaction progress parameter and k is the rate constant.

where ξ is the reaction rate progress parameter, ξ_∞ is the value of ξ at $t = \infty$ (i.e., as the reaction reaches completion or equilibrium). An equivalent definition is $\tau = -dt/d\ln|\xi - \xi_\infty|$. The longer the mean reaction time, the slower the reaction is. By this definition, the mean reaction time may vary as the reaction goes on. Because the mean reaction time is defined using the reaction rate, the rate law can be directly compared with the definition of τ so as to find τ. Hence, even though the definition is more complicated than the half-life, obtaining the mean reaction time is often simpler. For a first-order reaction, τ is independent of the initial concentration. Still using the decay of ^{147}Sm as an example,

$$d\xi/dt = -d[^{147}Sm]/dt = \lambda_{147}[^{147}Sm].$$

Because $\xi_\infty = [^{147}Sm]_0$ and $\xi = [^{147}Sm]_0 - [^{147}Sm]$, we have $(\xi_\infty - \xi) = [^{147}Sm]$. Hence,

$$d\xi/dt = \lambda_{147}(\xi_\infty - \xi). \tag{1-61}$$

By definition $d\xi/dt = (\xi_\infty - \xi)/\tau$; therefore,

$$\tau = 1/\lambda_{147}. \tag{1-62}$$

The above simple formula is one of the reasons why some authors prefer the use of the mean reaction time (or relaxation timescale) instead of the half-life. The mean reaction time is longer than the half-life.

For a second-order reaction, the mean reaction time is not so simple. For example, for reaction $2A \rightarrow$ product, $d\xi/dt = k[A]^2$. Because $\xi_\infty = [A]_0/2$, and $\xi = ([A]_0 - [A])/2$, then $(\xi_\infty - \xi) = [A]/2$. Therefore, $d\xi/dt = k[A]^2 = [A]/(2\tau)$. Simplification leads to

$$\tau = 1/(2k[A]). \tag{1-63}$$

The mean reaction time during a reaction varies as the concentration varies if the reaction is not a first-order reaction. Expressions of mean reaction time of various types of reactions are listed in Table 1-2. In practice, half-lives are often used in treating radioactive decay reactions, and mean reaction times are often used in treating reversible chemical reactions.

1.3.6 Dependence of reaction rate constant on temperature; Arrhenius equation

Experimental data show that the reaction rate constant depends on temperature, and often in the following form:

$$k = A\exp[-E/(RT)], \tag{1-64}$$

where k is the reaction rate constant for a reaction, T is temperature (always in kelvins), R is the universal gas constant ($8.314\,J\,mol^{-1}\,K^{-1}$; in older books and

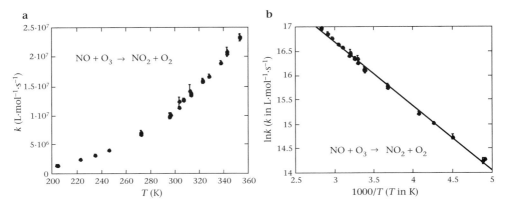

Figure 1-2 Rate coefficients (Borders and Birks, 1982) for gas-phase reaction $NO + O_3 = NO_2 + O_2$. Two data points with large errors are excluded. (a) k versus T; (b) $\ln k$ versus $1000/T$ to linearize the plot.

journal articles, the value $1.9872 \text{ cal mol}^{-1} \text{K}^{-1}$ is used), E is called the activation energy (in J or cal), and A is called the preexponential parameter (for lack of a better name) and is also the value of k as T approaches ∞. The relation was first discovered by the Swedish chemist Svante August Arrhenius (1859–1927), and hence bears the name *Arrhenius equation*. Because the activation energy E and the gas constant R often occur together as E/R (which has the dimension of temperature), (E/R) is often grouped together in this book, and the Arrhenius equation is hence in the form of $k = A \exp(-B/T)$.

Given the Arrhenius equation for a reaction, i.e., given the preexponential factor A and the activation energy E as well as the applicable temperature range, k can be found at any temperature within the range. The calculation is not complicated but one must (i) maintain consistency between units, and (ii) be especially careful about the unit of temperature (which must be converted to kelvins).

On the other hand, given experimental data of k versus T at several temperatures (either you made the measurements or there are literature data), one can use the data to obtain the Arrhenius equation by regression. Then the Arrhenius equation can be used for both interpolation (which is usually reliable) and extrapolation. Caution must be exercised for extrapolation because if an equation is extrapolated too far outside the data coverage (in this case, the temperature range), the error might be greatly amplified, and the activation energy E might change with temperature (see Figure 1-17 in a later section; see also Lasaga, 1998).

> **Example 1.1** If $E = 250$ kJ/mol, $A = 10^{10} \text{ s}^{-1}$, and the applicable temperature range is 270 to 1000 K, find k at 500°C.

Solution: First, convert temperature to Kelvin scale: $T = 773.15$ K. Because it is within the range of 270 to 1000 K, the formula $k = A \exp[-E/(RT)]$ can be used. Use $R = 8.314$ J mol^{-1} K^{-1}. To make sure all units are consistent, E must be in J/mol (instead of kJ/mol). That is, $E = 250,000$ J/mol. Using the formula, one finds that $k = 1.288 \times 10^{-7}$ s$^{-1} = 4.065$ yr^{-1}.

For the purpose of viewing data and for regression, the Arrhenius equation is often rewritten in the following form:

$$\ln k = \ln A + [-E/(RT)]. \tag{1-65}$$

By letting $y = \ln k$, $x = 1/T$, constant $\ln A = a$, and constant $(-E/R) = b$, then the above equation becomes

$$y = a + bx. \tag{1-66}$$

Hence, the exponential Arrhenius equation has been transformed to a linear equation. Figure 1-2 shows kinetic data in k versus T (Figure 1-2a) and in $\ln k$ versus $1/T$ (Figure 1-2b). Actually, $1000/T$ instead of $1/T$ is often used so that the numbers on the horizontal axis are of order 1, which is the same relation except now the slope is $0.001E/R$. Because the linear relation is so much simpler and more visual, geochemists and many other scientists love linear equations because data can be visually examined for any deviation or scatter from a linear trend. Hence, they take extra effort to transform a relation to a linear equation. As will be seen later, many other equations encountered in geochemistry are also transformed into linear equations.

Linear regression Given experimental data (x_i, y_i), where $i = 1, 2, \ldots, n$, fitting the data to an equation, such as $y = a + bx$, where a and b are parameters to be obtained by fitting, is not as trivial as one might first think. By plotting the data, one can always draw by hand a straight line that fits the data. Nowadays with help from graphing or spreadsheet programs, the task is simple if one does not want to pay much attention to data uncertainties. However, experimental data always have uncertainties. For example, every temperature measurement may have some error, or there may be temperature fluctuations during an experiment, and each estimate of the reaction rate coefficient may have a large error. Furthermore, the error for one experiment may differ from that of another. To treat errors in a rigorous fashion, more advanced linear regression algorithms must be used (see below).

For a given data set of (x_i, y_i), where $i = 1, 2, \ldots, n$, the simplest non-eyeball fit of the data, which is usually what graphing programs and spreadsheet programs use, can be obtained as follows. First calculate the average of x_i's and y_i's (the averages are denoted as $\bar{x}$ and $\bar{y}$):

$$\bar{x} = (x_1 + x_2 + \cdots + x_n)/n \equiv \sum_{i=1}^{n} x_i/n, \tag{1-67}$$

$$\bar{y} = \sum_{i=1}^{n} y_i/n. \tag{1-68}$$

Then a and b can be calculated:

$$b = \frac{\sum_{i=1}^{n} (x_i - \bar{x})(y_i - \bar{y})}{\sum_{i=1}^{n} (x_i - \bar{x})^2}, \tag{1-69a}$$

$$a = \bar{y} - b\bar{x}. \tag{1-69b}$$

In the above simple fit, the implicit assumptions are that (i) x_i values have no errors, and (ii) all y_i values have identical error. The above fit cannot account for different errors in y_i, nor errors in x_i, nor correlations in the errors. Therefore, data with high accuracy would not be emphasized as they deserve, and data with large uncertainty would not be deemphasized. To account for data uncertainties, more advanced programs must be used, and the best is by York (1969). Most radiogenic isotope geochemists use such a program for fitting isochrons, but other geochemists do not necessarily do that. If a program treats all errors perfectly and the fit equation is $y = a + bx$, and x is switched with y and the data are refitted, the equation would be exactly $x = (y - a)/b = -a/b + (1/b)y$.

Example 1.2 The following data are from Besancon (1981) for Fe–Mg disordering reaction between M1 and M2 sites of an orthopyroxene. The errors are estimated from the number of experimental data points for each determination and whether there are enough points between the initial state and the final equilibrium state.

T (K)	k (s^{-1})	$1000/T$	$\ln k$
873 ± 3	$1.24 \times 10^{-5} \overset{\times}{\div} 1.6$	1.1453 ± 0.0039	-11.30 ± 0.47
973 ± 3	$2.45 \times 10^{-4} \overset{\times}{\div} 1.4$	1.0276 ± 0.0032	-8.31 ± 0.34
1023 ± 3	$0.00295 \overset{\times}{\div} 3$	0.9774 ± 0.0029	-5.83 ± 1.10
1073 ± 3	$0.00677 \overset{\times}{\div} 2$	0.9318 ± 0.0026	-5.00 ± 0.69

Solution: If simple linear least-squares fitting is used, the result is $\ln k = 23.576 - 30{,}559/T$ (dashed line in Figure 1-3). If York's linear least-squares fitting program is used, the resulting equation is $\ln k = 21.762 - 29.029/T$ (solid line in Figure 1-3). The more advanced fit emphasizes and passes through data with smaller error bars, as expected.

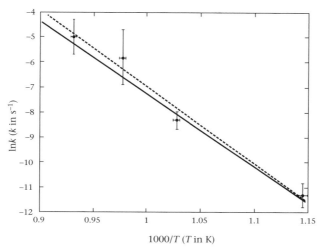

Figure 1-3 Comparison of simple least-squares fitting (dashed line) versus weighted least-squares fitting (solid line) that accounts for all individual errors and correlations. The data are Fe–Mg disordering rate coefficients in an orthopyroxene from Besancon (1981). Errors are estimated from the paper. The equation of the dashed line is $y = 23.576 - 30.559x$. The equation of the solid line is $y = 21.762 - 29.029x$. The solid line goes through the data point with the smallest error within its errors, but the simple fit does not.

1.3.7 Nonisothermal reaction kinetics

Except for radioactive decays, other reaction rate coefficients depend on temperature. Hence, for nonisothermal reaction with temperature history of $T(t)$, the reaction rate coefficient is a function of time $k(T(t)) = k(t)$. The concentration evolution as a function of time would differ from that of isothermal reactions. For unidirectional elementary reactions, it is not difficult to find how the concentration would evolve with time as long as the temperature history and hence the function of $k(t)$ is known. To illustrate the method of treatment, use Reaction $2A \rightarrow C$ as an example. The reaction rate law is (Equation 1-51)

$$d\xi/dt = k[A]^2 = k([A]_0 - 2\xi)^2,$$

where $x|_{t=0} = 0$ and k is a function of t. Rearranging the above leads to

$$d\xi/(k\ dt) = ([A]_0 - 2\xi)^2, \tag{1-70a}$$

Define

$$\alpha = \int_0^t k\ dt. \tag{1-70b}$$

Hence, $\alpha|_{t=0} = 0$, and $d\alpha = k\ dt$. Therefore, Equation 1-70a becomes

$$d\xi/(d\alpha) = ([A]_0 - 2\xi)^2. \tag{1-70c}$$

The above equation is equivalent to Equation 1-51 by making α equivalent to kt. Hence, the solution can be obtained similar to Equation 1-53 as follows:

$$[A] = [A]_0/(1 + 2[A]_0\alpha). \tag{1-70d}$$

In general, for unidirectional elementary reactions, it is easy to handle non-isothermal reaction kinetics. The solutions listed in Table 1-2 for the concentration evolution of elementary reactions can be readily extended to nonisothermal reactions by replacing kt with $\alpha = \int k\,dt$. The concepts of half-life and mean reaction time are not useful anymore for nonisothermal reactions.

The most often encountered thermal history by geologists is continuous cooling from a high temperature to room temperature (such as cooling of volcanic rocks, plutonic rocks, and metamorphic rocks). One of the many ways to approximate the cooling history is as follows:

$$T = T_0/(1 + t/\tau_c), \tag{1-70e}$$

where T_0 is the initial temperature and τ_c is the time for temperature to cool from the initial temperature to half of the initial temperature. (Other temperature versus time functions are discussed in later chapters.) Because k depends on temperature as $A\exp[-E/(RT)]$, the dependence of k on time may be expressed as follows:

$$K = A\exp[-E(1 + t/\tau_c)/(RT_0)] = A\exp[-E/(RT_0)]\exp[-Et/\tau_c RT_0)]. \tag{1-70f}$$

Let $k_0 = A\exp[-E/(RT_0)]$, meaning the initial value of the rate coefficient, and $\tau = \tau_c(RT_0/E)$. Then the expression of k becomes

$$k = k_0\,e^{-t/\tau}. \tag{1-70f}$$

That is, k decreases with time exponentially with a timescale of τ. Therefore, α can be found to be

$$\alpha = \int_0^t k\,dt = k_0\tau(1 - e^{-t/\tau}). \tag{1-70g}$$

If the reaction occurs at high temperature, but the rate at room temperature is negligible (i.e., negligible reaction even for 4 billion years), the integration can be carried out to $t = \infty$, leading to

$$\alpha = k_0\tau. \tag{1-70h}$$

Therefore, under the conditions of continuous cooling and negligible reaction rate at room temperatures, the degree of the reaction is equivalent to that at the high temperature T_0 (where the rate coefficient is k_0) for a finite duration of $\tau = \tau_c(RT_0/E)$. An example of calculations is shown below.

Example 1.3 For a first-order reaction $A \rightarrow B$, $k = \exp(-6.00 - 25{,}000/T)$ s^{-1}. Suppose the temperature history is given by $T = 1400/(1 + t/1000)$, where T is in K and t is in years. The initial concentration of A is $[A]_0 = 0.1$ mol/L. Find the final concentration of A after cooling to room temperature.

Solution: The initial temperature $T_0 = 1400$ K. The rate coefficient at T_0 can be found to be

$$k_0 = \exp(-6.00 - 25{,}000/T_0)\,\text{s}^{-1} = 4.35 \times 10^{-11}\,\text{s}^{-1} = 0.00137, y^{-1}.$$

The cooling timescale $\tau_c = 1000$ years. The activation energy is 25,000R. Hence, the timescale for k to decrease is

$$\tau = 1000 \cdot 1400/25{,}000 = 56 \text{ years}.$$

Hence,

$$\alpha = k_0\tau = 0.077.$$

From Table 1-2, replacing kt by α, it can be found that the concentration

$$[A] = [A]_0 e^{-\alpha} = 0.1 \times 0.926 = 0.0926 \text{ mol/L}.$$

This concludes the solution.

   Additionally, we can also estimate whether the reaction at room temperature is significant. At 298 K, $k = \exp(-6.00 - 25{,}000/298)\,\text{s}^{-1} = 9.1 \times 10^{-40}\,\text{s}^{-1} = 2.88 \times 10^{-32}\,\text{yr}^{-1}$. Even if the sample has been at this temperature for the whole age of the Earth, kt would be of the order 1.3×10^{-22}, and $e^{-kt} = 1$. The extent of the reaction at room temperature is negligible.

The above analyses show that it is fairly easy to deal with temperature variation for unidirectional elementary reaction kinetics containing only one reaction rate coefficient. Analyses similar to the above will be encountered often and are very useful. However, if readers get the impression that it is easy to treat temperature variation in kinetics in geology, they would be wrong. Most reactions in geology are complicated, either because they go both directions to approach equilibrium, or because there are two or more paths or steps. Therefore, there are two or more reaction rate coefficients involved. Because the coefficients almost never have the same activation energy, the above method would not simplify the reaction kinetic equations enough to obtain simple analytical solutions.

1.3.8 More complicated homogeneous reactions

A *reversible reaction* can go both forward and backward, depending on the initial concentrations of the species. Most chemical reactions are reversible. For example, Reactions 1-6 to 1-12 are all reversible.

   A *chain reaction* is accomplished by several sequential steps. Chain reactions are also known as consecutive reactions or sequential reactions. For a chain reaction, the slowest step is the rate-limiting step. For example, the decay of ^{238}U to ^{206}Pb is a chain reaction. Nuclear hydrogen burning by the PP I chain process is also a chain reaction.

If a reaction can be accomplished by two or more paths, the paths are called *parallel paths* and the reaction is called a *parallel reaction*. The overall reaction rate is the sum of the rates of all the reaction paths. The fastest reaction path is the rate-determining path. For nuclear hydrogen burning, the PP I chain is one path, the PP II chain is another path, and the CNO cycle is yet another path.

A *branch reaction* is when the reactants may form different products. It is similar to a parallel reaction in that there are different paths, but unlike a parallel re-action in that the different paths lead to different products for a branch reaction but to the same product (eventually) for a parallel reaction. For example, ^{40}K undergoes a branch reaction, one branch to ^{40}Ar and the other to ^{40}Ca.

For bimolecular second-order reactions and for trimolecular reactions, if the reaction rate is very high compared to the rate to bring particles together by diffusion (for gas-phase and liquid-phase reactions), or if diffusion is slow com-pared to the reaction rate (for homogenous reaction in a glass or mineral), or if the concentrations of the reactants are very low, then the reaction may be lim-ited by diffusion, and is called an *encounter-controlled reaction*.

An example of branch reactions is discussed in Section 1.7.2. The quantitative treatment of the kinetics of other reactions is complicated, and is the subject of Chapter 2.

1.3.9 Determination of reaction rate laws, rate constants, and mechanisms

Reaction rate laws are determined experimentally. For reactions known to be elementary reactions, it is necessary to experimentally determine the rate con-stant. For other reactions that may or may not be elementary, it is necessary to experimentally determine the reaction rate law and the rate constant. If the reaction rate law conforms to that of an elementary reaction, i.e., for reaction $\alpha A + \beta B \rightarrow$ products, the reaction rate law is $d\xi/dt = k[A]^{\alpha}[B]^{\beta}$, then the reaction is considered consistent with an elementary reaction, but other information to confirm that no other steps occur is necessary to demonstrate that a reaction is elementary. It is possible that a reaction has the "right" reaction rate law, but is shown later to be nonelementary based on other information.

1.3.9.1 Determination of the reaction rate constant

^{7}Be is the lightest nuclide that decays by electron capture and hence is the best nuclide to demonstrate whether or not the decay constant for electron capture depends on the chemical environment and pressure. (The decay constants for α-decay and β-decay through emission of an electron or a positron from the

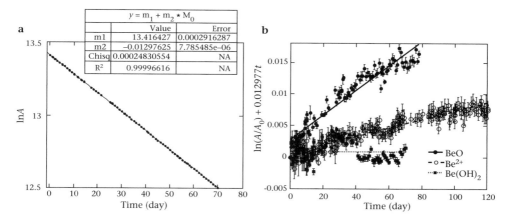

Figure 1-4 (a) Logarithm of the activity of ^{7}Be (counts per minute) in Be(OH)$_2$ versus time to demonstrate that the decay is a first-order reaction and that the decay constant is 0.012977 day^{-1} (or half-life is 53.41 days). Error is about the size of the points. (b) $\ln(A/A_0) + 0.012977t$ versus t to compare the decay constant of ^{7}Be in different compounds. The error bars are at the 2σ level. The data indicate that the decay constant of ^{7}Be depends on the chemical environment. Data from Huh (1999).

nucleus do not depend on temperature, pressure and chemical environment because these are processes inside the nucleus. The decay constant for β-decay through electron capture may depend on these factors because they may affect the behavior of K-shell electrons and, hence, their probability to be captured.) Because the variation of decay constant for electron capture may impact on the accuracy of the ^{40}K–^{40}Ar dating method, the accurate determination of the decay constant of ^{7}Be is of special interest. Huh (1999) determined the rate constant for the decay of ^{7}Be to ^{7}Li. The reaction is known to be a first-order reaction with the rate law of d[^{7}Be]/d$t = -k$[^{7}Be], where [^{7}Be] can be the concentration of ^{7}Be, or the activity of ^{7}Be, or the total number of ^{7}Be atoms. Hence, the task is to determine the value of decay constant k. Huh (1999) measured the variation of ^{7}Be activity (proportional to concentration) with time. The data can be fit by the exponential decay equation similar to Equation 1-47a:

$$[^7\text{Be}] = [^7\text{Be}]_0 e^{-\lambda t},$$

or

$$A = A_0 e^{-\lambda t},$$

where A is activity of ^{7}Be and equals $\lambda[^7$Be]. Figure 1-4a shows the decay data in the compound Be(OH)$_2$ plotted as $\ln A$ versus t. The decay constant of ^{7}Be in Be(OH)$_2$ is found to be 0.012977 day^{-1}.

To examine whether there are small differences in the decay constant of ^{7}Be in different compounds, Figure 1-4b plots $\ln(A/A_0) + 0.012977t$ versus t for ^{7}Be decay in Be(OH)$_2$, BeO, and dissolved Be^{2+}. If the decay constant of ^{7}Be in

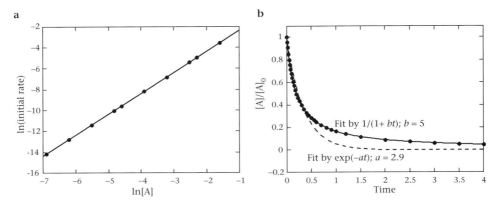

Figure 1-5 Determination of the order of hypothetical reactions with respect to species A. (a) The initial reaction rate method is used. The initial rate versus the initial concentration of A is plotted on a log–log diagram. The slope 2 is the order of the reaction with respect to A. The intercept is related to k. (b) The concentration evolution method is used. Because the exponential function (dashed curve) does not fit the data (points) well, the order is not 1. The solution for the second-order reaction equation (solid curve) fits the data well. Hence, the order of the reaction is 2.

Be(OH)$_2$ were the same as that in other compounds, they would all follow the same horizontal trend within error. Figure 1-4b shows that $\ln(A/A_0) + 0.012977t$ versus t follows different trend for different compounds, indicating that the decay constant of ^{7}Be depends on the kind of compound Be is in. The variation of the decay constant amounts to about 1.5%. Tossell (2002) raised doubt about the experimental results based on calculated electron density of Be in various compounds, but the reliability of the theoretical calculation has not been verified.

1.3.9.2 Determination of the reaction rate law

For a reaction $\alpha A + \beta B \rightarrow$ products, to determine the reaction rate law, one often-used method is to vary the concentration of one species at a time and keep the concentration of the other species constant. This requires that each of the reactants (A and B) can be prepared in the pure form, and that the concentration can be varied freely and independently of the other species. For example, one may first fix the concentration of B at very high concentration (such as 1 M), and vary the concentration of A at low concentration levels but with a large concentration range (such as 0.001–0.01 M). A large range of concentration for A is necessary to develop an accurate reaction rate law. During the reaction, the concentration of B may be regarded as constant. The order of the reaction with respect to A can be determined using either of the following methods:

(1) *Initial rate method.* If the initial production rate of the product can be determined directly as a function of the initial concentration of A, such as rate $\propto [A]^n$,

one finds the order of the reaction with respect to A to be n. Practically, this can be done by plotting ln(initial rate) versus ln[A], and the slope would be n, the order of the reaction. If the relation is not linear (which means the order is not defined), or if the slope is not an integer, then the reaction is not an elementary reaction. This method is called the initial rate method. Figure 1-5a shows an example of determining the order of a reaction using the initial rate method.

(2) *Concentration evolution method.* If the rate cannot be directly measured, but the concentration evolution of A as a function of time can be measured, then the order of the reaction can be compared with theoretical solutions (Table 1-2). For example, if $[A] = [A]_0 - kt$, i.e., if [A] decreases linearly with t, then the order with respect to A is zero. If $[A]/[A]_0 = \exp(-kt)$, i.e., if ln[A] is linear to t, then the order with respect to A is 1. If $[A]/[A]_0 = 1/(1 + 2k[A]_0 t)$, i.e., if 1/[A] is linear to t, then the order with respect to A is 2. Figure 1-5b shows an example of determining the order of a reaction using the concentration evolution method.

After obtaining the order of the reaction with respect to A, one can fix the concentration of A at a very high concentration, and examine the order of the reaction with respect to B. In this way the complete reaction rate law can be developed.

The above methods of investigating the order of the reaction with respect to each species independently, although simple and practical for many reactions (such as atmospheric reactions and aqueous reactions) studied by chemists and geochemists, is often difficult to apply to homogeneous reactions in a silicate melt or mineral because the concentration of each species may not be varied freely and independently. This will become clear later when the kinetics for the Fe–Mg order–disorder reaction in orthopyroxene and the interconversion reaction between molecular H_2O and OH groups in silicate melt are discussed.

For very rapid reactions such as the ionization of H_2O, it is difficult to determine the rate constants using conventional methods. One often-used method is the relaxation method. The system is initially at equilibrium under a given set of conditions. The conditions are then suddenly changed so that the system is no longer at equilibrium. The system then relaxes to a new equilibrium state. The speed of relaxation is measured, usually by spectrophotometry, and the rate constants can be obtained. One technique to change the conditions is to increase temperature suddenly by the rapid discharge from a capacitor. This technique is called temperature-jump technique.

1.3.9.3 Reaction mechanisms

After obtaining the reaction rate law, if it does not conform to an elementary reaction, then the next step is to try to understand the reaction mechanism, i.e., to write down the steps of elementary reactions to accomplish the overall reaction. This task is complicated and requires experience. Establishing the mechanism for a homogeneous reaction is, in general, more like arguing a case in court, than a mathematical proof. A few general rules are as follows. First, elementary reactions are usually *monomolecular* or *bimolecular*. Only very rarely are they *termolecular*. Secondly, rate constants for elementary reactions increase with temperature roughly according to Arrhenius law, i.e., the activation energy E is positive and not a strong function of temperature. Thirdly, it might be necessary to examine whether the reaction rate is affected by light, whether there are color changes or other indications that might point to the formation of some intermediate species, etc. It is from synthesizing all the experimental observations that a reaction mechanism may emerge. The proposed mechanism should have testable consequences, such as measurable intermediate species, which can be investigated further to verify or reject the proposed mechanism.

1.4 Mass and Heat Transfer

The physical transport of mass from one position to another plays a significant to dominant role in many kinetic processes. For example, a zoned crystal becomes homogeneous through diffusion, and magma erupts through fluid flow. Diffusion and fluid flow are two ways to accomplish the physical transport of masses, referred to as mass transfer. Besides pure mass transfer problems, mass transfer also plays an important role in many heterogeneous reactions (reactions involving two or more phases). The following are some examples of mass transfer problems encountered in geology:

- Homogenization of a zoned crystal through diffusion

- The change of melt inclusion composition by diffusion through the host mineral

- Diffusive loss of Ar from a mineral, affecting age determination (closure temperature)

- Diffusive exchange of isotopes, affecting age determination (closure temperature)

- Diffusive exchange of isotopes, affecting temperature determination (thermometry)

- Diffusion in a temperature gradient (Soret diffusion)

- Exchange of components between phases (heterogeneous reactions)

- Spinodal decomposition of a phase into two phases (heterogeneous reactions)

- Mass transfer during bubble growth in volcanic eruptions (heterogeneous reactions)

a 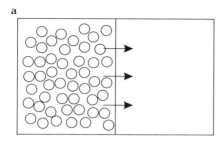 b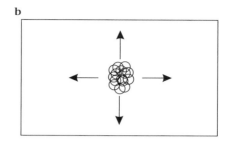

Figure 1-6 Two examples of random motion of particles, which will lead to net flux (a) to the right-hand side, and (b) from the center to all directions.

- The transport of pollutant in river or in ground water: both diffusion and flow

- Mass transfer during crystal growth or dissolution (heterogeneous reactions)

In this section, we focus on diffusive mass transfer. The mathematical description of mass transfer is similar to that of heat transfer. Furthermore, heat transfer may also play a role in heterogeneous reactions such as crystal growth and melting. Heat transfer, therefore, will be discussed together with mass transfer and examples may be taken from either mass transfer or heat transfer.

1.4.1 Diffusion

Diffusion is due to random particle motion in a phase. The random motion leads to a net mass flux when the concentration of a component is not uniform (more strictly speaking, when the chemical potential is not uniform). Hence, a zoned crystal can be homogenized through diffusion. Some examples of diffusion are shown in Figure 1–6.

The phenomenological law that describes diffusion is

$$\mathbf{J} = -D\, \partial C/\partial x, \tag{1-71}$$

where $\mathbf{J}$ is the diffusive flux, D is the diffusion coefficient (also referred to as diffusivity), C is the concentration, x is distance, $\partial C/\partial x$ is the concentration gradient (a vector), and the negative sign means that the direction of diffusive flux is opposite to the direction of concentration gradient (i.e., diffusive flux goes from high to low concentration, but the gradient is from low to high concentration). The above equation was first proposed by the German physiologist Adolf Fick (1829–1901) and hence bears the name *Fick's law* (sometimes called Fick's first law). The unit of D is length^2/time, such as m^2/s, mm^2/s, and μm^2/s $(1\ \mu\text{m}^2/\text{s} = 10^{-6}\ \text{mm}^2/\text{s} = 10^{-12}\ \text{m}^2/\text{s})$. The value of the diffusion coefficient is a

Table 1-3a Diffusion coefficients in aqueous solutions

Dissolved gas molecules	D (m^2/s) at 25°C	Ions	D (m^2/s) at 25°C
Ar	2.00×10^{-9}	H$^+$	9.1×10^{-9}
Air	2.00×10^{-9}	Li$^+$	1.0×10^{-9}
CO$_2$	1.92×10^{-9}	Na$^+$	$1.3 - 10^{-9}$
CO	2.03×10^{-9}	OH$^-$	5.2×10^{-9}
He	6.28×10^{-9}	Cl$^-$	2.0×10^{-9}
N$_2$	1.88×10^{-9}	Br$^-$	2.1×10^{-9}
O$_2$	2.10×10^{-9}		

Note. Molecular diffusivities from Cussler (1997, p. 112); ionic diffusivities from Pilling and Seakins (1995, p. 148).

characterization of the "rate" of diffusion and, hence, is very important in quantifying diffusion. Many experimental studies have been carried out to determine diffusivity. Typical values of diffusion coefficients are as follows:

In gas, D is large, about 10 mm^2/s in air at 298 K;
In aqueous solution, D is small, about 10^{-3} mm^2/s in water at 298 K;
In silicate melts, D is small, about 10^{-5} mm^2/s at 1600 K;
In a solid, D is very small, about 10^{-11} mm^2/s in silicate mineral at 1600 K.

Diffusion coefficients of some ionic and molecular species are listed in Table 1–3. The diffusivities in Table 1–3 are molecular, ionic, or atomic diffusivities due to the random motion of particles excited by thermal energy. If a system is disturbed randomly, such as fish swimming in a lake, wave motion, boating, as well as other random disturbances, then mass transport may also be described by diffusion (*eddy diffusion*) on a scale larger than each perturbation, but the diffusivity may be significantly larger. If not much activity happens in water, eddy diffusivity (or turbulent diffusivity) would be only slightly higher than molecular diffusivity. In seawater, vertical turbulent diffusivity is about 10^{-5} m^2/s (e.g., Gregg et al., 2003), 4 orders of magnitude greater than molecular diffusivity. If the water is a main waterway for shipping, eddy diffusivity can be many orders of greater. For example, eddy diffusivity may be as high as 70 m^2/s. This and other concepts of diffusion, such as *self-diffusion*, *tracer diffusion*, *chemical diffusion*, *grain boundary diffusion*, and *effective diffusion through a porous medium*, are examined in Chapter 3.

Table 1-3b ^{18}O diffusion coefficients in some minerals under hydrothermal conditions ($P_{H_2O} = 100$ MPa)

Mineral	Direction	D (m^2/s)	T range (K)	D along other directions
Albite	Bulk	$\exp(-29.10 - 10{,}719/T)$	623–1073	
Anorthite	Bulk	$\exp(-25.00 - 13{,}184/T)$	623–1073	
Orthoclase	Bulk	$\exp(-26.12 - 12{,}882/T)$	623–1073	
Almandine	Isotropic	$\exp(-18.93 - 36{,}202/T)$	1073–1273	
Apatite	//**c**	$\exp(-18.53 - 24{,}658/T)$	823–1473	⊥**c** 3 orders slower
Biotite	⊥**c**	$\exp(-20.82 - 17{,}110/T)$	773–1073	//**c** 4 orders slower
Muscovite	⊥**c**	$\exp(-18.68 - 19{,}626/T)$	785–973	//**c** 4 orders slower
Phlogopite	⊥**c**	$\exp(-18.08 - 21{,}135/T)$	873–1173	//**c** 4 orders slower
Calcite	Bulk	$\exp(-18.78 - 24{,}658/T)$	673–1073	
Diopside	//**c**	$\exp(-22.62 - 27{,}174/T)$	973–1523	⊥**c** 2 orders slower
Hornblende	//**c**	$\exp(-25.33 - 20{,}632/T)$	923–1073	⊥**c** 1 to 1.3 orders slower
Richterite	//**c**	$\exp(-17.32 - 28{,}684/T)$	923–1073	
Tremolite	//**c**	$\exp(-26.94 - 19{,}626/T)$	923–1073	
Magnetite	Isotropic	$\exp(-21.77 - 22{,}645/T)$	773–1073	
Quartz(α)	//**c**	$\exp(-3.96 - 34{,}158/T)$	773–823	⊥**c** 2 orders slower
Quartz(β)	//**c**	$\exp(-23.94 - 17{,}079/T)$	873–1073	⊥**c** 2 orders slower

Reference: Brady (1995).
Note. The direction of diffusion is the direction of fastest diffusion.

Although the diffusion coefficient is related to the diffusion "rate," it is difficult to define a single diffusion "rate" during diffusion because the diffusion distance is proportional not to duration, but to the square root of duration:

$$x \propto t^{1/2}. \tag{1-72}$$

Table 1-3c Ar diffusion coefficients in some minerals

Mineral	Orientation	Shape model	T range (K)	D (m^2/s)	T_c (K)
Hornblende	Powder	Sphere	773–1173	$\exp(-12.94 - 32{,}257/T)$	770
Phlogopite	Powder	Cylinder	873–1173	$\exp(-9.50 - 29{,}106/T)$	646
Biotite	Powder	Cylinder	873–1023	$\exp(-11.77 - 23{,}694/T)$	554
Orthoclase	Powder	Sphere	773–1073	$\exp(-13.48 - 21{,}685/T)$	529

Note. The closure temperature T_c (see later discussion) depends on grain size and cooling rate; here it is calculated for a radius of 0.1 mm and a cooling rate of 5 K/Myr (Brady, 1995). Cylinder shape model means that the grains are treated as infinitely long cylinders with diffusion along the cross section (in the plane $\perp\mathbf{c}$).

The diffusion coefficient increases rapidly with temperature. The dependence of diffusivity on temperature also follows the Arrhenius relation,

$$D = Ae^{-E/(RT)}, \tag{1-73}$$

where A is the preexponential factor and equals the value of D as T approaches ∞, E is the activation energy, and R is the universal gas constant.

Fick's first law relates the diffusive flux to the concentration gradient but does not provide an equation to solve for the evolution of concentration. In general, diffusion treats problems in which the concentration of a component or species may change with both spatial position and time, i.e., $C = C(x, t)$, where x describes the position along one direction. Therefore, a differential equation for $C(x, t)$ must include differentials with respect to both t and x. That is, a partial differential equation is necessary to describe how C would vary with x and t. It can be shown that under simple conditions, the flux equation and mass conservation can be transformed to the following equation:

$$\frac{\partial C}{\partial t} = D \frac{\partial^2 C}{\partial x^2}. \tag{1-74}$$

where C is a function of x and t, and D is assumed to be independent of C. This equation is called the *diffusion equation*, and is also referred to as Fick's second law. The mathematics of diffusion is complicated and is discussed in Chapter 3. In this section, some results are presented and explained but not derived.

From the diffusion equation, it can be seen that if $\partial^2 C/\partial x^2 = 0$, then $\partial C/\partial t = 0$, meaning that the concentration at the position would not vary with time. Hence, if the initial concentration is uniform, the concentration would not change with time. If the initial concentration profile is linear and the concentrations at the two ends are not changed from linear distribution, because

$\partial^2 C/\partial x^2 = 0$ for a linear profile, the concentration profile would remain as a linear profile. A linear profile is often the steady state for a diffusion-controlled profile.

Some examples of concentration evolution (Crank, 1975) are shown and explained below. These are all often-encountered problems in diffusion. The purpose of the examples and the qualitative discussion is to help readers develop familiarity and gain experience in treating diffusion in a qualitative fashion.

1.4.1.1 Point-source diffusion

Initially, a substance is concentrated at one point along a straight line that extends to infinity to both sides. For convenience, the position of the point is defined as $x = 0$. One real-world problem is the spill of toxic substance into a narrow lake. This problem is called the one-dimensional (or 1-D) point-source problem. With time, the substance would diffuse out and be diluted. The concentration variation as a function of time is shown in Figure 1–7a. The mathematical description of the concentration of the substance as a function of x and t is

$$C = \frac{M}{(4\pi Dt)^{1/2}} e^{-x^2/4Dt}, \tag{1-75}$$

where M is the total initial mass at the point source. For the unit of C to be mass/volume, the unit of M must be mass/area (such as kg/m^2). Therefore, even though this is a 1-D diffusion problem, one must know the cross-section area of the spill (and the cross-section area is assumed to be constant for 1-D diffusion) so that mass per unit area of the cross section can be calculated. The concentration at the center ($x = 0$) is $C = M/(4\pi Dt)^{1/2}$. That is, the center concentration is infinity at zero time, and decreases as $(t)^{-1/2}$.

If a substance is initially concentrated at one point and then diffuses into three dimensions (along a spherical radius r) such as Figure 1–6b, then,

$$C = \frac{M}{(4\pi Dt)^{3/2}} e^{-r^2/4Dt}, \tag{1-76}$$

where M is the total initial mass (in kg) at the point source. The concentration variation along r is shown in Figure 1–7b. This problem is called the three-dimensional (or 3-D) point-source problem. The dispersion of the mass is more rapid compared to 1-D point-source diffusion.

1.4.1.2 Half-space diffusion

One example of half-space diffusion is the cooling of an oceanic plate. The oceanic plate when created at the mid-ocean ridge is hot, with a roughly uniform temperature of about 1600 K. It is cooled at the surface (quenched by ocean water) as it moves away from the ocean ridge. For simplicity, ignore complexities

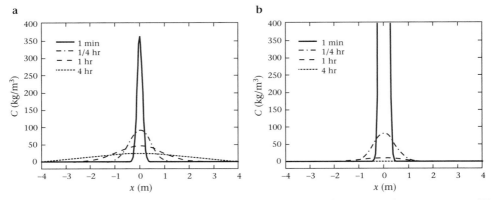

Figure 1-7 Evolution of concentration profiles for (a) one-dimensional point-source diffusion and (b) 3-D point-source diffusion. This calculation is made for $M = 100\,\text{kg/m}^2$, and $D_{\text{eddy}} = 10^{-4}\,\text{m}^2/\text{s}$ (eddy diffusivity). In the 3-D case, C was much higher at the center at smaller times because initial mass distribution was at a point instead of a plane; but at greater times, the concentration dissipates much more rapidly.

such as the intrusion of sheeted dikes, crystallization of the magma, and hydrothermal circulation. The surface temperature may be regarded as constant at ocean floor temperature (about 275 K). The plate thickness may be regarded as infinite. This problem with uniform initial temperature and a constant surface temperature is referred to as the (1-D) half-space diffusion problem. The evolution of temperature with time is diagrammed in Figure 1–8a. The low surface temperature gradually propagates into the interior of the oceanic plate.

Another example of half-space diffusion problem is as follows. A thin-crystal wafer initially contains some ^{40}Ar (e.g., due to decay of ^{40}K). When the crystal is heated up (metamorphism), ^{40}Ar will diffuse out and the surface concentration of ^{40}Ar is zero. If the temperature is constant and, hence, the diffusivity is constant, the problem is also a half-space diffusion problem.

A third example is as follows. Initially a crystal has a uniform $\delta^{18}\text{O}$. Then the crystal is in contact with a fluid with a higher $\delta^{18}\text{O}$. Ignore the dissolution of the crystal in the fluid (e.g., the fluid is already saturated with the crystal). Then ^{18}O would diffuse into the crystal. Because fluid is a large reservoir and mass transport in the fluid is rapid, $\delta^{18}\text{O}$ at the crystal surface would be maintained constant. Hence, this is again a half-space diffusion problem with uniform initial concentration and constant surface concentration. The evolution of $\delta^{18}\text{O}$ with time is shown in Figure 1–8b.

1.4.1.3 Diffusion couple

If two samples of different composition are placed together in contact, then diffusion will occur to bridge the compositional difference and homogenize the

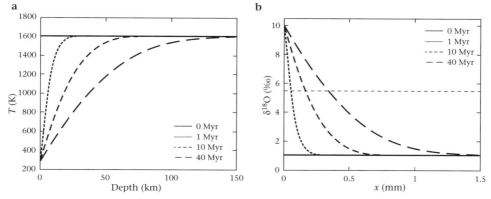

Figure 1-8 Heat and mass diffusion in a semi-infinite medium in which the diffusion profile propagates according to square root of time. (a) The evolution of temperature profile of oceanic plate. The initial temperature is 1600 K. The surface temperature (at depth = 0) is 275 K. Heat diffusivity is 1 mm^2/s. (b) The evolution of δ^{18}O profile in a mineral. Initial δ^{18}O in the mineral is 1‰. The surface δ^{18}O is 10‰. $D = 10^{-22}$ m^2/s.

two samples. The concentration evolution is plotted in Figure 1–9. The diffusion-couple profile may be regarded to consist of two profiles, one to the left ($x < 0$) with a surface concentration that is the arithmetic average of the two initial concentrations, and one to the right ($x > 0$), with the same surface concentration. The center composition gradually propagates into both sides.

1.4.1.4 Homogenization of an oscillatorily zoned crystal

Initially the concentration of a component in a crystal is a periodic function such as a sine (or cosine) function as follows:

$$C|_{t=0} = C_0 + A \sin(2\pi x/l), \tag{1-77}$$

where A is the amplitude of the concentration fluctuation and l is the periodicity. As time progresses, the concentration profile would stay as a periodic function, but the amplitude would decrease with time:

$$C(x, t) = C_0 + A e^{-4\pi^2 Dt/l^2} \sin(2\pi x/l). \tag{1-78}$$

The new amplitude is $A e^{-4\pi^2 Dt/l^2}$, decreasing with time exponentially. The concentration evolution is diagrammed in Figure 1–10.

1.4.1.5 Diffusion distance

The diffusion profile is a smooth profile. Even if the initial concentration distribution is not smooth, diffusion smoothes out any initial discontinuities. Therefore, there is no well-defined diffusion front (except for some special cases),

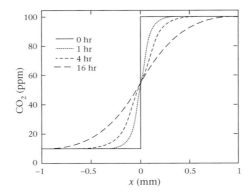

Figure 1-9 The evolution of concentration profile in a diffusion couple of silicate melt. $D = 1 \ \mu m^2/s$. Initially, the concentration profile is a discontinuous step function. The profile then becomes smooth.

and it is difficult to define a diffusion distance. Nonetheless, as duration of diffusion increases, the diffusion profile becomes longer and its length is proportional to square root of time (Figures 1–7 to 1–10). Hence, it is useful to define a characteristic diffusion distance similar to the definition of half-life. Unfortunately, unlike the case for radioactive decay or first-order reactions where there is a unique definition of half-life, in diffusion, there is no unique definition of *characteristic diffusion distance*. Some authors define the characteristic distance as $x_c = (Dt)^{1/2}$, where x_c means the characteristic diffusion distance (if it would cause no confusion, the subscript "c" is often dropped from the notation). This definition roughly corresponds to the *midconcentration distance* (see definition below). Others define $x_c = (4Dt)^{1/2}$, and still others define $x_c = (\pi Dt)^{1/2}$. All formulas state that the characteristic "diffusion distance" is proportional to $(Dt)^{1/2}$.

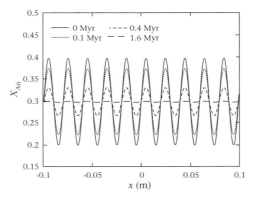

Figure 1-10 The evolution of concentration profile of an oscillatorily zoned plagioclase crystal. $D = 10^{-24} \ m^2/s$. The width of each cycle is 0.02 mm. In 1.6 Myr, the crystal is nearly homogeneous.

To avoid confusion and maintain consistency, we need a unique definition of diffusion distance, similar to the concept of half-life. In this book, a midconcentration distance for diffusion (x_{mid}) is defined. The *midconcentration distance* is the distance at which the concentration is halfway between the maximum and minimum along the profile at a specific time. For the half-space diffusion problem where x_{mid} is best defined because there is a definite surface where diffusion commences and the surface concentration is constant,

$$x_{mid} = 0.95387(Dt)^{1/2}. \tag{1-79}$$

For the point source diffusion problem,

$$x_{mid} = (4Dt \ln 2)^{1/2} = 1.6651(Dt)^{1/2}. \tag{1-80}$$

For a diffusion couple, the definition of x_{mid} requires some thinking because the mid-concentration of the whole diffusion couple is right at the interface, which does not move with time. This is because for a diffusion couple every side is diffusing to the other side. On the other hand, if a diffusion couple is viewed as two half-space diffusion problems with the interface concentration viewed as the fixed surface concentration, then, x_{mid} equals $0.95387(Dt)^{1/2}$, the same as the half-space diffusion problem.

Knowing that $x_{mid} = \alpha(Dt)^{1/2}$, one can also estimate the time required for diffusion to reach some depth x (that is, for the concentration at that depth to be halfway between the initial and final concentrations) using

$$t_{1/2} = x^2/(\alpha^2 D). \tag{1-81}$$

Example 1.4 Knowing $D = 10^{-22}$ m^2/s, for half-space diffusion, estimate the time for the depth of 1 μm to reach the midconcentration.

Solution: Because $\alpha = 0.95387$ for half-space diffusion,

$$t_{1/2} = x^2/(\alpha^2 D) = (10^{-6})^2/(0.95387^2 \times 10^{-22}) \text{ s} = 348 \text{ yr}.$$

1.4.1.6 Microscopic view of diffusion

Statistically, diffusion can be viewed as random walk of atoms or molecules. Consider diffusion in an isotropic crystal, such as Mg and Mn exchange in spinel for a case in which the only concentration gradient is along the x direction. Consider now two adjacent lattice planes (left and right) at distance l apart. If the jumping distance of Mg is l and the frequency of Mg ions jumping away from the original position is f, then the number of Mg ions jumping from left to right is $\frac{1}{6}n_L f \, dt$, and the number of Mg ions jumping from right to left is $\frac{1}{6}n_R f \, dt$, where n_L and n_R are the number of Mg ions per unit area on the left-hand side plane and on the right-hand side plane. The factor $\frac{1}{6}$ in the expressions is due to the fact that every ion jump can be decomposed to motion in six directions on three orthogonal axes (left, right, up, down, front, back), and we are considering only one direction (from left to right or from right to left). The jumping frequency f is assumed to be the same from left to right or from right to left, i.e., random walk is assumed. Therefore, the net flux from the left plane to the right plane is

$$\mathbf{J} = \frac{1}{6}(n_L - n_R)f. \tag{1-82a}$$

Since $n_L = lC_L$ and $n_R = lC_R$, where C_L and C_R are the concentrations of Mg on the two planes, then

$$\mathbf{J} = \frac{1}{6}l(C_L - C_R)f. \tag{1-82b}$$

Now because

$$C_L - C_R = -l\frac{\partial C}{\partial x} \tag{1-82c}$$

we have

$$\mathbf{J} = -\frac{1}{6}l^2 f\frac{\partial C}{\partial x}. \tag{1-82d}$$

Comparing this with Fick's law (Equation 1-71), we have

$$D = \frac{1}{6}l^2 f. \tag{1-82e}$$

Thus, microscopically, the diffusion coefficient may be interpreted as one-sixth of the jumping distance squared times the overall jumping frequency. Since l is of the order 3×10^{-10} m (interatomic distance in a lattice), the jumping frequency can be roughly estimated from D. For $D \approx 10^{-17}$ m^2/s such as Mg diffusion in spinel at 1400°C, the jumping frequency is $6D/l^2 \sim 700$ per second. Because ion jumping requires a site to accept the ion, the jumping frequency in solids depends on the concentration of vacancies and other defects. In liquids, the jumping frequency depends on the flexibility of the liquid structure, and is hence related to viscosity.

For an anisotropic crystal, jumping frequencies in different directions may not be the same, and, hence, D along each crystallographic direction may be different. The relation between D and jumping frequency may be written as follows:

$$D_i = l_i^2 f_i, \tag{1-82f}$$

where D_i is diffusivity along a crystallographic direction i, l_i is the jumping distance along the direction, and f_i is the jumping frequency along the direction.

1.4.2 Convection

Diffusion is one mode of mass transfer. If the phase is fluid or if the temperature is high for the solid to show fluid behavior, there is a second way to transfer mass, which is fluid flow, sometimes referred to as convection, or convective mass transfer, or advection, depending on the authors and the problems at hand. (More strictly, especially in meteorology, advection means horizontal motion of gas and convection means vertical motion. However, the distinction is not always made.) There is mass flux due to fluid flow. In the case of unidirectional laminar flow along the x direction, the concentration variation as a function of x and t can be related through the following partial differential equation:

$$\frac{\partial C}{\partial t} = -u\frac{\partial C}{\partial x}, \tag{1-83}$$

where u is the flow rate along the x-direction. Because diffusion is always present, the full equation in the presence of flow is called the convective diffusion equation and is as follows:

$$\frac{\partial C}{\partial t} = D\frac{\partial^2 C}{\partial x^2} - u\frac{\partial C}{\partial x}, \tag{1-84}$$

Convection enhances mass transfer. The general convective diffusion equation is not trivial to solve and will be dealt with later.

1.5 Kinetics of Heterogeneous Reactions

There are a variety of heterogeneous reactions. Most reactions encountered by geologists are heterogeneous reactions. The kinetic aspects of various heterogeneous reactions have been reviewed by a number of authors (Kirkpatrick, 1975, 1981; Berner, 1978). Heterogeneous reactions may be classified as at least three different types.

(1) The first and the simplest type is *component exchange* between phases without growth or dissolution of either phase. Examples include oxygen isotope exchange between two minerals, and Fe^{2+}–Mg^{2+} exchange between olivine and garnet or between olivine and melt. This is essentially a diffusion problem.

(2) The second type is *simple phase transitions* in which one phase transforms into another of identical composition, e.g., diamond $\rightarrow$ graphite, quartz $\rightarrow$ coesite, and water $\rightarrow$ ice. This type sounds simple, but it involves most steps of heterogeneous reactions, including nucleation, interface reaction, and coarsening.

(3) The third and the most common type is *complex phase transformations*, including the following: (i) some components in a phase combine to form a new phase (e.g., H_2O exsolution from a magma to drive a volcanic eruption; the precipitation of calcite from an aqueous solution, $Ca^{2+} + CO_3^{2-} \rightarrow$ calcite; the condensation of corundum from solar nebular gas; and the crystallization of olivine from a basaltic magma), (ii) one phase decomposes into several phases (e.g., spinodal decomposition, or albite $\rightarrow$ jadeite + quartz), (iii) several phases combine into one phase (e.g., melting at the eutectic point, or jadeite +

quartz → albite), and (iv) several phases react to form several new phases (e.g., spinel + enstatite → forsterite + pyrope).

At the fundamental level, it may be said that volcanology, igneous petrology, and metamorphic petrology are studies of different heterogeneous reactions. Volcanic eruption is powered by a heterogeneous reaction, the exsolution of H_2O from magma. The crystallization of igneous rocks consists of a series of heterogeneous reaction. Furthermore, igneous rocks display a variety of fabrics due to the dependence of nucleation and crystal growth rates on thermal history. Metamorphism is characterized by heterogeneous reactions in the solid state, although usually involving release or incorporation of a fluid phase. Many sedimentary rocks are due to chemical precipitation from water, again, heterogeneous reactions. Hence, understanding the kinetics of heterogeneous reactions is essential to volcanology, and igneous, metamorphic, and sedimentary petrology. For materials scientists, producing high-quality crystals also requires an understanding of crystal growth.

The processes that may be encountered during a heterogeneous reaction include nucleation, interface reaction, and mass transfer (or heat transfer). *Nucleation* forms the embryo from which the new phase can grow. Nucleation may be either homogeneous (nucleation of a new phase inside an existing phase) or heterogeneous (nucleation of a new phase at the interface of two existing phases). Nucleation is not well understood. The classical theory for *homogeneous nucleation* based on the atomic scale fluctuation, though well developed, often predicts nucleation rates many orders of magnitude lower than experimental data (Kirkpatrick, 1981). *Heterogeneous nucleation* does not require as much energy as homogeneous nucleation. Therefore, nucleation in most natural cases is probably heterogeneous. *Interface reaction* is the attachment or detachment of atoms to or from the surface of a phase, allowing the new phase to grow or shrink. The theoretical basis for understanding interface reaction rates is available but more experimental data are needed. *Heat transfer* carries the extra heat (such as latent heat released by crystal growth) away from the crystal surface (otherwise the crystal growth would stop due to the local increase in temperature), or brings the necessary heat to the interface (such as during melting). *Mass transfer* (including diffusion and convection) brings the necessary ingredients to the growing phase and excess components away from the phase. All heterogeneous reactions except for simple phase transitions require mass transfer. Binary diffusion can be described very well, as long as the diffusion coefficient for a specific application is known, but diffusion in a multicomponent system can be mathematically complex.

1.5.1 Controlling factors and "reaction laws"

The rates of heterogeneous reactions do not usually follow rate laws of homogeneous reactions; instead, they involve multisteps and are controlled by the slowest step in the process. (Sometimes "linear laws" and "parabolic laws" are mentioned, but they differ from the rate laws of homogeneous reactions.) Some gas phase reactions occur on the surface of catalysts or the container. For these reactions, the reaction rate depends on surface area. For a fixed surface area, the reaction rate may depend on the power of the concentration of the gas component, and hence a reaction order may be defined.

Component exchange between phases is controlled by mass transfer. Between solid phases, mass transfer is through diffusion where the exchange of components may be used as a geospeedometer (Lasaga, 1983). Convection rather than diffusion may play a dominant role if fluid phases are involved. In reactions between solid and fluid phases, diffusion in the solid phase is usually the slowest step. However, dissolution and reprecipitation may occur and may accomplish the exchange more rapidly than diffusion through the solid phase.

Simple phase transitions require no mass transfer (but they may require heat transfer). For example, the glass transition (where a liquid cools to a glass) does not require mass transfer but many homogeneous reactions may be involved. Simple phase transitions can be classified as first order and second order (not to be confused with first order and second order reactions). In all phase transitions, the Gibbs free energy is continuous. *First-order phase transitions* are those in which there is a discontinuity in the enthalpy and first derivatives of the Gibbs free energy. They always involve nucleation and interface reaction and hence can be very slow. For example, the transition from diamond to graphite requires the breaking of C–C bonds. Because these bonds in diamond are very strong, the transition is very slow at room temperature and pressure. (The melting temperature of diamond is lower than that of graphite because diamond has lower entropy. The melting temperature is an equilibrium property, depending on both bond strength and entropy, and is not purely determined by bond strength.) First-order phase transitions release or absorb heat and hence require heat transfer. *Second-order transitions* are those in which the enthalpies of the two phases are the same at the transition but there is a discontinuity in heat capacity (and second derivatives of the Gibbs free energy). They do not require heat transfer, and usually they do not require nucleation. The heat capacity versus temperature curve determined by heating at a constant rate often shows λ-shaped pattern in the vicinity of the transition, and therefore these phase transitions are also called lambda transitions. The glass transition and magnetic transitions are examples of second-order phase transitions. Quantitative understanding of simple phase transitions from nucleation to growth is not available yet.

Complex phase transformations require nucleation, interface reaction, and mass transfer, with the slowest step controlling the reaction rate. Nucleation and growth are usually followed by *coarsening*, also known as *Ostwald ripening*, in which fewer and larger crystals replace many small crystals to minimize the interface area and total free energy. Complete quantitative description or modeling of complex phase transformations from nucleation to growth to coarsening is not available yet. However, some processes have been studied in detail, one of which is the growth or dissolution of a single existing crystal from a large reservoir of melt or aqueous solution. The growth and dissolution of a single bubble can be treated similarly.

Because interface reaction and mass/heat transfer are sequential steps, crystal growth rate is controlled by the slowest step of interface reactions and mass/heat transfer. For a crystal growing from its own melt, the growth rate may be controlled either by interface reaction or heat transfer because mass transfer is not necessary. For a crystal growing from a melt or an aqueous solution of different composition, the growth rate may be controlled either by interface reaction or mass transfer because heat transfer is much more rapid than mass transfer. Different controls lead to different consequences, including the following cases:

(1) If the rate is controlled by interface reaction (Figure 1-11a) and if other conditions (such as temperature and pressure) are kept constant, then (i) the growth rate is independent of time (that is, the growth distance is proportional to time) leading to *linear growth law*, (ii) the concentration in the melt is uniform (no concentration gradient), and (iii) stirring the solution would not affect the growth or dissolution rate.

(2) If the rate is controlled by diffusive mass transfer (Figure 1-11b) and if other conditions are kept constant, then (i) the growth (or dissolution) distance is proportional to the square root of time, referred to as the *parabolic growth law* (an application of the famous square root law for diffusion), (ii) the concentration in the melt is not uniform, (iii) the concentration profile propagates into the melt according to square root of time, and (iv) the interface concentration is near saturation. For the rate to be controlled by diffusion in the fluid, it cannot be stirred.

(3) If the rate is controlled by convective mass transfer and steady state is reached (Figure 1-11c), then (i) the growth rate is independent of time, leading to the *linear growth law* (similar to the case of interface control, but different from diffusion control), (ii) the concentration in the melt is not uniform (different from interface control, but similar to diffusion control), (iii) the concentration profile stays roughly the same with time (i.e., the concentration profile does not become longer with time, different from diffusion control), (iv) the interface concentration is near saturation (similar to diffusion control but different from interface control). The length of the concentration profile is called the *boundary layer thickness* (to be defined later). Convection can be induced by stirring (more rapid stirring would lead to stronger convection and thinner boundary layer thickness), or natural motion in the fluid (such as a crystal freely falling in water or melt). In the case of a crystal moving through a fluid, the boundary layer thickness depends on whether one is measuring the profile in the leading side or the trailing side of the crystal (e.g., Levich, 1956; Kerr, 1995).

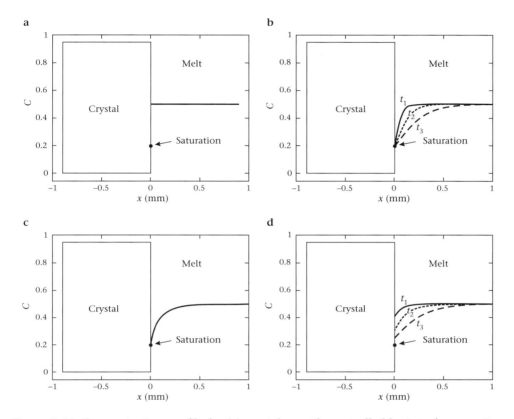

Figure 1-11 Concentration profile for (a) crystal growth controlled by interface reaction (the concentration profile is flat and does not change with time), (b) diffusive crystal growth with $t_2 = 4t_1$ and $t_3 = 4t_2$ (the profile is an error function and propagates according to $t^{1/2}$), (c) convective crystal growth (the profile is an exponential function and does not change with time), and (d) crystal growth controlled by both interface reaction and diffusion (both the interface concentration and the length of the profile vary).

(4) If the rate is controlled by convective mass transfer but steady state is not reached, then there may not be a simple relation between the growth distance and time.

(5) If the growth rate is controlled by both interface reaction and diffusion (Figure 1-11d), then (i) the concentration profile is not flat, (ii) the interface concentration changes with time toward the saturation concentration, (iii) the diffusion profile propagates into the melt, and (iv) the growth rate is not constant, nor does it obey the parabolic growth law.

The concentration profiles for crystal growth under different controls and their evolution with time are shown in Figure 1-11. Whether crystal growth (or dissolution) is controlled by interface reaction or mass transfer can be determined experimentally using these criteria. Theoretically, when departure from equilibrium (i.e., degree of oversaturation or undercooling) is small (e.g., undercooling of 0.01 K), the interface reaction rate approaches zero and growth should be controlled by interface reaction rate. As departure from equilibrium increases, interface reaction rate increases and mass transport may become the rate-determining step. Furthermore, because heat transport is much more rapid than mass transport, melting or growth of a crystal in its own melt is often controlled by interface reaction, but dissolution or growth of a crystal in a different melt is often controlled by mass transfer.

Experimental study shows that crystal melting in its own melt is extremely fast for diopside (e.g., Kuo and Kirkpatrick, 1985), indicating very high interface reaction rates. On the other hand, crystal growth or dissolution in a melt of different composition is slow (e.g., Harrison and Watson, 1983; Donaldson, 1985; Zhang et al., 1989), indicating control by mass transfer. Theoretical modeling shows that crystal growth or dissolution in a melt of different composition is often controlled by mass transfer (Zhang et al., 1989). Some recent studies indicate that the growth or dissolution of some minerals may be controlled by both interface reaction and mass transfer (e.g., Acosta-Vigil et al., 2002; Shaw, 2004).

1.5.1.1 Mineral dissolution in aqueous solutions

Crystal growth or dissolution in aqueous solutions may be controlled either by interface reaction or by mass transfer. Berner (1978) summarized some of the dissolution results and suggested a rule of thumb for rate control in aqueous solutions: For minerals with low solubility ($<10^{-4}$ mol/L), the control is interface reaction; for minerals with high solubility ($>10^{-3}$ mol/L), the control is mass transfer (Table 1-4). However, there are exceptions: opal has a high solubility (0.002 mol/L) but its dissolution rate is controlled by interface reaction, whereas AgCl has a low solubility (10^{-5} mol/L) but the dissolution rate is controlled by mass transport. One possible explanation is proposed here. AgCl is a univalent ionic compound, whose bonds are relatively easy to break, leading to high interface reaction rate, and hence to mass transfer control of the dissolution rate. On the other hand, bonds in $SiO_2 \cdot nH_2O$ are difficult to break, leading to low interface reaction rate and hence to interface control. That is, whether dissolution is controlled by interface reaction or mass transfer is determined by not only solubility but also bond strength (which influences the interface reaction rate). For complicated compounds, there is no easy quantitative measure of bond strength. In a simple approach, valence product $(z_+z_-)_{max}$ may be used to characterize the bond strength, where z_+ is the valence of the cation and z_- is the valence of the anion involved in the actual bond breakage. For example, during the dissolution of $BaSO_4$, S–O bonds are not broken. Hence, $(z_+z_-)_{max} = 2 \cdot 2 = 4$ (not $6 \cdot 2 = 12$). For the dissolution of $KAlSi_3O_8$, Si–O bonds are broken. Hence, $(z_+z_-)_{max} = 4 \cdot 2 = 8$. Figure 1-12 (see page 54) shows the same data of Berner (1978)

Table 1-4 Dissolution mechanism for some substances

Mineral	Solubility (mol/L)	$(z_+z_-)_{max}$	Mineral	Solubility (mol/L)	$(z_+z_-)_{max}$
Interface reaction control			**Transport control**		
$KAlSi_3O_8$ (KFs)	3×10^{-7}	8	AgCl (chlorargyrite)	1×10^{-5}	1
$NaAlSi_3O_8$ (NaFs)	6×10^{-7}	8	$Ba(IO_3)_2$	8×10^{-4}	2
$BaSO_4$ (barite)	1×10^{-5}	4	$CaSO_4 \cdot 2H_2O$ (gypsum)	0.005	4
$SrCO_3$ (strontianite)	3×10^{-5}	4	$Na_2SO_4 \cdot 10H_2O$	0.2	2
$CaCO_3$ (calcite)	6×10^{-5}	4	$MgSO_4 \cdot 7H_2O$ (epsomite)	3	4
Ag_2CrO_4	1×10^{-4}	2	$Na_2CO_3 \cdot 10H_2O$	3	1
$SrSO_4$ (celestine)	9×10^{-4}	4	KCl (sylvite)	4	1
SiO_2 (opal)	0.002	8	NaCl (halite)	5	1
Mixed control			$MgCl_2 \cdot 6H_2O$ (bischofite)	5	2
$PbSO_4$ (anglesite)	1×10^{-4}	4			

Note. From Berner, 1978.

in a plot of solubility (log scale) versus $(z_+z_-)_{max}$ and a line is drawn to separate the region of transport control and that of interface reaction control. The new proposal accounts the observations well.

1.5.1.2 Linear versus parabolic laws

Sometimes crystal growth, dissolution, or oxidation is said to follow a linear growth law or a parabolic growth law. The *linear law* means that the thickness of the crystal depends linearly on time,

$$\Delta x = At, \tag{1-85}$$

where Δx is the thickness of growth or dissolution or oxidation, and A is a constant. The linear growth law means a constant growth or dissolution or oxidation rate. (Because this is for heterogeneous reaction instead of homogeneous reaction, constant growth rate does not mean zeroth-order reaction.) As shown

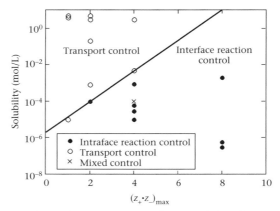

Figure 1-12 Control mechanisms of mineral dissolution in aqueous solutions. Data are from Berner (1978). A straight line is drawn to separate transport control and interface reaction control although there is no theoretical basis for whether the boundary should be linear. Almost without exception, those with transport control lie above a straight line, and those with interface reaction control lie below the line. The only significant departure from the rule is the dissolution of $PbSO_4$ (cross in the figure) that lies inside the region for the interface reaction control, but is actually controlled by both interface reaction and mass transport.

above, the linear growth law may result from either interface-controlled growth, or convective crystal growth reaching steady state.

The *parabolic law* means that the growth (or dissolution or oxidation) distance is proportional to the square root of time,

$$\Delta x = At^{1/2}. \tag{1-86}$$

Because diffusion distance is proportional to $t^{1/2}$, the above relation indicates diffusion control.

An example of parabolic versus linear reaction is the oxidation of pure fayalite in the gas phase (Mackwell, 1992). Mackwell found that fayalite oxidation in air at 1043 K follows a parabolic law, whereas oxidation at 1303 K in a mixture of CO and CO_2 follows a linear law (Figure 1-13, see page 55). Fayalite oxidation is a complicated heterogeneous reaction. The parabolic law indicates that the oxidation process is controlled by diffusion—not diffusion in air (which is extremely rapid), but diffusion of O_2 through the already oxidized layer to reach the inside pristine fayalite. The linear reaction law indicates control by the oxidation rate at the interface (which is the interface reaction rate). It is not clear why there is a change in the controlling mechanisms. Notwithstanding interface or diffusion control, as well as the change in the gas medium, the oxidation is much more rapid at the higher temperature, consistent with expectations.

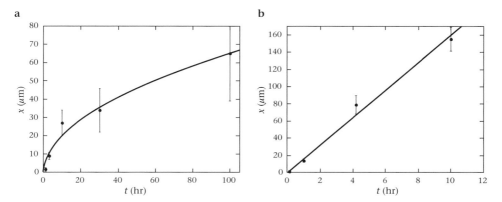

Figure 1-13 Comparison of (a) parabolic oxidation law for fayalite oxidation at 1043 K in air and (b) linear oxidation law for fayalite oxidation at 1303 K in a mixture of CO and CO_2 (Mackwell, 1992). The vertical axis x is the thickness of the oxidized layer.

1.5.2 Steps in heterogeneous reactions

1.5.2.1 Nucleation

Nucleation is the formation of a new phase, either inside a pre-existing phase (*homogeneous nucleation*), or at the interface of two or more preexisting phases (*heterogeneous nucleation*). The classical nucleation theory of homogeneous nucleation is well developed but does not seem to work. Without a better theory, a calculated result from the theory is illustrated in Figure 1-14 (dashed curve, see page 56). Homogeneous nucleation is very difficult and the rate is very small unless the degree of supersaturation or undercooling is huge. As a melt cools down and when a mineral becomes exactly saturated, the nucleation rate is zero. As the melt cools further, the nucleation rate increases, but very slowly with decreasing temperature. For very slowly cooled rocks (such as plutonic rocks), slow nucleation over very long time can still result in a significant number of nuclei for growth into a coarsely crystalline rock. Only when undercooling is huge (such as 800 K below the melting temperature) would the nucleation rate be high. The nucleation rate then reaches a maximum and decreases as temperature is further decreased. The reason for the decrease of the nucleation rate is due to the slowdown of mass transfer in the melt as temperature continues to cool. Heterogeneous nucleation is much easier. Hence, nucleation in nature is believed to be almost always heterogeneous. The heterogeneous nucleation rate of phase C at the interface of phases A and B depends on the interfacial properties between A and B, B and C, and C and A.

1.5.2.2 Interface reaction rate

The interface reaction rate depends on temperature and on the degree of saturation. Right at saturation, the interface reaction rate is zero. If the melt (or

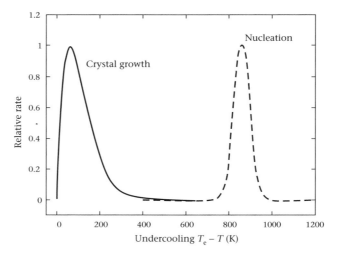

Figure 1-14 Homogeneous nucleation rate (dashed curve) and interface-reaction-controlled crystal growth rate (solid curve) plotted against the undercooling. Undercooling ($T_e - T$, where T_e is the equilibrium or saturation temperature between the crystal and the melt, and T is the system temperature) is an often-used alternative term to describe supersaturation. Crystal–liquid equilibrium is at an undercooling of zero, at which both the growth and nucleation rates are zero. The vertical axis plots the relative rates, that is, the nucleation and growth rates divided by their respective maximum rate. In this example, maximum crystal growth rate (42 μm/s) occurs at an undercooling of 60°C and maximum nucleation rate (6×10^7 m^{-3} s^{-1}) occurs at an undercooling of 860°C.

solution) is supersaturated with respect to a crystal, the crystal will grow. If the melt is undersaturated with respect to a crystal, the crystal will dissolve. The degree of saturation can vary with temperature or composition or pressure. If temperature is the only control, then crystal growth rate is zero at the saturation temperature, crystal dissolution or melting rate will increase rapidly with increasing temperature, and crystal growth rate will increase first as temperature decreases from the saturation temperature, but will reach a maximum, and then decrease and approach zero as the temperature decreases further. The reason for decreasing crystal growth rate as temperature decreases further is because mass transfer (either diffusion or convection) is slowed down as temperature decreases. The maximum crystal growth rate is usually reached when undercooling is on the order of several tens of kelvins. The dependence of interface reaction rate on temperature is shown as the solid curve in Figure 1-14. The comparison in Figure 1-14 shows that to reach maximum nucleation rate requires much larger undercooling than to reach maximum crystal growth rate.

1.5.2.3 Volume growth and coarsening

Growth of a single crystal in a large fluid reservoir is an idealized simple case. In nature, often many crystals of the same phase nucleate and then grow in a fluid

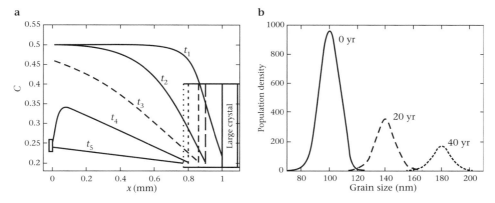

Figure 1-15 (a) Schematic drawing of crystal growth, interaction of diffusion profiles, and coarsening. From t_1 to t_3, the large crystal (on the right-hand side) grows by itself (the dashed lines that touch the respective concentration profiles). A small crystal (small rectangle near $x=0$) appears at t_4 and grows slowly. At t_5, there is roughly a linear concentration profile between the small and large crystals. The small crystal dissolves and the large one grows. Finally, the small crystal is completely consumed. (b) Simulated results of Ostwald ripening of gibbsite crystals. Initial average size was 100 nm. The crystal size distribution is assumed to be Gaussian. As the average crystal size increases linearly with time from 0 to 40 years, the number of crystals decreases. From Steefel and van Cappellen (1990).

phase simultaneously and compete for nutrients. Hence, their diffusion profiles overlap, which reduces the growth rate. As the volume fraction of crystals increases, the bulk melt composition moves toward the equilibrium composition with the crystals and the growth rate gradually decreases to zero. For example, at a given temperature, a magma may contain only one mineral phase (such as olivine), and the volume fraction of the mineral phase at equilibrium may be 10%. Under this scenario, as the volume fraction of olivine reaches 10%, there would be no more volume growth of olivine. In a crystallizing silicate melt, not only can many crystals of a single mineral phase grow simultaneously, many crystals of several mineral phases can also grow simultaneously.

When the volume fraction of the crystals of one phase (or many phases) reaches the equilibrium value, there would be no more volume growth. Nonetheless, the size of crystals can still increase. This process is called coarsening, also referred to as *Ostwald ripening*. Coarsening occurs because for a given volume of a phase, a smaller number of larger crystals have smaller surface area compared with a larger number of smaller crystals. Reduction of surface area reduces surface energy, leading to a thermodynamically more stable system. The process occurs by the growth of larger crystals and resorption of smaller crystals (Figure 1-15a). A schematic example of crystal size distribution due to Ostwald Ripening is shown in Figure 1-15b.

The above is for isothermal crystal growth. In nature, crystallization occurs in a continuously cooled magma. The cooling rate plays a main role in controlling crystallization, and the nucleation and interface-reaction rates shown in Figure 1-14 are instructive in understanding crystallization under various cooling rates.

If the cooling rate is extremely high, there would be little time for crystal nucleation and growth, and the melt would be cooled into a glass. If the cooling rate is slowed a little, then there would be some growth, mostly in the form of *dendritic growth* (see later). As cooling rate decreases further, some crystals would grow, and the rest would be glass. The result is a phenocrystic glassy rock. If the cooling rate is slowed more, there would be phenocrysts and crystallized matrix, a porphyritic rock. If the cooling rate is extremely slow (such as plutonic rock), the coarsening step would wipe out the small grains, and lead to a rock with roughly uniform grain size (such as gabbro and granite). If slow cooling is coupled with a volatile-rich melt, huge crystals may grow, leading to a pegmatite.

Aspects of nucleation, growth of a single crystal, volume growth, and coarsening processes are discussed in greater depth in Chapter 4. It will be seen that although some of these processes can be quantified well, it is not possible yet to quantitatively predict the kinetics of many of these processes.

1.6 Temperature and Pressure Effect on Reaction Rate Coefficients and Diffusivities

The rate constants of elementary reactions, the diffusion coefficient, and other rate constants (such as attachment rate and detachment rate) all increase with increasing temperature, roughly according to the Arrhenius equation. This equation is one of the most fundamental relations in kinetics. Although the pressure effect is usually small for chemists, it can be large for geochemists because the pressure range encountered by geochemists can be very large. Experimental data show that diffusion coefficient D often depends on pressure exponentially. Hence, when both temperature and pressure effects are considered, the diffusion coefficient depends on temperature and pressure (P) as follows:

$$D = A \exp[-(E + P\Delta V)/(RT)], \tag{1-87}$$

where ΔV is a constant often referred to as the activation volume (the meaning of ΔV and E will be clearer later). The units of $P\Delta V$ must be the same as E and RT. Experimentally determined ΔV values may be either positive or negative. That is, D may decrease or increase with increasing P. At pressures <4 GPa negative activation volume (D increasing with pressure) is usually associated with fully or nearly fully polymerized melts such as jadeite and silica melt (e.g., Shimizu and Kushiro, 1984; Kubicki and Lasaga, 1988), and positive activation volume is associated with other melts (e.g., Watson, 1979a, 1981; Behrens and Zhang, 2001). In a larger pressure range, the sign of ΔV may change. For example, molecular dynamics simulations (Kubicki and Lasaga, 1988) show that ΔV for silica melt changes from negative (D increases with P) at low pressures to positive (D decreases with P) at pressures > 13 Gpa. Experimental data of Reid et al. (2001) show that oxygen and silicon self-diffusivities in diopside liquid decrease with pressure at $P < 9$ Gpa, and then increase with pressure; and those of Tinker and Lesher (2001) show that oxygen and silicon self-diffusivities in dacitic melt increase with pressure at $P < 4$ Gpa, and then decrease with pressure.

Similarly, the reaction rate coefficient for an elementary reaction takes the following form:

$$k = A \exp[-(E + P\Delta V)/(RT)]. \tag{1-88}$$

At constant pressure, if the logarithm of the reaction rate constant or the diffusion constant is plotted against $1/T$ (often $1000/T$ so that the numbers are nice numbers around 1), we would get a straight line. At constant temperature, if the logarithm of the reaction rate constant or the diffusion constant is plotted against P, we should also get a straight line. In a small temperature range (a couple of hundred degrees at high T), the activation energy can be regarded as roughly a constant. However, over a large temperature range, the reaction (or diffusion) mechanism may change with temperature, which would cause a change in the slope of the straight-line relation.

The Arrhenius relation means that the rate constant or the diffusivity increases with temperature. Typically, at low temperatures (0–60°C), a 10-degree increase in temperature results in a doubling of reaction rates. In this section, two theories are introduced to account for the Arrhenius relation and reaction rate laws. Collision theory is a classical theory, whereas transition state theory is related to quantum chemistry and is often referred to as one of the most significant advances in chemistry.

1.6.1 Collision theory

At the molecular level, a bimolecular elementary reaction is accomplished by bringing the two molecules together to collide. If the orientation is right and the molecules have enough energy, the collision would turn the reactants to the product. The energy must be high enough to overcome the activation energy for the reaction to occur. *Collision theory* is best developed for gas-phase reactions. Consider the following elementary reaction

$$A + B \rightarrow C, \tag{1-89}$$

in a gas phase. Let f_{AB} be the collision rate of A and B with the right orientation per unit volume. The reaction rate is the number of such collisions with energy high enough to overcome the activation energy (E) of the reaction. Based on Boltzmann's distribution law, the probability of an event with energy greater than E is proportional to $e^{-E/(RT)}$. Hence, the reaction rate per unit volume is

$$dN_A/(V dt) = -f_{AB} e^{-E/(RT)}, \tag{1-90}$$

where N_A is the number of molecules and V is the volume of the gas system. Dividing both sides by the Avogadro's number ($N_{av} = 6.0221 \times 10^{23}$) leads to

$$d\xi/dt = (f_{AB}/N_{av})e^{-E/(RT)}. \tag{1-91}$$

The collision frequency f_{AB} can be calculated from the kinetic theory of gases. The result is (cf., Atkins, 1982, p. 872)

$$f_{AB} = \sigma^*\left(\frac{8k_BT}{\pi m_{AB}}\right)^{1/2}\frac{N_A}{V}\frac{N_B}{V} = \sigma^*\left(\frac{8k_BT}{\pi m_{AB}}\right)^{1/2}N_{av}^2[A][B], \tag{1-92}$$

where σ^* is the effective cross section for the reaction, k_B is the Boltzmann constant ($\equiv R/N_{av}$), and m_{AB} is the reduced mass of the two colliding species ($1/m_{AB} = 1/m_A + 1/m_B$). The collision cross section σ for hard sphere species of radii r_A and r_B is $\pi(r_A + r_B)^2$ based on the collision theory. But for two molecules to react during a collision, the orientations of the two molecules have to be right, so the cross section for reactions (σ^*) should be the "effective" cross section at the right orientation, which is most likely smaller than the kinetic collision cross section. The more complicated the molecules, the smaller the ratio σ^*/σ. Combining Equations 1-91 and 1-92 leads to

$$\frac{d\xi}{dt} = N_{av}\sigma^*\left(\frac{8k_BT}{\pi m_{AB}}\right)^{1/2}[A][B]e^{-E/(RT)}, \tag{1-93}$$

which can be rewritten as

$$d\xi/dt = k[A][B], \tag{1-94}$$

where

$$k = N_{av}\sigma^*\left(\frac{8k_BT}{\pi m_{AB}}\right)^{1/2}e^{-E/(RT)}. \tag{1-95}$$

Hence, collision theory gives the correct reaction rate law (Equation 1-94) and a reaction rate constant of the form

$$k = A'T^{1/2}\,e^{-E/(RT)}. \tag{1-96}$$

Although this form differs from the Arrhenius equation in that the pre-exponential term $A'T^{1/2}$ depends slightly on T, because the exponential dependence usually dominates, the weak dependence of the pre-exponential term on T may be regarded as negligible and the whole term $A'T^{1/2}$ regarded as a constant A. Hence, it is possible to roughly derive the Arrhenius relation from the collision theory.

Because σ^* is a parameter that cannot be calculated from first principles, Equation 1-95 cannot be used to calculate reaction rate constant k from first principles. Furthermore, the collision theory applies best to bimolecular reactions. For monomolecular reactions, the collision theory does not apply. Trying to calculate reaction rates from first principles for all kinds of reactions, chemists developed the transition state theory.

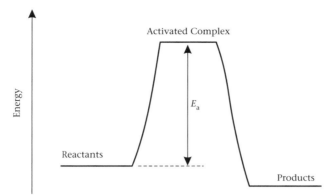

Figure 1-16 Transition state and activation energy (referred to as either E_a or E, or ΔE_a). The energy state of the activated complex is the transition state. Energy (or enthalpy) for reactants, activated complex, and products is plotted against reaction progress.

1.6.2 Transition state theory

The Arrhenius equation can also be roughly derived using the transition state theory. In the context of the transition state theory, for reactants to react to form the products, reactants must first pass through a high-energy state (Figure 1-16), called the transition state. Similarly, for a particle to diffuse from one site to another, the particle goes through a high-energy state, such as the orifice through which the particle must pass. This state can be regarded as the transition state. The unstable species at the high-energy transition state (or the atom at the orifice during diffusion) is called the activated complex (typically signified using the symbol ‡ to highlight that it is not a stable or normal species). Hence, the reaction mechanism for a one-step (i.e., elementary) reaction is

reactants (low energy) $\rightarrow$ activated complex‡(highest energy)
$$\rightarrow \text{products (low energy)}$$

diffusion species in one site $\rightarrow$ activated complex‡
$$\rightarrow \text{diffusion species in the new site}$$

The necessity to go through a high-energy transition state for a reaction or diffusion is the origin of the activation energy. The backward reaction would go through the same transition state. The transition state theory and the concept of the activated complex are also applicable to other kinetic processes, such as the attachment and detachment reactions (which can be regarded as the forward and backward reactions) for heterogeneous reactions.

The activated complex of the transition state is at the highest energy along the reaction coordinate, i.e., its energy is higher than that of the reactants by the activation energy. In the transition state theory, it is further assumed that the activated complex has a concentration in equilibrium with the reactants. For example, for a bimolecular elementary reaction $A + B \rightarrow C$, transition state theory presumes the following process:

$$A + B \underset{K^{\ddagger}}{\rightleftharpoons} AB^{\ddagger} \xrightarrow{k^{\ddagger}} C \tag{1-97}$$

where $K^{\ddagger} = [AB^{\ddagger}]/\{[A][B]\}$ is the "equilibrium" constant between the activated complex and the reactants, and $k^{\ddagger}$ is the reaction rate constant from the activated complex to the product. Therefore,

$$d\xi/dt = k^{\ddagger}[AB^{\ddagger}] = k^{\ddagger}K^{\ddagger}[A][B]. \tag{1-98}$$

Because the reaction rate law for the reaction is $d\xi/dt = k[A][B]$, the rate constant k for the elementary reaction $A + B \rightarrow C$ is

$$k = k^{\ddagger}K^{\ddagger}. \tag{1-99}$$

Since the activated complex is highly unstable, it may not survive the vibration of the complex. Hence, the rate coefficient $k^{\ddagger}$ can be thought to be proportional to the fundamental frequency ν; i.e., we may write

$$k^{\ddagger} = \kappa\nu \tag{1-100}$$

where κ is called the transmission coefficient. The fundamental frequency ν of the activated complex can be estimated based on quantum mechanics to be

$$\nu = k_B T/h, \tag{1-101}$$

where k_B is the Boltzmann constant and h is the Planck constant. At 1000 K, the fundamental frequency ν is about 2×10^{13} per second.

Based on thermodynamic considerations, the "equilibrium" constant $K^{\ddagger}$ can be written as

$$K^{\ddagger} = \frac{[AB^{\ddagger}]}{[A][B]} = \frac{\gamma_A \gamma_B}{\gamma^{\ddagger}} \frac{a^{\ddagger}}{a_A a_B} = \frac{\gamma_A \gamma_B}{\gamma^{\ddagger}} e^{-\Delta G^{\ddagger}/(RT)} = \frac{\gamma_A \gamma_B}{\gamma^{\ddagger}} e^{\Delta S^{\ddagger}/R} e^{-\Delta H^{\ddagger}/(RT)} \tag{1-102}$$

Therefore, we have

$$k = \kappa \frac{k_B T}{h} \frac{\gamma_A \gamma_B}{\gamma^{\ddagger}} e^{\Delta S^{\ddagger}/R} e^{-\Delta H^{\ddagger}/(RT)} = A'T \, e^{-E/(RT)} \tag{1-103}$$

where A' collects various parameters. This equation differs from the Arrhenius equation and that derived from the collision theory in the power of T in the pre-exponential factor (zeroth power in the Arrhenius equation, ½ power in the collision theory, and first power in the transition state theory). Nonetheless, as temperature varies, the exponential dependence dominates and one may view $A'T$ roughly as a constant pre-exponential factor, and Equation 1-103 is roughly the same as the Arrhenius equation. Experimental data show that in a small temperature range (when $1000/T$ varies by ≤ 0.5), the Arrhenius equation is indeed a good approximation.

Both the collision and transition state theories lead to a nonlinear dependence of $\ln k$ versus $1/T$, which is theoretically resolvable with data over a large temperature range. High-quality experimental data show that over a large temperature range, $\ln k$ versus $1/T$ can be nonlinear, which is referred to as non-Arrhenian behavior. For example, the data plotted in Figure 1-2b show a slight nonlinearity. Nonetheless, the nonlinear behavior matches neither the transition state nor the collision theory. It is hence difficult to use experimental data to verify a theory. In addition to the theoretical dependence of the pre-exponential factor on T, the activation energy may also vary with temperature because of a change in the reaction mechanism. The data in Figure 1-2b with slight nonlinearity (replotted in Figure 1-17) may be fit well by the following three-parameter equation:

$$k = 3.61 \cdot T^{3.32} \exp(-428.7/T) \text{ L mol}^{-1}\text{s}^{-1},$$

which is perfectly consistent neither with the transition state theory, nor the collision theory. Other three-parameter equations may also fit the data well (see caption of Figure 1-17).

In the context of the transition state theory, the activation energy for the reaction can be approximately identified as the enthalpy necessary to form the activated complex:

$$\Delta H^{\ddagger} = H^{\ddagger} - (H_A + H_B). \tag{1-104}$$

This understanding can be used to infer the pressure dependence of k because enthalpy depends on pressure. It can be shown that

$$\left(\frac{\partial H}{\partial P}\right)_T = V(1 - \alpha_P T). \tag{1-105}$$

For reactions between ideal gases, $1 - \alpha_P T = 0$, and $\Delta H^{\ddagger}$ and the activation energy are independent of pressure. For reactions in the condensed phases, thermal expansivity is small and $1 - \alpha_P T \approx 1$. Hence,

$$\left(\frac{\partial \Delta H^{\ddagger}}{\partial P}\right)_T = \Delta V^{\ddagger}, \tag{1-106}$$

and

$$\Delta H^{\ddagger}|_P \approx \Delta H^{\ddagger}|_{P_0} + (P - P_0)\Delta V^{\ddagger}, \tag{1-107}$$

where V is partial molar volume, $\Delta V^{\ddagger} = V^{\ddagger} - V_A - V_B$ and is roughly constant for reactions in a condensed phase. The standard state of one atmosphere is small enough and may be approximated as zero ($P_0 = 0$), leading to

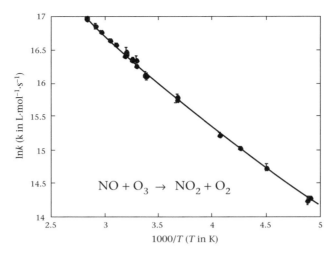

Figure 1-17 Rate coefficients (Borders and Birks, 1982) for gas-phase reaction $NO + O_3 = NO_2 + O_2$ in a $\ln k$ versus $1/T$ (K) plot. Two data points with significantly larger errors are excluded. Although the Arrhenius relation (linear relation) is a good approximation, there is a small nonlinearity. The data can be fit well by $k = 3.61T^{3.32}\exp(-428.7/T)$ $L\,mol^{-1}\,s^{-1}$. Two other relations that can fit the data equally well are $\ln k = 23.18 - 3047.6/(T + 138.09)$, and $\ln k = 30.90 - (1196.4/T)^{0.324}$.

$$\Delta H^{\ddagger} = \Delta H^{\ddagger}_0 + P\Delta V^{\ddagger}, \tag{1-108}$$

$$k = AT\exp[-(\Delta H^{\ddagger}_0 + P\Delta V^{\ddagger})/(RT)], \tag{1-109}$$

where $\Delta V^{\ddagger} = V^{\ddagger} - (V_A + V_B)$ and is referred to as the activation volume. Because it is the difference of volumes of different species in the condensed state, $\Delta V^{\ddagger}$ is usually small and can be either positive or negative. Therefore, k (and diffusion coefficient D) may either decrease or increase with pressure. That is, if you have the intuition that the pressure effect is opposite to the temperature effect and k and D must decrease with pressure, this is a misconception. Furthermore, $\Delta V^{\ddagger}$ is usually small, which leads to two consequences: (i) k and D do not depend on pressure as strongly as on temperature, and (ii) $\Delta V^{\ddagger}$ may change signs as pressure increases. That is, k and D may first increase (or decrease) and then decrease (or increase) with pressure. Specifically, when k or D increases with pressure, it means only that $\Delta V^{\ddagger}$ is negative, and does not mean that the volume of the activated complex is negative.

In theory, one can use statistical thermodynamics to calculate the partition functions of all the species from first principles, $\Delta S^{\ddagger}$, $\Delta H^{\ddagger}$, and hence k. For simple systems, the calculation results are in good agreement with experimental data (e.g., Chapter 3 in Laidler, 1987). For complicated geological systems, however, it is not possible to calculate k from first principles, but the concept of activated complexes is very useful for a microscopic understanding of the reaction processes. Lasaga (1998) provided a more thorough account of the transition state theory. Felipe et al. (2001) discussed some applications in geochemistry.

Even if the reaction rate coefficient or diffusion coefficient cannot be calculated from first principles, it is often possible to make a very rough estimate of the activation energy, which might be helpful. Given a reaction, the activation energy can be intuitively thought to be related to bond strengths. Some examples are given below.

(1) The activation energy for diffusion increases from gas, to liquid, to solid. For diffusion in gas, not much energy is needed for gas molecules to go between other gas molecules. The activation energy is small and diffusion coefficient is large. For diffusion in a liquid, particles are more tightly packed and hence it is more difficult for a particle to move through a neck of other particles. Hence, the activation energy is greater. For solids, the activation energy is even greater.

(2) It is relatively easier for a small neutral molecule (such as He or H_2) to move through a liquid or solid structure, than for a large molecule (such as Xe). Hence, the activation energy for diffusion of small molecules is small and the diffusion coefficient is large.

(3) For ionic species (such as Mg^{2+}, O^{2-}, Si^{4+}) in solid or liquid, moving about requires the breakage of ionic bonds of the species themselves, as well as disrupting other ionic bonds when a species is squeezing through the structure. The size of the ion is not as important as the charge. For example, Si^{4+} is a small cation, but it forms strong bonds with oxygen in silicate minerals and melts. Hence, the activation energy for Si diffusion is large.

(4) For nuclear fusion reactions, the activation energy is the energy barrier and can be roughly estimated from

$$E = E_0 \frac{Z_1 Z_2}{A_1^{1/3} + A_2^{1/3}},$$
(1-110)

where E is the activation energy, $E_0 = 1.2 \, \text{MeV}$, Z_1 and Z_2 are the numbers of protons in nuclei 1 and 2, and A_1 and A_2 are the mass numbers of the two nuclei. Because activation energy for chemical reaction is in the order of 1 eV (or 96.5 kJ/mol) and that for nuclear reactions is of the order 1 MeV, nuclear reactions require a temperature of 100 million kelvins compared to several hundreds of kelvins for geochemical reactions.

The transition-state theory may be applied to component processes of heterogeneous reactions. Examples of heterogeneous reaction include the oxidation of metal sodium to Na_2O versus diamond to CO_2. To oxidize sodium metal, relatively weak sodium bonds must be broken as well as the O–O bond of O_2. To oxidize diamond, the very strong C–C bond in diamond must be broken as well as the O–O bond. Hence, burning diamond is more difficult than burning sodium metal. Similarly, converting diamond to graphite requires high activation energy for the interface reaction. Therefore, reaction rate at room temperature is basically zero, and diamond, although thermodynamically unstable, can survive almost forever under these conditions.

1.7 Inverse Problems

Whereas chemical kineticists are lucky to treat mostly reaction kinetics given initial conditions, one of the critical tasks of geochemical kineticists is to infer past conditions from present-day observations using kinetics. Such inferences are called *inverse problems*. Furthermore, whereas chemical kineticists usually only have to treat reactions under isothermal conditions, geochemical kineticists are required to deal with reactions under variable temperatures, especially under cooling because igneous and metamorphic rocks often experienced high temperature and then cooled down. Inverse problems are often closely related to reaction kinetics during cooling. Below we introduce the basics of these problems.

1.7.1 Reactions and diffusion during cooling

Rocks and their minerals often experienced complicated temperature and pressure history. The temperature and pressure history of volcanic, plutonic, and ultra-high-pressure metamorphic rocks are schematically shown in Figure 1-18. Most of the rocks geologists study now are surface rocks and have cooled to surface temperatures (except for rocks recovered from deep drilling). Hence, the endpoint of the temperature–pressure history is surface conditions. Volcanic rocks have the simplest thermal history. A completely molten or partially crystallized magma rapidly rises to the surface, often adiabatically, and then cools down rapidly on the surface (Figure 1-18a). For submarine eruptions, the quench is extremely rapid, and cooling rate at the surface of the rock may exceed 1000 K/s. For subaerial eruptions, cooling rate of pyroclasts in an eruption column may be a few to a few tens of kelvins per second (Xu and Zhang, 2002). Cooling of lava flows may take days to years depending on the thickness of the flows.

The P-T-t path of a plutonic igneous rock (Figure 1-18b) depends on the size of the magma chamber, the depth of emplacement, the time uplift began, and the uplift rate. The cooling rate may be as low as 1 K/Myr for large magma bodies intruding deep in the crust.

The P-T-t paths of metamorphic rocks may be much more complicated. They are usually heated first to a peak temperature and then cooled to room temperature. There are a variety of metamorphic rocks. Figure 1-18c shows a hypothetical cooling history of an ultra-high-pressure metamorphic rock, which was subducted to great depth and then returned to the surface. Before subduction, the premetamorphic rock could be a basalt or a sedimentary rock. From 0 to 2 Myr, the slab is modeled as being subducted at 0.08 m/yr and an angle of 45°. At 2 Myr, the slab is at the depth of 113 km. As the slab subducts, it is heated up but the temperature is much lower than the normal geotherm (as marked in Figure 1-18c). At this depth, the crustal part of the slab detaches from the rest of the slab, and returns to the surface. During the return, the rock is heated up until its temperature becomes the same as the geotherm. (There is slight overshoot in temperature because the rock is rising from hotter to colder mantle.)

The above P-T-t paths do not cover all temperature–pressure histories. For example, all deep-seated rocks may be brought up rapidly by volcanic eruptions as xenoliths and hence may cool from a high temperature to room temperature rapidly. Furthermore, through drilling it is possible to obtain samples at high temperatures and pressures.

The Arrhenius equation indicates that the reaction rate constant depends strongly on temperature. At high temperature, the reaction rate constant is large and the reaction rate is high. Hence, equilibrium is much easier to reach at high temperatures. As temperature decreases, the reaction rate also decreases. Therefore, at low enough temperature, the reaction may effectively stop. The same is true for diffusion. At high temperatures, diffusive loss or exchange is much more significant and may be able to maintain equilibrium. However, at sufficiently low temperatures, diffusion may effectively stop. Hence, the final species concentrations in a mineral or in a glass, as well as the overall composition of minerals depend on the thermal history, especially the cooling rate. If the cooling is rapid, the extent of reaction progress during cooling is less, and the final assemblage would reflect "equilibrium" at a higher temperature. On the other hand, if the cooling is slow, the extent of the reaction during cooling would be greater, and the final assemblage would reflect "equilibrium" at a lower temperature.

Because of the complex thermal history, quantitative treatment of reaction kinetics, or the extent of the reaction as a function of time, must incorporate the dependence of the rate coefficient(s) with time. The treatment may be simple, as shown in Section 1.3.7, but more often is complicated for real geologic reactions. Some of the concepts are discussed below but the rigorous derivations are presented in later chapters.

Consider a homogeneous reaction such as Fe–Mg order–disorder reaction in an orthopyroxene (Reaction 1-12), $Fe^{2+}_{M2}(opx) + Fe^{2+}_{M1}(opx) \rightleftharpoons Fe^{2+}_{M1}(opx) + Mg^{2+}_{M2}$ (opx). The equilibrium constant K of the above reaction depends on temperature and the composition. Use simple exchange coefficients $K_D = (Fe/Mg)_{M1}/(Fe/Mg)_{M2}$ to describe the distribution at equilibrium. Wang et al. (2005) showed that for $Fe/(Fe + Mg) \leq 0.6$,

$$K_D = \exp(0.391 - 2205/T), \tag{1-111}$$

where T is in K. For a plutonic or metamorphic rock, the cooling rate is low. Hence, the reaction continues to relatively low temperatures, leading to a very small K_D, meaning almost all Fe^{2+} is in the M2 site, and very little in M1 site (i.e., orthopyroxene is highly ordered). For a volcanic rock, the cooling rate is very

a

b

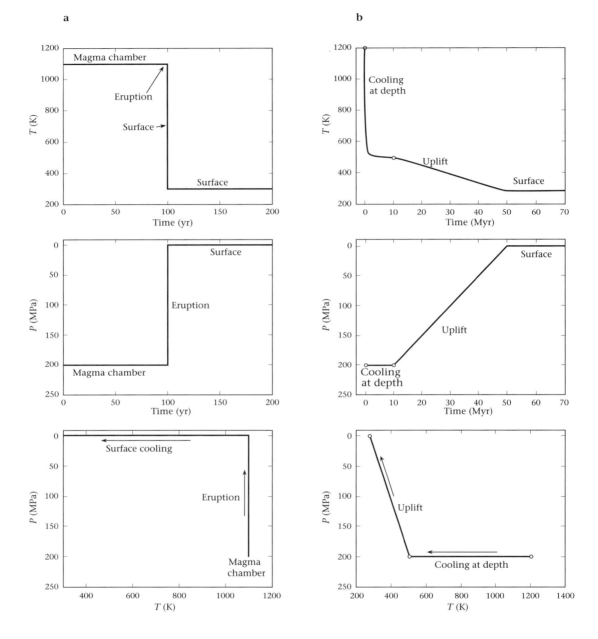

c

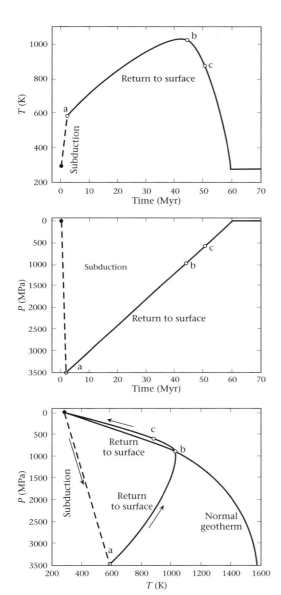

Figure 1-18 Schematic temperature and pressure history for (a) a volcanic (rhyolitic) flow, (b) a plutonic rock, and (c) a metamorphic rock. In each of (a), (b), and (c), there are three panels: (top) T versus time, (middle) P versus time (note that high pressure is pointing down to mimic depth in the so-called geophysical plot), and (bottom) P versus T. Note that unit of time is different in (a). (a) For a volcanic flow, both temperature and pressure are high in the magma chamber. During the eruption, the pressure decreases rapidly (using an eruption velocity of 0.02 m/s), but temperature variation is small (using an adiabatic $\partial T/\partial P$ of 1.46×10^{-8} K/Pa). Upon reaching the surface, the magma cools rapidly and pressure stays at 0.1 MPa. (b) For a plutonic rock, from 0 to 10 Myr, magma cools at 200 MPa to reach the ambient temperature of 500 K (calculated assuming conductive cooling). At $t = 10$ Myr, there was uplift (without it we would not be able to collect the rocks) at 5 Pa/yr. The temperature changes with pressure at 1.1 K/MPa (geotherm). At $t = 50$ Myr, the rock reached the surface (280 K and 0.1 MPa). (c) The starting point (the solid point) is at the surface ($P = 0.1$ MPa and $T = 280$ K). The rock first subducts and then rebounds. Point a indicates the peak pressure, point b indicates the peak temperature, and point c indicates a possible closure temperature (T_c) or apparent equilibrium temperature (T_{ae}). The concept of T_c and T_{ae} can be found in the next section.

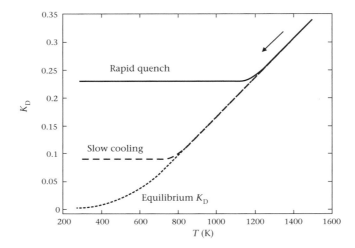

Figure 1-19 Schematic diagram showing how K_D for the Fe–Mg order–disorder reaction varies during cooling. The arrow indicates the progression of time. The thin dashed curve shows how the equilibrium K_D varies with temperature as the system cools. The solid curve shows how K_D varies with temperature during rapid quench in a volcanic rock. The thick and long dashed curve shows how K_D varies during slow cooling in a plutonic rock.

high, and there is more Fe^{2+} in the M1 site (although still not as much as in the M2 site). The variation of instantaneous K_D with temperature during cooling is shown schematically in Figure 1-19 for two cases: rapid quench in a volcanic rock and slow cooling in a plutonic rock.

The relationship shown in Figure 1-19 is the basis for using kinetics to infer the cooling rate. Quantitative models have been developed to understand the behavior of various reactions and diffusion during cooling (Dodson, 1973; Ganguly, 1982; Giletti, 1986; Eiler et al., 1992, 1994; Zhang, 1994; Ganguly and Tirone, 1999; Ni and Zhang, 2003). With such models and with careful experimental investigation of the reaction rates, reaction kinetics can be used as an indicator of cooling rate. In theory, if the kinetic data are available, not only the cooling rate, but the full thermal history (including heating and cooling) can be modeled. However, inferring complicated thermal history from observed phase compositions and speciation is rarely possible (Zhang et al., 1995).

The reaction rate constant and the diffusivity may depend weakly on pressure (see previous section). Because the temperature dependence is much more pronounced and temperature and pressure often co-vary, the temperature effect usually overwhelms the pressure effect. Therefore, there are various cooling rate indicators, but few direct decompression rate indicators have been developed based on geochemical kinetics. Rutherford and Hill (1993) developed a method to estimate the decompression (ascent) rate based on the width of the breakdown rim of amphibole phenocryst due to dehydration. Indirectly, decompression rate (e.g., uplift rate due to erosion) may be also estimated from temperature history if the temperature–pressure relation follows a geotherm. However, high erosion rate may result in substantial deviation from a steady-state geotherm. In such cases, thermokinematic models are necessary to relate uplift and thermal history and to estimate the decompression rate (Reiners and Ehlers, 2005).

1.7.2 *Geochronology, closure age, and thermochronology*

One of the most important geochemical tools is isotope geochronology, i.e., the use of radiogenic isotope systems to determine the age of a mineral or rock. Isotope geochronology is the inverse problem of the kinetics of first-order reactions (radioactive decay). In geochronology, the amount of radioactive parent and radiogenic daughter is measured, and the age is then determined using, e.g., Equations 1-49. The difference between the forward and inverse problems is on which are known and which are to be found. In forward problems, we know the initial conditions and want to predict the future. In inverse problems, we know the current state and want to infer the past (age and the initial conditions). Much of geochemistry is on developing methods to solve inverse problems.

In geochronology, the present concentrations of the parent and daughter nuclides are measured. The age will be determined by either the isochron method (Equation 1-49), or by knowing the initial concentration of the daughter (Equation 1-48c) or the initial concentration of the parent. Use the ^{40}K–^{40}Ar method as an example. ^{40}K undergoes branch decay as follows:

$$^{40}K \Bigg\langle {\substack{^{40}Ar \\ ^{40}Ca}} . \tag{1-112}$$

That is, the decay of ^{40}K (a branch reaction) produces both ^{40}Ar and ^{40}Ca. The fraction of ^{40}K decaying to ^{40}Ar is 0.1048 (10.48%), and that to ^{40}Ca is 0.8952. The total decay constant of ^{40}K is referred to as $\lambda_{40} = 5.543 \times 10^{-10}$ yr^{-1}. The branch decay constant from ^{40}K to ^{40}Ar is $\lambda_e = 0.1048\lambda_{40} = 5.81 \times 10^{-11}$ yr^{-1}. In ^{40}K–^{40}Ar dating, the growth of ^{40}Ar in a mineral is measured. In the simplest treatment, initial ^{40}Ar in a mineral is assumed to be negligible because Ar does not like any crystal structure. If the mineral formed instantaneously and was closed since its formation, the concentration of ^{40}Ar in the mineral is related to the parent concentration as follows (e.g., Equation 1-47c with initial daughter nuclide being 0):

$$^{40}Ar = (\lambda_e/\lambda_{40})\,^{40}K[\exp(\lambda_{40}t) - 1], \tag{1-113}$$

where ^{40}Ar and ^{40}K are molar concentrations of ^{40}Ar and ^{40}K. After measuring $^{40}Ar/^{40}K$, an age can be calculated by solving the above equation for t:

$$t = (1/\lambda_{40})\ln[1 + (^{40}Ar/^{40}K)/(\lambda_e/\lambda_{40})]. \tag{1-114}$$

An example of this calculation is shown below.

Example 1.5 The ^{40}K–^{40}Ar system is applied to determine the age of hornblende. The decay constant of ^{40}K is $\lambda_{40} = 5.543 \times 10^{-10}$ yr^{-1} (half-life of ^{40}K is 1.25 billion years). The branch decay constant from ^{40}K to ^{40}Ar is $\lambda_e = 0.581 \times 10^{-10}$ yr^{-1}. Assume there was no initial ^{40}Ar in the mineral. The measured $^{40}Ar/^{40}K$ ratio at present is 0.1048. Find the age of the mineral.

Solution: Use Equation 1-114, $t = \ln(1 + 1)/\lambda_{40} = 1.25 \times 10^9$ yr, the same as the half-life.

Comment: This result can be understood because when $^{40}Ar/^{40}K = 0.1048$, which is the fraction of the branch decay, it means that the amount of ^{40}K that has decayed away is the same as the present amount. That is, exactly half of ^{40}K decayed away. Thus, the age is the same as the half-life.

Accurate dating of the formation age (or age at the peak temperature) of a rock requires that the system be closed to both the parent and the daughter nuclides right after the formation. Hence, age determination for unaltered volcanic rocks is relatively straightforward, because volcanic rocks cooled and solidified rapidly from magmas (high temperature with homogeneous radiogenic isotopic compositions). (Alteration and weathering may reset ages of older volcanic rocks.) Many igneous rocks formed at high temperatures and metamorphic rocks reached high peak temperatures, but they cooled down slowly and their minerals might have experienced continuous loss of the daughter nuclide. Hence, it is difficult to determine the formation age of slowly cooled igneous rocks or the peak temperature age of metamorphic rocks. Because some isotopic systems (such as U–Pb in zircon) are more retentive than others, they may constrain the formation or peak temperature ages better. Moreover, by using several isotopic systems with different retention abilities, it may be possible to determine the whole cooling history.

The parent nuclide is structurally incorporated in a mineral and hence is less likely lost from the mineral. For the K–Ar method (based on the branch decay of ^{40}K to ^{40}Ar), the concentration of K in the host mineral (such as hornblende) may not change much because it is part of the mineral formula, and diffusion of the parent (K) is expected to be slower than that of Ar because K is charged. The daughter Ar, on the other hand, does not fit into the site of the parent and hence diffuses out. For the Sm–Nd method, the daughter nuclide fits well into the mineral structure, but isotopic exchanges with the surroundings can still occur through diffusion. Even though the decay constant does not depend on temperature, because the loss and isotopic exchange depend on temperature, the age determined from isotopic dating may still depend on thermal history. That is, the age may not be the true age, but an apparent age (called *closure age* t_c). Although the diffusive loss of the daughter nuclide makes isotope geochronology more complicated, geochemists are resourceful in turning complexities into advantages. In this case, the complexity of closure age is used to obtain the full thermal history (temperature versus time), that is, to obtain much more information of the rock than a single formation age.

Therefore, to kineticists and informed geochronologists, the age obtain from an isochron equation or from Example 1-5 is an apparent age, and is called the closure age (Dodson, 1973) because it means the age since the closure of the mineral, not necessarily since the formation of the mineral. The closure age may differ from the true age or formation age because of diffusive loss (or exchange) of the daughter nuclide. For the closure age to be the same as the formation age, the mineral must have cooled down rapidly (for volcanic rocks) or formed at not-so-high a temperature (for metamorphic rocks) so that diffusive loss from the mineral is negligible.

Consider the mineral hornblende, which is formed at a high temperature. There is potassium in hornblende. As ^{40}K decays, 10.48% of it goes to ^{40}Ar. At high temperatures, diffusion rate is high (Arrhenius equation) and the daughter isotope in the system does not necessarily build up according to the decay law. Therefore, if cooling rate is slow, ^{40}Ar produced from the decay of ^{40}K in hornblende grain may be completely or partially lost to the surroundings (which can be the grain boundaries, from where ^{40}Ar can be rapidly transported away due to fast grain-boundary diffusion). For example, at 850°C (a typical rhyolitic or granitic magma temperature), the diffusion coefficient of Ar in hornblende is $7.5 \times 10^{-19}\,m^2/s$ (Harrison, 1981). A volcanic rock may cool down in 1 month, with a diffusion distance $<1.5\,\mu m$. Hence, volcanic hornblende may be regarded to completely retain its radiogenic Ar except for slowly cooled interior of thick lava flows. However, a typical cooling timescale for a plutonic rock is about 1 Myr. A hornblende with radius $\leq 0.5\,mm$ would completely lose its Ar at 850°C, and would only begin to partially accumulate when T is low enough (such as T_1 in Figure 1-20, corresponding to time t_1). When the temperature is very low (such as T_2 in Figure 1-20, corresponding to time t_2), then Ar would be completely retained, and $^{40}Ar/^{40}K$ increases with t roughly linearly with a slope of λ_e, to the present day. The closure age lies between t_1 and t_2 as shown in Figure 1-20. Shallow level intrusions often form sheets and may cool faster, and the retention of Ar in their minerals is between that of volcanic and plutonic rocks.

A geologist who collects the rock now (150 Myr after the formation of the rock) measures $^{40}Ar/^{40}K$, and uses Equation 1-114 to calculate the age. This is equivalent to starting from the endpoint of the solid curve in Figure 1-20b and extrapolating using the slope of the dashed line until $^{40}Ar/^{40}K = 0$, leading to a closure age t_c of 130 Ma. The temperature at the time of t_c is defined to be the closure temperature T_c, which lies between T_1 and T_2. That is, the closure temperature is not the temperature at which the system became closed (which is T_2), nor the temperature at which Ar began to accumulate in the mineral (which is T_1). It is just some apparent temperature corresponding to the closure age. Nonetheless, T_c is uniquely defined but T_1 and T_2 are not. The definition of

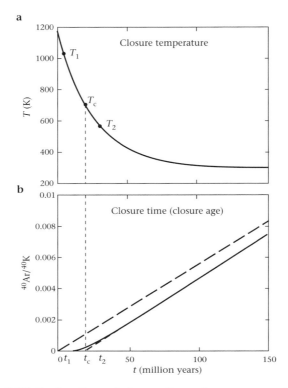

Figure 1-20 Explanation of closure time, closure age, and closure temperature (T_c). (a) The cooling history. (b) ^{40}Ar accumulation history. The long dashed line shows the accumulation of ^{40}Ar if there were no Ar loss since formation. The thin solid curve shows a real accumulation history. The short dashed line shows how the age is obtained from the present-day ^{40}Ar/^{40}K ratio. For a mineral grain cooling down from 1200 K, when the temperature is between 1200 and 1037 K (at 5 Myr), all ^{40}Ar is lost once produced. Then from 1037 to 571 K (at 30 Myr), there is partial loss, and the loss becomes smaller and smaller. Below 571 K, essentially all newly produced ^{40}Ar is retained. When ^{40}Ar/^{40}K ratio is determined, one calculates the age based on the present-day ^{40}Ar/^{40}K ratio and the age corresponds to the time of closure (t_c, about 20 Myr). That is, the age is 130 Ma, although the mineral formed at 150 Ma. The temperature at $t = 20$ Myr is the closure temperature ($\sim$704 K). Adapted from Dodson (1973).

T_1 and T_2 depends on how well we can quantify (e.g., to 1% relative, or 0.01% relative) the beginning of Ar accumulation or cessation of Ar loss.

The closure temperature defined above is related to the diffusion property, the grain size of the mineral, and the "shape" of the crystals as follows (Dodson, 1973):

$$T_c = \frac{E/R}{\ln\left[\dfrac{GAT_c^2}{a^2 qE/R}\right]}, \tag{1-115}$$

or

$$q = \frac{GT_c^2 D_{T_c}}{a^2 E/R},$$
(1-116)

where E and A are the activation energy and pre-exponential factor for diffusion, R is the gas constant, q is the cooling rate when the temperature was at T_c, a is the grain size, and G is the shape factor. Both of the above equations may be referred to as *Dodson's equation*. The shape factor G takes the value of 55, 27, or 8.7 for a sphere (a is the radius), long cylinder (a is the radius) or plane sheet (a is the half thickness), respectively, for isotropic diffusion and slow decay of the parent (i.e., for age much less than the half-life of the parent). The above formulations apply to whole mineral grains. If mineral grains are broken and only the center part is used, the value of G would be different (see Chapter 5). Because minerals are often anisotropic in terms of diffusion, the "shape" of mineral grains means the effective shape accounting for the fastest diffusion direction. For magnetite and garnet, diffusion is isotropic and the grains are roughly spherical, hence the "shape" is sphere. For apatite, the physical shape is often a cylinder, with length greater than the base diameter. Hence, one might think that the effective "shape" would correspond to a cylinder (which means infinitely long and diffusive loss is perpendicular to the **c**-axis). However, for oxygen diffusion in apatite, diffusivity parallel to the **c**-axis is about 3 orders of magnitude larger than that perpendicular to the **c**-axis (Farver and Giletti, 1989). Unless the length is more than 1.5 orders of magnitude larger than the diameter (i.e., length to diameter ratio is greater than 32:1), diffusive loss would be mostly by diffusion parallel to the **c**-axis, and the *effective shape* should be a plane sheet. (Remember that diffusion distance is proportional to square root of diffusivity.) For mica, the effective shape is not the physical shape of a plane sheet, but an infinitely long cylinder for both oxygen and Ar diffusion because oxygen diffusivity perpendicular to the **c**-axis is about 4 orders of magnitude greater than that parallel to the **c**-axis (Fortier and Giletti, 1991), and because Ar diffusivities are obtained assuming this shape (Giletti, 1974; Harrison et al., 1985). For oxygen diffusion in hornblende (Farver and Giletti, 1985), if hornblende length to base diameter ratio is much less than 3.7:1, then it should be treated as a plane sheet; if the ratio is 3.7:1, it can be treated as sphere; and if the ratio is much larger than 3.7:1, it should be treated as cylinder. For Ar diffusion in hornblende, the effective shape is a sphere because Ar diffusivity was obtained assuming this shape (Harrison, 1981).

In summary, when using an isotopic system to determine the age, the meaning of the age is the closure age, as defined in Figure 1-20. The temperature at the time of closure is referred to as the closure temperature (T_c), which varies from one mineral to another. T_c decreases as diffusivity in the mineral increases, or activation energy decreases, or grain size decreases. T_c of some isotopic systems (such as U–Pb in zircon) is high and the isotopic systems hence record an older age that may approach the formation age of igneous rocks, whereas other isotopic systems (such as K–Ar) record younger age corresponding to lower temperatures.

Because the closure temperature depends on the diffusion properties of a mineral and on the isotopic systems, by determining the closure ages of several minerals (sometimes using different isotopic systems), the closure temperature can be found to be a function of closure age. Figure 1-21 shows an example. That is, with the concept of closure temperature and closure age, it is possible to reveal the full temperature–time history of a rock (thermochronology), more than simply the formation age.

Example 1.6 Ar diffusivity in hornblende depends on temperature:

$$D = \exp(-12.94 - 32,257/T)$$

where D is in m^2/s and T is in K. Ar diffusion in hornblende grains is treated as isotropic diffusion in spheres. If radius of hornblende is 0.5 mm and the cooling rate is about 10 K/Myr, find the closure temperature T_c of hornblende.

Solution: Use Equation 1-115. From the conditions given, we can obtain

$$G = 55 \text{(for spheres)}; \ E/R = 32257 \text{ K}; \ A = \exp(-12.94) \text{ m}^2/\text{s} = 2.4 \times 10^{-6} \text{ m}^2/\text{s};$$

$$a^2 = (0.5 \text{ mm})^2 = 2.5 \times 10^{-7} \text{ m}^2; \ q = 10 \text{ K/Myr} = 3.17 \times 10^{-13} \text{ K/s}.$$

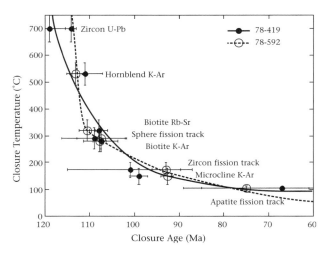

Figure 1-21 Cooling history of two granitoid samples (78–419 and 78–592) from the Separation Point Batholith of New Zealand. Data are from Harrison and McDougal (1980). The emplacement age of the batholith is >115 Ma. The curves are plotted to guide the eyes. It can be seen that cooling rate was high initially and decreases gradually as the temperature approaches surface temperature, as expected. The temperature versus time relation is roughly exponential.

Now T_c can be found by iteration using Equation 1–115. For example, let the first guess of T_c be 800 K, then the first iteration gives $T_c = 848$ K. Use this as the guess to calculate T_c, then $T_c = 845.5$ K. Use this as the guess to calculate T_c again, $T_c = 845.6$ K. This is good enough precision. (If you would like higher precision, you may carry out more iterations.)

1.7.3 Geothermometry, apparent equilibrium temperature, and geospeedometry

In *geothermometry*, geologists use information in a rock to infer the temperature history of a rock. For example, if a rock contains sillimanite, it can be inferred that it must have experienced temperatures above the triple point of sillimanite, kyanite, and andalusite. If a rock contains sanidine, it must have experienced high temperatures and have cooled down rapidly. The appearance of specific minerals only gives a limit on the temperature. To infer the specific temperature, homogeneous or heterogeneous reactions with continuous variation of K on temperature are required.

Use the Fe–Mg order–disorder reaction in an orthopyroxene (Reaction 1-12 with equilibrium constant given by Equation 1-111) again as an example. In a basalt, K in orthopyroxene might be 0.2, and the inferred temperature is 1102 K. In a gabbro, $K = 0.04$ would lead to $T = 611$ K. Although temperature quantification is relatively easy once the dependence of K on T is known, the meaning of the inferred temperature requires clarification. Clearly, they do not mean the present-day temperature of the rocks, which are at room temperature. They mean some temperature that the rock has experienced in the history of the rock. Although thermodynamicists often refer to temperatures as equilibrium temperature or peak temperature or formation temperature, kineticists refer such inferred temperatures as *apparent equilibrium temperatures* (denoted as T_{ae}). In literature, the apparent equilibrium temperature is sometimes referred to as the closure temperature. In this book, the two (apparent equilibrium temperature and closure temperature) are different, with T_{ae} being calculated from a chemical reaction equilibrium, and T_c being inferred from closure to diffusion. The distinction between them will become clearer later.

In the terminology of kinetics, temperatures inferred from all geothermometers are apparent equilibrium temperatures, which become true equilibrium temperatures only when two conditions are satisfied: (i) equilibrium was reached at a high initial temperature or peak temperature, and (ii) subsequent changes to the species concentrations involved were negligible. The first condition is often realized for igneous rocks and for relatively high-grade metamorphic rocks. The second condition is rarely satisfied for homogeneous reactions, except for rapidly quenched volcanic rocks, but is much more often satisfied for heterogeneous reactions. Hence, equilibrium thermometry based on homogeneous reactions is best applied to volcanic rocks. And even then, there may still be some quench effects. For example, even for rapidly quenched volcanic rocks, the apparent equilibrium temperature inferred from the Fe–Mg order–disorder reaction is lower than the magmatic temperatures because of the extreme high reaction rate at temperatures $>1000°C$. For metamorphic and plutonic rocks, application of thermobarometry using homogeneous reactions must be treated with extreme caution. Rather, the inferred apparent equilibrium temperature is an indicator of cooling rate as shown in Figure 1-19.

Despite the complexities, experienced metamorphic petrologists are able to piece together the metamorphic P-T history. One reason is that heterogeneous reactions are usually better in preserving peak temperature and pressure conditions because they are less likely to reset during cooling. Without melting, metamorphic rocks almost never completely re-equilibrate. Some reactions even preserve information on the formation temperature and pressure that predate the peak temperature, allowing inference of prograde metamorphic history. The likelihood to re-equilibrate during thermal evolution depends on the rates of the reaction steps involved. For example, the Fe–Mg exchange reaction between olivine and garnet (Reaction 1-15) is accomplished by diffusion of Fe and Mg between the two phases, and Fe–Mg diffusivities are relatively high. Hence, the reaction may continue during cooling. That means, the thermobarometer based on Fe–Mg exchange between olivine and garnet (Reaction 1-15) (this reaction is mostly a thermometer because its dependence on pressure is weak) may record an apparent temperature different from the formation or peak temperature depending on the peak temperature and cooling rate. On the other hand, the reaction to convert $Mg_2Si_2O_6$ and $MgAl_2SiO_6$ components in pyroxene to garnet (Reaction 1-16) is much slower under dry conditions because it requires the breakdown of pyroxene components to reorganize into garnet. (In the presence of fluid, the reaction can be accelerated through dissolution and reprecipitation.) Hence, the thermobarometer based on Al-in-opx coexisting with garnet (Reaction 1-16) (this reaction is mostly a barometer but also depends on temperature) may record the peak pressure and temperature more faithfully. If two or more phases are involved on one side of the reaction (e.g., exchange reactions, and Reaction 1-17), one kinetic control for the heterogeneous reaction is the phases involved must be able to transfer mass to one another. Hence, in metamorphic rocks, if no fluid is present and two potentially reactive phases are not adjacent to (touching) each other, they would not react and would be able to record a relatively high temperature. Another way to isolate a phase is to include it in a nonreactive mineral, such as inclusions in diamond. By selecting minerals or assemblages, some would record P-T conditions at relatively low retrograde temperature, some would record P-T conditions at relatively high temperature, and some would preserve prograde P-T conditions. It may hence be possible to piece together the whole P-T history. For partially reacted metamorphic reactions such as exchange reactions, petrologists developed clever ways to remove the effect of resetting. For example, reintegration of exsolution lamellae will reveal the composition before exsolution, integration of the compositional profile may reveal the total mass gained by one phase, which can then be added back to the other phase to obtain the "original" composition. The use of mineral inclusions in a noninteractive host may further reveal critical information on rock history. This discussion also suggests that temperatures recorded by different thermobarometers are not necessarily identical, and an understanding of kinetics is necessary to interpret the P-T data.

The concept of apparent equilibrium temperature is simple and is best quantified for homogeneous reactions by following species concentration evolution as a function of temperature and cooling rates through numerical simulations (Figure 1-22). For more complicated reactions such as oxygen isotope exchange between two minerals or Fe–Mg exchange between two ferromagnesian minerals, the meaning of the apparent equilibrium temperature is less clear. Hence, subsequent discussion in this section will focus on homogeneous reactions, and one kind of heterogeneous reaction will be discussed in the next section.

As a phase cools, a homogeneous reaction in the phase proceeds continuously and tries to maintain instantaneous equilibrium. Since reaction rates decrease with temperature, it is impossible to maintain instantaneous equilibrium at lower temperatures and several stages may be distinguished based on the reaction rates. If the initial temperature (T_0) is high enough, the homogeneous reaction reaches equilibrium instantaneously. At some lower temperature, however, the reaction rate is not rapid enough to maintain equilibrium, so the reaction begins to deviate noticeably from equilibrium. This temperature is defined as T_d (deviation temperature) here (Figure 1-22b). At a still lower temperature, the reaction effectively stops and is "quenched." This temperature may be defined as T_q (quench temperature). The phase continues to cool until the final temperature (T_∞) is reached, but no more reaction takes place. The final speciation in the "quenched" phase reflects an apparent equilibrium temperature (T_{ae}), which is defined as the temperature one obtains by calculating the "equilibrium" temperature of the quenched speciation. An *instantaneous apparent equilibrium temperature* is defined as the temperature one obtains by calculating the equilibrium temperature from the instantaneous species concentrations. Figure 1-22a shows a hypothetical cooling history, and Figure 1-22b shows how T_d, T_{ae}, and T_q can be obtained for the given cooling history. The latter figure may be compared with Figure 1-19 to follow the evolution of species concentrations and the apparent equilibrium temperature.

In detail, three stages of the reaction can be distinguished based on Figure 1-22. In the first stage (from T_0 to T_d), the homogeneous reaction rate is so rapid that instantaneous equilibrium is maintained (i.e., departure from equilibrium is negligible) as the temperature decreases. In the second stage (from T_d to T_q), the reaction continues but instantaneous equilibrium is not maintained; departure from instantaneous equilibrium increases with time. In the third stage (from T_q to T_∞), effectively no reaction takes place. Note that T_{ae} is uniquely defined but

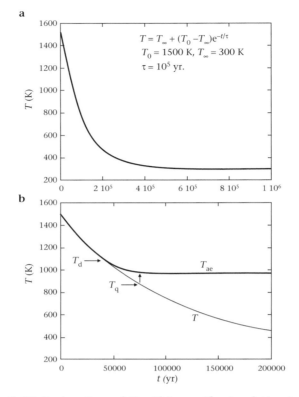

Figure 1-22 Explanation of T_{ae}. This specific simulation is for reaction A$\rightleftharpoons$B with $k_f = 10^9 e^{-30,000/T}$ and $k_b = 10^7 e^{-25,000/T}$. (a) An exponential cooling history. (b) Comparison of actual temperature T and the instantaneous apparent equilibrium temperature (T_{ae}). Note change in timescale of the horizontal axis from (a) to (b). For this specific case, $T_d \approx 1070$ K, $T_q \approx 870$ K, and $T_{ae} = 973.7$ K. To understand the calculation algorithm, it is necessary to quantitatively understand reactions during cooling, discussed in the next chapter. Adapted from Zhang (1994).

T_d and T_q are not; they depend on how precise the measurements are and how "noticeably" and "effectively" are defined. Therefore, even though T_d and T_q are conceptually useful, they do not contain important information and cannot be used in a quantitative fashion. Note also that T_{ae} is neither the colloquially referred "last equilibrium temperature" (the last temperature where equilibrium was maintained, which is T_d), nor "the temperature at which the reaction stopped" (which is T_q). Although T_{ae} is uniquely defined, this temperature does not correspond to any specific event in the rock.

   If a rock cooled down rapidly, it would spend a short time at each temperature interval and hence would be able to reach equilibrium only at relatively high temperatures. Therefore, T_{ae} would be higher. If a rock cooled down slowly, it would have a lower T_{ae}. That is, T_{ae} is an indication of and is positively correlated with cooling rate (q). Although the principle for the relation between T_{ae} and cooling rate q is simple, the quantification of the relation is not. An approximate relation is obtained by Zhang (1994):

$$q \approx \frac{2T_{ae}^2}{\tau_r E/R},$$
(1-117)

where T_{ae} is the apparent equilibrium temperature determined from the homogeneous reaction, R is the gas constant, E is the greater of the forward reaction activation energy and the backward reaction activation energy, q is the cooling rate when the temperature of the rock was T_{ae}, and τ_r is the mean reaction time (see Table 1-2) of the reaction at $T = T_{ae}$.

The concepts of T_{ae} and T_c are similar, although one is for a chemical reaction and the other is for diffusive transport. The T_{ae} is the temperature calculated from species concentrations of a reaction after a cooling history by assuming that the equilibrium was reached under isothermal condition and was never perturbed afterward. The T_c is the temperature at the apparent age that is calculated from a radiogenic system after a cooling history by assuming that the system was always closed. Furthermore, the expressions for T_{ae} (Equation 1-117) and T_c (Equation 1-116) are also similar. Recognizing that a^2/D_{Tc} in Equation 1-116 is a diffusion timescale, one sees that the two equations have the same form, with G in Equation 1-116 equivalent to the factor of 2 in Equation 1-117. The closure temperature is always a calculated property from diffusion parameters and grain size and shape by assuming a cooling rate, but the apparent equilibrium temperature is obtained from measured species concentrations and the expression of the equilibrium constant as a function of temperature without assuming a cooling rate.

1.7.4 Geospeedometry using exchange reactions between two or more phases

The section above deals with homogeneous reactions. Many geothermometers are based on heterogeneous reactions. The kinetics of these reactions are much more difficult to quantify. Hence, the quantitative meaning of the apparent equilibrium temperature is not necessarily known for many of these reactions. For example, one class of geothermometers is based on oxygen isotope exchange between two minerals such as quartz and magnetite, or Fe–Mg exchange between two minerals such as biotite and garnet, which are relatively simple heterogeneous reactions. The significance of the apparent equilibrium temperature for the exchange reactions has been investigated but only recently did it begin to emerge. Below, oxygen isotope fractionation is employed as an example to illustrate the meaning of apparent temperatures inferred from such heterogeneous reactions.

The oxygen isotope fractionation factor α between the two phases depends on temperature as follows (Chiba et al., 1989):

$$\alpha_{qz/mt} = \exp(6290/T^2). \tag{1-118}$$

For San Jose tonalite, $\delta^{18}O_{qz} = 9.7‰$ and $\delta^{18}O_{mt} = 2.0‰$ (Giletti, 1986). Hence, $\alpha_{qz/mt} = (1 + 9.7‰)/(1 + 2.0‰) = 1.0077$ and the apparent equilibrium temperature can be easily calculated to be $T_{ae} = 906$ K $= 633°C$. The expression of the exchange coefficient based on Fe–Mg exchange reaction between biotite and garnet is more complicated than Equation 1-118 due to the effect of other components on the exchange (Gessmann et al., 1997).

As a rock cools down, α varies. On the surfaces of two minerals that are in contact, the isotopic ratios vary in an effort to maintain equilibrium. Diffusion in each phase tries to keep the interior isotopic ratio the same as the surface ratio. However, when temperature is sufficiently low, diffusion becomes too slow to be effective and the mineral may be regarded as closed. Furthermore, at some temperature, interface reaction rate may also be too slow. Therefore, considering two minerals, there are two Dodson closure temperatures (one for each mineral characterizing diffusion in the mineral), and there is also an apparent equilibrium temperature between the two minerals characterizing isotopic equilibrium. Each of the two closure temperatures can be calculated using Equation 1-115, given the diffusion property, cooling rate, grain shape and size of a mineral.

However, when studying a rock, the cooling rate is not given, but is one of the most important parameters to be found. The apparent equilibrium temperature can be calculated from measured isotopic ratios in both minerals given the temperature dependence of the isotopic fractionation factor, such as $\alpha_{A/B} = \exp(A/T^2)$. The meaning of this apparent equilibrium temperature has been investigated in a sequence of papers: an earlier attempt to elucidate the meaning (Giletti, 1986) turned out to be incorrect; and only recently do the meaning of T_{ae} and its application begin to emerge (Eiler et al., 1992, 1993, 1994; Ni and Zhang, 2003). For a bimineralic rock, the apparent equilibrium temperature lies between the two closure temperatures (Eiler et al., 1993). For a multimineralic rock, T_{ae} between most mineral pairs may not have any meaning at all (and in some cases T_{ae} may not even be defined because the fractionation may be reversed), but *the mineral pair with the largest isotopic fractionation always yields a T_{ae} between Dodson closure temperatures of the two minerals* (Ni and Zhang, 2003). If these two minerals happen to have similar diffusion properties such that the two closure temperatures are similar, then T_{ae} would be a good proxy of both closure temperatures. Thus, two cooling rates can be calculated using Equation 1-116 using T_{ae} as T_c, each for one mineral. The true cooling rate would lie between the two cooling rates (Ni and Zhang, 2003). If the metamorphic and plutonic rocks underwent late heating events, there may be resetting, which would complicate the treatment. These conclusions may be applied to other exchange reactions, but may not be applied to some types of heterogeneous reactions such as the barometer reaction of Al-in-opx coexisting with garnet (Reaction 1-16).

1.7.5 Concluding remarks

In summary, because rocks experienced continuous cooling history (i.e., did not equilibrate at a single temperature and quench), whenever the equilibrium constant of a homogeneous or heterogeneous reaction depends on temperature, the equilibrium constant calculated from the measured concentrations in the cooled rock reflects an apparent equilibrium temperature. This temperature may or may not have much temperature meaning. Under favorable conditions, the apparent equilibrium temperature may be used to infer the cooling rate (geo-speedometry). Whenever diffusion in a phase is of concern, there is a closure temperature and the diffusion profile may be used as a geospeedometer. If the diffusion is for a radiogenic species that can be used to date minerals, age and temperature determinations can be coupled as thermochronology.

The concepts of T_{ae} and T_c may be confusing to students. The following brief comparison may help to further clarify the difference between T_{ae} and T_c. The concept of T_{ae} is encountered in treating equilibrium of a reaction (homogeneous or heterogeneous), and that of T_c is encountered in treating diffusion. In a single phase, T_{ae} is encountered in treating homogeneous reactions and describes roughly the temperature at which the equilibrium was reached. T_c is encountered in diffusion (and more generally mass transfer) in a single phase when the species can be lost to the surroundings, and indicates roughly the closure of the phase. Both T_{ae} and T_c are encountered for heterogeneous exchange reactions between two phases, because for both phases the species can go out or come in (and hence there are two closure temperatures), and for the whole reaction there is an equilibrium constant (and hence there is an apparent equilibrium temperature).

1.8 Some Additional Notes

1.8.1 Mathematics encountered in kinetics

The mathematical difficulty increases from homogeneous reactions, to mass transfer, and to heterogeneous reactions. To quantify the kinetics of homogeneous reactions, ordinary differential equations must be solved. To quantify diffusion, the diffusion equation (a partial differential equation) must be solved. To quantify mass transport including both convection and diffusion, the combined equation of flow and diffusion (a more complicated partial differential equation than the simple diffusion equation) must be solved. To understand kinetics of heterogeneous reactions, the equations for mass or heat transfer must be solved under other constraints (such as interface equilibrium or reaction), often with very complicated boundary conditions because of many particles.

Hence, the following chapters will start from homogeneous reactions and proceed to diffusion and mass transfer, and then to heterogeneous reactions.

1.8.2 Demystifying some processes that seem to violate thermodynamics

Some natural processes seem to violate thermodynamics. Thermodynamics states that any spontaneous reaction should result in a lower Gibbs free energy, and the opposite cannot happen. However, some natural processes seem to do just the opposite. There are two kinds of processes that seem to violate thermodynamics. One kind is the formation of a phase that is not the most stable. For example, opal often precipitates from seawater even though quartz is more stable; aragonite sometimes precipitates from seawater even though calcite is more stable. When opal or aragonite forms, each is more stable than the components in seawater, though less stable than quartz or calcite. Hence, this process does not violate thermodynamics at all because thermodynamics states only that the process must lead to lower Gibbs free energy, and it does not state that the product must have the lowest Gibbs free energy. Although the equilibrium state must have the lowest free energy, thermodynamics does not say when the equilibrium state would be reached or whether it will be reached at all.

The second kind is more enigmatic in that the product has higher free energy. For example, photosynthesis produces carbohydrate and O_2, which have higher free energy than the reactants H_2O and CO_2. Another example is the production of ozone in the Earth's stratosphere. These seem to be impossible from a thermodynamic point of view. A closer look at the processes indicates that thermodynamics is not violated (and indeed, thermodynamics cannot be violated), but we may say that thermodynamics is "manipulated" by a good understanding of kinetics. That is, kinetics can make what seems impossible happen. Both nature and humans know how to manipulate thermodynamics and kinetics without violating thermodynamics but achieving the opposite effect to thermodynamics. One mechanism to achieve the feat of creating a less stable species B from a stable species A is as follows: (i) starting from the stable species A, add a lot of energy to make a transient species C that is less stable than both A and B; and (ii) starting from C, it is possible to form either A or B. By controlling some conditions, create B (instead of or in addition to A) from C. Three examples are given below, all based on the above mechanism. The first example is easy to understand, and is discussed to make it easier to understand photosynthesis and the photochemical production of ozone.

1.8.2.1 Conversion of quartz crystals to SiO$_2$ glass

At room temperatures, crystalline quartz is thermodynamically more stable than silica glass. From ancient times, humans have converted quartz to silica glass as follows. First, by adding energy, quartz is heated up and melted. As the melt cools down, both quartz and silica glass are more stable than the melt. Hence, which phase to produce can be manipulated. If the cooling is extremely slow, the more stable form, quartz, would be produced. To produce silica glass, one needs only to cool the melt rapidly. Here the end product may be controlled by varying the cooling rate. Thermodynamics is not violated but quartz is converted to glass. Nature has accomplished similar feats. For example, heating by meteoritic impact/shock or lightening may produce melt that cools rapidly to glass (Taylor, 1967; Essene and Fisher, 1986). Another example is the amorphization of quartz by meteorite impact, which may be altered later at room temperature to tridymite (less stable than quartz at room temperature).

1.8.2.2 Photosynthesis

The most famous example of nature manipulating thermodynamics is probably the photosynthesis reaction, without which most life forms would not be present on the Earth. Thermodynamically, $CO_2 + H_2O$ are much more stable than carbohydrate $+ O_2$. However, nature can convert $CO_2 + H_2O$ to carbohydrate $+ O_2$ through the photosynthesis reaction:

$$CO_2 + H_2O + h\nu \rightarrow (CH_2O) + O_2, \tag{1-119}$$

where (CH_2O) denotes carbohydrate, and $h\nu$ denotes energy from photons. The above reaction is an overall reaction with complicated steps aided by organic and inorganic catalysts. The key to the above reaction is the additional energy $h\nu$ from photons. By analogy to the case of converting quartz to silica glass, $h\nu$ in the photosynthesis reaction plays the same role as heat in converting quartz to silica glass. However, nature accomplishes photosynthesis more elegantly and efficiently than human's conversion of quartz to silica glass. In highly simplified terms and ignoring the various intermediate steps and catalysts, the reaction is accomplished as follows. First, by adding energy ($h\nu$) to $CO_2 + H_2O$, an intermediate with higher energy than both $CO_2 + H_2O$ and carbohydrate $+ O_2$ is produced. Which end product to produce can then be manipulated. Nature manipulates the reaction path so that carbohydrate and O_2 are produced (Figure 1-23). Thermodynamics is not violated although the opposite is achieved.

1.8.2.3 Ozone in the stratosphere

Ozone production in the stratosphere is another example. Ozone concentration in the atmosphere is not the equilibrium concentration (which is negligible). The mechanism of ozone production is as follows:

$$O_2 + h\nu \rightarrow O + O, \tag{1-120}$$

$$O + O_2 \rightarrow O_3. \tag{1-121}$$

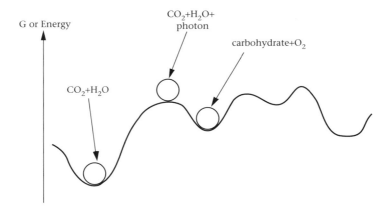

Figure 1-23 Energetics of photosynthesis.

The first step (Reaction 1-120) produces the highly reactive O radical, which can either recombine to form O_2, or react with O_2 to form O_3. Somehow a significant fraction (more than the equilibrium fraction) goes to O_3, often with the help of molecules such as N_2. Hence, the photochemical production of ozone is another example of nature manipulating thermodynamics and kinetics to produce something that "should not be there," similar to photo-synthesis.

1.8.3 Some other myths

To understand kinetics, one must first understand thermodynamics. However, overemphasizing the relation between thermodynamics and kinetics sometimes leads to some misleading statements (which are called myths here). One is that reaction rate is driven by Gibbs free energy difference (ΔG) between reactants and products, and some authors would assume that the reaction rate is proportional to ΔG. Although a reaction happens because of ΔG between reactants and products, the word "driven" should not be interpreted to imply that reaction rate is proportional to ΔG. For a given reaction near equilibrium (i.e., ΔG is small), there may be a rough proportionality between reaction rate and ΔG. However, this cannot be extended to cases when ΔG is large for the same reaction (even at a given temperature; Oelkers, 2001), and cannot be extended to compare rate coefficients of different reactions. For example, ΔG for isotopic exchange reactions is extremely small, orders of magnitude smaller than that for chemical reactions, but isotopic exchange reaction rate coefficients are not orders of magnitude smaller. A second example is that ΔG for nuclear reactions (such as hydrogen burning to helium) is huge, but the reaction is nonexistent at room temperature. The third example is crystal growth rate from a melt as a function of temperature. Just slightly below the melting temperature, the growth rate in-creases as the temperature decreases because of increasing ΔG (rough proportionality between reaction rate and ΔG). However, as the temperature decreases further, ΔG continues to increase, but the crystal growth rate decreases (Figure 1-14). The reaction rate is largely controlled by the activation energy, and may or may not be related to Gibbs free energy difference.

Another myth is that there is a driving force for diffusion. The truth is that diffusion is due to the random motion of atoms. There is no driving force to move atoms along a direction, but random motion is able to produce a directional change of concentrations if there is an initial concentration gradient.

Another myth arises from the intuition that pressure effect is opposite to the temperature effect. This is not true in kinetics. Therefore, kinetic constants (reaction rate constants, diffusion coefficients, etc.) almost always increase with increasing temperature, but they may decrease or increase with increasing pressure. Both positive and negative pressure dependences are well accounted for by the transition-state theory and are not strange.

1.8.4 Future research

Geochemical kinetics is still in its infancy, and much research is necessary. One task is the accumulation of kinetic data, such as experimental determination of reaction rate laws and rate coefficients for homogeneous reactions, diffusion coefficients of various components in various phases under various conditions (temperature, pressure, fluid compositions, and phase compositions), interface reaction rates as a function of supersaturation, crystal growth and dissolution rates, and bubble growth and dissolution rates. These data are critical to geological applications of kinetics. Data collection requires increasingly more sophisticated experimental apparatus and analytical instruments, and often new progresses arise from new instrumentation or methods.

Few homogeneous reactions in minerals and silicate liquids have been investigated in enough detail so that the reaction rate law is obtained, the reaction mechanism is understood, and the rate coefficient as a function of temperature is known accurately enough to allow inference of cooling rates. The difficulty often lies in the measurement of various species concentrations in minerals and melts, such as $^{18}O-^{16}O$ exchange among OH^-, SiO_4^{4-}, PO_4^{3-}, SO_4^{2-}, and other oxygen species in a mineral.

For experimental determination of diffusion coefficients, a large database is already available. Nonetheless, data for specific applications are often difficult to find because the data may not cover the right temperature range, mineral compositions, or fluid conditions. In geospeedometry applications, data often must be extrapolated to much lower temperatures and the accuracy of such extrapolation is difficult to assess. Because the timescale of geological processes is often in the order of Myr, and that of experiments is at most years, instrumental methods to measure very short profile are the key for the determination of diffusion coefficients that are applicable to geologic problems.

For heterogeneous reaction kinetics, both data accumulation and theoretical development are essential. Because of the complexities of kinetics, the theoretical basis for many kinetic problems is not understood yet. One famous example is the nucleation theory. Although the classical nucleation theory is beautiful and it can provide a scheme to fit experimental data, it does not have predictive power. No other nucleation theory that has predictive power is available. Much work is necessary to understand nucleation from first principles. Furthermore, because a nucleus is usually at nanometer scale, it is critical to understand nanomaterial properties. Another famous class of problems is to determine for a given condition (temperature, pressure, and composition) and system, what phase would form first. Sometimes, that phase is not the most stable phase. Examples include the formation of opal (instead of quartz) from seawater, the famous dolomite problem (dolomite is more stable in seawater but calcite forms), the formation of aragonite (calcite is more stable), etc. Many kinetic problems in low-temperature geochemistry (such as weathering, reactions in sediment during diagenesis, reactions in water and oceans) remain to be solved. A major effort is necessary to understand the kinetics of solid-state phase transformations (metamorphic reactions).

Even when the principles of interface reaction and diffusion are thought to be understood, the integrated results may still require major new work. For example, the growth rate of an individual crystal in an infinite melt can be predicted if parameters are known, but the growth rates of many crystals (and different minerals), i.e., the kinetics of crystallization of a magma, is not quantitatively understood.

The many exciting and rich problems call for new talents into geochemical kinetics. It is hoped that this book will encourage students to become geochemical kineticists.

Problems

1.1 Which of the following reactions are homogeneous reactions and which are heterogeneous? Among the homogeneous reactions, which reactions can be guessed to be an overall reaction? Assuming that all other homogeneous reactions are elementary, write down the reaction rate law and specify the molecularity and order.

a. Nuclear hydrogen burning: $4\,^1H \rightarrow {}^4He$.

b. Chemical hydrogen burning: $2H_2(gas) + O_2(gas) \rightarrow 2H_2O(gas)$.

c. Ionization of H_2O: $2H_2O(aq) \rightarrow H_3O^+(aq) + OH^-(aq)$.

d. Alpha decay of ^{147}Sm: $^{147}Sm \rightarrow {}^{143}Nd + {}^4He$.

e. Beta-decay of ^{87}Rb: $^{87}Rb \rightarrow {}^{87}Sr$.

f. Decay chain of ^{238}U: $^{238}U \rightarrow {}^{206}Pb + 8\,^4He$.

g. Part of the decay chain of ^{238}U: $^{238}U \rightarrow {}^{234}U + {}^{4}He$.

h. Irradiation of K-bearing minerals by neutrons: $^{39}K + {}^{1}n \rightarrow {}^{39}Ar + {}^{1}H$.

i. $H_2O(aq) + CO_2(gas) \rightarrow H_2CO_3(aq)$

j. $H_2O(aq) + CO_2(aq) \rightarrow H_2CO_3(aq)$.

k. $H_2CO_3(aq) \rightarrow H_2O(aq) + CO_2(aq)$.

l. $H^+(aq) + HCO_3^-(aq) \rightarrow H_2CO_3(aq)$.

m. $C(diamond) \rightarrow C(graphite)$.

n. $H_3O^+(aq) + OH^-(aq) \rightarrow 2H_2O(aq)$.

o. $2Fe_{0.947}O(wüstite) + 0.4205O_2(gas) \rightarrow 0.947Fe_2O_3(hematite)$.

1.2 Name some kinetic problems that you have already encountered or you may encounter in the future in your research. Then decide whether they belong to (i) kinetics of homogeneous reactions; (ii) mass transfer, or (iii) kinetics of heterogeneous reactions (including phase transformation).

1.3 Calculate ΔG_r° (standard state Gibbs free energy change of the reaction) for the following reactions. Remember that $\Delta G_r^\circ = -RT \ln K$, where K is the equilibrium constant for a given reaction, R is the gas constant, and T is temperature in K. For example, if $K = 0.1$ and $T = 1000$ K, then $\Delta G_r^\circ = 19{,}145$ J. Please be careful; the equilibrium constant for the given reaction may not be the same as α or K_D that are given to you below.

a. Reaction: $H_2O(liquid) + HDO(vapor) \rightleftharpoons H_2O(vapor) + HDO(liquid)$, at $25°C$, $\ln(\alpha_{D/H,\ liq/vapor}) = 24840/T^2 - 76.25/T + 0.0526$, where T is in kelvins.

b. Reaction: $H_2^{18}O(liq) + H_2^{16}O(vapor) \rightleftharpoons H_2^{18}O(vapor) + H_2^{16}O(liq)$ at $25°C$, $\ln(\alpha_{18O/16O, liq/vapor}) = 766/T^2 + 1.205/T - 0.0035$, where T is in kelvins.

c. Reaction: $Mg_2SiO_4(olivine) + 2FeO(melt) \rightleftharpoons Fe_2SiO_4(olivine) + 2MgO(melt)$ at $1300°C$, equilibrium $K_D = (Fe/Mg)_{oliv}/(Fe/Mg)_{melt} = 0.3$. Note that for ideal mixing, $a_{Mg_2SiO_4}^{oliv} = (X_{Mg_2SiO_4}^{oliv})^2$, $a_{Fe_2SiO_4}^{oliv} = (X_{Fe_2SiO_4}^{oliv})^2$, where a is activity and X is mole fraction.

(*Comment*: After you have solved the problem, you should find that ΔG_r° is much smaller for isotopic exchange reactions than for "normal" chemical reactions. Sometimes ΔG_r° for a reaction is called the driving force for the reaction, and the reaction rate is assumed to be proportional to ΔG_r°. Because isotopic reactions are not any slower than chemical reactions, you can see that the driving force concept defined this way is not very helpful.)

1.4 Given a homogeneous reaction $2A + B \rightarrow C$.

   a. Suppose the rate of consumption of A is 0.001 M/s, what is the rate of consumption of B, and the rate of production of C?

   b. Suppose the reaction rate law is $d\xi/dt = k[A][B]^2$. What is the order of the reaction? Is this reaction an elementary reaction or an overall reaction?

   c. Suppose the reaction rate law is $d\xi/dt = k[A][B]$. If $[A] = 0.01$ M, $[B] = 0.005$ M, and $k = 10$ L·mol^{-1}·s^{-1}. Find $d[A]/dt$.

1.5. For a hypothetical homogeneous reaction $2A + B \rightarrow C$, the following data have been obtained (units are arbitrary).

  (i) Dependence of $d[C]/dt$ on [A] at a fixed [B]

[A]	0.01	0.03	0.1	0.3	1	2
$d[C]/dt$	0.0209	0.0631	0.209	0.628	2.10	4.19

  (ii) Dependence of $d[C]/dt$ on [B] at a fixed [A]

[B]	0.01	0.03	0.1	0.3	1	2
$d[C]/dt$	0.0069	0.0209	0.0703	0.208	0.696	1.42

  (iii) $d[C]/dt$ does not depend on concentrations of C or other species.

Answer the following questions.

   a. Find the reaction rate law. Is this reaction elementary?

   b. From the above reaction rate law, given initial concentrations $[A]_0$ and $[B]_0$ with $[A]_0 = 2[B]_0$, find how the concentration would evolve with time.

   c. Using the equation you derive above, calculated and plot how the concentration of A evolves with time for $[A]_0 = 0.01$ M, $[B]_0 = 0.005$ M, and $k = 10$ L mol^{-1} s^{-1}.

   d. What is the half-life of the above reaction?

1.6 For reaction $NO_2(gas) + O_3(gas) \rightarrow NO_3(gas) + O_2(gas)$, the rate constant $k = \exp(18.10 - 2450/T)$ L mol^{-1} s^{-1}, where T is between 230 and 360 K. Calculate k at 25°C.

1.7 The forward reaction rate coefficient for reaction $CO_2(aq) + H_2O(aq) \rightleftharpoons$ $H_2CO_3(aq)$ is 0.002 s^{-1} at $0°C$ and 0.043 s^{-1} at $25°C$. Estimate the activation energy of the forward reaction. (In research, you need more data points to do a linear regression to get the activation energy.)

1.8 The reaction $Fe_{M2}^{2+}(opx) + Mg_{M1}^{2+}(opx) \rightleftharpoons Fe_{M1}^{2+}(opx) + Mg_{M2}^{2+}(opx)$ has been investigated by Wang et al. (2005) for an orthopyroxene sample with $Fe/(Fe + Mg) = 0.011$. Some data (with 2σ errors) are shown in the data table below. Use the data to find the activation energy of the reaction and the Arrhenius expression of k. Use (i) simple linear regression and (ii) the best linear regression method that you can find (the best is the York program).

T (°C)	600 ± 3	700 ± 3	800 ± 3	900 ± 3
$\ln k_f$ (s^{-1})	-15.40 ± 0.32	-10.65 ± 0.04	-7.04 ± 0.46	-4.03 ± 0.22

1.9 The following are some hypothetical data for the decay of ^7Be ($^7Be \rightarrow {}^7Li$) in BeO. Find the reaction rate law and the rate coefficient. What is the order of the reaction? Is the reaction an elementary reaction?

t (day)	0	3	10	20	40	60	80	100	120
$^7Be/^7Be_0$	1	0.9624	0.8800	0.7744	0.5997	0.4644	0.3597	0.2785	0.2157

1.10 Heat conduction distance can be calculated in a similar way as diffusion distance. For example, mean heat conduction distance can be calculated as $(\kappa t)^{1/2}$, where κ is the heat diffusivity. Hence, mean time for heat conduction to reach a distance x (for one-dimensional heat conduction) is x^2/κ. Assume that $\kappa = 0.8 \text{ mm}^2/\text{s}$ for the Earth and for a meteorite. (It is nice that κ does not vary much with temperature or phase.)

 a. Calculate the mid-distance of heat conduction in the Earth over the age of the Earth. Is heat conduction an efficient way for the whole Earth to lose heat?

 b. Assume that the lithosphere thickness is roughly the mid-distance of heat conduction. Estimate the lithosphere thickness for a craton whose age is (i) 1 Ga; (ii) 3 Ga.

 c. When a meteorite passes through the Earth's atmosphere, the surface is heated up by friction with the atmosphere (or due to high-speed collision of atmospheric molecules with the surface of the meteorite). For the purpose of this calculation, assume that the atmosphere is 60 km thick. (The thickness of the atmosphere is not well defined. The atmospheric pressure decreases by 50% every 5-km increase in height.) Assume that a typical meteorite falls to the Earth at 30 km/s. Estimate the timescale for a meteorite to go through the atmosphere. Then estimate the thickness of the surface layer that is melted during its flight through the atmosphere.

1.11 This problem illustrates the applications of heat conduction in cooking. Assume that $\kappa = 0.8 \, mm^2/s$ for an egg (both the white and the yolk)

a. Get a large chicken egg (which actually means a medium chicken egg since in supermarkets all chicken eggs are at or above "medium") and measure its shortest radius.

b. For heat conduction into a sphere (three-dimensional) with a radius of r, when $t = 0.5r^2/\kappa$, the center temperature of the sphere is roughly the same as the surface temperature (the temperature would have changed by 98.6% toward the final temperature). Estimate the time for the center of the egg to be well cooked (to $\geq 98°C$) when you boil a whole large chicken egg.

c. When making a scrambled egg in a frying pan, if you want to cook the egg well but you want to cook it rapidly, what would you do?

1.12 An initially homogeneous garnet crystal with a radius of 10 mm is surrounded by a homogeneous metamorphic fluid that is in equilibrium with the surface of garnet. The fluid can be viewed as an infinite reservoir and is uniform in composition through convection. The garnet crystal is isotropic. Ignore the dissolution of garnet in the fluid. All the problems below can be treated as one-dimensional and semi-infinite diffusion because diffusion distances are much smaller than the radius of the given garnet.

a. Estimate the mean (or "half") diffusion distance into garnet at $t = 1, 10^3, 10^6$ yr for a constant $D = 10^{-14} \, mm^2/s$ at a high temperature.

b. Estimate the time required for the concentration at a position r away from the surface to be roughly midway between the initial and the surface concentration for $r = 1, 10$, and $100 \, \mu m$ for a constant $D = 10^{-14} \, mm^2/s$ at a high temperature.

c. The garnet crystal is cooled down to room temperature. D for Fe–Mg exchange in garnet at room temperature is not well known, but it is roughly estimated to be about $10^{-40} \, mm^2/s$. Calculate mean diffusion distance if $t = 4.5$ billion years (age of the Earth). Compare this with typical atomic layer distance. Is diffusion at room temperature significant?

1.13. You are conducting high-temperature experiments to determine the controlling factor(s) for the dissolution of a mineral in a melt. If there are two or more possibilities, list all of them and discuss how to distinguish the possibilities.

a. If the dissolution distance of a mineral is proportional to the square root of experimental duration, what controls the dissolution rate?

b. If the dissolution distance is proportional to the experimental duration, what controls the dissolution rate?

1.14 For the following substances dissolving in water, estimate whether the dissolution is controlled by interface reaction or by mass transfer.

a. NaBr

b. Olivine

c. Orthopyroxene

1.15 An investigator studied the oxidation kinetics of fayalite (Mackwell, 1992). He measured how the thickness of the oxidized layer depends on time.

a. At $770°C$ in air, the thickness of the oxidized layer changes with time as follows:

Time (h)	1	3	10	30	100
Thickness (μm)	6	11	21	35	65

Please give an approximate expression relating the thickness with time. What do you think controls the oxidation kinetics?

b. At $1030°C$ in CO–CO_2 mixture, the thickness of the oxidized layer changes with time as follows:

Time (h)	0.1	1	4.2	10	69.2
Thickness (μm)	1	14	67	150	all oxidized

Please give an approximate expression relating the thickness with time. What do you think controls the oxidation kinetics?

1.16 For the Fe–Mg order–disorder reaction between M1 and M2 sites in orthopyroxene, $Fe^{2+}_{M2}(opx) + Mg^{2+}_{M1}(opx) \rightleftharpoons Fe^{2+}_{M1}(opx) + Mg^{2+}_{M2}(opx)$, the equilibrium constant $K_D = (Fe/Mg)_{M1}/(Fe/Mg)_{M2} = \exp(0.391 - 2205/T)$.

a. Suppose $K_D = 0.06$. Find the apparent equilibrium temperature.

b. Suppose the relaxation timescale τ at the above equilibrium temperature is 910 yr, and the activation energy for the kinetics of the Fe–Mg order–

disorder reaction is 317.56 kJ/mol. Find the cooling rate of the host rock of orthopyroxene.

c. At which temperature of the rock was the cooling rate the same as calculated above?

1.17 Suppose Ar diffusivity in hornblende is $D = \exp(-12.94 - 32{,}257/T)$ m²/s, and treat hornblende crystals as spheres.

a. Estimate the closure temperature of hornblende if the radius of hornblende is 0.5 mm and the cooling rate is 5°C/Myr.

b. Estimate the closure temperature of hornblende if the radius of hornblende is still 0.5 mm but the cooling rate is 1°C/Myr.

c. Estimate the closure temperature of hornblende if the radius of hornblende is 0.2 mm and the cooling rate is 10°C/Myr.

1.18 Zheng et al. (1998) reported $\delta^{18}O$ measurement of individual minerals in eclogites. The following table shows results for one rock (sample 94M45):

Mineral	Quartz	Phengite	Omphacite	Garnet	Rutile
$\delta^{18}O$	+8.2‰	+6.1‰	+5.3	+4.8	+3.3

Oxygen isotope fractionation factors between mineral pairs may be expressed as $\ln \alpha = A/T^2$ with values of A for various mineral pairs given below:

Mineral pair	Quartz–magnetite	Quartz–rutile	Quartz–garnet	Quartz–diopside	Quartz–mica
A	6290	4690	3150	2750	1370

a. Find the apparent equilibrium temperature between quartz–phengite, quartz–omphacite, quartz–garnet, and quartz–rutile.

b. Which temperature is the most reliable?

2 Kinetics of Homogeneous Reactions

The basic aspects on kinetics of homogeneous reactions were covered in Chapter 1. A one-directional homogeneous reaction such as

$$aA + bB \rightarrow \text{product},$$

may be either an *elementary reaction* if the reaction occurs as written in one step, or an *overall reaction* involving two or more steps or paths. If the reaction is an elementary reaction, which must be verified experimentally, the reaction rate law can be written directly by using the stoichiometric coefficients as exponents on respective species concentrations as

$$d\xi/dt = k[A]^a[B]^b,$$

where ξ is the reaction progress parameter $\{=-d[A]/(a\,dt) = -d[B]/(b\,dt)\}$, t is time, k is the reaction rate constant (or reaction rate coefficient), and [A] and [B] are concentrations of A and B. The order (or the overall order) of the reaction is $a + b$. The order of the reaction with respect to A is a, and that with respect to B is b. If the concentration of A (or B) is constant (e.g., if A is the solvent), then the reaction rate law can be simplified as

$$d\xi/dt = k'[B]^b,$$

where k' is the conditional rate coefficient. Obviously, $k' = k[A]^a$.

The reaction rate constant increases with increasing temperature (except for rare gas-phase reactions with negative activation energies), and may either increase or decrease with increasing pressure. The equation for such dependence is

$$k = A \exp[-(E+P\Delta V)/(RT)],$$

where A is the pre-exponential factor, E is the activation energy of the reaction (which is almost always positive; see Lasaga, 1998, p. 63, for a discussion of negative activation energy), and ΔV is the activation volume of the reaction (which may be either positive or negative).

If reaction $a\mathrm{A} + b\mathrm{B} \rightarrow$ product is an overall reaction, the rate law may differ from $d\xi/dt = k[\mathrm{A}]^a[\mathrm{B}]^b$. It is even possible that the order of the reaction is a fractional number, or there may not be an order at all (e.g., Reaction 1-34). On the other hand, even if the reaction rate law is $d\xi/dt = k[\mathrm{A}]^a[\mathrm{B}]^b$ for reaction $a\mathrm{A} + b\mathrm{B} \rightarrow$ product, the reaction is not necessarily an elementary reaction.

An overall reaction may be accomplished by parallel paths or sequential steps. If it is realized by *sequential steps*, it is called a chain reaction and the slowest step determines the reaction rate. If it is achieved by *parallel paths*, it is called a parallel reaction and the fastest path (more accurately, the sum of the rates of all paths) determines the reaction rate. To distinguish parallel paths versus sequential steps, each of the parallel paths can accomplish the whole reaction alone; but each of the sequential steps cannot.

The half-life of a reactant and the mean reaction time of a reaction are two measures of the time to reach equilibrium. The half-life $t_{1/2}$ is the time for the reactant to decrease to half of its initial concentration, or more generally, the time for it to decrease to halfway between the initial and the final equilibrium concentration. The mean reaction time τ is roughly the time it takes for the reactant concentration to change from the initial value to $1/e$ toward the final (equilibrium) value. The rigorous definition of the mean reaction time τ is through the following equation (Equation 1–60):

$$\frac{d\xi}{dt} = \frac{\xi_\infty - \xi}{\tau},$$

where ξ is the reaction rate progress parameter, and ξ_∞ is the value of ξ at $t = \infty$ (i.e., the reaction has reached equilibrium).

Experiments at a constant temperature are often carried out to investigate the kinetics of a reaction at a high temperature. The rate coefficient is a constant and the rate equation can be solved relatively easily. By varying the temperature of isothermal experiments, the dependence of the rate coefficient on temperature may be obtained.

If a system experienced a complicated thermal history, the rate coefficient would depend on time and the solution to the rate equation would be more complicated. For the special case of reaction kinetics described by one single rate coefficient, the concentration evolution with time can be solved relatively easily.

For continuous cooling, the concept of *apparent equilibrium temperature* (T_{ae}) has been developed, which is the temperature calculated assuming that the species concentrations after cooling down reflect an equilibrium under some isothermal condition. That is, from the species concentrations, an *apparent equilibrium constant* K_{ae} is calculated (it is called the apparent equilibrium constant because the system did not really reach equilibrium at any temperature). Based on $K_{ae} = \exp(A - B/T_{ae})$, T_{ae} is defined as $B/(A - \ln K_{ae})$. The value of T_{ae} increases with increasing cooling rate. If this relation between T_{ae} (or K_{ae}) and cooling rate is quantified, the reaction can be used to be a cooling rate indicator (geospeedometer).

In this chapter, we examine in depth the kinetics of reversible reactions, chain reactions, parallel reactions, and other reactions.

2.1 Reversible Reactions

Many homogeneous reactions, known as reversible reactions, go both ways toward equilibrium. The common exception is the decay of radioactive nuclides. Consider the following reversible reaction

$$A \rightleftharpoons B, \tag{2-1}$$

with equilibrium constant $K = [B]_e/[A]_e$, where subscript "e" means at equilibrium. Based on the principle of microscopic reversibility, in a reversible reaction, the mechanism of the backward reaction is exactly the reverse of the mechanism in the forward reaction. Assume that both the forward and backward reactions are elementary (and hence first-order) reactions. Denote the rate constant for the forward reaction as k_f and that for the backward reaction as k_b. Because the net reaction rate at equilibrium is zero, we have,

$$(d\xi/dt)_e = (d[B]/dt)_e = -(d[A]/dt)_e = k_f[A]_e - k_b[B]_e = 0, \tag{2-2}$$

where subscript "e" means at equilibrium. Therefore,

$$k_f/k_b = K. \tag{2-3}$$

The above relation is a fundamental relation and holds for other types of reactions although it is specifically derived for Reaction 2-1. Hence, when two of the three parameters K, k_f, and k_b are known, the third can be calculated.

2.1.1 Concentration evolution for first-order reversible reactions

The evolution of the concentration of either A or B with time can be obtained by solving an ordinary differential equation. Let the initial concentrations of A and B be $[A]_0$ and $[B]_0$. Using the reaction progress parameter ξ, $[A] = [A]_0 - \xi$, and $[B] = [B]_0 + \xi$. By mass balance $[A] + [B] = [A]_0 + [B]_0$. Hence,

$$d\xi/dt = k_f[A] - k_b[B] = -(k_f + k_b)\xi + (k_f[A]_0 - k_b[B]_0). \tag{2-4}$$

This is a first-order ordinary differential equation and its solution is

$$\xi = \frac{k_f[A]_0 - k_b[B]_0}{k_f + k_b}\{1 - \exp[-(k_f + k_b)t]\}. \tag{2-5a}$$

Hence, the variations of [A] and [B] with time are

$$[A] = [A]_0\,e^{-(k_f+k_b)t} + k_b([A]_0 + [B]_0)\{1 - e^{-(k_f+k_b)t}\}/(k_f + k_b), \tag{2-5b}$$

$$[B] = [B]_0\,e^{-(k_f+k_b)t} + k_f([A]_0 + [B]_0)\{1 - e^{-(k_f+k_b)t}\}/(k_f + k_b). \tag{2-5c}$$

Therefore, [A] approaches exponentially the equilibrium concentration of $([A]_0 + [B]_0)k_b/(k_f + k_b)$ as $t \to \infty$. Similarly, $[B] \to ([A]_0 + [B]_0)k_f/(k_f + k_b)$ as $t \to \infty$. Hence, $[B]_\infty/[A]_\infty = k_f/k_b = K$. That is, the equilibrium condition is satisfied by the solution, as is necessary. Equations 2–5a to 2–5c may be expressed in terms of departure from equilibrium in a general form as

$$\Delta C = \Delta C_0 e^{-(k_f+k_b)t}, \tag{2-6}$$

where C means concentration (either ξ, or [A], or [B]), ΔC means departure from the equilibrium concentration $C - C|_{t\to\infty}$, and ΔC_0 means $C|_{t=0} - C|_{t\to\infty}$.

The time to reach half-equilibrium is $\ln(2)/(k_f + k_b)$. The mean reaction time (or relaxation timescale, or reaction timescale, see Chapter 1) is

$$\tau_r = 1/(k_f + k_b). \tag{2-7}$$

One example of first-order reversible reactions is Reaction 1-7,

$$CO_2(aq) + H_2O(aq) \rightleftharpoons H_2CO_3(aq). \tag{2-8}$$

The reaction rate law is

$$d\xi/dt = k_f[CO_2] - k_b[H_2CO_3].$$

The concentration of CO_2 and H_2CO_3 would change as

$$[CO_2] = [CO_2]_0 e^{-(k_f+k_b)t} + k_b([CO_2]_0 \\ + [H_2CO_3]_0)\{1 - e^{-(k_f+k_b)t}\}/(k_f + k_b). \tag{2-9a}$$

$$[H_2CO_3] = [H_2CO_3]_0 e^{-(k_f+k_b)t} + k_f([CO_2]_0 \\ + [H_2CO_3]_0)\{1 - e^{-(k_f+k_b)t}\}/(k_f + k_b). \tag{2-9b}$$

At 25°C, $k_f = 10^{-1.37}$ s^{-1}, and $K = 10^{-2.54}$ (Table 1-1). The mean time to reach equilibrium is $1/(k_f + k_b) = 1/(10^{-1.37} + 15) = 0.066$ s. Hence, equilibrium is reached rapidly and human eyes will not be able to see the equilibrium in progress (such as gradual color change if a pH indicator is used).

If the reaction occurs in a solution with high concentration of OH$^-$ (high pH), such as bubbling CO_2 into an alkaline solution, $H_2CO_3(aq)$ would dissociate to

HCO_3^- and H^+ (which is rapidly consumed by reaction with OH^-). The dissociation reaction is much more rapid (the rate constant is $10^{6.9}$ s^{-1}) than k_b above, and consumes newly produced H_2CO_3. Because $[H_2CO_3]$ is almost negligible, the reaction may be viewed as unidirectional from CO_2 to HCO_3^-, with a mean reaction time of $1/k_f = 23$ s. This is a long enough time for human eyes to perceive the progress of the reaction (Jones et al., 1964).

2.1.2 Concentration evolution for second-order reversible reactions

For second-order elementary reactions, we similarly have $K = k_f/k_b$. The evolution of concentrations is, however, much more complicated. One specific case, the ionization of water (Reaction 1-9),

$$H_2O \rightleftharpoons H^+ + OH^-, \tag{2-10}$$

is used here as an example to show how the problem can be solved. The forward reaction is pseudo-zeroth order, and the backward reaction is second order. The degree of mathematical complexity of a reversible reaction is determined by the highest order of the forward and backward reactions. The reaction rate law is

$$d\xi/dt = k_f - k_b[H^+][OH^-]. \tag{2-11}$$

At equilibrium at 298 K, $[H^+]_e[OH^-]_e = k_f/k_b = K_w = 10^{-14} M^2$. Knowing $k_f = 0.0015$ M/s, one can find that $k_b = 1.5 \times 10^{11}$ $M^{-1}s^{-1}$ (Table 1-1).

To solve the above equation, it is necessary to express $[H^+]$ and $[OH^-]$ in terms of ξ:

$$[H^+] = [H^+]_0 + \xi, \tag{2-12a}$$

$$[OH^-] = [OH^-]_0 + \xi. \tag{2-12b}$$

Hence, Equation 2-11 can be written as

$$d\xi/dt = k_f - k_b([H^+]_0 + \xi)([OH^-]_0 + \xi). \tag{2-13}$$

That is,

$$d\xi/dt = -k_b\{([H^+]_0 + \xi)([OH^-]_0 + \xi) - K_w\}. \tag{2-14}$$

Before presenting the general method to solve the above ordinary differential equation, let's first consider a simple case with special initial condition.

2.1.2.1 Solution for a special initial condition

For the special case $[H^+]_0 = 0$ and $[OH^-]_0 = 0$,

$$-k_b dt = \frac{d\xi}{(\xi - \sqrt{K_w})(\xi + \sqrt{K_w})} \tag{2-15}$$

Integrating the above equation gives the solution:

$$[H^+] = [OH^-] = \xi = K_w^{1/2}(1 - e^{-2k_b K_w^{1/2}t})/(1 + e^{-2k_b K_w^{1/2}t}).\qquad(2\text{-}16)$$

The equilibrium $[H^+]$ concentration (i.e., as $t \to \infty$) is $K_w^{1/2} = 10^{-7}$ mol/L. The concentration evolution for this special initial condition is shown in Figure 2-1a.

2.1.2.2 General solution

The method for solving Equation 2-14 for the general initial condition is instructive because it is the method to solve all second-order reversible reactions. The first step is to note that the terms inside the braces in the right-hand side of Equation 2-14 is a quadratic form $\xi^2 + ([H^+]_0 + [OH^-]_0)\xi + ([H^+]_0[OH^-]_0 - K_w)$, which can be rewritten as $(\xi - \xi_1)(\xi - \xi_2)$, with ξ_1 and ξ_2 being the two zeros of the quadratic. That is, Equation 2-14 can be rewritten as

$$d\xi/dt = -k_b(\xi - \xi_1)(\xi - \xi_2),\qquad(2\text{-}17)$$

where ξ_1 and ξ_2 are solved from

$$\xi^2 + ([H^+]_0 + [OH^-]_0)\xi + ([H^+]_0[OH^-]_0 - K_w) = 0.\qquad(2\text{-}18)$$

The explicit expression for ξ_1 and ξ_2 is

$$\xi_{1,2}$$
$$= \frac{-([H^+]_0 + [OH^-]_0) \pm \sqrt{([H^+]_0 + [OH^-]_0)^2 + 4(K_w - [H^+]_0[OH^-]_0)}}{2}.\qquad(2\text{-}19)$$

One of ξ_1 and ξ_2 is the value of ξ at equilibrium (referred to either ξ_∞ or ξ_e or ξ_f, where the subscript means "$t = \infty$," "equilibrium," or "final"). Here, ξ_∞ is used because the subscript "f" is used to denote forward reaction, and because "$t = \infty$" may include both equilibrium and complete reactions. Choose ξ_1 to be ξ_∞. Rearrange Equation 2-17 as

$$-k_b dt = \frac{d\xi}{(\xi - \xi_1)(\xi - \xi_2)}.\qquad(2\text{-}20)$$

Integration leads to

$$\ln\frac{\xi_2(\xi - \xi_1)}{\xi_1(\xi - \xi_2)} = -k_b(\xi_1 - \xi_2)t.\qquad(2\text{-}21)$$

Letting $x = \exp[-k_b(\xi_1 - \xi_2)t]$, then

$$\xi = \frac{(x - 1)\xi_1\xi_2}{x\xi_1 - \xi_2}.\qquad(2\text{-}22)$$

Therefore, for a given initial distribution, one needs to find ξ_1 and ξ_2 and then solve for the evolution of concentrations. In numerical calculations one must be

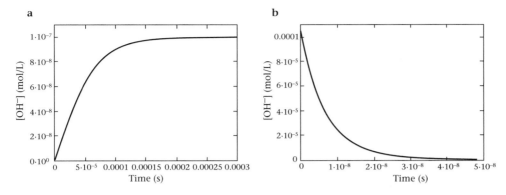

Figure 2-1 The concentration evolution of $[OH^-]$ for two initial conditions. (a) $[H^+]_0 = 0$ and $[OH^-]_0 = 0$. At $t = 3.7 \times 10^{-5}$ s, $[H^+] = 0.5 \times 10^{-7}$ M. That is, the time to reach half-equilibrium is 37 μs. At $t = 2 \times 10^{-4}$ s, $[H^+] = 0.995 \times 10^{-7}$ M. Complete equilibrium of this reaction is reached in 0.2 ms. (b) $[H^+]_0 = 0.001$ M and $[OH^-]_0 = 0.0001$ M. The time to reach half-equilibrium is 4.76 ns, shorter than the half-time (37 μs) for (a) (zero initial concentrations of H^+ and OH^-) by almost 4 orders of magnitude.

careful with truncation errors. If $[H^+]_0 = 0.001$ M $= 1$ mM, and $[OH^-]_0 = 0.0001$ M $= 0.1$ mM, then $\xi_1 = -0.0999999889$ mM, and $\xi_2 = -1$ mM. Calculated $[OH^-]$ concentration evolution is shown in Figure 2-1b.

The concentration evolution curves of Figures 2-1a and 2-1b may be used to estimate the half-life or mean reaction time. When Figures 2-1a and 2-1b are compared, the mean reaction time is found to differ by four orders of magnitude! Hence, for second-order reactions, the timescale to reach equilibrium in general depends on the initial conditions. This is in contrast to the case of first-order reactions, in which the timescale to reach equilibrium is independent of the initial conditions.

The method to find the mean time to reach equilibrium without calculating the full concentration evolution curve is as follows. By comparing the definition of mean reaction time (Equation 1-60) $d\xi/dt = -(\xi - \xi_\infty)/\tau$, and Equation 2-17, the following is obtained:

$$1/\tau = k_b(\xi - \xi_2). \tag{2-23}$$

Therefore, given the initial conditions, first find ξ_2, then the mean reaction time as a function of ξ can be calculated. According to Vieta's formulas, the two roots of Equation 2-18 satisfy

$$\xi_1 + \xi_2 = -([H^+]_0 + [OH^-]_0), \tag{2-24}$$

where ξ_1 is ξ_∞ (ξ at equilibrium). That is,

$$\begin{aligned} \xi_2 &= -(\xi_1 + [H^+]_0 + [OH^-]_0) = -([H^+]_\infty + [OH^-]_0) \\ &= -([H^+]_0 + [OH^-]_\infty). \end{aligned} \tag{2-25}$$

Hence,

$$
\begin{aligned}
1/\tau &= k_b(\xi + [H^+]_\infty + [OH^-]_0) = k_b([H^+]_\infty + [OH^-]) \\
&= k_b([H^+] + [OH^-]_\infty).
\end{aligned}
\tag{2-26}
$$

That is,

$$
\tau = \frac{1}{k_b([H^+]_\infty + [OH^-])} = \frac{1}{k_b([H^+] + [OH^-]_\infty)}.
\tag{2-22}
$$

Hence, as long as the equilibrium concentration is known, the mean reaction time can be calculated at any given species concentration. Although k_f does not appear in the above equation, it does not indicate that the mean reaction time is independent of k_f because the equilibrium concentration depends on the equilibrium constant that equals k_f/k_b.

2.1.2.3 Other second-order reversible reactions

Even though the above example is for a specific reaction, the procedures for solving the differential equation are general for all other kinds of second-order reversible reactions. Table 2-1 lists solutions for concentration evolution of all

Table 2-1 Relaxation timescale and concentration evolution for reversible reactions

Reaction	Concentration expressed by ξ	τ	$\xi(t)$
1. $A \rightleftharpoons B$	$A = A_0 - \xi$ $B = B_0 + \xi$	$\dfrac{1}{k_f + k_b}$	$\xi = \xi_\infty\{1 - \exp[-(k_f + k_b)t]\}$
2. $A + B \rightleftharpoons C + D$	$A = A_0 - \xi;$ $B = B_0 - \xi;$ $C = C_0 + \xi;$ $D = D_0 + \xi$	$\dfrac{1}{k_f(A + B_\infty) + k_b(C + D_\infty)}$	$\ln\dfrac{\xi_2(\xi - \xi_\infty)}{\xi_\infty(\xi - \xi_2)}$ $= (k_f - k_b)(\xi_\infty - \xi_2)t$
3. $A + B \rightleftharpoons 2C$	$A = A_0 - \xi;$ $B = B_0 - \xi;$ $C = C_0 + 2\xi$	$\dfrac{1}{k_f(A + B_\infty) + 2k_b(C + C_\infty)}$	$\ln\dfrac{\xi_2(\xi - \xi_\infty)}{\xi_\infty(\xi - \xi_2)}$ $= (k_f - 4k_b)(\xi_\infty - \xi_2)t$
4. $A + B \rightleftharpoons C$	$A = A_0 - \xi;$ $B = B_0 - \xi;$ $C = C_0 + \xi$	$\dfrac{1}{k_f(A + B_\infty) + k_b}$	$\ln\dfrac{\xi_2(\xi - \xi_\infty)}{\xi_\infty(\xi - \xi_2)} = k_f(\xi_\infty - \xi_2)t$
5. $2A \rightleftharpoons C$	$A = A_0 - 2\xi;$ $C = C_0 + \xi$	$\dfrac{1}{2k_f(A + A_\infty) + k_b}$	$\ln\dfrac{\xi_2(\xi - \xi_\infty)}{\xi_\infty(\xi - \xi_2)} = 4k_f(\xi_\infty - \xi_2)t$

Notes. A, B, C, and *D* are the concentrations of A, B, C, and D, respectively. The subscripts ∞ and 0 mean $t = \infty$ (i.e., at equilibrium) and $t = 0$, respectively. From Zhang (2004).

first- and second-order reactions and the expression of the mean reaction time (Zhang, 1994).

Most first- and second-order reactions can be handled using formulations in Table 2-1. One example is to use results in Table 2-1 to obtain solution for Reaction 2-10 $H_2O \rightleftharpoons H^+ + OH^-$ with $k_f = 0.0015$ M/s and $k_b = 1.5 \times 10^{11}$ M^{-1}+s^{-1}. Reaction 4 in Table 2-1 is of similar type and the solutions may be used with a few transformations. First, rewrite Reaction 2-10 in the same form as Reaction 4 in Table 2-1: $H^+ + OH^- \rightleftharpoons H_2O$ with $k'_f = 1.5 \times 10^{11}$ M^{-1} s^{-1} and $k''_b = 0.0015$ M/s. Second, rewrite the rate law from $d\xi/dt = k'_f[H^+][OH^-] - k''_b$, to $d\xi/dt = k'_f[H^+][OH^-] - k'_b[H_2O]$, where $k'_b[H_2O] = k''_b$, meaning that $k'_b = k''_b/55.51 = 2.7 \times 10^{-5}$ s^{-1}. With these transformations, H^+ corresponds to A, OH^- corresponds to B, H_2O corresponds to species C, and the reaction law of $H^+ + OH^- \rightleftharpoons H_2O$ corresponds to that of $A + B \rightleftharpoons C$ in Table 2-1. Hence, the mean reaction time can be found using the solution in Table 2-1:

$$\tau = \frac{1}{k'_f([H^+] + [OH^-]_\infty) + k'_b} \approx \frac{1}{k'_f([H^+] + [OH^-]_\infty)}, \tag{2-28}$$

where $k'_f = 1.5 \times 10^{11}$ M^{-1}s^{-1} and $k'_b = 2.7 \times 10^{-5}$ s^{-1}. The extra term k'_b in the above equation compared to Equation 2-27 arises from the consideration of H_2O concentration variation (which is negligible) in using the treatment of Reaction 4 in Table 2-1. Because k'_f in the above equation is equivalent to k_b in Equation 2-27, the above equation is the same as the solution of Equation 2-27. Because of symmetry, the above solution can also be written as

$$\tau \approx \frac{1}{k'_f([H^+]_\infty + [OH^-])}. \tag{2-29}$$

For example, if the initial condition is $[H^+]_0 = 0$ and $[OH^-]_0 = 0$, the equilibrium $[OH^-]_\infty$ is easily estimated to be 10^{-7} M. Hence, $\tau = 6.7 \times 10^{-5}$ s at $t = 0$, and decreases to $\tau = 3.3 \times 10^{-5}$ s at $t = \infty$. If the initial condition is $[H^+]_0 = 0.001$ M, and $[OH^-]_0 = 0.0001$ M, the equilibrium $[H^+]_\infty$ can be easily found to be about 0.0009 M. Hence, $\tau = 6.7 \times 10^{-9}$ s at $t = 0$, and increases to $\tau = 7.4 \times 10^{-9}$ s at $t = \infty$. From Equations 2-28 and 2-29, it can be seen that τ can be reduced by increasing $[H^+]_0$ or $[OH^-]_0$. For this second-order reaction, τ depends strongly on the initial conditions, but only weakly on time.

Using the solution for concentration evolution for reaction $A + B \rightleftharpoons C$ in Table 2-1, we obtain the concentration evolution for reaction $H^+ + OH^- \rightleftharpoons H_2O$ to be

$$\ln \frac{\xi_2(\xi - \xi_\infty)}{\xi_\infty(\xi - \xi_2)} = k'_f(\xi_\infty - \xi_2)t. \tag{2-30}$$

Because k'_f in the above equation is equivalent to k_b in Equation 2-21, the above solution is identical to the solution Equation 2-21 except for a negative sign, which is because ξ in the above equation equals $-\xi$ in Equation 2-21.

2.1.2.4 Isotopic exchange reactions

One special type of second-order reactions is isotopic exchange reactions, such as exchange between Fe^{2+} and Fe^{3+} in aqueous solution:

$$^{56}Fe^{2+} + {}^{55}Fe^{3+} \rightleftharpoons {}^{56}Fe^{3+} + {}^{55}Fe^{2+}. \tag{2-31}$$

The forward reaction rate constant is 0.87 $M^{-1}s^{-1}$ (Lasaga, 1981). The backward reaction constant is roughly the same as the forward reaction constant because the isotopic fractionation factor is roughly 1 (the difference from 1 is in the fourth decimal place; e.g., Anbar et al., 2000). The net rate is

$$\frac{d\xi}{dt} = k_f([{}^{56}Fe^{2+}]_0 - \xi)([{}^{55}Fe^{3+}]_0 - \xi)$$
$$- k_b([{}^{56}Fe^{3+}]_0 + \xi)([{}^{55}Fe^{2+}]_0 + \xi). \tag{2-32}$$

Recognizing that $k_f = k_b$, then,

$$\frac{d\xi}{dt} = k_f A - k_f B\xi, \tag{2-33}$$

where $A = [{}^{56}Fe^{2+}]_0[{}^{55}Fe^{3+}]_0 - [{}^{56}Fe^{3+}]_0[{}^{55}Fe^{2+}]_0$, and $B = [{}^{56}Fe^{2+}]_0 + [{}^{55}Fe^{3+}]_0 + [{}^{56}Fe^{3+}]_0 + [{}^{55}Fe^{2+}]_0$. Hence, the kinetics of a second-order isotopic exchange reaction reduces to a first-order kinetics because the forward and backward reaction coefficients are almost identical (but relaxation timescale depends on initial conditions). The solution to the rate equation is

$$\xi = (A/B)[1 - \exp(-k_f Bt)]. \tag{2-34}$$

The above solution is similar to that for first-order reversible reactions (Reaction 1) in Table 2-1, if $k_f([{}^{56}Fe^{2+}]_0 + [{}^{55}Fe^{3+}]_0 + [{}^{56}Fe^{3+}]_0 + [{}^{55}Fe^{2+}]_0)$ in the above equation is identified as $(k_f + k_b)$ in Table 2-1, but the mean reaction time $1/(k_f B)$ depends on the total concentration of ^{55}Fe and ^{56}Fe, which is a characteristic of second-order reactions.

In summary, because the forward reaction rate constant is similar to the back reaction rate constant, the concentration evolution of a second-order isotopic exchange reaction often reduces to that of a first-order reaction (exponential evolution) but the rate "constants" and mean reaction time for the reduced reaction depend on total concentrations.

2.1.3 Reversible reactions during cooling

Geologists often must deal with chemical reactions during cooling. The quantitative aspects for a simple case of reaction kinetics during cooling, and the qualitative aspects for more complicated reactions during cooling, were presented in Chapter 1. In this section, the quantitative aspects of reversible reactions are presented. A simple first-order reversible reaction is used as an example to derive analytical solutions. For second-order reversible reactions under cooling, calculations of the concentration evolution with time, the final species concentrations, and T_{ae} require numerical solutions (Ganguly, 1982; Zhang, 1994).

2.1.3.1 Analytical solution for a first-order reversible reaction

Consider the following first-order reversible reaction (Reaction 2-1):

$$A \rightleftharpoons B. \tag{2-1}$$

From mass balance, $[A] + [B] = [A]_0 + [B]_0$. The reaction rate law is as follows:

$$d[A]/dt = -k_f[A] + k_b[B],$$

$$d[A]/dt = -(k_f + k_b)[A] + k_b[A]_0 + [B]_0), \tag{2-35}$$

where k_f and k_b depend on T and hence on t. Assume that the temperature dependence of k_f and k_b is

$$k_f = A_f e^{-E_f/(RT)}, \tag{2-36a}$$

$$k_b = A_b e^{-E_b/(RT)}. \tag{2-36b}$$

The equilibrium constant is

$$K = [B]_e/[A]_e = A_K e^{-\Delta H/(RT)}. \tag{2-37}$$

Because T depends on t, k_f and k_b depend on t. The analytical solution of Equation 2-35 is complicated:

$$[A] = \left\{ [A]_0 + ([A]_0 + [B]_0) \int_0^t k_b(t') e^{\int_0^{t'} [k_f(t'') + k_b(t'')]dt''} dt' \right\}$$
$$\times e^{-\int_0^t [(k_f(t') + k_b(t')]dt'}. \tag{2-38}$$

To integrate the above equation, the functional form of the dependence of T on t, $T(t)$, must be assumed. Usually, two functions are used, one is referred to as asymptotic cooling, and the other as exponential cooling. The two functions are

$$\text{Asymptotic}: T = T_\infty + (T_0 - T_\infty)/(1 + t/\tau_c), \tag{2-39}$$

$$\text{Exponential}: T = T_\infty + (T_0 - T_\infty)\exp(-t/\tau_c), \tag{2-40}$$

where T_0 and T_∞ are the initial and final temperatures respectively, and τ_c characterizes the cooling timescale. The meaning of τ_c is the time for the temperature to reach the midtemperature of $(T_0 + T_\infty)/2$ in the case of asymptotic cooling (similar to the half-time of cooling), and that for the temperature to reach e^{-1} toward the final temperature (similar to the mean time of cooling). In asymptotic cooling, temperature approaches T_∞ asymptotically. In exponential cooling, temperature approaches T_∞ exponentially. Both lead to smaller cooling rate as the system cools down, which is expected. A third function, $T = T_0 - qt$, where

q is the cooling rate (defined as $-dT/dt$), referred to as linear cooling, was used in earlier years but is not used in this book because (i) constant cooling rate is less realistic than either asymptotic or exponential cooling; (ii) it does not lead to simpler mathematical solutions of kinetic problems (although the function of temperature versus time is simple); and (iii) there is no minimum in temperature (at large t, T would be negative; calculation should be ended before T is negative).

The exponential thermal history is probably the best description of the cooling history of a volcanic or an igneous rock; the asymptotic cooling history is the second best description. Nonetheless, for modeling cooling history, either one would suffice. This is because the most interesting part of the thermal history is a small region near T_c or T_{ae} (such as between T_d and T_q in Figure 1-22), and either Equation 2-39 or 2-40 would fit the small region well. Since reaction progress for most solid-state homogeneous reactions at low temperature (e.g., below 500 K) is negligible (e.g., Example 1-3 in Chapter 1) over the age of the Earth, the assumed value of T_∞ does not affect the calculated final species concentrations significantly. The asymptotic cooling history with $T_\infty = 0$,

$$T = T_0/(1 + t/\tau_c), \tag{2-41}$$

is often used because it leads to simple expressions of k_f and k_b and sometimes easy analytical results.

Combining Equations 2-36a and 2-41, k_f depends on time as

$$k_f = A_f e^{-E_f/(RT)} = A_f e^{-E_f(1+t/\tau_c)/(RT_0)} = k_{f0} e^{-E_f t/(RT_0 \tau_c)} = k_{f0} e^{-t/\tau_f}, \tag{2-42a}$$

where k_{f0} is k_f at $t = 0$ and equals $A_f e^{-E_f/(RT_0)}$, and $\tau_f = RT_0 \tau_c/E_f$ is a time constant for k_f to decrease to $1/e$ of the initial value. Similarly,

$$k_b = k_{b0} e^{-t/\tau_b}, \tag{2-42b}$$

where k_{b0} is k_b at $t = 0$ and equals $A_b e^{-E_b/(RT_0)}$, and $\tau_b = RT_0 \tau_c/E_b$ is a time constant for k_b.

The two parameters, τ_f and τ_b, are very useful in inferring cooling rates and it is helpful to give their expression under a more general cooling history. The expressions of τ_f and τ_b can be obtained using the definition of mean times (Equation 1-60):

$$\frac{dk_f}{dt} = \frac{k_{f\infty} - k_f}{\tau_f}, \tag{2-43}$$

leading to

$$\tau_f = -\frac{k_f}{dk_f/dt} = -\frac{A_f e^{-E_f/(RT)}}{A_f e^{-E_f/(RT)} \frac{E_f}{RT^2} \frac{dT}{dt}} = \frac{RT^2}{E_f q}, \tag{2-43a}$$

where $q = -dT/dt$ and is the cooling rate. Similarly,

$$\tau_b = \frac{RT^2}{E_b q}. \tag{2-43b}$$

The general solution of Equation 2-35, as well as a simple solution for the special case of $E_f = 2E_b$, can be found in Box 2-1 (Zhang, 1994). For the special case of $E_f = 2E_b$, if the initial temperature is high so that the final speciation does not depend on the initial temperature, the final concentration of species A after complete cooling down of the system is

---

Box 2.1 Kinetics of reversible reaction A ⇌ B for cooling history of
  $T = T_0/(1 + t/\tau_c)$.

Using $k_f = k_{f0}e^{-t/\tau_f}$ and $k_b = k_{b0}e^{-t/\tau_b}$, the two integrals in Equation 2–38 may be written as

$$\int_0^t [(k_f(t') + k_b(t')]dt' = \int_0^t [(k_{f0}e^{-t'/\tau_f} + k_{b0}e^{-t'/\tau_b}]dt$$
$$= k_{f0}\tau_f(1 - e^{-t/\tau_f}) + k_{b0}\tau_b(1 - e^{-t/\tau_b})$$

and

$$\int_0^t k_b(t')e^{\int_0^{t'} [k_f(t'') + k_b(t'')]dt''} dt' = \int_0^t k_{b0}e^{-t'/\tau_b}e^{k_{f0}\tau_f(1 - e^{-t'/\tau_f}) + k_{b0}\tau_b(1 - e^{-t'/\tau_b})}dt'.$$

Inserting the above into Equation 2–38, and letting $u = e^{-t/\tau_b}$, $\gamma = \tau_b/\tau_f = E_f/E_b$, $\mu = k_{b0}/k_{f0} = 1/K_0$, $\alpha = [A]_0 e^{-k_{f0}\tau_f - k_{b0}\tau_b}$, and $\beta = k_{b0}\tau_b([A]_0 + [B]_0)$, then u is between 0 and 1 with $u|_{t=0} = 1$, $u|_{t=\infty} = 0$, and $du = -u\,dt/\tau_b$. Equation 2-38 becomes

$$[A] = \left\{ \alpha + \beta \int_u^1 e^{-k_{f0}\tau_f(u'^\gamma + \gamma\mu u')}du' \right\} e^{k_{f0}\tau_f(u^\gamma + \gamma\mu u)}.$$

For the special case of $\gamma = 2$, i.e., if $E_f = 2E_b$ and $\tau_b = 2\tau_f$, the above equation becomes

$$[A] = \left[\alpha + \beta \int_u^1 e^{-k_{f0}\tau_f(u'^2 + 2\mu u')}du' \right] e^{k_{f0}\tau_f(u^2 + 2\mu u)}.$$

Let $\eta^2 = k_{f0}\tau_f(u + \mu)^2$, then $\eta = (k_{f0}\tau_f)^{1/2}(u + \mu)$, $\eta_0 = \eta|_{t=0} = (k_{f0}\tau_f)^{1/2}(1 + \mu)$, $\eta_\infty = \eta|_{t=\infty} = (k_{f0}\tau_f)^{1/2}\mu$, $k_{f0}\tau_f(u^2 + 2\mu u) = \eta^2 - k_{f0}\tau_f\mu^2$, and $d\eta' = (k_{f0}\tau_f)^{1/2}du'$. The above equation becomes

(continued on next page)

(continued from previous page)

$$[A] = \alpha e^{\eta^2 - k_{f0}\tau_f \mu^2} + \beta(k_{f0}\tau_f)^{-1/2} e^{\eta^2} \int_{\eta}^{\eta_0} e^{-\eta'2} d\eta'.$$

Hence,

$$[A] = \alpha e^{\eta^2 - k_{f0}\tau_f \mu^2} + \frac{1}{2}\sqrt{\pi}\beta(k_{f0}\tau_f)^{-1/2} e^{\eta^2}[\mathrm{erf}(\eta_0) - \mathrm{erf}(\eta)],$$

where $\mathrm{erf}(\eta)$ is an error function (Appendix 2), defined as

$$\mathrm{erf}(\eta) = \frac{2}{\sqrt{\pi}} \int_0^{\eta} e^{-u^2} du.$$

Further simplification leads to

$$[A] = \{[A]_0 e^{\eta^2 - \eta_0^2} + \sqrt{\pi}\eta_{\infty}([A]_0 + [B]_0)e^{\eta^2}[\mathrm{erf}(\eta_0) - \mathrm{erf}(\eta)]\}.$$

Often we are only interested in the concentration of A as time approaches infinity (i.e., the freezing-in concentration in the rock that we collect and investigate), which is

$$[A]_{\infty} = \{[A]_0 e^{\eta_{\infty}^2 - \eta_0^2} + \sqrt{\pi}\eta_{\infty}([A]_0 + [B]_0)e^{\eta_{\infty}^2}[\mathrm{erf}(\eta_0) - \mathrm{erf}(\eta_{\infty})]\}.$$

If T_0 is high enough, equilibrium is maintained at the high temperature during cooling, meaning that the final A_{∞} does not depend on the initial species concentration $[A]_0$, leading to $[A]_0\exp(\eta_{\infty}^2 - \eta_0^2) \approx 0$ and $\mathrm{erf}(\eta_0) \approx 1$. Hence,

$$[A]_{\infty} \approx ([A]_0 + [B]_0)\sqrt{\pi}\eta_{\infty}\exp(\eta_{\infty}^2)\mathrm{erfc}(\eta_{\infty}),$$

where $\mathrm{erfc}(\eta) = 1 - \mathrm{erf}(\eta)$.

$$[A]_{\infty} \approx ([A]_0 + [B]_0)\sqrt{\pi}\eta_{\infty}\exp(\eta_{\infty}^2)\mathrm{erfc}(\eta_{\infty}), \tag{2-44}$$

where

$$\eta^2 = k_{f0}\tau_f(e^{-t/\tau_b} + k_{b0}/k_{f0})^2, \tag{2-45a}$$

$$\eta_0^2 = k_{f0}\tau_f(1 + k_{b0}/k_{f0})^2, \tag{2-45b}$$

$$\eta_{\infty}^2 = k_{f0}\tau_f(k_{b0}/k_{f0})^2. \tag{2-45c}$$

Equation 2-44 holds if η_0 is large so that $\mathrm{erf}(\eta_0) = 1$ and $\exp(\eta_{\infty}^2 - \eta_0^2) \approx 0$. The erfc function in Equation 2-44 is summarized in Appendix 2.

The parameter η can also be expressed as

$$\eta = \sqrt{\frac{A_f \tau_c R T_0}{E_f}} \left\{ \frac{1}{A_K} + \exp\left[-\frac{E_b}{RT_0}\left(1 + \frac{t}{\tau_c}\right)\right]\right\}, \tag{2-46}$$

$$\eta_0 = \sqrt{\frac{A_f \tau_c R T_0}{E_f}} \left\{ \frac{1}{A_K} + \exp\left[-\frac{E_b}{RT_0}\right]\right\}, \tag{2-47}$$

and

$$\eta_\infty = \frac{1}{A_K} \sqrt{\frac{A_f \tau_c R T_0}{E_f}}. \tag{2-48}$$

The above analytic solution has two applications: (i) to investigate the concentration evolution under the special conditions given above, and (ii) to check the accuracy of numerical programs. One application is given in Example 2-1.

Example 2.1 For reaction $A \rightleftharpoons B$, if $T_0 = 1500\,K$, $K = 2000\,\exp(-7500/T)$, $k_f = 40\,\exp(-15{,}000/T)$, $k_b = 0.02\,\exp(-7500/T)$, $T = 1500/(1 + t/10^6)$ with t in years and k_f and k_b in year^{-1}, find T_{ae}.

Solution: From the conditions given, the cooling history follows asymptotic cooling with $T_\infty = 0$, and $E_f/R = 15{,}000\,K = 2E_b/R$. Hence, the conditions for using the solution of Equations 2-46 to 2-48 are satisfied. First find η_∞ and η_0 using Equations 2-47 and 2-48 with $A_f = 20\,y^{-1}$, $A_K = 2000$, $\tau_c = 10^6\,$yr, $T_0 = 1500\,K$, and $E_f/R = 15{,}000\,K$. The results are

$$\eta_\infty = 1.00 \quad \text{and} \quad \eta_0 = 14.476$$

Hence, $\mathrm{erf}(\eta_0) = 1$, and $\exp(\eta_\infty^2 - \eta_0^2) \approx 0$. Using Equation 2-44, $[A]_\infty \approx 0.7579([A]_0 + [B]_0)$. Hence, $K_\infty = [B]_\infty/[A]_\infty = (1 - 0.7579)/0.7579 = 0.3195$. Therefore,

$$T_{ae} = 7500/\ln(2000/K_\infty) = 858\ K.$$

The mean times for the kinetic constants to decrease (τ_f and τ_b) may be thought to be roughly the same as the mean time for the reaction at T_{ae}. That is, one may intuitively guess that

$$\tau_r|_{T_{ae}} \approx \max(\tau_f, \tau_b). \tag{2-49}$$

In the above relation, if τ_f and τ_b are not independent of temperature, it is understood that they mean τ_f and τ_b at $T = T_{ae}$. Using the analytical solutions for the case of $E_f = 2E_b$ it may be shown that this is roughly the case. Because the kinetics of a reaction does not depend on which direction is called the forward reaction, the relation also holds for $E_b = 2E_f$. For other conditions, the relation is not accurate. Based on numerical simulations (Zhang, 1994), the following expression, which is a modification of the above equation by a small factor of $2 \min(E_f, E_b)/\max(E_f, E_b)$, is a good approximation:

$$\tau_r|_{T_{ae}} \approx \max(\tau_f, \tau_b) \frac{2 \min(E_f, E_b)}{\max(E_f, E_b)} = \frac{2 T_{ae}^2}{qE/R}, \tag{2-50}$$

where E is understood to be the greater of the activation energies for the forward and backward reactions. The above is Equation 1-117 for the inference of cooling rates.

2.1.3.2 Numerical solutions for second-order reversible reactions

For the kinetics of second-order reversible reactions (Reactions 2 to 5 in Table 2-1) under variable temperature, an analytical solution is not available. The evolution of species concentrations may be calculated through numerical methods. Consider the following second-order reversible reaction as an example:

$$A + B \rightleftharpoons C + D. \tag{2-51}$$

Given the initial concentrations $[A]_0$, $[B]_0$, $[C]_0$, and $[D]_0$, the species concentrations are related to the reaction progress parameter as

$$[A] = [A]_0 - \xi, \tag{2-52a}$$

$$[B] = [B]_0 - \xi, \tag{2-52b}$$

$$[C] = [C]_0 + \xi, \tag{2-52c}$$

$$[D] = [D]_0 + \xi. \tag{2-52d}$$

The reaction rate law is

$$d\xi/dt = k_f[A][B] - k_b[C][D]. \tag{2-53}$$

That is,

$$d\xi/dt = k_f([A]_0 - \xi)([B]_0 - \xi) - k_b([C]_0 + \xi)([D]_0 + \xi), \tag{2-54}$$

with the initial condition of $\xi = 0$. Given the values of $[A]_0$, $[B]_0$, $[C]_0$, and $[D]_0$, a temperature history such as Equation 2-39 or Equation 2-40, and the dependence of k_f and k_b on temperature (such as Equations 2-36a and 2-36b), Equation 2-54 can be solved numerically using, e.g., the Runge-Kutta method. For other reactions, the relevant reaction laws can also be integrated numerically from the initial concentrations, the initial temperature, the cooling history and the dependence of k_f and k_b on temperature.

In numerical solutions, the calculation is usually ended when the temperature is about 280 K (surface temperature of the Earth), or when the temperature is $\leq T_\infty + 1$ K because the temperature of T_∞ would take an infinite amount of time to reach. Another built-in control might be to end the calculation when t is the same as the age of the rock. When the age is not available, the calculation should

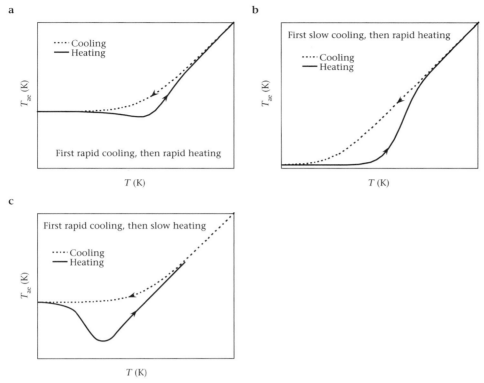

Figure 2-2 Calculated cooling–heating behavior for a second-order reversible reaction. The equilibrium and kinetic parameters are from Reaction 2-55. From Zhang (unpublished).

continue no more than $t = 4.5$ billion years. Some numerical results may be found in Zhang (1994).

From numerical solutions, results shown in Figures 1-19 and 1-22 may be obtained, which illustrates the behavior of kinetics during cooling and the concept of the *apparent equilibrium temperature* (T_{ae}). It is also possible to investigate the behavior of a reaction during a cooling and heating cycle. To gain familiarity with reaction kinetics during cooling and heating, Figure 2-2 shows some calculated results expressed as T_{ae} at any instant as a function of real system temperature. Recall that T_{ae} is defined to be the apparent temperature calculated from the instantaneous species concentrations as if the concentrations were equilibrium concentrations at the temperature. For example, from the calculated species concentrations of Reaction 2-51 using the above procedure, the apparent equilibrium constant K_{ae} is calculated as [C][D]/{[A][B]}. If reaction is rapid and equilibrium is reached at the temperature, this apparent equilibrium constant is the true equilibrium constant. Otherwise, it is simply the reaction quotient. Because the equilibrium constant depends on temperature $K = \exp(a - b/T)$, meaning by definition $K_{ae} = \exp(a - b/T_{ae})$, the apparent equilibrium temperature at any instant may be calculated.

All three figures in Figure 2-2 show hysteresis in a cooling–heating cycle. The cooling curves (dashed curves in Figure 2-2) depend on cooling rate but do not depend on thermal history of the sample because the initial temperature is made high enough so that equilibrium is reached instantaneously. For example, the curves for rapid cooling in Figures 2-2a and 2-2c are identical. On the other hand, the heating curves (the solid curves in Figure 2-2) depend on both the heating rate and the thermal history because the starting temperature is low and there is no equilibrium at the beginning. Hence, the beginning T_{ae} for a given heating rate may differ due to difference in prior cooling rate. For example, the rapid heating curve in Figure 2-2a differs from that in Figure 2-2b. The large decrease in T_{ae} during slow heating in Figure 2-2c might appear strange and is hence explained here. Because the sample was cooled rapidly, the quenched species concentrations correspond to a high apparent equilibrium temperature. For clarity of discussion, suppose K for Reaction 2-51 increases with temperature, meaning the concentrations of C and D increase with temperature. Because of rapid prior cooling, at the beginning of heating, T_{ae} is high (also for clarity of discussion, suppose this high T_{ae} is 1000 K), meaning high concentrations of C and D. When the sample is heated up slowly, at each temperature, there is more time to react compared with the reaction time during prior cooling. Hence, even though during prior cooling the reaction roughly stopped at 950 K, during heating up, the reaction would be noticeable at 850 K. The equilibrium concentrations of C and D at 850 K are much smaller than the initial concentrations of C and D (i.e., those at $T_{ae} = 1000$ K). Therefore, Reaction 2-51 goes to the left-hand side, leading to a decrease in T_{ae}. Given a prior cooling rate, the slower the heating rate, the larger drop in T_{ae} there would be. These results will also be helpful for us to understand glass transition and inferences of cooling rates in nature.

2.1.3.3 Homogeneous reactions that have been investigated

Only two high-temperature homogeneous reactions have been investigated in detail for their kinetics by geochemists. One is the Fe–Mg order–disorder reaction in orthopyroxene, and the other is the hydrous species interconversion reaction in rhyolitic melt. The two reactions have been applied as *geospeedometers* in various geochemical and meteoritic problems. Because they are often encountered in geochemical kinetics literature, the two reactions are discussed in depth below.

Some other intracrystalline exchange reactions have also been investigated to some extent, such as Fe, Ni, and Mg exchange between M1 and M2 sites in olivine (Ottonello et al., 1990; Henderson et al., 1996; Redfern et al., 1996; Heinemann et al., 1999; Merli et al., 2001), Fe and Mg exchange between M1 + M2 + M3 and M4 sites in amphibole (Ghiorso et al., 1995), order–disorder reaction for Mg and Al, or for Mg and Fe^{3+}, between the tetrahedral and octahedral sites (O'Neill, 1994; Harrison and Putnis, 1999; Andreozzi and Princivalle, 2002), and ^{18}O and ^{16}O exchange reaction between OH^- groups and SO_4^{2-} groups in alunite,

$KAl_3(SO_4)_2(OH)_6$ (Stoffregen et al., 1994a,b). However, the understanding of these reactions is not enough yet for geospeedometry applications of natural samples. Many other intracrystalline exchange reactions and silicate melt reactions have not been investigated because the analytical techniques are not available.

2.1.4 Fe–Mg order–disorder reaction in orthopyroxene

2.1.4.1 General consideration

Fe^{2+}–Mg^{2+} order–disorder reaction in orthopyroxene (opx), i.e., Reaction 1-12,

$$Fe^{(M2)}Mg^{(M1)}Si_2O_6(opx) \rightleftharpoons Mg^{(M2)}Fe^{(M1)}Si_2O_6(opx), \tag{2-55}$$

is a reversible homogeneous reaction. In the above distribution reaction of Fe^{2+} and Mg^{2+} between M1 and M2 sites, because Fe^{2+} prefers M2 site, the forward reaction (increasing Fe in M1 site) is the disordering reaction and the backward reaction (increasing Fe in M2 site) is the ordering reaction. Another notation of the above reaction is

$$Fe^{(M2)}(opx) + Mg^{(M1)}(opx) \rightleftharpoons Mg^{(M2)}(opx) + Fe^{(M1)}(opx). \tag{2-55a}$$

This reaction has been investigated extensively by high-temperature geochemists. Maintaining the consistency to use capital K for equilibrium constants and lower-case k for kinetic properties, the exchange coefficient is denoted as K_D:

$$K_D = \frac{X_{Mg}^{M2} X_{Fe}^{M1}}{X_{Fe}^{M2} X_{Mg}^{M1}} = \frac{(Fe/Mg)_{M1}}{(Fe/Mg)_{M2}}. \tag{2-56}$$

Some authors (e.g., Stimpfl et al., 1999) suggested denoting the above equilibrium coefficient as k_D (lowercase k) because it is an intracrystalline reaction (as opposed to intercrystalline reactions), and K as reaction rate constants of intracrystalline reactions. The suggestion is not adopted in this book because k is used to denote kinetic coefficients.

If mixing in each site is not ideal, K_D would differ from the real equilibrium constant by the quotient of activity coefficients and hence may depend on composition. The measurement of the site occupancy (the fraction of Fe and Mg in each of M1 and M2 sites) is not trivial. There are two methods to determine the intracrystalline site distribution. One is by Mössbauer spectroscopy (MS), in which there are a pair of outer and smaller peaks, which are due to Fe in M1 site, and a pair of inner and larger peaks, which are due to Fe in M2 site (Figure 2-3). The ratio of Fe in M1 site to Fe in M2 site is assumed to be the area ratio of the pair of M1 peaks to the pair of M2 peaks. Using total Fe content from electron microprobe analysis, and the ratio from Mössbauer spectroscopy, Fe(M1) and Fe(M2) concentrations can be obtained.

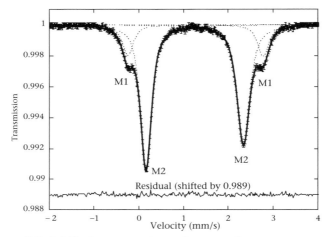

Figure 2-3 A Mössbauer spectrum for an orthopyroxene sample with ferrosilite mole fraction of 1.1% equilibrated at 900°C. The solid curve is the fit to the spectrum, and the dashed curves are individual fit lines. The residual is also shown (shifted upward by 0.989 units so that it can be shown clearly). The area ratio of two M1 peaks to two M2 peaks is 0.2367 ± 0.0055. Fe^{3+} is negligible. The cation distribution (or formula) of this orthopyroxene (4 cation and 6 oxygen basis) is as follows: tetrahedral site, $Al_{0.018}Si_{1.982}$; M1, $Fe_{0.0041}Mg_{0.9776}Ti_{0.0007}Al_{0.0176}$; M2, $Ca_{0.0016}Fe_{0.0172}Mg_{0.9803}$. There is also minor amount of Na (0.0004) and Mn (0.0005) in M2 site. From Wang et al. (2005).

The second method of measuring the site occupancy of Fe and Mg in M1 and M2 sites in orthopyroxene is by X-ray diffraction (XRD). In this method, single-crystal X-ray reflection intensity data are collected. The positions (angles) of X-ray reflections are determined by the structure, but the intensities are influenced by concentrations of the individual elements in each site. Hence, with structure refinement, information on elemental distribution between M1 and M2 sites may be obtained.

In both the MS and XRD methods, the sites of other elements must be assigned. The calculation procedure is typically as follows. Given analytical data, calculate the orthopyroxene formula on a four-cation basis. If the number of oxygen is not 6, then some Fe is allowed to be ferric so that the total number of oxygen is 6. Si is assigned to the tetrahedral site. If there is not enough Si to fill the tetrahedral site, then Al is used to fill the site. The rest of Al is assigned to M1 site. Cr, Fe^{3+}, and Ti are also assigned to M1 site. Ca, Na, and Mn are assigned to M2 site.[1] In the MS method, the ratio of Fe^{2+} in the M1 site to that in the M2 site is assumed to be the same as the area ratio of M1 peaks to M2 peaks (Figure 2-3)

---

[1] Sometimes Mn is assigned to both M1 and M2 sites similar to Fe^{2+}. However, Stimpfl (2005) showed experimentally that Mn should be assigned to the M2 site, as is the case for the orthopyroxene mineral donpeacorite, $(Mn, Mg)(Mg, Fe)Si_2O_6$ (Petersen et al., 1985).

and the distribution of Fe^{2+} and Mg is calculated from the ratio. In the XRD method, Fe and Mg are allowed to distribute between M1 and M2 sites and the fractions of Fe^{2+} and Mg in both sites are varied by maintaining mass balance so that calculated X-ray intensities can match the observed X-ray intensities. Unfortunately, the MS measurement and XRD measurement of the same sample do not give the same concentrations (such as concentration of Fe^{2+} in M1 site), although each technique is self-consistent.

The equilibrium and kinetics of the Fe–Mg order–disorder reaction have been investigated experimentally and theoretically by a number of authors (Mueller, 1969; Virgo and Hafner, 1969; Saxena and Ghose, 1971; Besancon, 1981; Ganguly, 1982; Saxena, 1983b; Anovitz et al., 1988; Molin et al., 1991; Yang and Ghose, 1994; Kroll et al., 1997; Stimpfl et al., 1999; Schlenz et al., 2001; Zema et al., 2003; Wang et al., 2005). The experimental results are complicated and confusing. One source of confusion is due to inconsistency between MS data and XRD data. For example, K_D value obtained from MS measurement is 10 to 25% greater than that obtained from XRD measurements. Hence, comparisons and applications must be based on one or the other technique. Another source of confusion is due to large scatters of K_D even with the same technique.

Early data using the MS method seemed to indicate strong compositional dependence of K_D, especially at low Fe or low Mg contents, which is strange because at low concentrations of one component it is expected that the mixing approaches constant activity coefficient for the component (Henry's law) and ideality for the other component (Raoult's law), leading to composition-independent K_D. Recent work removed this confusion and showed that K_D for this reaction is almost independent of orthopyroxene composition from X_{Fs} of 0.01 to at least 0.6 (Wang et al., 2005), where X_{Fs} is the mole fraction of ferrosilite component and equals $(X_{Fe}^{M1} + X_{Fe}^{M2})/2$. Previously observed compositional dependence of K_D is attributed to low accuracy at low Fe content or to instability of ferrosilite at ambient pressures at low Mg content. Using the MS data, K_D depends on temperature as follows (Wang et al., 2005):

$$K_D = \exp(0.391 - 2205/T). \tag{2-57}$$

Figure 2-4 shows some of the new and old data. Uncertainty in K_D at the 2σ level (σ is standard deviation) is about 10% relative.

Data obtained by the XRD method were equally confusing. Many equations for the expression of K_D have been reported in the literature. The most recent results apparently still indicate that K_D is a step function of X_{Fs} (Stimpfl et al., 1999): For X_{Fs} between 0.19 and 0.75, K_D depends on temperature as follows:

$$K_D = \exp(0.547 - 2557/T). \tag{2-58a}$$

For X_{Fs} between 0.11 and 0.17, K_D depends on temperature as follows:

$$K_D = \exp(0.603 - 2854/T). \tag{2-58b}$$

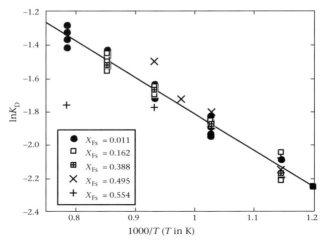

Figure 2-4 K_D values for the Fe–Mg order–disorder reaction in orthopyroxene as a function of temperature for selected compositions. The line is a fit to the data for $X_{Fs} = 0.011$ and 0.162 (Equation 2-57). Data sources are as follows: $X_{Fs} = 0.011$ and 0.162 (Wang et al., 2005); $X_{Fs} = 0.388$ (Anovitz et al., 1988); $X_{Fs} = 0.495$ (Besancon, 1981); $X_{Fs} = 0.554$ (Virgo and Hafner, 1969). Data for $X_{Fs} > 0.60$ are not included to avoid complexities due to the decomposition of orthopyroxene. From Wang et al. (2005).

The values of K_D from the two expressions differ by about 20%. Most likely, this sudden jump of K_D at $X_{Fs} \approx 0.18$ is an artifact. When applying the orthopyroxene geospeedometer, it is necessary to choose the appropriate relation of K_D.

Given K_D or temperature, the calculation of equilibrium concentrations from initial concentrations or from bulk compositions, which is prerequisite for any kinetic calculation, is shown in Box 2-2.

We now turn to the reaction kinetics. The reaction is usually assumed to be an elementary reaction, and the reaction rate law is written as

$$\frac{dX_{Fe}^{M1}}{dt} = k_f X_{Fe}^{M2} X_{Mg}^{M1} - k_b X_{Fe}^{M1} X_{Mg}^{M2}. \tag{2-59}$$

Because X_i^j's are dimensionless (mole fractions), the unit of k_f and k_b for this second-order reaction is s^{-1} (not $M^{-1} s^{-1}$). Some recent experimental kinetic data are shown in Figure 2-5 on how equilibrium is reached from both sides (Wang et al., 2005). The data show that for the ordering reaction (solid circles in Figure 2-5), the concentration variation toward the equilibrium value is monotonic. However, for the disordering reaction (open symbols in Figure 2-5), the concentration variation toward equilibrium is not monotonic. If real, the nonmonotonic behavior indicates that Reaction 2-55 is not an elementary reaction, or Fe^{3+} participates in the reaction. Ignoring the nonmonotonic behavior, the reaction rate coefficient can be obtained at a fixed temperature and a given bulk composition. The reaction rate coefficient depends not only on temperature, but

Box 2.2 Calculation of the equilibrium species concentrations of the Fe–Mg order–disorder reaction in orthopyroxene

This reaction is of type 2 in Table 2-1. For simplicity, use the simple notation of

$$A = X_{Fe}^{M2}; \quad B = X_{Mg}^{M1}; \quad C = X_{Fe}^{M1}; \quad \text{and} \quad D = X_{Mg}^{M2}.$$

Two different sets of concentration conditions may be given. If the initial concentrations of all four species are given, the concentrations after any degree of reaction may be expressed as

$$A = A_0 - \xi; \quad B = B_0 - \xi; \quad C = C_0 + \xi; \quad \text{and} \quad D = D_0 + \xi,$$

where ξ is the reaction progress parameter. At equilibrium,

$$K_D = \frac{CD}{AB} = \frac{(C_0 + \xi)(D_0 + \xi)}{(A_0 - \xi)(B_0 - \xi)},$$

where subscript 0 means the initial concentration. Only one unknown (ξ) is in the above equation. Rearrange the above to the standard quadratic form:

$$(1 - K_D)\xi^2 + [K_D(A_0 + B_0) + (C_0 + D_0)]\xi + (C_0D_0 - K_DA_0B_0) = 0.$$

By solving ξ from the above, $X_{Fe}^{M2}, X_{Mg}^{M1}, X_{Fe}^{M1}$, and X_{Mg}^{M2} can be obtained.
 On the other hand, if the conditions are given such that

$$X_{Fe}^{M1} + X_{Mg}^{M1} = C_1; \quad X_{Fe}^{M2} + X_{Mg}^{M2} = C_2; \quad \text{and} \quad X_{Fe}^{M1} + X_{Fe}^{M2} = C_3,$$

where C_1, C_2, and C_3 are constants ($C_3 = 2X_{Fs}$, C_1 and C_2 might be 1 if there are no Mn, Ca, Cr, Al, etc.), then using $x = X_{Fe}^{M2}$ as the independent variable to express all other species concentrations, the equilibrium equation is

$$K_D = \frac{CD}{AB} = \frac{(C_3 - x)(C_2 - x)}{x(C_1 - C_3 + x)}.$$

Rearrange:

$$(1 - K_D)x^2 + [K_D(C_3 - C_1) - (C_2 + C_3)]x + C_2C_3 = 0.$$

Hence, x can be solved and concentrations of other species can then be calculated.

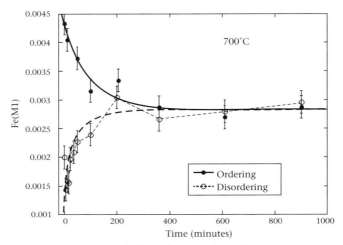

Figure 2-5 Experimental data showing how Fe concentration in the M1 site approaches equilibrium value from both above and below the equilibrium concentration. Figures like this are used in both equilibrium studies (to show that equilibrium is indeed reached since the same final state is reached from opposite directions, which is called a pair of reversals) and kinetic studies (to infer the reaction rate constants). From Wang et al. (2005).

also on the composition (Ganguly, 1982; Kroll et al., 1997). The evaluated reaction rate coefficient is independent of MS versus XRD method, and takes the following form for $X_{Fs} < 0.55$ (Kroll et al., 1997):

$$\ln k_f = 23.33 - (32,241 - 6016X_{Fs}^2)/T, \tag{2-60}$$

where k_f is in s^{-1} and T is in K. The value of k_b can be calculated from k_f/K_D. The two-sigma uncertainty in calculated k_f and k_b is about a factor of 3. The above equation means that the activation energy is roughly linear to the square of X_{Fs}. The dependence of k_f on the bulk composition of orthopyroxene according to the above equation may be seen from the following example: at 973.15 K, $k_f = 5.5 \times 10^{-5}$ s^{-1} at $X_{Fs} = 0$; and $k_f = 2.6 \times 10^{-4}$ s^{-1} at $X_{Fs} = 0.5$.

The intracrystalline exchange is controlled by Fe–Mg interdiffusion in the lattice and the rate coefficients are related to the Fe–Mg interdiffusion coefficients (Ganguly and Tazzoli, 1994). Hence, the dependence of k_f on the composition of orthopyroxene is related to the dependence of diffusion coefficient on the composition. For interdiffusion in olivine, there is a relatively large database, and $\ln D$ is found to depend linearly on the mole fraction of the forsterite component of olivine (Morioka and Nagasawa, 1991). It has also been suggested that the reaction rate coefficients k_f and k_b are proportional to $f_{O_2}^{1/6}$ (Ganguly and Tazzoli, 1994; Stimpfl et al., 2005) because such a relation holds for Fe–Mg interdiffusion in olivine (Buening and Buseck, 1973). The dependence of rate coefficients on the overall composition (such as X_{Fs}) and on conditions such as T, P, and f_{O_2} does not affect whether the reaction is second order or not.

Knowing K_D (Equation 2-57) and k_f (Equation 2-60), and assuming that Reaction 2-55 is an elementary reaction, the kinetics of the reaction can be treated using the methods and formulations presented earlier in this section. For example, the mean reaction time and the concentration evolution with time at a constant temperature can be calculated, and isothermal experimental kinetic data can be fitted to obtain the reaction rate constants, using the formulas in Table 2-1. Reaction during cooling may be solved using the Runge-Kutta method. For a natural orthopyroxene, by measuring Fe and Mg mole fractions in M1 and M2 sites, T_{ae} can be calculated from Equation 2-57 or 2-58. Furthermore, q at T_{ae} can also be calculated. The calculation of cooling rate is the main application for investigating this reaction, and this application (as a geospeedometer) is discussed in Chapter 5.

In the literature on intracrystalline reactions, another formulation, which is more general than that shown in Table 2-1, has been advanced to treat the kinetics of order–disorder reactions (Mueller, 1969; Ganguly, 1982). The method is outlined below to help readers follow the literature. Those who are not interested in such details may jump to Section 2.1.5.

2.1.4.2 Ganguly's treatment of the kinetics of order–disorder reactions

This method is developed explicitly for crystalline sites that contain different numbers of ions, such as Fe–Mg order–disorder between M1/M2/M3 (these three sites have been treated as roughly identical sites in terms of Fe–Mg distribution) and M4 sites of amphibole. The outline in this section is based on the treatment of Ganguly (1982).

Assume that there are two nonequivalent lattice sites in a mineral, referred to as α and β, and two ions, referred to as i and j, may partition between the two sites. The intracrystalline reaction may be written as

$$i(\alpha) + j(\beta) \rightleftharpoons i(\beta) + j(\alpha), \tag{2-61}$$

where $i(\alpha)$ means ion i on site α. Let C_α and C_β be the total number of α and β sites per unit volume of the mineral, respectively; $C_i^\alpha, C_i^\beta, C_j^\alpha$, and C_j^β be the numbers of i and j in α and β sites, respectively; and $X_i^\alpha, X_i^\beta, X_j^\alpha$, and X_j^β and be the mole fractions of i and j in α and β sites, respectively. Let $C_0 = C_\alpha + C_\beta$, $p = C_\alpha/C_0$, and $q = C_\beta/C_0$. Note that $p + q = 1$. Thus,

$$C_i^\alpha = X_i^\alpha C_\alpha = p X_i^\alpha C_0, \tag{2-62a}$$

$$C_i^\beta = X_i^\beta C_\beta = q X_i^\beta C_0, \tag{2-62b}$$

$$C_j^\alpha = X_j^\alpha C_\alpha = p X_j^\alpha C_0, \tag{2-62c}$$

$$C_j^\beta = X_j^\beta C_\beta = q X_j^\beta C_0. \tag{2-62d}$$

The exchange coefficient can be written as

$$K_D = \frac{X_i^\beta X_j^\alpha}{X_i^\alpha X_j^\beta} = \frac{(i/j)_\beta}{(i/j)_\alpha}. \tag{2-63}$$

Based on mass balance, $C_i^\alpha + C_i^\beta = pX_i^\alpha C_0 + qX_i^\beta C_0 = \text{constant}$, and $C_j^\alpha + C_j^\beta = pX_j^\alpha C_0 + qX_j^\beta C_0 = \text{constant}$. That is,

$$pX_i^\alpha + qX_i^\beta = pX_{i0}^\alpha + qX_{i0}^\beta = \overline{X}_i, \tag{2-64a}$$

$$pX_j^\alpha + qX_j^\beta = pX_{j0}^\alpha + qX_{j0}^\beta = \overline{X}_j, \tag{2-64b}$$

$$X_i^\alpha + X_j^\alpha = X_{i0}^\alpha + X_{j0}^\alpha, \tag{2-65a}$$

$$X_i^\beta + X_j^\beta = X_{i0}^\beta + X_{j0}^\beta, \tag{2-65b}$$

where subscript 0 means the initial concentration, and overbar means average. The value of $X_i^\alpha + X_j^\alpha$ does not necessarily equal 1 because there may be other elements in site α. From Equations 2-64a to 2-65b, X_i^β, X_j^α, and X_j^β can be expressed as

$$X_j^\alpha = X_{i0}^\alpha + X_{j0}^\alpha - X_i^\alpha, \tag{2-66a}$$

$$X_i^\beta = (\overline{X}_i - pX_i^\alpha)/q, \tag{2-66b}$$

$$X_j^\beta = X_{i0}^\beta + X_{j0}^\beta - (\overline{X}_i - pX_i^\alpha)q. \tag{2-66c}$$

To solve the equilibrium mole fractions from initial mole fractions, insert the above into Equation 2-63, leading to

$$K_D = \frac{(X_{i0}^\alpha + X_{j0}^\alpha - X_i^\alpha)(\overline{X}_i - pX_i^\alpha)}{X_i^\alpha[qX_{i0}^\beta + qX_{j0}^\beta - (\overline{X}_i - pX_i^\alpha)]}. \tag{2-67}$$

Let $y = X_i^\alpha$, and rearrange,

$$Ay^2 + By + C = 0, \tag{2-68}$$

where

$$A = p(K_D - 1), \tag{2-69a}$$

$$B = K_D(qX_{j0}^\beta - pX_{i0}^\alpha) + \overline{X}_i + p(X_{i0}^\alpha + X_{j0}^\alpha), \tag{2-69b}$$

$$C = -\overline{X}_i(X_{i0}^\alpha + X_{j0}^\alpha). \tag{2-69c}$$

Therefore, the equilibrium X_i^α must lie between 0 and the smaller of $\overline{X}_i/p$ and $X_{i0}^\alpha + X_{j0}^\alpha$, and is one of the following two values:

$$\frac{-B \pm \sqrt{B^2 - 4AC}}{2A}. \tag{2-70}$$

Next we focus on the kinetics. Assuming that both the forward and backward reactions of Reaction 2-61 are elementary reactions, then the reaction law can be written as

$$-\frac{dC_i^\alpha}{dt} = k_{f1}C_i^\alpha C_j^\beta - k_{b1}C_i^\beta C_j^\alpha, \tag{2-71}$$

where k_{f1} and k_{b1} are the reaction rate coefficients. Note that α and β above are not exponents, but refer to two different crystalline sites. The above can be written as

$$-\frac{dX_i^\alpha}{dt} = qC_0(k_{f1}X_i^\alpha X_j^\beta - k_{b1}X_i^\beta X_j^\alpha). \tag{2-72}$$

That is,

$$-\frac{dX_i^\alpha}{dt} = qk_{b1}C_0(K_D X_i^\alpha X_j^\beta - X_i^\beta X_j^\alpha). \tag{2-73}$$

Let $y = X_i^\alpha$, and insert Equations 2-66a to 2-66c into the above differential equation; then

$$-\frac{dy}{dt} = k_{b1}C_0[K_D y(qX_{i0}^\beta + qX_{j0}^\beta - \overline{X}_i + py) - (\overline{X}_i - py)(X_{i0}^\alpha + X_{j0}^\alpha - y)] \tag{2-74}$$

That is,

$$-\frac{dy}{dt} = k_{b1}C_0(Ay^2 + By + C). \tag{2-75}$$

Hence,

$$-C_0 \int_{t_1}^{t_2} k_{b1}dt = \int_{X_i^\alpha(t_1)}^{X_i^\alpha(t_2)} \frac{dy}{(Ay^2 + By + C)}. \tag{2-76}$$

For isothermal reaction kinetics, and hence constant K_D and k_{b1}, the above can be integrated to obtain

$$-k_{b1}C_0 t = \frac{1}{\sqrt{B^2 - 4AC}} \left[\ln \left| \frac{(2Ay + B) - \sqrt{B^2 - 4AC}}{(2Ay + B) + \sqrt{B^2 - 4AC}} \right| \right]_{y=X_i^\alpha(0)}^{y=X_i^\alpha(t)}. \tag{2-77}$$

Then, the concentration evolution may be calculated.

For nonisothermal reaction kinetics, Ganguly (1982) applied the above solution for a small time interval $\Delta t \to 0$. In this time interval, K_D and k_{b1} may be regarded as constant. Hence,

$$-k_{b1}C_0\Delta t = \frac{1}{\sqrt{B^2 - 4AC}} \left[\ln \left| \frac{(2Ay + B) - \sqrt{B^2 4AC}}{(2Ay + B) + \sqrt{B^2 - 4AC}} \right| \right]_{X_i^\alpha(t_1)}^{X_i^\alpha(t_2)} \tag{2-78}$$

Box 2.3 Fe–Mg order–disorder in orthopyroxene; comparison of different formulations

For Fe–Mg order–disorder in orthopyroxene, let $\alpha = M2$, $\beta = M1$, $i = Fe$, and $j = Mg$. Then, Reaction 2-61 is the same as Reaction 2-55, and $p = 1/2$ and $q = 1/2$. The reaction law becomes

$$\frac{dX_{Fe}^{M1}}{dt} = pC_0(k_{f1}X_{Fe}^{M2}X_{Mg}^{M1} - k_{b1}X_{Fe}^{M1}X_{Mg}^{M2}).$$

Comparing with Equation 2-59, it can be seen that

$$k_f = pk_{f1}C_0 = 0.5k_{f1}C_0;$$

$$k_b = pk_{b1}C_0 = 0.5k_{b1}C_0.$$

Hence, there is a difference of a factor of 2 between k_f in Equation 2-59 and $k_{f1}C_0$ in Equation 2-71 to 2-78.

The whole calculation procedure is as follows (Ganguly, 1982). Starting from $t_1 = 0$, choosing a Δt that is small, calculate T at $(t_1 + \Delta t/2)$ and then K_D and k_{b1} at that T, and find new $X_i^?$ using the above expressions. Repeat this process until the required temperature or the required time is reached.

The definitions of the reaction rate coefficients by Ganguly (1982) differ from those in Table 2-1. The difference is explained in Box 2-3.

Some other similar reactions have also been investigated, such as Fe–Mg order–disorder in cummingtonite and olivine. A brief description on how to treat intracrystalline Fe–Mg exchange reaction in cummingtonite can be found in Box 2-4. Because mass balance for intracrystalline reactions with many sites is confusing, it is discussed in Box 2-5.

2.1.5 Hydrous species reaction in rhyolitic melt

Another geochemical reaction that has been investigated extensively as a geospeedometer by high-temperature geochemists is the hydrous species reaction in rhyolitic melt. As the H_2O component dissolves in silicate melt, it partially reacts with oxygen in the melt to form OH groups (Reaction 1-10):

$$H_2O_m(melt) + O(melt) \rightleftharpoons 2OH(melt). \tag{2-79}$$

Hence, there are at least two hydrous species in silicate melt, H_2O molecules (referred to as H_2O_m in this section) and OH groups (referred to as OH). Total H_2O will be referred to as H_2O_t. The equilibrium constant for the interconvert reaction is

Box 2.4 Fe–Mg order–disorder in cummingtonite

For Fe-Mg order-disorder in cummingtonite, let $\alpha = M4$, $\beta = M1 + M2 + M3$, $i = Fe$, and $j = Mg$. There are 2 moles of M4 and 5 moles of $M1 + M2 + M3$ per formula unit of cummingtonite. Hence, $p = 2/7$ and $q = 5/7$. The reaction law becomes

$$-\frac{dC_{Fe}^{M4}}{dt} = -\frac{dC_{Mg}^{M1+M2+M3}}{dt} = \frac{dC_{Fe}^{M1+M2+M3}}{dt} = \frac{dC_{Mg}^{M4}}{dt}.$$
$$= k_{f1} C_{Fe}^{M4} C_{Mg}^{M1+M2+M3} - k_{b1} C_{Fe}^{M1+M2+M3} C_{Mg}^{M4}$$

The above takes the following form when expressed in terms of mole fractions:

$$-p\frac{dX_{Fe}^{M4}}{dt} = -q\frac{dX_{Mg}^{M1+M2+M3}}{dt} = q\frac{dX_{Fe}^{M1+M2+M3}}{dt} = p\frac{dX_{Mg}^{M4}}{dt}$$
$$= pqC_0 k_{b1} (k_D X_{Fe}^{M4} X_{Mg}^{M1+M2+M3} - X_{Fe}^{M1+M2+M3} X_{Mg}^{M4}),$$

where $pq = 10/49$.

Box 2.5 Mass balance for intracrystalline reactions

Mass balance for intracrystalline reactions must consider the number of moles of the site per formula unit, as shown in Box 2-4. To clarify this point further, here is an example. Consider $^{18}O-^{16}O$ exchange reaction between OH^- groups and SO_4^{2-} units in alunite, $KAl_3(SO_4)_2(OH)_6$. The equilibrium constant is

$$\alpha = (^{18}O/^{16}O)_{SO_4}/(^{18}O/^{16}O)_{OH} = \exp(0.00096 + 8 \times 10^2/T^2),$$

where T is in K (Stoffregen et al., 1994a,b). In each formula unit of alunite, there are six moles of oxygen in the OH^- site, and two moles of SO_4^{2-}, leading to 8 moles of oxygen in the SO_4^{2-} site. Assuming alunite is a closed system, the mass balance equation is hence

$$6(^{18}O/^{16}O)_{OH} + 8(^{18}O/^{16}O)_{SO_4} = \text{constant}$$
$$= [6(^{18}O/^{16}O)_{OH} + 8(^{18}O/^{16}O)_{SO_4}]_{initial}.$$

Knowing the initial condition and α (or temperature), the two unknowns $(^{18}O/^{16}O)_{OH}$ and $(^{18}O/^{16}O)_{SO_4}$ at equilibrium can hence be solved from the above two equations.

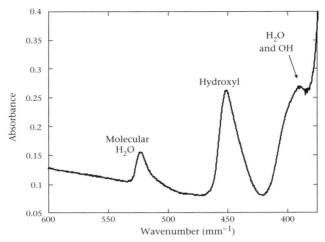

Figure 2-6 An FTIR spectrum of hydrous rhyolitic glass in the near-IR region of 600- to 375-mm^{-1} wavenumbers. The absorbance A is defined as $-\log(I/I_0)$, where I_0 is the infrared beam intensity without the sample, and I is the intensity with the sample in the beam path. This sample contains about 0.8 wt% H_2O_t. Both 523- and 452-mm^{-1} peaks are combination modes. There are more hydrous peaks outside the region: The fundamental stretch of OH (the strongest peak) is at 355 mm^{-1} and the absorbance is usually off the scale. The overtone of the fundamental stretch is at 710 mm^{-1}, which is weak. As H_2O_t content increases, the ratio of 523-mm^{-1} peak height to 452-mm^{-1} increases.

$$K = \frac{[OH]^2}{[H_2O_m][O]},\qquad\qquad (2\text{-}80)$$

where brackets mean activities approximated by mole fractions. The two species may be directly seen by well-resolved peaks in infrared (IR) or Raman spectra; Figure 2-6 shows an IR spectrum, with clearly separated H_2O_m and OH peaks. Calibrations for infrared spectroscopy have been carried out for rhyolitic, basaltic, dacitic, and andesitic melts. The equilibrium and the kinetics of the hydrous species reaction have been investigated in rhyolitic melt and have been applied as a geospeedometer. For other melts, the work is limited for of various reasons. Reviews of several aspects on dissolved water in rhyolitic melt can be found in Zhang (1999b) and Zhang et al. (2007).

2.1.5.1 Definitions of H₂O species and contents

To nonspecialists and beginners, the definitions of H_2O_t, H_2O_m, and OH weight percent and mole fractions may be confusing. The definition of mass fraction (or weight percent) of H_2O_m and H_2O_t is straightforward. The definition of mass fraction (or weight percent) of OH is not the actual mass fraction of OH per se, but the mass fraction of extracted H_2O that was present in the glass in the form of

OH. That is, it is the mass fraction of the species of two OH groups minus one oxygen. This is because in Reaction 2-79 two OH groups minus one oxygen would form one molecular H_2O. This definition of OH would lead to $(H_2O_t) = (H_2O_m) + (OH)$ in terms of mass fraction or wt%. The definition of mole fraction of OH is, however, the mole fraction of OH per se, not the mole fraction of $2OH - O$. In terms of mole fraction, $[H_2O_t] = [H_2O_m] + [OH]/2$.

Three definitions of H_2O mole fractions are encountered in the literature. They are summarized below (Zhang, 1999b):

(1) In this book, mole fractions on a single oxygen basis are used, following the work of Stolper (1982b). The calculation of the mole fractions of H_2O_t, H_2O_m, OH, and O for hydrous rhyolitic melts/glasses is

$$[H_2O_t] = (C/18.015)/\{C/18.015 + (1-C)/W\}, \tag{2-81}$$

$$[H_2O_m] = [H_2O_t](H_2O_m)/C, \tag{2-82}$$

$$[OH] = 2\{[H_2O_t] - [H_2O_m]\}, \tag{2-83}$$

$$[O] = 1 - [H_2O_m] - [OH], \tag{2-84}$$

where parentheses indicate mass fraction, C is the mass fraction of H_2O_t, and W is the mass of dry rhyolite (from Mono Craters, California) per mole of oxygen and is 32.49 g/mol. For a haplogranitic melt AOQ (Qz28Ab38Or34, where Qz means quartz, SiO_2, Ab means albite, $NaAlSi_3O_8$, and Or means orthoclase, $KAlSi_3O_8$) composition, $W = 32.6$ g/mol. For albite ($NaAlSi_3O_8$), $W = 32.778$ g/mol.

(2) Some authors (e.g., Moore et al., 1998a) defined the H_2O oxide mole fraction by treating each oxide (e.g., SiO_2) as one unit (whereas the definition on a single oxygen basis treats SiO_2 as two units and Al_2O_3 as three units). In this definition, $X_{H_2O_t} = (C/18.015)/\Sigma(C_i/W_i)$, where C_i is the mass fraction of oxide component i (including H_2O) and W_i is the molar mass of the oxide.

(3) In the H_2O–$NaAlSi_3O_8$ system, Burnham (1975) and other authors following him treated $NaAlSi_3O_8$ as one unit (whereas the definition on a single oxygen basis treats $NaAlSi_3O_8$ as eight units). In this definition, $X_{H_2O_t} = (C/18.015)/\{C/18.015 + (1-C)/262.22\}$, where 262.22 is the molar mass of $NaAlSi_3O_8$.

The three definitions above result in very different mole fractions. For example, 5.0 wt% H_2O_t in albite melt translates into an H_2O_t mole fraction of 0.0874 on a single-oxygen basis, 0.161 using oxide moles, and 0.434 using $NaAlSi_3O_8$ as one unit. The third definition is the basis of the statement "a small weight percent of H_2O leads to a large mole fraction."

2.1.5.2 Measurement of H₂O species concentrations

The absolute methods (meaning no independent calibration is necessary) for determining H_2O_t include manometry (Newman et al., 1986), Karl-Fischer titration (Behrens et al., 1996), and nuclear reaction analyses (NRA). NRA is a bulk method for H_2O_t and requires a density estimate to convert data to weight percent. Secondary ion mass spectrometry (SIMS) can measure H_2O_t, but it requires independent calibration using an absolute method. The absolute methods and SIMS method cannot distinguish the individual species. The best tool currently available on quantitative measurement of hydrous species concentrations is Fourier transform infrared spectrometry (FTIR) (Stolper, 1982a; Newman et al., 1986; Zhang et al., 1997a). Raman spectrometry can also distinguish the two hydrous species, and hence in theory may be developed as a quantitative tool to measure species concentrations. In practice, however, quantitative Raman seems to be more complicated than IR. Contrasting conclusions about whether hydrous species concentrations can be determined accurately by Raman have been reached by different groups (e.g., Thomas, 2000; Arredondo and Rossman, 2002).

The infrared spectrum in Figure 2-6 shows three peaks in this near-IR region: the 523-mm^{-1} peak for H_2O_m, the 452-mm^{-1} peak for OH, and the $\sim$390-mm^{-1} peak for $H_2O_m + OH$. The $\sim$390-mm^{-1} peak is not very useful because it is not well resolved. The intensity (either peak height or area) of each peak is roughly proportional to the concentration of the corresponding species according to Beer's law. Hence, mass fraction of H_2O_t may be expressed as

$$H_2O_t = \frac{18.015A_{523}}{d\rho\varepsilon_{523}} + \frac{18.015A_{452}}{d\rho\varepsilon_{452}}, \tag{2-85}$$

where 18.015 is the molar mass of H_2O, A_{523} and A_{452} are absorbances of the 523- and 452-mm^{-1} peaks, d and ρ are the thickness and density of the sample, and ε_{523} and ε_{452} are the extinction coefficients (or molar absorptivities) of the 523- and 452-mm^{-1} peaks. The first term on the right-hand side of the above equation means H_2O_m mass fraction, and the second terms means OH mass fraction. Using the above relation, for a glass sample of a given anhydrous melt composition, if an absolute method is used to determine H_2O_t content, IR spectrum is taken to determine A_{523} and A_{452}, and the thickness and densities are known, we would have one equation. For many samples with the same anhydrous composition but with different H_2O_t, we would have many equations, from which the two unknowns ε_{523} and ε_{452} can be determined by linear regression (e.g., Newman et al., 1986). Another way to present calibration data is by modifying Equation 2-85:

$$\frac{18.015A_{523}}{\rho dC} = \varepsilon_{523} - \frac{\varepsilon_{523}}{\varepsilon_{452}}\frac{18.015A_{452}}{\rho dC}, \tag{2-86}$$

where C is H_2O_t content (mass fraction). Let $y = 18.015A_{523}/(\rho dC)$ and $x = 18.015A_{452}/(\rho dC)$. Plotting y versus x would yield a straight line, with the intercept of ε_{523}, and the slope of $\varepsilon_{523}/\varepsilon_{452}$. Hence, both extinction coefficients may be obtained. Figure 2-7 shows such a plot.

Knowing ε_{523} and ε_{452} (i.e., after calibration), H_2O_t, H_2O_m, and OH contents of an unknown glass of the same anhydrous composition can be obtained from IR

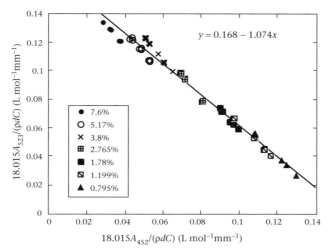

Figure 2-7 A diagram for IR calibration (Equation 2-86). The data for each given H_2O_t content is obtained by heating the sample to different temperatures to vary the species concentrations. For a perfect calibration, all the trends would lie on a single straight line. However, there is some scatter. Furthermore, the slope defined by data for one fixed H_2O_t content does not equal that for another. Hence, the calibration results ($\varepsilon_{523} = 0.168$; $\varepsilon_{452} = 0.156$) shown in this diagram have a relative precision of only about 10%, whereas the relative precision of IR band intensity data is about 1%.

spectrum. If ε_{523} and ε_{452} are not constants, then a more advanced method must be used (Zhang et al., 1997a). Although the above sounds simple and although band intensities determined from FTIR spectra are reproducible to 1% relative, because of (i) large uncertainties in the absolute methods of determining H_2O_t content, and (ii) possible variations of ε_{523} and ε_{452} with H_2O_t, the accuracy of FTIR determination of H_2O_t and species concentrations is of the order 5 to 10%, much worse than the measurement precision of FTIR band intensities.

2.1.5.3 Species equilibrium

The equilibrium constant of Reaction 2-79 as a function of temperature and H_2O content has been investigated extensively. Earlier results were confusing, because of the handling of the effect of quenching and the interpretation of in situ results. The conclusions from the debates (see review by Zhang (1999b) and reconciliation by Withers et al. (1999) are (i) very early (1980s) speciation data obtained by quenching from temperatures of >800°C were affected by quenching, and the strong dependence of the equilibrium constant of Reaction 2-79 on H_2O_t from such early data was invalid; (ii) later speciation data (1991 and later) obtained by quenching from intermediate temperatures (400–600°C) do not have a quench problem; (iii) early (1995) in situ speciation data were not interpreted

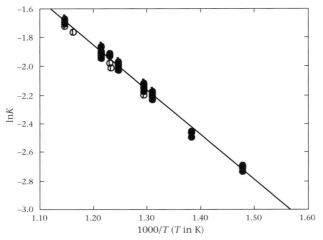

Figure 2-8 The equilibrium constant of Reaction 2-79 as a function of temperature in $\ln K$ versus $1000/T$ plot. The rough straight line means that the standard state enthalpy change of Reaction 2-79 is constant. Solid circles are 1-atm data from Zhang et al. (1997a) and open circles are 500-MPa data from Zhang (unpublished data).

correctly because the temperature dependence of the extinction coefficients was not taken into account; and (iv) later in situ speciation data (2001 and later) do not have such problems. All data still have some uncertainties because of uncertainties in the extinction coefficients (of the order 10% relative, and larger for the in situ data). There are two expressions for species equilibrium constant K for Reaction 2-79. One is based on experimental data measured on glasses quenched from intermediate-temperature (400–600°C) melts (Zhang et al., 1997a; Ihinger et al., 1999):

$$K = \exp(1.876 - 3110/T). \tag{2-87}$$

Figure 2-8 shows some experimental data. The above formulation is applicable to a temperature range of 400–600°C. Whether it can be extrapolated reliably to higher temperatures is debated.

Another expression is based on the in situ data of Nowak and Behrens (2001) for a haplogranitic melt (roughly the same as rhyolitic melt of Zhang et al., 1997a and Ihinger et al., 1999) at 500–800°C:

$$K = \exp(3.33 - 4210/T). \tag{2-88}$$

Given K and total H_2O mole fraction, the species mole fractions can be calculated as follows (Zhang, 1999b):

$$[H_2O_m] = \frac{8X^2}{8X + K(1 - 2X) + \sqrt{\{K(1 - 2X)\}^2 + 16KX(1 - X)}}, \tag{2-89}$$

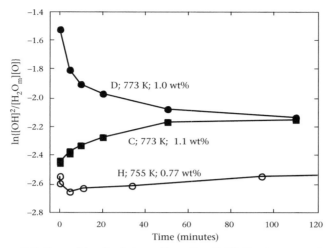

Figure 2-9 Some kinetic data on Reaction 2-79. For two samples (C and D at 773 K), equilibrium is reached monotonically. In the third sample (H), the species concentrations do not evolve monotonically with time: OH content first decreases away from equilibrium and then increases toward equilibrium. The reaction rate for sample H is significantly slower because of both low H_2O_t and low temperature (equilibrium would require about 1000 minutes). Data are from Zhang et al. (1995) but recalculated using the calibrations of Zhang et al. (1997a).

and

$$[OH] = 2\{X - [H_2O_m]\}, \tag{2-90}$$

where $X = [H_2O_t]$.

Understanding the speciation reaction equilibrium of Reaction 2-79, i.e., how the species concentration depends on H_2O_t and T, is critical in understanding H_2O diffusion. More on the speciation is presented in Chapter 3 when H_2O diffusion is discussed.

2.1.5.4 Kinetics of the reaction

The kinetics of Reaction 2-79 has been investigated through isothermal experiments but the reaction law is not well understood. If the reaction is assumed to be an elementary reaction, the reaction rate law would be

$$d\xi/dt = k_f[H_2O_m][O] - k_b[OH]^2, \tag{2-91}$$

where brackets mean mole fractions. The assumed reaction rate law would encounter the following difficulties. (i) Starting from an initial [OH], the assumed reaction law means that [OH] should approach the equilibrium concentration monotonically. However, experimental data show that this is not the case (Zhang et al., 1995). Some data can be found in Figure 2-9. (ii) Even if the nonmonotonic behavior is ignored, the rate constants using the assumed reaction rate law would depend roughly to the 7th power of total H_2O content (Zhang et al., 1997b). Better modeling can be achieved by considering subspecies (Zhang et al., 1995) such as OH pairs versus singletons, different OH groups (e.g., SiOH or AlOH), or different kinds of O, but these are complicated and unconstrained. Because of the difficulties in modeling the reaction kinetics quantitatively, and because the main application for understanding the kinetics of this reaction is geospeedometry, Zhang et al. (2000) empirically calibrated the geospeedometer. The application to geospeedometry is discussed in Chapter 5.

2.2 Chain Reactions

Chain reactions are a type of overall reactions, which require two or more steps to accomplish. They are also known as consecutive reactions or sequential reactions. Examples of chain reactions include nuclear hydrogen burning, nuclear decay chains, ozone production, and ozone decomposition. Some steps of a chain reaction may be rapid and some may be slow. The slowest step is the *rate-determining step*. During a chain reaction, some intermediate and unstable species may be produced and consumed continuously.

Chain reactions may lead to either steady state (in which concentration of intermediate species reaches a constant) or explosion (in which the reaction rate increases exponentially with time). If the concentrations of intermediate catalyst species reach a constant quickly, there would be a steady state. If the concentrations of intermediate catalyst species grow exponentially, there would be explosion.

In treating chain reactions, two concepts are often used: (i) the concept of rate-determining step, in which the slowest step is the rate-determining step; and (ii) the concept of *steady state*, which assumes that the concentration of a trace level intermediate is constant ($dC/dt = 0$) because it is rapidly produced and consumed, leading to steady state. The difference between the steady state and the thermodynamic equilibrium state of a reaction is that at the steady state only for some species in a reaction are the concentrations constant, whereas at equilibrium all species concentrations involved in the reaction are constant. Thermodynamic equilibrium is reached when Gibbs free energy is minimized. A third concept often referred to is that of *quasi-equilibrium*, which is not an independent concept. With this concept, the fast reaction steps are treated as being able to maintain equilibrium. This concept may be viewed as a special case of the concept of steady state, and is easy to apply. Below, some chain reactions are discussed to illustrate the different methods of treating chain reactions. The radioactive decay series are discussed in detail because they are powerful dating and tracing tools in geochemistry and because they illustrate the principles well.

2.2.1 Radioactive decay series

Probably the most important chain reactions in geochemistry are the decay series of ^{238}U, ^{235}U, and ^{232}Th (Table 2-2). For chemical chain reactions, the intermediate steps are not necessarily directly observed, and the reaction steps are inferred from experimental reaction rate laws and other information. For the radioactive decay series, the intermediate steps of the chain reactions can be directly observed through the measurement of α-decay and β-decay particles of the intermediate species. The consideration of the decay series also clearly illustrates the concepts of (i) rate-determining step and (ii) steady state. Understanding of the series has been applied to date geologic samples and to investigate the dynamics of partial melting and melt transport.

For chemical chain reactions, the goal of kinetic studies is to infer the reaction steps and to obtain the overall reaction rate law, rather than how the concentrations of the intermediate species evolve with time. For the decay series, the intermediate steps are observable and the overall reaction rate laws are also known. Our treatment below is to solve for the behavior of some intermediate species because these species are often of geological significance, such as Rn as an environmental hazard, ^{234}Th as a dating tool, and ^{234}U as a dating and tracing method. Hence, the mathematical treatment of the decay series is much more thorough than that of chemical chain reactions. In other words, the investigation of chemical chain reactions is at the basic level, but that of decay series is at a more advanced level.

2.2.1.1 Secular equilibrium

Three decay chains are shown in Table 2-2, with all the reaction steps and the decay constants (rate coefficients). The decay chain of ^{238}U is used as an example for detailed discussion below. The ^{238}U-series is a long and complicated decay chain: starting from ^{238}U, after 8 α-decays and 6 β-decays (in places there are different branches and paths), the final stable product is ^{206}Pb.

Every decay reaction in each decay chain is a first-order elementary reaction. To solve the concentration of each species in the ^{238}U decay series, the reaction rate laws for every species (ignoring the minor effect of different states of ^{234}Pa) are written below:

$$\mathrm{d}^{238}\mathrm{U}/\mathrm{d}t = -\lambda_{238_U}{}^{238}\mathrm{U} \tag{2-92a}$$

$$\mathrm{d}^{234}\mathrm{Th}/\mathrm{d}t = \lambda_{238_U}{}^{238}\mathrm{U} - \lambda_{234_{Th}}{}^{234}\mathrm{Th} \tag{2-92b}$$

$$\mathrm{d}^{234}\mathrm{Pa}/\mathrm{d}t = \lambda_{234_{Th}}{}^{234}\mathrm{Th} - \lambda_{234_{Pa}(2)}{}^{234}\mathrm{Pa} \tag{2-92c}$$

$$\mathrm{d}^{234}\mathrm{U}/\mathrm{d}t = \lambda_{234_{Pa}(2)}{}^{234}\mathrm{Pa} - \lambda_{234_U}{}^{234}\mathrm{U} \tag{2-92d}$$

$$\mathrm{d}^{230}\mathrm{Th}/\mathrm{d}t = \lambda_{234_U}{}^{234}\mathrm{U} - \lambda_{230_{Th}}{}^{230}\mathrm{Th} \tag{2-92e}$$

Table 2-2a Decay steps in the decay chain of ^{238}U

Reaction	Decay constant	Half-life
$^{238}U \rightarrow {}^{234}Th + {}^{4}He$	$\lambda_{238} = 1.55125 \times 10^{-10} \text{ yr}^{-1}$	4.4683 Byr
$^{234}Th \rightarrow \underline{^{234}Pa}(2)$	$\lambda_{234_{Th}} = 10.5 \text{ yr}^{-1}$	24.10 d
$\underline{^{234}Pa}(2)$ to ^{234}U and $^{234}Pa(1)$	$\lambda_{234_{Pa}(2)} = 3.1 \times 10^{5} \text{ yr}^{-1}$	1.17 min
$\quad \underline{^{234}Pa}(2) \rightarrow {}^{234}U$ (99.84%)	$\lambda_{234_{Pa}(2),1} = 3.1 \times 10^{5} \text{ yr}^{-1}$	6.69 h
$\quad \underline{^{234}Pa}(2) \rightarrow {}^{234}Pa(1)$ (0.16%)	$\lambda_{234_{Pa}(2),2} = 5 \times 10^{2} \text{ yr}^{-1}$	
$\quad \underline{^{234}Pa}(1) \rightarrow {}^{234}U$	$\lambda_{234_{Pa}(1)} = 908 \text{ yr}^{-1}$	
$^{234}U \rightarrow {}^{230}Th + {}^{4}He$	$\lambda_{234_U} = 2.835 \times 10^{-6} \text{ yr}^{-1}$	244 kyr
$^{230}Th \rightarrow {}^{226}Ra + {}^{4}He$	$\lambda_{230_{Pa}Th} = 9.195 \times 10^{-6} \text{ yr}^{-1}$	75.4 kyr
$^{226}Ra \rightarrow {}^{222}Rn + {}^{4}He$	$\lambda_{226_{Ra}} = 4.33 \times 10^{-4} \text{ yr}^{-1}$	1599 yr
$^{222}Rn \rightarrow {}^{\mathbf{218}}\mathbf{Po} + {}^{4}He$	$\lambda_{222_{Ra}} = 66.21 \text{ yr}^{-1}$	3.8235 d
$^{\mathbf{218}}\mathbf{Po}$ to (i) ^{218}At and (ii) $^{214}Pb + {}^{4}He$	$\lambda_{218_{Po},\text{total}} = 1.18 \times 10^{5} \text{ yr}^{-1}$	3.10 min
$\quad ^{\mathbf{218}}\mathbf{Po} \rightarrow {}^{218}At$ (branch 1; 0.02%)	$\lambda_{218_{Po},1} = 23.6 \text{ yr}^{-1}$	
$\quad ^{\mathbf{218}}\mathbf{Po} \rightarrow {}^{214}Pb + {}^{4}He$ (branch 2; 99.98%)	$\lambda_{218_{Po},2} = 1.18 \times 10^{5} \text{ yr}^{-1}$	
$^{\mathbf{218}}\mathbf{At}$ to (i) ^{218}Rn and (ii) $^{214}Bi + {}^{4}He$	$\lambda_{218_{At},\text{total}} = 1.5 \times 10^{7} \text{ yr}^{-1}$	1.5 s
$\quad ^{\mathbf{218}}\mathbf{At} \rightarrow {}^{218}Rn$ (branch 1; 0.1%)	$\lambda_{218_{At},1} = 1.5 \times 10^{4} \text{ yr}^{-1}$	
$\quad ^{\mathbf{218}}\mathbf{At} \rightarrow {}^{\mathbf{214}}\mathbf{Bi} + {}^{4}He$ (branch 2; 99.9%)	$\lambda_{218_{At},2} = 1.5 \times 10^{7} \text{ yr}^{-1}$	
$^{218}Rn \rightarrow {}^{\mathit{214}}\mathit{Po} + {}^{4}He$	$\lambda_{218_{Rn}} = 6.2 \times 10^{8} \text{ yr}^{-1}$	0.035 s
$^{214}Pb \rightarrow {}^{\mathbf{214}}\mathbf{Bi}$	$\lambda_{214_{Pb}} = 1.35 \text{ x}10^{4} \text{ yr}^{-1}$	27 min
$^{\mathbf{214}}\mathbf{Bi}$ to (i) ^{214}Po and (ii) $^{210}Tl + {}^{4}He$	$\lambda_{214_{Bi},\text{total}} = 1.83 \text{ x}10^{4} \text{ yr}^{-1}$	19.9 min
$\quad ^{\mathbf{214}}\mathbf{Bi} \rightarrow {}^{\mathit{214}}\mathit{Po}$ (branch 1; 99.979%)	$\lambda_{214_{Bi},1} = 1.83 \text{ x}10^{4} \text{ yr}^{-1}$	
$\quad ^{\mathbf{214}}\mathbf{Bi} \rightarrow {}^{210}Tl + {}^{4}He$ (branch 2; 0.021%)	$\lambda_{214_{Bi},2} = 3.8 \text{ yr}^{-1}$	
$^{214}Po \rightarrow {}^{\mathbf{210}}\mathbf{Pb} + {}^{4}He$	$\lambda_{214_{Po}} = 1.336 \times 10^{11} \text{ yr}^{-1}$	163.7 μs
$^{210}Tl \rightarrow {}^{\mathbf{210}}\mathbf{Pb}$	$\lambda_{210_{Tl}} = 2.80 \times 10^{5} \text{ yr}^{-1}$	78 s
$^{\mathbf{210}}\mathbf{Pb}$ to (i) ^{210}Bi and (ii) $^{206}Hg + {}^{4}He$	$\lambda_{210_{Pb},\text{total}} = 0.0307 \text{ yr}^{-1}$	22.6 yr
$\quad ^{\mathbf{210}}\mathbf{Pb} \rightarrow {}^{210}Bi$ (branch 1; 100%)	$\lambda_{210_{Pb},1} = 0.0307 \text{ yr}^{-1}$	
$\quad ^{\mathbf{210}}\mathbf{Pb} \rightarrow {}^{206}Hg + {}^{4}He$ (branch 2; 19 ppb)	$\lambda_{210_{Pb},2} = 5.8 \times 10^{-10} \text{ yr}^{-1}$	
$^{\mathbf{210}}\mathbf{Bi}$ to (i) ^{210}Po and (ii) $^{206}Tl + {}^{4}He$	$\lambda_{210_{Bi},\text{total}} = 50.5 \text{ yr}^{-1}$	5.01 d
$\quad ^{\mathbf{210}}\mathbf{Bi} \rightarrow {}^{210}Po$ (branch 1; 100%)	$\lambda_{210_{Bi},1} = 50.5 \text{ yr}^{-1}$	
$\quad ^{\mathbf{210}}\mathbf{Bi} \rightarrow {}^{\mathit{206}}\mathit{Tl} + {}^{4}He$ (branch 2; 1.32 ppm)	$\lambda_{210_{Bi},2} = 6.7 \times 10^{-5} \text{ yr}^{-1}$	
$^{206}Hg \rightarrow {}^{\mathit{206}}\mathit{Tl}$	$\lambda_{206_{Hg}} = 4.4 \times 10^{4} \text{ yr}^{-1}$	8.2 min

Table 2-2a (*continued*)

Reaction	Decay constant	Half-life
$^{210}Po \rightarrow {}^{206}Pb + {}^4He$	$\lambda_{210Po} = 1.83 \ yr^{-1}$	138.38 d
$^{206}Tl \rightarrow {}^{206}Pb$	$\lambda_{206Tl} = 8.68 \times 10^4 \ yr^{-1}$	4.20 min

Note. The steps from ^{238}U to ^{206}Pb are 8 α-decays and 6 β-decays. The net reaction is $^{238}U \rightarrow {}^{206}Pb + 8{}^4He$. Data are from Lockheed Martin (2002) and Firestone and Shirley (1996). The nuclide in bold undergoes branch decays. The convention adopted here is that branch 1 is β-decay and branch 2 is α-decay. The nuclide in italics receives radiogenic contribution from two sources. The underlined nuclide has two different states that can both decay to other nuclides (if one state undergoes internal transition to another state, it is not listed). ^4He receives multiple sources and is not highlighted.

$$d^{226}Ra/dt = \lambda_{230Th}{}^{230}Th - \lambda_{226Ra}{}^{226}Ra \tag{2-92f}$$

$$d^{222}Rn/dt = \lambda_{226Ra}{}^{226}Ra - \lambda_{222Rn}{}^{222}Rn \tag{2-92g}$$

$$d^{218}Po/dt = \lambda_{222Rn}{}^{222}Rn - (\lambda_{218Po,1} + \lambda_{218Po,2}){}^{218}Po \tag{2-92h}$$

$$d^{218}At/dt = \lambda_{218Po,1}{}^{218}Po - (\lambda_{218At,1} + \lambda_{218At,2}){}^{218}At \tag{2-92i}$$

$$d^{214}Pb/dt = \lambda_{218Po,2}{}^{218}Po - \lambda_{214Pb}{}^{214}Pb \tag{2-92j}$$

$$d^{218}Rn/dt = \lambda_{218At,1}{}^{218}At - \lambda_{218Rn}{}^{218}Rn \tag{2-92k}$$

$$d^{214}Bi/dt = \lambda_{214Pb}{}^{214}Pb + \lambda_{218At,2}{}^{218}At - (\lambda_{214Bi,1} + \lambda_{214Bi,2}){}^{214}Bi \tag{2-92l}$$

$$d^{214}Po/dt = \lambda_{218Rn}{}^{218}Rn + \lambda_{214Bi,1}{}^{214}Bi - \lambda_{210Pb214}Po \tag{2-92m}$$

$$d^{210}Tl/dt = \lambda_{214Bi,2}{}^{214}Bi - \lambda_{210Tl}{}^{210}Tl \tag{2-92n}$$

$$d^{210}Pb/dt = \lambda_{214Po}{}^{214}Po + \lambda_{210Tl}{}^{210}Tl - (\lambda_{210Pb,1} + \lambda_{210Pb,2}){}^{210}Pb \tag{2-92o}$$

$$d^{210}Bi/dt = \lambda_{210Pb,1}{}^{210}Pb - (\lambda_{210Bi,1} + \lambda_{210Bi,2}){}^{210}Bi \tag{2-92p}$$

$$d^{206}Hg/dt = \lambda_{210Pb,2}{}^{210}Pb - \lambda_{206Hg}{}^{206}Hg \tag{2-92q}$$

$$d^{210}Po/dt = \lambda_{210Bi,1}{}^{210}Bi - \lambda_{210Po}{}^{210}Po \tag{2-92r}$$

$$d^{206}Tl/dt = \lambda_{210Bi,2}{}^{210}Bi + \lambda_{206Hg}{}^{206}Hg - \lambda_{206Tl}{}^{206}Tl \tag{2-92s}$$

$$d^{206}Pb/dt = \lambda_{210Po}{}^{210}Po + \lambda_{206Tl}{}^{206}Tl \tag{2-92t}$$

The above set of differential equations (20 equations) can be solved with the help of linear algebra (matrix operation). Even though the math is not particularly

Table 2-2b Individual decay steps in the decay chain of ^{235}U

Reaction	Decay constant	Half-life
$^{235}U \rightarrow {}^{231}Th + {}^{4}He$	$\lambda_{235} = 9.8485 \times 10^{-10}$ yr^{-1}	703.81 Myr
$^{231}Th \rightarrow {}^{231}Pa$	$\lambda_{234_{Th}} = 238.1$ yr^{-1}	1.063 d
$^{231}Pa \rightarrow {}^{227}Ac + {}^{4}He$	$\lambda_{231_{Pa}} = 2.11 \times 10^{-5}$ yr^{-1}	32.8 kyr
^{227}Ac to (i) ^{227}Th and (ii) ^{223}Fr	$\lambda_{227_{Ac,total}} = 0.0318$ yr^{-1}	21.774 yr
$\quad$**^{227}Ac** $\rightarrow {}^{227}Th$ (branch 1; 98.62%)	$\quad\lambda_{227_{Ac,1}} = 0.0314$ yr^{-1}	
$\quad$**^{227}Ac** $\rightarrow {}^{223}Fr + {}^{4}He$ (branch 2; 1.38%)	$\quad\lambda_{227_{Ac,2}} = 4.4 \times 10^{-4}$ yr^{-1}	
$^{227}Th \rightarrow {}^{223}Ra + {}^{4}He$	$\lambda_{227_{Th}} = 13.5$ yr^{-1}	18.72 d
^{223}Fr to (i) ^{223}Ra and (ii) **^{219}At**	$\lambda_{223_{Fr,total}} = 1.67 \times 10^{4}$ yr^{-1}	21.8 min
$\quad$**^{223}Fr** $\rightarrow {}^{223}Ra$ (branch 1; 99.994%)	$\quad\lambda_{223_{Fr,1}} = 1.67 \times 10^{4}$ yr^{-1}	
$\quad$**^{223}Fr** $\rightarrow$ **^{219}At** $+ {}^{4}He$ (branch 2; 0.006%)	$\quad\lambda_{223_{Fr,2}} = 1.0$ yr^{-1}	
$^{223}Ra \rightarrow {}^{219}Rn + {}^{4}He$	$\lambda_{223_{Ra}} = 22.15$ yr^{-1}	11.435 d
^{219}At to (i) ^{219}Rn and (ii) $^{215}Bi + {}^{4}He$	$\lambda_{219_{At,total}} = 3.9 \times 10^{5}$ yr^{-1}	56 s
$\quad$**^{219}At** $\rightarrow {}^{219}Rn$ (branch 1; 3%)	$\quad\lambda_{219_{At,1}} = 1.2 \times 10^{4}$ yr^{-1}	
$\quad$**^{219}At** $\rightarrow {}^{215}Bi + {}^{4}He$ (branch 2; 97%)	$\quad\lambda_{219_{At,2}} = 3.9 \times 10^{5}$ yr^{-1}	
$^{219}Rn \rightarrow$ **^{215}Po** $+ {}^{4}He$	$\lambda_{219_{Rn}} = 5.52 \times 10^{6}$ yr^{-1}	3.96 s
$^{215}Bi \rightarrow {}^{215}Po$	$\lambda_{215_{Bi}} = 4.8 \times 10^{4}$ yr^{-1}	7.6 min
^{215}Po to (i) ^{215}At and (ii) $^{211}Pb + {}^{4}He$	$\lambda_{215_{Po,total}} = 1.23 \times 10^{10}$ yr^{-1}	0.00178 s
$\quad$**^{215}Po** $\rightarrow {}^{215}At$ (branch 1; 2.3 ppm)	$\quad\lambda_{215_{Po,1}} = 2.8 \times 10^{4}$ yr^{-1}	
$\quad$**^{215}Po** $\rightarrow {}^{211}Pb + {}^{4}He$ (branch 2; 99.99977%)	$\quad\lambda_{215_{Po,2}} = 1.23 \times 10^{10}$ yr^{-1}	
$^{215}At \rightarrow$ **^{211}Bi** $+ {}^{4}He$	$\lambda_{215_{At}} = 2 \times 10^{11}$ yr^{-1}	0.1 ms
$^{211}Pb \rightarrow$ **^{211}Bi**	$\lambda_{211_{Pb}} = 1.01 \times 10^{4}$ yr^{-1}	36.1 min
^{211}Bi to (i) ^{211}Po and (ii) $^{207}Tl + {}^{4}He$	$\lambda_{211_{Bi,total}} = 1.70 \times 10^{5}$ yr^{-1}	128 s
$\quad$**^{211}Bi** $\rightarrow {}^{211}Po$ (branch 1; 0.276%)	$\quad\lambda_{211_{Bi,1}} = 469$ yr^{-1}	
$\quad$**^{211}Bi** $\rightarrow {}^{207}Tl + {}^{4}He$ (branch 2; 99.724%)	$\quad\lambda_{211_{Bi,2}} = 1.70 \times 10^{5}$ yr^{-1}	
$^{211}Po \rightarrow {}^{207}Pb + {}^{4}He$	$\lambda_{211_{Po}} = 4.24 \times 10^{7}$ yr^{-1}	0.516 s
$^{207}Tl \rightarrow {}^{207}Pb$	$\lambda_{207_{Tl}} = 7.64 \times 10^{4}$ yr^{-1}	4.77 min

Note. The steps from ^{235}U to ^{207}Pb are 7 α-decays and 4 β-decays. The net reaction is $^{235}U \rightarrow {}^{207}Pb + 7{}^{4}He$. See Table 2-2a for notation.

Table 2-2c Individual decay steps in the decay chain of ^{232}Th

Reaction	Decay constant	Half-life
^{232}Th → ^{228}Ra + ^{4}He	$\lambda_{232} = 4.948 \times 10^{-11}$ yr^{-1}	14.01 Byr
^{228}Ra → ^{228}Ac	$\lambda_{228_{Ra}} = 0.1203$ yr^{-1}	5.76 yr
^{228}Ac → ^{228}Th	$\lambda_{228_{Ac}} = 988$ yr^{-1}	6.15 h
^{228}Th → ^{224}Ra + ^{4}He	$\lambda_{228_{Th}} = 0.362$ yr^{-1}	1.913 yr
^{224}Ra → ^{220}Rn + ^{4}He	$\lambda_{224_{Ra}} = 69.2$ yr^{-1}	3.66 d
^{220}Rn → ^{216}Po + ^{4}He	$\lambda_{220_{Rn}} = 3.93 \times 10^5$ yr^{-1}	55.6 s
^{216}Po → ^{212}Pb + ^{4}He	$\lambda_{216_{Po}} = 1.51 \times 10^8$ yr^{-1}	0.145 s
^{212}Pb → ^{212}Bi	$\lambda_{212_{Pb}} = 571$ yr^{-1}	10.64 hr

^{212}Bi to (i) ^{212}Po and (ii) ^{208}Tl + ^{4}He
  ^{212}Bi → ^{212}Po (branch 1; 64.06%)
  ^{212}Bi → ^{208}Tl + ^{4}He (branch 2; 35.94%)
	$\lambda_{212_{Bi,total}} = 6.02 \times 10^3$ yr^{-1}
	$\lambda_{212_{Bi,1}} = 3.86 \times 10^3$ yr^{-1}
	$\lambda_{212_{Bi,1}} = 2.16 \times 10^3$ yr^{-1}
		1.009 hr

^{212}Po → ^{208}Pb + ^{4}He	$\lambda_{212_{Po}} = 7.34 \times 10^{13}$ yr^{-1}	0.298 μs
^{208}Tl → ^{208}Pb	$\lambda_{212_{Po}} = 1.194 \times 10^5$ yr^{-1}	3.053 min

Note. The steps from ^{232}Th to ^{208}Pb are 6 α-decays and 4 β-decays. The net reaction is ^{232}Th → ^{208}Pb + 6^{4}He. See Table 2-2a for notation.

difficult once you know how to obtain eigenvalues of a matrix, the final result is very messy. Another way to solve it is to go step by step. The slowest step (rate-determining step) in this overall reaction is the first step, the decay of ^{238}U to ^{234}Th with $\lambda_{238} = 1.55125 \times 10^{-10}$ yr^{-1} (half-life = 4.468 Byr). This is the *rate-determining step* of the reaction and controls the production of ^{206}Pb. The second slowest step is the decay of ^{234}U with $\lambda_{238_U} = 2.84 \times 10^{-6}$ yr^{-1} (half-life = 0.244 Myr). The difference in the rate constants between the slowest and the second slowest decays is more than four orders of magnitude. The large differences in the decay constants are important in simplifying the treatment of this series, and in developing the concept of *secular equilibrium* below. The first equation in the set of equations, which is for ^{238}U, can be solved easily:

$$^{238}U = {^{238}U_0} e^{-\lambda_{238}t}, \tag{2-93}$$

where subscript 0 means initial. Define the *activity* of a radioactive nuclide as the decay rate (number of decays per unit time): $A_{238_U} = \lambda_{238}{}^{238}U$. Note this activity (radioactivity) differs from the activity of a component in thermodynamics. Multiplying the above equation by λ_{238} transforms it to

$$A_{238_U} = A^0_{238_U} e^{-\lambda_{238} t}, \tag{2-94}$$

where superscript 0 means initial. The second in the set of equations (Equation 2-92b) can be solved next:

$$d^{234}Th/dt = \lambda_{238_U}{}^{238}U - \lambda_{234_{Th}}{}^{234}Th. \tag{2-95}$$

$$d^{234}Th/dt + \lambda_{234_{Th}}{}^{234}Th = \lambda_{238}{}^{238}U_0 e^{-\lambda_{238} t}. \tag{2-96}$$

Or

$$d(^{234}Th e^{\lambda_{234Th} t})/dt = \lambda_{238}{}^{238}U_0 e^{(\lambda_{234Th} - \lambda_{238}) t}. \tag{2-97}$$

Hence,

$$^{234}Th = {}^{234}Th_0 e^{-\lambda_{234Th} t} + \frac{\lambda_{238}{}^{238}U_0}{\lambda_{234_{Th}} - \lambda_{238}} (e^{-\lambda_{238} t} - e^{-\lambda_{234Th} t}). \tag{2-98}$$

Multiplying the above equation by $\lambda_{234_{Th}}$, then

$$A_{234_{Th}} = \left(A^0_{234_{Th}} - \frac{\lambda_{234_{Th}}}{\lambda_{234_{Th}} - \lambda_{238_U}} A^0_{238_U} \right) e^{-\lambda_{234Th} t} + \frac{\lambda_{234_{Th}}}{\lambda_{234_{Th}} - \lambda_{238_U}} A_{238_U}. \tag{2-99}$$

Because $\lambda_{234_{Th}} - \lambda_{238} \approx \lambda_{234_{Th}}$, the above equation becomes

$$A_{234_{Th}} \approx (A^0_{234_{Th}} - A^0_{238_U}) e^{-\lambda_{234Th} t} + A_{238_U}. \tag{2-100}$$

The above equation may be written in terms of excess activity $\Delta A = A_{234Th} - A_{238U}$ as

$$\Delta A = \Delta A^0 e^{-\lambda_{234} t}. \tag{2-101}$$

That is, the initial extra or deficient ^{234}Th activity would decay away exponentially. When $\lambda_{234Th} t \gg 1$ (for example, if the system has been closed for 20 years, $\lambda_{234Th} t = 21$; and $\lambda_{238U} t \approx 3.1 \times 10^{-10}$), $\Delta A = 0$, meaning that ^{234}Th activity equals ^{238}U activity. That is,

$$A_{234_{Th}} \approx A_{238_U}. \tag{2-102}$$

In other words, $\lambda_{238_U}{}^{238}U \approx \lambda_{234Th}{}^{234}Th$. In summary, because $\lambda_{234Th} \gg \lambda_{238U}$, when $\lambda_{234Th} t \gg 1$, the activity of ^{234}Th is the same as that of ^{238}U. This is referred to as *secular equilibrium*, meaning *steady state*.[2] Because of the simplicity of Equation 2-102 compared to the equation of $\lambda_{238_U}{}^{238}U \approx \lambda_{234_{Th}}{}^{234}Th$, activity is often used in treating decay series. This example also demonstrates that, for an unstable (i.e., a rapidly reacted) transient species such as ^{234}Th, a steady state is reached when the timescale under consideration is much longer than the half-life of the transient. Under steady state, the net reaction rate of the unstable transient is (Equation 2-95) $A_{238_U} - A_{234_{Th}} = 0$. The accuracy of the steady-state assumption can be found as follows:

$$\begin{aligned}
\mathrm{d}^{234}\mathrm{Th}/\mathrm{d}t &= A_{238_U} - A_{234_{Th}} \\
&= \frac{-\lambda_{238_U}}{\lambda_{234_{Th}} - \lambda_{238_U}} A_{238_U} - \left(A^0_{234_{Th}} - \frac{\lambda_{234_{Th}}}{\lambda_{234_{Th}} - \lambda_{238_U}} A^0_{238_U} \right) e^{-\lambda_{234_{Th}}t}.
\end{aligned} \tag{2-103}$$

As $e^{-\lambda_{234}t}$ approaches zero, then $\mathrm{d}^{234}\mathrm{Th}/\mathrm{d}t \approx -(\lambda_{238_U}/\lambda_{234_{Th}})A_{238_U}$. That is, the relative error in the steady-state assumption for the ^{234}Th species, characterized by $(\mathrm{d}^{234}\mathrm{Th}/\mathrm{d}t)/A_{238_U}$ is $(\lambda_{238_U}/\lambda_{234_{Th}}) = 1.48 \times 10^{-12}$, the ratio of the two reaction rate coefficients. For other chain reactions, the relative error in assuming the second species reached steady-state may be estimated similarly. Although the relative error in the steady-state assumption for subsequent species in the decay chain is more complicated to estimate, it suffices to say that the error is small.

Because λ_{238_U} is much smaller than the other decay constants (smaller than the second slowest decay in the chain by 4 orders of magnitude), the intermediate steps can reach steady state with low and roughly constant concentration. If the time interval we are interested in is longer than a couple of Myr, which is much longer than $1/\lambda_{234_U}$ (the decay of ^{234}U is the second slowest in the chain), then steady state is reached:

$$\begin{aligned}
A_{238_U} &= A_{234_{Th}} = A_{234_{Pa}} = A_{234_U} = A_{230_{Th}} = A_{226_{Ra}} = A_{222_{Rn}} = A_{218_{Po}} \\
&= A_{218_{At}} + A_{214_{Pb}} = A_{218_{Rn}} + A_{214_{Bi}} = A_{214_{Po}} + A_{210_{Tl}} = A_{210_{Pb}} \\
&= A_{210_{Bi}} + A_{206_{Hg}} = A_{210_{Po}} + A_{206_{Tl}}.
\end{aligned} \tag{2-104}$$

The above condition of equal activity of all radioactive nuclides in a decay chain (except for branch decays) is known as *secular equilibrium*. More detailed solutions for the concentration evolution of intermediate species can be found in Box 2-6.

---

[2] True thermodynamic equilibrium is reached when Gibbs free energy is minimized for a closed system maintained at constant temperature and constant pressure, or when entropy is maximized for an isolated system. For radioactive decays, the true equilibrium state is when all the radioactive nuclides are gone. For other nuclear reactions discussed later, there is no real equilibrium except when the reactants are all gone. Nuclear physicists refer to steady states as "equilibrium" (e.g., for nuclear reactions such as nuclear hydrogen burning, discussed later), or secular "equilibrium" for radioactive decays. The term "secular equilibrium" is retained here, but the term "equilibrium" is not used when discussing nuclear hydrogen burning.

Box 2.6 More solutions and discussion of the decay chain

To solve for the concentration evolution of the intermediate species, following the same procedure in the text but using simple notation of $N_1 = {}^{238}U$, $N_2 = {}^{234}Th$, $N_3 = {}^{234}Pa$, $N_4 = {}^{234}U$, $N_5 = {}^{230}Th$, etc., the evolution of the concentration for the first four nuclides in the decay chain can be found as

$$N_1 = N_1^0 e^{-\lambda_1 t},$$

$$N_2 = N_2^0 e^{-\lambda_2 t} + \lambda_1 N_1^0 \left[\frac{e^{-\lambda_1 t}}{\lambda_2 - \lambda_1} + \frac{e^{-\lambda_2 t}}{\lambda_1 - \lambda_2} \right], \text{ or } A_2 \approx (A_2^0 - A_1^0) e^{-\lambda_2 t} + A_1,$$

$$N_3 = N_3^0 e^{-\lambda_3 t} + \lambda_2 N_2^0 \left[\frac{e^{-\lambda_2 t}}{\lambda_3 - \lambda_2} + \frac{e^{-\lambda_3 t}}{\lambda_2 - \lambda_3} \right]$$

$$+ \lambda_1 \lambda_2 N_1^0 \left[\frac{e^{-\lambda_1 t}}{(\lambda_2 - \lambda_1)(\lambda_3 - \lambda_1)} + \frac{e^{-\lambda_2 t}}{(\lambda_1 - \lambda_2)(\lambda_3 - \lambda_2)} \right.$$

$$\left. + \frac{e^{-\lambda_3 t}}{(\lambda_1 - \lambda_3)(\lambda_2 - \lambda_3)} \right],$$

$$N_4 = N_4^0 e^{-\lambda_4 t} + \lambda_3 N_3^0 \left[\frac{e^{-\lambda_3 t}}{\lambda_4 - \lambda_3} + \frac{e^{-\lambda_4 t}}{\lambda_3 - \lambda_4} \right]$$

$$+ \lambda_2 \lambda_3 N_2^0 \left[\frac{e^{-\lambda_2 t}}{(\lambda_3 - \lambda_2)(\lambda_4 - \lambda_2)} + \frac{e^{-\lambda_3 t}}{(\lambda_2 - \lambda_3)(\lambda_4 - \lambda_3)} \right.$$

$$\left. + \frac{e^{-\lambda_4 t}}{(\lambda_2 - \lambda_4)(\lambda_3 - \lambda_4)} \right]$$

$$+ \lambda_1 \lambda_2 \lambda_3 N_1^0 \left[\frac{e^{-\lambda_1 t}}{(\lambda_2 - \lambda_1)(\lambda_3 - \lambda_1)(\lambda_4 - \lambda_1)} + \frac{e^{-\lambda_2 t}}{(\lambda_1 - \lambda_2)(\lambda_3 - \lambda_2)(\lambda_4 - \lambda_2)} \right.$$

$$\left. + \frac{e^{-\lambda_3 t}}{(\lambda_1 - \lambda_3)(\lambda_2 - \lambda_3)(\lambda_4 - \lambda_3)} + \frac{e^{-\lambda_4 t}}{(\lambda_1 - \lambda_4)(\lambda_2 - \lambda_4)(\lambda_3 - \lambda_4)} \right].$$

One can notice the symmetry of the above equations, so that solution for N_5, N_6, N_7, and N_8 can also be written down, although the equation is very long. However, starting from N_9, parallel decay paths are encountered, and the solutions are more complicated and cannot be written down from the symmetry of the above equations.

When $\lambda_{234\text{Th}} t \gg 1$ and when $\lambda_{234\text{Pa}} t \gg 1$, then

$$N_2 = \frac{\lambda_1 N_1}{\lambda_2 - \lambda_1} \quad \text{and} \quad N_3 = \frac{\lambda_1 \lambda_2 N_1}{(\lambda_2 - \lambda_1)(\lambda_3 - \lambda_1)}.$$

Hence,

(continued on next page)

(continued from previous page)

$$\lambda_1 N_1 = (\lambda_2 - \lambda_1)N_2 = \frac{\lambda_2 - \lambda_1}{\lambda_2}(\lambda_3 - \lambda_1)N_3 = \frac{(\lambda_2 - \lambda_1)(\lambda_3 - \lambda_1)}{\lambda_2 \lambda_3}(\lambda_4 - \lambda_1)N_4$$

$$= \frac{(\lambda_2 - \lambda_1)(\lambda_3 - \lambda_1)(\lambda_4 - \lambda_1)}{\lambda_2 \lambda_3 \lambda_4}(\lambda_5 - \lambda_1)N_5$$

$$= \frac{(\lambda_2 - \lambda_1)(\lambda_3 - \lambda_1)(\lambda_4 - \lambda_1)(\lambda_5 - \lambda_1)}{\lambda_2 \lambda_3 \lambda_4 \lambda_5}(\lambda_6 - \lambda_1)N_6$$

$$= \frac{(\lambda_2 - \lambda_1)(\lambda_3 - \lambda_1)(\lambda_4 - \lambda_1)(\lambda_5 - \lambda_1)(\lambda_6 - \lambda_1)}{\lambda_2 \lambda_3 \lambda_4 \lambda_5 \lambda_6}(\lambda_7 - \lambda_1)N_7$$

For example, secular equilibrium between N_4 (^{234}U) and N_5 (^{234}Th) is

$$\lambda_4 N_4 = \lambda_5 N_5, \quad \text{or} \quad A_4 = A_5.$$

Because λ_{238_U} is much smaller than the other decay constants, the approximate secular equilibrium equations have a precision of better than 0.01%.

If a system initially contains only ^{238}U but no other daughters of ^{238}U in the decay chain, in the first hundred thousand years, the decay of ^{238}U would mostly produce the intermediates ^{234}U, ^{230}Th, and ^{226}Ra, and the final stable product ^{206}Pb. Figure 2-10 shows the activity evolution of selected species with time.

The concentrations of intermediate species may affect ^{238}U–^{206}Pb dating if the age is small. Ignoring the concentrations of intermediate species, then

$$^{206}\text{Pb} = {}^{206}\text{Pb}_0 + ({}^{238}\text{U}_0 - {}^{238}\text{U}) = {}^{206}\text{Pb}_0 + {}^{238}\text{U}(e^{\lambda_{238_U} t} - 1). \tag{2-105}$$

And the ^{238}U–^{206}Pb isochron equation (Section 1.3.5.2) is

$$^{206}\text{Pb}/^{204}\text{Pb} = ({}^{206}\text{Pb}/^{204}\text{Pb})_0 + ({}^{238}\text{U}/^{204}\text{Pb})(e^{\lambda_{238_U} t} - 1). \tag{2-106}$$

However, because intermediate species also consume ^{238}U, Equation 2-105 does not hold for short times (or low radiogenic $^{206}\text{Pb}^* = {}^{206}\text{Pb} - {}^{206}\text{Pb}_0$), that is, $^{206}\text{Pb}^* \neq ({}^{238}\text{U}_0 - {}^{238}\text{U})$. The calculated concentration of $^{206}\text{Pb}^*$ is plotted in Figure 2-11, and compared with that of $\Delta^{238}\text{U} = ({}^{238}\text{U}_0 - {}^{238}\text{U})$. If the two are identical within error, then the isochron method above is accurate. If the age is young, one has to take special care.

One simple way to estimate the effect of intermediate species is as follows. After steady state (secular equilibrium) is reached, the concentrations of the intermediate species do not vary much, and the decay of ^{238}U would basically produce ^{206}Pb. Hence, at the timescale longer than 2 Myr, we have

$$d^{238}\text{U}/dt = -\lambda_{238_U}{}^{238}\text{U} \tag{2-107}$$

$$^{234}\text{Th}/^{238}\text{U} \approx \lambda_{238_U}/\lambda_{234_{Th}} = 1.48 \times 10^{-11} \ll 1 \tag{2-108a}$$

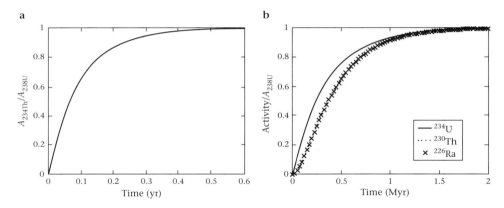

Figure 2-10 Evolution of the activity of (a) ^{234}Th, and (b) ^{234}U, ^{230}Th, and ^{226}Ra with time when there was initially no ^{234}Th, ^{234}U, ^{230}Th, and ^{226}Ra.

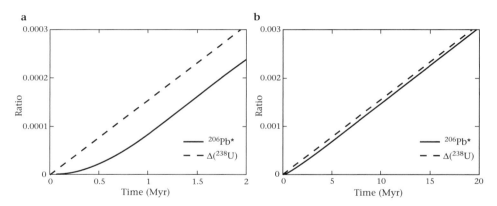

Figure 2-11 Evolution of ^{206}Pb*/^{238}U and Δ^{238}U/^{238}U atomic ratio with time. Initially there was only ^{238}U. ^{206}Pb* means radiogenic ^{206}Pb. Simple isochron dating assumes ^{206}Pb$^\star = \Delta^{238}$U. (a) The first 2 Myr, and (b) over the first 20 Myr.

$$^{234}\text{Pa}/^{238}\text{U} \approx \lambda_{238_\text{U}}/\lambda_{234_\text{Pa}} = 5.0 \times 10^{-16} \ll 1 \qquad (2\text{-}108\text{b})$$

$$^{234}\text{U}/^{238}\text{U} \approx \lambda_{238_\text{U}}/\lambda_{234_\text{U}} = 5.46 \times 10^{-5} \ll 1 \qquad (2\text{-}108\text{c})$$

$$^{230}\text{Th}/^{238}\text{U} \approx \lambda_{238_\text{U}}/\lambda_{230_\text{Th}} = 1.72 \times 10^{-5} \ll 1 \qquad (2\text{-}108\text{d})$$

$$^{226}\text{Ra}/^{238}\text{U} \approx \lambda_{238_\text{U}}/\lambda_{226_\text{Ra}} = 3.58 \times 10^{-7} \ll 1 \qquad (2\text{-}108\text{e})$$

$$\vdots$$

$$\text{d}^{206}\text{Pb}/\text{d}t \approx \lambda_{238\text{U}}{}^{238}\text{U} \qquad (2\text{-}109)$$

That is, the production of ^{206}Pb is mainly determined by the rate-determining first step. The total number of atoms of intermediate species is about 72 ppm of the amount of ^{238}U, and about 3/4 of the intermediates (in terms of concentrations or the number of atoms) is ^{234}U.

On the basis of the above discussion, if a relative precision of 1% in age is required, the application of the ^{238}U–^{206}Pb geochronometer to ages younger than 47 Ma (if instrumental analytical accuracy allows such determination) would require a careful account of the intermediate species, where 47 Myr is the time required for 7200 ppm of all ^{238}U to decay and $72/7200 = 0.01$. If both ^{238}U and ^{234}U are incorporated with equal activity, then for ages younger than 11 Ma it would be necessary to consider the intermediate species.

For the ^{235}U decay series to ^{207}Pb, the slowest step (and hence the rate-determining step) is the decay of ^{235}U with decay constant of 9.8485×10^{-10} yr^{-1} (half-life 703.81 Myr). The second slowest step is the decay of ^{231}Pa (i.e., ^{231}Pa is the most stable intermediate) with half-life of 32.8 kyr. The third slowest step is the decay of ^{227}Ac (half-life 21.8 yr). That is, intermediates with long half-lives are less abundant compared to the decay series of ^{238}U. At secular equilibrium, the total number of atoms of all intermediate species is 47 ppm of that of ^{235}U. If relative precision of 1% in age is needed, application of the ^{235}U–^{207}Pb geochronometer to ages younger than 4.8 Ma requires careful evaluation of the intermediate species, where 4.8 Myr is the time required for 4700 ppm of total ^{235}U to decay.

For the ^{232}Th decay series to ^{208}Pb, the slowest step is the decay of ^{232}Th with decay constant of 4.948×10^{-11} yr^{-1} (half-life 14.01 Byr). The longest-lived intermediate is ^{228}Ra with a half-life of only 5.75 yr. That is, there are basically no intermediates with long enough half-lives of geologic interest. At secular equilibrium, the total number of atoms of intermediate species is 0.55 ppb of that of ^{232}Th. Hence, there is no need to account for the concentrations of the intermediate species when applying the ^{232}Th–^{208}Pb geochronometer.

2.2.1.2 Disturbed decay chain and applications

In the long history of the Earth, the secular equilibrium of a decay chain is often reached but may be disturbed by geological processes because the elements in the decay chain have different chemical properties. That is, fractionation between different phases due to chemical equilibrium (or even disequilibrium process) produces secular disequilibrium in each phase if the whole system is assumed to be at secular equilibrium. To clarify, consider a two-phase system such as water and a clay mineral. Suppose the first phase takes U preferentially and the second phase takes Th preferentially. The whole system is at secular equilibrium at $A_{238_U} = A_{234_{Th}} = A_{234_{Pa}} = A_{234_U} = A_{230_{Th}} = \cdots$. However, in the first phase, $A_{238_U} > A_{230_{Th}}$, and in the second phase, $A_{234_U} < A_{230_{Th}}$. If the two phases are separated (for example, by sinking of solid particles) rapidly, each phase would be out of secular equilibrium.

Such disturbance and the subsequent return of a disturbed system back to secular equilibrium may provide powerful tools to study a variety of geological processes, including the determination of recent age, the estimation of sedimentation rate, and the investigation of the dynamics of mantle partial melting. Because of the practical applications, the theories have been well developed.

Other conditions being equal, the intermediate species with longer half-lives in a decay series have more opportunities to be fractionated from their parents. Hence, in the decay series of ^{238}U, two nuclides ^{230}Th and ^{226}Ra have a greater chance to be fractionated. In the ^{235}U decay series, ^{231}Pa (half-life 32.8 kyr) has the greatest chance to be fractionated. In the ^{232}Th decay series, all the intermediate species have short half-lives (the longest half-life of intermediates is 5.75 yr for ^{228}Ra ($\lambda = 0.1205$ yr^{-1}) and the disturbance of this decay system does not have much utility. That is, the U-series (including ^{238}U and ^{235}U series) disequilibrium is much more often applied. Some examples of disturbed decay chain (i.e., fractionation of the intermediate species) are given below:

(1) Under oxidized conditions on the Earth's surface, U solubility in water is high in the form of UO_2^{2+} (where the oxidation state of U is +6), but Th (as Th^{4+}) solubility is extremely low (Broecker and Peng, 1982). Hence, U stays in water, whereas Th isotopes (^{232}Th, plus ^{238}U decay products of ^{230}Th with half-life of 75,400 years and ^{234}Th with half-life of 24.1 days) would precipitate into sediment. For example, $^{232}Th/^{238}U$ ratio is about 4 in the crust, but is 5 orders of magnitude lower in seawater (about 4×10^{-5}, Chen et al., 1986). In the decay chain of ^{238}U, ^{234}Th activity is almost the same as ^{238}U activity, but ^{230}Th activity is much smaller, indicating that the timescale for removing Th from water by sedimentation is much longer than 24.1 days (the half-life of ^{234}Th) and much shorter than 75,400 years (the half-life of ^{230}Th). If young sediment is measured, there would be overabundance of ^{230}Th, compared to ^{238}U. The decay of the extra ^{230}Th activity may be used to date sedimentation rate (Chapter 5). In the ^{235}U series, Pa solubility is extremely small. Hence, Pa is rapidly removed from water. There is an underabundance of ^{231}Pa in water and overabundance of ^{231}Pa in young sediment.

(2) Because Th/U ratio is low in seawater, it is also low in corals grown from seawater, with $^{232}Th/^{238}U$ of 0.8×10^{-5} to 12×10^{-5} (Edwards et al., 1986/87). This huge deficiency in Th means that initial ^{230}Th (the fifth nuclide in the ^{238}U decay series) in coral may be ignored. By measuring the activity ratio of $^{230}Th/^{238}U$, it is possible to estimate the age of coral and also to calibrate the ^{14}C geochronometer (Chapter 5).

(3) During mantle partial melting, the partition coefficients of Th, Pa, and Ra are different from that of U. Assuming the melt and the mantle residue as a whole maintains secular equilibrium, if the melting process is slow, there is chemical equilibrium between the phases, which means each phase (such as the melt phase) is out of secular equilibrium because of different partition coefficients (McKenzie, 1985).

If the melt is then extracted rapidly at the midocean ridges, young midocean ridge basalt would show disequilibrium in terms of activity ratios. Hence, U-series disequilibrium measured on recent mantle-derived basalt may provide information on the dynamics of mantle partial melting and melt extraction processes. For example, consider the partial melting of garnet peridotite. Both Th and U are incompatible elements and Th is more incompatible than U, meaning that Th and U are strongly enriched in the melt phase but Th/U ratio in the melt is greater than in the solid residue. To find the concentration of a trace element in the melt, it is necessary to assume equilibrium (hence, the melting process should be slow). If the batch-melting model is adopted, then the concentration of an element (or radioactive nuclide) and nuclide ratio may be expressed as (Gast, 1968; Shaw, 1970; Zou, 2007)

$$C_{i,\text{melt}} = C_{i,0}/[F + D_i(1 - F)], \tag{2-110}$$

and

$$(C_i/C_j)_{\text{melt}} = (C_i/C_j)_0[F + D_j(1 - F)]/[F + D_i(1 - F)], \tag{2-111}$$

where $C_{i,\text{melt}}$ is the concentration of element i in the melt, $C_{i,0}$ is the concentration of element i in the whole system (melt plus solid phases), D_i is partition coefficient (note D more often refers to the diffusion coefficient) of element i, and F is the degree of partial melting. If $F \gg D_i$ and $F \gg D_j$, then $[F + D_j(1 - F)] \approx F$ and $[F + D_j(1 - F)] \approx F$, leading to $(C_i/C_j)_{\text{melt}} = (C_i/C_j)_0$. To produce $^{230}\text{Th}/^{238}\text{U}$ disequilibrium in the melt, the Th/U ratio in the melt must be significantly different from the whole system in secular equilibrium. Hence, the degree of partial melting must be small, of the order of the partition coefficients (about 10^{-3} for Th and U, and about 10^{-4} for Ra and Pa). If such a melt is rapidly (meaning a timescale not much longer than the half-life of the intermediate nuclide) extracted, erupted, solidified, sampled, and analyzed, $^{230}\text{Th}/^{238}\text{U}$ disequilibrium would be observed. If the melt is extracted slowly but there is continuous reaction with the matrix mantle (i.e., the mantle continues to contribute/consume Th and U in the melt through partial melting and diffusive exchange), U-series disequilibrium would also be preserved. The extra ^{230}Th activity begins to decay away once the melt is isolated from the mantle. McKenzie (1985) was the first to apply U-series disequilibrium to model the dynamics of mantle partial melting. Peate and Hawkesworth (2005) and Zou (2007) reviewed the applications of U-series disequilibrium to mantle melting and magma differentiation.

(4) Crystallization of magma can also fractionate the elements. For example, the Ra/Th ratio may be high in plagioclase and especially in potassium feldspar (Ra can enter potassium feldspar structure through RaAl substitution of KSi) (e.g., Cooper and Reid, 2003). Therefore, there may be U-series disequilibrium in phenocrysts. By measuring activities of U-series species, the timescale of magma differentiation may be constrained. In general, the intermediate with longer

half-lives has greater opportunity to be fractionated from its parent because there is more time for fractionation before its decay.

(5) The gaseous species Rn (^{222}Rn, ^{220}Rn, ^{219}Rn, and ^{218}Rn) may escape from rocks into groundwater and the atmosphere and is a health hazard. Because Rn is denser than air, it tends to stay near the ground, such as in the basement of a house. Radon or its airborne radiogenic daughters may be inhaled. Inhaled Rn would decay to Po, sticking to the lung and undergoing a series of further α-decays in the lung. These energetic α-particles can disrupt DNA in lung cells, potentially causing lung cancer. According to the U.S. Environmental Protection Agency, radon is the number one cause of lung cancer among nonsmokers, responsible for about 21,000 deaths a year. Hence, a radon test is often requested by potential home buyers.

Below, the return of a disturbed system to secular equilibrium is examined. Suppose a decay system is disturbed so that ^{234}Th activity differs from ^{238}U activity. Based on Equation 2-100, the excess ^{234}Th activity, i.e., $A^0_{234_{Th}} - A^0_{238_U}$, would decay away with the decay constant of $\lambda_{234_{Th}}$, and after about 5 half-lives (or 10 half-lives depending on the measurement precision) of ^{234}Th, $A_{234_{Th}}$ would be the same as A_{238_U}.

A more useful and also more difficult derivation is the return of $A_{230_{Th}}$ to secular equilibrium because of the many terms of the intermediate species. Starting from the general solution in Box 2-6, using the magnitudes of values of λ_i's to simplify,

$$A_5 = A_1 + \left[\frac{\lambda_5 A^0_4}{\lambda_5 - \lambda_4} - \frac{\lambda_5 A^0_1}{(\lambda_5 - \lambda_4)}\right] e^{-\lambda_4 t} + \left[A^0_5 - \frac{\lambda_5 A^0_4}{\lambda_5 - \lambda_4} + \frac{\lambda_4 A^0_1}{(\lambda_5 - \lambda_4)}\right] e^{-\lambda_5 t}. \quad (2\text{-}112)$$

If the activity of ^{234}U equals that of ^{238}U (these are two isotopes of the same element), then the evolution of ^{230}Th activity would be

$$A_{230_{Th}} - A_{238_U} = (A^0_{238_{Th}} - A^0_{238_U}) e^{-\lambda_{230_{Th}} t} . \quad (2\text{-}113)$$

That is, the excess activity would decay away with the decay constant of $\lambda_{230_{Th}}$.

Curiously, in water, activity of ^{234}U (A_4) is usually greater than that of ^{238}U, roughly about 1.144 times the activity of ^{238}U (Chen et al., 1986). Furthermore, the activity of ^{234}Th (A_5) in water is negligible. Hence, for corals grown from seawater, Equation 2-112 becomes

$$A_5 = A_1 + \frac{0.144\lambda_5}{(\lambda_5 - \lambda_4)} A^0_1 e^{-\lambda_4 t} + \frac{(\lambda_4 - 1.144\lambda_5)}{(\lambda_5 - \lambda_4)} A^0_1 e^{-\lambda_5 t} . \quad (2\text{-}114)$$

These equations may be applied to dating of corals (Chapter 5).

2.2.2 Chain reactions leading to negative activation energy

In treating the radioactive decay series of ^{238}U, we explored the concepts of rate-determining step and steady state, and learned how they are applied to treat the reaction kinetics. In this section we use an example to explore the concept of quasi-equilibrium and apply it to treat reaction kinetics. This example comes from Bamford and Tipper (1972, pp. 169–170) and also shows how a chain reaction may lead to an apparent negative activation energy.

The oxidation of NO to NO_2 may be written as

$$2NO(gas) + O_2(gas) \rightarrow 2NO_2(gas), \tag{2-115}$$

Experimentally, the reaction is found to be third order with reaction law of $d\xi/dt = k_{115}[NO]^2[O_2]$. From the reaction law, it seems that the reaction is an elementary reaction. However, it was found (Bamford and Tipper, 1972, p. 169) that $k_{115} = T \exp(-0.187 + 1000/T)$ for $293 < T < 500$ K, decreasing with temperature, in contrast with the normal Arrhenian behavior of reactions. A reaction mechanism that involves chain reactions that can explain the apparent negative activation energy is as follows. Suppose the above reaction is accomplished by the following two elementary reactions:

$$\text{Fast reaction: } NO(gas) + NO(gas) \rightleftharpoons N_2O_2(gas), \tag{2-116}$$

$$\text{Slow reaction: } N_2O_2(gas) + O_2(gas) \rightarrow 2NO_2(gas), \tag{2-117}$$

where the first step is the fast step and equilibrium is roughly reached with equilibrium constant:

$$K_{116} = [N_2O_2]/[NO]^2, \tag{2-118}$$

and the second step is slow with rate constant k_{117}. N_2O_2 is not a stable species but has been detected spectroscopically. Therefore, the rate of the formation of NO_2 is

$$d[NO_2]/dt = 2 \, d\xi/dt = 2k_{117}[N_2O_2][O_2] = 2k_{117}K_{116}[NO]^2[O_2]. \tag{2-119}$$

One can therefore see that $k_{115} = k_{117}K_{116}$. K_{116} decreases strongly with temperature because ΔH for the fast reaction (first step) is about -172 kJ. Hence, even though k_{117} behaves normally (i.e., increases with T), k_{115} still decreases with T.

2.2.3 Thermal decomposition of ozone

In this section, we use another chain reaction to show the relation between the steady-state treatment and the quasi-equilibrium treatment. The former is more general than the latter, and leads to more complete but also more complicated results. Ozone, O_3, is present in the stratosphere as the ozone layer, and in the troposphere as a pollutant. Ozone production and destruction in the atmosphere is primarily controlled by photochemical reactions, which are discussed in a later section. Ozone may also be thermally decomposed into oxygen, O_2, although this is not the primary process in the atmosphere. The net (overall) reaction for the thermal decomposition of ozone is

$$2O_3(gas) \rightarrow 3O_2(gas). \tag{2-120}$$

In experimental studies, thermal decomposition occurs partly on the surface of the container and partly in the gas phase. The part occurring on the surface of the container is a heterogeneous reaction, and the part occurring in the gas phase is a homogeneous reaction. The homogeneous and heterogeneous parts can be separated by varying the surface/volume ratio. The surface reaction rate is proportional to $[O_3]$, but the homogeneous reaction rate law depends on the concentration of O_2: when $[O_2]$ is very high, the reaction is second order with respect to O_3; when $[O_2]$ concentration is very low, the reaction is first order with respect to O_3. Only the homogeneous reaction is considered here. The inferred homogeneous reaction mechanism is

$$O_3(gas) + M \rightleftharpoons O(gas) + O_2(gas) + M, \tag{2-121}$$

$$O(gas) + O_3(gas) \rightarrow 2O_2(gas), \tag{2-122}$$

where M is a catalyst species in the gas phase (such as Ar), and O is an intermediate species with low concentration. The first step above is rapid and reaches quasi-equilibrium with equilibrium constant K_{121} and rate constants k_{121f} and k_{121b}. The second reaction is slow with rate constant k_{122}. Application of the steady-state treatment to the concentration of O leads to

$$d[O]/dt = k_{121f}[O_3][M] - k_{121b}[O][O_2][M] - k_{122}[O][O_3] = 0. \tag{2-123}$$

Therefore,

$$[O] = \frac{k_{121f}[O_3][M]}{k_{121b}[O_2][M] + k_{122}[O_3]}. \tag{2-124}$$

Hence, the decomposition rate law is

$$-d[O_3]/dt = k_{121f}[O_3][M] - k_{121b}[O][O_2][M] + k_{122}[O][O_3]. \tag{2-125}$$

Inserting the expression for [O] into the above equation gives

$$-\frac{d[O_3]}{dt} = \frac{k_{121f}k_{122}[O_3]^2[M]}{k_{121b}[O_2][M] + k_{122}[O_3]}. \tag{2-126}$$

Therefore, when $[O_2]$ is very high (and hence does not change with time during the reaction), the reaction is second order with respect to O_3. When $[O_2]$ is high and varied from one experiment to another, the reaction rate is inversely proportional to $[O_2]$. When $[O_2]$ concentration is very low, the reaction is first order with respect to O_3. All these are consistent with observations (Benson and Axworthy, 1957).

If the quasi-equilibrium treatment is used to treat Reactions 2-121 and 2-122, the result would be a simpler rate law: $-d[O_3]/dt = k_{122}[O][O_3] = k_{122}K_{121}[O_3]^2/$

$[O_2]$, which is the same as one special case of Equation 2-126 when $[O_2]$ is high, but does not cover the situation when $[O_2]$ is low. Hence, the quasi-equilibrium treatment may be viewed as a special case of the steady-state treatment.

Readers might have noticed that the two chain reactions, (i) Reactions 2-121 and 2-122 and (ii) Reactions 2-116 and 2-117, are similar, but were treated differently. Reactions 2-116 and 2-117 were treated using the quasi-equilibrium assumption, but may also be treated using the steady-state concept. The result is a more complicated expression, which would reduce to the experimental reaction rate law if $k_{117}[O_2] \ll k_{116b}$. Readers can work this problem out as an exercise. Therefore, the quasi-equilibrium treatment is a special case of the steady-state treatment.

2.3 Parallel Reactions

Some net (overall) reactions may be accomplished by several paths, with each path leading to the same end result. Such reactions are called parallel reactions. In a parallel reaction, the overall reaction rate is the sum of all paths:

$$r_{\text{overall}} = r_1 + r_2 + r_3 + \cdots \tag{2-127}$$

where subscripts 1, 2, 3, . . . indicate the individual paths. Hence, the fastest path determines the overall reaction rate, instead of the slowest path. This is in contrast to chain reactions, in which the slowest step determines the overall reaction rate. Each path of a parallel reaction may be a simple elementary reaction or a complicated chain reaction. Parallel reactions have been encountered in the discussion of decay chains and are discussed in more depth in this section. Below, three examples of parallel reactions are discussed to elucidate the principles of treating them.

In treating parallel reaction, two concepts are often used: (i) the concept of rate-determining path, in which the fastest path is the rate-determining path, and (ii) the concept of *steady state*, also called the concept of quasi-stationary states of trace-level intermediates.

2.3.1 Electron transfer between Fe²⁺ and Fe³⁺ in aqueous solution

One example of parallel reactions is the electron transfer between Fe^{2+} and Fe^{3+} in an aqueous solution. One path is through Reaction 2-31:

$$^{56}Fe^{2+}(aq) + {}^{55}Fe^{3+}(aq) \rightleftharpoons {}^{56}Fe^{3+}(aq) + {}^{55}Fe^{2+}(aq), \tag{2-31}$$

The forward reaction rate constant is $k_{31f} = k_{31b} = 0.87 \text{ M}^{-1}\text{s}^{-1}$ (Table 1-1a) for the above reaction, and the backward reaction rate constant is about the same (isotopic fractionation between Fe^{2+} and Fe^{3+} is very small).

In the presence of Cl^- anion, Fe^{3+} may be complexed with Cl^- as $FeCl^{2+}$ and there is a second path for the electron transfer:

$$^{56}Fe^{2+}(aq) + {}^{55}FeCl^{2+}(aq) \rightleftharpoons {}^{56}FeCl^{2+}(aq) + {}^{55}Fe^{2+}(aq), \tag{2-128}$$

where Cl^- is a ligand to Fe^{3+} in $FeCl^{2+}$. Both the forward and backward reaction rate coefficients of Reaction 2-128 are $k_{128f} = k_{128b} = 5.4$ $M^{-1}s^{-1}$ (Table 1-1a), 6.2 times those of Reaction 2-31. The overall forward rate for the electron transfer between the two isotopes of Fe is the sum of the reaction rates of the two paths:

$$\frac{d\xi}{dt} = k_{31f}[^{55}Fe^{3+}][^{56}Fe^{2+}] - k_{31b}[^{56}Fe^{3+}][^{55}Fe^{2+}]$$
$$+ k_{128f}[^{55}FeCl^{2+}][^{56}Fe^{2+}] - k_{128b}[^{56}FeCl^{2+}][^{55}Fe^{2+}].$$

That is,

$$\frac{d\xi}{dt} = k_f[^{55}Fe^{3+}][^{56}Fe^{2+}] - k_b[^{56}Fe^{3+}][^{55}Fe^{2+}], \tag{2-129}$$

where $k_f = k_{31f} + k_{128f}[FeCl^{2+}]/[Fe^{3+}]$ and is the overall forward (and backward) reaction rate coefficient. Solution in Section 2.1.2.4 can be applied to the parallel reactions by letting $k_f = k_{31f} + k_{128f}[FeCl^{2+}]/[Fe^{3+}]$. Furthermore, the importance of the two reaction paths can be easily evaluated: If $[FeCl^{2+}]/[Fe^{3+}] > k_{31f}/k_{128f}$, i.e., if $[FeCl^{2+}]/[Fe^{3+}] > 0.161$, then path 2 contributes more to the overall reaction. If $[FeCl^{2+}]/[Fe^{3+}] < 0.161$, then path 1 contributes more to the overall reaction.

2.3.2 From dissolved CO₂ to bicarbonate ion

Another example of parallel reactions that require more complicated treatment than the above example is the reaction from dissolved CO_2 to form HCO_3^-. The following accounts are based on Lasaga (1998). Let's first consider the case of very low HCO_3^- concentration so that the backward reaction does not have to be considered. One path is

$$CO_2(aq) + OH^-(aq) \rightarrow HCO_3^-(aq). \tag{2-130}$$

The equilibrium constant is $K_{130} = 10^{7.9}$ M^{-1} and the reaction rate law is

$$d[HCO_3^-]/dt = k_{130f}[CO_2][OH^-], \tag{2-131}$$

where the reaction rate constant $k_{130f} = 10^{3.924}$ $s^{-1}M^{-1}$. The rate of HCO_3^- formation depends on the concentration of OH^-. For example, for a pH of 7, $d[HCO_3^-]/dt = 10^{-3.076}[CO_2]$.

A second path that involves a chain reaction is

$$CO_2(aq) + H_2O(aq) \rightarrow H_2CO_3(aq); \tag{2-8}$$

$$H_2CO_3(aq) \rightarrow H^+(aq) + HCO_3^-(aq). \tag{2-132}$$

In the above chain reaction, the first step is the slow step and the second is the rapid step. The equilibrium constant for Reaction 2-8 is $K_8 = 0.00287$, and the forward rate constant is $k_{8f} = 0.043$ s^{-1}. The equilibrium constant for Reaction 2-132 is $K_{132} = 10^{-3.77}$ and the rate constant is $k_{132f} = 10^{6.9}$ s^{-1}. Hence, the first step is the rate-determining step. Using the steady-state concept, the reaction rate law is (ignoring the backward reaction of Reaction 2-132):

$$d\{[HCO_3^-] + [H_2CO_3]\}/dt = k_{8f}[CO_2]. \tag{2-133}$$

Assuming $[H^+]$ is constant (e.g., it is buffered), then

$$d[HCO_3^-]/dt = k_{8f}[CO_2]/(1 + [H^+]/K_{132}). \tag{2-134}$$

Yet a third path is the following chain reactions:

$$CO_2(aq) + H_2O(aq) \rightarrow H_2CO_3(aq); \tag{2-8}$$

$$H_2CO_3(aq) + OH^-(aq) \rightarrow H_2O(aq) + HCO_3^-(aq). \tag{2-135}$$

This path differs from the second path only in the second step. For simplicity of considerations below, only the first two paths are considered.

The parallel paths lead to the same net result of converting CO_2 into HCO_3^-. For equilibrium considerations, it does not matter which reactions one writes to calculate the equilibrium species concentrations. However, in kinetics, one has to consider the kinetics of all paths. To evaluate the relative importance of path 1 and path 2, we compare $\{d[HCO_3^-]/dt\}_{path1}$ and $\{d[HCO_3^-]/dt\}_{path2}$:

$$\{d[HCO_3^-]/dt\}_{path1} = k_{130f}[CO_2][OH^-], \tag{2-136}$$

$$\{d[HCO_3^-]/dt\}_{path2} = k_{8f}[CO_2]/(1 + [H^+]/K_{132}). \tag{2-137}$$

Hence, if $[OH^-] > (k_{8f}/k_{130f})/(1 + [H^+]/K_{132})$, i.e., if $[OH^-] > 10^{-5.29}/(1 + [H^+]/K_{132})$, the first step would be more important. For example, if pH > 8.71, then the first path is more important. Otherwise, the second path is more important.

Because Reactions 2-130, 2-8, and 2-132 are reversible, the backward reaction should be considered for more quantitative analyses. Hence, for the first reaction path, the rate is

$$\{d[HCO_3^-]/dt\}_{path1} = k_{130f}[CO_2][OH^-] - k_{130b}[HCO_3^-]. \tag{2-138}$$

The mean reaction time is (Table 2-1)

$$\tau_{path1} = \frac{1}{k_{130f}([CO_2]_\infty + [OH^-]) + k_{130b}}, \tag{2-139}$$

with $k_{130f} = 10^{3.924}$ s^{-1}M^{-1} and $k_{130b} = 10^{-3.976}$ s^{-1}.

For the second reaction path, we have

$$\{d[HCO_3^-]/dt\}_{path2} = k_{132f}[H_2CO_3] - k_{132b}[H^+][HCO_3^-]. \tag{2-140}$$

Using the steady-state assumption,

$$d[H_2CO_3]/dt = k_{8f}[CO_2] - k_{8b}[H_2CO_3] - k_{132f}[H_2CO_3]$$
$$+ k_{132b}[H^+][HCO_3^-] \approx 0. \tag{2-141}$$

That is,

$$k_{8f}[CO_2] - k_{8b}[H_2CO_3] \approx k_{132f}[H_2CO_3] - k_{132b}[H^+][HCO_3^-]. \tag{2-142}$$

Hence,

$$\{d[HCO_3^-]/dt\}_{path2} \approx k_{8f}[CO_2] - k_{8b}[H_2CO_3]. \tag{2-143}$$

Because Reaction 2-8 is the slow step and Reaction 2-132 is the rapid step, H_2CO_3 formed by Reaction 2-8 would react away rapidly through Reaction 2-132 and $[H_2CO_3]$ would be much smaller than the equilibrium concentration. That is,

$$\{d[HCO_3^-]/dt\}_{path2} \approx k_{8f}[CO_2]. \tag{2-144}$$

The mean reaction time for path 2 is hence $1/k_{8f} \approx 23$ s. The total rate for HCO_3^- production is the combination of the two paths. By comparing the rates of two paths, the dominant path can be inferred.

2.3.3 Nuclear hydrogen burning

Nuclear hydrogen burning might be said to be the most important reaction in the solar system because it powers the Sun, and hence indirectly surface processes on planets, including life cycles on the Earth. Hydrogen burning also powers all main sequence stars. The reaction consists of complicated parallel and chain reactions. There are several paths (parallel reactions), with each path being a chain reaction. The following accounts are based on Fowler et al. (1975), Harris et al., (1983), Zeilik et al. (1992), and Lodders and Fegley (1998). In the core of the Sun, the temperature is 10 to 15.5 million kelvins, and at least five parallel paths are present. There are three PP chains (beginning with proton–proton reaction) called *PP I, PP II, and PP III chains*, among which the PP III chain does not contribute significantly. Furthermore, because there are heavy nuclides in the Sun, there are also other paths involving heavier nuclides. One of these paths involves carbon, nitrogen, and oxygen, and is called the *CNO cycle*. Another is the *Ne–Na cycle*. Among the cycles involving heavy nuclides, the CNO cycle is the most important. Because the CNO cycle requires a higher activation energy (or energy barrier, as it is called in nuclear physics), the importance of the CNO cycle increases with temperature. At a temperature of about 18 million kelvins, the CNO cycle and the PP chains generate roughly the same amount of energy. At 15 million kelvins (the temperature at the center of the Sun is 15.51 million kelvins), the energy from the CNO cycle is about an order of magnitude less than that from the PP chains. At about 12 million kelvins, energy from the CNO cycle is about two orders of magnitude less. For simplicity, only the PP chains are considered below.

All PP chains start with the following two reactions:

$$2\,^1\text{H} \rightarrow\, ^2\text{H} \;(1.442\,\text{MeV}), \tag{2-145}$$

$$^2\text{H} +\, ^1\text{H} \rightarrow\, ^3\text{He} \;(5.493\,\text{MeV}). \tag{2-146}$$

After these two steps, the reaction becomes branched. Reaction 2-146 is often followed by the following reaction:

$$2\,^3\text{He} \rightarrow\, ^4\text{He} + 2\,^1\text{H} \;(12.86\,\text{MeV}). \tag{2-147}$$

Reactions 2-145, 2-146, and 2-147 (three steps) comprise the *PP I chain*. In the presence of ^4He, PP II and PP III chains also operate. Because ^4He is the product of hydrogen burning, its concentration increases as the reaction continues, which leads to a rise in the reaction rate of PP II and PP III chains if concentrations of other species are kept constant. That is, PP II and PP III chains are *auto-catalyzed*. In the core of the Sun, among the PP chains, the PP I chain accounts for 69% of ^4He production (the fraction varies with temperature and hence radial position in the Sun), and PP II and PP III chains account for 31%. That is, about 31% of the time, Reaction 2-146 is followed by the following reaction:

$$^3\text{He} +\, ^4\text{He} \rightarrow\, ^7\text{Be} \;(1.586\,\text{MeV}). \tag{2-148}$$

About 99.7% of the resultant ^7Be will react as follows:

$$^7\text{Be} \rightarrow\, ^7\text{Li} \;(0.862\,\text{MeV}), \tag{2-149}$$

$$^7\text{Li} +\, ^1\text{H} \rightarrow 2\,^4\text{He} \;(17.348\,\text{MeV}). \tag{2-150}$$

Reactions 2-145, 2-146, 2-148, 2-149, and 2-150 are called the *PP II chain* (5 steps). About 0.3% of ^7Be from Reaction 2-148 will react as follows:

$$^7\text{Be} +\, ^1\text{H} \rightarrow\, ^8\text{B} \;(0.137\,\text{MeV}), \tag{2-151}$$

$$^8\text{B} \rightarrow\, ^8\text{Be} \;(17.979\,\text{MeV}), \tag{2-152}$$

$$^8\text{Be} \rightarrow 2\,^4\text{He} \;(0.0918\,\text{MeV}), \tag{2-153}$$

Reactions 2-145, 2-146, 2-148, 2-151, 2-152, and 2-153 are called the *PP III chain* (6 steps). Reactions 2-149 and 2-152 are β-decays, but the former is through electron capture, and the latter is through the emission of a positron. Because PP III chain accounts for only about 0.1% of the three PP chains, only PP I and PP II chains (i.e., from Reaction 2-145 to Reaction 2-150) are considered below. The first step (Reaction 2-145) in the PP chains is the slowest step and controls the overall rate of the reaction. The intermediate species have low concentrations.

Hence, the steady-state assumption may be applied. Because the backward reactions are negligible, the rate equations may be written as

$$\frac{d[^1H]}{dt} = -2k_{145}[^1H]^2 - k_{146}[^1H][^2H] + 2k_{147}[^3He]^2 - k_{150}[^1H][^7Li], \qquad (2\text{-}154)$$

$$\frac{d[^2H]}{dt} = k_{145}[^1H]^2 - k_{146}[^1H][^2H] = 0, \qquad (2\text{-}155)$$

$$\frac{d[^3He]}{dt} = k_{146}[^1H][^2H] - 2k_{147}[^3He]^2 - k_{148}[^3He][^4He] = 0, \qquad (2\text{-}156)$$

$$\frac{d[^4He]}{dt} = k_{147}[^3He]^2 - k_{148}[^3He][^4He] + 2k_{150}[^1H][^7Li], \qquad (2\text{-}157)$$

$$\frac{d[^7Be]}{dt} = k_{148}[^3He][^4He] - k_{149}[^7Be] = 0. \qquad (2\text{-}158)$$

$$\frac{d[^7Li]}{dt} = k_{149}[^7Be] - k_{150}[^1H][^7Li] = 0. \qquad (2\text{-}159)$$

The reaction rate coefficients in the above equations may be related to reaction rates per pair of particles λ_{ij} in nuclear physics (e.g., Fowler et al., 1975; Harris et al., 1983) by $k = \lambda_{ij}/(1 + \delta_{ij})$, where $\delta_{ij} = 0$ except for $i = j$, for which $\delta_{ij} = 1$. That is, for Reactions 2-145 and 2-147 in which two identical particles collide to react, the definition of k is half of λ_{ii} defined by nuclear physicists; and for reactions in which different particles collide, the definition of k is the same as λ_{ij}. The reaction rate coefficients depend on temperature in a complicated way (Table 2-3) and may be calculated as the average value of the product of relative velocity times cross section. The concentrations of the intermediate species can be derived as follows. From Equation 2-155, $k_{145}[^1H]^2 = k_{146}[^1H][^2H]$. That is,

$$[^2H] = k_{145}[^1H]/k_{146}. \qquad (2\text{-}160)$$

Combining Equations 2-156 and 2-155 leads to

$$2k_{147}[^3He]^2 + k_{148}[^4He][^3He] - k_{145}[^1H]^2 = 0, \qquad (2\text{-}161)$$

from which $[^3He]$ can be solved to obtain

$$[^3He] = \frac{2k_{145}[^1H]^2}{k_{148}[^4He] + \sqrt{(k_{148}[^4He])^2 + 8k_{145}k_{147}[^1H]^2}}. \qquad (2\text{-}162)$$

Equations 2-158 and 2-159 lead to $k_{150}[^1H][^7Li] = k_{149}[^7Be] = k_{148}[^3He][^4He]$. Hence,

$$[^7Be] = k_{148}[^3He][^4He]/k_{149}; \qquad (2\text{-}163)$$

$$[^7Li] = k_{148}[^3He][^4He]/(k_{150}[^1H]). \qquad (2\text{-}164)$$

Table 2-3 Rate coefficients of some nuclear reactions

Reaction	Rate coefficient

2-145
$^1\text{H} + {}^1\text{H} \rightarrow {}^2\text{H}$

$$k_{111} = 1.91 \times 10^{-15} \left(\frac{10^9}{T}\right)^{2/3} \left\{ 1 + 0.123 \left(\frac{T}{10^9}\right)^{1/3} \right.$$
$$\left. + 1.09 \left(\frac{T}{10^9}\right)^{2/3} + \frac{0.938T}{10^9} \right\} \exp\left[-3.38 \left(\frac{10^9}{T}\right)^{1/3} \right]$$

2-146
$^2\text{H} + {}^1\text{H} \rightarrow {}^3\text{He}$

$$k_{112} = 2240 \left(\frac{10^9}{T}\right)^{2/3} \left\{ 1 + 0.112 \left(\frac{T}{10^9}\right)^{1/3} + 3.38 \left(\frac{T}{10^9}\right)^{2/3} \right.$$
$$\left. + \frac{2.65T}{10^9} \right\} \exp\left[-3.72 \left(\frac{10^9}{T}\right)^{1/3} \right]$$

2-147
$^3\text{He} + {}^3\text{He} \rightarrow$
$^4\text{He} + 2\,{}^1\text{H}$

$$k_{113} = 2.98 \times 10^{10} \left(\frac{10^9}{T}\right)^{2/3} \exp\left[-12.276 \left(\frac{10^9}{T}\right)^{1/3} \right]$$
$$\left\{ 1 + 0.034 \left(\frac{T}{10^9}\right)^{1/3} - 0.199 \left(\frac{T}{10^9}\right)^{2/3} - \frac{0.047T}{10^9} \right.$$
$$\left. + 0.162 \left(\frac{T}{10^9}\right)^{4/3} + 0.032 \left(\frac{T}{10^9}\right)^{4/3} + 0.019 \left(\frac{T}{10^9}\right)^{5/3} \right\}$$

2-148
$^3\text{He} + {}^4\text{He} \rightarrow {}^7\text{Be}$

$$k_{114} = 5.79 \times 10^6 \left[\frac{T}{10^9 (1 + 0.0495T/10^9)} \right]^{5/6} \left(\frac{10^9}{T}\right)^{3/2}$$
$$\exp\left\{ -12.826 \Big/ \left[\frac{T}{10^9 (1 + 0.0495T/10^9)} \right]^{1/3} \right\}$$

2-149
$^7\text{Be} \rightarrow {}^7\text{Li}$

$$k_{115} = 1.34 \times 10^{-10} \left(\frac{10^9}{T}\right)^{1/2} \exp\left[-\frac{2.515 \times 10^6}{T} \right]$$
$$\times \left\{ 1 - 0.537 \left(\frac{T}{10^9}\right)^{1/3} + 3.86 \left(\frac{T}{10^9}\right)^{2/3} + \frac{1.2T}{10^9} \right.$$
$$\left. + \frac{2.7 \times 10^6}{T} \right\}$$

Table 2-3 (continued)

Reaction	Rate coefficient
2-150 $^7\text{Li} + {}^1\text{H} \rightarrow {}^4\text{He} + {}^4\text{He}$	

$$k_{116} = 8.04 \times 10^8 \left(\frac{10^9}{T}\right)^{2/3} \exp\left[-8.471 \left(\frac{10^9}{T}\right)^{1/3}\right.$$

$$\left. - \left(\frac{T}{30.068 \times 10^9}\right)^2\right] \left\{1 + 0.049 \left(\frac{T}{10^9}\right)^{1/3}\right.$$

$$+ 0.23 \left(\frac{T}{10^9}\right)^{2/3} + \frac{0.079T}{10^9} - 0.027 \left(\frac{T}{10^9}\right)^{4/3}$$

$$\left. - 0.023 \left(\frac{T}{10^9}\right)^{5/3}\right\}$$

Note. The unit of k is based on time (s) and concentration (mol/cm^3). The reaction rate coefficients as a function of temperature are from Fowler et al. (1975) and Harris et al. (1983). Note that for Reactions 2-145 and 2-147, the definition of k is consistent with chemists' definition used in this book and is half of λ_{ij} defined by nuclear physicists. That is, $k = \lambda_{ij}/(1 + \delta_{ij})$, where λ_{ij} is the reaction rates per pair of particles, and $\delta_{ij} = 0$ except for $i = j$ for which $\delta_{ij} = 1$. The concentration unit is not converted to mol/L.

Therefore, the net production rate of ^4He from PP I and PP II chains is

$$\frac{d[^4\text{He}]}{dt} = k_{147}[^3\text{He}]^2 + k_{148}[^3\text{He}][^4\text{He}]. \tag{2-165}$$

The first term in the right-hand side of the above equation represents contribution from the PP I chain and the second term represents contribution from the PP II chain. The relative importance of each chain depends on the kinetic constants (which depend on temperature) and the concentrations of ^3He and ^4He. Because the concentration of ^3He can be solved from the quadratic equation above, the relative importance of PP I and PP II chains can be evaluated numerically at any given temperature. Figure 2-12 shows a calculated example of reaction rate of PP I and PP II chains. For the Sun, the PP I chain is more important.

To find the reaction rate of ^1H, recognizing that the net reaction is $4{}^1\text{H} \rightarrow {}^4\text{He}$, the net consumption rate of ^1H is

$$-\frac{d[^1\text{H}]}{dt} = 4\frac{d[^4\text{He}]}{dt} = 4(k_{147}[^3\text{He}] + k_{148}[^4\text{He}])[^3\text{He}]. \tag{2-166}$$

By solving for the reaction rate, the energy production rate can then be calculated. The calculation at each temperature is not difficult, but the application to the whole Sun is time-consuming because it is necessary to model the temperature as a function of radius, and to integrate over the whole Sun (the core) to obtain the luminosity.

Figure 2-12 Nuclear reaction rates $d[^4He]/dt$ by PP I and PP II chains as a function of temperature. The unit of temperature is megakelvins (MK). The unit of the reaction rate is somewhat arbitrary. The highest temperature in this calculation is 15.6 MK, roughly corresponding to the center temperature of the Sun. The concentrations of species used in the calculation of the reaction rates are the modeled species concentrations in the standard solar model (Bahcall, 1989).

2.4 Some Special Topics

2.4.1 Photochemical production and decomposition of ozone, and the ozone hole

Many chemical reactions in the atmosphere, such as those related to ozone production and destruction, are often complicated chain and parallel reactions, plus another complication in which some steps are initiated or controlled by photon fluxes from the Sun. Such reactions are called *photochemical reactions*. Kinetics of photochemical reactions differ from that of thermal reactions in that the reaction rate coefficients for thermal reactions depend on temperature, whereas the reaction rate coefficients of photochemical reactions depend on the photon flux and wavelength. The effect of different wavelengths is handled by integration of absorption cross sections and relative solar intensity with respect to wavelength. In the atmosphere, the photon flux depends on the altitude, latitude, season, and time of the day. The handling of photochemical reaction kinetics may be simplified by considering (i) only daily averages, (ii) only yearly averages (by integration with respect to time), (iii) only global averages as a function of altitude (by integration with respect to latitude for a given altitude), or (iv) only daily or yearly global averages (by integration with respect to time and latitude). It is hence necessary to understand the absorption of sunlight in the atmosphere and the fraction that penetrates to a specific altitude as a function of wavelength. These considerations are beyond the scope of this book.

Ozone in the atmosphere is a good example of photochemical reactions. Atmospheric ozone is not due to equilibrium. The production and decomposition of ozone are largely by *photochemical* process, and the concentration of ozone in the stratosphere is at steady state, controlled by the kinetics of photochemical production and decomposition.

2.4.1.1 Photochemical production and consumption of ozone

The ozone layer in the atmosphere is an important protective layer for life on the Earth. Ozone is photochemically produced from O_2 in the atmosphere. The following account is from Pilling and Seakins (1995). First, oxygen atoms are generated by short-wavelength UV photolysis (at wavelengths below 242 nm) in the stratosphere. That is, UV photons split the oxygen molecule as follows:

$$O_2 + h\nu \rightarrow 2O \tag{2-167}$$

The active oxygen atom may then combine with oxygen molecules to generate ozone:

$$O + O_2 + M \rightarrow O_3 + M \tag{2-168}$$

The above two reactions account for the layered structure of ozone in the stratosphere. (i) At lower altitudes, the requisite short wavelengths for oxygen photolysis are absent because they are already absorbed by oxygen molecules higher up. Hence, O_3 concentration is low at lower altitudes. (ii) At altitudes above the ozone layer, because of the decrease in $[O_2]$ due to the general pressure reduction with altitude, the concentration of O_2 is low, reducing the efficiency for the termolecular combination of Reaction 2-168. Hence, O_3 concentration is also low.

The ozone concentration is limited by two further reactions that destroy ozone:

$$O_3 + h\nu \rightarrow O_2 + O \tag{2-169}$$

$$O + O_3 \rightarrow 2O_2 \tag{2-170}$$

Reactions 2-167 to 2-170 constitute the *Chapman mechanism* for the creation and destruction of ozone in the unpolluted stratosphere.

To obtain the concentration of $[O_3]$, it is necessary to solve for $[O]$ and $[O_3]$ from two equations $d[O]/dt = 0$ and $d[O_3]/dt = 0$. The resulting equation for $[O_3]$ is a quadratic equation:

$$k_{169}k_{170}[O_3]^2 + k_{167}k_{170}[O_2][O_3] - k_{167}k_{168}[M][O_2]^2 = 0. \tag{2-171}$$

The solution for $[O_3]$ is

$$\frac{[O_3]}{[O_2]} = \frac{k_{167}\left\{\sqrt{1 + 4k_{168}k_{169}[M]/(k_{167}k_{170})} - 1\right\}}{2k_{169}}. \tag{2-172}$$

where k_{167} and k_{169} depend on the UV photon flux and hence the altitude.

The maximum concentration of ozone in the stratosphere (or the ozone layer) is about 9 ppm at an altitude of about 35 km. That is, the concentration of ozone in the so-called ozone layer is still very low. Transport of ozone in the atmosphere modifies ozone concentration levels at each altitude and latitude. It is emphasized that the steady-state concentration of O_3 in the stratosphere is not the thermodynamic equilibrium concentration, but is established by kinetics of photochemical reactions.

2.4.1.2 Ozone hole

The substantial concentration of ozone in the stratosphere can be significantly depleted by comparatively small amounts of other substances. The significantly depleted ozone level in polar regions (mostly over Antarctica) is referred to as the *ozone hole*.

Anthropogenic ozone depletion is through catalyst reactions of the type

$$O_3 + X \rightarrow XO + O_2, \tag{2-173}$$

$$XO + O \rightarrow O_2 + X, \tag{2-174}$$

with the net effect of $O + O_3 \rightarrow 2O_2$. In the above, X is a free radical (such as photochemically formed Cl or Br from anthropogenic CFCs and halons) acting as a catalyst; it participates in the reaction but is not consumed. Reaction 2-173 is a parallel path to destroy O_3, in addition to the natural paths of Reactions 2-169 (photochemical reaction) and 2-170. The effect of the catalyst reactions is to increase the decomposition rate of ozone, but this does not affect the production rate, resulting a shift of the balance of ozone concentration to a lower value.

With the addition of Reactions 2-173 and 2-174, the production and consumption of ozone include both chain and parallel reactions. The method of solution is nonetheless similar to the case without anthropogenic ozone destruction. To solve for the concentration of $[O_3]$, it is necessary to solve for $[XO]$, $[O]$, and $[O_3]$ from three equations: $d[XO]/dt = 0$, $d[O]/dt = 0$, and $d[O_3]/dt = 0$.

2.4.2 Diffusion control of homogeneous reactions

In a homogeneous phase in which particles are randomly distributed, the rate of a reaction, especially when the concentrations of the reacting species are low, must be influenced by the rate at which the reactants diffuse into each other. This effect is known as encounter control or microscopic diffusion control. In contrast, macroscopic diffusion control means the case when the liquid phase is heterogeneous and is mixed together, such as the mixing of milk and coffee. Macroscopic diffusion control is not considered under homogeneous reactions. If, upon encounter, the reaction rate is very fast compared to the rate to bring the species together, then the reaction is said to be fully controlled by encounter.

If, upon encounter, the reaction rate is much slower than the rate to bring the species together, then the reaction is not controlled by encounter. If the two rates are comparable, then the reaction is partially controlled by encounter (or diffusion).

For a fully diffusion-controlled (or encounter-controlled) reaction,

$$A + B \rightarrow \text{product},\tag{2-175}$$

the rate constant is (Atkins, 1982)

$$k_D = 4\pi(D_A + D_B)d_{AB}N_{av}\frac{\beta}{(e^\beta - 1)},\tag{2-176}$$

where k_D is the rate constant for a fully diffusion-controlled reaction, D_A and D_B are the diffusivity of A and B, respectively, d_{AB} is the critical distance between A and B within which A and B would react immediately, N_{av} is Avogadro's number, and $\beta = z_A z_B e^2/(4\pi\varepsilon_0\varepsilon d_{AB}k_B T)$, in which z_A and z_B are the electric charges of the ions (with negative or positive signs), e is proton charge, ε_0 is permittivity of vacuum (8.8542×10^{-12} $C^2 N^{-1} m^{-2}$), ε is the relative dielectric constant of the solvent (78.54 for water at 298.15 K), k_B is the Boltzmann constant, and T is temperature in kelvins. For aqueous solutions at 298.15 K, if $d_{AB} = 3$ Å, the term $\beta/(e^\beta - 1)$ takes the value of 1, 2.6, 4.8, 7.1, 9.5, 0.24, 0.041, 0.0057, 0.00070 for $z_A z_B$ of 0, −1, −2, −3, −4, 1, 2, 3, and 4. For k_D to have the unit of $M^{-1}s^{-1}$ (the normal unit for k of a second-order reaction), D_A and D_B must be in the unit of dm^2/s, and d_{AB} must be in the unit of dm. For neutral molecules,

$$k_D = 4\pi(D_A + D_B)d_{AB}N_A, \quad \text{if } A \neq B, \text{neutral},\tag{2-177a}$$

$$k_D = 2\pi(D_A + D_B)d_{AB}N_A, \quad \text{if } A = B, \text{neutral}.\tag{2-177b}$$

Three reactions are discussed below.

Consider Reaction 2-10b:

$$H^+ + OH^- \rightarrow H_2O.$$

Given $D_{H^+} = 9.1 \times 10^{-7}$ dm^2/s, $D_{OH^-} = 5.2 \times 10^{-7}$ dm^2/s (Table 1-3a), and $\beta/(e^\beta - 1) = 2.6$, if $d_{AB} = 5$ Å $= 5 \times 10^{-9}$ dm (which is a little too large), then k_D would match the observed value of 1.4×10^{11} $M^{-1}s^{-1}$. Hence, this reaction seems to be fully diffusion controlled.

All reactions in aqueous solutions should have rate constants smaller than that calculated from fully diffusion-controlled reactions because diffusion must play a role and the reaction rate may not be faster than the rate to bring the reactants together.

The Fe–Mg order–disorder reaction (Reaction 2-55),

$$Fe^{(M2)}Mg^{(M1)}Si_2O_6(\text{opx}) \rightleftharpoons Mg^{(M2)}Fe^{(M1)}Si_2O_6(\text{opx}),\tag{2-55}$$

is assumed to be controlled by Fe–Mg interdiffusion, or more specifically, the jumping of Fe^{2+} and Mg^{2+} along neighboring M1 and M2 sites that form continuous chains in some crystallographic directions (Ganguly and Tazzoli, 1994). With the assumption, they derived the diffusivity in orthopyroxene from the reaction rate coefficients using Equation 1-82b:

$$D_i = l_i^2 f_i, \tag{2-178}$$

where D_i is diffusivity along a crystallographic direction i, l_i is the jumping distance along the direction, and f_i is the jumping frequency along the direction. By examining the crystal structure of orthopyroxene (Figure 1 in Ganguly and Tazzoli, 1994), along crystallographic direction **c**, M1 and M2 sites alternate to form a closely packed continuous zigzag chain. Along crystallographic direction **b**, there is a small gap after each pair of M1 and M2 sites, making the jumping exchange more difficult. Along crystallographic direction **a**, neighboring layers of octahedral sites (M1 and M2) are separated by a tetrahedral layer. Based on such information, Ganguly and Tazzoli (1994) assumed that exchange along direction **a** is negligible, and diffusive exchange along both **c** and **b** directions contribute to the average of the forward and backward reaction rate coefficients. From the average reaction rate coefficient, only the average diffusivity along **c** and **b** directions can be found, although crystallographic consideration suggests that diffusion along **c** is faster than that along **b**. Hence, Ganguly and Tazzoli (1994) assumed that

$$D_{\text{Fe-Mg}} = l^2 f \approx l^2 (k_f + k_b), \tag{2-179}$$

where $D_{\text{Fe-Mg}}$, l and f are all for **c** and **b** directions, and k_f and k_b are the forward and backward reaction rate coefficients of Reaction 2-55. In the formulation of Ganguly and Tazzoli (1994), there was a factor of 1/2 for the term $(k_f + k_b)$. The factor does not appear in the above equation because Ganguly's definition of the reaction rate coefficient for this reaction differs from that adopted in this book by a factor of 2. The above equation can be expressed as

$$D_{\text{Fe-Mg}} \approx l^2 k_f (1 + 1/K_D). \tag{2-180}$$

Using Equations (2-60) and (2-57) for k_f and K_D expressions (which are newer versions and differ slightly from the expressions used by Ganguly and Tazzoli, 1994), and letting $l = (3.692 + 0.1 X_{\text{Fs}})$ Å (the average distance between the centers of M1 and M2 sites along **c** and **b** directions), then it can be found that

$$\ln D_{\text{Fe-Mg}} \approx 20.02 - (30,357 - 6106 X_{\text{Fs}}^2)/T. \tag{2-181}$$

The above method follows that of Ganguly and Tazzoli (1994) but the expression of D is slightly different because of the use of newer expressions of k_f and K_D. For example, at 800°C and $X_{\text{Fs}} = 0.2$, the above expression gives $D = 1.32 \times 10^{-21}$ m²/ s, and the expression of Ganguly and Tazzoli (1994) gives $D = 2.02 \times 10^{-21}$ m²/s.

In the investigation of the kinetics of Reaction 2-79, $H_2O(melt) + O(melt) \rightleftharpoons$ $2OH(melt)$, it was found that the concentration of a given species may initially deviate from equilibrium further, and then gradually approach equilibrium. One explanation suggested by Zhang et al. (1995) is that the backward reaction of Reaction 2-79 may be controlled by diffusion because OH groups are on average separated by several oxygen atoms. Zhang et al. (1995) modeled the backward reaction as a diffusion-controlled process. Because the model is fairly complicated and there is no direct observational evidence yet, the model is not discussed here.

2.4.3 Glass transition

2.4.3.1 General

The structure of a silicate melt depends on temperature and pressure. Above the liquidus, the structure changes rapidly in response of temperature and pressure changes. Below the liquidus of the melt, crystallization should occur. If the cooling is slow, crystallization does occur, resulting in a crystalline rock. However, if cooling is rapid, crystallization may be suppressed, resulting in a glass. The transition from liquid to a glass is called *glass transition*, which is a region of temperature in which molecular rearrangements occur on a timescale of seconds to months (Scherer, 1986), similar to the cooling timescale. The temperature at which glass transition occurs is called the *glass transition temperature* T_g. The temperature is also referred to as the *fictive temperature* T_f because the glass property is related to the fictive liquid at this temperature. More accurately, the glass transition temperature is the terminal fictive temperature after the glass is cooled down. The fictive temperature is a more general concept than the glass transition temperature. For example, the fictive temperature T_f is also defined at every temperature (or any instant) during cooling or heating, similar to the variation of T_{ae} during cooling (Figure 1-22b, Figure 2-2). The fictive temperature (T_f) concept is essentially identical to the apparent equilibrium temperature (T_{ae}) concept; the former is used in glass literature, and the latter is used in geochemical kinetics literature. Because the concepts of apparent equilibrium temperature and fictive temperature are similar, one may use reaction kinetics to understand glass transition (Zhang, 1994). The behavior of the fictive temperature as a function of temperature may be obtained similarly from the behavior of apparent equilibrium temperature, and there is hysteresis between cooling and heating. In Figure 2-2, one may substitute T_{ae} by T_f, and understand how the fictive temperature depends on cooling rate during cooling, and on both heating rate and thermal history during heating. From the point of view of reaction kinetics, many homogeneous reactions occur and are near equilibrium in the liquid state. These reactions rearrange the particles in the liquid and hence are part of the structure of the liquid.

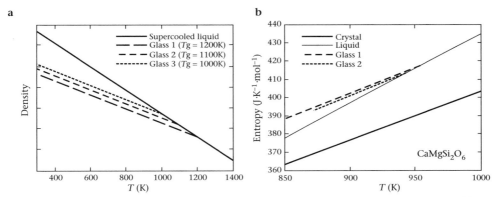

Figure 2-13 Schematic drawing of (a) density as a function of temperature, and (b) entropy as a function of temperature for glasses with different cooling rates and hence different glass transition temperature (Martens et al., 1987). The entropy of the undercooled liquid is estimated assuming constant heat capacity.

As the liquid is cooling down, the reactions try to maintain equilibrium. At some temperature T_1, equilibrium cannot be maintained anymore. At a lower temperature T_2, the reactions essentially stop. Glass transition occurs between T_1 and T_2. The apparent equilibrium constant of a reaction would roughly follow the behavior diagramed in Figure 1-17 and the dashed curves in Figure 2-2. Hence, glass transition may be viewed as reaction kinetics during cooling although during glass transition there might be many undefined reactions.

The structure of a glass is similar to that of a liquid at the fictive temperature, with short-range order but long-range disorder. For a given composition, the glass transition temperature is not fixed, but depends on the cooling rate, especially the cooling rate near the glass transition temperature. Sometimes, the glass transition temperature is reported or discussed without specifying a cooling rate. In such cases, either the viscosity at T_g is understood to be 10^{12} Pa·s or the cooling rate is understood to be about 10 K/min.

Although the glass transition is sometimes referred to as a second-order phase transition, more accurately it is a kinetic process responding to cooling. The properties of a glass at room temperature or other temperatures are not state functions. (Hence, when you see thermodynamic properties of a glass such as enthalpy listed in a handbook, they are approximate values.) They depend not only on temperature, pressure, and composition, but on the thermal history such as cooling rate as well. In other words, glass properties also depend on the fictive temperature. For example, the density of a glass at room temperature is lower if it was quenched more rapidly from high temperature compared to a glass of the same composition but quenched more slowly (Figure 2-13a); the entropy of a glass at room temperature is higher if it was quenched down more rapidly (Figure 2-13b). The structure of a glass at room temperature corresponds to the structure of a liquid at T_f. The physical properties of a glass depend on the liquid property at T_f and other elastic modifications in the glass state.

In terms of mechanical properties, the liquid state behaves as a viscous fluid, and the glass state behaves as an elastic solid. Given a noncrystalline material, whether it is in the glass state or liquid state depends on the timescale of the process of interest. For example, a melt at a given T-P condition (such as albite melt at 1200 K and 0.1 MPa), if it is rapidly smashed or dropped to the ground, the melt would behave as a glass and shatter into sharp-angled pieces; if it is stressed slowly, it would flow. Such a behavior is called *viscoelastic*. The mean time for stress relaxation of a *viscoelastic material* is roughly

$$\tau = \eta/G, \tag{2-182}$$

where η is shear viscosity, G is shear modulus and is roughly constant (≈ 10 GPa; Dingwell and Webb, 1990), and τ is the relaxation timescale. The above relation is called the *Maxwell relation*. For example, the viscosity of a dry albite melt at 1200 K and 0.1 MPa is about 6×10^9 Pa·s. Hence, the relaxation timescale is about 0.6 s. For a timescale longer than this, the material behaves as a liquid. For a timescale shorter than 0.6 s, the material behaves as a glass.

In summary, differences between the liquid and glass include the following: (i) Liquid is an equilibrium state structurally (although some reactions nonessential to the structure may not be at equilibrium, such as oxidation of Fe^{2+} to Fe^{3+} by dissolved oxygen in water) but glass is a disequilibrium state, with structural reactions frozen at the fictive temperature. (ii) Liquid is viscous (Newtonian liquid) and glass is elastic. In the glass transition region, the glass or liquid is a viscoelastic material, behaving partially elastically and partially viscously (not necessarily Newtonian). Whether something is in the liquid state or the glass state depends on the timescale of consideration. A silicate melt at 1000 K in an eruption is able to flow and hence is a liquid on the timescale of days, but during magma fragmentation (timescale of seconds or less) it fragments into angular pieces (after cooling down rapidly these angular pieces are very sharp and must be handled with caution). That is, on the timescale of seconds, the melt behaves as an elastic glass. For a glass heated to 900 K, on the timescale of seconds it is still a glass, but on the timescale of hours or longer it can flow under a pressure load. The latter is the basis of parallel-plate viscometry. Therefore, it is important to note the timescale of consideration when determining whether a material is glass or liquid. Similarly, it is important to specify the cooling rate when discussing the glass transition temperature.

2.4.3.2 Different definitions of glass transition temperatures

Glass transition temperature or the fictive temperature may be investigated or diagrammed using different methods, resulting in different definitions. These definitions are all similar and can be made identical. The rheological definition of glass transition, in its simplest form, is that glass transition occurs when the viscosity is 10^{12} Pa·s. The Maxwell timescale at $\eta = 10^{12}$ Pa·s is $\tau = \eta/G \approx 100$ s. Hence, with this definition, the timescale of interest is a few minutes for observable changes to occur. However, because glass transition is a kinetic property, the glass transition temperature depends on cooling rate. Hence, the usual definition is for a "normal" cooling rate in glass studies, 10 K/min. When the cooling rate increases, glass transition would occur at a higher temperature; when the cooling rate decreases, glass transition would occur at a lower temperature. Quantitatively, the mean reaction or relaxation time τ is proportional to viscosity η but inversely proportional to cooling rate q. Hence, viscosity at the glass transition temperature is inversely proportional to q. Therefore, for a given cooling rate q, the glass transition occurs at a viscosity of $10^{11.22}/q$, where the unit of $10^{11.22}$ is Pa·K and the unit of q is K/s. Clearly there is some arbitrariness in this definition, especially in the value of $10^{11.22}$ Pa·K, which does not correspond to any physically significant property. Hence, some authors have adjusted this parameter slightly to make the rheological definition to be the same as other definitions.

Another definition is based on the measurement of a property as a liquid cools down or as a glass is heated up at a given rate. If density (or volume) is measured, its variation with temperature follows a curve that has two linear segments, one linear trend at low temperatures with a shallower slope corresponding to the glass state, and one linear trend at high temperatures with a steeper slope corresponding to the liquid state. The temperature corresponding to the intersection of the two straight lines is the glass transition temperature. Similarly, heat capacity ($C_p = \partial H/\partial T$, where H is enthalpy) may be measured, or its integrated equivalent, heat content, may be measured as a function of temperature. The intersection of the two enthalpy segments would correspond to the glass transition temperature.

The equivalence of these T_g definitions, or the equivalence of volume, enthalpy, viscosity, and reaction relaxation has been verified provided that the exact values such as $10^{11.22}$ Pa·K can be varied by a small amount (e.g., Toplis et al., 2001; Sipp and Richet, 2002). For example, Toplis et al. (2001) adopted the constant to be $10^{11.5}$ Pa·K to match fictive temperature obtained from heat capacity curves and the rheologically defined T_g.

For hydrous silicate melts, the behavior of Reaction 2-79, $H_2O(melt) + O(melt) \rightleftharpoons 2OH(melt)$, upon cooling has been investigated. For a given cooling rate, the OH and H_2O concentrations in the quenched glass correspond to an apparent equilibrium temperature T_{ae}. This T_{ae} has also been found to be similar to the rheological T_g (Zhang et al., 2003). If the constant $10^{11.22}$ is changed to $10^{11.45}$ Pa·K, the T_{ae} of the reaction is in quantitative agreement with the rheological T_g (Zhang et al., 2003). This example demonstrates that glass transition is related to the cessation of homogeneous reactions.

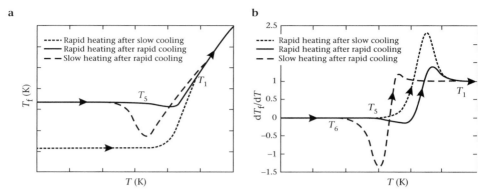

Figure 2-14 Schematic curves of (a) T_f versus T and (b) dT_f/dT versus T upon heating for samples that had different prior cooling rates.

2.4.3.3 Fictive temperature as a function of temperature and heating/cooling rate

Figure 2-2 shows how the apparent equilibrium temperature varies with temperature and cooling rate during cooling, and with temperature, heating rate, and prior cooling history during heating. The fictive temperature varies in a similar fashion. (Figures 1-19 and 1-22 may also be referred to in order to review reaction kinetics during cooling.) The behavior of the glass during cooling is relatively straightforward, with higher cooling rate leading to less reaction during cooling and higher terminal fictive temperature (that is, T_g). The behavior of glass during heating is more complicated because the fictive temperature depends not only on temperature and heating rate, but also on the prior history (which highlights that properties of glass are not state functions). The behavior of glass properties during heating is an important tool to characterize glass properties.

Figure 2-14a shows how T_f varies with T during heating at the same heating rate for glass with different cooling history. To show the variation of T_f with T more clearly, the variation of dT_f/dT with T is shown in Figure 2-14b, which highlights rapid kinetic changes of glass properties in the glass transition region. The explanation of Figure 2-14b is as follows. (Figure 2-14a can be understood by comparison with Figure 2-2. Furthermore, because Figure 2-14a represents the integrated form of Figure 2-14b, an understanding of the latter means an understanding of the former.) For clarity of explanation, some values will be used even though the diagrams are schematic. Consider the solid curve for rapid heating after rapid cooling (same absolute values of dT/dt). Suppose prior rapid cooling led to a terminal fictive temperature (T_g) of 920 K. It means that during cooling, the fictive temperature was about 940 K when the system temperature was 920 K, and the reaction continued (although slowly) to about 800 K to reach the T_g of 920 K. When heating up, the reaction rate began to be noticeable also at about 800 K, which is below T_f of 920 K, leading to the glass property to move to lower fictive temperature. This is why there is a significant decrease in T_f in the solid curve. As temperature increases, reaction rate increases, $T_f - T$ decreases. At some temperature, when the temperature increases by 1 K, T_f would increase by more than 1 K, leading to the maximum in the solid curve. At T_1, the glass transition is over, and the glass becomes a fully equilibrium liquid. Between T_5 and T_1, the glass property would change to reach equilibrium at T_1. Hence, if the starting glass has a lower T_f (meaning it experienced a slower cooling history), more reaction would occur, leading to a larger peak in the dT_f/dT curve (short-dashed curve in Figure 2-14b).

2.4.3.4 Kinetic heat capacity curve as a function of temperature

During glass transition, many homogeneous reactions are happening. Heat is absorbed or released with reactions. For clarity in explaining the concepts, the hydrous species reaction (Reaction 2-79) is used as an example. As temperature increases, the equilibrium goes to the right-hand side (producing more OH). For convenience, the side that is favored at higher temperatures is referred to as the higher temperature side. Reaction toward the higher temperature side requires addition of heat (enthalpy of the reaction), and reaction toward the lower temperature side releases heat. Hence, every homogeneous reaction would affect the heat capacity curve because of the release or absorption of heat. At equilibrium, as temperature increases, the extent of the reaction to the high-temperature side increases, meaning the reaction absorbs heat, contributing to C_p. If the reaction is infinitely slow at the given temperature, it would not contribute to C_p. If heating and cooling are infinitely slow, the C_p versus T curve would be the equilibrium curve and would be independent of whether it is heating or cooling, and independent of the heating or cooling rate (i.e., no hysteresis). However, heat capacity measurements are carried out at a finite heating or cooling rate, such as 10 or 5 K/min. There are, hence, nonequilibrium (kinetic) effects, which generate specific shapes of heat capacity curves.

First consider the heat capacity curve measured during the cooling path of the heating–cooling cycle (Figure 2-15a). At high enough temperatures ($T > T_1$), the reaction rate is very rapid and there is equilibrium at every temperature. Hence, every reaction contributes fully to the heat capacity, and C_p is independent of cooling rate. As temperature decreases, the reaction begins to deviate from equilibrium. If the cooling rate is high (solid curve in Figure 2-15a), this deviation occurs at a higher temperature (T_1), and the reaction effectively stops at T_3, below which the reaction does not contribute to C_p anymore. If the cooling is slow (dashed curve in Figure 2-15a), this deviation occurs at a lower temperature (T_4), and the reaction effectively stops at T_5. Below T_5, the two heat capacity curves are about the same. The heat capacity curve below T_5 may be referred to as reaction-free heat capacity (or glass heat capacity), and that above T_1 may be referred to as fully reactive heat capacity (or liquid heat capacity).

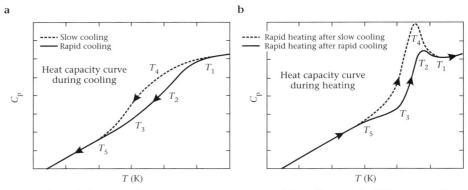

Figure 2-15 Schematic heat capacity curve upon (a) cooling with different cooling rates, and (b) heating with the same heating rate for samples with different prior cooling rates. Temperature increases to the right-hand side. (Note that the two curves do not represent cooling–heating cycles.) From Zhang (unpublished).

Next consider the heat capacity curve measured during the heating path of the heating–cooling cycle (Figure 2-15b). Suppose both samples from Figure 2-15a are now heated up using a single heating rate of rapid heating (with rate about the same as rapid cooling in Figure 2-15a). Before heating, the OH concentration is high for the rapidly cooled sample, corresponding to a high T_{ae}, such as 920 K. For the slowly cooled sample, the OH concentration is low, corresponding to a low T_{ae}, such as 810 K. Upon heating from low temperature, both samples have higher OH concentration than the equilibrium concentration, but reaction kinetics is too slow at low temperature. As temperature increases to high enough (such as 850 K, or T_5 in Figure 2-15b), the reaction rate begins to be noticeable. Because the rapidly cooled sample (with $T_{ae} = 920$ K) contains more OH than the equilibrium concentration at 850 K, the reaction goes to the lower temperature side and releases heat, thus reducing the heat capacity (heat absorbed per degree of heating) compared to the "normal" linear trend. On the other hand, the slowly cooled sample (with $T_{ae} = 810$ K), contains less OH, reacts toward the higher temperature side, and absorbs heat, thus increasing the heat capacity. As temperature increases further, reaction rate increases. At T_1, complete equilibrium is reached, meaning high OH concentration. In a small temperature range (such as T_3 to T_1), the OH concentration changes from the initial to the equilibrium concentration. Because the slowly cooled sample contains much less initial OH, to reach the equilibrium OH concentration means formation of much more OH, leading to a large peak in the heat capacity curve (dashed curve in Figure 2-15b). For the rapidly cooled sample, it contains more OH to begin with, and reaction to reach equilibrium would cause a C_p increase, but smaller than that for the slowly cooled sample. In the heating path, after the sample reached equilibrium, the heat capacity drops back to the fully reactive heat capacity (above T_1 in Figure 2-15b).

The temperature range of T_5 to T_1 is referred as the *glass transition* region. The calorimetric glass transition temperature (T_g) may be defined as the temperature at which C_p is at maximum in the heating path. Comparison of Figure 2-15b and Figure 2-14b shows that the C_p versus T curve is similar to the dT_f/dT versus T curve: By adding a linear baseline (accounting for heat capacity due to vibrational, rotational, and translational motion) to the dT_f/dT versus T curve, the resulting curve would mimic the C_p versus T curve. This is understandable because both characterize how $d\xi/dt$ depends on temperature.

In glass–liquid, there are many homogeneous reactions, and hence the heat capacity curve is the integrated effect of all these reactions. Furthermore, some factors that contribute to heat capacity are not necessarily reactions, but vibrational, rotational, and translational motion (e.g., the linear part of the glass heat capacity curve). The purpose of using Reaction 2-79 is to facilitate the explanation of the concepts, and not to mean that this reaction alone accounts for the full heat capacity curve of glass transition. The above discussion explains that the λ-shape of the C_p versus T curve upon heating is a kinetic phenomenon. Furthermore, it shows that the heat capacity curve upon heating depends on the thermal history of the glass being heated up, which may be applied to infer cooling rate of the glass (Wilding et al., 1995, 1996a,b).

Glass scientists have investigated glass properties largely empirically and developed empirical relations by summarizing observations of cooling and heating behaviors but without much theoretical basis (e.g., Scherer, 1986). Because the concept of fictive temperature is similar to that of the apparent equilibrium temperature, and because glass transition likely involves homogeneous reactions, it may be productive to use the concept of homogeneous reaction kinetics under variable temperatures to guide the study of glass transition so as to gain a deeper and more quantitative and predictive understanding of glass transition (Zhang, 1994). In fact, all figures on glass transition in this section are generated using homogeneous reaction kinetics. Nonetheless, due to the complexity of structural relaxation (e.g., there are likely many homogeneous reactions), a single fictive temperature is not enough to completely characterize the property of a glass (Scherer, 1986), suggesting that multiple homogeneous reactions are needed to model the glass transition and glass properties.

Problems

2.1 Half-life versus half-time to reach equilibrium.

a. The half-life of a reactant is the time interval after which half of it has been turned into the product. Find the relationship between the half-life of A and the rate constant for a first-order reaction A → B, where $d\xi/dt = k[A]$.

b. For radioactive decay of A → B, the final concentration of A is zero. Hence, the half-life of A is well defined. For many chemical reactions of A ⇌ B, when it reaches equilibrium, the concentration of A is not zero (and in fact it may still be pretty high). Hence, instead of half-life, on can define a half reaction time to be the time when [A] changes from the initial concentration to halfway between the initial and the final equilibrium concentrations. Given that the rate constant for the forward reaction is k_f and that of the backward reaction is k_b, find the relation between half reaction time and the rate constants.

2.2 Treat both the forward and backward reactions of $H_2O(aq) + CO_2(aq) \rightleftharpoons H_2CO_3(aq)$ as elementary reactions. The forward reaction rate coefficient is 0.002 s^{-1} at $0°C$ and 0.043 s^{-1} at $25°C$.

a. Write down the reaction rate law accounting for both the forward and backward reaction.

b. The equilibrium constant for the above reaction is 0.00287 at $25°C$. Find the backward reaction rate coefficient at $25°C$.

c. At $25°C$, if initially $[CO_2] = 0.1$ mM and $[H_2CO_3] = 0$, calculate and plot how $[CO_2]$ and $[H_2CO_3]$ evolve with time.

d. At $25°C$, if initially $[CO_2] = 0.1$ mM and $[H_2CO_3] = 0$, and if the solution contains high $[OH^-]$ concentration (e.g., $pH = 11$) so that H_2CO_3 would immediately react with OH^- to become HCO_3^- and H_2O, calculate and plot how $[CO_2]$ evolves with time.

2.3 Assume (i) that both the forward and backward reactions of the following electron transfer reaction are elementary: $^{56}Fe^{2+} + {}^{55}Fe^{3+} \rightleftharpoons {}^{56}Fe^{3+} + {}^{55}Fe^{2+}$, and (ii) that the forward reaction constant ($k = 0.87 \text{ M}^{-1}\text{s}^{-1}$) equals the backward reaction constant.

a. Is assumption (ii) reasonable? Why?

b. Calculate the evolution of concentrations of $^{56}Fe^{2+}$, $^{56}Fe^{3+}$, $^{55}Fe^{2+}$, $^{55}Fe^{3+}$ as a function of time (using kt as the horizontal axis) for an initial condition of $[^{56}Fe^{2+}]_0 = 0.1$ mM, $[^{55}Fe^{3+}]_0 = 0.01$ mM, $[^{56}Fe^{3+}]_0 = 0.01$ mM, and $[^{55}Fe^{2+}]_0 = 0.2$ mM.

c. Calculate the half-time to reach equilibrium and the relaxation timescale for case b.

2.4 This problem explores the concept of relaxation timescale (τ_r) for a first-order reaction. It is simplest to use formula in Table 2-1 but you might have to do some conversion. Consider a first-order reaction $H_2CO_3 \rightleftharpoons H_2O + CO_2$ with $k_f \approx 15 \text{ s}^{-1}$ and $k_b \approx 0.043 \text{ s}^{-1}$ at $25°C$. Determine τ_r for

a. $[H_2CO_3]_0 = 1.02$ M; $[CO_2]_0 = 0$ M;

b. $[H_2CO_3]_0 = 0$ M; $[CO_2]_0 = 1.02$ M.

What general conclusion do you get for the relaxation timescale for first-order reactions?

2.5 This problem explores the concept of relaxation timescale (τ_r) for a second-order reaction. It is simplest to use formula in Table 2-1 but you might have to do some conversion. Consider a second-order reaction $2H_2O \rightleftharpoons H_3O^+ + OH^-$ with $k_f \approx 0.0015$ M s^{-1} and $k_b \approx 1.5 \times 10^{11}$ M^{-1} s^{-1} at 25°C. Find τ_r for

a. $[H_3O^+]_0 = [OH^-]_0 = 0$ M;

b. $[H_3O^+]_0 = [OH^-]_0 = 10^{-3}$ M;

c. $[H_3O^+]_0 = 10^{-6}$ M; $[OH^-]_0 = 10^{-7}$ M.

What general conclusion do you get for the relaxation timescale for second-order reactions?

2.6 Water dissolves into a silicate melt or glass in at least two forms: H_2O molecules (denoted as H_2O_m) and OH groups (denoted as OH). H_2O molecules are free and neutral. OH groups are associated with either Al or Si or some other cation. Total water concentration (denoted as H_2O_t) can be expressed as $[H_2O_t] = [H_2O_m] + 0.5[OH]$ in terms of mole fractions. H_2O_m and OH interconvert in the melt structure according to the following reaction:

$H_2O_m(melt) + O(melt) \rightleftharpoons 2OH(melt)$,

where O is a bridging oxygen. Let $K = [OH]^2/\{[H_2O_m][O]\}$ at equilibrium. The concentrations of H_2O_m, OH, and O are often expressed as mole fractions on a single oxygen basis so that $[H_2O_m] + [OH] + [O] = 1$. (For this homework problem, each 1 wt% of total water is roughly equivalent to $[H_2O_t] = 0.0178$. If you use the reaction in your research, this approximation would not be good enough. Use Equation 2-81.)

a. K for the reaction is 0.1 at 470°C and 0.3 at 740°C. Calculate and plot the mole fractions of H_2O_m and OH as a function of H_2O_t at each temperature. Let H_2O_t (mole fraction) vary from 0 to 0.1.

b. Assume that the reaction is elementary. Assume that the forward reaction rate coefficient is 0.002 s^{-1} and $K = 0.1$ at 470°C. Initially, $[H_2O] = 0.01$ and $[OH] = 0.01$. Calculate and plot how $Q = [OH]^2/\{[H_2O][O]\}$ approaches K.

2.7 The following are some real experimental data for the Fe–Mg order–disorder reaction in orthopyroxene at 600°C (Wang et al., 2005):

$Fe(M2) + Mg(M1) \rightleftharpoons Fe(M1) + Mg(M2)$.

The unit of the concentrations is the mole fraction on each site (either M1 or M2 site).

a. Find the equilibrium constant at this temperature.

b. Assume that the reaction is an elementary reaction and find the reaction rate coefficient.

c. What is the unit of the rate coefficient? Is it the same as the unit for second-order reactions in aqueous reactions?

Data table:

t (min)	Fe(M1)	Mg(M1)	Fe(M2)	Mg(M2)
0	0.00450	0.9769	0.0174	0.9807
600	0.00425	0.9771	0.0176	0.9804
1,920	0.00380	0.9774	0.0179	0.9801
3,720	0.00361	0.9778	0.0183	0.9798
6,000	0.00335	0.9780	0.0185	0.9795
11,760	0.00281	0.9786	0.0191	0.9790
20,300	0.00261	0.9788	0.0193	0.9788
29,700	0.00233	0.9790	0.0195	0.9785
48,165	0.00232	0.9790	0.0195	0.9785

2.8 Fe^{2+} and Mg in orthopyroxene can partition between M1 and M2 sites through the reaction $Fe(M2) + Mg(M1) \rightleftharpoons Fe(M1) + Mg(M2)$. Ganguly et al. (1994) expressed K_D for the intracrystalline exchange reaction as $\exp(0.888 - 3062/T)$, where T is in K. (New data have led to a new expression, but we will use the old expression in this homework problem.)

a. At 1000 K, calculate K and then calculate Fe and Mg concentrations in M1 and M2 for an orthopyroxene of the following composition:

$$(Mg_{1.6221}Fe_{0.3309}Ca_{0.026}Cr_{0.021})(Al_{0.021}Si_{1.979})O_6.$$

You can assume that all Ca, Mn, and Na are in the M2 site and all Cr, Fe^{3+}, Ni, and Al^{3+} are in the M1 site.

b. Ganguly et al. (1994) showed that the reaction rate coefficient for the forward reaction can be written as

$$\ln k_f = (26.2 + 6.0 X_{Fs}) - 31{,}589/T,$$

where k_f is in min^{-1} and T is in K. Note that k_f is in min^{-1} and not in s^{-1}. Note also that $k_f = C_0 k_1$, where C_0 is the total concentration of the M1 and M2 sites (this definition of C_0 by Ganguly is 2 times the definition of C_0 in Zhang, 1994) and k_1 is the rate coefficient defined by

$$d\xi/dt = k_1[\text{Fe(M2)}][\text{Mg(M1)}] - k_2[\text{Fe(M1)}][\text{Mg(M2)}].$$

For a pyroxene with the above overall composition, initially, [Fe(M1)] = 0.1, and [Mg(M1)] = 0.879. At 1000 K, calculate and plot how Fe(M1) approaches the equilibrium concentration.

2.9 The half-lives of ^{234}U and ^{238}U are 2.45×10^5 yr and 4.468×10^9 yr.

   a. Calculate the natural ^{234}U/^{238}U ratio, and compare it with the observed ratio 5.5×10^{-5}. What have you assumed?

   b. Estimate the ratio at 1.0 Ma.

   c. Evaluate the ratio at 4.0 Ga.

   d. The present day ^{238}U/^{235}U ratio is 137.88. Calculate the ratio at 4.0 Ga.

   e. Calculate the isotopic abundances of ^{234}U, ^{235}U, and ^{238}U at 4.0 Ga.

2.10 ^{14}C is a cosmogenic nuclide produced in the atmosphere. ^{14}C is unstable and decays into ^{14}N with a half-life of 5730 years ($\lambda = 0.00012097$ yr^{-1}). Assume that the concentration (or activity) of ^{14}C in the atmosphere is a steady-state concentration that did not vary with time. If the activity of ^{14}C in a plant tissue 13.56 dpm per gram of carbon, calculate the atomic ratio of ^{14}C/C.

2.11 In the decay chain of ^{238}U, let $N_1 = {}^{238}$U, $N_2 = {}^{234}$Th, and $N_3 = {}^{234}$Pa, derive how the concentration of N_3 would change with time. You may ignore different excitation states of ^{234}Pa, and simplify the decay of ^{234}Pa as one-step decay to ^{234}U with a half-life of 1.17 min.

2.12 At the center of the Sun, the temperature is about 15.6 million kelvins, the concentration of ^{1}H at the center of the Sun is about 50 mol/cm^3, and that of ^{4}He is about 23 mol/cm^3. Use Table 2-3 and the relevant equations to solve for ^{3}He concentration, and evaluate the relative importance of the PP I chain and PP II chain. (The relative contribution of each chain does not have to be the same as stated in the text for the core of the Sun because here only the center of the Sun is considered.)

2.13 Consider ozone production and decomposition in the atmosphere, including anthropogenic contribution to the decomposition. Use the steady-state treatment.

Solve for the concentration of [XO], [O], and [O_3] (expressed as a function of [O_2], [X], and [M]). The concentration of [O_3] should be the same as Equation 2-176.

2.14 Estimate the relaxation time of a melt with a viscosity of 100 Pa s.

2.15 Mader et al. (1996) carried out experiments to simulate volcanic eruptions. In the experiments, 4 mL of K_2CO_3 solution (K_2CO_3 concentration is 6 M) is injected in a few milliseconds through 96 holes into a 100-mL HCl solution (HCl concentration is 6 M). Knowing that the equilibrium constants for the following reactions are

Reaction 0: $H_2CO_3 \rightleftharpoons H_2O + CO_2$; $K_0 = [CO_2]/[H_2CO_3] \approx 500$

Reaction 1: $H_2CO_3 \rightleftharpoons H^+ + HCO_3^-$; $K_1 = [H^+][HCO_3^-]/[H_2CO_3]$
$$\approx 2.2 \times 10^{-4}$$

Reaction 2: $HCO_3^- \rightleftharpoons H^+ + CO_3^{2-}$; $K_2 = [H^+][CO_3^{2-}]/[HCO_3^-]$
$$\approx 4.7 \times 10^{-11}$$

a. Calculate equilibrium concentration of H^+, CO_2, H_2CO_3, HCO_3^-, and CO_3^{2-} in the 6 M K_2CO_3 solution.

b. As the K_2CO_3 solution and the HCl solution mix, the concentrations of the above species change. Calculate the species concentrations when the mixture has an equivalent volume ratio (volume of the HCl solution to that of the K_2CO_3 solution) of 0.5, 1, 2, 4, 25 (the ratio of 25 is the final mixture). Ignore degassing of CO_2 (i.e., assume that all CO_2 is dissolved in water). Plot how the concentration of CO_2 changes with this volume ratio.

c. Now consider the kinetics of the reaction. Assume that reactions 1 and 2 are rapid (μs timescale) but reaction 0 is slow. The reaction rate constant for the forward reaction of reaction 0 is $k_0 \approx 15$ s^{-1}. Assuming that the mixing timescale is ~1 ms, discuss how the concentrations of CO_2, H_2CO_3, HCO_3^-, CO_3^{2-}, and H^+ change with time after the injection.

3 Mass Transfer: Diffusion and Flow

The physical transport of mass is essential to many kinetic and dynamic processes. For example, bubble growth in magma or beer requires mass transfer to bring the gas components to the bubbles; radiogenic Ar in a mineral can be lost due to diffusion; pollutants in rivers are transported by river flow and diluted by eddy diffusion. Although fluid flow is also important or more important in mass transfer, in this book, we will not deal with fluid flow much because it is the realm of fluid dynamics, not of kinetics. We will focus on diffusive mass transfer, and discuss fluid flow only in relation to diffusion.

The basic aspects of diffusion were discussed in Chapter 1. Diffusion is due to random particle motion in a phase (Figure 1-6). Hence, solutions to diffusion problems are often related to statistics, such as Gaussian distribution and error function. If there is no gradient in chemical potential, random motion still occurs, although it does not lead to detectable changes. Given a gradient in chemical potential (or concentration gradient), random motion would lead to a net mass flux. The mass flux $\mathbf{J}$ is roughly proportional to the concentration gradient (Equation 1-71) according to Fick's first law:

$$\mathbf{J} = -D\partial C/\partial x,$$

where D is the diffusivity, $\partial C/\partial x$ is the concentration gradient (a vector), and the negative sign means that the direction of diffusive flux is opposite to the direction of the concentration gradient (i.e., diffusive flux goes from high to low concentration, but the gradient is from low to high concentration). The above flux law may also be written in terms of chemical potential (Appendix 1). The diffusivity increases with increasing temperature, and may decrease or increase with increasing pressure (Equation 1-88):

$$D = A \exp[-(E + P\Delta V)/(RT)],$$

where A is the pre-exponential factor, E is the activation energy and is positive, and ΔV is the volume difference between the activated complex and the diffusing species and may be either positive or negative. In a single phase, the diffusivity varies from one species to another, usually by a few orders of magnitude. In different phases, the diffusivity may vary by many orders of magnitude. Typical diffusivities in the gas phase at 298 K, in aqueous solution at 298 K, in silicate melt at 1600 K and in silicate mineral at 1600 K are 10 mm^2/s, 10^{-3} mm^2/s, 10^{-5} mm^2/s, and 10^{-11} mm^2/s, respectively.

From the phenomenological flux equation, the diffusion equation may be derived. For a constant diffusivity, the diffusion equation for one-dimensional diffusion takes the following form (Equation 1-74):

$$\frac{\partial C}{\partial t} = D \frac{\partial^2 C}{\partial x^2}.$$

Given initial and boundary conditions, the concentration variation as a function of x and t can be solved from the diffusion equation. The above diffusion equation was given without derivation in Chapter 1. In this chapter, the diffusion equation is derived and solved.

Diffusion does not proceed at a constant rate. The length of a diffusion profile, as characterized by the *mid-diffusion distance*, is not proportional to time. Rather, the mid-diffusion distance is proportional to the square root of time (Equation 1-79):

$$x_{mid} = \gamma(Dt)^{1/2},$$

where γ is of order 1, but depends slightly on the boundary and initial conditions. For example, γ equals 0.95387 for half-space diffusion, and 1.6651 for point-source diffusion.

Diffusion is ubiquitous in nature: whenever there is heterogeneity, there is diffusion. In liquid and gas, flow or convection is often present, which might be the dominant means of mass transfer. However, inside solid phases (minerals and glass), diffusion is the only way of mass transfer. Diffusion often plays a major role in solid-state reactions, but in the presence of a fluid dissolution and recrystallization may dominate.

Diffusion plays an important role in many geological processes, including homogenization of an originally zoned crystals, loss of radiogenic daughter (such as Ar from decay of K, and Pb from decay of U and Th) from a mineral, crystal growth from a liquid or gas or other solid phases, crystal dissolution, xenolith digestion, fluid–rock interaction, and magma mixing. The study of diffusion not only helps us to understand these processes, but also provides a tool to infer thermal history of rocks (inverse problems). Understanding diffusion is critical to thermochronology (and the concept of closure temperature). Several geospeedometry techniques are based on diffusion.

Mathematically, studies of diffusion often require solving a diffusion equation, which is a partial differential equation. The book of Crank (1975), *The Mathematics of Diffusion*, provides solutions to various diffusion problems. The book of Carslaw and Jaeger (1959), *Conduction of Heat in Solids*, provides solutions to various heat conduction problems. Because the heat conduction equation and the diffusion equation are mathematically identical, solutions to heat conduction problems can be adapted for diffusion problems. For even more complicated problems, including many geological problems, numerical solution using a computer is the only or best approach. The solutions are important and some will be discussed in detail, but the emphasis will be placed on the concepts, on how to transform a geological problem into a mathematical problem, how to study diffusion by experiments, and how to interpret experimental data.

In addition to the similarity between the heat conduction equation and the diffusion equation, erosion is often described by an equation similar to the diffusion equation (Culling, 1960; Roering et al., 1999; Zhang, 2005a). Flow in a porous medium (*Darcy's law*) often leads to an equation (Turcotte and Schubert, 1982) similar to the diffusion equation with a concentration-dependent diffusivity. Hence, these problems can be treated similarly as mass transfer problems.

3.1 Basic Theories and Concepts

3.1.1 Mass conservation and transfer

In dealing with mass transfer problems, one of the basic equations is mass conservation, meaning that mass is almost always conserved. The modifier "almost always" is needed because mass is not conserved in nuclear reactions. For example, when ^{238}U decays to ^{206}Pb and 8 α-particles, a mass of 0.0555 atomic mass units is lost per ^{238}U atom, or 0.0233% mass loss. Even though such mass loss is measurable with a mass spectrometer, it is still negligible for most other applications (e.g., gravimetric analyses of concentrations often are no better than 1% relative precision). In chemical reactions, although some mass loss or gain is also involved (see Problem 3.1), the amount is so miniscule that it cannot be measured with current technology. Hence, mass conservation is an excellent approximation.

For a closed system, the total mass of the system is conserved. For a component that is made of nonradioactive and nonradiogenic nuclides, the concentration of the component in the whole system can increase or decrease only through chemical reactions. The mass of a radioactive component decreases with time due to decay, whereas that of a radiogenic component increases with time (nuclear reaction). On the basis of mass conservation, some relations can be derived that are used in understanding diffusion and mass transfer. The relations take different forms for total mass versus the mass of a component or species.

3.1.1.1 Mass conservation

We first derive a relation for total mass conservation. Consider an arbitrary volume V enclosed in a surface Ω. The mass inside the volume is $\int \rho dV$, where ρ is density (in kg/m^3) and dV is an infinitesimal volume in the volume V. The time derivative of the mass in the volume (i.e., the rate of the variation of the mass with time) is

$$\frac{\partial}{\partial t} \int_V \rho \, dV = \int_V \frac{\partial \rho}{\partial t} dV = -\int_\Omega \mathbf{J} \cdot d\mathbf{S} = -\int_V \nabla \cdot \mathbf{J} dV, \tag{3-1}$$

where $\mathbf{J}$ is the mass flux (a vector with unit of $\mathrm{kg\,m^{-2}\,s^{-1}}$), $d\mathbf{S} = \mathbf{n} dS$, where $\mathbf{n}$ is the unit vector normal to the surface and pointing outward, dS is an infinitesimal surface area (a scalar), $d\mathbf{S}$ is an infinitesimal surface vector, and ∇ is the divergent when it is applied to a vector $\mathbf{J}$ and is the gradient when applied to a scalar (i.e., ∇ operator turns a vector to a scalar and a scalar to a vector). The first equality is simply an exchange of the order of integration and differentiation. The second one is based on the law of mass conservation, which states that the total mass change in the volume equals the mass flux into the volume from the surface surrounding the volume. The third equality is based on Gauss' theorem. From the second and the fourth terms of Equation 3-1, since V is arbitrary, we obtain

$$\frac{\partial \rho}{\partial t} = -\nabla \cdot \mathbf{J}. \tag{3-2a}$$

This is the differential form of the mass balance equation in three dimensions. Since $\mathbf{J}$ can be written as $\rho \mathbf{u}$, where $\mathbf{u}$ is the flow velocity of the fluid, the above equation can also be written as

$$\frac{\partial \rho}{\partial t} = -\nabla \cdot (\rho \mathbf{u}), \tag{3-2b}$$

which is known as the continuity equation in fluid mechanics.

To gain an intuitive understanding of Equation 3-2a, the one-dimensional case of Equation 3-2a is derived using a more visual method. Consider a cubic volume with the lower-left-front corner at (x, y, z) and the lengths of sides being dx, dy, and dz. Assume that the flux is one-dimensional along the x-direction (Figure 3-1). Then the total mass variation in the volume ($dV = dx \, dy \, dz$) equals the flux into dV from the left-hand side (x) of the volume $J_x \, dy \, dz$ minus the flux out of dV from the right-hand side $(x + dx)$ of the volume $J_{x+dx} \, dy \, dz$:

$$\frac{\partial \rho}{\partial t} dV = J_x \, dy \, dz - J_x(x + dx) dy \, dz = -\frac{\partial J_x(x)}{\partial x} dV, \tag{3-3a}$$

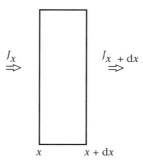

Figure 3-1 Relation between concentration increase in an element volume and fluxes into and out of the volume. The flux along x-axis points to the right (x-axis also points to the right). The flux at x is J_x, and that at $x+dx$ is J_{x+dx}. The net flux into the small volume is $(J_x - J_{x+dx})$, which causes the mass and density in the volume to vary.

where J_x (a scalar) is the flux along the x-direction, and $dy\,dz$ is the area across which the flux flows. Hence,

$$\frac{\partial \rho}{\partial t} = -\frac{\partial J_x(x)}{\partial x}, \tag{3-3b}$$

which is the one-dimensional form of Equation 3-2a.

Next, we treat the case of mass conservation of a species. The difference between the conservation of total mass and the conservation of the mass of a species is that other species may react to form the species under consideration. Hence, the reactions must be included. The conservation equation for a species k can be written as

$$\frac{\partial}{\partial t} \int_V \rho_k dV = \int_V \frac{\partial \rho_k}{\partial t} dV = -\int_\Omega \mathbf{J}_k \cdot d\mathbf{S} + \sum_{i=1}^{n} \int_V \nu_{ki} \frac{d\xi_i}{dt} dV$$

$$= -\int_V \nabla \cdot \mathbf{J}_k dV + \int_V \sum_{i=1}^{n} \nu_{ki} \frac{d\xi_i}{dt} dV, \tag{3-4}$$

where ρ_k is the mass of species k in a unit volume (i.e., concentration in kg/m^3 or mol/m^3), $d\xi_i/dt$ is the net chemical reaction rate (i.e., rate of forward reaction minus rate of backward reaction) of reaction i, ν_{kj} is the stoichiometric coefficient of species k in reaction i, and n is the total number of reactions. The value of ν_{kj} is positive when component k is a product and negative when component k is a reactant. All reactions should be included in Equation 3-4, not just the independent ones (this is different from equilibrium thermodynamics and is one of the reasons why kinetics is much more complicated than thermodynamics). From Equation 3-4, we have

$$\frac{\partial \rho_k}{\partial t} = -\nabla \cdot \mathbf{J}_k + \sum_{i=1}^{n} \nu_{ki} \frac{d\xi_i}{dt}. \tag{3-5}$$

One example is mass conservation of ^{40}Ar in a mineral (such as hornblende). Because ^{40}K decays to ^{40}Ar at a rate of $\lambda_e {}^{40}K = \lambda_e {}^{40}K_0\, e^{-\lambda t}$, where $^{40}K_0$ is the initial content of ^{40}K, λ is the overall decay constant of ^{40}K, and λ_e is the branch decay constant of ^{40}K to ^{40}Ar, the concentration of ^{40}Ar can be expressed as

$$\frac{\partial\, ^{40}\mathrm{Ar}}{\partial t} = -\nabla \cdot \mathbf{J}_{40_{Ar}} + \lambda_e {}^{40}\mathrm{K}. \tag{3-5a}$$

Another example for mass conservation of a species is the conservation of molecular H_2O concentration in a silicate melt. Because OH groups can convert to molecular H_2O (Reaction 1-10, $H_2O(\text{melt}) + O(\text{melt}) \rightleftharpoons 2OH(\text{melt})$), assuming the reaction is elementary (which may not be correct), the concentration of molecular H_2O may be expressed as

$$\frac{\partial C_{H_2O_m}}{\partial t} = -\nabla \cdot \mathbf{J}_{H_2O_m} + k_b C_{OH}^2 - k_f C_{H_2O_m} C_O, \tag{3-5b}$$

where $C_{H_2O_m}$, C_{OH}, and C_O are concentrations (mol/m^3) of molecular H_2O, hydroxyl, and anhydrous oxygen, k_f and k_b are the forward and backward reaction rate coefficients, and $\mathbf{J}_{H_2O_m}$ is the diffusive flux of molecular H_2O.

For any given system, it is possible to choose a set of components whose concentrations are independent of chemical reactions even though the choice is not unique. For example, if chemical elements are chosen as components, the concentrations are conservative with respect to chemical reactions (but not with respect to nuclear reactions). If oxide components are chosen, they are conservative except for redox (shorthand for reduction/oxidation) reactions. If conservative components are used, then Equation 3-5 reduces to

$$\frac{\partial \rho_k}{\partial t} = -\nabla \cdot \mathbf{J}_k. \tag{3-5c}$$

In the above equation, $\mathbf{J}_k$ may include the flux of several species because there may be several species for the component. For example, the MgO component in a rock can be in the form of Mg_2SiO_4 and $MgSiO_3$, and the H_2O component in a melt can be in the form of molecular H_2O and OH groups. For the H_2O component, because of two species, the variation of total H_2O concentration ($C_{H_2O_t}$) with time may be expressed as

$$\frac{\partial C_{H_2O_t}}{\partial t} = -\nabla \cdot \mathbf{J}_{H_2O_m} - \frac{1}{2}\nabla \cdot \mathbf{J}_{OH}, \tag{3-5d}$$

where $C_{H_2O_t}$ is the concentration of total H_2O (in mol/m^3), $\mathbf{J}_{H_2O_m}$ is the flux of molecular H_2O, $\mathbf{J}_{OH}$ is flux of hydroxyl group, and the factor $\frac{1}{2}$ is because two OH groups convert to one H_2O molecule. Comparing Equations 3-5b and 3-5d, if the concentration variation of a conservative component is considered, species interconversion reactions do not enter the expression but there may be extra flux terms; on the other hand, if the concentration variation of a nonconservative species is considered, there is only one flux term but there may be reactive terms.

3.1.1.2 Diffusion

Diffusion is due to the random motion of particles (atoms, ions, molecules). The random motion is excited by thermal energy. In the case of pure diffusion, there is no bulk flow, only the redistribution of the components. Nonetheless, exchange of components may result in a shift of the mass center if a heavier particle such as Fe^{2+} exchanges with a lighter particle such as Mg^{2+}; it may result in a volume shift if a larger particle exchanges with a smaller particle. If the two particles have the same volume, then there is no volume shift, even if there is mass shift. This shows that it is necessary to carefully account for the *reference frame*. For Fe^{2+}–Mg^{2+} exchange, it may be said that there is no bulk motion in a volume-fixed reference frame, but there is bulk shift of the gravity center. Reference frame is a subtle issue (Brady, 1975a), and more discussion will be found later using examples for which knowing the reference frame is critical (Section 4.2.1).

In one single phase that is stable with respect to *spinodal decomposition*, the effect of random motion of atoms is to homogenize the phase if it is initially inhomogeneous. This process is similar to the following process (sometimes referred to as the drunkard's walk): In a large room many people are initially on one side of the room; but after getting drunk they start to walk randomly. After some time, people will be randomly dispersed in the room. That is, at any instant, the number of people per unit area (as long as the area is much greater than the area occupied by one person) is uniform. "Random" is the key in this process.

Now, consider particle motion in a phase (solid, liquid, gas). If the phase is initially inhomogeneous, random motion of atoms tends to homogenize the phase. If several phases are present and there are exchanges between the phases, the interphase reaction or exchange tends to make the chemical potential of all exchangeable components the same in all phases and diffusion again works to homogenize each phase. Hence, at equilibrium, the chemical potential of a component is constant.

In the most general sense, diffusion responds to a chemical potential gradient. When the phase is stable (would not separate into two phases), the chemical potential and concentration of a component are positively correlated. Hence, in a stable phase diffusion responds to a concentration gradient and tends to homogenize the phase to minimize Gibbs free energy. However, if the phase is not stable (that is, if it undergoes spontaneous decomposition), diffusion would help to create two phases from one single phase (that is, to create compositional heterogeneity to minimize Gibbs free energy), and will help to homogenize each phase. One example is spinodal decomposition of alkali feldspar: at high temperature, sodium and potassium feldspar mix more extensively. As temperature is lowered, a homogeneous feldspar phase is not stable and separates into two phases, forming perthite. The direction of diffusion (from a less homogeneous to more homogeneous phase in terms of chemical potential) is dictated by the second law of thermodynamics (see Appendix 1).

The mass conservation equation only relates concentration variation with flux, and hence cannot be used to solve for the concentration. To describe how the concentrations evolve with time in a nonuniform system, in addition to the mass balance equations, another equation describing how the flux is related to concentration is necessary. This equation is called the constitutive equation. In a binary system, if the phase (diffusion medium) is stable and isotropic, the diffusion equation is based on the constitutive equation of Fick's law:

$$\mathbf{J}_2 = -D\nabla C_2, \tag{3-6}$$

where $\mathbf{J}_2$ and C_2 are diffusive flux and concentration (in mol/m^3) of component 2. If diffusion is one-dimensional, the above equation reduces to Equation 1-71. D is diffusivity (which can be *self-diffusivity*, *tracer diffusivity*, or *interdiffusivity*, etc.). The diffusivity in general depends on the composition of the system (i.e., mole fraction of component 2, X_2). D can be regarded as constant if (i) the two components are two isotopes of the same element, (ii) the composition range is small, or (iii) mixing between the two components is ideal and the two components have identical intrinsic diffusivity. Using the appropriate reference frame, the diffusive flux for component 1 is opposite to that for component 2 (so that there is no bulk flow): $\mathbf{J}_1 = -\mathbf{J}_2$.

Because $\partial C_2/\partial t = -\nabla \mathbf{J}_2$ from mass balance, combining with Equation 3-6, we have

$$\partial C_2/\partial t = \nabla \cdot (D\nabla C_2). \tag{3-7}$$

The above equation is known as the three-dimensional diffusion equation. One can also write the diffusion in terms of the first component. Hence, C_2 is replaced by C (concentration of either component 1 or component 2) below. The above equation is general and accounts for the case when D depends on concentration (such as chemical diffusion to be discussed later).

If the bulk molar density, i.e., $\rho = C_1 + C_2$ in mol/m^3 is constant, then

$$\partial X/\partial t = \nabla \cdot (D\nabla X), \tag{3-7a}$$

where $X =$ mole fraction. One rough example is Fe^{2+}–Mg^{2+} exchange in olivine. If the bulk mass density, i.e., $\rho = C_1 + C_2$ in kg/m^3 is constant, then

$$\partial w/\partial t = \nabla \cdot (D\nabla w), \tag{3-7b}$$

where $w =$ mass fraction (or wt%). One approximate example is diffusion in silicate melt. In reality, neither molar density nor mass density is constant in a system, but the approximations are often made in literature for simplicity. Although Equations 3-7, 3-7a, and 3-7b are different, the difference is small in most cases and no attempt will be made to distinguish them. The choice of the equations will be based on convenience instead of rigorousness. In some cases, there is large variation in density (such as from a mineral to a melt). Then, concentrations in mol/m^3 or kg/m^3 will be used.

If D is constant, Equation 3-7 becomes

$$\partial C/\partial t = D\nabla^2 C, \tag{3-8}$$

If diffusion is one-dimensional, Equation 3-7 becomes

$$\frac{\partial C}{\partial t} = \frac{\partial}{\partial x}\left(D\frac{\partial C}{\partial x}\right). \tag{3-9}$$

If diffusion is one-dimensional and D is independent of C and x, the above equation becomes

$$\frac{\partial C}{\partial t} = D\frac{\partial^2 C}{\partial x^2}, \tag{3-10}$$

which is Equation 1-74. In a multicomponent system, diffusion of one component is affected by all other components. For an anisotropic diffusion medium, the diffusion coefficient depends on the direction. The diffusion equations for these two situations are more complex and are discussed later.

Mathematically, the diffusion equation is identical to the heat conduction equation:

$$\rho c \partial T/\partial t = \nabla \cdot (k\nabla T), \tag{3-11a}$$

where ρ is density, c is heat capacity per unit mass, k is the *heat conductivity*, and T is temperature. With constant heat conductivity, the above can be written as

$$\partial T/\partial t = \kappa\nabla^2 T, \tag{3-11b}$$

where $\kappa = k/(\rho c)$ and is called *heat diffusivity*. So one can apply solutions for heat conduction problems to similar diffusion problems by letting $\kappa = D$.

3.1.1.3 Fluid flow

Diffusion is one mode of mass transfer. Fluid flow is another mode of mass transfer and may be the dominant mode. The total mass flux due to fluid flow is

$$\mathbf{J} = \rho\mathbf{u}, \tag{3-12}$$

where ρ is the density (total mass per unit volume), and $\mathbf{u}$ is the flow velocity vector. The mass conservation law (also referred to as the continuity equation) for the case of fluid flow is

$$\partial\rho/\partial t = -\nabla \cdot \mathbf{J} = -\nabla \cdot (\rho\mathbf{u}) = -\rho\nabla \cdot \mathbf{u} - \mathbf{u} \cdot \nabla\rho. \tag{3-13}$$

If the fluid density is uniform (meaning $\nabla\rho = 0$) and does not change with time (meaning $\partial\rho/\partial t = 0$), then

$$\nabla \cdot \mathbf{u} = 0. \tag{3-14}$$

The result that the divergence of flow velocity ($\nabla \cdot \mathbf{u}$) is zero is general as long as the fluid is uniform and incompressible. Liquid and "solid" fluids are approximately incompressible. However, gas is compressible and hence $\nabla \cdot \mathbf{u} \neq 0$ for gas.

Next we consider the flux of a component instead of total mass flux. Similar to Equation 3-12, the flux of a conserved component (no source nor sink) due to fluid flow is

$$\mathbf{J}_k = C_k \mathbf{u}, \tag{3-15}$$

where C_k is the mass of component k per unit volume (mol/m^3 or kg/m^3). The mass conservation law for component k is

$$\partial C_k/\partial t = -\nabla \cdot \mathbf{J}_k = -\nabla \cdot (C_k \mathbf{u}). \tag{3-16}$$

If the fluid density is uniform and invariant with time (hence, $\nabla \cdot \mathbf{u} = 0$), then

$$\partial C_k/\partial t = -\mathbf{u} \cdot \nabla C_k. \tag{3-17}$$

3.1.1.4 General mass transfer

The general case of mass transfer includes both diffusion and convection. Hence, there are both diffusive flux and convective flux for a component. Therefore, the total flux is the sum of the two fluxes. For a given component in a binary and isotropic system, the total flux is

$$\mathbf{J} = C\mathbf{u} - D\nabla C, \tag{3-18}$$

where D is the *interdiffusivity*. Hence, the general mass transfer equation for a binary and isotropic system with constant D is

$$\partial C/\partial t = -\nabla \cdot \mathbf{J} = \nabla \cdot (D\nabla C) - \nabla \cdot (C\mathbf{u}) = D\nabla^2 C - \mathbf{u} \cdot \nabla C. \tag{3-19}$$

In deriving the above equation, the condition of $\nabla \cdot \mathbf{u} = 0$ is assumed. The above equation takes the following form in three dimensions in the *Cartesian coordinate system* (x, y, z) if D is independent of C, x, y, and z:

$$\frac{\partial C}{\partial t} = D\left(\frac{\partial^2 C}{\partial x^2} + \frac{\partial^2 C}{\partial y^2} + \frac{\partial^2 C}{\partial z^2}\right) - u_x\frac{\partial C}{\partial x} - u_y\frac{\partial C}{\partial y} - u_z\frac{\partial C}{\partial z}, \tag{3-19a}$$

where u_x, u_y and u_z are the three components of $\mathbf{u}$. The above equation takes the following form in one dimension:

$$\frac{\partial C}{\partial t} = D\frac{\partial^2 C}{\partial x^2} - u_x\frac{\partial C}{\partial x}. \tag{3-19b}$$

3.1.2 Conservation of energy

The equation for the conservation of energy is similar to that for mass conservation. The equation is obtained following similar steps as the diffusion equation: starting from the equation for the conservation of energy, combining it with the constitutive heat conduction law (Fourier's law), which is similar to Fick's law (in fact, Fick's law was proposed by analogy to Fourier's law), the following heat conduction equation (Equation 3-11b) is derived:

$$\partial T/\partial t = \kappa \nabla^2 T, \tag{3-20}$$

which is of the same form as the diffusion equation. If convection is included, a similar heat transfer equation dealing with both heat conduction and convection can be derived. The mathematical similarity between heat transfer and mass transfer equations is a blessing because many solutions to heat conduction problems (Carslaw and Jaeger, 1959) may be applied to diffusion problems, and vice versa.

3.1.3 Conservation of momentum

The momentum equation (the *Navier-Stokes equation*) for fluid flow (De Groot and Mazur, 1962) is complicated and difficult to solve. It is the subject of fluid mechanics and dynamics and is not covered in this book. When fluid flow is discussed in this book, the focus is on the effect of the flow (such as a flow of constant velocity, or boundary flow) on mass transfer, not the dynamics of the flow itself.

For flow in a porous medium, Darcy's law describes the flow rate:

$$\mathbf{u} = -\frac{k}{\eta}\nabla P, \tag{3-21}$$

where $\mathbf{u}$ is the volumetric flow rate per unit area (or average flow velocity over a given area), k is the permeability (unit is m^2) of the porous medium, η is the viscosity of the fluid, and ∇P is the pressure gradient. When the flow rate is known, the effect of flow on mass transport may be accounted for by Equation 3-15.

3.1.4 Various kinds of diffusion

Before going into detailed treatments of diffusion problems, the concepts of various kinds of diffusion are summarized here. These concepts will be encountered in later parts of this chapter, which will enhance your understanding of them.

Although random motion of particles is always present, even in a single-component system, the process cannot be studied experimentally in a one-component system because there is no way to label the species of a single component. A binary system is a system with two distinguishable components (such as two different isotopes, or two chemical components). Diffusion in a binary system will be discussed extensively because it is the simplest system and because more complicated diffusion problems are often converted into these simpler problems.

If the two components are two different isotopes of the same element, the diffusion is called *self-diffusion*. An example of self-diffusion is ^{18}O diffusion between two olivine crystals of identical chemical composition but one is made of normal oxygen and the other is made of ^{18}O-enriched oxygen. For self-diffusion, the number of chemical components may be more than two, but the concentration difference is in two isotopes of the same element only. For example, oxygen self-diffusion in basalt (there are many components, including SiO_2, TiO_2, Al_2O_3, Fe_2O_3, FeO, MgO, CaO, Na_2O, and K_2O) is for cases when there is no chemical composition difference across the diffusion medium, but there is ^{18}O and ^{16}O concentration difference. (If there are concentration differences in all three stable isotopes of oxygen, ^{18}O, ^{17}O, and ^{16}O, then strictly this is a multicomponent diffusion problem, but the multicomponent effect is considered negligible.) Self-diffusivity is constant across the whole profile because the only variation along the profile is in isotopic ratio, which is not expected to affect diffusion coefficient.

If one component is at a trace level but with variable concentrations (e.g., from 1 to 10 ppb) and concentrations of other components are uniform, the diffusion is called *tracer diffusion*. An example of tracer diffusion is ^{14}C diffusion into a melt of uniform composition (Watson, 1991b) when the concentration of ^{14}C is below ppb level. Usually only for a radioactive nuclide such as ^{14}C or ^{45}Ca, can such low concentrations be measured accurately to obtain concentration profiles. If a radioactive nuclide diffuses into a melt that contains the element (such as ^{45}Ca diffusion into a Ca-bearing melt), it is still called tracer diffusion although it may be through isotopic exchange.

When the concentration levels are higher, such as Ni diffusion between two olivine crystals with the same compositions except for Ni content (e.g., one contains 100 ppm and the other 2000 ppm), it may be referred to as either tracer diffusion or *chemical diffusion*. Tracer diffusivity is constant across the whole profile because the only variation along the profile is the concentration of a trace element that is not expected to affect the diffusion coefficient.

Other general cases in binary systems are referred to as *interdiffusion* or *binary diffusion*. For example, Fe–Mg diffusion between two olivine crystals of different X_{Fo} (mole fraction of forsterite Mg_2SiO_4) is called Fe–Mg interdiffusion. Interdiffusivity often varies across the profile because there are major concentration changes, and diffusivity usually depends on composition.

Diffusion in a system with three or more components is called *multicomponent diffusion*. One example is diffusion of Ca, Fe, Mn, and Mg in a zoned garnet (Ganguly et al., 1998a). Another example is diffusion between an andesitic melt and a basaltic melt (Kress and Ghiorso, 1995). The rigorous description of multicomponent diffusion is complicated. Numerous diffusivities (as elements of a diffusivity matrix) may be necessary to describe the diffusion, and each of these may vary in a complex way. One simple method of treating multicomponent diffusion is to consider only the component of interest and treat all other components as a combined "component," which is called *effective binary diffusion*. The diffusion coefficient is referred to as the *effective binary diffusion coefficient* (*EBDC*) or *effective binary diffusivity*.

If both chemical concentration gradients and isotopic ratio gradients are present (e.g., basaltic melt with $^{87}Sr/^{86}Sr$ ratio of 0.705 and andesitic melt with $^{87}Sr/^{86}Sr$ ratio of 0.720), the homogenization of isotopic ratio is referred to as *isotopic diffusion* (Lesher, 1990; Van Der Laan et al., 1994), although some prefer to call it *isotopic homogenization*. If there are concentration gradients in both major and trace elements, the diffusion of the trace elements is referred to as *trace element diffusion* (Baker, 1989). Isotopic diffusion and trace element diffusion are really part of multicomponent diffusion, which is complicated to handle. Isotopic diffusion should not be confused with self-diffusion, and trace element diffusion should not be confused with tracer diffusion.

Interdiffusion, effective binary diffusion, and multicomponent diffusion may be referred to as *chemical diffusion, meaning there are major chemical concentration gradients*. Chemical diffusion is defined relative to self diffusion and tracer diffusion, for which there are no major chemical concentration gradients.

If a diffusion component is present as two or more different species, the diffusion of the component is often referred to as *multispecies diffusion* (Zhang et al., 1991a,b). Multispecies diffusion is distinguished from multicomponent diffusion in that in the former case, the multiple species are from one component.

If the diffusion medium is isotropic in terms of diffusion, meaning that diffusion coefficient does not depend on direction in the medium, it is called *diffusion in an isotropic medium*. Otherwise, it is referred to as *diffusion in an anisotropic medium*. Isotropic diffusion medium includes gas, liquid (such as aqueous solution and silicate melts), glass, and crystalline phases with isometric symmetry (such as spinel and garnet). Anisotropic diffusion medium includes crystalline phases with lower than isometric symmetry. That is, most minerals are diffusionally anisotropic. An isotropic medium in terms of diffusion may not be an isotropic medium in terms of other properties. For example, cubic crystals are not isotropic in terms of elastic properties. The diffusion equations that have been presented so far (Equations 3-7 to 3-10) are all for isotropic diffusion medium.

Self-diffusion and tracer diffusion are described by Equation 3-10 in one dimension, and Equation 3-8 in three dimensions. For interdiffusion, because D may vary along a diffusion profile, the applicable diffusion equation is Equation 3-9 in one dimension, or Equation 3-7 in three dimensions. The descriptions of multispecies diffusion, multicomponent diffusion, and diffusion in anisotropic systems are briefly outlined below and are discussed in more detail later.

3.1.4.1 Multispecies diffusion

In a silicate melt or aqueous solution, a component may be present in several species. The species may interconvert and diffuse simultaneously. For example, the H_2O component in silicate melt can be present as at least two species, molecular H_2O (referred to as H_2O_m) and hydroxyl groups (referred to as OH) (Stolper, 1982a). The diffusion of such a multispecies component is referred to as multispecies diffusion (Zhang et al., 1991a,b). Starting from Equation 3-5d, the one-dimensional diffusion equation for this multispecies component can be written as

$$\frac{\partial C_{H_2O_t}}{\partial t} = \frac{\partial}{\partial x}\left(D_{H_2O_m}\frac{\partial C_{H_2O_m}}{\partial x}\right) + \frac{1}{2}\frac{\partial}{\partial x}\left(D_{OH}\frac{\partial C_{OH}}{\partial x}\right). \tag{3-22a}$$

In the above equation, there are three unknowns: $C_{H_2O_t}$, $C_{H_2O_m}$, and C_{OH} (all in mol/L or mole fractions). Hence, two more equations are needed for a solution:

$$\text{Mass balance:} \quad C_{H_2O_t} = C_{H_2O_m} + \frac{1}{2}C_{OH}; \tag{3-22b}$$

$$\text{Equilibrium constant:} \quad K = \frac{X_{OH}^2}{X_{H_2O_m}(1 - X_{H_2O_m} - X_{OH})}, \tag{3-22c}$$

where X means mole fractions on a single oxygen basis (Stolper, 1982b; Zhang, 1999b). To predict how concentration profiles evolve with time, it is necessary to solve simultaneously the above three equations numerically (e.g., Zhang et al., 1991a,b).

3.1.4.2 Multicomponent diffusion

The full treatment of multicomponent diffusion requires a diffusion matrix because the diffusive flux of one component is affected by the concentration gradient of all other components. For an N-component system, there are $N-1$ independent components (because the concentrations of all components add up to 100% if mass fraction or molar fraction is used). Choose the Nth component as the dependent component and let $n = N - 1$. The diffusive flux of the components can hence be written as (De Groot and Mazur, 1962)

$$\frac{\partial C_1}{\partial t} = \frac{\partial}{\partial x}\left(D_{11}\frac{\partial C_1}{\partial x}\right) + \frac{\partial}{\partial x}\left(D_{12}\frac{\partial C_2}{\partial x}\right) + \cdots + \frac{\partial}{\partial x}\left(D_{1n}\frac{\partial C_n}{\partial x}\right), \tag{3-23a}$$

$$\frac{\partial C_2}{\partial t} = \frac{\partial}{\partial x}\left(D_{21}\frac{\partial C_1}{\partial x}\right) + \frac{\partial}{\partial x}\left(D_{22}\frac{\partial C_2}{\partial x}\right) + \cdots + \frac{\partial}{\partial x}\left(D_{2n}\frac{\partial C_n}{\partial x}\right), \tag{3-23b}$$

$$\vdots$$

$$\frac{\partial C_n}{\partial t} = \frac{\partial}{\partial x}\left(D_{n1}\frac{\partial C_1}{\partial x}\right) + \frac{\partial}{\partial x}\left(D_{n2}\frac{\partial C_2}{\partial x}\right) + \cdots + \frac{\partial}{\partial x}\left(D_{nn}\frac{\partial C_n}{\partial x}\right), \tag{3-23n}$$

where D_{ij} is the diffusivity of component i due to the concentration gradient of component j. In matrix notation, the above equation may be written as

$$\frac{\partial}{\partial t} \begin{pmatrix} C_1 \\ C_2 \\ \vdots \\ C_n \end{pmatrix} = \frac{\partial}{\partial x} \begin{pmatrix} D_{11} & D_{12} & \cdots & D_{1n} \\ D_{21} & D_{22} & \cdots & D_{2n} \\ \vdots & \vdots & \vdots & \vdots \\ D_{n1} & D_{n2} & \cdots & D_{nn} \end{pmatrix} \frac{\partial}{\partial x} \begin{pmatrix} C_1 \\ C_2 \\ \vdots \\ C_n \end{pmatrix}. \tag{3-24a}$$

If every D_{ij} above is independent of C and x and can hence be taken out of the differential, then the above set of equations can be written in the matrix form as

$$\frac{\partial}{\partial t} \begin{pmatrix} C_1 \\ C_2 \\ \vdots \\ C_n \end{pmatrix} = \begin{pmatrix} D_{11} & D_{12} & \cdots & D_{1n} \\ D_{21} & D_{22} & \cdots & D_{2n} \\ \vdots & \vdots & \vdots & \vdots \\ D_{n1} & D_{n2} & \cdots & D_{nn} \end{pmatrix} \frac{\partial^2}{\partial x^2} \begin{pmatrix} C_1 \\ C_2 \\ \vdots \\ C_n \end{pmatrix}. \tag{3-24b}$$

3.1.4.3 Diffusion in an anisotropic medium

For binary diffusion in an isotropic medium, one diffusion coefficient describes the diffusion. For binary diffusion in an anisotropic medium, the diffusion coefficient is replaced by a diffusion tensor, denoted as $\underline{\mathbf{D}}$. The diffusion tensor is a second-rank symmetric tensor representable by a 3×3 matrix:

$$\underline{\mathbf{D}} = \begin{pmatrix} \underline{D}_{11} & \underline{D}_{12} & \underline{D}_{13} \\ \underline{D}_{12} & \underline{D}_{22} & \underline{D}_{23} \\ \underline{D}_{13} & \underline{D}_{23} & \underline{D}_{33} \end{pmatrix}. \tag{3-25a}$$

The diffusion coefficient along any direction (l, m, n) may be obtained as

$$D_{(l, m, n)} = (l, m, n) \begin{pmatrix} \underline{D}_{11} & \underline{D}_{12} & \underline{D}_{13} \\ \underline{D}_{12} & \underline{D}_{22} & \underline{D}_{23} \\ \underline{D}_{13} & \underline{D}_{23} & \underline{D}_{33} \end{pmatrix} \begin{pmatrix} l \\ m \\ n \end{pmatrix}, \tag{3-25b}$$

where (l, m, n) is a unit vector, i.e., $l^2 + m^2 + n^2 = 1$. Treatment of anisotropic diffusion is discussed later, where it will be simplified to Equation 3-8. The diffusion tensor describes binary diffusion in an anisotropic medium, and differs from the diffusion matrix, which describes multicomponent diffusion in an isotropic medium.

In the general case of three-dimensional multicomponent diffusion in an anisotropic medium (such as Ca–Fe–Mg diffusion in pyroxene), the mathematical description of diffusion is really complicated: it requires a diffusion matrix in which every element is a second-rank tensor, and every element in the tensor may depend on composition. Such a diffusion equation has not been solved. Because rigorous and complete treatment of diffusion is often too complicated, and because instrumental analytical errors are often too large to distinguish exact solutions from approximate solutions, one would get nowhere by considering all these real complexities. Hence, simplification based on the question at hand is necessary to make the treatment of diffusion manageable and useful.

3.1.4.4 Volume diffusion versus grain-boundary diffusion

All of the above discussion of diffusion involves mass transfer inside a fluid phase or solid grains, and hence is called *volume diffusion*, meaning diffusion through a volume. (Volume diffusion is not to be confused with volume shift, which is the shift of total volume if a small particle exchanges with a larger particle.) A single solid phase may consist of many grains, and a rock contains many mineral grains. Diffusive mass transport along grain boundaries is referred to as *grain-boundary diffusion*. Atoms on surfaces are underbonded. The atomic structure at grain boundaries is deformed with more defects. Both lead to more available jumping sites and lower activation energies for diffusion, meaning grain boundaries are more reactive. Hence, grain-boundary diffusion coefficients may be orders of magnitude greater than volume diffusion (Nagy and Giletti, 1986; Farver and Yund, 2000; Milke et al., 2001). Nagy and Giletti (1986) concluded that grain-boundary diffusion of oxygen is at least four orders of magnitude faster than volume diffusion. Farver and Yund (2000) showed that grain-boundary diffusion of ^{30}Si dominates Si transfer through fine-grained (about 4.5 μm size) forsterite aggregates and derived that grain-boundary diffusivity is 10^9 times greater than volume diffusivity of Si. Hence, mass transport across a rock may be through grain-boundary diffusion, especially for a fine-grained rock with abundant grain boundaries. If our interest is to understand the bulk effect of mass transfer, such as how rapidly mass may be transferred from one location to another in a rock, and if the space scale of our interest is much larger than grain sizes in the rock, these diffusion problems may be treated mathematically the same as volume diffusion, but the diffusivity would not be volume diffusivities in single grains, but some effective diffusivity due to the multiple grain-boundary diffusion paths. Even though grain-boundary diffusion may transfer masses across a rock medium, it cannot modify the interior composition of grains. Therefore, when concentration profiles are measured in a crystal, volume diffusion is the cause. In this book, diffusion means volume diffusion in an isotropic medium unless otherwise specified.

3.1.4.5 Eddy diffusion

All of the above discussion of diffusion considers physical motion of particles excited by thermal energy of the system (because the system is not at 0 K), rather than by outside factors. Eddy diffusion is different. It is due to random disturbance in water by outside factors, such as fish swimming, wave motion, ship cruising, and turbulence in water. On a small length scale (similar to the length scale of disturbance), the disturbances are considered explicitly as convection or flow in the mass transfer equation (Equation 3-19). On a length scale much larger than the individual disturbances, the collective effect of all of the disturbances leads to random dispersion of particles. If the intensities of these disturbances are not too variable with time or from one part of the lake or river to another part, the collective dispersion effect can be considered random, and hence can also be described by the diffusion equation. This diffusion is referred to as *eddy diffusion*, and the diffusivity is referred to as *eddy diffusivity*. The eddy diffusivities depend on the many disturbances in the water body and vary from one water body to another. The values are typically orders of magnitude larger than molecular diffusivities. In deep ocean water, eddy diffusivity may be only a few orders of magnitude greater than molecular diffusivity.

3.2 Diffusion in a Binary System

3.2.1 Diffusion equation

When one refers to the diffusion equation, it is usually the binary diffusion equation. Although theories for multicomponent diffusion have been exten- sively developed, experimental studies of multicomponent diffusion are limited because of instrumental analytical error and theoretical complexity, and there are yet no reliable diffusivity matrix data for practical applications in geology. Multicomponent diffusion is hence often treated as effective binary diffusion by treating the component under consideration as one component and combining all the other components as the second component.

The one-dimensional binary diffusion equation with constant diffusion coef- ficient is (Equation 3-10)

$$\frac{\partial C}{\partial t} = D \frac{\partial^2 C}{\partial x^2}. \tag{3-10}$$

From this equation, the rate of change of the concentration at a given point is proportional to the curvature (the second derivative) of the concentration curve as a function of x, as diagrammed in Figure 3-2. If the curvature is zero, e.g., if the concentration is a linear function of distance, the concentration at the point does not change with time because the inflow balances the outflow. When the curvature (second derivative) is positive (Figure 3-2a), i.e., C at the position is lower than its neighbors, diffusion increases the concentration. When the cur- vature (second derivative) is negative (Figure 3-2b), i.e., C at the position is higher than its neighbors, diffusion decreases the concentration. Therefore, diffusion acts to reduce the curvature and to reduce the wrinkles in the con- centration distribution, which is what we would intuitively expect.

Equation 3-10 is the most basic diffusion equation to be solved, and has been solved analytically for many different initial and boundary conditions. Many other more complicated diffusion problems (such as three-dimensional diffu- sion with spherical symmetry, diffusion for time-dependent diffusivity, and

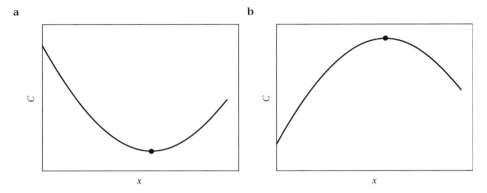

Figure 3-2 Concentration curves for (a) positive curvature and (b) negative curvature. Diffusion will increase the concentration along the curve for case (a), and will decrease the concentration for case (b).

multicomponent diffusion) may be transformed to the form of Equation 3-10. Therefore, analytical solutions of Equation 3-10 may also be applied to these more complicated diffusion problems with some transformation and adaptation.

The next several sections discuss initial and boundary conditions, and methods to solve Equation 3-10 given such conditions. In the process of learning the methods, typical solutions to the diffusion equation will be presented. These solutions will be encountered in later discussions. More solutions are presented in Appendix 3.

3.2.2 Initial and boundary conditions

Many solutions exist for a partial differential equation such as Equation 3-10. For example,

$$C = constant, \tag{3-26a}$$

$$C = a + bx, \tag{3-26b}$$

and

$$C = (A/t^{1/2})e^{-x^2/4Dt} \tag{3-26c}$$

plus many more are all solutions to Equation 3-10. You may verify that each of the above function satisfies Equation 3-10. Therefore, one cannot simply solve the diffusion equation to get the general solution. To solve a partial differential equation such as the diffusion equation, it is necessary to specify the *initial and boundary conditions*, and then to solve for the specific solution.

The initial condition for Equation 3-10 generally takes the form $C|_{t=0} = f(x)$. That is, the concentration distribution is given at $t=0$. The simplest initial condition is that $C|_{t=0} = $ constant.

The boundary conditions are more complicated and three cases may be distinguished:

(1) For one-dimensional diffusion, there are two ends. If both ends participate in diffusion, the diffusion medium is called a *finite medium* with two boundaries. The two boundaries may be defined differently, such as (i) $x = -a$ and $x = +a$, or (ii) $x = 0$ and $x = a$, or (iii) $x = a$ and $x = b$, whichever is more convenient to solve the problem. Hence, there are two boundary conditions. Each boundary condition may specify either the concentration at the boundary, such as $C|_{x=0} = g(t)$, or the concentration gradient at the boundary, such as $(\partial C/\partial x)|_{x=0} = g(t)$, or a mixed condition, such as $(\partial C/\partial x)|_{x=0} + gC|_{x=0} = g(t)$, where g is a constant.

(2) For one-dimensional diffusion, if diffusion starts in the interior and has not reached either of the two ends yet, the diffusion medium is called an *infinite medium*. An infinite diffusion medium does not mean that we consider the whole universe as the diffusion medium. One example is the diffusion couple of only a few millimeters long (discussed later). In an infinite medium, there is no boundary, but one often specifies the values of $C|_{x=-\infty}$ and $C|_{x=\infty}$ as constraints that must be satisfied by the solution. These constraints mean that the concentration must be finite as x approaches $-\infty$ or $+\infty$, and the concentrations at $+\infty$ or $-\infty$ must be the same as the respective initial concentrations. These obvious conditions often help in simplifying the solutions.

(3) If diffusion starts from one end (surface) and has not reached the other end yet in one-dimensional diffusion, the diffusion medium is called a *semi-infinite medium* (also called *half-space*). There is, hence, only one boundary, which is often defined to be at $x = 0$. This boundary condition usually takes the form of $C|_{x=0} = g(t)$, $(\partial C/\partial x)|_{x=0} = g(t)$, or $(\partial C/\partial x)|_{x=0} + aC|_{x=0} = g(t)$, where a is a constant. Similar to the case of infinite diffusion medium, one often also writes the condition $C|_{x=\infty}$ as a constraint.

Because an "infinite" or a "semi-infinite" reservoir merely means that the medium at the two ends or at one end is not affected by diffusion, whether a medium may be treated as infinite or semi-infinite depends on the timescale of our consideration. For example, at room temperature, if water diffuses into an obsidian glass from one surface and the diffusion distance is about $5\,\mu m$ in 1000 years, an obsidian glass of 50 μm thick can be viewed as a semi-infinite medium on a thousand-year timescale because $5\,\mu m$ is much smaller than 50 μm. However, if we want to treat diffusion into obsidian on a million-year timescale, then an obsidian glass of 50 μm thick cannot be viewed as a semi-infinite medium.

In three-dimensional diffusion, the boundary itself can be complicated, and boundary conditions may also be complicated except for some simple geometry, sphere, cube, long-cylinder, etc. The meaning of initial and boundary conditions will be clearer after some examples below.

3.2.3 Some simple solutions to the diffusion equation at steady state

Steady state may be reached in a diffusion problem proceeding for a long time in a finite medium. Steady state means that the concentration at any point does not change with time any more, i.e.,

$$\partial C/\partial t = 0. \tag{3-27}$$

For steady-state solution, the initial condition does not matter because the steady state does not depend on the initial condition. Only the boundary condition is necessary for solving the steady-state diffusion equation. The three-dimensional diffusion equation at steady state is

$$\nabla(D\nabla C) = 0. \tag{3-28}$$

If D is constant, the equation is known as the Laplace equation:

$$\nabla^2 C = 0. \tag{3-29a}$$

In Cartesian coordinates, the equation takes the following form:

$$\frac{\partial^2 C}{\partial x^2} + \frac{\partial^2 C}{\partial y^2} + \frac{\partial^2 C}{\partial z^2} = 0. \tag{3-29b}$$

3.2.3.1 One-dimensional steady-state diffusion

Steady state in one dimension is described by

$$\frac{\mathrm{d}}{\mathrm{d}x}\left(D\frac{\mathrm{d}C}{\mathrm{d}x}\right) = 0. \tag{3-30}$$

Integration leads to

$$D\,\mathrm{d}C/\mathrm{d}x = A, \tag{3-30a}$$

where A is a constant. How to integrate the above depends on whether and how D varies. If D is constant, then

$$C = a + bx, \tag{3-30b}$$

where a and b are two constants. That is, C is a linear function of x. Knowing boundary conditions at $x = x_1$ and $x = x_2$, the two constants can be determined. For example, if $C = C_1$ at $x = x_1$ and $C = C_2$ at $x = x_2$, then

$$C = C_1 + (C_2 - C_1)(x - x_1)/(x_2 - x_1). \tag{3-30c}$$

Figure 3-3a shows the one-dimensional steady-state diffusion profile with constant D. D cannot depend on time for steady state. If D is a function of x, then

$$C = \int \mathrm{d}C = \int (A/D)\mathrm{d}x. \tag{3-30d}$$

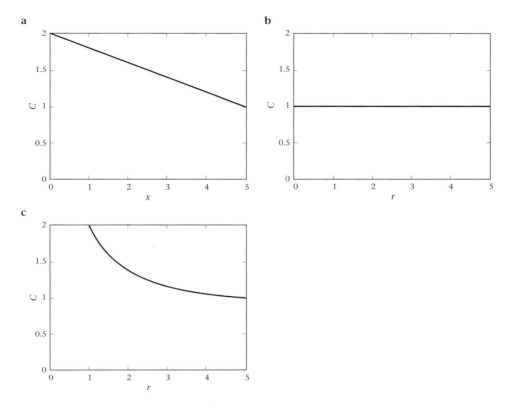

Figure 3-3 Steady-state diffusion profile in (a) one dimension with concentrations at the two ends fixed, (b) a solid sphere with constant concentration on the surface (at $r=5$), and (c) a spherical shell (radius from 1 to 5) with concentrations at the two surfaces fixed.

If D is a function of C, then

$$\int D\,dC = f(C) = \int A\,dx = Ax + B,\tag{3-30e}$$

C can be solved from $f(C) = Ax + B$. One application of this solution is for an insulated metal rod with one end in ice water (0°C) and the other end in boiling water (100°C).

3.2.3.2 Radial steady-state diffusion

Steady-state diffusion in three dimensions with spherical symmetry (i.e., the concentration is a function of r only) is described by an ordinary differential equation (which is Equation 3-28 simplified for spherical symmetry, cf. Equation 3-66b later):

$$\frac{\partial}{\partial r}\left(Dr^2\frac{\partial C}{\partial r}\right) = 0.\tag{3-31}$$

Integrating once, we have

$$Dr^2 \partial C / \partial r = A, \tag{3-31a}$$

where A is a constant. Integrating again, we have,

$$\int D\,dC = \int (A/r^2)dr = B - A/r, \tag{3-31b}$$

where B is another constant. If D is constant, then

$$C = B' - A'/r, \tag{3-31c}$$

where A' and B' are two constants to be determined by boundary conditions. One natural "boundary condition" is that C must be finite at $r=0$ and therefore, $A' = 0$. Hence, the steady-state solution for diffusion inside a solid sphere with a constant surface concentration is uniform concentration for the whole sphere, i.e.,

$$C = C_0. \tag{3-31d}$$

Figure 3-3b shows the profile. Note that the condition that C must be finite at $r=0$ is a very important natural requirement. This requirement is also applied in many other problems, such as heat conduction and elastic equilibrium.

Now consider steady state for a spherical shell from R_1 to R_2 with $R_1 < R_2$ (instead of a solid sphere). For constant D, if the boundary conditions are

$$C|_{r=R_1} = C_1, \tag{3-31e}$$

and

$$C|_{r=R_2} = C_2, \tag{3-31f}$$

then using the above two conditions to solve for A' and B' in Equation 3-31c, the steady-state solution is

$$C = C_1 + (C_2 - C_1)\frac{1 - R_1/r}{1 - R_1/R_2}. \tag{3-31g}$$

Figure 3-3c displays the steady-state concentration profile for a spherical shell. One application of this solution is for a spinel crystal inside a magma chamber, where the spinel contains a large melt inclusion at its core. The diffusion profile in the spinel (which is a spherical shell) in equilibrium with the melt inclusion and the outside melt reservoir would follow Equation 3-31g.

3.2.4 One-dimensional diffusion in infinite or semi-infinite medium with constant diffusivity

Several general methods are available for solving the diffusion equation, including Boltzmann transformation, principle of superposition, separation of variables, Laplace transform, Fourier transform, and numerical methods. In this and the next several sections, four methods are introduced briefly, in the context of deriving solutions for often encountered diffusion problems. Laplace transform and Fourier transform are two powerful methods but they are not covered in this book because more mathematical background is required (Crank, 1975; Carslaw and Jaeger, 1959).

This section introduces the method of Boltzmann transformation to solve one-dimensional diffusion equation in infinite or semi-infinite medium with constant diffusivity. For such media, if some conditions are satisfied, Boltzmann transformation converts the two-variable diffusion equation (partial differential equation) into a one-variable ordinary differential equation.

3.2.4.1 Diffusion couple

In experimental studies of diffusion, the diffusion-couple technique is often used. A diffusion couple consists of two halves of material; each is initially uniform, but the two have different compositions. They are joined together and heated up. Diffusive flux across the interface tries to homogenize the couple. If the duration is not long, the concentrations at both ends would still be the same as the initial concentrations. Under such conditions, the diffusion medium may be treated as infinite and the diffusion problem can be solved using Boltzmann transformation. If the diffusion duration is long (this will be quantified later), the concentrations at the ends would be affected, and the diffusion medium must be treated as finite. Diffusion in such a finite medium cannot be solved by the Boltzmann method, but can be solved using methods such as separation of variables (Section 3.2.7) if the conditions at the two boundaries are known. Below, the concentrations at the two ends are assumed to be unaffected by diffusion.

Define the interface between the two halves as $x=0$. Define the initial concentration at $x<0$ as C_L, and that in the $x>0$ half as C_R. Assume constant D. The diffusion-couple problem may be written as the following mathematical problem:

Diffusion equation: $\quad \dfrac{\partial C}{\partial t} = D\dfrac{\partial^2 C}{\partial x^2}$ \hfill (3-10)

Initial condition: $\quad C|_{t=0} = \begin{cases} C_L & x<0 \\ C_R & x>0 \end{cases}$, \hfill (3-32a)

Boundary condition 1: $\quad C|_{x=-\infty} = C_L$, \hfill (3-32b)

Boundary condition 2: $\quad C|_{x=\infty} = C_R$. \hfill (3-32c)

Even though the boundary conditions are not necessary, writing them out would avoid confusion.

3.2.4.2 Boltzmann transformation

In the above diffusion-couple problem (as in other diffusion problems), the concentration C depends on two independent variables, x and t. Briefly, the *Boltzmann transformation* uses one variable $\eta = x/\sqrt{4t}$ (some authors use $\eta = x/\sqrt{t}$; some others use $\eta = x/\sqrt{Dt}$ if D is constant; they are all equivalent) to replace two variables x and t. This works only under special conditions. Below, the method is described first and the conditions for its use are discussed afterward.

Assume that C depends on only one variable, $\eta = x/\sqrt{4t}$. This assumption will be verified later. Then, express the partial differentials in the diffusion equation in terms of the total differential with respect to η:

$$\frac{\partial C}{\partial t} = \frac{dC}{d\eta}\frac{\partial \eta}{\partial t} = -\frac{\eta}{2t}\frac{dC}{d\eta}, \tag{3-33a}$$

$$\frac{\partial C}{\partial x} = \frac{dC}{d\eta}\frac{\partial \eta}{\partial x} = \frac{1}{2t^{1/2}}\frac{dC}{d\eta}, \tag{3-33b}$$

$$\frac{\partial}{\partial x}\left(\frac{\partial C}{\partial x}\right) = \frac{1}{4t}\frac{d}{d\eta}\left(\frac{dC}{d\eta}\right). \tag{3-33c}$$

The partial differentials have been replaced by the total differentials because it is assumed that C depends on only one variable, η. Hence, Equation 3-10 can be written as

$$-\frac{\eta}{2t}\frac{dC}{d\eta} = \frac{D}{4t}\frac{d}{d\eta}\left(\frac{dC}{d\eta}\right), \tag{3-33d}$$

i.e.,

$$D\frac{d^2C}{d\eta^2} + 2\eta\frac{dC}{d\eta} = 0. \tag{3-34}$$

The above transformation from Equation 3-10 to 3-34 is called the Boltzmann transformation. The two variables x and t are replaced by a single variable η, and the partial differential equation becomes an ordinary differential equation. The transformation works *only if the initial and boundary conditions can also be written in terms of the single variable* η. To solve the transformed ordinary differential equation, define $w = dC/d\eta$. Hence, Equation 3-34 becomes

$$D\frac{dw}{d\eta} + 2\eta w = 0.$$

$$w = w_0 e^{-\eta^2/D}.$$

$$C = C_{\eta=0} + \int_0^\eta w\ d\eta' = C_{\eta=0} + \int_0^\eta w_0 e^{-\eta'^2/D}d\eta'.$$

Defining $\xi = \eta/\sqrt{D} = x/(\sqrt{4Dt})$, the solution can be expressed as

$$C = C_{\xi=0} + \sqrt{D}\, w_0 \int_0^{x/\sqrt{4Dt}} e^{-\xi^2}\,d\xi,$$

where ξ is a dummy variable. Because the integral in the above equation is related to the error function defined as (Appendix 2)

$$\mathrm{erf}(z) = \frac{2}{\sqrt{\pi}} \int_0^z e^{-\xi^2}\,d\xi, \tag{3-35}$$

the solution can be written as

$$C = a\,\mathrm{erf}\left(\frac{x}{\sqrt{4Dt}}\right) + b. \tag{3-36}$$

Note that neither initial nor boundary conditions have been applied yet. The above equation is the general solution for infinite and semi-infinite diffusion medium obtained from Boltzmann transformation. The parameters a and b can be determined by initial and boundary conditions as long as initial and boundary conditions are consistent with the assumption that C depends only on η (or ξ). Readers who are not familiar with the error function and related functions are encouraged to study Appendix 2 to gain a basic understanding.

3.2.4.3 Solution to the diffusion-couple problem

Return to the diffusion-couple problem. The initial condition (Equation 3-32a) can be rewritten using the variable ξ (or η) as follows. The conditions $t=0$ and $x<0$ are equivalent to $\xi = x/\sqrt{4Dt} = -\infty$, and $t=0$ and $x>0$ equivalent to $\xi = \infty$. Hence, the initial condition expressed using the single variable ξ is

$$\text{Initial condition:} \quad C = \begin{cases} C_L & \xi = -\infty \\ C_R & \xi = \infty. \end{cases} \tag{3-37a}$$

And the boundary conditions become

$$\text{Boundary condition 1:} \quad C|_{\xi=-\infty} = C_L, \tag{3-37b}$$

$$\text{Boundary condition 2:} \quad C|_{\xi=\infty} = C_R. \tag{3-37c}$$

Comparing Equation 3-37a with Equations 3-37b and 3-37c shows that Equation 3-37a is identical to Equations 3-37b and 3-37c. That is, both the initial condition and the boundary conditions are transformed to the same two equations using the variable ξ. Only when this happens can the problem be solved using Boltzmann transformation. If the diffusion duration is long, and C at $x=\infty$ changes with time, it would be impossible to write boundary condition 2 using a single variable ξ, and the boundary condition would be inconsistent with the expression of $C|_{\xi=\infty} = C_R$ derived from the initial condition. For such a case, Boltzmann transformation would not be applicable. That is, Boltzmann transformation applies only to infinite and semi-infinite diffusion problems.

For the diffusion couple, because $\mathrm{erf}(-\infty) = -1$ and $\mathrm{erf}(\infty) = 1$, applying Equations 3-37b and 3-37c to Equation 3-36 leads to

$$C_L = -a + b, \tag{3-37d}$$

and

$$C_R = a + b. \tag{3-37e}$$

Solving a and b leads to

$$a = (C_R - C_L)/2, \tag{3-37f}$$

and

$$b = (C_L + C_R)/2. \tag{3-37g}$$

Hence, the solution to the diffusion-couple problem in one dimension with constant D is

$$C = \frac{C_L + C_R}{2} + \frac{C_R - C_L}{2} \mathrm{erf}\left(\frac{x}{2\sqrt{Dt}}\right). \tag{3-38}$$

This solution is shown in Figure 1-9. It is a widely used solution in experiments and in modeling diffusion behavior in nature.

3.2.4.4 Diffusion in semi-infinite medium with constant surface concentration

Another experimental method to investigate diffusion is the so-called half-space method, in which the sample (e.g., rhyolitic glass with normal oxygen isotopes) is initially uniform with concentration C_∞, but one surface (or all surfaces, as explained below) is brought into contact with a large reservoir (e.g., water vapor in which oxygen is all ^{18}O). The surface concentration of the sample is fixed to be constant, referred to as C_0. The duration is short so that some distance away from the surface, the concentration is unaffected by diffusion. Define the surface to be $x = 0$ and the sample to be at $x \geq 0$. This diffusion problem is the so-called half-space or semi-infinite diffusion problem.

Assuming constant D, the diffusion equation is Equation 3-10:

$$\frac{\partial C}{\partial t} = D\frac{\partial^2 C}{\partial x^2} \tag{3-10}$$

The initial and boundary conditions are as follows:

Initial condition: $\qquad C|_{t=0, x \geq 0} = C_\infty,$ (3-39a)

Boundary condition 1: $\quad C|_{x=0, t>0} = C_0,$ (3-39b)

Boundary condition 2: $\quad C|_{x=\infty, t>0} = C_\infty,$ (3-39c)

Using Boltzmann transformation, the initial and boundary conditions become

Initial condition: $\qquad C|_{\eta=\infty} = C_\infty,$ (3-39d)

Boundary condition 1: $\quad C|_{\eta=0} = C_0,$ (3-39e)

Boundary condition 2: $\quad C|_{\eta=\infty} = C_\infty.$ (3-39f)

Note that boundary condition 2 is consistent with the initial condition (meaning the medium is semi-infinite). Applying these conditions to the general solution Equation 3-36 leads to

$$C_0 = b,$$ (3-39g)

and

$$C_\infty = a + b.$$ (3-39h)

Solving for a and b from the above two equations, and replacing them into Equation 3-36 leads to

$$C = C_0 + (C_\infty - C_0)\mathrm{erf}[x/(4Dt)^{1/2}].$$ (3-40a)

The above solution can also be written in terms of the complimentary error function as

$$C = C_\infty + (C_0 - C_\infty)\mathrm{erfc}[x/(4Dt)^{1/2}].$$ (3-40b)

The solution is plotted in Figure 1-8 and is symmetric with respect to $x = 0$.

Although the above example is specifically for diffusion from one surface to the interior of a sample, it can also be applied to the following:

(1) Diffusion from two opposite surfaces to the interior of a planar sample as long as diffusion has not reached the center of the sample, e.g., $(4Dt)^{1/2} <$ half-thickness. In this case, for each side, the surface is treated as $x = 0$, and the diffusion profile is calculated from the surface. When the two profiles are combined, the diffusion profile of the whole plane-sheet sample is shown in Figure 3-4.

(2) Diffusion from the surface of a three-dimensional sample, such as a sphere, as long as the diffusion distance is much smaller than the radius of the sample, e.g., $4(Dt)^{1/2} < 1\%$ of the radius. For larger diffusion distances, approximation using Equation 3-40 does not work well. For example, in three-dimensional diffusion, the center is more easily affected by diffusion than in one-dimensional

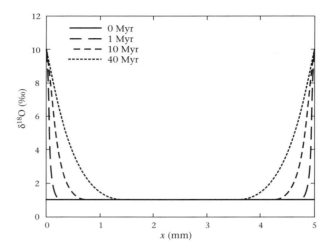

Figure 3-4 Calculated oxygen isotope diffusion profile from both sides of a plane-sheet mineral. Initial $\delta^{18}O$ in the mineral is 1‰. The surface $\delta^{18}O$ is 10‰. The diffusivity $D = 10^{-22}$ m^2/s. Compare this with Figure 1-8b.

diffusion because mass goes toward the center from all directions instead of just two opposite directions.

Example 3.1 *Cooling of an oceanic plate* (Parsons and Sclater, 1977). At mid-ocean ridges, mantle upwelling and melting produce new oceanic crust. The temperature of the newly formed plate at mid-ocean ridges is roughly uniform at $T_0 \approx 1300°C$ throughout except at the surface. The surface is quenched in water and has a temperature of roughly $T_s \approx 0°C$. As the newly created plate moves away, it is cooled further by heat loss from the upper surface in contact with ocean water. Heat loss to the sides is negligible because the temperature gradient is small. If convective heat loss (due to, for example, hydrothermal fluid) is ignored, solve the heat conduction equation to obtain the temperature profile as a function of time.

Solution: Heat conduction during aging of the plate (that is, as it moves away from the ocean ridge) can be described by the heat-diffusion problem in a semi-infinite medium. The solution is

$$T = T_s + (T_0 - T_s)\mathrm{erf}\left(\frac{z}{2\sqrt{\kappa t}}\right), \tag{3-41a}$$

where κ is heat diffusivity, t is age of the part of the oceanic plate, and z is depth below the ocean floor. For a constant half spreading velocity u, the age $t = x/u$, where x is the distance from the ridge. Hence, the above solution may be written as

$$T = T_s + (T_0 - T_s)\mathrm{erf}\left(\frac{z}{2\sqrt{\kappa x/u}}\right). \tag{3-41b}$$

The temperature profile is plotted in Figure 1-8a. The heat flow at the ocean floor ($z = 0$) is

$$q = -\left(-k\frac{\partial T}{\partial z}\Big|_{z=0}\right) = k\frac{T_0 - T_s}{\sqrt{\pi\kappa t}}, \tag{3-41c}$$

where k is heat conductivity. Because of cooling, oceanic plate density increases with time or with distance away from the ridge. Hence, the plate shrinks, and the top of the plate sinks. Ocean floor depth hence can be calculated and is expected to increase linearly with $t^{1/2}$, which is confirmed by observation (Parsons and Sclater, 1977).

3.2.4.5 Diffusion distance and square root of time dependence

Estimation of diffusion distance or diffusion time is one of the most common applications of diffusion. For example, if the diffusion distance of a species (such as ^{40}Ar in hornblende or Pb in monazite) is negligible compared to the size of a crystal, it would mean that diffusive loss or gain of the species is negligible and the isotopic age of the crystal reflects the formation age. Otherwise, the calculated age from parent and daughter nuclide concentrations would be an apparent age, which is not the formation age, but is defined as the closure age. This has important implications in geochronology. Another example is to evaluate whether equilibrium between two mineral phases (or mineral and melt) is reached: if the diffusion distances in the two phases are larger than the size of the respective phases, then equilibrium is likely reached.

The diffusion distance concept is best defined for infinite and semi-infinite media diffusion problems. In these cases, C depends on $x/(4Dt)^{1/2}$, so if at time t_1 the concentration is C_1 at x_1, then at time $t_2 = 4t_1$ the concentration is C_1 at $x_2 = 2x_1$ (because $\xi_1 = x_1/(4Dt_1)^{1/2} = \xi_2 = x_2/(4Dt_2)^{1/2}$). This fact is often referred as the square root of time dependence. That is, the distance of penetration of a diffusing species is proportional to the square root of time. In other words, the concentration profile propagates into the diffusion medium according to square root of time. It can also be shown that the amount of diffusing substance entering the medium per unit area increases with square root of time. The square root dependence is often expressed as

$$x_{\text{diffusion}} \approx (Dt)^{1/2} \tag{3-42a}$$

The square root of time dependence also holds for concentration-dependent diffusivity and the D value in the above equation would be a kind of average D across the profile.

Because a diffusion profile does not end abruptly (except for some special cases), it is necessary to quantify the meaning of diffusion distance. To do so, examine Equation 3-40a. Define the distance at which the concentration is halfway between C_0 and C_∞ to be the *mid-distance of diffusion*, x_{mid}. The concept of x_{mid} is similar to that of half-life $t_{1/2}$ for radioactive decay. From the definition, x_{mid} can be solved from the following:

$$(C_\infty + C_0)/2 = C_0 + (C_\infty - C_0)\text{erf}[x_{mid}/(4Dt)^{1/2}].$$

Therefore,

$$\text{erf}[x_{mid}/(4Dt)^{1/2}] = 0.5,$$

and

$$x_{mid}/(4Dt)^{1/2} = \text{erf}^{-1}(0.5) = 0.476936.$$

That is,

$$x_{mid} = 0.953872(Dt)^{1/2}. \tag{3-42b}$$

The above is Equation 1-79, which was presented in Chapter 1 without derivation.

If one is interested in the diffusion time instead of the diffusion distance, for a given x, the time (t_{mid}) required for the concentration at this x to reach mid-concentration $(C_\infty + C_0)/2$ is proportional to the square of its distance from the surface (derived from Equation 3-42b):

$$t_{mid} = (1.099056)(x^2/D). \tag{3-42c}$$

The above equation can be used to estimate the half-time to reach equilibrium. Because the coefficients of 0.953872 and of 1.099056 are not much different from 1, for many applications and colloquial referencing, the mid-diffusion distance is often simplified as $(Dt)^{1/2}$, and the mid-diffusion time simplified as x^2/D.

Example 3.2 *Monazite dating.* The diffusion coefficient of Pb in monazite is given by $D = \exp(-0.06 - 71{,}200/T)$ m^2/s where T is in K (Cherniak et al., 2004). A monazite crystal in a metamorphic rock is about 100 μm across. It is estimated that the peak metamorphic temperature was 700°C and monazite formed near peak metamorphism. The metamorphic event lasted for about 20 Myr. Find the diffusion distance and evaluate whether the monazite grain lost a significant amount of Pb. That is, evaluate whether monazite grain can be used to determine the age of peak metamorphism.

Solution: The diffusivity at 700°C can be calculated as

$$D = \exp(-0.06 - 71{,}200/973.15) = 1.58 \times 10^{-32} \text{m}^2/\text{s}.$$

Mid-distance of diffusion is $x_{mid} = (0.953872)(Dt)^{1/2} = 3 \times 10^{-9}\,m = 0.003$ μm. Hence, diffusion distance is negligible compared to the size of the grain. In fact, even if the grain were held at 700°C for the duration of the age of the Earth, the diffusion distance would still be negligible. Hence, loss of Pb from monazite is negligible, and the grain can be used to determine the age of peak metamorphism.

Comments: A complexity that may affect the above conclusion is recrystallization or internal annealing of monazite either due to influx of fluids or following radiation damage, which resets the age, sometimes such that the core is apparently younger than the rim (DeWolf et al., 1993). One indication of resetting is the crystallization of Th oxides or silicates in the vicinity of the monazite grain. The quantification of the diffusion distance during cooling or for a given temperature history, which would allow the quantification of the closure temperature of monazite with respect to Pb is discussed later.

Because the diffusion distance is proportional to the square root of time, instead of the first power of time, diffusion rate is a less well-defined concept. The rate of the diffusion front moving into the diffusion medium would be dx/dt, which is not a constant, but is proportional to $1/t^{1/2}$. Hence, there is no fixed diffusion rate or velocity. A diffusion-controlled reaction is said to follow the *parabolic law* because the square of the reaction thickness is proportional to the duration of the reaction. If the reaction thickness is proportional to duration, then the reaction is said to follow a *linear law*, and the controlling mechanism would be different from diffusion. Therefore, one way to examine the control mechanism of a reaction such as mineral dissolution or growth or oxidation is to plot the thickness of the reaction against time.

Another consequence of the square-root-of-time dependence of diffusion distance is that diffusion profile in a short duration might be a significant fraction of the diffusion profile in a long duration. For example, if the mid-distance of diffusion is $0.2\,\mu m$ in one year, it would be $0.63\,\mu m$ in 10 years, $2\,\mu m$ in 100 years, and $6.3\,\mu m$ in 1000 years. That is, the diffusion distance increases very slowly with time. Looking at it from another angle, a diffusion profile might be affected in a short duration if the boundary condition changes suddenly. For example, say diffusion proceeded for 1000 years and the midlength of the profile is about $6.3\,\mu m$. Then the surface condition suddenly changed since last year. In one year, the mid-diffusion distance is $0.2\,\mu m$, meaning the concentration in a surface layer of more than $0.2\,\mu m$ thick would have changed. Hence, measurement of the near-surface layer would reflect the condition established recently, instead of the conditions of the last 1000 years. To the uninitiated, it might be surprising that a profile established by 1000 years of diffusion would be significantly altered by diffusion in one year, but this is simply due to the property of diffusion and the square root dependence of the diffusion distance.

3.2.4.6 Diffusion and partitioning between two phases

The diffusion couple discussed above consists of two halves of the same phase. If the two halves are two minerals, such as Mn–Mg exchange between spinel and garnet (Figure 3-5), there would be both partitioning and diffusion. Define the diffusivity in one half ($x < 0$) to be D^L, and in the other half ($x > 0$) to be D^R. Both D^L and D^R are constant. Let w be the concentration (mass fraction) of a minor element (such as Mn). The initial condition is

$$w|_{t=0, x<0} = w_\infty^L, \tag{3-43a}$$

and

$$w|_{t=0, x>0} = w_\infty^R. \tag{3-43b}$$

At the boundary $x = 0$, there is partitioning of Mn between the two phases, and the partition coefficient is K. Find the solution to the diffusion problem.

Each side satisfies separately the conditions for applying Boltzmann transformation ($\eta = x/\sqrt{4Dt}$); hence, the solution is

$$w^L = a^L + b^L \mathrm{erfc}\frac{|x|}{2\sqrt{D^L t}}, \quad \text{for } x < 0, \tag{3-43c}$$

$$w^R = a^R + b^R \mathrm{erfc}\frac{x}{2\sqrt{D^R t}}, \quad \text{for } x > 0. \tag{3-43d}$$

The initial condition and boundary conditions are used to solve for the four constants a^L, b^L, a^R, and b^R. There are two equations from the initial conditions:

$$w^L|_{\eta=-\infty} = w_\infty^L = a^L, \tag{3-43e}$$

$$w_R|_{\eta=\infty} = w_\infty^R = a_R, \tag{3-43f}$$

and one partitioning condition at the boundary,

$$(w^R/w^L)|_{x=0} = (a^R + b^R)/(a^L + b^L) = K, \tag{3-43g}$$

but we need another condition to determine the four constants. This is a boundary condition that was not explicitly given, but mass balance requires that the flux from the left-hand side equals the flux into the right-hand side:

$$\rho^L D^L \frac{\partial w^L}{\partial x}\Big|_{x=-0} = \rho^R D^R \frac{\partial w^R}{\partial x}\Big|_{x=+0}. \tag{3-43h}$$

That is,

$$\rho^L b^L \sqrt{D^L} = -\rho^R b^R \sqrt{D^R}. \tag{3-43i}$$

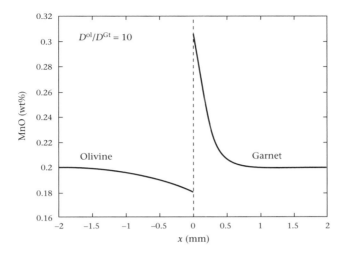

Figure 3-5 MnO partition between and diffusion in two minerals, olivine and garnet. Diffusional anisotropy of olivine is ignored. Initially, MnO in both phases were 0.2 wt%. As the two minerals come into contact, there will be diffusion to try to reach the equilibrium state. The partition coefficient $K = (M_n)_{oliv}/(M_n)_{gt}$ is assumed to be 0.59. The diffusivity in olivine is assumed to be 10 times that in garnet, resulting in a wider diffusion profile with a smaller slope in olivine.

With a^L and a^R given in Equations 3-43e and 3-43f, and b^L and b^R solved from Equations 3-43g and 3-43i, the four constants in Equations 3-43c and 3-43d are found. The solution is hence

$$w^L = w^L_\infty + \frac{\gamma(w^R_\infty - Kw^L_\infty)}{1 + K\gamma} \operatorname{erfc} \frac{|x|}{2\sqrt{D^L t}}, \qquad \text{for } x < 0, \tag{3-44a}$$

$$w^R = w^R_\infty + \frac{Kw^L_\infty - w^R_\infty}{1 + K\gamma} \operatorname{erfc} \frac{x}{2\sqrt{D^R t}}, \qquad \text{for } x > 0, \tag{3-44b}$$

where $\gamma = (\rho^R/\rho^L)(D^R/D^L)^{1/2}$. Figure 3-5 shows a calculated example. This solution is useful in studying diffusion in geological problems.

3.2.5 Instantaneous plane, line, or point source

Another widely used solution is for an *instantaneous plane/line/point source* (such as spill of a toxic pollutant, or diffusion of rare Earth elements from a tiny inclusion of monazite or xenotime into a garnet host) to diffuse away in either one dimension, two dimensions, or three dimensions (Figure 1-6b). If the source is initially in a plane, which may be defined as $x = 0$ (note that $x = 0$ represents a plane in three-dimensional space), then diffusion is one dimensional. If the source is initially a line, which may be defined as $x = 0$ and $y = 0$, then diffusion is two dimensional. If the source is initially a point, which may be defined as $x=0$, $y=0$, and $z=0$, then diffusion is three dimensional.

The mathematical translation of the plane-source problem is as follows. Initially, there is a finite amount of mass M but very high concentration at $x=0$, i.e., the density or concentration at $x=0$ is defined to be infinite (which is unrealistic but merely an abstraction for the case in which initially the mass is concentrated in a very small region around $x=0$). The initial condition is not consistent with that required for Boltzmann transformation. Hence, other methods must be used to solve the case of plane-source diffusion. Because this is the classical random walk problem, the solution can be found by statistical treatment as the following Gaussian distribution:[1]

$$C = \frac{M}{(4\pi Dt)^{1/2}} e^{-x^2/(4Dt)}. \tag{3-45a}$$

The concentration evolution with time is shown in Figure 1-7a. This solution is symmetric with respect to $x=0$ and approaches zero as x approaches $-\infty$ or ∞. The concentration at $x=0$ is proportional to $1/t^{1/2}$. At $t=0$, the concentration is infinity at $x=0$ and zero elsewhere. The integration of C from $x=-\infty$ to $x=\infty$ equals M, the initial total mass, satisfying the initial condition.

If the plane source is on the surface of a semi-infinite medium, the problem is said to be a thin-film problem. The diffusion distance stays the same, but the same mass is distributed in half of the volume. Hence, the concentration must be twice that of Equation 3-45a:

$$C = \frac{M}{(\pi Dt)^{1/2}} e^{-x^2/(4Dt)}. \tag{3-45b}$$

The solution to a line-source (i.e., along the z-axis at $x=0$ and $y=0$) problem with total mass M is

$$C = \frac{M}{4\pi Dt} e^{-(x^2+y^2)/(4Dt)}. \tag{3-45c}$$

The solution to a point-source (i.e., at $x=0$, $y=0$, and $z=0$) problem with total mass M is

$$C = \frac{M}{(4\pi Dt)^{3/2}} e^{-(x^2+y^2+z^2)/(4Dt)}. \tag{3-45d}$$

These sets of equations also describe the classic case of Brownian motion or random walk. The initial condition is that all M particles were at the central point, and then spread in one dimension (along a line), two dimensions (along a

---

[1] To verify the equation is the solution to the diffusion problem, you may verify that the expression satisfies the diffusion equation, the initial condition and the boundary conditions.

plane), or three dimensions. After some time of random walk, most would still be near at the center, and some would be far away. The mean square displacement of the particles from the center may be found as follows for the case of one-dimensional (plane-source) diffusion:

$$\langle x^2 \rangle = \frac{1}{M} \int_{-\infty}^{\infty} x^2 \frac{M}{\sqrt{4\pi Dt}} e^{-x^2/(4Dt)} dx = 2Dt. \tag{3-45e}$$

For two-dimensional (line-source) diffusion, the mean square displacement is $4Dt$. For three-dimensional (point-source) diffusion, the mean square distance is $6Dt$.

Equations 3-45a to 3-45d, in conjunction with the following superposition principle, are powerful in deriving solutions for the diffusion equation with infinite medium.

3.2.6 Principle of superposition

The diffusion equation with constant diffusivity (Equation 3-8) is said to be linear, which means that if f and g are solutions to the equation, then any linear combination of f and g, i.e., $u = af + bg$, where a and b are constants, is also a solution. To show this, we can write

$$\frac{\partial u}{\partial t} = a \frac{\partial f}{\partial t} + b \frac{\partial g}{\partial t}. \tag{3-46a}$$

Since it is assumed that f and g are solutions to the diffusion equation, then $\partial f/\partial t = D\nabla^2 f$ and $\partial g/\partial t = D\nabla^2 g$. Therefore,

$$\frac{\partial u}{\partial t} = a \frac{\partial f}{\partial t} + b \frac{\partial g}{\partial t} = aD\nabla^2 f + bD\nabla^2 g = D\nabla^2(af + bg) = D\nabla^2 u. \tag{3-46b}$$

Hence, u is also a solution to the diffusion equation.

This result is known as the *principle of superposition*. The principle is useful in solving diffusion equations with the same boundary conditions, but different initial conditions, or with the same initial conditions but different boundary conditions, or other more general cases. Suppose we want to find the solution to the diffusion equation for the following initial condition:

$$C|_{t=0} = af + bg. \tag{3-46c}$$

If the diffusion problem with the same boundary conditions but the initial condition of $C|_{t=0} = f$ has been solved to be C_1, and the problem with the same boundary conditions but the initial condition of $C|_{t=0} = g$ has been solved to be C_2, then the solution to the diffusion problem for the initial condition of $C|_{t=0} = af + bg$ is

$$C = aC_1 + bC_2. \tag{3-46d}$$

Two applications of the principle are given below.

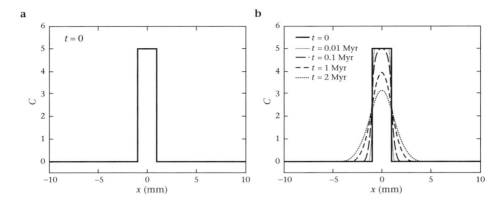

Figure 3-6 Diffusion in an infinite medium with an extended source. (a) The extended source with width $\delta = 1$ mm; (b) the solution.

3.2.6.1 Diffusion in an infinite medium with an extended source

For one-dimensional diffusion in an infinite medium with constant D, if the initial condition is an extended source, meaning C is finite in a region $(-\delta, \delta)$, and 0 outside the region (Figure 3-6a):

$$C|_{t=0, -\delta < x < \delta} = C_0, \tag{3-47a}$$

$$C|_{t=0, x<-\delta} = C|_{t=0, x>\delta} = 0, \tag{3-47b}$$

the problem can be solved using the principle of superposition.

The extended source can be viewed as a summation (or integral) of point plane sources. The mass density at each plane $\xi \in (-\delta, \delta)$ is $C_0 \, d\xi$. At position x, which is distance $|x - \xi|$ away from this plane, according to Equation 3-45a, the concentration due to this plane source is

$$\frac{C_0 d\xi}{(4\pi Dt)^{1/2}} e^{-(x-\xi)^2/4Dt}, \tag{3-47c}$$

Therefore, the concentration at x due to the extended instantaneous source can be found by summing all the plane sources:

$$C(x, t) = \sum \frac{C_0 d\xi}{(4\pi Dt)^{1/2}} e^{-(x-\xi)^2/4Dt} = \int_{-\delta}^{\delta} \frac{C_0}{(4\pi Dt)^{1/2}} e^{-(x-\xi)^2/4Dt} d\xi. \tag{3-48a}$$

Carrying out the integration leads to

$$C(x, t) = \frac{C_0}{2} \left\{ \mathrm{erf}\, \frac{x+\delta}{2\sqrt{Dt}} - \mathrm{erf}\, \frac{x-\delta}{2\sqrt{Dt}} \right\}. \tag{3-48b}$$

This is the solution to the problem and is plotted in Figure 3-6b. To compare the above with Equation 3-45a, note that $M = 2\delta C_0$. As δ approaches zero, the above solution approaches Equation 3-45a, as it should.

3.2.6.2 Diffusion in an infinite medium with arbitrary initial distribution

Using the principle of superposition, following the same procedure above, several other general solutions can be derived. For example, the solution for arbitrary initial distribution $C|_{t=0} = f(x)$ for one-dimensional diffusion in an infinite medium with constant D can be found by integration:

$$C(x, t) = \sum \frac{C_0 d\xi}{(4\pi Dt)^{1/2}} e^{-(x-\xi)^2/(4Dt)} = \int_{-\infty}^{\infty} \frac{f(\xi)}{(4\pi Dt)^{1/2}} e^{-(x-\xi)^2/(4Dt)} d\xi. \tag{3-49}$$

For one-dimensional half-space diffusion with constant D and an initial distribution of $C|_{t=0} = f(x)$ as well as other conditions, the solutions can be found in Appendix 3.

3.2.7 One-dimensional finite medium and constant D, separation of variables

When the medium is finite, there will be two boundaries in the case of one-dimensional diffusion. This finite one-dimensional diffusion medium will also be referred as plate sheet (bounded by two parallel planes) or slab. The standard method of solving for such a diffusion problem is to separate variables x and t when the boundary conditions are zero. This method is called *separation of variables*. As will be clear later, the method is applicable only when the boundary conditions are zero.

Starting from Equation 3-10, assume that the function $C(x, t)$ may be separated into the product of a function that depends only on x, $\xi(x)$, and a function that depends only on t, $\tau(t)$. Then

$$C(x, t) = \xi(x) \cdot \tau(t), \tag{3-50a}$$

where ξ is a function of x only and τ is a function of t only. Substituting the above into Equation 3-10 leads to

$$\xi \frac{d\tau}{dt} = D\tau \frac{d^2\xi}{dx^2}, \tag{3-50b}$$

which may be written as

$$\frac{1}{D\tau} \frac{d\tau}{dt} = \frac{1}{\xi} \frac{d^2\xi}{dx^2}. \tag{3-50c}$$

Because by assumption the left-hand side is a function of t only (independent of x) and the right-hand side is a function of x only (independent of t), they cannot be equal unless they equal the same constant. It will be seen later that the constant must be negative for the solution to be meaningful. Let this negative constant be $-\lambda^2$. That is,

$$\frac{1}{\tau}\frac{d\tau}{dt} = -\lambda^2 D, \tag{3-50d}$$

and

$$\frac{1}{\xi}\frac{d^2\xi}{dx^2} = -\lambda^2. \tag{3-50e}$$

The solutions are

$$\tau(t) = e^{-\lambda^2 Dt}, \tag{3-50f}$$

and

$$\xi(x) = A\sin(\lambda x) + B\cos(\lambda x). \tag{3-50g}$$

From Equation 3-50f, it can be seen that the constant defined as $-\lambda^2$ must be negative (which is why it is defined as $-\lambda^2$); otherwise, τ and hence C would increase with time exponentially, which violates the condition that diffusion leads to approach to equilibrium (and hence more uniform concentration).

The above derivation has not made use of the initial and boundary conditions yet, and shows only that λ may take any constant value. The value of λ can be constrained by boundary conditions to be discrete: $\lambda_1, \lambda_2, \ldots$, as can be seen in the specific problem below. Because each function corresponding to given λ_n is a solution to the diffusion equation, based on the principle of superposition, any linear combination of these functions is also a solution. Hence, the general solution for the given boundary conditions is

$$C = \sum_{n=0}^{\infty} [A_n\sin(\lambda_n x) + B_n\cos(\lambda_n x)]e^{-\lambda_n^2 Dt} \tag{3-51}$$

To find the specific solution for the given initial condition, A_n and B_n must be determined from the initial condition (some might have already been determined from the boundary conditions). Since the solution is expressed as a Fourier series, this method is also called the *Fourier series method*.

We now apply the method to a specific problem of one-dimensional diffusion in $0 \le x \le L$ with constant D, with the following conditions:

Initial condition: $\quad\quad C|_{t=0} = C_0$.

Boundary conditions: $\quad C|_{x=0} = C|_{x=L} = C_1$.

First, we note that the boundary condition is not zero. To use the method of separation of variables, changes must be made so that the boundary condition is zero. Let $u = C - C_1$; then u satisfies the diffusion equation and the following initial and boundary conditions:

$$u|_{t=0} = C_0 - C_1.$$

$$u|_{x=0} = u|_{x=L} = 0.$$

The solution is of the type of Equation 3-51. To satisfy $u|_{x=0} = 0$, all the B_n in Equation 3-51 must be zero (this is because not only collectively but also individually $A_n \sin(\lambda_n x) + B_n \cos(\lambda_n x)$ must be zero). To satisfy $u|_{x=L} = 0$, $\lambda_n L$ must equal $n\pi$, where $n = 1, 2, \ldots$. That is, $\lambda_n = n\pi/L$. This example shows why the boundary conditions must be zero for the Fourier series method, because otherwise λ_n cannot be constrained. Replacing B_n and λ_n into Equation 3-51 leads to

$$u = \sum_{n=1}^{\infty} A_n \sin \frac{n\pi x}{L} e^{-n^2 \pi^2 Dt/L^2},$$

where A_n is to be determined by the initial condition:

$$C_0 - C_1 = \sum_{n=1}^{\infty} A_n \sin \frac{n\pi x}{L}, \quad \text{for } 0 < x < L.$$

Hence, A_n's are the coefficients of the Fourier sine series expansion of $C_0 - C_1$. The values of A_n's can be obtained by multiplying both sides by $\sin(m\pi x/L)$ and integrating from 0 to L using the following relationship:

$$\int_0^L (C_0 - C_1) \sin \frac{m\pi x}{L} dx = \sum_{n=1}^{\infty} A_n \int_0^L \sin \frac{n\pi x}{L} \sin \frac{m\pi x}{L} dx.$$

Because

$$\int_0^L \sin \frac{n\pi x}{L} \sin \frac{m\pi x}{L} dx = \begin{cases} 0, & n \neq m, \\ L/2, & n = m, \end{cases}$$

and

$$\int_0^L \sin \frac{m\pi x}{L} dx = \begin{cases} 0, & m = 2, 4, 6, \ldots, \\ 2L/(m\pi), & m = 1, 3, 5, \ldots, \end{cases}$$

hence,

$$A_n = 4(C_0 - C_1)/(n\pi), \quad \text{for } n = 1, 3, 5, \ldots.$$

Therefore, the solution is

$$C = C_1 + \frac{4(C_0 - C_1)}{\pi} \sum_{n=0}^{\infty} \frac{1}{2n+1} \sin \frac{(2n+1)\pi x}{L} e^{-(2n+1)^2 \pi^2 Dt/L^2}. \tag{3-52a}$$

With a computer, the concentration profile can be calculated easily using the above formula. The mass loss or gain from the sheet is $M_t = \int C \, dx - C_0 L$ and may be expressed as follows:

$$\frac{M_t}{M_\infty} = 1 - \frac{8}{\pi^2} \sum_{n=0}^{\infty} \frac{e^{-(2n+1)^2 \pi^2 Dt/L^2}}{(2n+1)^2}, \tag{3-52b}$$

where $M_\infty = (C_1 - C_0)L$ is the mass loss or gain at time of infinity. The above two equations converge rapidly for large Dt/L^2 but slowly when $Dt/L^2 \ll 1$. For small t, the following two equations (obtained using other methods, Crank, 1975) converge rapidly:

$$C = C_0 + (C_1 - C_0) \sum_{n=0}^{\infty} (-1)^n \left[\text{erfc} \frac{nL + x}{2\sqrt{Dt}} + \text{erfc} \frac{(n+1)L - x}{2\sqrt{Dt}} \right], \tag{3-52c}$$

and

$$\frac{M_t}{M_\infty} = \frac{M_t}{(C_1 - C_0)L} = \frac{4\sqrt{Dt}}{L} \left[\frac{1}{\sqrt{\pi}} + 2 \sum_{n=1}^{\infty} (-1)^n \text{ierfc} \frac{nL}{2\sqrt{Dt}} \right]. \tag{3-52d}$$

More solutions in finite diffusion medium may be found in Appendix 3.

3.2.8 Variable diffusion coefficient

Diffusion equation 3-10 is for constant diffusivity. When diffusivity varies for one-dimensional diffusion, then Equation 3-9 must be used. Diffusivity may vary as a function of t (e.g., when temperature varies with time), or as a function of C (diffusivity in general depends on composition), and less often, as a function of x.

3.2.8.1 Time-dependent D and diffusion during cooling

Because D increases with increasing temperature (the Arrhenius equation 1-73), time-dependent D is often encountered in geology because an igneous rock may have cooled down from a high temperature, or metamorphic rock may have experienced a complicated thermal history. If the initial and boundary conditions are simple and if D depends only on time, the diffusion problem is easy to deal with. Because D is independent of x, Equation 3-9 can be written as

$$\frac{\partial C}{D \partial t} = \frac{\partial^2 C}{\partial x^2}. \tag{3-53a}$$

Define

$$\alpha = \int_0^t D \, dt. \tag{3-53b}$$

Hence, $\alpha|_{t=0} = 0$, and $d\alpha = D dt$. The unit of α is length^2. Therefore, Equation 3-53a becomes

$$\frac{\partial C}{\partial \alpha} = \frac{\partial^2 C}{\partial x^2}. \tag{3-53c}$$

The above equation is equivalent to Equation 3-10 by making α equivalent to t and D equal 1. Hence, solutions obtained for constant D may be applied to time-dependent D. For example, by analogy to Equation 3-38, the solution to the diffusion-couple problem for time-dependent D is

$$C = \frac{C_1 + C_2}{2} + \frac{C_2 - C_1}{2} \operatorname{erf}\left(\frac{x}{2\sqrt{\alpha}}\right), \tag{3-54a}$$

or explicitly,

$$C = \frac{C_1 + C_2}{2} + \frac{C_2 - C_1}{2} \operatorname{erf}\left(\frac{x}{2\sqrt{\int_0^t D \, dt}}\right). \tag{3-54b}$$

Similarly, by analogy to Equation 3-40b, the solution for semi-infinite medium with uniform initial concentration and constant surface concentration is

$$C = C_\infty + (C_0 - C_\infty) \operatorname{erfc}[x/(4\alpha)^{1/2}]. \tag{3-54c}$$

These solutions are used often in treating diffusion during cooling.

Estimation of diffusion distance The most common application of time-dependent D is to evaluate the effect of diffusion for a given temperature history. If D as a function of temperature is known, the mid-distance of diffusion can be expressed as

$$x_{\text{mid}} = (0.953872)\alpha^{1/2} = (0.953872)(\int_0^t D \, dt)^{1/2}. \tag{3-54d}$$

In particular, consider a thermal history of monotonic cooling (such as an igneous rock, especially a volcanic rock) represented by the asymptotic cooling model (Equation 2-41):

$$T = T_0/(1 + t/\tau_c),$$

where τ_c is the cooling timescale for temperature to decrease from T_0 to $T_0/2$. In asymptotic cooling, temperature approaches 0 K asymptotically. (A discussion of other cooling models can be found in Section 2.1.3.1.) Combining the above cooling function with the Arrhenius relation, the diffusion coefficient depends on time as follows:

$$D = A \, e^{-E/(RT)} = A \, e^{-E(1+t/\tau_c)/(RT_0)} = D_0 \, e^{-Et/(\tau_c RT_0)} = D_0 \, e^{-t/\tau}, \tag{3-55a}$$

where A is the preexponential factor, E is the activation energy, R is the universal gas constant, $D_0 = Ae^{-E/(RT_0)}$ is the diffusivity at $t = 0$, and $\tau = \tau_c(RT_0/E)$ is another time constant (for diffusivity to decrease to 1/e of the initial value). For typical activation energy and initial temperature encountered in geology, $\tau < \tau_c$. Because diffusivity depends strongly on temperature, diffusion near room temperature is negligible. Therefore, the value of the integration from initial time to the time when the rock reached surface (so that scientists can collect it) may be treated as from $t = 0$ to $t = \infty$. Hence,

$$\alpha = \int D\,dt = \int_0^\infty D_0\,e^{-t/\tau}dt = D_0\tau. \tag{3-55b}$$

$$x_{mid} = (0.953872)\left(\int_0^\infty D_0,\,e^{-t/\tau}dt\right)^{1/2} = (0.953872)(D_0\tau)^{1/2}. \tag{3-55c}$$

Estimation of cooling rate Another application is to estimate cooling rate from a diffusion profile in a mineral. For this application, it is necessary to know that the concentration profile in a mineral is due to diffusion. For example, if a garnet crystal has a core, a mantle, and a thin crust, it is possible that initially the transition from the core to the mantle is sharp and then diffusion produces an Fe–Mg profile between the core and the mantle. (Whether there was an initial sharp compositional difference may be evaluated from an element that diffuses very slowly, such as Zr or P.) If the profile is short compared to the radius of the core and thickness of the mantle, the diffusion profile may be treated as a roughly one-dimensional diffusion couple. Based on measured profiles, the distance x_{mid} from the interface may be determined, or $\alpha = \int D dt$ may be obtained from fitting the profile by Equation 3-54a. Given the diffusion property (the activation energy and the pre-exponential factor), if, furthermore, the initial temperature T_0 (i.e., the growth temperature of the mantle) is known, then D_0 can be estimated. Then the parameters τ and τ_c may be obtained from Equations 3-55a,b,c:

$$\tau = \int D\,dt/D_0. \tag{3-56a}$$

$$\tau = x_{mid}^2/(0.909872D_0). \tag{3-56b}$$

$$\tau_c = \tau E/(RT_0). \tag{3-56c}$$

Although τ_c characterizes the cooling history, sometimes one would like to know the cooling rate q rather than τ_c. For asymptotic cooling, the cooling rate q is

$$q = -dT/dt|_{t=0} = T_0/\tau_c = D_0RT_0^2/(E\alpha), \tag{3-57a}$$

By combining Equations 3-55b and 3-57a, q may be expressed as

$$q = RT_0^2/(\tau E). \tag{3-57b}$$

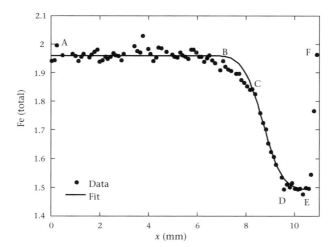

Figure 3-7 Measured Fe concentration (moles of Fe in the garnet formula) profile in a large garnet grain. The position $x = 0$ roughly corresponds to the center of garnet. The profile from the center to point E ($x = 10.5$ mm) is interpreted to be due to prograde garnet growth, with relatively low temperature at the beginning of garnet growth (such as 500°C) at $x = 0$, and peak temperature at point E. The part of the profile from E to F corresponds to retrograde garnet growth, which is not considered in the error function fit. The part of the profile between points B and C is not well fit, which might be related to the growth part of the profile. From Zhang and Chen (2007).

The above equations are rigorously derived geospeedometry equations. Other geospeedometry equations, such as Equations 1-113 and 1-117, which were presented without proof, often take a similar form.

Example 3.4. Diffusion across two zones of garnet. Figure 3-7 shows a concentration profile in garnet, which has a core (from A to B) with roughly uniform Fe content (Zhang and Chen, 2007). The core is surrounded by a "mantle" layer (from C to E; the part with uniform Fe concentration is >0.5 mm wide) that is also roughly homogeneous. The part from A to E grew during prograde metamorphism. The thin rim (from E to F) is due to retrograde growth. Between the "core" and "mantle" zones, there is a profile. Part of the profile (B to C) might be due to growth, and part to diffusion. The part due to growth may be investigated using a slowly diffusing element such as P. For example, if the phosphorus profile is a step function, then one may reasonably assume that the two zones grew under different conditions with a sharp transition in between, and hence the compositional zoning between the two zones is due to diffusion after peak metamorphism. If so, an error function may be fit to the profile to obtain $\int D \, dt = 0.2 \, \text{mm}^2$ (solid curve in

Figure 3-7). Using diffusion data of Ganguly et al. (1998), at 900°C and 1.2 GPa (assumed to be the peak condition), tracer diffusivity of $D(\text{Fe}) = 1.08 \times 10^{-21}$ m^2/s, and $D(\text{Mg}) = 1.16 \times 10^{-20}$ m^2/s. Garnet composition near the center of the diffusion couple is $\text{Fe}/(\text{Fe} + \text{Mg}) = 0.64$ and $\text{Mg}/(\text{Fe} + \text{Mg}) = 0.36$. Hence, the interdiffusivity of $D(\text{Fe–Mg}) \approx 2.58 \times 10^{-21}$ m^2/s (ignoring the nonideality of garnet) at 900°C and 1.2 GPa (see Section 3.6.2.4 for estimation of interdiffusivity from tracer diffusivities). Hence, if it is assumed that diffusion occurred at the peak condition without cooling, the diffusion timescale would be no more than

$$\tau = \int D\,dt/D_0 = (2 \times 10^{-7} \text{ m}^2)/(2.58 \times 10^{-21} \text{ m}^2/\text{s}) = 2.5 \text{ Myr}.$$

Assuming asymptotic cooling $T = T_0/(1 + t/\tau_c)$, then $q = -dT/dt|_{t=0} = T_0/\tau_c = T_0/[E/(RT_0) \cdot 2.5 \text{ Myr}] = 19$ K/Myr. Because part of the profile is likely due to growth, the cooling rate estimated above is the lower limit. Actual cooling rate must be greater than this.

Careful readers might notice that diffusion in garnet is three-dimensional with spherical geometry, and should not be treated as one-dimensional diffusion. Section 5.3.2.1 addresses this concern.

3.2.8.2 Concentration-dependent D and Boltzmann analysis

When D depends on concentration (which implies that D depends on x and t because C depends on x and t), Equation 3-9

$$\frac{\partial C}{\partial t} = \frac{\partial}{\partial x}\left(D(C)\frac{\partial C}{\partial x}\right) \tag{3-9}$$

cannot be simplified to Equation 3-10. Because Equation 3-9 is nonlinear, the principle of superposition cannot be applied, and the equation usually can be solved only by numerical methods, either in an infinite, semi-infinite, or finite medium. Nonetheless, as long as the initial and boundary conditions allow, Boltzmann transformation can still be applied, leading to the following equation:

$$\frac{d}{d\eta}\left(D\frac{dC}{d\eta}\right) = -2\eta\frac{dC}{d\eta}, \tag{3-58a}$$

where $\eta = x/(4t)^{1/2}$. Because this equation indicates that C is a function of η ($=x/(4t)^{1/2}$) for the right initial and boundary conditions, the square root of time dependence (x_{mid} is proportional to square root of time) still rigorously holds for concentration-dependent D for diffusion in infinite or semi-infinite medium.

The above ordinary differential equation may be solved numerically to obtain C as a function of $x/(4t)^{1/2}$, but the numerical solution is not much easier than directly solving Equation 3-9. That is, Boltzmann transformation from Equation 3-9 to 3-58a does not help much in solving the equation. The real usefulness of Boltzmann transformation in the case of concentration-dependent D is to extract concentration-dependent diffusivity from experimental diffusion profiles.

If D is constant, an experimental diffusion profile can be fit to the analytical solution (such as an error function) to obtain D. If it depends on concentration and the functional dependence is known, Equation 3-9 can be solved numerically, and the numerical solution may be fit to obtain D (e.g., Zhang et al., 1991a; Zhang and Behrens, 2000). However, if D depends on concentration but the functional dependence is not known a priori, other methods do not work, and Boltzmann transformation provides a powerful way (and the only way) to obtain D at every concentration along the diffusion profile if the diffusion medium is infinite or semi-infinite. Starting from Equation 3-58a, integrate the above from η_0 to ∞, leading to

$$D\frac{dC}{d\eta}\bigg|_{\eta=\infty} - D\frac{dC}{d\eta}\bigg|_{\eta=\eta_0} = -2\int_{C(\eta_0)}^{C(\infty)} \eta\, dC. \tag{3-58b}$$

For a diffusion couple, or for half-space diffusion, $\partial C/\partial x = 0$ at $x = \infty$. That is, $(dC/d\eta)|_{\eta=\infty} = 0$. Therefore, the above can be written as

$$D\frac{dC}{d\eta}\bigg|_{\eta=\eta_0} = 2\int_{C(\eta_0)}^{C(\infty)} \eta\, dC. \tag{3-58c}$$

Hence,

$$D = \frac{2\int_{C(\eta_0)}^{C(\infty)} \eta\, dC}{(dC/d\eta)|_{\eta=\eta_0}}. \tag{3-58d}$$

The above equation is the basic equation to estimate D at every concentration by numerically evaluating the integral in the numerator and the derivative in the denominator from C versus η relation, obtainable from a measured profile and experimental duration. Because the duration t for a given experimental profile is a constant, Equation 3-58d can also be written as

$$D = \frac{\int_{C(x_0)}^{C(\infty)} x\, dC}{2t(dC/dx)|_{x=x_0}}. \tag{3-58e}$$

This method of extracting concentration-dependent D is usually referred to as Boltzmann analysis.

To use Equation 3-58d or 3-58e, it is necessary to know the interface position $x = 0$ (i.e., $\eta = 0$) because the value of the integration depends on the exact position of $x = 0$. For a semi-infinite diffusion medium with fixed interface, this is easy ($x = 0$ is the surface). However, for a diffusion couple, the location of the

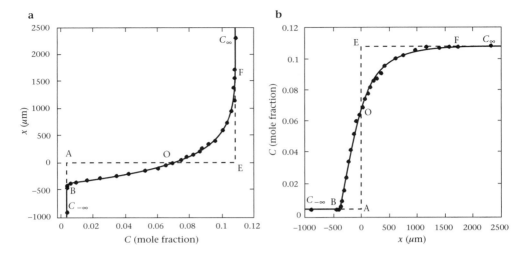

Figure 3-8 Concentration plots in Boltzmann-Matano analysis of an experimental diffusion-couple profile. (a) Plot of x versus C for the calculation of the integral $\int x\,dC$. (b) C versus x. The slope can be evaluated using this plot. As x approaches $-\infty$ (that is, for large negative $x < -500$ μm), C is roughly 0.003. As x approaches ∞ (that is, for large $x > 2000$ μm), C approaches 0.108. The data and the fit (using $D_{OH} = 0$ and $D_{H_2Om} = D_0\exp(aX_{H_2Ot})$, see Section 3.3.1) are for exp# Rhy-DC9 from Zhang and Behrens (2000).

interface must be found. Let $C_\infty > C_{-\infty}$. For Equations 3-58d and 3-58e to be applicable, the interface of the diffusion couple must satisfy

$$\int_{C(-\infty)}^{C(\infty)} x\,dC = \int_0^\infty (C_\infty - C)dx - \int_{-\infty}^0 (C - C_{-\infty})dx = 0, \tag{3-59}$$

otherwise the Boltzmann analysis would give infinite D at $x = -\infty$ because $(dC/dx)|_{x=-\infty} = 0$. The interface satisfying the above condition is called the *Matano interface* (Matano, 1933), and may differ from the real (marked) interface because of volume shift during diffusive exchange.

In summary, Boltzmann analysis for a diffusion-couple profile involves the following steps:

(1) Starting from measured C versus x profile, roughly estimate the interface position (e.g., at midconcentration). Plot x (as vertical axis) versus C (as horizontal axis) as in Figure 3-8a.

(2) It is important to smooth the profile in an objective way. The smoothing could be done by hand or by fitting a function to the data. A given profile may be fit by a high-order polynomial (because polynomials cannot produce constant concentrations at the ends, such a fit should not be applied to calculate concentration the the ends), a piecewise polynomial, or functions such as $a_0 + a_1[1 - 1/(1 + a_2e^{x/a_3})] + a_4[1 - 1/(1 + a_5e^{x/a_6})]$. The function does not have to be expressed by a single formula, but may be continuous stepwise segments. With the function, the concentration at any given x can be calculated.

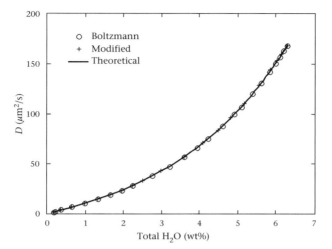

Figure 3-9 Calculated D by applying the Boltzmann-Matano method (open circles) and the algorithm of Sauer and Freise (1962) (pluses) to the experimental data shown in Figure 3-8. The profile was fit by Zhang and Behrens (2000) assuming an analytical function of D as a function of H_2O content (curve). The fit is shown in Figure 3-8, and the fit is used to calculate the integral and derivative to obtain D. Hence, the method is circular, but intended to show that as long as data precision is high, accurate D can be obtained.

(3) Carry out the integration $\int_{C(-\infty)}^{C(\infty)} x\,dC$. If it is not zero, then add a constant (positive or negative) to x until the integration is zero. Graphically, it is equivalent to the condition that the area of "triangle" AOB is the same as the area of EOF in Figure 3-8.

(4) Replot x versus C using new x values, which differs from old x values by a constant. Make another plot of C versus x (Figure 3-8b). For any given value of x_0, find C. Then find $\int_{C(x_0)}^{C(\infty)} x\,dC$ and $(dC/dx)_{x=x_0}$. Then find D at this concentration using Equation 3-58e.

(5) Because D is inversely proportional to the slope of the concentration profile, it is important to minimize the error in determining the slope. Specifically, D should not be determined near the two ends where the slope is almost zero and hence cannot be determined with precision. The best place to determine D is near the interface of the diffusion couple where both the slope and the integral are significantly different from zero. Diffusivity should not be determined at

$$C < C_{-\infty} + 4\sigma,$$

or at

$$C > C_\infty - 4\sigma,$$

where σ is standard deviation of the measurement.

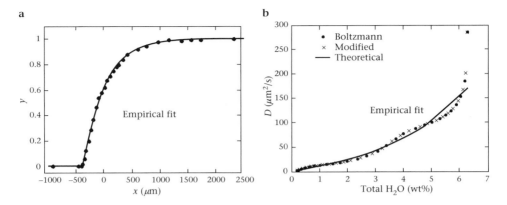

Figure 3-10 (a) Another fit of the same concentration profile shown in Figure 3-8, and (b) diffusivity (points and crosses) obtained from the fit. Although the empirical fit appears to match the data well, the D values oscillate around the theoretical solid curve obtained in Figure 3-9.

In the above application of the Boltzmann method to a diffusion couple, it is necessary to find the position of the interface accurately. A modified method that makes this step unnecessary is proposed by Sauer and Freise (1962):

$$D = \frac{1}{2t(dy/dx)|_{x=x_0}} \left[y|_x \int_x^{+\infty} (1-y)dx + (1-y|_x) \int_{-\infty}^x y\,dx \right], \quad (3\text{-}60a)$$

where D is diffusivity at x, $y = (C - C_{min})/(C_{max} - C_{min})$ so that $y = 0$ at $x = -\infty$ and $y = 1$ at $x = \infty$, with subscripts "min" and "max" meaning minimum concentration (i.e., mean concentration at one undisturbed end) and maximum concentration (i.e., mean concentration at the other end). If the curve is plotted so that $y = 1$ at $x = -\infty$ and $y = 0$ at $x = \infty$, then

$$D = \frac{-1}{2t(dy/dx)|_{x=x_0}} \left[(1-y|_x) \int_x^{+\infty} y\,dx + y|_x \int_{-\infty}^x (1-y)dx \right], \quad (3\text{-}60b)$$

In this method, to obtain D accurately, the key is to smooth or empirically fit the data well so that the integral and especially the slope at every point can be found accurately. For example, for the curve shown in Figure 3-8, using the fit in the figure to smooth the data and to evaluate the integral and slope, calculated D as a function of concentration is shown in Figure 3-9.

Another empirical function that fits the concentration profile in Figure 3-8 well is shown in Figure 3-10a. However, although the fit appears "perfect," the D values extracted as a function of concentration (Figure 3-10b) are significantly different from those in Figure 3-9. This result occurs because the slope is not necessarily well fit by any empirical function. Furthermore, D at C near C_∞ (e.g., within 5% of C_∞) cannot be obtained with any accuracy using the Boltzmann method. Therefore, although the Boltzmann method may be used to investigate the overall trend of the dependence of D on C, small variations (<30% relative) should not be interpreted as significant. My personal preference is that if there is a theoretical basis to assume a functional relation between D and C, it is better to test whether such a relation fits the diffusion profile, rather than trying different empirical functions to fit the diffusion profile and then obtain how D varies with C.

3.2.8.3 D that depends on x

When D depends on space coordinates, the diffusion equation is in general difficult to solve because D cannot be taken out of the differential in Equation 3-9, and Boltzmann transformation cannot be applied. The solution given in Equation 3-44 may be viewed as a special case of D depending on x: D takes the value of D_L at $x < 0$ and D_R at $x > 0$.

3.2.9 Uphill diffusion in binary systems and spinodal decomposition

Fick's first law $\mathbf{J} = -D\nabla C$ (Equation 3-6) applies only to ideal or nearly ideal mixtures. Many liquid and solid solutions are nonideal. Extreme nonideality in a binary system results in a miscibility gap (usually at low temperatures, although some solvi persist up to the melting point). For example, the alkali feldspar $(Na,K)AlSi_3O_8$ system exhibits a large miscibility gap: at high temperature, the entropy effect becomes more dominant with increasing temperature, allowing more mixing; as the temperature is decreased, the large difference in the ionic radii of Na^+ and K^+ results in more favorable energetics by separating into two phases. As a phase in such a binary system is cooled from high temperature, it passes through a critical point, below which the phase decomposes into two phases at equilibrium. In this process, the concentration profile evolves from initial homogeneity to the final two-phase coexistence (i.e., heterogeneity), accomplished by random perturbation and diffusion. In the case of ideal and nearly ideal solutions, mass flux goes from high concentration to low concentration, and diffusion homogenizes the sample. However, in the case of a phase decomposing into two phases, diffusion produces heterogeneity from homogeneity. In the latter case, diffusion is able to transfer mass from low concentration to high concentration. The diffusion of mass from low to high concentration, which makes a sample more heterogeneous, is called *uphill diffusion*.

Uphill diffusion occurs in binary systems because, strictly speaking, diffusion brings mass from high chemical potential to low chemical potential (De Groot and Mazur, 1962), or from high activity to low activity. Hence, in a binary system, a more rigorous flux law is (Zhang, 1993):

$$\mathbf{J} = -(\mathcal{D}/\gamma)\nabla a, \tag{3-61}$$

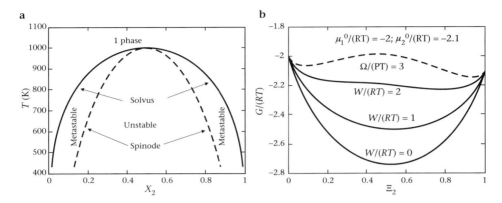

Figure 3-11 (a) Calculated phase diagram for a binary system described by a symmetric regular solution with $W = 2000R$, where R is the gas constant. Outside the solid curve is the single-phase region where one phase is stable. Inside the dashed curve (spinode) a single phase is unstable and decomposes spontaneously to two phases. Between the solid and dashed curves, one phase is metastable. (b) The plot of G (normalized to RT) versus composition for four $W/(RT)$ values. $W/(RT) = 0$ means an ideal solution, $W/(RT) = 2$ means the critical point, corresponding to $T = 1000$ K for $W = 2000R$ in (a), and $W/(RT) = 3$ corresponds to $T = 667$ K in (a). The solvus is obtained by the common tangent, and the spinode by $\partial^2 G/\partial X^2 = 0$.

where $\mathcal{D}$ is "intrinsic" diffusivity and is always positive, γ is the activity coefficient, and a is the activity of the diffusing component. The above formulation includes γ so that $\mathcal{D} = D$ for ideal and nearly ideal solutions with constant activity coefficient. Simulations show that $\mathcal{D}$ is similar to self-diffusivity (Zhang, 1993). From $a = \gamma C$, and comparing Equations 3-61 and 3-6, the relation between $\mathcal{D}$ and D is

$$D = \frac{\mathcal{D}}{\gamma} \frac{\mathrm{d}a}{\mathrm{d}C} = \mathcal{D}\left[1 + \frac{\mathrm{d}(\ln \gamma)}{\mathrm{d}(\ln C)}\right] = \mathcal{D}\left(1 + \frac{C}{\gamma} \frac{\mathrm{d}\gamma}{\mathrm{d}C}\right). \tag{3-62}$$

To illustrate how a and γ depend on composition in a nonideal binary system, Figure 3-11 shows the solvus and its spinode in a temperature versus concentration phase diagram, as well as the Gibbs energy variation with concentration at several temperatures assuming a symmetric regular solution model in which $\Delta G_{\mathrm{mix}} = X_1 X_2 W + RT(X_1 \ln X_1 + X_2 \ln X_2)$, where X_1 and X_2 are the mole fractions of components 1 and 2, and W is the interaction parameter.

If the composition of a phase falls inside the spinode, the phase would undergo spontaneous decomposition into two phases. That is, heterogeneity would arise from homogeneity. From an energetic point of view, the homogeneous phase is at an unstable equilibrium state, meaning that without disturbance at all, the equilibrium could persist, but the tiniest microscopic perturbation (there would always be perturbation due to thermal motion) would grow to produce hetero-

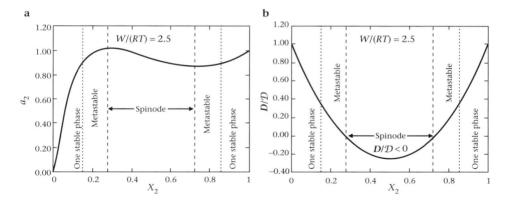

Figure 3-12 Calculated relation between (a) activity, and (b) D/D, and mole fraction of component 2 for a symmetric regular solution model at $W/(RT) = 2.5$. Inside the spinode, the activity decreases as mole fraction increases, leading to negative D/D (which is why diffusion goes uphill in terms of concentration).

geneity, from one phase to two phases. There are many examples of unstable equilibrium. For example, later we will encounter the equilibrium between critical-sized nuclei of a new phase and the old phase, which is also an unstable equilibrium. Any perturbation in the size of the nucleus would lead to its further growth or consumption. Another example is a perfectly linear river. When perturbation leads to a trivially small dent on one riverbank, the river flow would erode the dent further to increase the curvature, leading to a curved river path. There is no stable equilibrium shape of a river. Hence, the river path wanders around in geologic time.

Figure 3-12 shows how the activity and $D/\mathcal{D}$ depend on mole fraction for the specific case of $W/(RT) = 2.5$ (corresponding to $T = 800$ K in Figure 3-11a). It can be seen that $D = 0$ on the spinode and $D < 0$ inside the spinode.

To treat spinodal decomposition (and uphill diffusion), one may assume that $\mathcal{D}$ is a positive constant and calculate D according to Equation 3-62 given a thermodynamic model for the binary system. From the concentration-dependent D, solving the diffusion equation would require numerical method because D is a complicated function of C and changes from positive outside the spinode to negative inside the spinode. A simple treatment of spinodal decomposition to explain how a small fluctuation grows spontaneously to a different phase is as follows: Consider an initially homogeneous phase that has a composition inside the spinode and hence with $D < 0$. The phase is bounded by $x = 0$ to $x = L$. Because of thermal motion of atoms, there will always be small heterogeneities. Suppose the concentration profile of component 2 shows a small initial periodic fluctuation so that

$$C|_{t=0} = C_0 + \varepsilon \sin (2n\pi x/L), \tag{3-63a}$$

where C_0 is the initial homogeneous concentration of component 2 in the phase, ε is a very small positive number (such as $10^{-15}C_0$), and n is an integer. Let's consider only small concentration changes so that D can be viewed as constant. Assuming that $C|_{x=0} = C|_{x=L} = C_0$, the solution of the diffusion equation for this problem is

$$C(x, t) = C_0 + \varepsilon e^{-4n^2\pi^2 Dt/L^2} \sin(2n\pi x/L). \tag{3-63b}$$

Hence, the fluctuation would change with time with the same periodicity of L/n, but the amplitude would change as $\varepsilon e^{-4n^2\pi^2 Dt/L^2}$.

When the phase is stable with respect to spinodal decomposition, D is positive, the amplitude of the fluctuation ($\varepsilon e^{-4\,n^2\pi^2 Dt/L^2}$) decreases exponentially with time. That is, any fluctuation would rapidly disappear due to diffusion and the system returns to uniform composition. However, inside the spinode, D is negative, the amplitude increases with time, eventually leading to the decomposition of the phase. This simple treatment ignores the compositional dependence of D (which actually changes from negative to positive as the spinode is crossed), surface energy and strain energy effects. A fuller explanation of spinodal decomposition can be found in the literature (see Kirkaldy and Young, 1987 for a review).

From the above analysis, it can be seen that D in Fick's first law $\mathbf{J} = -D\nabla C$ (Equation 3-6) may be either positive or negative (accounting for uphill diffusion), and it can vary from positive to negative along a spinodal decomposition diffusion profile. If, on the other hand, Fick's law is modified as $\mathbf{J} = -(\mathcal{D}/\gamma)\nabla a$ (Equation 3-61), then $\mathcal{D}$ is always positive in a binary system.

Uphill diffusion in a binary system is rare and occurs only when the phase undergoes spinodal decomposition. In multicomponent systems, uphill diffusion occurs often, even when the phase is stable. The cause for uphill diffusion in multicomponent systems is different from that in binary systems and will be discussed later.

3.2.10 Diffusion in three dimensions; different coordinates

3.2.10.1 Three-dimensional diffusion in Cartesian coordinates

The three-dimensional diffusion equation in Cartesian coordinates is

$$\frac{\partial C}{\partial t} = \nabla(D\nabla C) = \frac{\partial}{\partial x}\left(D\frac{\partial C}{\partial x}\right) + \frac{\partial}{\partial y}\left(D\frac{\partial C}{\partial y}\right) + \frac{\partial}{\partial z}\left(D\frac{\partial C}{\partial z}\right). \tag{3-64a}$$

If D is constant, then,

$$\frac{\partial C}{\partial t} = D\nabla^2 C = D\left(\frac{\partial^2 C}{\partial x^2} + \frac{\partial^2 C}{\partial y^2} + \frac{\partial^2 C}{\partial z^2}\right). \tag{3-64b}$$

3.2.10.2 Three-dimensional diffusion in cylindrical coordinates

In cylindrical coordinates, using $x = r \cos \theta$ and $y = r \sin \theta$, Equation 3-64a can be transformed into (Crank, 1975)

$$\frac{\partial C}{\partial t} = \frac{1}{r} \frac{\partial}{\partial r} \left(Dr \frac{\partial C}{\partial r} \right) + \frac{1}{r^2} \frac{\partial}{\partial \theta} \left(D \frac{\partial C}{\partial \theta} \right) + \frac{\partial}{\partial z} \left(D \frac{\partial C}{\partial z} \right). \tag{3-65}$$

3.2.10.3 Three-dimensional diffusion in spherical coordinates

In spherical coordinates, using $x = r \sin \theta \cos \phi$, $x = r \sin \theta \sin \phi$, and $z = r \cos \theta$, then

$$\frac{\partial C}{\partial t} = \frac{1}{r^2} \left\{ \frac{\partial}{\partial r} \left(Dr^2 \frac{\partial C}{\partial r} \right) + \frac{1}{\sin \theta} \frac{\partial}{\partial \theta} \left(D \sin \theta \frac{\partial C}{\partial \theta} \right) + \frac{1}{\sin^2 \theta} \frac{\partial}{\partial \phi} \left(D \frac{\partial C}{\partial \phi} \right) \right\}. \tag{3-66a}$$

If C is independent of θ and ϕ, and D is constant, then

$$\frac{\partial C}{\partial t} = \frac{D}{r^2} \frac{\partial}{\partial r} \left(r^2 \frac{\partial C}{\partial r} \right), \tag{3-66b}$$

which can also be written as

$$\frac{\partial (rC)}{\partial t} = D \frac{\partial^2 (rC)}{\partial r^2}. \tag{3-66c}$$

Define $w = rC$ in Equation 3-66c; then the above equation for three-dimensional diffusion with spherical symmetry is reduced to the same form as Equation 3-10:

$$\frac{\partial w}{\partial t} = D \frac{\partial^2 w}{\partial r^2}. \tag{3-66d}$$

The above equation is important and very useful because many solutions obtained for the simple one-dimensional diffusion equation can be applied to the spherical diffusion problem. Below is an example.

Example 3.5 Consider three-dimensional diffusion in a solid sphere of radius a with spherical geometry (meaning concentration depends only on r with $0 \leq r \leq a$). The initial and boundary conditions are

Initial condition: $C|_{t=0} = f(r)$.

Boundary conditions: $C|_{r=a} = C_0$.

Find the solution.

Solution: Let $w = rC$. Then w satisfies the following one-dimensional diffusion equation

$$\frac{\partial w}{\partial t} = D \frac{\partial^2 w}{\partial r^2},$$

and the following initial and boundary conditions:

$$w|_{t=0} = rf(r),$$

$$w|_{r=0} = 0,$$

$$w|_{r=a} = aC_0.$$

Note that $w|_{r=0} = 0$ is a natural boundary condition based on the other natural boundary condition that $C|_{r=0}$ is finite. This one-dimensional diffusion problem can be solved using the method of separation of variables and Fourier series discussed in Section 3.2.7, and the specific solution can be found in Appendix A3.2.4c as

$$w = C_0 r + \frac{2aC_0}{\pi} \sum_{n=1}^{\infty} \frac{(-1)^n}{n} \sin \frac{n\pi r}{a} e^{-Dn^2\pi^2 t/a^2}$$

$$+ \frac{2}{a} \sum_{n=1}^{\infty} \sin \frac{n\pi r}{a} e^{-Dn^2\pi^2 t/a^2} \int_0^a yf(y) \sin \frac{n\pi y}{a} dy. \tag{3-67a}$$

The solution for C is

$$C = C_0 + \frac{2aC_0}{\pi r} \sum_{n=1}^{\infty} \frac{(-1)^n}{n} \sin \frac{n\pi r}{a} e^{-Dn^2\pi^2 t/a^2}$$

$$+ \frac{2}{ar} \sum_{n=1}^{\infty} \sin \frac{n\pi r}{a} e^{-Dn^2\pi^2 t/a^2} \int_0^a yf(y) \sin \frac{n\pi y}{a} dy. \tag{3-67b}$$

Example 3.5a If $C|_{t=0} = f(r) = C_1$ and $C|_{r=a} = C_0$, the above general solution can be integrated to obtain

$$C = C_0 + (C_0 - C_1) \frac{2a}{\pi r} \sum_{n=1}^{\infty} \frac{(-1)^n}{n} \sin \frac{n\pi r}{a} e^{-Dn^2\pi^2 t/a^2}. \tag{3-68a}$$

The concentration at the center ($r = 0$) can be found as

$$C = C_0 + 2(C_0 - C_1) \sum_{n=1}^{\infty} (-1)^n e^{-Dn^2\pi^2 t/a^2}. \tag{3-68b}$$

The total amount of diffusing substance entering or leaving the sphere is

$$\frac{M_t}{M_\infty} = 1 - \frac{6}{\pi^2} \sum_{n=1}^{\infty} \frac{1}{n^2} e^{-Dn^2\pi^2 t/a^2}, \tag{3-68c}$$

where M_∞ is the final mass gain or loss as t approaches ∞, and equals $4\pi a^3 (C_0 - C_1)/3$.

The above three equations converge rapidly for large Dt/a^2, but slowly for small Dt/a^2. To improve the convergence rate, the following three equations may be used for small Dt/a^2:

$$C = C_1 + (C_0 - C_1)\frac{a}{r}\sum_{n=0}^{\infty}\left\{\mathrm{erfc}\,\frac{(2n+1)a - r}{\sqrt{4Dt}} - \mathrm{erfc}\,\frac{(2n+1)a + r}{\sqrt{4Dt}}\right\}, \tag{3-68d}$$

$$C = C_1 + (C_0 - C_1)\frac{2a}{\sqrt{\pi Dt}}\sum_{n=0}^{\infty}e^{-(2n+1)^2 a^2/(4Dt)}. \tag{3-68e}$$

$$\frac{M_t}{M_\infty} = 6\frac{\sqrt{Dt}}{a}\left\{\frac{1}{\sqrt{\pi}} + 2\sum_{n=1}^{\infty}\mathrm{ierfc}\,\frac{na}{\sqrt{Dt}}\right\} - 3\frac{Dt}{a^2}. \tag{3-68f}$$

The solutions for a spherical shell ($a \leq r \leq b$) or an infinite sphere but with a hole within ($r \geq a$) can be found in Appendix 3.

Example 3.5b If $C|_{t=0} = f(r) = C_1$ and $C|_{r=a} = C_0 = 0$ (meaning uniform initial concentration and zero surface concentration), the above solution can be simplified to obtain

$$C = \frac{2aC_1}{\pi r}\sum_{n=1}^{\infty}\frac{(-1)^{n+1}}{n}\sin\frac{n\pi r}{a}\,e^{-Dn^2\pi^2 t/a^2}. \tag{3-68g}$$

For rapid convergence with small Dt/a^2, the following may be used:

$$C = C_1 - C_1\frac{a}{r}\sum_{n=0}^{\infty}\left\{\mathrm{erfc}\,\frac{(2n+1)a - r}{\sqrt{4Dt}} - \mathrm{erfc}\,\frac{(2n+1)a + r}{\sqrt{4Dt}}\right\}. \tag{3-68h}$$

3.2.11 Diffusion in an anisotropic medium; diffusion tensor

In an isotropic medium, D is a scalar, which may be constant or dependent on time, space coordinates, and/or concentration. In anisotropic media (such as crystals other than cubic symmetry, i.e., most minerals), however, diffusivity also depends on the diffusion direction. The diffusivity in an anisotropic medium is a second-rank symmetric tensor $\underline{\mathbf{D}}$ that can be represented by a 3×3 matrix (Equation 3-25a). The tensor is called the *diffusivity tensor*. Diffusivity along any given direction can be calculated from the diffusivity tensor (Equation 3-25b). Each element in the tensor may be constant, or dependent on time, space coordinates and/or concentration.

The diffusion equation in an anisotropic medium is complicated. Based on the definition of the diffusivity tensor, the diffusive flux along a given direction (except along a principal axis) depends not only on the concentration gradient along this direction, but also along other directions. The flux equation is written as $\mathbf{F} = -\underline{\mathbf{D}}\nabla C$ (similar to Fick's law $\mathbf{F} = -D\nabla C$ but the scalar D is replaced by the tensor $\underline{\mathbf{D}}$), i.e.,

$$\begin{pmatrix} F_x \\ F_y \\ F_z \end{pmatrix} = -\begin{pmatrix} \underline{D}_{11} & \underline{D}_{12} & \underline{D}_{13} \\ \underline{D}_{12} & \underline{D}_{22} & \underline{D}_{23} \\ \underline{D}_{13} & \underline{D}_{23} & \underline{D}_{33} \end{pmatrix}\begin{pmatrix} \partial C/\partial x \\ \partial C/\partial y \\ \partial C/\partial z \end{pmatrix}. \tag{3-69a}$$

The derivative of concentration with time is related to the flux:

$$\frac{\partial C}{\partial t} = -\nabla \cdot \mathbf{F} = -\frac{\partial F_x}{\partial x} - \frac{\partial F_y}{\partial y} - \frac{\partial F_z}{\partial z}.$$

(3-69b)

Therefore, for constant $\underline{\mathbf{D}}$, combining Equations 3-69a and 3-69b leads to

$$\frac{\partial C}{\partial t} = \underline{D}_{11}\frac{\partial^2 C}{\partial x^2} + \underline{D}_{22}\frac{\partial^2 C}{\partial y^2} + \underline{D}_{33}\frac{\partial^2 C}{\partial z^2}$$
$$+ 2\underline{D}_{23}\frac{\partial^2 C}{\partial y \partial z} + 2\underline{D}_{13}\frac{\partial^2 C}{\partial x \partial z} + 2\underline{D}_{12}\frac{\partial^2 C}{\partial x \partial y}.$$

(3-69c)

Using a coordinate transform, the above equation can be reduced to

$$\frac{\partial C}{\partial t} = D_1\frac{\partial^2 C}{\partial \xi^2} + D_2\frac{\partial^2 C}{\partial \eta^2} + D_3\frac{\partial^2 C}{\partial \xi^2}.$$

(3-69d)

This coordinate transformation is the same as that by which the ellipsoid

$$\underline{D}_{11}x^2 + \underline{D}_{22}y^2 + \underline{D}_{33}z^2 + 2\underline{D}_{12}xy + 2\underline{D}_{23}yz + 2\underline{D}_{13}xz = \text{const.}$$

(3-69e)

is transformed to

$$D_1\xi^2 + D_2\eta^2 + \mathcal{D}_3\xi\zeta^2 = 1.$$

(3-69f)

The new axes ξ, η, ζ are known as the principal axes of diffusion and D_1, D_2, and D_3 are principal diffusion coefficients (i.e., diffusivity along the directions of the principal axes). Diffusion along a principal axis direction is not affected by concentration gradient along other directions. That is,

$$F_\xi = -D_1 \partial C/\partial \xi,$$

(3-70a)

$$F_\eta = -D_2 \partial C/\partial \eta,$$

(3-70b)

$$F_\zeta = -D_3 \partial C/\partial \zeta.$$

(3-70c)

Treating diffusion along each principal axis is hence relatively simple. The principal axes coincide with the crystallographic axes for crystals with at least orthorhombic symmetry (Nye, 1985). To further simplify, define

$$\xi' = \xi/\sqrt{D_1}, \quad \eta' = \eta/\sqrt{D_2}, \quad \zeta' = \zeta/\sqrt{D_3}.$$

(3-70d)

Then the diffusion equation becomes

$$\frac{\partial C}{\partial t} = \frac{\partial^2 C}{\partial \xi'^2} + \frac{\partial^2 C}{\partial \eta'^2} + \frac{\partial^2 C}{\partial \zeta'^2}.$$

(3-70e)

This equation is the same as the diffusion equation in isotropic media with $D = 1$. Hence, theoretically, solutions in isotropic media can be applied to diffusion problems in anisotropic media after these transformations. However, it must be realized that the transformed coordinates may correspond to strange and unin-

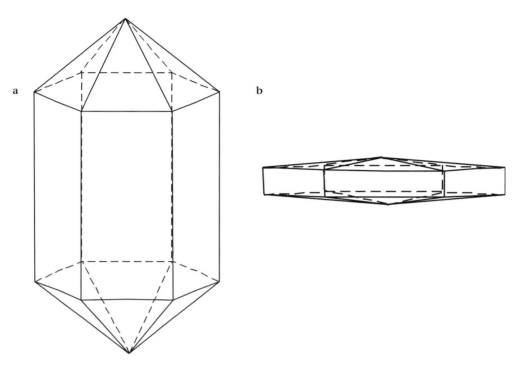

Figure 3-13 (a) The shape of a "perfectly" shaped quartz crystal, and (b) the effective "shape" of quartz crystal with respect to oxygen diffusion (after coordinate transformation using Equation 3-70d). The length (thickness) to diameter ratio is about 2:1 in (a) and 1:5 in (b).

tuitive shapes. If the original crystal is roughly spherical in shape, after transformation, the shape may not be spherical. A specific example is as follows. Oxygen diffusivity in quartz along the **c**-axis is about 100 times larger than that perpendicular to the **c**-axis (Giletti and Yund, 1984). Suppose the length of a quartz crystal is about 2 times the diameter (Figure 3-13a). Use Equation 3-70d to transform the coordinates. Let the direction of ζ be along the **c**-axis; then the axis of ζ would be compressed by a factor of 10 compared to the axes of ξ and η. That is, in the transformed coordinates, the effective shape of quartz would be a thin disk (Figure 3-13b), and the diffusive loss would be mostly along the **c**-axis. For oxygen diffusion in mica, D perpendicular to the **c**-axis is about four orders of magnitude greater than that parallel to the **c**-axis (Fortier and Giletti, 1991). Hence, the distortion of the shape after transformation is even greater.

In geologic applications, treating anisotropic diffusion in its full mathematical rigor is usually avoided because of mathematical complexities. Usually the diffusivities along the crystallographic axes are obtained. If the diffusivities along three principal crystallographic directions are similar, the mineral may be treated as roughly isotropic in terms of diffusion. If the diffusivity along one direction is much larger, then diffusion along other directions is usually ignored even if diffusion in the mineral proceeds along all directions. Hence, the grain is considered to be a thin slab or plane sheet (Figure 3-13b) bounded by two parallel planes. If the diffusivity in a plane is much larger compared to diffusivity perpendicular to the plane, the grain would be considered to be an infinitely long cylinder. For example, for oxygen diffusion in mica, because oxygen diffusivity in the plane perpendicular to the **c**-axis is about four orders of magnitude greater than that parallel to the **c**-axis (Fortier and Giletti, 1991), diffusion parallel to the **c**-axis is often ignored, and diffusion in the plane perpendicular to the **c**-axis is considered. Hence, even though mica has a shape of plane sheet, it is treated as an infinitely long cylinder in terms of oxygen diffusion.

Sometimes, an anisotropic mineral is in diffusive exchange with another phase only along one direction, e.g., an anisotropic mineral such as biotite is in contact with another ferromagnesian mineral such as garnet only along one surface and there is no contact with ferromagnesian mineral along other surfaces (Usuki, 2002). If no fluid is present and grain boundary diffusion is not rapid enough (otherwise diffusion along other directions may still occur), diffusion along this direction instead of the fastest diffusion direction must be considered. If the direction is parallel to a crystallographic axis, then diffusivity along this axis may be used and the concentration gradient is perpendicular to the contact surface. Otherwise, to decipher the direction of diffusion and diffusivity along the direction, it is necessary to use Equation 3-25b, or to transform the shape of the crystal using the method in Figure 3-13 into an effective shape. In the effective shape, diffusion may be treated as isotropic, and the diffusive gradient is perpendicular to the contact surface.

Quantifying the diffusional anisotropy requires measurement of diffusion profiles along different crystallographic directions, which is not always easy. For some species with low concentrations, such as Ar diffusion in minerals, the diffusivities are obtained using the bulk extraction or bulk equilibrium method using whole mineral grains (Section 3.6.1.2), instead of the profiling method. It is necessary to assume the effective shape of the mineral grains to estimate diffusivity. In such a case, later applications of the data must assume the same effective shape. For example, Harrison (1981) reported experimental Ar diffusion data on hornblende, assuming hornblende grains are spheres. When such data are applied, then the effective shape of hornblende grains must also be assumed to be spherical. Farver and Giletti (1985) showed that in terms of oxygen isotope diffusion in hornblende under hydrothermal conditions, the diffusivity along the **c**-axis is at least ten times greater than that along the **a**-axis, and the latter is slightly larger than that along the **b**-axis. Therefore, in terms of oxygen diffusion, hornblende is usually treated as a plane sheet. Nonetheless, in terms of Ar diffusion, hornblende must still be treated as spherical unless new data addressing the anisotropy are available.

3.2.12 Summary of analytical methods to obtain solution to the diffusion equation

The one-dimensional diffusion equation in isotropic medium for a binary system with a constant diffusivity is the most treated diffusion equation. In infinite and semi-infinite media with simple initial and boundary conditions, the diffusion equation is solved using the Boltzmann transformation and the solution is often an error function, such as Equation 3-44. In infinite and semi-infinite media with complicated initial and boundary conditions, the solution may be obtained using the superposition principle by integration, such as Equation 3-48a and solutions in Appendix 3. In a finite medium, the solution is often obtained by the separation of variables using Fourier series.

If the diffusion coefficient depends on time, the diffusion equation can be transformed to the above type of constant D by defining a new time variable $\alpha = \int D dt$ (Equation 3-53b). If the diffusion coefficient depends on concentration or x, the diffusion equation in general cannot be transformed to the simple type of constant D and cannot be solved analytically. For the case of concentration-dependent diffusivity, the Boltzmann transformation may be applied to numerically extract diffusivity as a function of concentration.

For three-dimensional diffusion, if there is spherical symmetry (i.e., concentration depends only on radius), the diffusion equation can be transformed to a one-dimensional type by redefining the concentration variable $w = rC$. This transformation would work for a solid finite sphere, a spherical shell, an infinite sphere with a spherical hole in the center, or an infinite sphere.

For three-dimensional diffusion in an anisotropic medium, theoretically it is possible to transform the diffusion equation to a form similar to that in an isotropic system. However, in practice, the transformed equation is rarely used, and diffusion is often simplified to be along the fastest diffusion direction.

In addition to the solution of various diffusion problems in this chapter, Appendix 3 summarizes some often-encountered solutions. Furthermore, Crank (1975) and Carslaw and Jaeger (1959) provide solutions to many other problems related to diffusion.

3.2.13 Numerical solutions

Many diffusion problems cannot be solved analytically, such as concentration-dependent D, complicated initial and boundary conditions, and irregular boundary shape. In these cases, numerical methods can be used to solve the diffusion equation (Press et al., 1992). There are many different numerical algorithms to solve a diffusion equation. This section gives a very brief introduction to the finite difference method. In this method, the differentials are replaced by the finite differences:

$$\partial C/\partial t \sim \Delta C/\Delta t, \tag{3-71a}$$

$$\partial C/\partial x \sim \Delta C/\Delta x, \tag{3-71b}$$

$$\partial^2 C/\partial x^2 \sim \Delta(\Delta C/\Delta x)/\Delta x. \tag{3-71c}$$

3.2.13.1 Nondimensionalizing the diffusion equation

Diffusion in a finite medium $(0, L)$ with a constant D can be nondimensionalized by letting

$$\xi = x/L, \tag{3-72a}$$

$$\tau = Dt/L^2, \tag{3-72b}$$

where ξ and τ are dimensionless "distance" and "time." The diffusion equation becomes

$$\frac{\partial C}{\partial \xi} = \frac{\partial^2 C}{\partial \tau^2}, \qquad \text{with } 0 < \xi < 1, \tau > 0. \tag{3-73}$$

If the concentrations at the two boundaries are given as C_0 and C_L, the concentration can also be normalized as $u = (C - C_0)/(C_L - C_0)$ so that $u = 0$ at $\xi = 0$ and $u = 1$ at $\xi = 1$. For other boundary conditions, there may or may not be a simple way to normalize. One of the advantages to nondimensionalize a diffusion equation is that the solution is independent of D and L. Hence, one solution can be applied to other situations with different D and L but similar boundary and initial conditions.

3.2.13.2 Stability

An algorithm for a partial differential equation is said to be stable if the truncation error introduced in a step is not amplified in the latter calculation steps. Unstable algorithms cannot be used in solving a diffusion equation because the errors would explode and overwhelm the values of C.

3.2.13.3 Three often-used algorithms

Divide medium $(0,1)$ into N equally spaced divisions (Figure 3-14). Let $\xi_i = i\,\Delta\xi$, with $\xi_0 = 0$ and $\xi_N = 1$ being the two boundaries. Let $\tau_j = j\,\Delta\tau$ with equally spaced time interval. (In more advanced programming, one may also divide the time and space into unequal parts.) Three algorithms are discussed below. Other algorithms may be numerically unstable.

Explicit method This is the simplest method but it is stable only for small time steps. Let

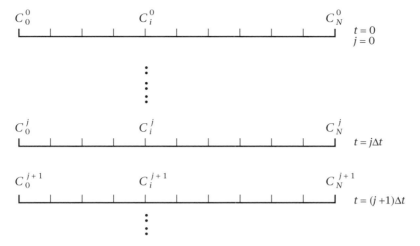

Figure 3-14 Schematics of dividing the diffusion medium into N equally spaced divisions. Starting from the initial condition (concentration at every nodes at $t=0$), C of the interior node at the next time step ($t=\Delta t$) can be calculated using the explicit method, whereas C at the two ends can be obtained from the boundary condition.

$$\left(\frac{\partial C}{\partial \tau}\right)_{i,j} = \frac{C_i^{j+1} - C_i^j}{\Delta \tau}, \qquad (3\text{-}74a)$$

$$\left(\frac{\partial C}{\partial \xi}\right)_{i,j} = \frac{C_{i+1/2}^j - C_{i-1/2}^j}{\Delta \xi}, \qquad (3\text{-}74b)$$

$$\left(\frac{\partial^2 C}{\partial \xi^2}\right)_{i,j} = \frac{(\partial C/\partial \xi)_{i+1/2}^j - (\partial C/\partial \xi)_{i-1/2}^j}{\Delta \xi} = \frac{C_{i+1}^j + C_{i-1}^j - 2C_i^j}{(\Delta \xi)^2}. \qquad (3\text{-}74c)$$

Then Equation 3-73 becomes

$$C_i^{j+1} = C_i^j + \alpha(C_{i+1}^j + C_{i-1}^j - 2C_i^j), \qquad (3\text{-}74d)$$

where $\alpha = \Delta \tau/(\Delta \xi)^2$. Starting from the initial concentration profile (or from the concentration profile at the jth time step at $\tau = j\Delta\tau$), the concentration C_i^{j+1} at the next step ($\tau = (j+1)\Delta\tau$) at space points $i = 1, 2, \ldots, N-1$ can be calculated directly using the above equation. C_0^{j+1} and C_N^{j+1} are calculated using the boundary conditions. For boundary condition $C|_{\xi=0} = f_0(t)$ and $C|_{\xi=1} = f_1(t)$, the use of the boundary condition is straightforward. The no flux boundary condition $\partial C/\partial \xi = 0$ at $\xi = 0$ means that the concentration profile is symmetric with respect to $\xi = 0$. That is, $C_{-1}^j = C_1^j$. Hence,

$$C_0^{j+1} = C_0^j + \alpha(C_1^j + C_{-1}^j - 2C_0^j) = C_0^j + \alpha(C_1^j - C_0^j). \qquad (3\text{-}74e)$$

Similarly, for no flux boundary condition $\partial C/\partial \xi = 0$ at $\xi = 1$,

$$C_N^{j+1} = C_N^j + \alpha(C_{N+1}^j + C_{N-1}^j - 2C_N^j) = C_N^j + \alpha(C_{N-1}^j - C_N^j). \tag{3-74f}$$

Because the diffusion profiles at the next time step are calculated directly from initial and boundary conditions, this method is called the *explicit method*. The method is stable only when $\alpha < 0.5$, i.e., $\Delta\tau/(\Delta\xi)^2 < 0.5$, and has only first-order precision because the expression for $(\partial C/\partial\tau)$ has only first-order precision. Hence, given $\Delta\xi$, it is necessary to choose a small $\Delta\tau$.

Implicit method This method is slightly more complicated, but it offers unconditional stability. Let

$$\left(\frac{\partial C}{\partial\tau}\right)_{i,j} = \frac{C_i^{j+1} - C_i^j}{\Delta\tau}. \tag{3-75a}$$

and

$$\left(\frac{\partial^2 C}{\partial\xi^2}\right)_{i,j} = \frac{C_{i+1}^{j+1} + C_{i-1}^{j+1} - 2C_i^{j+1}}{(\Delta\xi)^2} \tag{3-75b}$$

Then Equation 3-73 becomes

$$-\alpha C_{i-1}^{j+1} + (1 + 2\alpha)C_i^{j+1} - \alpha C_{i+1}^{j+1} = C_i^j, \tag{3-75c}$$

where $\alpha = \Delta\tau/(\Delta\xi)^2$. The left-hand side of the above equation contains three unknowns on the time level $j+1$, and the quantity on the right-hand side is known on the jth time level. If there are $N+1$ space points, then there will be $N-1$ equations of the type of Equation 3-75c. Coupled with two boundary conditions, the concentration profile at time step $j+1$ can be solved from $N+1$ linear equations. Because the concentrations at the next time step cannot be explicitly calculated but must be solved from the set of equations, this method is known as the *implicit method*. The advantage of this method is that it is stable for all α even though higher precision is obtained with smaller α (smaller $\Delta\tau$ steps), but it has only first-order precision because the expression for $(\partial C/\partial\tau)$ has only first-order precision.

Crank-Nicolson implicit method This method is a little more complicated but it offers high precision and unconditional stability. Let

$$\left(\frac{\partial C}{\partial\tau}\right)_{i,j+1/2} = \frac{C_i^{j+1} - C_i^j}{\Delta\tau}, \tag{3-76a}$$

and

$$\left(\frac{\partial^2 C}{\partial\xi^2}\right)_{i,j+1/2} = \frac{1}{2}\left(\frac{C_{i+1}^j + C_{i-1}^j - 2C_i^j}{(\Delta\xi)^2} + \frac{C_{i+1}^{j+1} + C_{i-1}^{j+1} - 2C_i^{j+1}}{(\Delta\xi)^2}\right) \tag{3-76b}$$

Then Equation 3-73 becomes

$$-\alpha C_{i-1}^{j+1} + (2 + 2\alpha)C_i^{j+1} - \alpha C_{i+1}^{j+1} = \alpha C_{i-1}^j + (2 - 2\alpha)C_i^j + \alpha C_{i+1}^j, \tag{3-76c}$$

where $\alpha = \Delta\tau/(\Delta\xi)^2$. This method is again implicit and the concentration profile at time step $j+1$ has to be solved from a set of linear equations. The method is unconditionally stable and has second-order precision in both time and space because the time derivative is now a central derivative that has second-order precision. Therefore, it is the preferred method for the numerical solution of a diffusion equation if D is constant.

3.2.13.4 Concentration-dependent D

If D depends on concentration, the explicit method is easy to adapt but the implicit methods are more difficult. Let

$$D = D_0 f(C) \tag{3-77}$$

where D_0 has the dimension of diffusivity and $f(C)$ is dimensionless. The diffusion equation can be nondimensionalized as follows. Let

$$\xi = x/L, \tag{3-77a}$$

and

$$\tau = D_0 t/L^2, \tag{3-77b}$$

where ξ and τ are dimensionless. The diffusion equation becomes

$$\frac{\partial C}{\partial \tau} = \frac{\partial}{\partial \xi}\left[f(C)\frac{\partial C}{\partial \xi}\right], \quad \text{with } 0 < \xi < 1, \ \tau > 0. \tag{3-77c}$$

Let

$$\left(\frac{\partial C}{\partial \tau}\right)_{i,j} = \frac{C_i^{j+1} - C_i^j}{\Delta\tau}, \tag{3-77d}$$

$$\left(\frac{\partial C}{\partial \xi}\right)_{i,j} = \frac{C_{i+1/2}^j - C_{i-1/2}^j}{\Delta\xi}, \tag{3-77e}$$

and

$$\frac{\partial}{\partial \xi}\left[f(C)\frac{\partial C}{\partial \xi}\right]_{i,j} = \frac{f(C_{i+1/2}^j)(C_{i+1}^j - C_i^j) - f(C_{i-1/2}^j)(C_i^j - C_{i-1}^j)}{(\Delta\xi)^2} \tag{3-77f}$$

Then Equation 3-77c becomes

$$C_i^{j+1} = C_i^j + \frac{\Delta\tau}{(\Delta\xi)^2}[f(C_{i+1/2}^j)(C_{i+1}^j - C_i^j) - f(C_{i-1/2}^j)(C_i^j - C_{i-1}^j)]. \tag{3-77g}$$

In the above equation, $C_{i+1/2}^j = (C_{i+1}^j + C_i^j)/2$, and $C_{i-1/2}^j = (C_i^j + C_{i-1}^j)/2$. For the explicit method to be stable, the condition $f_{max}\Delta\tau/(\Delta\xi)^2 < 0.5$ must be satisfied, where f_{max} is the maximum value of $f(C)$ in the range of C.

3.3 Diffusion of a Multispecies Component

Some components exist as only one species in any system, such as Ar in silicate melt. However, for many other components, there may be reactions involving the component, producing two or more species. This is often so in a silicate melt or an aqueous solution. For example, H_2O dissolves in a silicate melt as at least two species: molecular H_2O (denoted as H_2O_m and hydroxyl (denoted as OH).[2] CO_2 may be present in a silicate melt as molecular CO_2 and carbonate ion (CO_3^{2-}). Oxygen may be present in a silicate melt as molecular H_2O, hydroxyl, and anhydrous oxygen (which may be present as bridging oxygen, nonbridging oxygen, and free O^{2-}). A multispecies component may show specific diffusion behavior, depending on the kinds of reactions. The diffusion of a multispecies component is hence a theoretically interesting problem. Furthermore, the problem is of great practical importance because oxygen is the most abundant element in a silicate melt, and the volatile component H_2O plays a major role in volcanic eruptions and in affecting the properties of melts and minerals. Recall that multispecies diffusion (the focus of this section) is different from multicomponent diffusion (the focus of the next section).

There are two methods to write the diffusion equation for a multispecies component. One is to write the diffusion equation for the conserved component, and then relate the species concentrations by the reaction(s). Using one-dimensional H_2O diffusion as an example, the diffusion equation is Equation 3-22a:

$$\frac{\partial C_{H_2O_t}}{\partial t} = \frac{\partial}{\partial x}\left(D_{H_2O_m}\frac{\partial C_{H_2O_m}}{\partial x}\right) + \frac{\partial}{\partial x}\left(D_{OH}\frac{\partial C_{OH}/2}{\partial x}\right). \tag{3-78}$$

That is, the concentration change of the component with time is due to the diffusive flux of the two species. A factor of $\frac{1}{2}$ is required for the OH species because two OH groups convert into one H_2O molecule through the following reaction (Reaction 1-10):

$$H_2O(\text{melt}) + O(\text{melt}) \rightleftharpoons 2OH(\text{melt}), \tag{3-79}$$

---

[2] The hydroxyl may be bonded to Si, Al, Na, or other cations. In geochemistry, if the hydroxyl group is bonded to Si in an Si-centered tetrahedral, the bond is strong and the hydroxyl is not considered free (meaning free to move and free to react). If the hydroxyl is bonded to Na or K, the bond is weak and the hydroxyl is considered free. Free hydroxyl is often denoted as OH-, and strongly bonded hydroxyl is often denoted as OH (without the negative charge). For simplicity, all hydroxyls are denoted as OH here.

where O is an anhydrous oxygen. Define mole fractions on a single oxygen basis, $X_{H_2O_m} + X_{OH} + X_O = 1$. In the above diffusion equation, there are three unknowns: $C_{H_2O_t}$, $C_{H_2O_m}$, and C_{OH} (all in mol/L or mole fractions). Hence, two more equations are needed to solve Equation 3-78. One of the two equations is the mass balance equation (Equation 3-22b):

$$C_{H_2O_t} = C_{H_2O_m} + \left(\frac{1}{2}\right)C_{OH}. \tag{3-80a}$$

The second equation depends on whether equilibrium is rapidly reached locally between the species. If it is, then the equilibrium condition provides the third equation:

$$\text{Equilibrium constant:} \qquad K = \frac{X_{OH}^2}{X_{H_2O_m}(1 - X_{H_2O_m} - X_{OH})}. \tag{3-80b}$$

The equilibrium constant K depends on temperature and an approximate expression is $K = \exp(1.876 - 3110/T)$, where T is in K. Equations 3-78, 3-80a, and 3-80b can be solved simultaneous numerically (e.g., Zhang et al., 1991a,b) for three unknowns, $C_{H_2O_t}$, $C_{H_2O_m}$, and C_{OH}. If equilibrium is not reached, then Equation 3-80b would not be applicable and the third equation would involve the reaction and diffusion of a species, which is similar to the second way to mathematically describe the diffusion of a multispecies component.

The second way to write the diffusion equations for a multispecies component is to write the diffusion-reaction equation for each species. Starting from Equation 3-5b, the diffusion-reaction equation for each species is

$$\frac{\partial C_{H_2O_m}}{\partial t} = \frac{\partial}{\partial x}\left(D_{H_2O_m}\frac{\partial C_{H_2O_m}}{\partial x}\right) + k_b C_{OH}^2 - k_f C_{H_2O_m} C_O, \tag{3-80c}$$

$$\frac{\partial C_{OH}}{\partial t} = \frac{\partial}{\partial x}\left(D_{OH}\frac{\partial C_{OH}}{\partial x}\right) + 2k_f C_{H_2O_m} C_O - 2k_b C_{OH}^2, \tag{3-80d}$$

where $C_{H_2O_m}$, C_{OH}, and C_O are concentrations (mol/m^3) of molecular H_2O, hydroxyl, and anhydrous oxygen, and k_f and k_b are the forward and backward reaction rate coefficients of Reaction 3-79.

Because Equation 3-78 with two constraints (Equations 3-80a,b) is easier to solve than the diffusion-reaction equations (Equations 3-80c and 3-80d), if there is rough local equilibrium, the first method is the method of choice.

More generally, consider a species A that participates in a reaction of the following type:

$$n_A A + n_B B \rightleftharpoons n_C C. \tag{3-81}$$

Assuming that the above reaction is an elementary reaction, then a one-dimensional diffusion equation for such a species may be written as (the second method)

$$\frac{\partial [A]}{\partial t} = D_A \frac{\partial^2 [A]}{\partial x^2} - n_A k_f [A]^{n_A} [B]^{n_B} + n_A k_b [C]^{n_C}, \tag{3-82a}$$

where [A] is the concentration of A, D_A is the diffusion coefficient for component A, k_f is the forward reaction rate constant, and k_b is the backward reaction rate constant for Reaction 3-81. Since the concentrations of B and C will also change in a way similar to that of A, in this case, there will be three diffusion equations of the form of Equation 3-82a and they must be solved simultaneously.

The total concentration (w) of a multispecies component is independent of species interconversions of the type of Reaction 3-81, but is affected by the diffusion flux of individual species. Because each species may have a distinct diffusivity, the diffusion equation for w may be written as

$$\frac{\partial w}{\partial t} = \frac{\partial}{\partial x} \sum D_i \frac{\partial w_i}{\partial x}, \tag{3-82b}$$

where w_i and D_i are the concentration and diffusivity of species i. All D_i's are not equal. Since the species concentrations are related to one another by local reactions, if local equilibrium is reached (meaning high reaction rates), the equilibrium conditions will relate the species concentrations, and provide the additional constraints to solve Equation 3-82b. Due to these local reactions, Equation 3-82b is, in general, nonlinear and must be solved numerically. Therefore, use of components whose concentrations are independent of species concentrations has advantages here because only one partial differential equation needs to be solved in conjunction with some algebraic equations relating species concentrations.

3.3.1 Diffusion of water in silicate melts

The best-studied example of multispecies diffusion is that of dissolved water in silicate melt. Aspects of this problem have been mentioned before. In this section we thoroughly discuss this diffusion problem. This system played a critical role in the development of the concept and theory of multispecies diffusion and is also the first component for which the role of speciation in diffusion has been understood. In addition to the theoretical importance in the context of diffusion of a multispecies component, dissolved H_2O has a strong effect on melt and glass properties (such as strength of glasses and crystallization sequence of melts) and plays a critical role in explosive volcanic eruptions because exsolution of H_2O drives explosive volcanic eruptions. We use the example to illustrate the various aspects of diffusion of a multispecies component. For clarity, H_2O denotes the component, H_2O_t means total H_2O content, H_2O_m means H_2O molecules, and OH means hydroxyl groups.

To better understand diffusion in silicate melts, we first briefly review silicate melt structure. In natural silicate melts from basalt (about 50% SiO_2) to rhyolite (about 75% SiO_2), the structure is based on the aluminosilicate network made of SiO_4^{4-} and AlO_4^{5-} tetrahedra often connected to one other. Elements in the melt may be grouped as network formers (including Si, Al, and P) that form tetrahedra, and network modifiers (such as Mg, Fe, Ca, Na, and K) that form cations and connect or fill holes in the network tetrahedra. If two tetrahedra share one oxygen, this oxygen is said to be bridging oxygen. If the oxygen is used by one tetrahedron only, the oxygen is said to be nonbridging oxygen. If n of the four oxygens of a tetrahedron are bridging oxygens, the tetrahedron is denoted as Q_n. For example, Q_0 means isolated tetrahedra such as in an Mg_2SiO_4 melt, Q_1 means $Si_2O_7^{6-}$ units, Q_2 means chain silicate structure such as pyroxenes, and Q_4 is fully connected (such as pure SiO_2 or $NaAlSi_3O_8$ melt). The degree of polymerization means the degree of connectedness of the network. Fully polymerized melt means all the tetrahedra are connected to each other as Q_4 species. Rhyolitic melt is roughly fully polymerized. Basaltic melt is less polymerized than rhyolite and more polymerized than a pyroxene melt. Another characterization of the melt structure is NBO/T, the ratio of NBO (nonbridging oxygen) to T (tetrahedron). For Mg_2SiO_4 melt, all oxygens are NBO; hence, NBO/T = 4. In pyroxene melt, NBO/T = 2, and in SiO_2 or $NaAlSi_3O_8$ melt, NBO/T = 0. The general formula for calculating NBO/T is

$$\frac{NBO}{T} = 2\frac{\text{total oxygen}}{(Si + Al + P)} - 4,$$

where it is assumed that only Si, Al, and P form tetrahedron units, and Si, Al, P, and total oxygen are mole proportions. For rhyolitic melt with 77% SiO_2, NBO/T ≈ 0. For a basaltic melt with 50% SiO_2, NBO/T ≈ 0.9. Dissolved H_2O reacts with bridging oxygen to form OH groups (Reaction 1-10c), and hence depolymerizes the melt and increases NBO/T. Si–O (or Al–O) bonds in the network are strong, and the bonds connecting the network modifier to oxygen are relatively weak. Hence, it is difficult for network formers to diffuse, easier for charged network modifier (such as Na^+ or Mg^{2+}) to diffuse, and easiest for small neutral molecules (such as H_2O) or atoms (such as He) to diffuse through the melt structure.

Before going into the details and the mathematical aspects of H_2O diffusion, we first summarize the major features and developments of H_2O diffusion. It has been found experimentally that total H_2O diffusivity ($D_{H_2O_t}$) is a complicated function of total H_2O content (H_2O_t). Glass scientists first discovered that at low H_2O_t (<0.1 wt%), the diffusion profile has a particular shape that indicates that $D_{H_2O_t}$ is proportional to H_2O_t (e.g., Doremus, 1969, 2002). This was explained as follows. For the H_2O_m and OH interconversion reaction in the melt (Reaction 3-79), at equilibrium (Equation 3-80b), the $[H_2O_m]/[H_2O_t]$ ratio is proportional to H_2O_t. Assuming that H_2O_m is the diffusing species even though H_2O_m concentration is very small (often below detection limit), $D_{H_2O_t}$ would be proportional to H_2O_t. Later, geologists found that $D_{H_2O_t}$ depends on H_2O_t in a complicated fashion (Shaw, 1974; Delaney and Karsten, 1981). Subsequently, the microanalytical method to determine hydrous species concentrations was developed

(Stolper, 1982a,b). Afterward, in a study of H_2O diffusion for H_2O_t up to 1.7 wt%, the concentration profiles of the two species were measured, and the profiles were used to obtain both D_{OH} and $D_{H_2O_t}$ by fitting the profiles to the diffusion equation (Equation 3-78) by assuming that the two diffusivities are independent H_2O_t concentration (Zhang et al., 1991a). The results indicate that D_{OH} is negligible compared to $D_{H_2O_m}$. The OH profile is hence not due to OH diffusion itself, but due to H_2O_m diffusion and the interconversion reaction 3-79. By carrying out H_2O diffusion experiments to much higher H_2O_t, Nowak and Behrens (1997) showed that at high H_2O_t content, the diffusion profiles are inconsistent with constant $D_{H_2O_m}$. Zhang and Behrens (2000) inferred that H_2O diffusion profiles are consistent with $D_{H_2O_m}$ being an exponential function of H_2O_t. Consequently (i) $D_{H_2O_m}$ is roughly constant and $D_{H_2O_t}$ increases roughly proportionally to H_2O_t at low H_2O_t (≤ 2 wt% H_2O_t), and (ii) $D_{H_2O_t}$ increases exponentially with H_2O_t at higher H_2O_t. The exponential dependence of the diffusivity of a molecular species on H_2O_t has been confirmed by Ar diffusivity (Behrens and Zhang, 2001). The mathematical aspects are described below for the case of H_2O diffusion when local equilibrium of the species concentrations is reached.

As can be seen from the interconversion reaction between two hydrous species H_2O_m and OH (Reaction 3-79), the coefficient for OH is 2. This seemingly trivial point turns out to be the key in understanding H_2O diffusion. Because of this factor of 2, the equilibrium constant K involves the square of OH concentration but only the first power of H_2O_m concentration (Equation 3-80b). That is, the species concentrations of OH and H_2O_m are not proportional to each other, and neither concentration is proportional to total H_2O concentration (H_2O_t) in the entire concentration range. Solving Equations 3-80a,b for the two unknowns of H_2O_m and OH yields

$$[H_2O_m] = \frac{8X^2}{K(1-2X)+8X+\sqrt{K^2(1-2X)^2+16KX(1-X)}} \tag{3-83a}$$

and

$$[OH] = 2\{[H_2O_t] - [H_2O_m]\}, \tag{3-83b}$$

where $X = [H_2O_t]$, and the brackets mean mole fractions on a single-oxygen basis (Section 2.1.5.1). Figure 3-15 shows how the species concentrations vary with H_2O_t. At very low H_2O_t, H_2O_m concentration is very small, and OH is the dominant species, leading to H_2O_m/H_2O_t proportional to H_2O_t (Figure 3-15) and H_2O_m proportional to the square of H_2O_t. As H_2O_t concentration increases, the ratio H_2O_m/H_2O_t increases but deviates from the proportionality (the maximum value of the H_2O_m/H_2O_t ratio is 1).

Substituting Equations 3-83a and 3-83b into Equation 3-78 leads to a complicated nonlinear diffusion equation for $[H_2O_t]$, which must be solved numer-

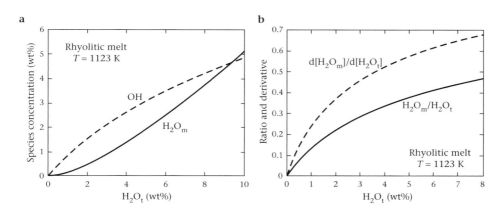

Figure 3-15 H_2O_m and OH species concentration, the ratio of H_2O_m/H_2O_t and the derivative $d[H_2O_m]/d[H_2O_t]$ versus H_2O_t concentration. The equilibrium constant at 1123 K is 0.41 from extrapolation of Zhang et al. (1997a).

ically (Zhang et al., 1991a; Zhang and Behrens, 2000). Comparing Equation 3-78 with the following "normal" diffusion equation

$$\frac{\partial[H_2O_t]}{\partial t} = \frac{\partial}{\partial x}\left\{D_{H_2O_t}\frac{\partial[H_2O_t]}{\partial x}\right\}, \tag{3-84}$$

we obtain

$$D_{H_2O_t} = D_{H_2O_m}\frac{d[H_2O_m]}{d[H_2O_t]} + D_{OH}\frac{d[OH]/2}{d[H_2O_t]}. \tag{3-85}$$

Because the derivatives are complicated functions of H_2O_t, $D_{H_2O_t}$ is a complicated function of H_2O_t.

Experiments show that when H_2O_t is low (≤ 1 wt%), the diffusion profile has a specific shape, which indicates that $D_{H_2O_t}$ is roughly proportional to H_2O_t. Figures 3-16a and 3-16b show two profiles (one for dehydration and one for hydration) at low H_2O_t. They cannot be well fit by the error function (dashed curves), meaning that $D_{H_2O_t}$ is not constant. But they can be fit well using $D_{H_2O_t}$ is proportional to H_2O_t (solid curves). One may also use Boltzmann analysis to obtain how $D_{H_2O_t}$ varies with H_2O_t, which also indicates that $D_{H_2O_t}$ is proportional to H_2O_t.

Concentration profiles of H_2O_m and OH after diffusion experiments can be measured by infrared spectroscopy. For the species profiles to reflect the concentrations at the experimental temperature and not altered during quench, the experimental temperature needs to be intermediate (usually 673 to 873 K) so that species reaction during quench is negligible. From such experiments, assuming D_{OH} and $D_{H_2O_m}$ are independent of H_2O_t, it is possible to fit all concentration

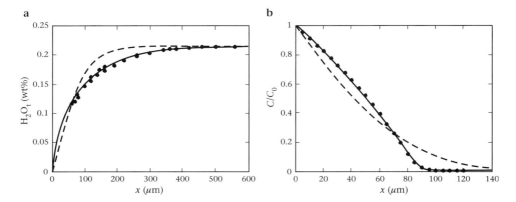

Figure 3-16 H_2O diffusion profile in (a) a dehydration experiment (Zhang et al., 1991a) and (b) a hydration experiment with data read from a figure in Roberts and Roberts (1964). Two outlier points are excluded in (a). The solid curves are fits to the data assuming D is proportional to C, and the dashed curves are fits assuming constant D (error function).

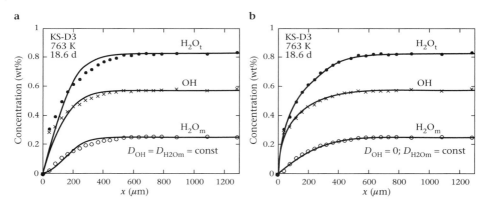

Figure 3-17 Concentration profiles of H_2O_m, OH, and H_2O_t from a dehydration experiment, fit by assuming (a) $D_{OH} = D_{H_2O_m}$, (b) $D_{OH}/D_{H_2O_m} = 0$.

profiles (H_2O_t, H_2O_m, and OH) to obtain both D_{OH} and $D_{H_2O_m}$ using numerically calculated profiles by solving Equations 3-78, 3-80a, and 3-80b. For example, if $D_{OH} = D_{H_2O_m}$, the H_2O_t profile would be an error function. Varying the $D_{OH}/D_{H_2O_m}$ ratio to another fixed value, the shape of the profiles would also change. Experimental profiles are best fit when $D_{OH}/D_{H_2O_m} = 0$ (Figure 3-17), demonstrating that H_2O_m is the diffusing species and OH diffusion is negligible. The OH concentration profile is due not to OH diffusion, but to H_2O_m diffusion and the local interconversion reaction between H_2O_m and OH (Reaction 3-79).

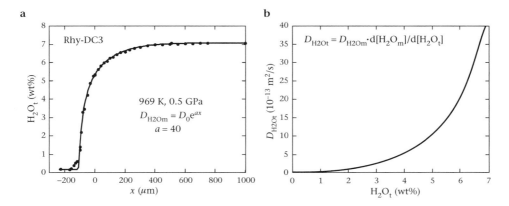

Figure 3-18 (a) Fit of an H_2O_t profile (from low to high H_2O_t) from a diffusion-couple experiment. The misfit at $x \approx -120$ μm is attributed to the convolution effect (limited spatial resolution of the measurements). (b) The dependence of $D_{H_2O_t}$ on H_2O_t from the fit in (a). From Zhang and Behrens (2000).

Because OH diffusion can be ignored compared to molecular H_2O diffusion, Equation 3-78 may be simplified to

$$\frac{\partial[H_2O_t]}{\partial t} = \frac{\partial}{\partial x}\left\{D_{H_2O_m}\frac{\partial[H_2O_m]}{\partial x}\right\}. \tag{3-86}$$

Total H_2O diffusivity $D_{H_2O_t}$ (Equation 3-85) may be expressed as follows:

$$D_{H_2O_t} = D_{H_2O_m}\,dX_{H_2O_m}dX_{H_2O_t}. \tag{3-86a}$$

For constant K, the derivative $dX_{H_2O_m}/dX_{H_2O_t}$ may be expressed as

$$\frac{dX_m}{dX} = \frac{16X}{b} - \frac{8X^2}{b^2}\left[8 - 2K + \frac{8K - 2K^2(1 - 2X) - 16KX}{\sqrt{[K(1-2X)]^2 + 16KX(1-X)}}\right], \tag{3-86b}$$

where $X_m = [H_2O_m]$, $X = [H_2O_t]$, and $b = 8X + K(1 - 2X) + \sqrt{[K(1-2X)]^2 + 16KX(1-X)}$. The derivative $dX_{H_2O_m}/dX_{H_2O_t}$ is roughly proportional to H_2O_t at low H_2O_t content and then gradually reaches a constant (Figure 3-15). Hence, at relatively low H_2O_t, fitting the H_2O_t profile by assuming that $D_{H_2O_t}$ is proportional to H_2O_t concentration is roughly equivalent to fitting the profile by assuming that $D_{OH}/D_{H_2O_m} = 0$ and $D_{H_2O_m}$ is independent of H_2O_t.

At significantly higher H_2O_t (up to 8 wt% H_2O_t), experimental work showed that $D_{H_2O_t}$ increases more rapidly at high H_2O_t, roughly exponentially. Because such high H_2O_t affects melt structure significantly, it is not surprising that $D_{H_2O_m}$ increases with H_2O_t. Based on the diffusivity of other molecular species (such as Ar), $D_{H_2O_m}$ is expected to increase with H_2O_t exponentially. The exponential dependence means that at low H_2O_t, $D_{H_2O_m}$ increases with H_2O_t slowly. Experimental concentration profiles at low and high H_2O_t are consistent with this understanding (Figure 3-18a). The variation of $D_{H_2O_t}$ with H_2O_t content hence comes from two factors: the dependence of $dX_{H_2O_m}/dX_{H_2O_t}$ on H_2O_t and the dependence of $D_{H_2O_m}$ on H_2O_t. The complete dependence of $D_{H_2O_t}$ on H_2O_t is hence shown in Figure 3-18b.

Some specific relations at low H_2O_t are presented below. When total water concentration is low, the mole fraction of molecular H_2O may be expressed as

$$[H_2O_m] \approx [OH]^2/K \approx 4[H_2O_t]^2/K. \tag{3-86c}$$

At low H_2O_t, $D_{H_2O_m}$ may be treated as roughly independent of H_2O content. Hence, the diffusion equation becomes

$$\frac{\partial[H_2O_t]}{\partial t} = \frac{\partial}{\partial x}\left\{ D_{H_2O_m} \frac{\partial}{\partial x} \frac{4[H_2O_t]^2}{K} \right\}, \tag{3-87a}$$

or

$$\frac{\partial[H_2O_t]}{\partial t} = \frac{\partial}{\partial x}\left\{ \frac{8}{K} D_{H_2O_m}[H_2O_t] \frac{\partial[H_2O_t]}{\partial x} \right\}, \tag{3-87b}$$

The total H_2O diffusion coefficient is, hence,

$$D_{H_2O_t} = D_{H_2O_m} \frac{8[H_2O_t]}{K} = D_0 \frac{X}{X_0}, \tag{3-87c}$$

where X_0 is concentration used for normalization (usually the highest H_2O_t along the profile), and D_0 is a constant equaling $D_{H_2O_t}$ at $X = X_0$ (that is, D_0 is the maximum diffusivity along the profile): $D_0 = 8D_{H_2O_m}X_0/K$. This derivation shows that at low H_2O_t, $D_{H_2O_t}$ is roughly proportional to $[H_2O_t]$. The solution of Equation 3-87a,b can be found in Crank (1975), and has been applied by Jambon et al. (1992) and Wang et al. (1996).

If the diffusivity of a component is proportional to its own concentration, when sorption (or hydration) experiments are carried out and when diffusivity is obtained by total mass gain rather than by diffusion profile, the obtained diffusivity (termed *diffusion-in diffusivity*, and denoted as D_{in}) is some average of the diffusivity along the profile, and is neither the diffusivity at the zero H_2O_t (which is zero) nor diffusivity at the maximum H_2O_t (which is D_0 in Equation 3-87c). The value of D_{in} lies between 0 and D_0. Similarly, from desorption experiments, if diffusivity is obtained by total mass loss, the obtained diffusivity (termed *diffusion-out diffusivity* and denoted as D_{out}) is also some average of the diffusivity along the profile. Experimental studies show that D_{in} and D_{out} differ by a factor of about 2 for H_2O diffusion. Suppose that during the sorption experiments the initial H_2O_t concentration in the material is zero, and during the desorption experiments the surface H_2O_t concentration in the material is zero. If D is proportional to concentration (Equation 3-87c), numerical solutions show that D_{in}

from sorption experiments and D_{out} from desorption experiments may be expressed as follows (Wang et al., 1996):

$$D_{in} = 0.619D_0, \tag{3-88a}$$

$$D_{out} = 0.347D_0, \tag{3-88b}$$

$$D_{in} = 0.78D_{out}. \tag{3-88c}$$

where D_0 is the diffusivity at the maximum concentration of the profile.

H_2O diffusivity in rhyolitic melt may be estimated as follows (Zhang and Behrens, 2000):

(1) At low H_2O_t (≤ 2 wt%),

$$D_{H_2O_t} = w\exp(10.49 - 10,661/T - 1.772P/T), \tag{3-88d}$$

where the unit of $D_{H_2O_t}$ is $\mu m^2/s$, T is in K, P is in MPa, and w is wt% of H_2O_t (for 1 wt% H_2O_t, $w = 1$).

(2) At low and high H_2O_t (up to 8 wt%),

$$D_{H_2O_m} = \exp[(14.08 - 13,128/T - 2.796P/T) + (-27.21 + 36,892/T + 57.23P/T)X], \tag{3-88e}$$

$$D_{H_2O_t} = X\exp(m)\left\{1 + \exp\left[56 + m + X\left(-34.1 + \frac{44620}{T} + \frac{57.3P}{T}\right) - \sqrt{X}\left(0.091 + \frac{4.77\times10^6}{T^2}\right)\right]\right\}, \tag{3-88f}$$

where the unit of D is $\mu m^2/s$, $m = -20.79 - 5030/T - 1.4P/T$, T is in K, P is in MPa, and X is the mole fraction of H_2O_t on a single oxygen basis and is calculated as $X = (w/18.015)/[w/18.015 + (100 - w)/32.49]$, where w is wt% of H_2O_t.

3.3.2 Diffusion of CO$_2$ component in silicate melts

The volatile component CO_2 dissolves in silicate melts and glasses in two forms, CO_2 molecule and CO_3^{2-} ion. Hereafter, CO_2 refers to the component of CO_2; molecular CO_2 or $CO_{2,molec}$ refers to the species. There is a homogeneous reaction between them:

$$CO_{2,molec} + O^{2-} \rightleftharpoons CO_3^{2-}, \tag{3-89}$$

where O^{2-} is free oxygen ion in the melt and is assumed to be abundant compared to concentrations of $CO_{2,molec}$ and CO_3^{2-} (hundreds of ppm). Total CO_2 concentration is $[CO_2]_{total} \equiv [CO_{2,molec}] + [CO_3^{2-}]$. The equilibrium constant K can hence be written as

$$K[O^{2-}] = [CO_3^{2-}]/[CO_{2,molec}], \tag{3-90a}$$

where $[O^{2-}]$ is roughly constant for a given melt because CO_2 concentration in a melt is usually low and Reaction 3-89 would not significantly change $[O^{2-}]$. In rhyolitic melt, free oxygen concentration $[O^{2-}]$ is low (oxygen is mostly bridging oxygen) and CO_2 is mostly present as $CO_{2,molec}$. In basaltic melt, $[O^{2-}]$ is high and most CO_2 is present as CO_3^{2-}. In intermediate melts such as dacite and andesite, quenched glasses show both $CO_{2,molec}$ and CO_3^{2-}.

Assuming local equilibrium between $CO_{2,molec}$ and CO_3^{2-}, because

$$[CO_{2,molec}] + [CO_3^{2-}] \equiv [CO_2]_{total}, \tag{3-90b}$$

and

$$[CO_3^{2-}] = K[O^{2-}][CO_{2,molec}], \tag{3-90c}$$

therefore,

$$[CO_{2,molec}] = [CO_2]_{total} / (1 + K[O^{2-}]), \tag{3-90d}$$

and

$$[CO_3^{2-}] = K[O^{2-}][CO_2]_{total} / (1 + K[O^{2-}]). \tag{3-90e}$$

Compared to the speciation reaction of H_2O, the stoichiometric coefficient is 1 for all carbon species in Reaction 3-89. Hence, carbon species concentrations are proportional to each other, meaning that the derivative of one species concentration with respect to the total concentration is a constant (Figure 3-19a). This simple point is the key for understanding the role of speciation in the diffusion of CO_2 component.

The diffusion coefficients for the two carbon species are almost certainly different but each may be assumed to be constant in a given melt. Because CO_2 is a linear molecule (long cylinder with about 1.4 Å radius, and 3 Å half-length) with a small radius on the base (only slightly larger than the radius of H_2O molecule), whereas CO_3^{2-} is a relatively large anion (thin disk with about 1.4 Å half-thickness, and 3 Å radius), and because diffusion of anions requires counter motion of other ions, it is expected that the diffusivity of CO_2 is much larger than that of CO_3^{2-}.

The diffusion equation for total CO_2 is

$$\frac{\partial [CO_2]_{total}}{\partial t} = D_{CO_{2,molec}} \frac{\partial^2 [CO_{2,molec}]}{\partial x^2} + D_{CO_3^{2-}} \frac{\partial^2 [CO_3^{2-}]}{\partial x^2} \tag{3-91}$$

Therefore, the diffusion coefficient of total CO_2 may be written as

$$D_{CO_{2,total}} = D_{CO_{2,molec}} \frac{d[CO_{2,molec}]}{d[CO_{2,total}]} + D_{CO_3^{2-}} \frac{d[CO_3^{2-}]}{d[CO_{2,total}]} \tag{3-91a}$$

That is, $D_{CO_{2,total}}$ is a weighted average of $D_{CO_{2,molec}}$ and $D_{CO_3^{2-}}$. Because $CO_{2,molec}$ concentration is proportional to that of CO_3^{2-}, and also proportional to that of

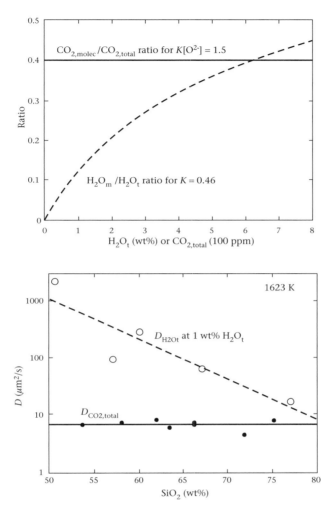

Figure 3-19 (a) Ratio of $CO_{2,molec}/CO_{2,total}$ as a function of $CO_{2,total}$ (in unit of 100 ppm), versus ratio of H_2O_m/H_2O_t as a function of H_2O_t (wt%). The former is a constant, and the latter is a curve passing through the origin. (b) The dependence of $D_{CO_{2,total}}$ and $D_{H_2O_t}$ on anhydrous melt composition. $D_{CO_{2,total}}$ is roughly independent of the anhydrous melt composition, but $D_{H_2O_t}$ depends strongly on the anhydrous melt composition. Data for $D_{CO_{2,total}}$ are from Nowak et al. (2004), and data for $D_{H_2O_t}$ are from Zhang and Behrens (2000), Freda et al. (2003), Behrens et al. (2004), and Liu et al. (2004). From Zhang et al. (2007).

$CO_{2,total}$, both derivatives in Equation 3-91a are independent of $CO_{2,total}$ concentration. That is, $D_{CO_{2,total}}$ is independent of $CO_{2,total}$, and the shape of the diffusion profiles of total CO_2 is similar to that of a simple nonreacting single species.

Assuming that CO_2 is the diffusing species, then

$$D_{CO_2, total} = D_{CO_2, molec} \frac{d[CO_{2, molec}]}{d[CO_{2, total}]}. \tag{3-91b}$$

If local equilibrium between $CO_{2,molec}$ and CO_3^{2-} is reached, then Equation 3-90d holds. Substituting Equation 3-90d into Equation 3-91b leads to

$$D_{CO_2, total} = D_{CO_2, molec} \frac{1}{1 + K[O^{2-}]}. \tag{3-91c}$$

Experimental data on CO_2 diffusion in rhyolitic to basaltic melts show that $D_{CO_2, total}$ is roughly independent of $CO_{2,total}$ concentration, consistent with the analyses above. However, experimental data also reveal a surprising result that $D_{CO_2, total}$ is also roughly independent of the anhydrous melt composition from rhyolite to basalt (Watson et al., 1982; Watson, 1991b; Nowak et al., 2004) (Figure 3-19b). CO_2 is the only component known to show this independence on anhydrous melt composition. The independence is even more interesting because CO_2 is mostly molecular CO_2 in rhyolite, but mostly CO_3^{2-} in basalt. Although $D_{CO_2, total}$ is almost independent of the anhydrous melt composition, it does depend on the dissolved H_2O content in the melt (Watson, 1991b).

The observed rough independence of $D_{CO_2, total}$ on anhydrous melt composition may be understood as follows. From Equation 3-91c, $D_{CO_2, total}$ increases as $D_{CO_2, molec}$ increases, but decreases as O^{2-} concentration increases. From rhyolite to basalt, diffusivity (such as Ar, H_2O, etc.) usually increases. Hence, we may expect that $D_{CO_2, molec}$ also increases from rhyolite to basalt, which would lead to an increase in $D_{CO_2, total}$. On the other hand, $[O^{2-}]$ increases from rhyolite to basalt, which would lead to a decrease in $D_{CO_2, total}$. Nowak et al. (2004) suggested that the rough independence of $D_{CO_2, total}$ on anhydrous melt composition is due to the coincidental cancellation of the two factors, one increasing and one decreasing $D_{CO_2, total}$.

In summary, the diffusion behavior of both H_2O and CO_2 demonstrates the importance of understanding the role of speciation in diffusion, and the very different consequences due to that role. Diffusion of a single-species component (such as Ar) usually does not depend on its own concentration (when the concentration is low), but depends on the melt composition. For a multispecies component, speciation affects the diffusion behavior. For H_2O, speciation makes the diffusion behavior very complicated: even at low H_2O concentrations, total H_2O diffusivity still depends on H_2O content (because the species concentrations are not proportional), in addition to the dependence on melt composition. If species concentrations are proportional to each other and hence to the total concentration of the component, then the diffusivity is independent of the concentration of the component, as in the case of CO_2 diffusion. Many multispecies components probably satisfy this condition that the concentrations of the major species are proportional to each other, leading to diffusivity that does not strongly depend on its own concentration. For the case of CO_2, in addition to concentration-independent diffusivity, there is another unique consequence: total CO_2 diffusivity is almost independent of the anhydrous melt composition. This simple diffusion behavior is explained to be a coincidence of both speciation and how species concentrations vary with anhydrous melt composition. The two cases (H_2O and CO_2) show the richness of the effect of speciation on diffusivity.

3.3.3 Diffusion of oxygen in melts and minerals

To investigate oxygen diffusion rate in minerals or melt, experiments of oxygen isotope diffusion have been carried under dry and hydrothermal conditions by using an ^{18}O-enriched fluid (such as $^{18}O_2$ or $C^{18}O_2$ under dry conditions or $H_2^{18}O$ under hydrothermal conditions). The surface of the mineral or melt is expected to be in isotopic equilibrium with the fluid phase, meaning high ^{18}O concentration at the surface. The high surface ^{18}O would diffuse into the mineral structure by exchanging with ^{16}O, constituting self-diffusion. However, water can also dissolve in the mineral (at the 100 ppm level) or melt (at the weight percent level) in hydrothermal experiments, and diffuse into the structure, carrying ^{18}O (because H_2O contains O). CO_2 or O_2 also dissolves in the mineral or melt, though at much smaller concentration, and diffuses into the structure. For the case of ^{18}O diffusion from a water vapor into a melt or mineral, as molecular H_2O (enriched in ^{18}O) diffuses into the structure, ^{18}O in molecular H_2O would exchange with structural oxygen, leading to more equitable $^{18}O/^{16}O$ ratio in different kinds of oxygen (oxygen in molecular H_2O, in OH, and in anhydrous structural oxygen). At equilibrium, $^{18}O/^{16}O$ ratio is roughly the same in all oxygen species (the difference is the isotopic fractionation factor, usually at per mil level). As molecular H_2O diffuses in further, it carries ^{18}O into the structure further. However, because of ^{18}O in H_2O exchanges with ^{16}O in other structural oxygen, $^{18}O/^{16}O$ is depleted rapidly in H_2O molecules diffusing further into the structure. Hence, the mean distance for H_2O diffusion and that for ^{18}O diffusion carried by H_2O can be very different (depending on H_2O concentration in the mineral or melt).

Therefore, oxygen diffusion may be through either ^{18}O–^{16}O exchange (self-diffusion), molecular H_2O diffusion, OH diffusion (both H_2O and OH contain oxygen and hence can carry ^{18}O), or molecular CO_2 or O_2 diffusion. That is, diffusion of oxygen in a mineral or melt may be pure self-diffusion, but may also be chemical diffusion of dissolved oxygen-bearing gas species. Oxygen isotopic diffusion is hence another example of multispecies diffusion.

It has been observed that water dramatically enhances oxygen diffusion in silicate melts as well as in many crystalline silicates. Furthermore, in systems studied in more detail, it has been observed that the diffusivity is roughly proportional to H_2O pressure (Farver and Yund, 1990), or H_2O content in the phase. This has been explained by the fact that the transport of ^{18}O is through the diffusion of molecular H_2O, rather than by self-diffusion of ^{18}O (meaning exchange of ^{18}O and ^{16}O). Zhang et al. (1991b) developed the theory for oxygen diffusion transported by a hydrous species such as H_2O_m, which has been experimentally confirmed (Behrens et al., 2007). The general diffusion equation for ^{18}O is

$$\frac{\partial C_{^{18}O}}{\partial t} = \frac{\partial}{\partial x}\left\{D_{O_{dry}}\frac{\partial C_{^{18}O_{dry}}}{\partial x} + D_{H_2O_m}\frac{\partial C_{H_2^{18}O_m}}{\partial x} + D_{OH}\frac{\partial C_{^{18}OH}}{\partial x}\right\} \tag{3-92}$$

$$D_{^{18}O} = D_{O_{dry}}X_{O_{dry}} + D_{OH}X_{OH} + D_{H_2O_m}X_{H_2O_m}. \tag{3-92a}$$

Assuming H_2O_m is the diffusing species, the diffusion equation for $R = {}^{18}O/({}^{16}O + {}^{17}O + {}^{18}O)$ is

$$\frac{\partial R}{\partial t} \approx \frac{\partial}{\partial x}\left(D_{H_2O_m}X_{H_2O_m}\frac{\partial R}{\partial x}\right). \tag{3-92b}$$

Therefore, given an expression of how $D_{H_2O_m}$ varies as a function of H_2O_t content, the H_2O_t profile can be solved from Equation 3-86 and ^{18}O profile can be solved from the above equation. The diffusivity of ^{18}O may be approximately expressed as

$$D_{^{18}O} \approx D_{H_2O_m}X_{H_2O_m}. \tag{3-92c}$$

If the same expression of $D_{H_2O_m}$ as a function of H_2O_t can match the profiles of both H_2O_t and ^{18}O in experiments simultaneously examining both ^{18}O and H_2O diffusion, it would be a verification of the hypothesis that H_2O_m is the diffusion species for both H_2O diffusion and ^{18}O diffusion. In Figure 3-20, simultaneous H_2O_t and R profiles from a single hydrothermal diffusion experiment in which $H_2^{18}O$ diffuses into a rhyolitic melt are shown and fit by the theory (solid curves). The almost perfect fits of both profiles and the agreement in D_0 and a obtained from fitting both profiles demonstrate that oxygen is indeed transported by molecular H_2O. The profile of R differs from the error function (dashed curve in Figure 3-20b).

In the above discussion, diffusion is considered to be the rate-limiting step, whereas the kinetics of the $^{18}O-{}^{16}O$ exchange reaction between different species or phases is assumed to be rapid. At relatively low temperatures, the kinetics of the exchange reactions may also limit the rate of the oxygen transport. Then both diffusion and reaction kinetics must be considered. Feng et al. (1993) conducted a detailed experimental study and developed theories to treat such a problem.

Many other components may also be present in silicate melts as two or more species. For example, although dry oxygen diffusivity is lumped together in Equation 3-92, oxygen diffusion in absolutely anhydrous systems may still depend on speciation: bridging oxygen, nonbridging oxygen, and free oxygen anion. Another more interesting example is the diffusion of sulfur. The component sulfur may be present in silicate melts as S^{2-}, SO_2, or SO_4^{2-}. These species

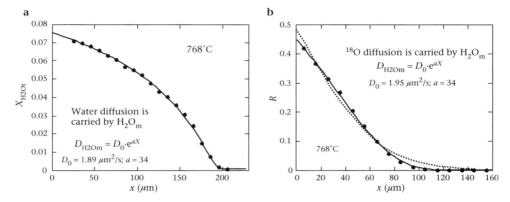

Figure 3-20 (a) H_2O profile and (b) ^{18}O profile during a simultaneous H_2O and ^{18}O diffusion experiment. The solid curves are fits by assuming that molecular H_2O is the diffusing species for both H_2O and oxygen with $D_{H_2O_m} = D_0 e^{aX}$, where $X = [H_2O_t]$ and $a = 34$. The high quality of the fits and similarity in D_0 values (1.89 $\mu m^2/s$ for H_2O profile and 1.95 $\mu m^2/s$ for ^{18}O profile) confirm the assumption. The dashed curve in (b) is an error function fit. From Behrens et al. (2007).

concentrations depend on oxygen fugacity but are proportional to one another. Hence, it is expected that the sulfur diffusion profile can be characterized by concentration-independent diffusion, but sulfur diffusivity depends strongly on f_{O_2}. Hence, understanding speciation and its role in diffusion will help us to understand the diffusion behavior.

3.4 Diffusion in a Multicomponent System

Most geological systems consist of more than two components. A natural silicate melt contains at least five major components (SiO_2, Al_2O_3, FeO, CaO, and Na_2O), and numerous minor and trace components. A natural garnet crystal may contain pyrope ($Mg_3Al_2Si_3O_{12}$), almandine ($Fe_3Al_2Si_3O_{12}$), spessartine ($Mn_3Al_2Si_3O_{12}$), grossular ($Ca_3Al_2Si_3O_{12}$), and/or andradite ($Ca_3Fe_2Si_3O_{12}$). In diffusion studies, systems with more than two components are known as multicomponent systems. Diffusion in such systems is considerably more complicated than that in a binary diffusion because every component also affects other components (Equations 3-23a to 3-23m). Although multicomponent diffusion theories are well developed (e.g., De Groot and Mazur, 1962; Kirkaldy and Young, 1987; Cussler, 1997), and although multicomponent diffusion effects in some synthetic ternary systems and a few quaternary systems have been experimentally and extensively investigated and hence are well understood, experiments to obtain multicomponent diffusion matrix for natural geological systems such as silicate melts is extremely difficult because of the many components involved and because of instrumental analytical errors (Trial and Spera, 1994). Without multicomponent diffusion matrices for treating such diffusion, multicomponent diffusion effects in silicate melts and minerals cannot be predicted. Hence, multicomponent diffusion in natural systems is usually treated using simplified approaches.

Uphill diffusion of some components is reported in silicate melts (e.g., Sato, 1975; Watson, 1982a; Zhang et al., 1989; Lesher, 1994; Van Der Laan et al., 1994). Recall that uphill diffusion in binary systems is rare and occurs only when the two-component phase undergoes spinodal decomposition. In multicomponent systems, uphill diffusion often occurs even when the phase is stable, and may be explained by cross-effects of diffusion by other components.

Several different ways have been developed to deal with diffusion in a multicomponent system. In order of increasing complexity and increasing accuracy, they are

- Effective binary approach,

- Activity-based effective binary approach,

- Concentration-based diffusivity matrix approach,

- Activity-based diffusivity matrix approach.

With each level of increased sophistication, the complexity of treating a diffusion problem increases tremendously, especially for natural silicate melts that have a large number of components. The simplest method (effective binary approach) is the most often used, but it cannot treat the diffusion of many components (such as those that show uphill diffusion). All the other methods have not been applied much to natural silicate melts.

Below, the effective binary approach and the concentration-based diffusivity matrix are introduced. The modified effective binary approach (Zhang, 1993) has not been followed up. The approach using the activity-based diffusivity matrix, similar to activity-based diffusivity $\mathcal{D}$ (Equations 3-61 and 3-62), is probably the best approach, but such diffusivities require systematic effort to obtain.

In this section, neutral oxide or endmember species are used as components. If ionic species are used as diffusion components, which as often the case for aqueous solutions, then one must also consider electroneutrality.

3.4.1 Effective binary approach

Fick's law is an empirical diffusion law for binary systems. For multicomponent systems, Fick's law must be generalized. There are several ways to generalize Fick's law to multicomponent systems. One simple treatment is called the effective binary treatment, in which Fick's law is generalized to a multicomponent system in the simplest way:

$$\mathbf{J}_i = -D_i \nabla C_i. \tag{3-93}$$

That is, diffusion of a component *i* in a multicomponent system is assumed to be independent of the presence of other components. In other words, all other components are treated as a combined "component" and the diffusion is viewed to be between the component itself and the combined "component." Therefore, this approach is known as the effective binary approach and the diffusivity obtained this way is known as effective binary diffusivity (or effective binary diffusion coefficient, EBDC). The diffusion equation using this approach is the same as the binary diffusion equation. Hence, solutions obtained before may be applied, and no specific mathematical development is necessary. Because of the simplicity of this approach and the complexity of the more rigorous approach discussed below, most authors have used this method in treating diffusion in geological liquids even though it is known to have limitations. One limitation is that uphill diffusion occurs often in multicomponent systems but is difficult to treat using the effective binary approach. If uphill diffusion profiles must be treated using the effective binary approach, one would have to allow the diffusivity to vary from positive to negative for these profiles. To avoid this, authors simply do not treat uphill diffusion profiles. Even though uphill diffusion in binary systems is easily predicted from thermodynamics (Figure 3-11), the conditions under which a component would display uphill diffusion cannot be predicted in multicomponent systems. Another limitation with the effective binary approach is that the effective binary diffusivities for a component extracted from experiments are overly sensitive to melt compositions and to the presence of concentration gradients of certain other components. For example, in a single uphill diffusion profile, effective binary diffusivities must vary by orders of magnitude and even to negative values. Even excluding uphill diffusion profiles, the extracted effective binary diffusivities may depend not only on the melt composition, but also on the directions and magnitudes of concentration gradients of other components (Equation 3-23). Because of such variations, monotonic profiles are usually fit by assuming a constant diffusivity and not worrying about the small details that are not well fit. One resulting drawback of this is that experimental diffusivities do not have general utility, and their use is limited to systems with similar compositions and similar concentration gradients.

Despite the various drawbacks, the effective binary approach is still widely used and will be widely applied to natural systems in the near future because of the difficulties of better approaches. For major components in a silicate melt, it is possible that multicomponent diffusivity matrices will be obtained as a function of temperature and melt composition in the not too distant future. For trace components, the effective binary approach (or the modified effective binary approach in the next section) will likely continue for a long time. The effective binary diffusion approach may be used under the following conditions (but is not limited to these conditions) with consistent and reliable results (Cooper, 1968):

(1) If the compositional difference along the diffusion direction is primarily in one component, and the difference in other components is due to the dilution effect, then diffusion of this component (not necessarily the other components) may be treated as effective binary, and the EBDC can be applied reliably to similar situations. Some examples are a diffusion couple made of dry rhyolite on one half and hydrous rhyolite on the other half, the hydration or dehydration of a silicate melt, and adsorption of a gas component by a glass.

(2) For a given set of conditions, e.g., for a basalt–rhyolite diffusion couple, or for the dissolution of a specific mineral in a specific melt, the direction and relative magnitude of concentration gradients of various components are roughly consistent, and the effective binary approach may be applied reliably to the component with the largest concentration gradient. For example, if the purpose is to estimate the MgO diffusion profile and olivine dissolution rate in an andesitic melt, EBDC extracted from olivine dissolution experiments in andesitic melt may be applied, but EBDC of MgO obtained from quartz dissolution in andesitic melt or from olivine dissolution in a basaltic melt may not be applicable to treat olivine dissolution in andesitic melt.

(3) If initial concentration gradients are in two complementing components, such as FeO and MgO (with the sum of FeO and MgO molar concentrations being a constant), and all other components have uniform concentration, the diffusion between the two components may be treated as interdiffusion, or effective binary diffusion.

In all of the three cases above, components other than the discussed major components often cannot be treated as effective binary diffusion, especially for a component whose concentration gradient is small or zero compared to other components. Such a component often shows uphill diffusion.

As long as care is taken so that effective binary diffusivity obtained from experiments under the same set of conditions is applied to a given problem, the approach works well. Although the limitations mean additional work, because of its simplicity and because of the unavailability of the diffusion matrices, the effective binary diffusion approach is the most often used in geological systems. Nonetheless, it is hoped that effort will be made in the future so that multi-component diffusion can be handled more accurately.

3.4.2 Modified effective binary approach

Zhang (1993) proposed the *modified effective binary approach* (also called *activity-based effective binary approach*). In this approach, the diffusive flux of a component is related to its activity gradient and all other components are treated as one combined component. The diffusive flux for any component i is expressed as (by analogy with Equation 3-61)

$$\mathbf{J}_i = - (\mathcal{D}_i/\gamma_i)\nabla a_i. \tag{3-94}$$

where $a_i = \gamma_i C_i$. The one-dimensional diffusion equation for component i is, hence,

$$\frac{\partial C_i}{\partial t} = \frac{\partial}{\partial x}\left(\frac{\mathcal{D}_i}{\gamma_i}\frac{\partial(\gamma_i C_i)}{\partial x}\right). \tag{3-95}$$

Zhang (1993) used this approach successfully to model concentration profiles from crystal dissolution experiments. The applicability needs to be investigated further. To establish this method, a major effort is necessary to extract and compile $\mathcal{D}$ values for geological applications.

3.4.3 Multicomponent diffusivity matrix (concentration-based)

A more rigorous way to generalize Fick's law is to use phenomenological equations based on linear irreversible thermodynamics. In this treatment of an N-component system, the diffusive flux of component i is (De Groot and Mazur, 1962)

$$\mathbf{J}_i = -\sum_{k=1}^{N} L_{ik}\nabla\mu_k = -\sum_{k=1}^{n}\left(L_{ik} - \frac{X_k}{X_n}L_{iN}\right)\nabla\mu_k = -\sum_{k=1}^{n}L'_{ik}\nabla\mu_k, \tag{3-96a}$$

where L_{ik} are the Onsager phenomenological coefficients, μ is the chemical potential, X is mole fraction, N is the number of components, and $n = N-1$. The second equality in Equation 3-96a is derived by using the Gibbs-Duhem relation:

$$\sum X_k \nabla\mu_k = 0,$$

where the summation is from 1 to N, which is used to cancel $\nabla\mu_N$. In an N-component system, only $n\ (=N-1)$ components are independent because

$$\sum \alpha_k \mathbf{J}_k = 0, \tag{3-96b}$$

in which $\alpha_k(k$ varies from 1 to N) are constants and depend on the choice of reference frame. If the diffusing species or components are ionic, then electroneutrality must also be considered. Using Equation 3-96b to eliminate $\mathbf{J}_N$, we can also write $\mathbf{J}_i$ as a function of the following independent "forces":

$$\mathbf{J}_i = -\sum_{k=1}^{m}L_{ik}\nabla\left(\mu_k - \frac{\alpha_k}{\alpha_N}\mu_N\right), \tag{3-96c}$$

where L_{ik} are the same as those in Equation 3-96a.

The entropy production during diffusion in a multicomponent system is (Appendix 1)

$$\sigma = \sum_{i=1}^{N}\sum_{j=1}^{N}L_{ij}\nabla\mu_i\nabla\mu_j = \sum_{i=1}^{n}\sum_{j=1}^{n}L_{ij}\nabla\left(\mu_i - \frac{\alpha_i}{\alpha_N}\mu_N\right)\nabla\left(\mu_j - \frac{\alpha_j}{\alpha_N}\mu_N\right), \tag{3-96d}$$

where the n by n matrix made of L_{ik} is called the *phenomenological coefficient matrix* **L**. Because $\sigma > 0$ if the gradients are not zero, **L** is positive definite if the phase is stable with respect to spinodal decomposition. Furthermore, the **L** matrix is symmetric based on Onsager's reciprocal principle (De Groot and Mazur, 1962).

Equation 3-96c can be written in terms of concentration gradients:

$$\mathbf{J}_i = -\sum_{k=1}^{n} D_{ik}\boldsymbol{\nabla}C_k \tag{3-96e}$$

where D_{ik} is the diffusion coefficient for component i due to concentration gradient in component k. The n by n matrix composed of D_{ik} is known as the *diffusivity matrix* **D**. D_{ii} is known as the *on-diagonal diffusivity* and D_{ik} when $i \neq k$ is known as *off-diagonal diffusivity* or *cross-diffusivity*. The diffusivity matrix is different from the second-rank tensor for diffusivity in an anisotropic medium. The former describes diffusion of all components in a multicomponent system, but the latter describes diffusion of one component in an anisotropic medium. The former is an n by n matrix, where n can be any positive integer (no more than the number of elements in a system), but the latter is always 3 by 3. The diffusivity matrix is not symmetric, but the second rank tensor is symmetric. Each element in the diffusivity matrix is itself a tensor if the medium is anisotropic.

Choosing an appropriate reference frame such as $\alpha_k = 1$ for all k, the diffusivity matrix can be written as

$$\mathbf{D} = \mathbf{LY} \tag{3-97a}$$

where **Y** is a symmetric matrix defined as

$$Y_{ij} = \frac{\partial^2 G}{\partial X_i \partial X_j}, \tag{3-97b}$$

where G is Gibbs free energy. If the phase is stable, then **Y** is positive definite. It can be shown that **D**, as the product of **L** and **Y** (both are symmetric and positive definite) is also positive definite but not necessarily symmetric. That is, all eigenvalues of **D** (denoted as $\lambda_1, \lambda_2, \ldots, \lambda_n$) are real and positive if the phase is stable.

Using the diffusivity matrix, the diffusion equation for component i can be written in the following form:

$$\frac{\partial C_i}{\partial t} = -\boldsymbol{\nabla}\mathbf{J}_i = \sum_{j=1}^{n} \boldsymbol{\nabla}(D_{ij}\boldsymbol{\nabla}C_j). \tag{3-98a}$$

Consider the case of one-dimensional diffusion and every element D_{ij} in **D** being a constant (otherwise the treatment would be much more complicated than below). The above equation can be simplified to

$$\frac{\partial C_i}{\partial t} = \sum_{j=1}^{n} D_{ij}\frac{\partial^2 C_j}{\partial x^2}. \tag{3-98b}$$

That is, the concentration profiles of all components are coupled. In matrix notation, Equation 3-98b can be written as

$$\frac{\partial \mathbf{C}}{\partial t} = \mathbf{D}\frac{\partial^2 \mathbf{C}}{\partial x^2},$$ (3-99)

in which $\mathbf{C}$ is the concentration vector and $\mathbf{D}$ is the diffusion coefficient matrix:

$$\mathbf{C} = \begin{pmatrix} C_1 \\ C_2 \\ \vdots \\ C_n \end{pmatrix}, \quad \mathbf{D} = \begin{pmatrix} D_{11} & D_{12} & \cdots & D_{1n} \\ D_{21} & D_{22} & \cdots & D_{2n} \\ \vdots & \vdots & \cdots & \vdots \\ D_{n1} & D_{n2} & \cdots & D_{nn} \end{pmatrix},$$ (3-99a)

Because the $\mathbf{D}$ matrix always has n real and positive eigenvalues, it can be diagonalized in the following fashion:

$$\mathbf{D} = \mathbf{T\Lambda T}^{-1}.$$ (3-99b)

where Λ is a diagonal matrix whose diagonal elements are the eigenvalues of matrix $\mathbf{D}$, and $\mathbf{T}$ is a matrix composed of eigenvectors of the $\mathbf{D}$ matrix. Each eigenvector, that is, each column in $\mathbf{T}$, can be made a unit vector. Replacing Equation 3-99b into Equation 3-99, and multiplying both sides by $\mathbf{T}^{-1}$ leads to

$$\frac{\partial \mathbf{T}^{-1}\mathbf{C}}{\partial t} = \Lambda\frac{\partial^2 \mathbf{T}^{-1}\mathbf{C}}{\partial x^2}.$$ (3-99c)

Let $\mathbf{w} = \mathbf{T}^{-1}\mathbf{C}$; then

$$\frac{\partial \mathbf{w}}{\partial t} = \Lambda\frac{\partial^2 \mathbf{w}}{\partial x^2}.$$ (3-99d)

Since Λ is a diagonal matrix, the above matrix equation is equivalent to

$$\frac{\partial w_i}{\partial t} = \lambda_i\frac{\partial^2 w_i}{\partial x^2} \quad (i = 1, 2, \ldots, n).$$ (3-99e)

Therefore, in the transformed components, the diffusion is decoupled, meaning that the diffusion of one component is independent of the diffusion of other components. The equation for each w_i can be obtained given initial and boundary conditions using the solutions for binary diffusion. The final solution for $\mathbf{C}$ is $\mathbf{C} = \mathbf{Tw}$. When the diffusivity matrix is not constant, the diffusion equation for a multicomponent system can only be solved numerically.

Values of a diffusion coefficient matrix, in principle, can be determined from multicomponent diffusion experiments. For ternary systems, the diffusivity matrix is 2 by 2, and there are four values to be determined for a matrix at each composition. For quaternary systems, there are nine unknowns to be determined. For natural silicate melts with many components, there are many unknowns to be determined from experimental data by fitting experimental diffusion profiles. When there are so many unknowns, the fitting of experimental concentration profiles is not a trivial task and there are many numerical problems. One serious numerical problem is that experimental errors in concentration profiles tend to amplify in the process of extracting diffusivity matrix by fitting concentration profiles. Special schemes have been proposed to experimentally extract the multicomponent diffusivity matrix of natural silicate melts (Trial and Spera, 1994), and bold efforts have been made to extract such a diffusivity matrix (Kress and Ghiorso, 1995; Mungall et al., 1998), but such matrices have not been verified. So far, the multicomponent diffusion theory is most successfully applied to ternary systems, as discussed below.

3.4.3.1 Solving the diffusion profiles given the diffusivity matrix (ternary systems)

In a ternary system, the **D** matrix is 2 by 2:

$$\mathbf{D} = \begin{pmatrix} D_{11} & D_{12} \\ D_{21} & D_{22} \end{pmatrix}. \tag{3-100a}$$

Suppose λ is an eigenvalue and $\mathbf{t}$ is an eigenvector; by definition,

$$\begin{pmatrix} D_{11} & D_{12} \\ D_{21} & D_{22} \end{pmatrix} \mathbf{t} = \lambda \mathbf{t}. \tag{3-100b}$$

Hence,

$$\begin{pmatrix} D_{11} - \lambda & D_{12} \\ D_{21} & D_{22} - \lambda \end{pmatrix} \mathbf{t} = \mathbf{0}. \tag{3-100c}$$

Therefore, the two eigenvalues λ_1 and λ_2 may be solved from the following quadratic equation:

$$\begin{vmatrix} D_{11} - \lambda & D_{12} \\ D_{21} & D_{22} - \lambda \end{vmatrix} = 0. \tag{3-100d}$$

The fact that the **D** matrix is positive definite means that λ_1 and λ_2 are real and positive. The above equation can be written as

$$\lambda^2 - (D_{11} + D_{22})\lambda + (D_{11}D_{22} - D_{12}D_{21}) = 0. \tag{3-100e}$$

Therefore,

$$\lambda = \frac{1}{2}\left[(D_{11} + D_{22}) \pm \sqrt{(D_{11} + D_{22})^2 - 4(D_{11}D_{22} - D_{12}D_{21})} \right]. \tag{3-100f}$$

For λ_1 and λ_2 both to be real and positive requires that

$$\lambda_1 + \lambda_2 = D_{11} + D_{22} > 0, \tag{3-100g}$$

$$\lambda_1 \lambda_2 = D_{11}D_{22} - D_{12}D_{21} > 0, \tag{3-100h}$$

$$(D_{11} + D_{22})^2 \geq 4(D_{11}D_{22} - D_{12}D_{21}), \tag{3-100i}$$

where the last condition insures that λ_1 and λ_2 are real and the first two conditions insure that λ_1 and λ_2 are positive. If one extracts a diffusion coefficient
matrix for a ternary system from experimental data, the first step is to check if the
matrix elements satisfy the above conditions. If they do not (some literature data
have this problem), the matrix should be discarded.

To solve a diffusion equation, one needs to diagonalize the **D** matrix. This is
best done with a computer program. For a ternary system, one can find the two
eigenvalues by solving the quadratic Equation 3-100e. The two vectors of matrix
T can then be found by solving

$$\begin{pmatrix} D_{11} & D_{12} \\ D_{21} & D_{22} \end{pmatrix} \begin{pmatrix} t_{11} \\ t_{21} \end{pmatrix} = \lambda_1 \begin{pmatrix} t_{11} \\ t_{21} \end{pmatrix} \tag{3-101a}$$

and

$$\begin{pmatrix} D_{11} & D_{12} \\ D_{21} & D_{22} \end{pmatrix} \begin{pmatrix} t_{12} \\ t_{22} \end{pmatrix} = \lambda_2 \begin{pmatrix} t_{12} \\ t_{22} \end{pmatrix}. \tag{3-101b}$$

The solution is

$$\begin{pmatrix} t_{11} \\ t_{21} \end{pmatrix} = \begin{pmatrix} D_{12} \\ \lambda_1 - D_{11} \end{pmatrix}, \tag{3-101c}$$

$$\begin{pmatrix} t_{12} \\ t_{22} \end{pmatrix} = \begin{pmatrix} \lambda_2 - D_{22} \\ D_{21} \end{pmatrix}. \tag{3-101d}$$

Because Equation 3-101a represents a set of homogeneous linear equations,
multiplying the solution by a positive or negative factor is still a solution.
Therefore, each column vector in Equation 3-101c and 3-101d can be made a unit
vector. Then the matrix **T** is obtained. With this matrix known, diffusion profiles
can be calculated by solving Equation 3-99c.

Below is a numerical example. Consider a three-component system MgO–
Al_2O_3–SiO_2 with MgO being component 1, Al_2O_3 being component 2, and SiO_2
being the dependent component. The diffusivity matrix (in arbitrary unit) is

$$\mathbf{D} = \begin{pmatrix} D_{11} & D_{12} \\ D_{21} & D_{22} \end{pmatrix} = \begin{pmatrix} 3 & 0.75 \\ 1 & 2 \end{pmatrix}. \tag{3-102a}$$

The two eigenvectors are solved from

$$\lambda^2 - 5\lambda + 5.25 = 0. \tag{3-102b}$$

The solution is

$$\lambda_1 = 3.5 \tag{3-102c}$$

and

$$\lambda_2 = 1.5. \tag{3-102d}$$

The eigenvector corresponding to eigenvalue λ_1 is

$$\begin{pmatrix} t_{11} \\ t_{21} \end{pmatrix} = \begin{pmatrix} D_{12} \\ \lambda_1 - D_{11} \end{pmatrix} = \begin{pmatrix} 0.75 \\ 0.5 \end{pmatrix} \sim \begin{pmatrix} 3/\sqrt{13} \\ 2/\sqrt{13} \end{pmatrix}. \tag{3-102e}$$

The eigenvector corresponding to eigenvalue λ_2 is

$$\begin{pmatrix} t_{12} \\ t_{22} \end{pmatrix} = \begin{pmatrix} \lambda_2 - D_{22} \\ D_{21} \end{pmatrix} = \begin{pmatrix} -0.5 \\ 1 \end{pmatrix} \sim \begin{pmatrix} -1/\sqrt{5} \\ 2/\sqrt{5} \end{pmatrix}. \tag{3-102f}$$

The **T** matrix is, hence,

$$\mathbf{T} = \begin{pmatrix} t_{11} & t_{12} \\ t_{21} & t_{22} \end{pmatrix} = \begin{pmatrix} 3/\sqrt{13} & -1/\sqrt{5} \\ 2/\sqrt{13} & 2/\sqrt{5} \end{pmatrix}. \tag{3-102g}$$

The inverse $\mathbf{T}^{-1}$ matrix is

$$\mathbf{T}^{-1} = \begin{pmatrix} q_{11} & q_{12} \\ q_{21} & q_{22} \end{pmatrix} = \begin{pmatrix} \sqrt{13}/4 & \sqrt{13}/8 \\ -\sqrt{5}/4 & 3\sqrt{5}/8 \end{pmatrix}. \tag{3-102h}$$

Defining vector $\mathbf{w} = \mathbf{T}^{-1}\mathbf{C}$, then

$$\begin{aligned} \mathbf{w} = \begin{pmatrix} w_1 \\ w_2 \end{pmatrix} = \mathbf{T}^{-1}\mathbf{C} &= \begin{pmatrix} q_{11} & q_{12} \\ q_{21} & q_{22} \end{pmatrix} \begin{pmatrix} \text{MgO} \\ \text{Al}_2\text{O}_3 \end{pmatrix} \\ &= \begin{pmatrix} \sqrt{13}/4 & \sqrt{13}/8 \\ -\sqrt{5}/4 & 3\sqrt{5}/8 \end{pmatrix} \begin{pmatrix} \text{MgO} \\ \text{Al}_2\text{O}_3 \end{pmatrix}. \end{aligned} \tag{3-103a}$$

Numerically,

$$\begin{aligned} \mathbf{w} = \begin{pmatrix} w_1 \\ w_2 \end{pmatrix} &= \begin{pmatrix} (\sqrt{13}/4)\text{MgO} + (\sqrt{13}/8)\text{Al}_2\text{O}_3 \\ -(\sqrt{5}/4)\text{MgO} + (3\sqrt{5}/8)\text{Al}_2\text{O}_3 \end{pmatrix} \\ &\approx \begin{pmatrix} 0.90\text{MgO} + 0.45\text{Al}_2\text{O}_3 \\ -0.56\text{MgO} + 0.84\text{Al}_2\text{O}_3 \end{pmatrix} \end{aligned} \tag{3-103b}$$

The first eigencomponent w_1 is mostly MgO, and the second eigencomponent w_2 is mostly Al_2O_3. Now we need the specific initial and boundary conditions to finish the calculation of the concentration profiles. Suppose this is a diffusion-couple problem. Initially (MgO, Al_2O_3) = (0.25, 0.35) at $x < 0$, and (MgO, Al_2O_3) = (0.35, 0.25) at $x > 0$. The solutions for w_1 and w_2 are

$$w_1 = 0.5(w_{1,-\infty} + w_{1,\infty}) + 0.5(w_{1,\infty} - w_{1,-\infty})\text{erf}[x/(4\lambda_1 t)^{1/2}], \tag{3-103c}$$

$$w_2 = 0.5(w_{2,-\infty} + w_{2,\infty}) + 0.5(w_{2,\infty} - w_{2,-\infty})\text{erf}[x/(4\lambda_2 t)^{1/2}], \tag{3-103d}$$

The final solution is

$$\begin{aligned} \mathbf{C} = \begin{pmatrix} \text{MgO} \\ \text{Al}_2\text{O}_3 \end{pmatrix} = \mathbf{Tw} &= \begin{pmatrix} t_{11} & t_{12} \\ t_{21} & t_{22} \end{pmatrix} \begin{pmatrix} w_1 \\ w_2 \end{pmatrix} \\ &= \begin{pmatrix} (3/\sqrt{13})w_1 + (-1/\sqrt{5})w_2 \\ (2/\sqrt{13})w_1 + (2/\sqrt{5})w_2 \end{pmatrix}. \end{aligned} \tag{3-103e}$$

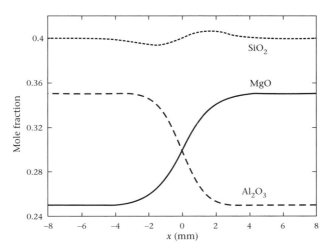

Figure 3-21 Calculated diffusion profiles for a diffusion couple in a ternary system. The diffusivity matrix is given in Equation 3-102a. The fraction of SiO_2 is calculated as $1 - MgO - Al_2O_3$. SiO_2 shows clear uphill diffusion. A component with initially uniform concentration (such as SiO_2 in this example) almost always shows uphill diffusion in a multi-component system.

The profiles are plotted in Figure 3-21. It can be seen that SiO_2 shows uphill diffusion.

3.4.3.2 Extracting the diffusivity matrix from diffusion profiles (ternary systems)

To extract diffusion matrix from experimental concentration profiles of multi-component diffusion is trickier. The best way is to do two diffusion-couple experiments under identical temperature, pressure, and average (or interface) composition but with very different, preferably orthogonal, concentration gradients. This can be done through choosing end-members of the diffusion couples as shown in Figure 3-22: in one diffusion couple, the end-member compositions are "a" and "b", whereas in the other diffusion couple, the end-member compositions are "c" and "d". The bulk (or interface) compositions of the two experiments are the same. Furthermore, the compositional difference across each diffusion couple should not be too large, otherwise another complexity (compositional dependence of diffusivity) would be introduced. On the other hand, the compositional difference should be much larger than the measurement uncertainty, otherwise the measured profiles would be too noisy. For example, the concentration difference between the two halves of each diffusion couple should be at least 10σ, where σ is the standard error of measurement. With two such diffusion-couple experiments, one may use Boltzmann

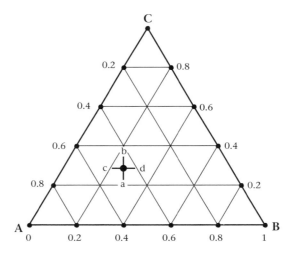

Figure 3-22 A diagram for the representation of compositions in a ternary system, with two hypothetical diffusion couples: a-b and c-d. The compositional gradient of the two diffusion couples are orthogonal to each other. For a given point inside the triangle, to find the fraction of a component (such as A), first draw a straight line parallel to BC, and then find where the straight line intersects the CA segment (with fraction indicated on the CA segment).

analysis (Section 3.2.4.2) to extract the four diffusion coefficients in the 2 by 2 diffusivity matrix as below.

Start from the diffusion equations in a ternary system:

$$\frac{\partial C_1}{\partial t} = \frac{\partial}{\partial x}\left(D_{11}\frac{\partial C_1}{\partial x} + D_{12}\frac{\partial C_2}{\partial x}\right), \tag{3-104a}$$

$$\frac{\partial C_2}{\partial t} = \frac{\partial}{\partial x}\left(D_{21}\frac{\partial C_1}{\partial x} + D_{22}\frac{\partial C_2}{\partial x}\right), \tag{3-104b}$$

Boltzmann transformation of the above leads to

$$D_{11}\frac{dC_1}{d\eta}\Big|_{\eta_0} + D_{12}\frac{\partial C_2}{d\eta}\Big|_{\eta_0} = 2\int_{C_1(\eta_0)}^{C_1(\infty)} \eta \, dC_1 \tag{3-104c}$$

$$D_{21}\frac{dC_1}{d\eta}\Big|_{\eta_0} + D_{22}\frac{\partial C_2}{d\eta}\Big|_{\eta_0} = 2\int_{C_2(\eta_0)}^{C_2(\infty)} \eta \, dC_2 \tag{3-104d}$$

Therefore, from each experiment, there are two equations relating diffusivities to integrals and slopes of the two concentration profiles. With two experiments (the compositions should be orthogonal or nearly so; Figure 3-22), there are four equations that can be used to solve for the four unknowns of the **D** matrix. More experiments along various directions would provide redundancy and constrain the **D** matrix better. The solution for the **D** matrix has to satisfy Equations 3-100g,h,i if the phase is stable with respect to spinodal decomposition.

Similar to binary diffusivities, each element in the diffusivity matrix is expected to depend on composition, sometimes strongly, especially for highly nonideal systems. If the nonideality is strong enough to cause a miscibility gap, the eigenvalues would vary from positive to zero and to negative. If there is no miscibility gap, the eigenvalues are positive but can still vary with composition.

3.4.4 Multicomponent diffusivity matrix (activity-based)

Writing the diffusive flux of a component *i* in terms of activity gradients of all independent components in an *N*-component system, the flux equation is

$$\mathbf{J}_i = -\sum_{k=1}^{n} \frac{\mathcal{D}_{ik}}{\gamma_k} \boldsymbol{\nabla} a_k, \tag{3-105a}$$

where $n = N - 1$, and $\mathcal{D}_{ik}$ is the "intrinsic" diffusion coefficient for component *i* due to activity gradient in component *k*. Using the activity-based diffusivity matrix, the diffusion equation may be written in the following form:

$$\frac{\partial C_i}{\partial t} = -\boldsymbol{\nabla}\mathbf{J}_i = \sum_{j=1}^{n} \boldsymbol{\nabla}\left(\frac{\mathcal{D}_{ij}}{\gamma_j} \boldsymbol{\nabla} a_j\right). \tag{3-105b}$$

Similar to binary systems, the intrinsic diffusivity matrix is expected to be much less dependent on the composition of the system, which is the main advantage of using the activity-based diffusivity matrix. To solve the above equation, it is necessary to know the relations between the activity of every component and all concentrations.

The set of diffusion equations based on the activity-based diffusivity matrix is expected to be complicated and solvable only numerically. Because of its complexity, no attempt has been made using this formulation to obtain the "intrinsic" diffusivity matrix.

3.4.5 Concluding remarks

Most diffusion processes encountered in Earth sciences are, strictly speaking, multicomponent diffusion. For example, even "self"-diffusion of oxygen isotopes from an ^{18}O-enriched hydrothermal fluid into a mineral is likely due to chemical diffusion of H_2O into the mineral (see Section 3.3.3). Because a natural melt contains at least five major components and many trace components, diffusion in nature is complicated to treat. For multicomponent and anisotropic minerals, the situation is even worse due to additional complexity of anisotropy. No reliable diffusivity matrix for natural silicate melts and minerals is available yet. If one insists on rigor in treating such diffusion problems, one would get nowhere in understanding diffusion in nature. Therefore, simplification is almost always made, most often using the effective binary approach.

The effective binary approach is sometimes excellent, and often not good enough. These points are summarized here again:

(1) If the difference in concentration is in one component only, e.g., one side contains a dry rhyolite, and the other side is prepared by adding H_2O to the rhyolite, then the main concentration gradient is in H_2O, and all other components have smaller concentration gradients. The diffusion of H_2O may be treated fairly accurately by effective binary diffusion. In other words, the diffusion of the component with the largest concentration gradient may be treated as effective binary, especially if the component also has high diffusivity. The diffusion of other components in the system may or may not be treated as effective binary diffusion.

(2) If the difference in concentration is in only two exchangeable components, such as FeO and MgO, the interdiffusion in a multicomponent system may be treated as effective binary. The diffusion of other components in the system may or may not be treated as effective binary diffusion.

(3) The diffusion of a component whose concentration gradient is the largest (i.e., ΔC between the ends is the largest) usually can be treated as binary diffusion. The diffusion of other components in the system may or may not be treated as effective binary diffusion.

(4) The components whose concentration gradient is small compared to other components cannot be treated as effective binary. They often show uphill diffusion.

For a complete description of the diffusion process, the diffusion matrix approach is necessary as in the following examples:

(1) Given a diffusion couple of different melts (e.g., basalt and andesite), if one wants to predict the diffusion behavior of all major components, the diffusion matrix approach is necessary.

(2) For the dissolution of a crystal into a melt, if one wants to predict the interface melt composition (that is, the composition of the melt that is saturated with the crystal), the dissolution rate, and the diffusion profiles of all major components, thermodynamic understanding coupled with the diffusion matrix approach is necessary (Liang, 1999). If the effective binary approach is used, it would be necessary to determine which is the principal equilibrium-determining component (such as MgO during forsterite dissolution in basaltic melt), estimate the concentration of the component at the interface melt, and then calculate the dissolution rate and diffusion profile. To estimate the interface concentration of the principal component from thermodynamic equilibrium, because the concentration depends somewhat on the concentrations of other components, only

with the multicomponent approach would it be possible to estimate the concentration accurately.

(3) Similarly, for the general problem of crystal growth in silicate melts, it is necessary to use the diffusion matrix approach.

Experimental determination of the diffusion coefficient matrix is time-consuming and labor-intensive. Nonetheless, diffusion studies have advanced significantly in recent years. Hence, with persistence and concerted effort, it is possible that reliable and reproducible diffusivity matrices for major components in some natural melts will become available in the near future.

In principle, the diffusion matrix approach can be extended to trace elements. My assessment, however, is that in the near future diffusion matrix involving 50 diffusing components will not be possible. Hence, simple treatment will still have to be used to roughly understand the diffusion behavior of trace elements: the effective binary diffusion model to handle monotonic profiles, the modified effective binary diffusion model to handle uphill diffusion, or some combination of the diffusion matrix and effective binary diffusion model.

3.5 Some Special Diffusion Problems

Many geological diffusion problems are complicated in one way or another. The diffusing component may participate in either homogeneous or heterogeneous reactions, such as the diffusion of a radioactive component, the absorption of a component that also reacts with the framework, crystal growth, and dissolution. This section covers some of these problems. The first class of problems to be discussed is diffusion involving homogeneous reactions, including the diffusion of a radioactive or radiogenic component. Diffusion of a multispecies component discussed in Section 3.3 also belongs to this class. The reaction may consume the component (that is, there is a sink for the component) or produce it (that is, there is a source for the component). The diffusion equation in such cases must include a term (or more terms if necessary) due to reactions. The most important case in this kind of diffusion problems is the diffusion of a radiogenic component, which forms the basis of thermochronology. The second class of problems is moving-boundary problems, often encountered in crystal dissolution and growth. The third class of problems involves diffusion and fluid flow such as pollutant transport along a river. Only the principles are briefly addressed in this section. Applications will be tackled in later respective sections, sometimes with great detail. For example, applications of the moving-boundary problem to crystal growth and dissolution, bubble growth and dissolution, including the possible presence of convection, are the subject of Chapter 4 on heterogeneous reaction kinetics, and applications of diffusive loss of radiogenic nuclide to thermochronology are the subject of Chapter 5 on inverse problems.

3.5.1 Diffusion of a radioactive component

The diffusion of a radioactive component is a relatively easy problem. It is discussed here to illustrate how coupled diffusion and homogeneous reaction can be treated, and to prepare for the more difficult problem of the diffusion of a radiogenic component. The diffusion of a radiogenic component, which is dealt with in Section 3.5.2, is an important geological problem because of its application in geochronology and thermochronology.

A radioactive component is consumed by its decay (homogeneous reaction). For example, one-dimensional diffusion of ^{238}U in zircon along the crystallographic axis **c** or along any direction in the **a-b** plane can be described by

$$\frac{\partial C}{\partial t} = D\frac{\partial^2 C}{\partial x^2} - \lambda C \tag{3-106}$$

where C is the concentration of ^{238}U in mol/m^3, D is the diffusion coefficient (assumed to be constant), and λ is the decay constant. The first term on the right-hand side of Equation 3-106 represents concentration change due to diffusion, and the second term represents concentration decrease due to decay. This diffusion equation can be converted to the simplest form of diffusion equation (and hence can be solved using methods introduced before) by the following procedure. Multiplying both sides of Equation 3-106 by $e^{\lambda t}$, we have

$$e^{\lambda t}\frac{\partial C}{\partial t} = D\frac{\partial^2 Ce^{\lambda t}}{\partial x^2} - \lambda Ce^{\lambda t}. \tag{3-106a}$$

Because

$$\frac{\partial(Ce^{\lambda t})}{\partial t} = e^{\lambda t}\frac{\partial C}{\partial t} + \lambda Ce^{\lambda t}, \tag{3-106b}$$

Equation 3-106a becomes

$$\frac{\partial(Ce^{\lambda t})}{\partial t} = D\frac{\partial^2 Ce^{\lambda t}}{\partial x^2}. \tag{3-106c}$$

Let $w = Ce^{\lambda t}$; then

$$\frac{\partial w}{\partial t} = D\frac{\partial^2 w}{\partial x^2}. \tag{3-106d}$$

Because the above equation is identical to the diffusion equation of a stable component, it can be solved the same way. After solving for w, then C can be found as $we^{-\lambda t}$. For diffusion of two isotopes, one stable and one radioactive, because they have the same diffusivity, the concentration profile for the radioactive nuclide is simply the concentration profile of the stable isotope multiplied by either (i) $F_0e^{-\lambda t}$, where F_0 is the initial isotopic ratio, or (ii) F, where F is the isotopic ratio at the time of measurement of the profiles.

3.5.2 Diffusion of a radiogenic component and thermochronology

Accurate dating of the formation age (or age at the peak temperature) of a rock requires that the system be closed to both the parent and the daughter nuclides right after the formation. Many igneous or metamorphic rocks formed at high temperatures underwent gradual cooling, during which there was often loss of the radiogenic daughter nuclide.

The diffusive loss of the daughter nuclide (and sometimes the parent nuclide) makes dating much more complicated, because the meaning of the age becomes obscure. Geochemists have turned this complexity to an advantage to infer more information from the isotopic systems. Because diffusion rate depends strongly on temperature, it is possible to infer the thermal history of the rock by detailed geochronologic studies. Because the isotopic systems both provide the age and constrain the temperature, they are the best geospeedometers to be applied to infer thermal history of rocks.

Consider the diffusion of ^{40}Ar in hornblende. Hornblende is anisotropic but the anisotropy is ignored here. Along a principal axis x, the one-dimensional diffusion can be described by

$$\frac{\partial(^{40}\text{Ar})}{\partial t} = D_{\text{Ar}} \frac{\partial^2(^{40}\text{Ar})}{\partial x^2} + \lambda_e(^{40}\text{K}), \tag{3-107a}$$

where D_{Ar} is the diffusivity of Ar in hornblende, and λ_e is the branch decay constant of ^{40}K to ^{40}Ar. The concentration profile of ^{40}K can be solved from Equation 3-106:

$$\frac{\partial(^{40}\text{K})}{\partial t} = D_K \nabla^2(^{40}\text{K}) - \lambda(^{40}\text{K}). \tag{3-107b}$$

where D_K is the diffusivity of K in hornblende and λ is the decay constant of ^{40}K. Note that D_{Ar} is expected to be much larger than D_K, and that $\lambda_e = 0.1048\lambda$. The first step is to solve for the concentration change of ^{40}K with time. Because K is structurally incorporated in the mineral, and because K diffusivity is smaller than Ar diffusivity, the loss or gain of ^{40}K may be ignored, though there could be exchange between Na and K in some cases.

Substituting the solution for ^{40}K into Equation 3-107a allows the ^{40}Ar profile to be solved, most likely numerically. Assume uniform initial K concentration and ignore its diffusion. Then $\lambda_e(^{40}\text{K}) = \lambda_e(^{40}\text{K}_0)e^{-\lambda t}$, and Equation 3-107a becomes

$$\frac{\partial(^{40}\text{Ar})}{\partial t} = D_{\text{Ar}} \frac{\partial^2(^{40}\text{Ar})}{\partial x^2} + \lambda_e {}^{40}\text{K}_0 e^{-\lambda t}, \tag{3-107c}$$

which has been solved by Dodson (1973) for the evolution of ^{40}Ar concentration as a function of time, from which the *closure temperature* and *closure age* are defined. In this section, these concepts are quantitatively elucidated using a simple approach.

As introduced in Section 1.7.3 and Figure 1-20, because of diffusive loss of Ar, the "age" obtained from the K–Ar system is not necessarily the real age (formation age or peak temperature age), but is an *apparent age* (t_a) as defined by Equation 1-114:

$$t_a = (1/\lambda_{40})\ln[1 + (^{40}\text{Ar}/^{40}\text{K})/(\lambda_e/\lambda_{40})], \tag{1-114}$$

where initial ^{40}Ar is assumed to be zero, and ^{40}Ar and ^{40}K are the present content of these species. If the concentrations of ^{40}Ar and ^{40}K are for the whole mineral, the apparent age and closure temperature are for the whole mineral. If the concentrations of ^{40}Ar and ^{40}K are measured at a single point in a mineral (such as the center of a mineral), then the apparent age and the closure temperature are for that point. The apparent age corresponds to the formation age if there was no argon loss subsequent to the formation of the mineral, as for volcanic rocks (which cool rapidly on the surface of the Earth). For plutonic and metamorphic rocks, cooling is slow (Figure 1-18) and Ar loss at high temperature may be significant. Hence, the apparent age does not mean the formation age, or the peak temperature age. Instead, the apparent age corresponds to the time when the system temperature was at a temperature called the *closure temperature*. Hence, the apparent age is also referred to as the *closure age* even though the system did not become completely closed at the time of closure age.

To estimate the closure temperature, it is necessary to estimate the diffusion distance. From earlier results (Section 3.2.8.1), for asymptotic cooling from the closure temperature (Equation 3-55),

$$T = T_c/(1 + t/\tau_c), \tag{3-108a}$$

where τ_c is the cooling timescale, and for D expressed as

$$D = Ae^{-E/(RT)} = D_c e^{-t/\tau}, \tag{3-108b}$$

where E and A are the activation energy and the preexponential factor for diffusion, R is the gas constant, D_c is the diffusivity at the temperature of T_c and equals $D|_{t=0} = D|_{T=T_c} = Ae^{-E/(RT_c)}$, and $\tau = \tau_c(RT_c/E)$ is the time constant for D to decrease, it can be found that the square of the diffusion distance is roughly

$$\int_0^\infty D\,dt = \int_0^\infty D_c\,e^{-t/\tau}\,dt = D_c\tau. \tag{3-108c}$$

Because the closure temperature (T_c) is low enough that diffusive loss below T_c is not major, the square of the diffusive distance $D_c\tau$ must be smaller than a^2, where a is the half-thickness of a plane sheet, or the radius of a sphere or an infinitely long cylinder. That is,

$$a^2 \gg D_c\tau. \tag{3-109a}$$

Let

$$a^2 = GD_c\tau, \tag{3-109b}$$

where $G > 1$ is a constant that depends on geometry. Substitution of various parameters into the above leads to

$$a^2 = G\tau A e^{-E/(RT_c)}, \tag{3-109c}$$

which can be rearranged as

$$\frac{E}{RT_c} = \ln\left(\frac{GA\tau}{a^2}\right). \tag{3-109d}$$

That is,

$$T_c = \frac{E}{R\ln\left(\frac{GA\tau}{a^2}\right)}. \tag{3-110a}$$

The parameter τ in the above equation may be replaced by the quench rate q, leading to

$$T_c = \frac{E}{R\ln\left(\frac{GART_c^2}{a^2Eq}\right)}, \tag{3-110b}$$

where q is the quench rate ($-dT/dt$) at the closure age (or closure temperature). Equation 3-110a or 3-110b is the equation for the closure temperature. The closure temperature is always a calculated property, not a directly measured property. The equations are referred to as *Dodson's equation* (Dodson, 1973). Dodson carried out a more detailed mathematical analysis by solving the diffusion equation and found that the constant G equals 55, 27, or 8.7 for a sphere, long cylinder, or plane sheet, respectively. The above analyses and equations are derived for whole grains. Another condition for the application of the above formulation is that the initial temperature was high so that at the initial high temperature there was essentially complete loss of Ar.

Because the diffusion properties differ for different minerals, by dating several minerals in a single rock, one would obtain different apparent ages. The curve of T_c versus t_a represents the cooling history (Figure 1-21).

Note that Equation 3-110b is concerned with the diffusive exchange with the surroundings and there is no specific requirement for radiogenic growth or radioactive decay. The closure temperature concept applies not only to radiogenic components, but also to diffusive exchange of any component. For example, closure temperature may be defined for oxygen exchange between a mineral and a fluid phase and this temperature may be calculated from the oxygen diffusivity parameters using Equation 3-110b. In a gradually cooled rock, there would be continuous exchange between various minerals, and one may also define the closure temperature of any mineral as if it were in equilibrium with an infinite fluid reservoir.

a   b

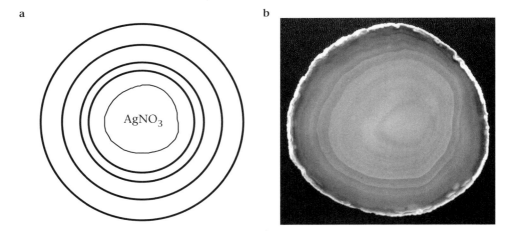

Figure 3-23 (a) Idealized sketch of Liesegang ring when a silver nitrate drop is added to a sodium chromate gel, and (b) a picture of agate.

3.5.3 Liesegang rings

If a silver salt (such as $AgNO_3$) diffuses into a gel containing chromate anion ($Na_2CrO_4 = 2Na^+ + CrO_4^{2-}$, yellow color), regularly spaced bands of precipitated silver chromate (Ag_2CrO_4, red color) develop in the gel (Stern, 1954). For example, by adding a drop of $AgNO_3$ solution to the center of a Na_2CrO_4 gel, red concentric layers of precipitated silver chromate will form (Figure 3-23a). This phenomenon is named *Liesegang rings*, after the German chemist Raphael E. Liesegang who discovered it. It is related to the diffusion of silver ion and the reaction of silver ion with chromate ion to form silver chromate when the ion product is greater than the equilibrium constant. There are many Liesegang-ring-like structures in igneous or sedimentary rocks, such as banded agate (Figure 3-23b), opal, orbicular structures in granites, and inch-scale layers in Stillwater Complex of Montana; whether any of these is caused by a diffusion–reaction process remains debatable.

To describe the process of the formation of such structures, it is necessary to write down the equations for a component that may be composed of several species and consider reactions among the species (Fisher and Lasaga, 1981). For example, for the case of diffusion of silver ions into a gel containing chromate ions, there are two species of Ag: one is Ag^+, which diffuses by interdiffusion with Na^+, and the other is Ag_2CrO_4 precipitation. The diffusivity of precipitated Ag_2CrO_4 is negligible. Therefore,

$$\frac{\partial [Ag]_{total}}{\partial t} = D_{Ag^+} \frac{\partial^2 [Ag^+]}{\partial x^2}.$$

(3-111a)

Assume that the initial CrO_4^{2-} concentration is uniform and CrO_4^{2-} diffusion is negligible. We have

$$[Ag^+]^2 \leq K_{sp}/[CrO_4^{2-}]. \tag{3-111b}$$

That is,

$$[Ag^+]^2 \leq K_{sp}/\{[CrO_4^{2-}]_0 - [Ag_2CrO_4]\} \tag{3-111c}$$

and

$$[Ag]_{total} = [Ag^+] + 2[Ag_2CrO_4], \tag{3-111d}$$

$$[CrO_4^{2-}]_0 = [CrO_4^{2-}] + [Ag_2CrO_4]. \tag{3-111e}$$

The above diffusion–reaction problem can be solved numerically. Fisher and Lasaga (1981) found that the solution to the above problem has bands of precipitated Ag_2CrO_4, similar to the observed Liesegang rings.

3.5.4 Isotopic ratio profiles versus elemental concentration profiles

During magma mixing, both chemical composition and isotopic ratios are heterogeneous. On a large scale, mixing is controlled by convection. On small scales (such as a 0.01-m scale, depending on other parameters), mixing or homogenization is controlled by diffusion. The homogenization of two different melts when there are gradients in both isotopic ratios and concentrations of all components is a complicated problem, and has been experimentally and theoretically treated by Baker (1989), Lesher (1990, 1994), Zhang (1993), and Van Der Laan et al. (1994). The theoretical treatment is approximate.

Experimental investigations were conducted using diffusion couples, such as basalt half with low $^{87}Sr/^{86}Sr$ isotopic ratio, and rhyolite half with high $^{87}Sr/^{86}Sr$ ratio. Both major element concentration gradients and trace element concentration gradients are present. Diffusion of the major elements is complicated and requires multicomponent approach. The focus of the studies, however, is often on the behavior of some trace elements (such as Sr and Nd in Baker (1989) and Lesher (1990, 1994)) or minor elements (such as K and Ca in Van Der Laan et al. (1994)) and isotopic ratios of the same elements. In the presence of major element concentration gradients, the diffusion of a trace element is not tracer diffusion. Trace or minor element diffusion in the presence of major concentration gradients display the following features:

(1) The effect of other concentration gradients sometimes leads to non-monotonic profile in the trace element, a typical indication of uphill diffusion.

(2) Even in the absence of uphill diffusion, a trace element concentration profile often does not match that for a constant diffusivity by using the effective binary diffusion treatment. Hence, the effective binary diffusivity depends on the chemical composition, which is expected.

(3) The effective binary diffusivities may also depend on the major element concentration gradients. That is, even with the same major element concentrations (e.g., at a point where $SiO_2 = 60$ wt%, and $Al_2O_3 = 15$ wt%), if the major element concentration gradients are different (e.g., in one diffusion couple, SiO_2 and Al_2O_3 both increase toward one direction, but in another diffusion couple, SiO_2 increases but Al_2O_3 decreases toward a direction), the effective binary diffusivity may differ.

(4) If the concentration gradients in the major elements are small, the length of the concentration profile of the trace element is similar to that of the isotopic fraction profile of the same element.

On the other hand, profiles of isotopic ratios, such as $^{87}Sr/^{86}Sr$, or the isotopic fractions, such as $^{87}Sr/(^{86}Sr + ^{87}Sr + ^{88}Sr)$, exhibit the following features:

(1) Isotopic fraction profiles are monotonic.

(2) The isotopic fraction profiles may be described by a roughly constant diffusion coefficient across major concentration gradients.

(3) If diffusivity is extracted from the profile of the isotopic fraction of an element, it may differ significantly from, and often greater than, the effective binary diffusivity obtained from the concentration profile of the trace or minor element.

(4) The interface position of the isotopic fraction profile is not necessarily the same as that of the concentration profile of the same element.

Based on these observations, the diffusivity extracted from isotopic fraction profiles is usually regarded to be similar to intrinsic diffusivity or self-diffusivity even in the presence of major element concentration gradients. That is, the multicomponent effect does not affect the length of isotopic fraction profiles (but it affects the isotopic fractions and the interface position). On the other hand, the diffusion of a trace or minor element is dominated by multicomponent effect in the presence of major element concentration gradients.

To quantify the diffusion profiles is a difficult multicomponent problem. The activity-based effective binary diffusion approach (i.e. modified effective binary approach) has been adopted to roughly treat the problem. In this approach,

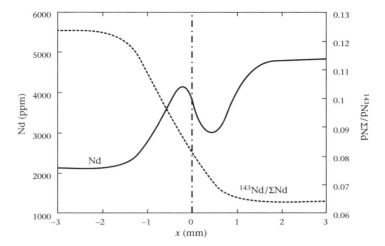

Figure 3-24 Calculated diffusion-couple profiles for trace element diffusion and isotopic diffusion in the presence of major element concentration gradients using the approximate approach of activity-based effective binary treatment. The vertical dot-dashed line indicates the interface. The solid curve is the Nd trace element diffusion profile (concentration indicated on the left-hand y-axis), which is nonmonotonic with a pair of maximum and minimum, indicating uphill diffusion. The dashed curve is the ^{143}Nd isotopic fraction profile. Note that the midisotopic fraction is not at the interface.

the concentration profile of the element is calculated by solving Equation 3-95. The concentration of each isotope of the element is also calculated by solving the same equation, from which the isotopic fraction can then be calculated. The calculations are in agreement with the main features of the experimental diffusion data on trace element concentrations and isotopic fractions. One example is shown in Figure 3-24.

3.5.5 Moving boundary problems

Moving boundary problems are a class of diffusion problems in which the boundary itself is moving. These are mostly encountered in crystal growth and dissolution, bubble growth and dissolution, solidification of a lava lake, freezing of a water lake, melting of ice, etc. That is, they are encountered in dealing with heterogeneous reaction kinetics. The methods of treating moving boundaries are summarized here, and specific problems are discussed in the appropriate sections dealing with the specific process.

For one-dimensional diffusion in a semi-infinite medium during crystal growth, define the crystal to be on the left-hand side and the melt on the right-

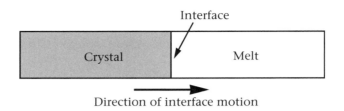

Direction of interface motion

Figure 3-25 Crystal growth in a melt.

hand side (Figure 3-25). The diffusion equation in the melt in the lab-fixed reference frame[3] is as follows (Equation 3-9):

$$\frac{\partial C}{\partial t} = \frac{\partial}{\partial x}\left(D\frac{\partial C}{\partial x}\right), \tag{3-112}$$

for $t > 0$ and $x > x_0$, where x_0 is the position of the interface between the crystal and melt. That is, x_0 is the position of the moving boundary. Define the initial boundary position (crystal-melt interface) to be $x = 0$. Assume a simple initial condition:

$$C|_{t=0} = C_\infty \qquad \text{for } x > 0. \tag{3-112a}$$

Denote boundary motion speed as u that may or may not depend on time. For crystal growth, the interface moves to the right with $x = x_0 > 0$. For crystal dissolution, the interface moves to the left with $x = x_0 < 0$. That is, u is positive during crystal growth and negative during crystal dissolution under our setup of the problem. The interface position can be found as

$$x_0 = \int u\,\mathrm{d}t. \tag{3-112b}$$

Because the crystal density usually differs significantly from that of the melt, it is necessary to distinguish the *crystal growth rate* and the *melt consumption rate* (or *melt dissolution rate*). The latter equals $\rho_{\text{cryst}}/\rho_{\text{melt}}$ times the crystal growth rate. Because we are interested in the melt phase, u in the above equation is specified as the melt consumption rate.

Assume the following boundary condition:

$$C|_{x=x_0} = C_0 \qquad \text{for } t > 0. \tag{3-112c}$$

The above equations (Equations 3-112, 3-112a,b,c) are the mathematical description for one-dimensional crystal growth or dissolution.

---

[3] The concept and subtleties of reference frames will be explored in more detail later in Section 4.2.1.1.

To solve the above problem, the usual method is to eliminate the moving boundary by adopting a reference frame that is fixed to the crystal–melt interface. That is, in the new reference frame and new coordinates y, the interface position is always at $y = 0$. Hence, we let

$$y = x - x_0, \tag{3-112d}$$

$$t_{\text{new}} = t. \tag{3-112e}$$

For example, for an experimental study of crystal growth, we measure the concentration as a function of distance away from the interface after the experiment. This would be a concentration profile in the interface-fixed reference frame. Using the new reference frame leads to

$$\frac{\partial C}{\partial t} = \frac{\partial C}{\partial t_{\text{new}}} \frac{\partial t_{\text{new}}}{\partial t} + \frac{\partial C}{\partial y} \frac{\partial y}{\partial t} = \frac{\partial C}{\partial t_{\text{new}}} - u \frac{\partial C}{\partial y}, \tag{3-113a}$$

$$\frac{\partial C}{\partial x} = \frac{\partial C}{\partial t_{\text{new}}} \frac{\partial t_{\text{new}}}{\partial x} + \frac{\partial C}{\partial y} \frac{\partial y}{\partial x} = \frac{\partial C}{\partial y}. \tag{3-113b}$$

Therefore, the diffusion equation in the new reference frame is

$$\frac{\partial C}{\partial t_{\text{new}}} = \frac{\partial}{\partial y} \left(D \frac{\partial C}{\partial y} \right) + u \frac{\partial C}{\partial y}. \tag{3-113c}$$

The initial condition becomes

$$C|_{t_{\text{new}} = 0} = C_\infty \quad \text{for } y > 0. \tag{3-113d}$$

The boundary condition becomes

$$C|_{y=0} = C_0 \quad \text{for } t_{\text{new}} > 0. \tag{3-113e}$$

To simplify notation, let's use x and t for y and t_{new} and remember that now x means the interface-fixed reference frame (and t still has the regular meaning because $t_{\text{new}} = t$). The above equations become

$$\frac{\partial C}{\partial t} = \frac{\partial}{\partial x} \left(D \frac{\partial C}{\partial x} \right) + u \frac{\partial C}{\partial x}, \tag{3-114a}$$

with initial condition of

$$C|_{t=0} = C_\infty \quad \textit{for } x > 0, \tag{3-114b}$$

and boundary condition of

$$C|_{x=0} = C_0 \quad \textit{for } t > 0. \tag{3-114c}$$

Other moving boundary problems such as crystal dissolution may be treated the same way. For example, for crystal dissolution, one way is to treat u as a negative parameter in the above equation. Alternatively, one may redefine u to

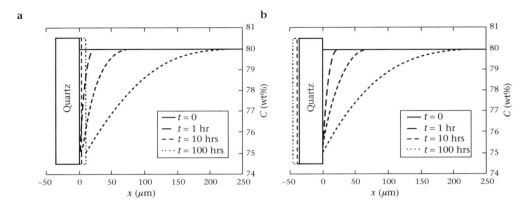

Figure 3-26 Quartz crystal growth and diffusion profile in (a) a laboratory-fixed reference frame and (b) an interface-fixed reference frame. At a given time, a given kind of curve is used to outline the crystal shape and plot the concentration profile.

be the melt growth rate (instead of the melt consumption rate), and the above equation would be changed simply by a negative sign in front of u.

If the boundary motion is controlled by the diffusion process itself so that the diffusion equation and the boundary motion velocity are coupled, the diffusion problem is called the *Stefan problem*. This can be the case if a magma is suddenly cooled by $100°C$ and is hence undercooled, and crystals grow under constant temperature and under the control of diffusion of nutrient chemicals (such as MgO during growth of olivine, or SiO_2 during growth of quartz). In this case of diffusion-controlled crystal growth or dissolution, the growth rate is related to square root of time and is not a constant.

If the boundary motion is controlled by an independent process, then the boundary motion velocity is independent of diffusion. This can happen if the magma is gradually cooling and crystal growth rate is controlled both by temperature change and mass diffusion. This problem does not have a name. In this case, u depends on time or may be constant. If the dependence of u on time is known, the problem can also be solved. The Stefan problem and the constant-u problem are covered below.

3.5.5.1. Stefan problem

In the Stephan problem, the boundary motion is controlled by diffusion itself. For example, the rate of a quartz crystal growth is related to how rapid mass can be diffused to the boundary (convection is not considered here, but will be considered in heterogeneous reactions). Let C be the concentration of SiO_2. The predicted concentration profile of SiO_2 as a function of time is shown in Figure 3-26. Suppose the SiO_2 concentration at saturation is 75 wt%. At $t=0$, the melt is uniformly supersaturated in SiO_2, e.g., 77 wt%. As quartz "magically" (as imposed by the initial and boundary conditions) begins to grow, SiO_2 in the interface melt suddenly drops to the equilibrium concentration at this temperature. Hence, initially the concentration gradient is infinite, leading to infinite crystal growth rate (which of course is not true due to limitation of crystal growth rate by the interface reaction rate, but the error introduced in terms of dissolution distance and concentration profile is small). As the diffusion profile gradually propagates into the melt, the concentration gradient becomes smaller, and the growth rate is reduced. The mathematical treatment is as follows.

For clarity, use w to denote mass fraction (dimensionless, i.e., the concentration unit is not kg/m^3 or mol/m^3) in the melt; w_{qtz} denotes mass fraction in quartz. The mass flux toward the interface (in the interface-fixed reference frame) is

$$J = -D(\rho_{melt} \partial w / \partial x)_{x=0},$$ (3-115a)

where ρ_{melt} is assumed to be a constant. This mass flux feeds crystal growth. Denote the melt consumption rate as u. The extra mass required for crystal growth is

$$\rho_{melt} u(w_{qtz} - w|_{x=0}),$$ (3-115b)

which must be equal to the mass flux, leading to

$$u(w_{qtz} - w|_{x=0}) = D(\partial w / \partial x)_{x=0}.$$ (3-115c)

Equations 3-114a, 114b, 114c, and 115c constitute the diffusive crystal growth problem, from which both $w(x, t)$ and u are to be solved. Because the diffusion profile propagates as square root of time, meaning that the total mass transported to the interface is proportional to the square root of Dt, it may be guessed that diffusive crystal growth distance is also proportional to the square root of Dt:

$$x_0 = 2\alpha(Dt)^{1/2},$$ (3-115d)

where α is a proportionality constant to be determined later. The crystal growth rate u is hence

$$u = \alpha(D/t)^{1/2}.$$ (3-115e)

The above equation would indicate an infinite growth rate at $t=0$, which is consistent with the diffusion equation (because the concentration gradient at $t=0$ is infinity), although in reality this would not happen. Because the growth rate quickly becomes finite, the initial infinite growth rate does not cause any numerical difficulty in solving the problem.

Applying Boltzmann transformation to Equation 3-114a with $u = \alpha(D/t)^{1/2}$ (similar to Section 3.2.4), after some steps (see Section 4.2.2.1), the solution is

$$w = w_\infty + (w_0 - w_\infty)\text{erfc}\left(\frac{x}{\sqrt{4Dt}} + \alpha\right)/\text{erfc}(\alpha).$$ (3-116)

Next, the parameter α must be determined. Using Equation 3-115c, because

$$(\partial w/\partial x)_{x=0} = -(2/\pi^{1/2})[(w_0 - w_\infty)/(4Dt)^{1/2}]\exp(-\alpha^2)/\mathrm{erfc}(\alpha), \tag{3-117a}$$

Equation 3-115c becomes

$$\begin{aligned}
\alpha(D/t)^{1/2}(w_{\mathrm{qtz}} - w|_{x=0}) = \\
-(D/t)^{1/2}(1/\pi^{1/2})(w_0 - w_\infty)\exp(-\alpha^2)/\mathrm{erfc}(\alpha).
\end{aligned} \tag{3-117b}$$

That is,

$$\pi^{1/2}\alpha\exp(\alpha^2)\,\mathrm{erfc}(\alpha) = (w_\infty - w_0)/(w_{\mathrm{qtz}} - w_0). \tag{3-117c}$$

As the parameter α is solved from the above equation, the dissolution rate u is uniquely solved (Equation 3-115e) and the concentration profile is also uniquely solved (Equation 3-116). Hence, this completes the solution of the diffusion and growth problem.

In the above treatment, diffusion in the melt is treated simply as effective binary diffusion. Rigorous treatment would have to consider multicomponent diffusion (Liang, 1999, 2000). The effective binary diffusion approach works roughly for the main diffusion component (that is, the component with the largest concentration difference between the crystal and melt). One example of quartz growth and diffusion is given below. The growth of quartz is considered because the composition of quartz is fixed and hence the problem is simple. For the growth of other minerals such as olivine, the composition of the crystal may vary with melt composition and hence with time. More examples and more detailed considerations are presented in the next chapter on heterogeneous reaction kinetics.

Example 3.6 Find the growth rate of a quartz crystal in a hydrous rhyolitic melt for the following set of conditions: $w_{\mathrm{qtz}} = 100$ wt%; $\rho_{\mathrm{qtz}}/\rho_{\mathrm{melt}} \approx 2.535/2.30 = 1.102$; initial SiO_2 concentration $w_\infty = 78.55$ wt%; the saturation SiO_2 concentration $w_0 = 75$ wt%; and $D = 0.01$ $\mu m^2/s$.

Solution: First find the right-hand side of Equation 3-117c:

$(w_\infty - w_0)/(w_{\mathrm{qtz}} - w_0) = 0.142.$

Solving parameter α (e.g., using a spreadsheet program) from

$\pi^{1/2}a\exp(a^2)\,\mathrm{erfc}(a) = 0.142,$

leads to

$\alpha = 0.0882.$

The dissolution distance is, hence,

$x_0 = 2\alpha(Dt)^{1/2} = 0.0176t^{1/2},$

where t is in s and x_0 is in μm. The concentration profile w is given by Equation 3-116:

$$C = 78.55 - 3.55 \, \text{erfc}\left(\frac{x}{\sqrt{4Dt}} + 0.0882\right)/\text{erfc}(0.0882).$$

Some profiles as a function of t are shown in Figure 3-26b.

3.5.5.2 Boundary motion with a constant velocity

Crystal growth rate may be constant, which could happen if temperature is decreasing or if there is convection. Smith et al. (1956) treated the problem of diffusion for constant crystal growth rate. In the interface-fixed reference frame, the diffusion equation in the melt is

$$\frac{\partial w}{\partial t} = D\frac{\partial^2 w}{\partial x^2} + u\frac{\partial w}{\partial x}, x > 0, t > 0. \tag{3-118}$$

where w is concentration in the melt, and u is melt consumption rate (density ratio times the crystal growth rate) and is a constant. For the initial condition of

$$w|_{t=0} = w_\infty, \tag{3-118a}$$

and boundary condition of

$$D\frac{\partial w}{\partial x}|_{x=0} = u(K-1)w_{x=0}, \tag{3-118b}$$

where K is the partition coefficient of the component between the crystal and the melt ($K = w_{c,0}/w_{L,0}$ and is assumed to be constant), the solution is

$$\frac{w}{w_\infty} = 1 + \frac{1-K}{2K}e^{-ux/D}\text{erfc}\frac{x-ut}{\sqrt{4Dt}}$$
$$- \frac{1}{2}\text{erfc}\frac{x+ut}{\sqrt{4Dt}} + \left(1 - \frac{1}{2K}\right)e^{-u(x+Kut)(1-K)/D}\text{erfc}\frac{x+(2K-1)ut}{\sqrt{4Dt}}. \tag{3-119}$$

At $x = 0$,

$$\frac{w_{x=0}}{w_\infty} = 1 + \frac{1-K}{2K}\text{erfc}\frac{-ut}{\sqrt{4Dt}} - \frac{1}{2}\text{erfc}\frac{ut}{\sqrt{4Dt}}$$
$$+ \left(1 - \frac{1}{2K}\right)e^{-u(Kut)(1-K)/D}\text{erfc}\frac{(2K-1)ut}{\sqrt{4Dt}}. \tag{3-119a}$$

If $K = 1$, then

$$w = w_\infty, \tag{3-119b}$$

as expected. The steady-state concentration profile is obtained by letting $t = \infty$:

$$\frac{w}{w_\infty} = 1 + \frac{1-K}{K}e^{-ux/D}. \tag{3-119c}$$

The concentration profile in the solid is found by multiplying $w_{x=0}$ by K and ignoring diffusion in the crystal:

$$\frac{w_c}{Kw_\infty} = 1 + \frac{1-K}{2K}\operatorname{erfc}\frac{-ut}{\sqrt{4Dt}} - \frac{1}{2}\operatorname{erfc}\frac{ut}{\sqrt{4Dt}}$$
$$+ \left(1 - \frac{1}{2K}\right)e^{-u(Kut)(1-K)/D}\operatorname{erfc}\frac{(2K-1)ut}{\sqrt{4Dt}}. \tag{3-120a}$$

Let $y = ut$, (i.e., $t = y/u$), where y means distance in the crystal measured from the beginning of the crystal growth; then

$$\frac{w_c}{w_\infty} = \frac{1}{2}\left[1 + \operatorname{erf}\sqrt{\frac{uy}{4D}}\right] + \left(K - \frac{1}{2}\right)e^{-uyK(1-K)/D}\operatorname{erfc}\left[\left(K - \frac{1}{2}\right)\sqrt{\frac{uy}{D}}\right]. \tag{3-120b}$$

If t approaches ∞, then y approaches ∞, leading to

$$w_c = w_\infty. \tag{3-120c}$$

That is, the concentration in the crystal is the same as that in the initial melt at steady state. Therefore, the growth of the crystal does not affect the mass excess or deficiency in the melt anymore, meaning that the concentration profile (in interface-fixed reference frame) in the melt is at steady state. Steady state may be reached only for elements whose concentration in a mineral can vary non-stoichiometrically.

3.5.6 Diffusion and flow

Diffusion is not a very effective way to transfer mass and to homogenize a fluid medium. For example, if you pour milk into coffee, you can patiently wait for the milk to diffuse through coffee. However, it would take a very long time. Natural convection (*free convection*) occurs even without stirring, and stirring with a straw or spoon can generate convection, called *forced convection*. Convection speeds up the homogenization process. By stirring, you produce and enhance convective motion in the fluid so that the fluid rapidly becomes homogeneous on a milli-meter scale. Homogenization to finer scale relies on diffusion. For example, for diffusion distance to reach 0.1 mm would take seconds.

Both diffusion and convection are modes of mass transfer. Typically, large-scale mass transfer is accomplished by convection, and small-scale mass transfer is accomplished by diffusion. Similarly, large-scale heat transfer in the Earth is through convection (mantle convection), and small-scale heat transfer is through heat conduction (e.g., through the lithosphere). To treat the compli-cated convection pattern and diffusion requires a large computational effort. Some simple problems can be treated analytically.

The general diffusion and flow equation is (Equation 3-19a)

$$\frac{\partial C}{\partial t} = D\left(\frac{\partial^2 C}{\partial x^2} + \frac{\partial^2 C}{\partial y^2} + \frac{\partial^2 C}{\partial z^2}\right) - u_x \frac{\partial C}{\partial x} - u_y \frac{\partial C}{\partial y} - u_z \frac{\partial C}{\partial z}. \tag{3-121a}$$

For one-dimensional diffusion and laminar flow with constant velocity along the direction, the above diffusion-flow equation can be written as

$$\frac{\partial C}{\partial t} = D\frac{\partial^2 C}{\partial x^2} - u\frac{\partial C}{\partial x}, \tag{3-121b}$$

where u is the flow velocity. The above equation is similar to the moving boundary diffusion problem for crystal dissolution, but the physical meaning is different. The reference frame is not moving.

Environmental scientists often have to investigate how pollutant dumped (intentionally or unintentionally) into a river is transported downstream (Boeker and van Grondelle, 1995). Suppose at a certain moment a toxic substance of mass M (in kg) spilled into river water at a certain location along the river. For simplicity, assume that the river flows at a constant velocity of u. In addition to flow, the toxic substance is dispersed by molecular diffusion and by many other processes in river water. These processes include turbulence in water due to irregularities on the bottom, or winding banks, fish swimming, boating, and other human and nonhuman activities on and in the river. At a small scale, these are convections that help disperse the toxic substance. On a larger scale, the disturbances due to the collective effect of these processes are almost random. Hence, the dispersion by these processes can be described mathematically by diffusion. This diffusion is different from molecular diffusion in previous sections. It has a special name: eddy diffusion. The eddy diffusivity, denoted as D_{eddy}, is not molecular diffusivity that may be measured in quiet water in lab settings. Instead, it is much larger because of all disturbances in a river and can be determined experimentally for specific river segments using the dispersion of nontoxic substances. The molecular or ionic diffusivity in water at room temperature is typically 2×10^{-9} m^2/s. Peter Schlosser (personal communication, 2003) found that an eddy diffusivity of 75 m^2/s (10 orders of magnitude larger than molecular diffusivity in water!) matches experimental data for the Hudson River. The eddy diffusivity along a river may not be constant and may depend on local flow conditions and other factors. In the following treatment, a constant eddy diffusivity is assumed for simplicity. To treat the problem with variable eddy diffusivity and flow rate, which is necessary in real calculations, it is necessary to use numerical methods. Also, a river has some finite width and depth, hence 3-D diffusion should be considered for accurate modeling. The simple calculations below nevertheless reveal the main features of pollutant transport in a river.

For the transport of a toxic substance, its concentration in river water at a given time and place is of great interest: if the concentration is greater than a threshold value, the water may not be used as drinking water; otherwise the water is fine (though one may still not want to drink it without treatment). To further simplify, it is assumed that the river is narrow and shallow, and concentration variation across the river width or depth is ignored. $\overline{C}$ is used to represent the average concentration of the toxic substance across the cross section of the river. Hence, instead of trying to solve for the concentration as a function of x, y, and z, we average the concentration along y and z directions so that the average concentration $\overline{C}$ depends only on x (along the river flow direction). Hence, the diffusion–flow equation is

$$\frac{\partial \overline{C}}{\partial t} = D_{\text{eddy}} \frac{\partial^2 \overline{C}}{\partial x^2} - u \frac{\partial \overline{C}}{\partial x}, \tag{3-121b}$$

with the initial condition

$$\overline{C}\big|_{t=0} = M\delta(x), \tag{3-121c}$$

and no boundary condition (infinitely long medium). Carry out the following coordinate transformation:

$$t' = t \tag{3-121d}$$

$$y = x - \int u \, dt. \tag{3-121e}$$

We obtain,

$$\frac{\partial \overline{C}}{\partial t'} = D_{\text{eddy}} \frac{\partial^2 \overline{C}}{\partial y^2}, \tag{3-121f}$$

and

$$\overline{C}\big|_{t'=0} = M\delta(y). \tag{3-121g}$$

The solution to the above problem is (Equation 3-45a)

$$\overline{C} = \frac{M}{2\sqrt{\pi D_{\text{eddy}} t'}} e^{-y^2/(4D_{\text{eddy}} t')} = \frac{M}{2\sqrt{\pi D_{\text{eddy}} t}} e^{-(x - \int u \, dt)^2/(4D_{\text{eddy}} t)}. \tag{3-122}$$

The above equation describes how the concentration of the toxic substance in river water varies as a function of time and distance x from the spill. The solution is simple and a spreadsheet program can carry out the calculations. Figure 3-27 shows the pollutant distribution in the river as a function of distance downstream after some specific times. The diagram indicates where the toxic substance has moved to at a specific time.

When we discussed moving boundary problems, we transformed the problem into boundary-fixed reference frame and converted the moving boundary to a

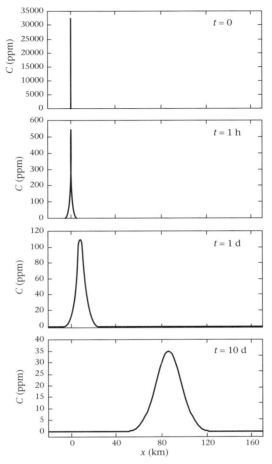

Figure 3-27 Evolution of pollutant concentration along a river as a function of time. Flow velocity is assumed to be constant $u = 0.1$ m/s. Eddy diffusivity is also assumed to be constant $D_{eddy} = 75$ m²/s. The concentration profile as a function of distance is smooth.

fixed boundary problem but made the diffusion equation more complicated. However, when we discussed diffusion and flow, we made the opposite transformation so that the diffusion equation is simplified. Nonetheless, in the latter case, the transformation meant that the moving center of the pollutant is fixed as the origin and hence in this sense it is similar to the transformation on the moving boundary problem. Because there is no interface or boundary in the diffusion and flow case (because of the infinite medium assumption), the transformation does not lead to a more complicated boundary condition. That is, the transformation of the moving boundary problem simplified the boundary condition but made the diffusion equation more complex. On the other hand, the transformation of the diffusion and flow problem simplified the diffusion equation without making the boundary condition more complex.

3.6 Diffusion Coefficients

Diffusion coefficients must be known to evaluate the rate of a specific diffusion process. Experimentalists have investigated diffusion of various components in different phases to provide diffusion data for understanding and quantifying the rate of diffusion, with applications ranging from estimating the time to reach equilibrium, to closure temperature, to bubble growth driving volcanic eruptions. Experimental study of diffusion is the only reliable method to obtain the diffusion coefficients and their dependence on temperature, pressure, phase, and composition. There are no generally applicable methods to calculate diffusivity in condensed phases. Nonetheless, various theoretical and empirical relations have been proposed, and they are useful under the conditions that were assumed to derive them. This section covers the experimental methods to obtain diffusion coefficients, and various theoretical and empirical relations between diffusivity and other parameters .

Diffusion coefficients vary widely, depending on temperature, pressure, the type of the phase, and the composition of the phase. The dependence on temperature and pressure can be described well by the Arrhenius relation including a pressure term (Equation 1-88):

$$D = A \exp[-(E + P\Delta V)/(RT)],$$

where A the a preexponential factor, E is the activation energy and is positive, and ΔV is the volume difference between the activated complex and the diffusing species and may be either positive or negative. The parameters A, E, and ΔV must be determined from experiments. The compositional dependence of diffusivity is addressed later in this section.

In the gas phase, typical D values at room temperature and pressure are of the order 10^{-5} to 10^{-4} m^2/s. To the first-order approximation, D is inversely proportional to pressure and is proportional to the absolute temperature raised to the 1.5 to 1.8 power (Cussler, 1997). There is not much of an activation energy for diffusion in the gas phase. Interdiffusivity in the gas phase may be found in Cussler (1997).

In aqueous solutions, typical D values at room temperature and pressure are about 2×10^{-9} m^2/s, and typical activation energy is 16 kJ/mol. D values at 25°C for selected species in aqueous solutions are given in Table 1-3a. The temperature dependence of some D values can be found in Appendix 4 (Table A4-1).

In silicate melts, typical D values at 1300°C are of the order 10^{-11} m^2/s, and typical activation energy is 250 kJ/mol. Highly charged cations (such as Si^{4+} and Zr^{4+}) or strongly bonded ions (such as bridging oxygen) diffuse more slowly and have higher activation energy than univalent or divalent cations (such as Fe^{2+}–Mg^{2+} interdiffusion). Appendix 4 (Table A4-2) lists selected diffusivities in silicate melts.

In ice, H_2O self-diffusivity at 273 K is of the order 5×10^{-15} m^2/s (Hobbs, 1974), and typical activation energy is 50–60 kJ/mol.

In silicate minerals, typical D values at 1200°C are of the order 10^{-16} m^2/s, and typical activation energy is about 300 kJ/mol. The diffusivity of cations depends on the charge of the cations. Highly charged cations (such as Si^{4+} and Zr^{4+}) diffuse more slowly and have higher activation energy than univalent or divalent cations (such as Fe^{2+}–Mg^{2+} interdiffusion). Tables 1-3b, 1-3c, and Appendix 4 list selected diffusivities in silicate minerals.

3.6.1 Experiments to obtain diffusivity

The purpose of most experimental studies of diffusion is to obtain accurate diffusion coefficients as a function of temperature, pressure, and composition of the phase. For this purpose, the best approach is to design the experiments so that the diffusion problem has a simple analytical solution. After the experiments, the experimental results are compared with (or fit by) the analytical solution to obtain the diffusivity. The method of choice depends on the problems. The often used methods include diffusion-couple method, thin-source method, desorption or sorption method, and crystal dissolution method.

After the experiment, the experimental charge is prepared for analysis of the diffusion component or species. The analytical methods include microbeam methods such as electron microprobe, ion microprobe, Rutherford backscattering, and infrared microscope to measure the concentration profile, as well as bulk methods (such as mass spectrometry, infrared spectrometry, or weighing) to determine the total gain or loss of the diffusion component or species. Often, the analysis of the diffusion profile is the most difficult step in obtaining diffusivity.

This section describes the experimental methods and focuses on the estimation of diffusivity after the experiment. The analytical methods are not described here. Estimation of diffusivity from homogeneous reaction kinetics (e.g., Ganguly and Tazzoli, 1994) is discussed in Chapter 2 and will not be covered here. Determination of diffusion coefficients is one kind of inverse problems in diffusion. This kind of inverse problem is relatively straightforward on the basis of solutions to forward diffusion problems. The second kind of inverse problem, inferring thermal history in thermochronology and geospeedometry, is discussed in Chapter 5.

3.6.1.1 Diffusion-couple method

In the diffusion-couple method, two cylinders of the same radius and roughly the same length are prepared. Each cylinder (called a half) is uniform in composition, but the two cylinders (called two halves) are different in composition, either in terms of chemical composition, or in terms of isotopic composition. The two halves are then placed together or pressed together in a high-pressure apparatus, making a long cylinder. To avoid convection, the denser half is placed at the bottom and the other half is on top. (Convection ensues easily for horizontally placed diffusion couple.) Then the assembly is brought to the desired pressure first and temperature next so that significant diffusion can occur (the sample may or may not be melted). After a designated duration, the sample is quenched to room temperature, and the long cylinder is sectioned. The concentration profile is then measured using a microbeam technique, or some other technique. The diffusion couple is one of the most commonly used experimental techniques in diffusion studies.

The experimental duration and the length of the diffusion couple are designed such that the length of the diffusion profile is short compared to the total length of the diffusion couple. Hence, the diffusion medium may be treated as infinite, meaning that at the ends of the two halves, the compositions are still the initial compositions. If D does not vary with concentration or distance, the concentration profile (C versus x) would be an error function. Hence, the first step to try to understand the profile is to fit an error function to the profile (Equation 3-38):

$$C = \frac{C_1 + C_2}{2} + \frac{C_2 - C_1}{2} \, \mathrm{erf}\left(\frac{x}{2\sqrt{Dt}}\right),$$

where C_1 and C_2 are the initial concentrations in the two halves; they are known either from the measurement of the initial compositions of the two halves, or from the measurement at the two ends. Hence, D is the only real unknown, although sometimes C_1 and C_2 may be allowed to vary in the fit. In Equation 3-38, x is defined such that $x = 0$ is the interface of the two halves, but the concentration measurement gives only an arbitrary x that is offset from the required interface-fixed coordinate by an unknown constant x_0. One may estimate x_0 from the data, especially if data quality is high, or obtain it from fitting using the following equation:

$$C = \frac{C_1 + C_2}{2} + \frac{C_2 - C_1}{2} \, \mathrm{erf}\left(\frac{x - x_0}{2\sqrt{Dt}}\right). \tag{3-123a}$$

Then the measured concentration profile can be fit by the above equation to obtain both D and x_0. This D is the diffusivity at the experimental temperature. One example can be found in Figure 3-28a. If the experimental duration is only slightly longer than the heating-up and cooling-down time, then correction of the heating-up and cooling-down time must be made to obtain D at the experimental temperature. This is usually done by treating t in Equation 3-38 to be the effective experimental duration t, which is the actual experimental duration at the experimental temperature, plus a Δt, which is the equivalent duration at the

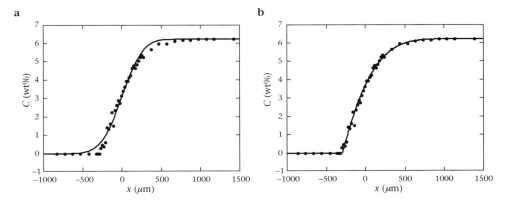

Figure 3-28 H_2O diffusion profile for a diffusion-couple experiment. Points are data, and the solid curve is fit of data by (a) error function (i.e., constant D) with $D = 167 \ \mu m^2/s$, which does not fit the data well; and (b) assuming $D = D_0(C/C_{max})$ with $D_0 = 409 \ \mu m^2/s$, which fits the data well, meaning that D ranges from $1 \ \mu m^2/s$ at minimum H_2O content (0.015 wt%) to $409 \ \mu m^2/s$ at maximum H_2O content (6.2 wt%). Interface position has been adjusted to optimize the fit. Data are adapted from Behrens et al. (2004), sample DacDC3.

experimental temperature based on the heating-up and cooling down thermal history. One method to find Δt is to carry out a zero-time experiment, meaning using the same heating-up and cooling-down rate and but let the experiment stay at the experimental temperature for zero time (e.g., Zhang and Stolper, 1991). Another way to estimate Δt is to use recorded thermal history and integrate the diffusion effect during heating up and cooling down if the activation energy is known (Zhang and Behrens, 2000).

If D depends on concentration, the first indication would come from the asymmetry of the diffusion-couple concentration profile. That is, there is no center symmetry with respect to $x = 0$, which means that one side approaches the end concentration more rapidly than the other side (Figure 3-28a). In such cases, D as a function of C can be obtained by Boltzmann analysis (Equation 3-58e):

$$D = \frac{\int_{C(x_0)}^{C(\infty)} x \ dC}{2t \cdot (dC/dx)|_{x = x_0}},$$ (3-123b)

where x is defined relative to the Matano interface, and D is at concentration of $C(x_0)$. Details can be found in Section 3.2.8.2.

The concentration dependence of D on C may be obtained by directly fitting the diffusion profile if the functional form of D as a function of C is known; for example,

$$D = D_0(C/C_0),$$ (3-123c)

where D_0 is a constant to be determined and C_0 is a normalizing concentration (usually the concentration at the high-concentration end of the couple). The diffusion profile for the above D can be calculated numerically and can then be fit to the experimental profile to obtain D_0. In this way, D as a function of C can also be determined. Figure 3-28b shows an example.

3.6.1.2 Desorption or sorption method

For a volatile component that can be absorbed or desorbed from a solid (glass or crystal), the desorption of sorption method can be used to determine diffusivity. In the desorption method, the glass or crystal initially contains the component (such as water or Ar) and upon heating in vacuum or in an atmosphere that is free of the gas component, the gas component would diffuse out so as to reach equilibrium with the atmosphere. For example, the dehydration experiments of Zhang et al. (1991a) and Wang et al. (1996) belong to this category. In sorption experiments, the opposite happens and the fluid component diffuses into the solid, such as oxygen diffusion into minerals (e.g., Giletti and Yund, 1984; Farver and Yund, 1990).

Exchange diffusion of nonvolatile components into a phase, such as ^{18}O or ^{87}Sr self-diffusion into a feldspar mineral, or Fe–Mg interdiffusion into olivine, is often conducted using a large fluid or powder reservoir to surround the phase to be investigated. The fluid or powder reservoir has a different chemical or isotopic composition from the phase, leading to isotopic or elemental exchange between the phase and the reservoir. Because the diffusion distance in the phase is typically small, meaning that the consumption of the diffusing component is trivial, the fluid or powder reservoir may be regarded to be an infinite reservoir with constant isotopic concentration. Hence, the surface concentration of the isotope in the phase may be regarded as constant. Therefore, these diffusion problems are similar to sorption. The advantages of using a fluid reservoir include its ability to maintain a surface concentration that is uniform through the whole surface area and is independent of time. However, the fluid may participate in the diffusion or may affect the diffusivity. A powder reservoir may not be able to maintain uniform surface concentration (e.g., some areas of the surface may be in contact with powder grains but other areas may not), nor constant surface concentration (diffusive transport in the powder, even along grain boundaries, may not be rapid enough).

With the desorption or sorption method, diffusivity may be extracted by measuring concentration profiles, or by measuring the total loss or gain of the component. The former is referred to as the *profiling technique*, in which a single grain is used, and concentration profile is measured after the experiment. In the other technique, many grains of similar shape and size are used, and the total mass gain or mass loss is determined after the experiment. This will be referred to as the *bulk technique*. The profiling technique is the preferred method for obtaining accurate diffusivity data, especially when diffusivity depends on concentration. The bulk technique may not produce accurate results because grain shapes are imperfect, because grain sizes are not the same, and especially because

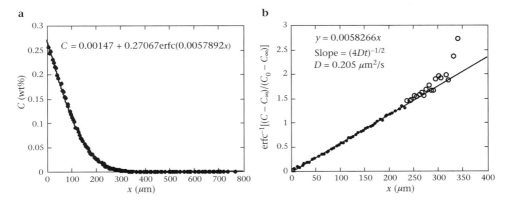

Figure 3-29 A half-space diffusion profile of Ar. $C_\infty = 0.00147$ wt% is obtained by averaging 45 points at 346 to 766 μm. Points are data, and the solid curve is a fit of (a) all data by the error function with $D = 0.207$ μm^2/s and $C_0 = 0.272$ wt%, and (b) data at $x \leq 230$ μm (solid dots) by the inverse error function. In (b), for larger x, evaluation of erfc$^{-1}[(C - C_\infty)/(C_0 - C_\infty)]$ becomes increasingly unreliable and even impossible as $(C - C_\infty)/(C_0 - C_\infty)$ becomes negative. Data are adapted from Behrens and Zhang (2001), sample AbDAr1.

the grains may be cleaved or cracked, which might lead to apparent diffusivities orders of magnitude greater than real diffusivities.

In the *profiling technique*, one grain is used and at least one side of the grain is prepared into a flat mirror surface (either by cleavage or by polishing). Diffusion from this surface into the sample is investigated. After the high-temperature and high-pressure experiment, a section is cut perpendicular to the polished surface. Concentration profile is measured as a function of distance away from this surface, from which the diffusivity is obtained. If D does not depend on C, then the diffusion profile would be an error function (Equation 3-40). For desorption experiments when the surface concentration is zero, the concentration profile would be

$$C = C_\infty \ \text{erf}[x/(4Dt)^{1/2}], \tag{3-124a}$$

where C_∞ is the initial concentration in the solid. For sorption experiments when the initial concentration is zero, the concentration profile would be

$$C = C_0 \ \text{erfc}[x/(4Dt)^{1/2}], \tag{3-124b}$$

where C_0 is the surface concentration in the solid. For sorption and desorption, if $C_\infty \neq 0$ and $C_0 \neq 0$, then,

$$C = C_\infty + (C_0 - C_\infty) \ \text{erfc}[x/(4Dt)^{1/2}]. \tag{3-124c}$$

Figure 3-29a shows an example. The surface position ($x = 0$) is in theory well known, and hence there is no need to allow x to vary. However, if the concentration profile is short and several micrometers in distance is important, there might still be a need to allow the real interface to differ from the observed because (i) the surface may be chipped, and/or (ii) the surface is not perfectly vertical for measurements using a transmission method (such as infrared spectroscopy).

In addition to the direct fit using error function above, another popular fitting method is to use the inverse error function to fit. For example, Equation 3-124c may be written as

$$\text{erfc}^{-1}[(C - C_\infty)/(C_0 - C_\infty)] = x/(4Dt)^{1/2}. \qquad (3\text{-}124\text{d})$$

Plotting $\text{erfc}^{-1}[(C - C_\infty)/(C_0 - C_\infty)]$ versus x would lead to a straight line with a slope of $1/(4Dt)^{1/2}$ if D is constant. Hence, D can be obtained (Figure 3-29b). There are some disadvantages in using the inverse error function fit: (i) C_0 must be guessed in calculating $\text{erfc}^{-1}[(C - C_\infty)/(C_0 - C_\infty)]$ and sometimes adjusted to optimize the fit; (ii) $\text{erfc}^{-1}[(C - C_\infty)/C_0 - C_\infty)]$ cannot be evaluated if $(C - C_\infty)/C_0 - C_\infty)$ is negative; and (iii) the error of calculating $\text{erfc}^{-1}[(C - C_\infty)/C_0 - C_\infty)]$ for a given C increases with x as $(C - C_\infty)$ approaches to 0, which means (a) the part of the profile with $(C - C_\infty) < 3\sigma$ cannot be used (σ is analytical error), and (b) the part of the profile with $(C - C_\infty) < 5\sigma$ must be used with caution.

In the profiling technique, the dependence of D on C may be obtained using either the Boltzmann method, or fitting the concentration profile with numerically calculated profile by assuming a specific relation between D and C, similar to the diffusion-couple method. For the Boltzmann method, the equation can be found by following steps in Section 3.2.8.2 and is as follows (Equation 3-58e):

$$D = \frac{\int_{C(x_0)}^{C(\infty)} x \, dC}{2t(dC/dx)_{x=x_0}},$$

where x is defined relative to the sorption or desorption surface and D is at concentration of $C(x_0)$. The Boltzmann method for the case of sorption or desorption is simpler than the diffusion couple because the surface is known.

The *bulk technique* is used when measurement of concentration profile is not available. In this technique, many grains of similar size and shape are heated to and held at the desired temperature for a given duration. After the experiment, the total mass loss or gain of the component by the grains is measured. From the mass loss or gain, the diffusion coefficient is calculated. To obtain diffusivity from mass loss experiments (most Ar and He diffusivities in minerals are obtained this way), it is necessary to assume that the initial concentration of the diffusion component is uniform. It is also necessary to assume the effective shape of the diffusing grains (cf. Section 3.2.11).

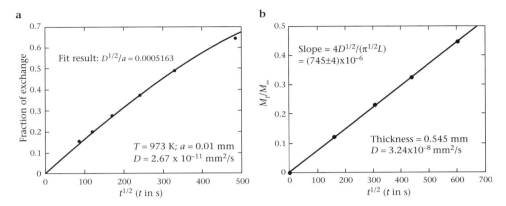

Figure 3-30 An example of obtaining diffusivity (a) from mass exchange data with spheres using Equation 3-125 (Gas/melilite oxygen isotope exchange data of Hayashi and Muehlenbachs (1986)) and (b) using mass loss data from a single thin wafer using Equation 3-126 (garnet dehydration data of Wang et al. (1996)).

If the grains are all equal-size spheres with radius a, from the mass loss or gain, the diffusivity may be calculated using Equation 3-68f of Section 3.2.10.3:

$$\frac{M_t}{M_\infty} = 6\frac{\sqrt{Dt}}{\sqrt{\pi}a}\left\{1 + 2\sqrt{\pi}\sum_{n=1}^{\infty} ierfc\frac{na}{\sqrt{Dt}}\right\} - 3\frac{Dt}{a^2} \approx \frac{6}{\sqrt{\pi}}\frac{\sqrt{Dt}}{a} - 3\frac{Dt}{a^2}, \tag{3-125}$$

where $M_\infty = 4\pi a^3 \Delta C/3$ with ΔC being the difference between the initial concentration and the surface concentration, and M_t is measured as the mass loss or gain from the spherical grains. The approximate relation (two terms) of Equation 3-125 has a relative accuracy of 0.1% if $M_t/M_\infty < 0.9$. When the equation is further simplified to only one term (square-root term), it does not have much applicability, with a relative accuracy of 1% only if $M_t/M_\infty < 0.04$. Figure 3-30a compares experimental data and fit to obtain D.

If a single grain of thin wafer of thickness L is used and total mass loss or gain is measured instead of the concentration profile, the diffusion coefficient may be obtained by fitting the data to Equation 3-52d:

$$\frac{M_t}{M_\infty} = \frac{4\sqrt{Dt}}{\sqrt{\pi}L}\left[1 + 2\sqrt{\pi}\sum_{n=1}^{\infty}(-1)^n ierfc\frac{nL}{2\sqrt{Dt}}\right] \approx \frac{4\sqrt{Dt}}{\sqrt{\pi}L}, \tag{3-126}$$

where the approximate relation has a relative precision of better than 0.7% when $M_t/M_\infty < 0.6$. Figure 3-30b shows experimental data and the fit to obtain D using this equation.

If it is possible to measure the diffusion profile, the profiling technique is preferred over the bulk technique. The disadvantage of the profiling technique is that it requires high spatial resolution in concentration measurement, as well as more tedious sample preparation. There are at least four disadvantages of the bulk technique compared to the profiling technique.

(1) One is possible overestimation of diffusivity if the grains contain cracks that would facilitate mass loss or gain, which are not accounted for in extracting diffusivity. For example, Pb diffusivity in zircon extracted using the bulk mass loss method may be many orders of magnitude greater than that obtained using the profiling method (Cherniak and Watson, 2000), and ^{18}O diffusivity extracted under dry conditions using the bulk extraction method by Connolly and Muehlenbachs (1988) is about 100 times greater than that obtained under similar conditions using the profiling method by Ryerson and McKeegan (1994). On the other hand, with the profiling method, cracks can be avoided and the diffusivity more likely reflects the true volume diffusivity.

(2) For minerals, if diffusion is anisotropic, the bulk method gives only an average diffusivity for an assumed effective shape, but cannot determine the diffusivity along different crystallographic directions. The profiling method is necessary to quantitatively resolve the anisotropy.

(3) If the diffusivity depends on concentration of the diffusing component, the measured diffusivity using the bulk technique is some average of the diffusivity, and diffusivity extracted from sorption experiments may differ from that from the desorption experiments (Zhang et al., 1991a; Wang et al., 1996). Hence, there is a need to distinguish the two: diffusion during sorption experiments is referred to as *in-diffusion*, and that during desorption experiments is referred to as *out-diffusion*. For one special case, the differences between *in-diffusivity* (D_{in}) and *out-diffusivity* (D_{out}) can be found in Sections 3.3.1 and 3.6.1.6. On the other hand, with the profiling method, the diffusivity as a function of concentration is independent of whether it is in-diffusion or out-diffusion.

(4) It is difficult to obtain how the diffusivity depends on concentration using the bulk mass loss or gain method, although it is possible to verify specific concentration dependence by conducting experiments from small degrees of mass loss to almost complete mass loss (Wang et al., 1996). On the other hand, the shape of diffusion profiles reveals the dependence of diffusivity on concentration.

3.6.1.3 Thin-source method

The thin-source method is also referred to as the *thin-film method*. One surface is cut into a plane surface and polished. A very thin layer is then sprayed or spread onto the surface. The thin layer contains the component of interest, which at high temperature diffuses into the interior of the sample from the polished surface. After the experiment, a section is cut perpendicular to the polished surface. Concentration profile is measured as a function of distance away from this surface. If the length of the concentration profile is much greater than (>100 times) the thickness of the thin layer on the surface, the problem may be treated as a

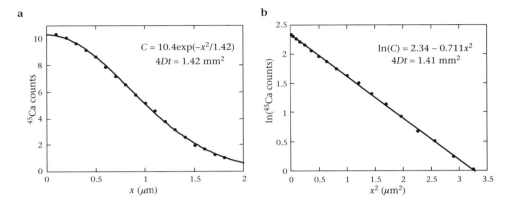

Figure 3-31 (a) A concentration profile from a thin-film experiment, and (b) a linearized plot of the same data. Data are read from Figure 2a of Watson (1979a).

thin-source problem (meaning the thin film is treated as an infinitely thin layer) and the solution is Equation 3-45b. If the length of the diffusion profile is shorter than the thickness of the thin film, the profile would approximate a diffusion couple. If the length of the concentration profile is in between the above two cases (e.g., the diffusion profile is about 5 times the thickness of the thin layer), then the problem must be treated using the general solution for a finite (or extended) source problem.

For the thin-source problem with constant D, the concentration profile is (Equation 3-45b):

$$C = \frac{M}{(\pi Dt)^{1/2}}\, e^{-x^2/(4Dt)} = C_0\, e^{-x^2/(4Dt)}, \tag{3-127a}$$

where $C_0 = [M/(\pi Dt)^{1/2}]$ is the concentration on the surface, which decreases with time. For a given concentration profile (which is measured at a given time, meaning a given t), one simply fits $C = C_0 e^{-x^2/(4Dt)}$ as a curve, or fits $\ln C$ versus x^2 as a straight line, to obtain D. A profile and two fits are shown in Figure 3-31.

Sometimes the concentration cannot be measured directly. One way to treat such a profile is to measure the total concentration by removing successively thin layers of the surface layers. Hence, the first measurement is the integrated total concentration of the species (such as β-counting) from 0 to ∞. Then a thin layer dx is removed and the second measurement is the integration from dx to ∞. The procedure is repeated until the whole profile is measured. Every measurement is hence the integral of the concentration profile $\int C\, dx$ from x to ∞, i.e.,

$$u = \int_x^\infty C_0 e^{-x^2/(4Dt)} dx = C_0\sqrt{\pi Dt}\,\, \mathrm{erfc}\left(\frac{x}{\sqrt{4Dt}}\right) = M\, \mathrm{erfc}\left(\frac{x}{\sqrt{4Dt}}\right). \tag{3-127}$$

That is, the resulting "profile" is an error function and can be fit in such a way to obtain D.

3.6.1.4 Crystal dissolution method

The dissolution of zircon into a melt (Harrison and Watson, 1983; Baker et al., 2002) is used as an example in the following discussion. As a zircon crystal dissolves into a melt, ZrO_2 concentration in the melt next to the crystal (the interface melt) is high, leading to ZrO_2 diffusion away from the crystal. After some time at high temperature, the charge is quenched. ZrO_2 concentration profile in the melt (now glass) is measured, which is fit to obtain the diffusion coefficient.

To use this method to obtain diffusivity, the dissolution must be diffusion controlled. The diffusion aspect was discussed in Section 3.5.5.1, and the heterogeneous reaction aspect is discussed later. The melt growth distance (L, which differs from the crystal dissolution distance by the factor of the density ratio of crystal to melt) may be expressed as (Equation 3-115d)

$$L = 2\alpha(Dt)^{1/2},$$

where α is a dimensionless parameter, and D is diffusivity. The constant α may be solved from the following equation (similar to Equation 3-117c but not identical because it is crystal dissolution here but crystal growth there):

$$\pi^{1/2}\alpha \ e^{\alpha^2} \mathrm{erfc}(-\alpha) = (w_0 - w_\infty)/(w_s - w_0), \tag{3-128}$$

where w_∞ is ZrO_2 mass fraction in the initial melt, w_0 is ZrO_2 mass fraction in the interface melt (in the melt right next to zircon crystal), and w_s is ZrO_2 mass fraction in the solid (zircon).

Because the melt growth distance is $L = 2\alpha(Dt)^{1/2}$, and melt growth rate is $u = \alpha(D/t)^{1/2} = L/2t$, the diffusion equation for mineral dissolution is (similar to Equation 3-114a)

$$\frac{\partial w}{\partial t} = \frac{\partial}{\partial x}\left(D\frac{\partial w}{\partial x}\right) - \alpha\sqrt{\frac{D}{t}}\frac{\partial w}{\partial x}, \tag{3-129}$$

where $\alpha(D/t)^{1/2}$ is the crystal dissolution rate, and there is a negative sign in front of the parameter α because we are dealing with crystal dissolution here and Equation 3-114a is for crystal growth. The above equation has been solved in Section 3.5.5, leading to the following equation for the concentration profile:

$$w = w_\infty + (w_0 - w_\infty)\frac{\mathrm{erfc}\left(\frac{x}{\sqrt{4Dt}} - \alpha\right)}{\mathrm{erfc}(-\alpha)} \tag{3-129a}$$

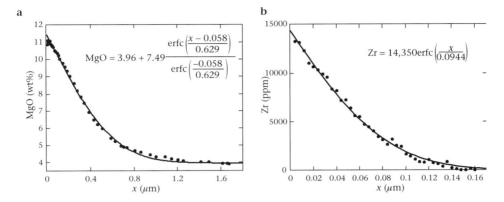

Figure 3-32 Diffusion profiles during mineral dissolution. (a) MgO diffusion profile during olivine dissolution and fit to the profile. Data from exp# 212 of Zhang et al. (1989). (b) Zr diffusion profile during zircon dissolution and fit to the profile ($L \approx 0.001$ mm). Data read from Figure 2a of Harrison and Watson (1983).

Because only L (not α) can be calculated from multiplying the measured crystal dissolution distance by the density ratio ρ^c/ρ (where ρ^c and ρ are crystal and melt density), for data fitting, the above equation needs to be recast into the following form:

$$w = w_\infty + (w_0 - w_\infty)\frac{\mathrm{erfc}\left(\dfrac{x - L}{\sqrt{4Dt}}\right)}{\mathrm{erfc}(-L/\sqrt{4Dt})}. \tag{3-129b}$$

When $w_0 < w_s$ (for example, when $w_0 < 0.005w_s$, such as dissolution of zircon), the dissolution rate is small, and α may be treated as roughly zero, $\mathrm{erfc}(-\alpha) \approx 1$, leading to the following simpler equation:

$$w = w_\infty + (w_0 - w_\infty)\mathrm{erfc}\left(\frac{x}{\sqrt{4Dt}}\right). \tag{3-129c}$$

Some concentration profiles are shown and fit in Figure 3-32.

Experimental concentration profiles from crystal dissolution experiments may show another complexity. The crystal dissolves at the high temperature of the experiment but as the sample is quenched, the temperature in the sample would drop below the saturation temperature, leading to crystal growth. This growth is only for a very short duration (a few seconds, depending on the quench rate), but the effect is to deplete MgO concentration for the case of olivine (Figure 3-32a); careful readers may notice that MgO concentration near the interface is slightly below the maximum. This part of the profile should not be used in diffusion profile fitting using Equation 3-129b.

If the diffusivity varies, the diffusivity as a function of composition may be obtained using the Boltzmann method. Starting from Equation 3-129 and applying Boltzmann transformation, $\eta = x/(2\sqrt{t})$, we obtain the following equation after some mathematical manipulation (similar to steps in Section 3.2.8.2):

$$D = \frac{\int_{w(x_0)}^{w(\infty)} (x - L)\mathrm{d}w}{2t(\mathrm{d}w/\mathrm{d}x)_{x=x_0}}, \tag{3-129d}$$

where x is distance to the crystal–melt interface (i.e., in the interface-fixed reference frame), and D is at concentration of $w(x_0)$.

3.6.1.5 Some comments about fitting diffusion profiles

High-precision and reliable concentration profiles are the key for a high-quality fit. In the profiling method, the diffusion distance determines the diffusivity. Hence, distance must be measured accurately. If the concentration is inaccurate by a constant factor, it would not affect the accuracy of diffusivity determination, unless one wants to relate diffusivity to real concentration. In fitting, it is important to exclude data points that are known to have problems, such as data points within 0.01 or 0.02 mm of the interface for the crystal dissolution method because of crystal growth during quench (Figure 3-32a), or microprobe and IR data points near cracks, or microprobe traverse data points with lower than expected totals, or data points at large x when the inverse error function is used (Figure 3-29b).

When a concentration profile is known to follow a theoretical equation and is fit by the equation, it is important to include "free data," which are natural constraints. For example, in desorption experiments, under the right conditions, the surface concentration is zero. Even if surface concentration cannot be directly measured, this free data point should be applied. Another example is that the fraction of mass loss or gain at time zero is zero. Hence, the linear fit between the fraction and square root of time should be forced through the (0, 0) point (Figure 3-30b). Although this seems a trivial issue, new practitioners may overlook it.

Another often-used technique in curve fitting is to first linearize the relation, and then to fit. Geochemists love straight lines. For example, the fraction of mass loss or gain may be plotted against the square root of time (Figure 3-30b), or the logarithm of concentration against distance squared (Figure 3-31b), or inverse error function against distance (Figure 3-29b). The linearization makes the figure simple and the verification of linearity is visual. In using the linearized form of an equation, it is necessary to understand how the absolute error in each value has been propagated and to account for these errors. One extreme example is the use of inverse error function against distance (Figure 3-29b). When the inverse error function $\mathrm{erfc}^{-1}[(C - C_{\infty})/(C_0 - C_{\infty})]$ is calculated from the original concentration data, for small values of $(C - C_{\infty})$, meaning at large distance x, the absolute error explodes. Sometimes the value of $\mathrm{erfc}^{-1}[(C - C_{\infty})/(C_0 - C_{\infty})]$ does not even exist because the measured C is slightly below C_{∞} due to analytical errors. Therefore, in such a fit, only when the absolute values of $(C - C_{\infty})$ are much larger than the analytical errors, would the calculated $erfc^{-1}[(C - C_\infty)/(C_0 - C_\infty)]$ be reliable for use in the linear fit. For other cases, such as the fraction of mass loss or gain plotted against the square root of time (Figure 3-30b), the linear relation holds under specific conditions (e.g., $M_t/M_\infty < 0.5$) and the issue of error propagation is not critical.

3.6.1.6 Values of diffusivity versus experimental methods

Tracer diffusivities are often determined using the thin-source method. Self-diffusivities are often obtained from the diffusion couple and the sorption methods. Chemical diffusivities (including interdiffusivity, effective binary diffusivity, and multicomponent diffusivity matrix) may be obtained from the diffusion-couple, sorption, desorption, or crystal dissolution method.

Diffusivity of a species in a phase is an intrinsic property and does not depend on the experimental method. If diffusivity does not depend on the concentration of the species, then diffusivity extracted from different methods has the same meaning and hence should all agree within experimental error. However, if diffusivity depends on the concentration of the species or component, the meaning of diffusivity extracted using different techniques may differ, leading to difference in diffusivity values.

If the mass loss technique is used to extract diffusivity, the diffusivity reflects an average of the diffusivity along the concentration profile during outward diffusion, which may be termed out-diffusivity D_{out}. If the mass gain technique is used, the diffusivity reflects another average along the concentration profile during inward diffusion, which may be termed in-diffusivity D_{in}. The average is weighted more heavily near the surface because this controls the diffusive flux out of or into the sample. If D is constant, then $D_{in} = D_{out}$. The two (D_{in} and D_{out}) may differ if D depends on concentration. Consider the specific case of diffusivity proportional to its own concentration ($D = D_0 C/C_0$) in a total concentration range 0 to C_0 (i.e., for mass loss experiment, the surface concentration is zero and the initial concentration is C_0; and for mass gain experiments, the surface concentration is C_0 and the initial concentration is zero). The expressions for D_{out} and D_{in} are as follows (Equations 3-88a,b,c; Zhang et al., 1991a; Wang et al., 1996):

$$D_{in} = 0.619D_0, \tag{3-88a}$$

$$D_{out} = 0.347D_0, \tag{3-88b}$$

$$D_{in} = 1.78D_{out}. \tag{3-88c}$$

The relations are shown in Figure 3-33. Because D_0 depends on C_0, D_{in} and D_{out} also vary with C_0. For other cases, D_{in} and D_{out} may be related differently.

When a diffusion-couple profile is fit by a constant diffusivity but the diffusivity actually changes with the concentration of the diffusing component, the extracted D is also an average. This average would differ from either D_{in} or D_{out}.

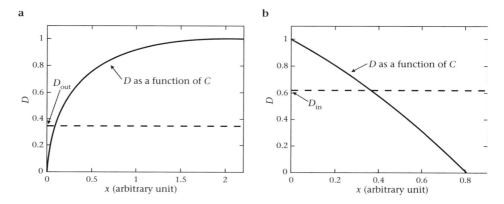

Figure 3-33 Comparison of (a) D_{out} during dehydration (outward H_2O diffusion) and (b) D_{in} during hydration (inward H_2O diffusion) and concentration-dependent D.

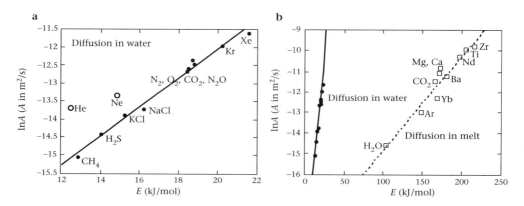

Figure 3-34 The compensation law for diffusion of some species in (a) water where $\ln A \approx -20.12 + 0.404E$, and (b) a silicate melt where $\ln A \approx -19.29 + 0.0453E$.

Hence, it is critical to understand the meaning of the extracted diffusivity in order to distinguish subtle differences between them.

3.6.2 Relations and models on diffusivity

3.6.2.1 Compensation law

The compensation law is a very rough empirical correlation between the activation energy and the pre-exponential factor of diffusion. Winchell (1969) showed that the logarithm of the pre-exponential factor (A) is roughly linear to the activation energy (E):

$$\ln A = a + bE, \tag{3-130}$$

where a and b are two constants, and called the relation the compensation law. Figures 3-34a and 3-34b show two compensation relations, one for diffusion in water and one for diffusion in silicate melts (data from Appendix 4). There is considerable scatter in both cases (sometimes more than a factor of 10), which suggests that the "law" is only approximate.

Given the compensation law $\ln A = a + bE$, it follows that (Lasaga, 1998)

$$D = \exp(\ln A - E/(RT)) = \exp(a + bE - E/(RT))$$
$$= \exp[a + E(b - 1/(RT))]. \tag{3-130a}$$

That is, if $b - 1/(RT) = 0$, meaning at a critical temperature of

$$T = 1/(bR), \tag{3-130b}$$

then all species would have the same diffusivity of

$$D = \exp(a). \tag{3-130c}$$

Therefore, the compensation law is equivalent to the statement that in a $\ln D$ versus $1/T$ plot, all lines for various species intersect at one common point.

The accuracy of the above relation can be addressed using Figures 3-34a and 3-34b. For diffusion in water, $\ln A \approx -20.12 + 0.40372E$. Hence, at $T = 297.9$ K, all species would have diffusivity of 1.8×10^{-9} m^2/s in water. Examination of Table 1-3a shows that this is only approximately so. For diffusion in silicate melts, $\ln A = -19.29 + 0.04526E$. Hence, at $T = 2657$ K, all species would have a diffusivity of 4.2×10^{-9} m^2/s in silicate melts. There are no data at such a high temperature to test this prediction.

In addition to applications to diffusion in the same phase, the compensation law has also been applied to the diffusion of a given species in many phases (Bejina and Jaoul, 1997). This is equivalent to the assumption that at some critical temperature, the diffusion coefficients of the species in all phases would be the same. The relation again is expected to be very approximate.

Not many practical uses have been found for the compensation law because it is not accurate enough. One potential use of the compensation law is that if one knows the diffusivity at one single temperature, then both the pre-exponential factor A and the activation energy E may be estimated. That is, the temperature dependence of the diffusivity may be inferred. In practice, however, because the compensation "law" itself is not accurate, the uncertainty of the approach is very large (intolerable in geologic applications). Hence, the approach is not recommended.

3.6.2.2 Diffusivity and ionic conductivity

Diffusion is due to random motion of particles. Conduction is due to motion of ions under an electric field. Ionic diffusivity and conductivity are hence related. Under an electric field, the velocity of an ion is proportional to the electric

Table 3-1 Molar conductivity of ions in infinitely dilute aqueous solutions at 298.15 K

Cations	Molar conductivity $\lambda_{+,0}$ (S·m^2 mol^{-1})	Anions	Molar conductivity $\lambda_{-,0}$ (S·m^2·mol^{-1})
H$^+$	0.03498	OH$^-$	0.01986
Li$^+$	0.00386	F$^-$	0.00555
Na$^+$	0.00501	Cl$^-$	0.00764
K$^+$	0.00735	Br$^-$	0.00781
Rb$^+$	0.00778	I$^-$	0.00768
Cs$^+$	0.00772	NO$_3^-$	0.00714
Ag$^+$	0.00619	HCO$_3^-$	0.00445
Mg^{2+}	0.01062	SO$_4^{2-}$	0.0160
Ca^{2+}	0.01190	CO$_3^{2-}$	0.01386
Sr^{2+}	0.01190		
Ba^{2+}	0.01272		

Note. To calculate mobility (m^2·s^{-1}·V^{-1}), use $\phi = \lambda_0/(zF)$; that is, divide the molar conductivity by the valence and then by the Faraday constant (96,485 C/mol). To calculate diffusivity, use $\mathcal{D} = RT\lambda_0/(zF)^2 = RT\phi/(zF)$.

potential gradient (similar to the diffusion flux proportional to the concentration gradient):

$$\mathbf{u}_+ = \phi_+ \nabla E, \tag{3-131a}$$

$$\mathbf{u}_- = \phi_- \nabla E, \tag{3-131b}$$

where the subscripts "+" and "−" mean cations and anions, respectively, E is the electric potential (in V), the unit of ∇E is V/m, and ϕ_+ and ϕ_- are proportionality constants, called *mobility*, whose unit is $(m \cdot s^{-1})/(V \cdot m^{-1}) = m^2 \cdot s^{-1} \cdot V^{-1}$. The mobility of an ion depends on the character of the ion (charge, size, etc.) and of the solution (viscosity, etc.). The concept of mobility is useful in linking various quantities. The mobilities of ions may be calculated from molar conductivity data listed in Table 3-1 (see footnotes to Table 3-1).

In the discussion below, a cation or anion is considered generally and the subscripts "+" and "−" are ignored. For ionic motion in solutions, the force experienced by the ion is $ze\nabla E$ (where z is valence and e is unit change), which must be balanced by the drag that equals the velocity times frictional coefficient f. That is,

$$ze\nabla E = f\mathbf{u}. \tag{3-131c}$$

A comparison with the definition of mobility leads to

$$\phi = ze/f. \tag{3-131d}$$

The conductance of an electrolyte solution characterizes the easiness of electric conduction; its unit is reciprocal ohm, $\Omega^{-1} = $ siemens $= S = A/V$. The electric conductivity is proportional to the cross-section area and inversely proportional to the length of the conductor. The unit of conductivity is S/m. The conductivity of an electrolyte solution depends on the concentration of the ions. *Molar conductivity*, denoted as λ, is when the concentration of the hypothetical ideal solution is 1 M $= 1000$ mol/m^3. Hence, the unit of molar conductivity is either $S\,m^{-1}\,M^{-1}$, or using SI units, $S\,m^2\,mol^{-1}$. For nonideal solutions, λ depends on concentration, and the value of λ at infinite dilution is denoted by subscript "0" (such as $\lambda_{+,0}$, and $\lambda_{-,0}$ for cation and anion molar conductivity). The conductivity is a directly measurable property. The molar conductivity at infinite dilution may be related to the mobility as follows:

$$\lambda_0 = \phi z F, \tag{3-131e}$$

where F is Faraday constant ($F = N_{av}e = 96{,}485$ C/mol).

According to *Kohlrausch's law* of the *independent migration of ions*, the total molar conductivity of an electrolyte (made of ν_+ cations and ν_- anions; e.g., $\nu_+ = 1$ and $\nu_- = 2$ for CaCl$_2$ in water) can be expressed as the summation of ionic molar conductivities:

$$\Lambda = \nu_+ \lambda_+ + \nu_- \lambda_-, \tag{3-132a}$$

where Λ, λ_+, and λ_- are molar conductivities of the electrolyte, the cation, and the anion. Molar conductivities of some ions at infinite dilution are listed in Table 3-1. When the concentration is low, the molar conductivity depends on the concentration as follows

$$\Lambda \approx \Lambda_0 - KC^{1/2}, \tag{3-132b}$$

where K is a constant and C is concentration. The above empirical equation is called *Kohlrausch's law*. The constant K may be approximated by $(b_1 + b_2 \Lambda_0)/ (1 + b_3 C^{1/2})$, where b_1, b_2, and b_3 are constants that depend on temperature and solvent properties.

Next the relations between diffusivity, mobility, and conductivity are considered. The flux of a cation may be expressed as

$$\mathbf{J} = C\mathbf{u} = C\phi \nabla E. \tag{3-133a}$$

The difference in chemical potential μ of an ion with valence z is related to the electric potential E as follows:

$$\Delta \mu = -zF\Delta E. \tag{3-133b}$$

Hence, $\nabla \mu = -zF\nabla E$, leading to

$$\nabla E = \nabla \mu / (z\mathbf{F}) \tag{3-133c}$$

For an ideal solution, $\mu = \mu_0 + RT \ln C$, and hence $\nabla \mu = RT \nabla C / C$. Therefore,

$$\mathbf{J} = C\phi \nabla E = - C\phi \nabla \mu / (zF) = -RT\phi \nabla C / (zF). \tag{3-133d}$$

In deriving the above relation, the solution is assumed to be ideal. Comparing the above with the diffusion flux equation leads to

$$\mathcal{D} = RT\phi / (zF), \tag{3-134a}$$

where $\mathcal{D}$ is tracer diffusivity at infinite dilution (or diffusivity in an ideal solution without considering cross-effects), which is equivalent to intrinsic diffusivity. Combining $\phi = ze/f$ (Equation 3-131d) with the above leads to

$$\mathcal{D} = k_B T / f, \tag{3-134b}$$

where k_B is Boltzmann constant (1.3807×10^{-23} J/K) and f is the friction coefficient. Both Equations 3-134a and 3-134b are referred to as the Einstein equation. Using Equation 3-134a, the diffusivity and conductivity may be related as follows:

$$\lambda_0 = \phi zF = (zF)^2 \mathcal{D} / (RT), \tag{3-134c}$$

where the subscript "0" means infinite dilution. Hence,

$$\mathcal{D} = RT\lambda_0 / (zF)^2, \tag{3-134d}$$

which applies to both cations and anions at infinite dilution. Therefore, the molar conductivity of an electrolyte at infinite dilution can be expressed as

$$\Lambda_0 = v_+ \lambda_{+,0} + v_- \lambda_{-,0} = (v_+ z_+^2 \mathcal{D}_+ + v_- z_-^2 \mathcal{D}_-)F^2 / (RT), \tag{3-134e}$$

which is known as the Nernst-Einstein relation, and relates ionic diffusivity and conductivity. The above relations are all derived for ideal (or infinitely dilute) solutions and for ionic species. Because an electrolyte solution (such as NaCl or $MgCl_2$) must be locally neutral, electroneutrality at every local region is a required condition. Considering electroneutrality and with the help of the concept of a self-consistent mean electrical potential (e.g., Lasaga, 1979), the diffusivity of the neutral electrolyte species is related to the ionic diffusivities as follows:

$$\mathcal{D} = \frac{\mathcal{D}_+ \mathcal{D}_- (z_+{}^2 C_+ + z_-{}^2 C_-).}{z_+{}^2 C_+ \mathcal{D}_+ + z_-{}^2 C_- \mathcal{D}_-.} \tag{3-135a}$$

If there is only one electrolyte of 1:1 type (such as NaCl, $ZnSO_4$), then $z_+ = z_-$ and $C_+ = C_-$, leading to,

$$\mathcal{D} = \frac{2\mathcal{D}_+ \mathcal{D}_-}{\mathcal{D}_+ + \mathcal{D}_-} \tag{3-135b}$$

Based on Kohlrausch's law and the relation between conductivity and diffusivity, electrolyte diffusivity at low concentrations decreases linearly with the square root of concentration. Using Equation 3-62, the chemical diffusivity of the electrolyte may be written as

$$D = \frac{\mathcal{D}_+ \mathcal{D}_- (z_+{}^2 C_+ + z_-{}^2 C_-)}{z_+{}^2 C_+ \mathcal{D}_+ + z_-{}^2 C_- \mathcal{D}_-} \left(1 + \frac{d\ln\gamma}{d\ln C}\right). \tag{3-135c}$$

Equations 3-131a,b to 3-134e are exact relations for infinite dilute electrolyte solutions and have been used to obtain diffusivity data from conductivity and vice versa.

Example 3.7 Use molar ionic conductivity data in Table 3-1 to calculate the mobility and diffusivity of Na^+, Cl^- and NaCl at infinite dilution and 298.15 K.

Solution: Molar ionic conductivity $\lambda_0 = \phi z F$, and diffusivity $\mathcal{D} = RT\phi/(zF)$, where $F = 96{,}485$ C/mol, and $RT/F = 0.025693$ V.

For Na^+, $\lambda_0 = 0.00501$ S m²/mol. Hence, $\phi = 5.19 \times 10^{-8}$ m² s⁻¹ V⁻¹; $\mathcal{D} = 1.33 \times 10^{-9}$ m²/s.

For Cl^-, $\lambda_0 = 0.00764$ S m²/mol. Hence, $\phi = 7.92 \times 10^{-8}$ m² s⁻¹ V⁻¹; $\mathcal{D} = 2.03 \times 10^{-9}$ m²/s.

NaCl diffusivity at infinite dilution at 298.15 K using Equation 3-135b is 1.61×10^{-9} m²/s. (Using the expression of Fell and Hutchison (1971), the diffusivity of NaCl at 298.15 K is 1.58×10^{-9} m²/s, in good agreement with the above calculation.)

3.6.2.3 Diffusivities, size, and viscosity

The above section is for diffusion of moving ionic species, whose diffusivity is related to conductivity instead of viscosity and particle size. Relations between diffusivity, size, and viscosity have also been developed, which usually apply to neutral particles. Einstein (1905) investigated Brownian motion and derived a relation for the diffusivity of a neutral particle. For a spherical particle moving in a fluid phase, assuming no-slip condition, the total drag force (including pressure drag and viscous drag) according to Stokes' law is $6\pi\eta a u$, where η is viscosity, a is radius of the sphere, and u is the velocity. That is, the frictional coefficient is $6\pi\eta a$. On the basis of Equation 3-134b, we obtain

$$\mathcal{D} = k_B T/(6\pi\eta a). \tag{3-136a}$$

This is called the *Stokes-Einstein equation*. Hence, the larger the particle, the smaller the tracer diffusivity. If the solution is nonideal, then $\nabla\mu = RT\nabla C(1 + \partial\ln\gamma/\partial\ln C)/C$. Hence, the Stokes-Einstein equation becomes

$$D = \frac{k_B T}{6\pi\eta a} \left(1 + \frac{d\ln\gamma}{d\ln C}\right). \tag{3-136b}$$

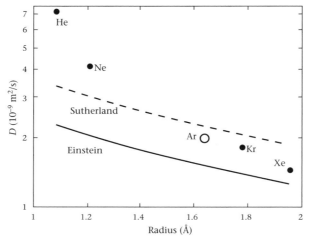

Figure 3-35 Comparison of calculated diffusivity and experimental diffusivity of noble gas elements in water. Noble gas radius from Zhang and Xu (1995). Molecular diffusivity data are from Jahne et al. (1987) except for Ar (Cussler, 1997). A different symbol for Ar is used because different sources for diffusion data may not be consistent. The solid curve is calculated from the Einstein equation, and the dashed curve is calculated from the Sutherland equation. The curve from Glasstone et al. (1941) is outside the scale.

In the derivation, Stokes flow is assumed for the particle, which assumes the liquid medium around the particle flows as a continuum. Hence, the particle size must be significantly larger than the molecules in the liquid matrix (such as H_2O molecules in water). The formulation is not necessarily valid for particles smaller than or about the same size as the matrix molecules themselves.

Many similar formulations have also been advanced (Cussler, 1997). One is by Sutherland (1905), predating Einstein's work, who used the slip condition, so that the total drag is $4\pi\eta au$ instead of $6\pi\eta au$. The result is a diffusivity that is 1.5 times the Einstéin diffusivity

$$\mathcal{D} = k_B T/(4\pi\eta a). \tag{3-136c}$$

Another is by Glasstone et al. (1941), which produces a diffusivity that is 3π times the Einstein diffusivity:

$$\mathcal{D} = k_B T/(2\eta a). \tag{3-136d}$$

Yet another is the Eyring equation (Glasstone et al., 1941):

$$\mathcal{D} = k_B T/(\eta l), \tag{3-136e}$$

where l is the effective jumping distance. Because the jumping distance is less well defined for a given particle, the Eyring equation cannot be directly compared with the Einstein equation.

Table 3-2 Diffusion coefficients in aqueous solutions at 25°C

Dissolved gas molecules	r (Å)	D (m²/s) experimental	D (m²/s) calc Einstein	D (m²/s) calc Sutherland	D (m²/s) calc Glasstone	l (Å) Eyring
He	1.08	7.22×10^{-9}	2.27×10^{-9}	3.42×10^{-9}	2.14×10^{-8}	6.4
Ne	1.21	4.16×10^{-9}	2.03×10^{-9}	3.04×10^{-9}	1.91×10^{-8}	11
Ar	1.64	2.00×10^{-9}	1.50×10^{-9}	2.24×10^{-9}	1.41×10^{-8}	23
Kr	1.78	1.84×10^{-9}	1.38×10^{-9}	2.07×10^{-9}	1.30×10^{-8}	25
Xe	1.96	1.47×10^{-9}	1.25×10^{-9}	1.88×10^{-9}	1.18×10^{-8}	31
SF_6	2.89	1.21×10^{-9}	0.85×10^{-9}	1.27×10^{-9}	8.0×10^{-9}	38

Note. Noble gas radius from Zhang and Xu (1995). Molecular diffusivity from Jahne et al. (1987) except for Ar (Cussler, 1997). For SF_6, the radius is based on S–F bond length of 1.56 Å plus the radius of F- (1.33 Å), and the diffusivity is from King and Saltzman (1995). The jumping distance is calculated from Equation 3-136e using pure water viscosity of 0.89 mPa·s at 25°C.

Some experimental diffusivity data of noble gases in water are shown in Figure 3-35 and Table 3-2 and compared with calculated diffusivities using different formula. The diffusivity clearly depends on the size of the diffusing species. It can be seen that as an order of magnitude approach, both Einstein and Sutherland equations work well for the case of water. However, the slope of the data is different from the slope of the calculated diffusivities. That is, experimental data on diffusivity of noble gases are not inversely proportional to the radius of noble gas atoms. For example, He diffusivity and the Einstein equation would imply a radius of 0.34 Å for He, which is clearly too small. Hence, the difference cannot be attributed to error in the radius estimation. The difference between the Einstein equation and data is best explained by the failure of Stokes flow for such small molecules. That is, the Einstein equation is expected to work better for larger particles, which is consistent with the trend in Figure 3-35. Although the Sutherland equation appears to work better, the fact that it intersects the experimental trend rather than approaching the experimental data as radius increases suggests that it may not work for larger particles. The easiest explanation is that the no-slip condition assumed by Einstein (1905) is better. For the Glasstone et al. equation, it predicts too high a diffusivity. For the Eyring equation, direct comparison is not possible but the jump distance can be calculated for He to Xe from the diffusion data (Table 3-2). The required jump distance of 6 to 31 Å to reproduce the experimental data is clearly too large. For a jump distance of 2.8 Å (diameter of an H_2O molecule), the calculated diffusivity may be a factor of 10 too large.

For silicate melts, the calculated diffusivity of noble gas elements using either of Equations 3-136a to 3-136d may deviate from experimental data by orders of magnitude. For example, in rhyolitic melt at $1100°C$ and 500 MPa and with 3 wt% water, the viscosity is 3.9 kPa s (Zhang et al., 2003), and Ar diffusivity is $1.6 \ 10^{-11} \ m^2/s$ (Behrens and Zhang, 2001). The calculated Einstein diffusivity is $1.6 \times 10^{-15} \ m^2/s$, 4 orders of magnitude less than the experimental data. Using other formulation does not significantly improve the agreement.

In summary, Einstein's equation is able to calculate diffusivity in water to within a factor of 3, and it works better for large neutral molecules. However, it does not work for more viscous silicate melts. Other equations do not work better. The Eyring equation does not work well for diffusion of noble gases in water or silicate melt. For silicate melts, much discussion is available on the applicability of the Eyring equation. In anhydrous melt, it seems that the Eyring equation relating oxygen diffusivity and viscosity is valid within a factor of 2 (Tinker et al., 2004). However, for hydrous melts, viscosity predicted from oxygen diffusivity using the Eyring equation is many orders of magnitude smaller than measured viscosity (Behrens et al., 2007). In short, the equations relating diffusivity, viscosity, and size (or jumping distance) are not accurate, but are useful as a rough guide of how diffusivity would vary.

3.6.2.4 Interdiffusivity and tracer diffusivity

The compositional dependence of interdiffusivity in binary systems has been investigated and a number of equations have been proposed. The models and resulting equations depend on whether the interdiffusing species are ions or neutral particles.

For interdiffusion between same-valence ions (ionic exchange) in an aqueous solution, or a melt, or a solid solution such as olivine $(Fe^{2+}, Mg^{2+})_2SiO_4$, an equation similar to Equation 3-135c has been derived from the Nernst-Planck equations first by Helfferich and Plesset (1958) and then with refinement by Barrer et al. (1963) with the assumption that (i) the matrix (or solvent) concentration does not vary and (ii) cross-coefficient L_{AB} (phenomenological coefficient in Equation 3-96a) is negligible, which is similar to the activity-based effective binary diffusion treatment. The equation takes the following form:

$$D_{AB} = \frac{\mathcal{D}_A \mathcal{D}_B (z_A^2 C_A + z_B^2 C_B)}{z_A^2 C_A \mathcal{D}_A + z_B^2 C_B \mathcal{D}_B} \left(1 + \frac{d \ln \gamma_B}{d \ln C_B}\right), \tag{3-137a}$$

where A and B are two components of the binary interdiffusion system (and not a cation–anion pair as in Equation 3-135c), z_A and z_B are valence charges of A and B, D_{AB} is the interdiffusivity, C_A and C_B are molar concentrations of A and B, $\mathcal{D}_A$ and $\mathcal{D}_B$ are self-diffusivities of A and B, and γ_B is the activity coefficient of B. D_{AB}, $\mathcal{D}_A$, and $\mathcal{D}_B$ all depend on concentration C_B. The term in parentheses accounts for the thermodynamic effect on diffusivity, and $d\ln\gamma_A/d\ln C_A = d\ln\gamma_B/d\ln C_B$. The above equation does not have a name. In case the valences of the two ions are the same, then

$$D_{AB} = \frac{\mathcal{D}_A \mathcal{D}_B (C_A + C_B)}{C_A \mathcal{D}_A + C_B \mathcal{D}_B} \left(1 + \frac{d \ln \gamma_B}{d \ln C_B} \right). \tag{3-137b}$$

If intrinsic interdiffusivity $\mathcal{D}$ (Equation 3-61) is used, then for ionic diffusion,

$$\mathcal{D}_{AB} = \frac{\mathcal{D}_A \mathcal{D}_B (C_A + C_B)}{C_A \mathcal{D}_A + C_B \mathcal{D}_B}. \tag{3-137c}$$

The above model for binary ionic diffusion has been extended to multicomponent ionic diffusion by Lasaga (1979).

For interdiffusion of neutral metal atoms in alloys, the following relation, referred to as the Darken-Hartley-Crank equation, has been derived (Darken, 1948; Shewmon, 1963; Kirkaldy and Young, 1987):

$$D_{AB} = \frac{C_B \mathcal{D}_A + C_A \mathcal{D}_B}{C_A + C_B} \left(1 + \frac{d \ln \gamma_B}{d \ln C_B} \right), \tag{3-138a}$$

where the symbols have the same meaning as in Equation 3-137a. If intrinsic interdiffusivity $\mathcal{D}$ (Equation 3-61) is used, then for interdiffusion of neutral atoms,

$$\mathcal{D}_{AB} = \frac{C_B \mathcal{D}_A + C_A \mathcal{D}_B}{C_A + C_B}. \tag{3-138b}$$

The above model for binary neutral species diffusion has been extended by Cooper (1965) and further extended by Richter (1993) to multicomponent systems.

The difference between Equations 3-137c and 3-138b can be substantial, and increases when the ratio of $\mathcal{D}_A / \mathcal{D}_B$ deviates more from 1. Figure 3-36 compares $\mathcal{D}_{AB}$ calculated from the two expressions assuming (i) ideal solutions and (ii) concentration-independent $\mathcal{D}_A$ and $\mathcal{D}_B$. Barrer et al. (1963) showed that there may be large errors in using Equation 3-137a to predict interdiffusivities. The extensions of these equations to multicomponent systems to predict diffusivity matrix from self-diffusivities (Lasaga, 1979; Richter, 1993) involve more assumptions and are not expected to be accurate.

For diffusion in minerals, it is possible to determine whether the diffusing species is ionic or neutral and hence to determine which model to use. For example, Fe–Mg interdiffusion in olivine is ionic diffusion, but Au–Ag interdiffusion in gold–silver alloy is neutral species diffusion. However, for silicate melts, many diffusing species are present, and it is often impossible to determine whether the diffusing species is ionic or neutral, leading to uncertainty on which model is correct. For example, Kress and Ghiorso (1995) obtained diffusion data in basaltic melt, tested the model of Richter (1993) that is an extension of the Darken model to multicomponent silicate melts, and found that the model failed. However, Kress and Ghiorso (1995) did not test the model of Lasaga (1979). Molecular dynamics simulations or first principles calculations may re-

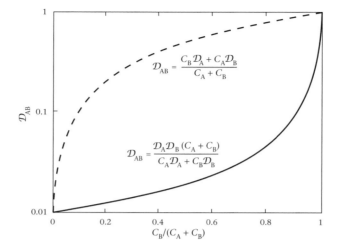

Figure 3-36 The dependence of interdiffusivity on composition for two models (Equations 3-137c versus 3-138b) for ideal solutions and concentration-independent $\mathcal{D}_A$ and $\mathcal{D}_B$. The solid curve is for interdiffusion of two ions of identical charge. The dashed curve is for interdiffusion of neutral atomic species such as in an alloy.

veal the diffusing species in silicate melts. For example, Kubicki and Lasaga (1993) investigated interdiffusion in $MgSiO_3$–Mg_2SiO_4 melts and concluded the diffusing species are Mg^{2+} ion, O^{2-} ion, and $[SiO_n]^{(4-2n)}$ complexes.

3.6.2.5 Diffusivity and ionic porosity

The diffusivity of a species in a phase depends on both the species and the phase (in addition to temperature and pressure). In this section, we examine relations on the diffusivity of a species in different phases, and the diffusivity of different species in a single phase.

It has been observed that in some phases, the diffusivity is smaller, and in other phases the diffusivity is larger. One explanation is that the diffusivity is larger if there is more "free" volume in a structure (Dowty, 1980b; Fortier and Giletti, 1989). The "free" volume in a structure is quantified by *ionic porosity*, defined as

$$IP = 1 - V_{ions}/V_0, \tag{3-139a}$$

where V_{ions} is the volume occupied by all ions in one mole of the substance, and V_0 is the molar volume of the structure. Given a mineral formula, V_{ions} can be calculated as follows:

$$V_{ions} = N_A \left(\frac{4}{3}\right) \pi \Sigma n_i r_i^3, \tag{3-139b}$$

where N_A is Avogadro's number (6.02214×10^{23}), i is an ion in the structure, n_i is the number of ion i in the mineral formula, r_i is the ionic radius, and the summation is over all ions (both cations and anions). In such calculations, the ionic radius of O^{2-} is taken to be 1.38 Å $= 1.38 \times 10^{-10}$ m. For the cations one must know the coordination number (CN) to know r_i from Shannon (1976). Therefore, ionic porosity can be calculated as

$$IP = 1 - 2.5225 \Sigma n_i r_i^3 / V_0, \tag{3-139c}$$

where r_i is in Å and V_0 is in cm^3/mol. For example, there are 2 moles of Mg^{2+} (CN $= 6$, $r = 0.720$ Å), 1 mole of Si^{4+} (CN $= 4$; $r = 0.26$ Å), and 4 moles of O^{2-} in 1 mole of forsterite Mg$_2$SiO$_4$. The molar volume of forsterite at 25°C and 0.1 MPa is 43.66×10^{-6} m^3/mol. Hence, the ionic porosity of forsterite Mg$_2$SiO$_4$ is

$$IP = 1 - N_A \left(\frac{4}{3}\right) \pi (2r_{Mg^{2+}}^3 + 1r_{Si^{4+}}^3 + 4r_{O^{2-}}^3) / V_0,$$

$$IP = 1 - 6.02214 \times 10^{23} \left(\frac{4}{3}\right) \pi 10^{-30} (2 \times 0.72^3 + 0.26^3 + 4 \times 1.38^3) / (43.66 \times 10^{-6})$$

$$= 0.348$$

or

$$IP = 1 - 2.5225 (2 \times 0.72^3 + 0.26^3 + 4 \times 1.38^3) / 43.66 = 0.348.$$

When comparing ionic porosity of different minerals, for self-consistency, the same set of ionic radii should be used, and the same temperature and pressure should be adopted to calculate the molar volume of the mineral. Table 3-3 lists the ionic porosity of some minerals. It can be seen that among the commonly encountered minerals, garnet and zircon have the lowest ionic porosity, and feldspars and quartz have the highest ionic porosity. More accurate calculation of IP may use actual X-ray data of average inter-ionic distance and determine the ionic radius in each structure.

There are some difficulties in using the above approach. One is that ionic radii are available at 0.1 MPa and 298 K, but not readily available at high temperatures and pressures. In the ionic porosity calculation, ionic radii are assumed to be independent of temperature and pressure as mineral phases vary with temperature and pressure. With this approximation, IP of stishovite is 0.042, which is clearly too small, and suggests that oxygen radius decreases with increasing pressure to reach stishovite stability. Another difficulty is that many minerals are anisotropic in terms of diffusion, but IP is defined for the whole mineral, not along individual crystallographic directions. Hence, it is necessary to decide whether the average diffusivity, such as the geometric average $(D_a D_b D_c)^{1/3}$, or diffusivity along the fastest diffusion direction is related to IP. Because the diffusivity along the fastest diffusion direction is often the most useful diffusivity (e.g., in estimating the closure temperature), this diffusivity has been related to IP

Table 3-3 Ionic porosity of some minerals at 0.1 MPa and 298.15 K

Mineral	Formula	Cation CN	Cation radii (Å)	V_0 $(10^{-6}$ m^3/mol)	IP
Stishovite	SiO_2	6	0.40	14.01	0.042
Pyrope	$Mg_3Al_2Si_3O_{12}$	8; 6; 4	0.890; 0.535; 0.26	113.16	0.242
Almandine	$Fe_3Al_2Si_3O_{12}$	8; 6; 4	0.92; 0.535; 0.26	115.11	0.250
Grossular	$Ca_3Al_2Si_3O_{12}$	8; 6; 4	1.12; 0.535; 0.26	125.38	0.273
Rutile	TiO_2	6	0.605	18.82	0.266
Zircon	$ZrSiO_4$	8; 4	0.84; 0.26	39.26	0.285
Monazite	$CePO_4$	9; 4	1.196; 0.17	44.66	0.309
Spinel	$MgAl_2O_4$	4; 6	0.57; 0.535	39.77	0.302
Magnetite	$Fe^{3+}(Fe^{2+}Fe^{3+})O_4$	4; 6; 6	0.49; 0.780; 0.645	44.52	0.356
Diopside	$CaMgSi_2O_6$	8; 6; 4	1.12; 0.720; 0.26	66.20	0.330
Hedenbergite	$CaFeSi_2O_6$	8; 6; 4	1.12; 0.780; 0.26	67.95	0.344
Enstatite	$MgSiO_3$	6; 4	0.720; 0.26	31.33	0.334
Ferrosilite	$FeSiO_3$	6; 4	0.780; 0.26	32.96	0.359
Forsterite	Mg_2SiO_4	6; 4	0.720; 0.26	43.63	0.348
Fayalite	Fe_2SiO_4	6; 4	0.780; 0.26	46.3	0.375
Coesite	SiO_2	4	0.26	20.64	0.355
Quartz(α)	SiO_2	4	0.26	22.69	0.414

Note. Ionic radii are from Shannon (1976). Molar volumes are from Berman (1988) except for a few minerals. If the minerals is not stable at 0.1 MPa and 298.15 K, the calculation is based on the volume of the metastable phase.

in the evaluation of Fortier and Giletti (1989). They showed that there is indeed a positive correlation between oxygen "self"-diffusivity (along the fastest diffusion direction) and ionic porosity, but the correlation is not perfect. One clear exception is for the mica group; the diffusivities are much greater than indicated by the trend of $\ln D$ versus IP. For eight minerals (anorthite, albite, potassium feldspar, quartz, hornblende, richterite, tremolite, and diopside), Fortier and Giletti (1989) presented the following equation for oxygen diffusivity at $P_{H_2O} = 100$ MPa:

$$\ln D = -13.8 - 78,288/T + \text{IP}(-29.9 + 147,345/T), \tag{3-139d}$$

where D is in m^2/s and the uncertainty in D is about a factor of 10.

Although the correlation between ionic porosity and diffusivity is imperfect, there is a rough trend that oxygen diffusivity in the minerals increases with increasing IP. The trend is useful in qualitative estimation of closure temperature (among other applications). Extending the relation to metallic systems, one prediction is that diffusion in face-centered cubic structure (25.95% free space) is slower that that in body-centered structure (31.98% free space) of the same metal composition. To avoid the issue of anisotropy, it would be worthwhile to reexamine the relations between diffusivity and ionic porosity using only isometric minerals.

The dependence of diffusivity of a given species in different phases may be applied to the dependence of diffusivity of different species in a single phase. The relation might be interpreted to be the dependence of diffusivity of a species on the size of the doorways for the species to pass through. When applied to diffusion of different species in a given mineral, we obtain the following: For a given doorway (that is, for species occupying the same lattice site), the diffusivity of different species is inversely related to the size of the species. For neutral molecules, their diffusion coefficient in liquid and glass decreases with the size of the molecule (e.g., Figure 3-35), consistent with the expectation. For diffusion in minerals, this effect is best examined for a given mineral composition and for species that have the same valence and occupy the same crystallographic sites so that the doorway diameter is fixed. For example, the diffusivity of the larger cation Ca^{2+} is smaller than that of the smaller cation Mg^{2+} in garnet. However, one may not conclude that in zircon the smaller cation Si^{4+} (ionic radius of 0.26 Å in tetrahedral site) diffuses more rapidly than the larger cation Zr^{4+} (ionic radius of 0.84 Å in octahedral site) because Si^{4+} and Zr^{4+} occupy different sites. Nor may one conclude that in feldspar Si^{4+} diffuses more rapidly than Al^{3+} because they have different valences. Nor may one conclude that in garnet or olivine Si^{4+} (ionic radius of 0.26 Å in tetrahedral site) diffuses more rapidly than Mg^{2+} (ionic radius of 0.72 Å in octahedral site) because Si^{4+} and Mg^{2+} have different valences and occupy different sites.

3.6.2.6 Point defects and diffusion; diffusivity and oxygen fugacity

Defects play a critical role in diffusion in a crystalline phase because diffusivity is roughly proportional to the concentration of vacancy defects. It is important to understand how defect concentration varies with other parameters.

In a lattice structure, if the periodicity is locally disturbed, then there is a defect. There are two types of defects: *point defects* and *extended defects*. A point defect may be any one of the following three types: (i) an atom or ion is absent from a site that normally would be occupied (*vacancies*), (ii) an atom or ion is present in an interstitial position that normally would be unoccupied (*interstitial defects*), or (iii) an atom or ion of an unexpected identity is occupying a site. If defects are due to impurity content, they are called *extrinsic defects*; otherwise, they are called *intrinsic defects*. For ionic compounds, the cation to anion ratio is fixed except for cations with multiple valences (otherwise, charge neutrality would be violated). With the production of defects, there are different ways to maintain charge neutrality for intrinsic defects. If stoichiometric proportions of vacancies are produced in cation and anion sites (e.g., for MgO, equal numbers of cation and anion vacancies), then it is called a *Schottky defect*. If equal numbers of vacancies and interstitials of one ion are produced (i.e., cations are removed from regular sites to interstitial sites leaving behind vacancies), then it is called a *Frenkel defect*. Most ionic crystals usually have one dominant type of intrinsic defects.

The equilibrium concentration of intrinsic defects in a structure depends on temperature. For the Schottky defect, the equilibrium constant K for the defect-generation reaction is

$$K = X_a X_c, \tag{3-140}$$

where X_a and X_c are the mole fractions of anion and cation vacancies. Therefore,

$$X_a = X_c = K^{1/2} = e^{-\Delta G_f/(2RT)}, \tag{3-141}$$

where ΔG_f is the Gibbs free energy for forming a pair of Schottky defects. The above dependence of defect concentration on temperature is similar to the dependence of diffusivity on temperature, with diffusion activation energy $\approx \Delta G_f/2$.

If an ionic structure contains ions with multiple valences, such as Fe, which can be either Fe^{2+} or Fe^{3+}, the mineral may not be stoichiometric. One example is wüstite, $Fe_{1-x}O$, in which most Fe has a valence of 2+, and some has a valence of 3+. The Fe/O ratio (that is, x in $Fe_{1-x}O$) depends on the oxygen fugacity according to the following reaction:

$$(1 - x)FeO + (x/2)O_2 \rightleftharpoons Fe_{1-x}O. \tag{3-142}$$

Assume that ionic diffusion in $Fe_{1-x}O$ occurs via cation vacancies. A defect reaction that conserves charge and atoms can be written as

$$2Fe^{2+} + \left(\frac{1}{2}\right)O_2 \rightleftharpoons 2Fe^{3+} + V + O^{2-}, \tag{3-143}$$

where V denotes a vacancy at the Fe^{2+} site. The equilibrium constant is

$$K = \frac{[Fe^{3+}]^2[V][O^{2-}]}{[Fe^{2+}]f_{O_2}^{1/2}}, \tag{3-144}$$

where brackets mean mole fractions. From the above reaction, the concentration of Fe^{3+} is two times the concentration of vacancy. Hence, the above equation becomes

$$K = \frac{4[V]^2[V][O^{2-}]}{[Fe^{2+}]f_{O_2}^{1/2}}, \tag{3-145}$$

Therefore,

$$[V] = \left(\frac{K[Fe^{2+}]}{4[O^{2-}]}\right)^{1/3} f_{O_2}^{1/6}. \tag{3-146}$$

Because $[O^{2-}]$ may be regarded as constant, the vacancy concentration is proportional to the 1/6 power of f_{O_2}, and to the 1/3 power of Fe^{2+} concentration (Lasaga, 1998). This relation has been applied to some minerals containing Fe^{2+} (such as olivine and pyroxene). If other vacancies are present, they must be considered too.

With a vacancy-dominated diffusion mechanism, diffusion coefficients are proportional to vacancy concentrations. In such a case, the diffusivity would be proportional to the 1/6 power of oxygen fugacity and 1/3 power of Fe^{2+} concentration. Experimental data have shown that diffusivity of ^{18}O, Fe–Mg, ^{30}Si, and Ni in olivine depends on the 0.2 to 0.3 power of f_{O_2} (e.g., Ryerson et al., 1989; Petry et al., 2004). For example, Fe–Mg interdiffusivity in olivine has been investigated by a number of authors. Buening and Buseck (1973) showed that Fe–Mg interdiffusivity is the greatest along the **c**-axis, and the least along the **b**-axis, with $D_c \sim 4D_a \sim 5D_b$. Even though the difference in diffusivity is not very large, for simplicity, diffusion in olivine is often treated to occur only along the **c**-axis. All authors showed that Fe–Mg interdiffusivity increases roughly exponentially with the concentration of the fayalite component, much more rapidly than the 1/3 power of Fe^{2+} concentration. Hence, the compositional variation in fayalite mole fraction must have another effect on the diffusivity. In terms of dependence on oxygen fugacity, Buening and Buseck (1973) inferred that D is proportional to oxygen fugacity to the 1/6 power, but Petry et al. (2004) inferred that D is proportional to oxygen fugacity to the 1/4.25 power. Furthermore, Chakraborty (1997) showed that the diffusivity values by Buening and Buseck (1973) are too high by two orders of magnitude, but did not reexamine the dependence on crystallographic orientation. Using the data of Chakraborty (1997) combined with f_{O_2} dependence of Petry et al. (2004), Fe–Mg interdiffusivity in olivine along the **c**-axis may be expressed as follows (at 1253–1573 K):

$$D_{//c} = (10^7 f_{O_2})^{1/4.25} \exp(-19.96 - 27,181/T + 6.56X_{Fa}), \tag{3-147}$$

where D is in m²/s, f_{O_2} is in Pa, T is temperature in K, and X_{Fa} is the mole fraction of the fayalite component. It is not clear whether the difference of the power of f_{O_2} from 1/6 can be attributed to experimental data uncertainty or to the presence of other defects.

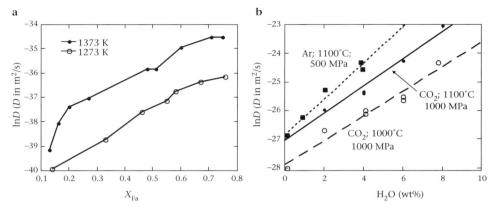

Figure 3-37 Compositional dependence of diffusivities. (a) Fe–Mg interdiffusivity along the **c**-axis in olivine as a function of fayalite content at $P = 0.1$ MPa and log $f_{O2} = -6.9 \pm 0.1$. Diffusion data are extracted using Boltzmann analysis. Some of the nonsmoothness is likely due to uncertainty in extracting interdiffusivity using the Boltzmann method. Data are from Chakraborty (1997). (b) Ar and CO_2 diffusivity in melt as a function of H_2O content. Data are from Watson (1991b) and Behrens and Zhang (2001).

3.6.2.7 Diffusivity and composition

The diffusivity of a species in one phase depends on the composition of the phase. For example, at a given temperature, pressure, and f_{O_2}, Fe–Mg inter-diffusivity in olivine increases rapidly with the concentration of Fa component. Figure 3–37a presents some data that shows that ln D is roughly linear to X_{Fa}, although there are also data showing that the relation is curved.

In silicate melts, the diffusivity is strongly affected by the H_2O concentration: the diffusivities of Ar, CO_2, and molecular H_2O all increase roughly exponentially with increasing H_2O content. That is, ln$D$ increases linearly with H_2O content. Figure 3-37b displays the dependence of CO_2 and Ar diffusivity on H_2O content.

The diffusivity of a species or component in silicate melts may also depend on the SiO_2 content. For H_2O diffusion in silicate melts, the diffusivity appears to decrease exponentially with increasing SiO_2 content (Behrens et al., 2004). For CO_2 diffusion in silicate melts, the diffusivity does not depend significantly on the dry melt composition from basalt to rhyolite (Watson et al., 1982; Watson, 1991b) but depends strongly on the H_2O content (Figure 3-37b). He diffusivity in silicate melts increases from basalt to rhyolite to silica (Shelby, 1972a, b; Jambon and Shelby, 1980; Kurz and Jenkins, 1981).

The dependence of diffusivity in silicate melts on composition is related to how melt structure (including degree of polymerization and ionic porosity) depends on composition. One the one hand, as SiO_2 concentration increases, the melt becomes more polymerized and the viscosity increases. Hence, diffusivity of most structural components, such as SiO_2 and Al_2O_3, decreases from basalt to rhyolite. On the other hand, as SiO_2 content increases, the ionic porosity increases. The increasing He diffusivity from basalt to rhyolite to silica, opposite to the viscosity trend, may be explained by ionic porosity increase because He is a small neutral molecule and can move through the holes without disrupting the structural units controlling viscosity. The effect of increasing H_2O content in silicate melts is to decrease the degree of polymerization (Burnham, 1975; Stolper, 1982a,b), and hence is opposite to that of SiO_2. Much more work is still necessary to understand and quantify the dependence of diffusivity on melt or mineral composition.

3.6.2.8 Diffusivity and radiation damage

When a radioactive nuclide decays, the smaller particles (β and α that is, electrons/positrons and ^{4}He nuclide) of the daughters are ejected at high speed, and the remaining daughter particle recoils. For simplicity of consideration, suppose the parent nuclide emits one particle and becomes the daughter. For example, the first step of ^{238}U decay emits an α-particle, and the daughter is ^{234}Th. The decay of ^{87}Rb is by emission of an electron, leaving the daughter of ^{87}Sr. Let the mass of the emitted particle be m_1, and that of the main daughter be m_2. Because of momentum conservation ($m_1v_1 = m_2v_2$, where v_1 is the velocity of the emitted particle and v_2 is the recoil velocity of the remaining daughter), the energy of the recoil is (m_1/m_2) times the energy of the emitted particle. That is, most of the energy is carried by the emitted particle. Emitted electrons are small in size and mass and hence can easily penetrate a crystal structure without causing much damage. Furthermore, the recoil of the main daughter from β-decay is low in energy because (m_1/m_2) is small. Hence, β-decay does not cause much damage to a crystal structure.[4] For α-decay, an α-particle is more massive and hence is able to displace atoms in a mineral structure. An α-particle may travel tens of micrometers in crystalline structure, knocking off electrons (ionizing effect) or displacing atoms. Furthermore, the recoil energy for α-decay is also greater so that the remaining daughter nuclide may recoil away from its original crystalline site. The most massive damage per decay is from fission, where the emitted nuclides are large and energetic enough to blaze a trail (called fission track) from the crystalline structure. Due to energy difference, α-particles predominantly deposit their energy by ionization, and the recoil particle and fission particles predominantly deposit their energy by displacing atoms. Some minerals, such as zircon and monazite, may contain high concentrations of U and/or Th. The damage from the decay of radioactive nuclides may render part of the structure amorphous. Such materials are referred to as metamict minerals. More on the radiation effect can be found in Ewing et al. (2000).

Radiation damage may cause at least two effects on diffusion. One is that radiation damage results in defects in crystalline structures, and they facilitate diffu-

---

[4]Radiation damage to life depends on whether the radioactive parent nuclides are already in the human body or outside the human body. If the radioactive nuclides are inside the human body, the damage effect is similar to that on crystal structures: more massive particles are more damaging. For radioactive nuclides not inside the human body, the more massive particles cannot penetrate much distance, and could be stopped by cloth or paper, and hence do not cause much damage to life tissues. The less massive β-particles and γ-rays are much more penetrating and can hence deliver energy to life tissues.

sion. Diffusion of all components in damaged minerals is faster than that in pristine minerals. The effect depends on the degree of amorphization, but has not been quantified. One possible way to quantify the relation between diffusivity and radiation damage (amorphization) may be through ionic porosity (Section 3.6.2.5). As the degree of radiation damage increases, the density of the mineral would decrease, and the ionic porosity would increase. Diffusivity would increase with ionic porosity. If amorphization is accompanied by hydration, diffusivity is expected to increase more.

The second effect is on the diffusive loss of daughters of radioactive nuclides. For example, ^{238}U decays into many daughters and finally becomes ^{206}Pb. All the daughters (including the intermediate ones) would have been knocked off their original site by about 10 nm due to recoil, and reside in slightly damaged environment. Hence, these daughter nuclides may diffuse more readily than the parent nuclides. One example is the comparison of the diffusivity of ^{238}U and ^{234}U. Because they are two isotopes of the same element and the mass difference is only 1.7%, the diffusivity difference is expected to be very small, no more than 0.85%. However, it has been found that ^{234}U is much easier to get out of mineral structures into water, as evidenced by secular disequilibrium between ^{234}U and ^{238}U, with $A_{234_U} \approx 1.144 A_{238_U}$ in seawater (Chen et al., 1986). The difference reflects recoil due to α-decay. A similar effect would apply to the diffusion of ^{206}Pb. This second effect applies especially to the diffusion of radiogenic ^4He, which would travel by tens of μm along a random direction. Therefore, the "diffusivity" of radiogenic ^4He consists of two parts: one is due to the initial random motion with an effective distance of tens of micrometers, and the second is normal diffusion.

3.6.2.9 Summary

Many relations have been proposed between diffusivity and other parameters, some theoretical and some empirical. Some of the relations are more accurate than others. For example, the equations relating conductivity and diffusivity for infinitely dilute solutions (hence, tracer diffusivities) are accurate, but the equations relating self-diffusivities and interdiffusivities are model dependent and not accurate, especially for concentrated solutions. The compensation law is empirical and very approximate, often with an uncertainty of a factor of ten or more. The relation between diffusivity and ionic porosity is useful for qualitative estimations. It has not been tested extensively for quantitative applications. There is no unique relation between diffusivity, viscosity, and size. Each of the Einstein, Sutherland, Glasstone, and Eyring equations is applicable when the appropriate assumptions are satisfied, but none is general and errors may be orders of magnitude. Nonetheless, there is a rough anti-correlation between diffusivity and viscosity. The compositional dependence of diffusivity has been examined in only a small number of systems. The logarithm of the diffusivity of a minor or trace component is often linear to the concentration of a major component, but much more work is necessary to examine whether the relation is general.

Problems

3.1 Calculate mass loss during nuclear and chemical reactions.

a. The nuclear hydrogen burning reaction may be written as follows: $4^1H \rightarrow {}^4He$. The mass of ^1H is 1.007825 atomic mass units (amu; 1 amu $= 1.6605 \times 10^{-27}$ kg), and that of ^4He is 4.002603 amu. Calculate the fractional mass loss during nuclear hydrogen burning.

b. The chemical hydrogen burning reaction may be written as follows: $H_2(g) + (\frac{1}{2})O_2(g) \rightarrow H_2O(g)$. The energy released is 242 kJ per mole of H_2O produced. Calculate the fractional mass loss during chemical hydrogen burning. Is this mass loss noticeable?

c. As H_2O vapor condenses to form H_2O liquid, 44 kJ/mol of energy is released. Calculate the fractional mass loss.

3.2 If the Fe–Mn interdiffusivity in a mineral is $1.0 \times 10^{-21}\,m^2/s$ at 800°C, $5.1 \times 10^{-20}\,m^2/s$ at 1000°C, and $9.1 \times 10^{-19}\,m^2/s$ at 1200°C, find the activation energy and the pre-exponential factor.

3.3 Calculate the following to the best precision possible of your calculator or computer (you are allowed to use a spreadsheet program):

a. erf(0.11), erfc(0.11), ierfc(0.11)

b. erfc(4.15) (this is a very small number but is not zero)

c. erfc(7.1) (this is a very small number but is not zero)

3.4 Following the steps below (also the steps on how the diffusion equation is derived in class), derive the heat conduction equation in one dimension. You should try to understand the concepts so that you can finish this problem without looking at the notes or book.

a. Consider thermal energy conservation in a small volume (Δx times the cross-section area). The thermal energy increase in the small volume is mass times heat capacity time the temperature increase. The mass equals density times volume.

b. The thermal energy flux is related to the temperature gradient according to Fourier's law:

$$J = -k\partial T/\partial x,$$

where J is the thermal energy flux and k is the thermal conductivity. Combine the energy conservation equation and Fourier's law to obtain the heat conduction equation.

3.5 The diffusion coefficient of Pb in monazite depends on temperature as $D = \exp(-0.06 - 71,200/T)$ m^2/s (Cherniak et al., 2004). You found a 100-μm-diameter monazite crystal in a metamorphic rock. Assume that the monazite crystal formed at peak metamorphic temperature.

   a. If the peak temperature of the metamorphic rock was estimated to be 600°C and the duration of metamorphism is 10 Myr, estimate how thick a layer of monazite has been affected by diffusive loss of Pb, and then determine whether it is possible to determine the peak metamorphism age.

   b. Do the same if the peak temperature of the metamorphic rock was 800°C.

   c. Do the same if the peak temperature of the metamorphic rock was 1000°C.

3.6 Use a spreadsheet program to do the calculations in this problem.

   a. Diffusion coefficient of molecular H$_2$O (H$_2$O$_m$) depends on T, P, and total H$_2$O (H$_2$O$_t$) concentration as follows (Zhang and Behrens, 2000):

$$D_{H_2O_m} = \exp[(14.08 - 13128/T - 2.796P/T) + (-27.21 + 36892/T + 57.23P/T)X],$$

where T is in kelvins, P is in MPa, X is the mole fraction of H$_2$O$_t$, and $D_{H_2O_m}$ is in μm^2/s. The mole fraction X in rhyolitic melt may be calculated from weight percent (w) as follows:

$$X = (w/18.015)/[(w/18.015) + (100 - w)/32.49].$$

Calculate molecular H$_2$O diffusivity ($D_{H_2O_m}$) under the following conditions:

T (K)	P (MPa)	X	$D_{H_2O_m}$ (μm^2/s)	K	dX_m/dX	$D_{H_2O_t}$ (μm2/s)
900	0.1	0.001				
900	0.1	0.01				
900	0.1	0.08				
1200	500	0.001				
1200	500	0.01				
1200	500	0.08				

b. Assume the equilibrium constant for the homogeneous reaction $H_2O_m + O =$ $2OH$ is $K = 6.53 \exp(-3110/T)$, independent of P and X. Calculate K at the above T, P, and X conditions.

c. Using (or rederiving) the relation between X_m (mole fraction of H_2O_m), X and K (i.e., X_m in terms of X and K), derive dX_m/dX. Show this expression. Then calculate dX_m/dX in the above table.

d. Calculate total H_2O diffusivity under the same conditions (in the above table) using the following relation: $D_{H_2O_t} = D_{H_2O_m} dX_m/dX$.

3.7 Fe–Mg interdiffusion in olivine along the **c**-axis (fastest diffusion direction) is

$$D_{//c} = (10^7 f_{O_2})^{1/4.25} \exp(-19.96 - 27,181/T + 6.56 X_{Fa})$$

where T is in K, f_{O_2} is in Pa, and X_{Fa} is the mole fraction of the fayalite component (Chakraborty, 1997; Petry et al., 2004). Estimate whether equilibrium is reached between 0.5-mm olivine (Fo88) and melt at 1300°C and $f_{O_2} = 0.01$ Pa in an experiment (2 days), and in a magma chamber (1000 yr). (That is, whether the mid-diffusion distance is much greater than the half-thickness of olivine.)

3.8 When you boil an egg, assuming that the boiling temperature is 100°C, how much time is necessary so that the temperature at the center of the egg reaches 90°C? Assume that the egg is a sphere, the initial temperature of the egg is 5°C, and heat diffusivity $\kappa = 0.8$ mm^2/s. Use the average radius of a chicken egg.

3.9 Watson (1979a) carried out tracer diffusion experiments by loading a small amount of ^{45}Ca tracer onto one surface of a cylinder. The cylinder was heated up and ^{45}Ca diffuses into the cylinder. Assume that diffusion is along the axis of the cylinder (i.e., there is no radial concentration gradient). Assume that $D = 10^{-11}$ m^2/s. The cylinder is 3 mm long. Calculate the diffusion profile (concentration normalized to the surface concentration) at $t = 2$ h and $t = 8$ h. How does the concentration profile at 8 h look like when compared to that at 2 h?

3.10 Air contains ~1% of Ar that can dissolve and diffuse into glass at high temperatures. For a glass cylinder heated to high temperature for 2 h with only one surface in contact with air (all other surfaces are welded to a metal capsule). The glass cylinder initially does not contain any Ar and can be viewed as semi-infinite. Ar diffusivity is 10^{-12} m^2/s. Calculate the diffusion profile.

3.11 Use the Boltzmann transformation to solve the following diffusion equation:

$$\frac{\partial C}{\partial t} = D \frac{\partial^2 C}{\partial x^2} - \frac{A}{\sqrt{t}} \frac{\partial C}{\partial x} \quad t > 0, x > 0,$$

where A is a constant. The initial condition is

$$C|_{t=0} = C_\infty \quad x > 0,$$

and the boundary condition is

$$C|_{x=0} = C_0 \quad t > 0.$$

3.12 An experiment is carried out to study oxygen isotope equilibrium between a spinel and a fluid. The equilibrium is reached at the experimental temperature. As the experimental charge is quenched, the ^{18}O concentration in the spinel in equilibrium with the fluid is assumed to vary as

$$C = C_0 + \alpha \sqrt{1 - \frac{\tau}{t + \tau}},$$

due to temperature decrease, where C_0 is the equilibrium concentration in spinel at the experimental temperature, and α is a constant depending on how the fractionation factor changes with temperature. The diffusion coefficient of ^{18}O in the spinel is assumed to vary according to

$$D = D_0/(1 + t/\tau)^2$$

due to temperature decrease, where D_0 is the diffusion coefficient at the experimental temperature, and τ is a characteristic time for quench. Assume that the diffusion can be viewed as through a one-dimensional semi-infinite medium (if you find that the center is also affected by diffusion using this simple approach, you can conclude that the experimental results are suspicious before you go to more sophisticated approach). Find:

  a. how ^{18}O concentration in spinel varies with distance away from the surface and time,

  b. the solution $C(x,t)$ as $t \to \infty$.

  c. total mass of ^{18}O that enters the spinel per unit area at $t \to \infty$.

3.13 Derive the explicit numerical algorithm for solving a diffusion equation for concentration-dependent D.

3.14 Write a computer program (either a spreadsheet, Fortran, Basic, or C++ program) to solve the following diffusion equation numerically:

$$\frac{\partial C}{\partial t} = D \frac{\partial^2 C}{\partial x^2} \quad t > 0, \ 0 < x < L,$$

where D is a constant. The initial condition is

$$C|_{t=0} = C_0 x/L, \quad x > 0,$$

and the boundary condition is

$$C|_{x=0} = C|_{x=L} = C_0/2 \quad t > 0.$$

Let $X = x/L$, $T = Dt/L^2$, and $w = C/C_0$. Use the explicit method with $\Delta X = 0.05$. Plot the result (w vs. X) at $T = 0.01$ using (i) $\Delta T = 0.001$ (stable); (ii) $\Delta T = 0.002$ (unstable). Compare the results. What can you conclude?

3.15 The following diffusion data are adapted from experimental diffusion data for water diffusion in a basaltic melt (Zhang and Stolper, 1991). The experiment was carried out at 1300°C and the duration of the experiment is 10 minutes. Using Boltzmann analysis to obtain diffusion coefficients or water as a function of water concentration. (*Hint:* You will probably need to use a spreadsheet program to do simple integration and differentiation. You may also try to write a simple program. You may fix the concentration at one end to be 0.410 and the other end to be 0.100.)

    a. Smooth the data in an objective way, either with a french curve or use your eye to draw a best fit curve through the data. You will appreciate the difficulties in Boltzmann analyses using real diffusion data because this data set is as good as any diffusion data one ever gets. You may also try curve fitting, but be careful to avoid *systematic* error, which may cause bias in your interpretation of the data. You may also try just the raw data, but it is difficult to calculate the differentials.

    b. Find the interface based on the smoothed data.

    c. Find the D at $C_{water} = 0.15$, 0.2, 0.25, and 0.3, 0.35 (wt%). Plot D vs. C_{water}.

z (mm)	C_{water}	z (mm)	C_{water}	z (mm)	C_{water}	z (mm)	C_{water}
7.0	0.1	5.95	0.2122	5.665	0.2795	5.175	0.3613
6.8	0.1	5.95	0.2105	5.65	0.2873	5.1	0.3707
6.6	0.1	5.925	0.2139	5.615	0.2891	5.075	0.3750
6.575	0.103	5.9	0.2250	5.6	0.2934	5	0.3863
6.5	0.1083	5.895	0.2259	5.575	0.3006	4.975	0.3856
6.5	0.1076	5.865	0.2346	5.55	0.3053	4.9	0.3832

z (mm)	C_{water}	z (mm)	C_{water}	z (mm)	C_{water}	z (mm)	C_{water}
6.475	0.1135	5.85	0.2353	5.525	0.3084	4.875	0.3905
6.4	0.1198	5.835	0.2422	5.5	0.3126	4.8	0.3918
6.375	0.1219	5.805	0.2493	5.475	0.3188	4.775	0.3991
6.3	0.1361	5.8	0.2516	5.45	0.3196	4.7	0.4022
6.275	0.1422	5.775	0.2551	5.4	0.3243	4.675	0.4104
6.2	0.1573	5.75	0.2625	5.3	0.3464	4.6	0.4090
6.195	0.1561	5.735	0.2664	5.275	0.3460	4.505	0.4156
6.1	0.1824	5.7	0.2745	5.225	0.3576	4.5	0.4103
6	0.1977	5.695	0.2741	5.2	0.3564		

3.16 Wang et al. (1996) studied diffusion of the hydrous component in pyrope. A natural pyrope wafer initially contains uniform OH content. The wafer was 1.636 mm thick. The total amount of OH in the wafer (average C below) was determined by an IR absorption band at 357 mm^{-1}. After a heating period, the amount of OH in the wafer was redetermined. The new OH content is less than the initial because some OH diffused out. Repeated heating and measurements yield a relation between average concentration and time. Assume that the surface concentration of OH is zero. Find the diffusivity of the hydrous component using the data below. Explain whether this diffusivity is diffusion-in or diffusion-out diffusivity. Under what conditions would the two differ?

Time (s)	0	300	900	2400	6000	11,400	18,600	27,600	38,400
Ave C	0.2018	0.1977	0.1953	0.1908	0.1852	0.1785	0.1733	0.1666	0.159

3.17 Assume that D is proportional to C (that is, $D = D_0 C/C_0$, where C_0 can be chosen as the highest concentration in a given profile, and D_0 is D at $C = C_0$).

  a. Calculate the diffusion couple profile numerically and plot C/C_0 against $x/\sqrt{4D_0 t}$.

  b. Does the profile match the experimental data in problem 3.15? If yes, find D_0.

  c. Calculate the mass loss from a plane sheet of thickness L and plot M_t/M_∞ vs. $\sqrt{t}$.

d. Does the profile match the experimental data in the above problem (3.16)? If yes, find D_0.

3.18 Consider Fe–Mg exchange between olivine (Fe, Mg)$_2$SiO$_4$ and spinel (Fe, Mg)Al$_2$O$_4$. Assume that D in olivine is 10^{-16} m^2/s and that D in spinel is 10^{-17} m^2/s. Initial composition of the two phases are Fe/(Fe + Mg) = 0.2 in olivine and Fe/(Fe + Mg) = 0.3 in spinel. Assume that the exchange occurred at constant temperature and that the crystals are very large so that you can treat each mineral as one-dimensional and semi-infinite. Ignore anisotropy of olivine. The exchange coefficient is $K_D = $ (Fe/Mg)$_{sp}$/(Fe/Mg)$_{ol}$ = 2.5.

a. Consult any book to find the concentration of Fe + Mg per unit volume of olivine and of spinel. Then find the respective Fe and Mg concentrations in mol/L in olivine and spinel.

b. Calculate the composition of olivine and spinel at their mutual interface.

c. Calculate (you can use a spreadsheet program) and plot the concentration profile after 100 years. Which profile (i.e., in which phase) is steeper?

3.19 Suppose there was a major spill of 600 kg of a toxic chemical (that can dissolve in water) in a river that is 20 m wide and 3 m deep. The local government of a city 180 km downstream from the spill site asks you to evaluate the water quality (whether it can be piped into the city water supply) in the river next to the city as a function of time. Suppose water flow rate is 2 m/s and width and depth of water of the river are constant. Assume an eddy diffusivity of 10 m^2/s. You find from EPA guidelines that the maximum tolerable concentration of the toxic substance for drinking water is 0.01 ppb.

a. Estimate the time interval for the toxic substance to spread across the river, so that for times much longer than this, one can treat the problem as a one-dimensional diffusion and flow problem.

b. Assuming that the concentration is uniform across the river, obtain the solution to the problem. Give values for each parameter in your solution. Check the units to make sure there is consistency. Then convert the concentration into ppb and rewrite the solution.

c. Plot the concentration of the toxic substance as a function of time at 180 km downstream.

d. Determine the time required for the toxic water to arrive at the city (180 km downstream). Use the EPR guideline to determine whether water is toxic or not (i.e., whether water can be piped into city water supply).

e. Because there are many uncertainties in your calculation, assume that the uncertainty in concentration is a factor of 2. Determine the time interval (starting from the time of spill) during which the city water supply should not take any water from the river.

3.20 Examine the applicability of the compensation law using the following examples (data can be found in the Appendix 4, plus your own search of data from literature).

a. Use diffusion data of various species in zircon.

b. Use diffusion data of Sr in various minerals.

c. Use diffusion data of O in various minerals.

4 Kinetics of Heterogeneous Reactions

Most reactions encountered by geologists are *heterogeneous reactions*, that is, reactions involving two or more phases. A heterogeneous reaction is a complicated process involving multiple steps and paths. The steps include nucleation, interface reaction, and mass/heat transport (which may be accomplished by diffusion and/or convection). To produce a new phase from an existing phase, the new phase must first form. The formation of tiny embryos of the new phase from another phase or phases is called *nucleation*. All heterogeneous reactions in which a new phase forms require the nucleation of stable embryos of the new phase, which serves as a template for the crystal to grow. For heterogeneous reactions in which all the phases are initially present (such as mineral dissolution), nucleation is not necessary. The growth of the new phase involves interface reaction and mass/heat transport. *Interface reaction* is the attachment and detachment of atoms, ions, and molecules to or from a phase. Hence, the growth of a new phase or the consumption of an old phase requires reactions at the interface. *Mass transport* brings the necessary ingredients to and excess components away from the new phase. For the growth or dissolution of a phase from another phase or other phases of different composition, such as olivine growth in a basaltic melt, mass transport is necessary. However, for the growth or melting of a mineral in its own melt, mass transfer is not necessary. *Heat transfer* brings the necessary heat to or excess heat away from the new phase. Because heterogeneous reactions always involve heat production or consumption, heat transfer is always present. Heat transfer is orders of magnitude faster than mass transfer in liquids and solids. Therefore, when mass transfer is necessary, heat transfer does not limit the reaction rate of heterogeneous reactions and is not considered. However, for the growth or melting of a mineral in its own melt, heat transfer may play a role in controlling the heterogeneous reaction rate. After the new phase completely replaces the old phase, or when the new phase is in equilibrium with the old phase (e.g., precipitation of crystals in an aqueous solution), there may be a *coarsening* step (also called *Ostwald ripening*) during which many small crystals are replaced by fewer larger crystals.

A given heterogeneous reaction or one of the steps may be accomplished by different paths. For example, nucleation may be realized by homogeneous and/or heterogeneous nucleation. *Homogeneous nucleation* means nucleation of a new phase inside an existing phase; whereas *heterogeneous nucleation* means nucleation of a new phase at the interface of two existing phases. Mass transfer may be achieved by diffusion and/or convection. Heat transfer may be attained by heat conduction and/or convection. An existing mineral out of equilibrium with a melt with respect to some exchange reactions (such as Fe–Mg exchange or isotopic exchange) may reach equilibrium by either (i) diffusion in the crystal or (ii) dissolution of the existing nonequilibrium crystals and reprecipitation of new crystals of the same mineral but equilibrium composition. These paths must be determined before quantitative understanding of heterogeneous reaction rates.

Among these steps and paths, diffusion is probably the best understood and can be quantified very well, though diffusion coefficients for a specific application may not be available and diffusion in a multicomponent system can be mathematically complex. Mass transport in the presence of convection is more complicated; however, some problems can be quantified. The theory of interface reaction rates is also available and seems to account for experimental data, but more reliable experimental data are needed on interface reaction rates of minerals in melts and in water. Nucleation is the least understood. The classical theory for homogeneous nucleation based on atomic scale fluctuation, though well developed, often predicts nucleation rates many orders of magnitude smaller than experimental data. Heterogeneous nucleation theory is also available, but the rate is inherently much more difficult to quantify because it depends on the type, number, and size distribution of heterogeneities. Nonetheless, it is widely thought that nucleation in natural systems is often heterogeneous.

The overall rates of heterogeneous reactions do not usually follow rate laws of homogeneous reactions. For example, there are no equivalents of first-, second-, or third-order reactions. Instead, a heterogeneous reaction may be limited by nucleation, interface reaction, or mass transport. For example, component exchange reactions are controlled by mass transfer. The rate of exsolution of gas from beer or champagne is first limited by bubble nucleation, and postnucleation bubble growth is controlled by mass transport. The dissolution of many silicate minerals in pure water is often controlled by interface reaction (Figure 1-12). Each of these controls leads to a relation between the extent of the reaction and time, which are some times called rate "laws," such as linear law (meaning that the extent of a reaction is proportional to time, or constant reaction rate), or parabolic law (meaning that the extent of a reaction is proportional to square root of time). However, such rate laws relating the extent of a reaction versus time differ from rate laws of homogeneous reactions. In the latter, the reaction rate depends on the concentration raised to a certain power, and that power is called the order of a reaction. For example, a linear law (i.e., constant reaction rate) for a heterogeneous reaction should not be called (or confused with) a zeroth-order homogeneous reaction.

There are numerous heterogeneous reactions. It might even be said that many branches of geological sciences are dealing with some specific heterogeneous reactions. For example, the main goal of volcano dynamicists is to understand the kinetics and dynamics of gas exsolution from magma, a relatively simple heterogeneous reaction, but with complicated kinetics and dynamics. Metamorphic petrologists aim to understand the metamorphic reactions (solid-state reactions) and "read" metamorphic rock for its temperature–pressure–time history. Igneous petrologists strive to understand equilibrium, kinetics, and dynamics of crystallization of magma (crystallization involves many heterogeneous reactions). For further treatment, heterogeneous reactions are grouped below.

(1) *Heterogeneous reactions that do not require nucleation of a new phase.* That is, all the phases involved in the reactions are initially present. Many of these reactions can be quantified well if the boundary conditions are simple. The following are some examples.

(1a) *Simple component exchange between phases without growth or dissolution of any phase.* Examples include oxygen isotope exchange between two minerals, such as quartz and magnetite; Fe^{2+}–Mg^{2+} exchange between ferromagnesian minerals, such as garnet and biotite; and hydrogen isotope exchange between hydrous minerals, such as apatite and mica. Nucleation is not necessary, and interface reaction is assumed to be rapid and hence not the rate-determining step. Component exchange between phases is controlled by mass transport. Between solid phases, mass transport is through diffusion. One simple case has been discussed in Section 3.2.4.6. Component exchange between minerals may be exploited as a geospeedometer (Lasaga, 1983; Lasaga and Jiang, 1995), which is covered in Chapter 5. Convection, instead of diffusion, may play a dominant role if at lease one of the phases is a fluid phase. Between solid and fluid, diffusion in the solid phase is usually the slowest and hence controls the reaction rate, but dissolution and reprecipitation may also accomplish the exchange, often more rapidly than diffusion through the solid phase. Hence, in the presence of a fluid phase, it is critical to determine the reaction path before quantitative modeling of the reaction rate. When dissolution and reprecipitation occur, the kinetics is more complicated and more difficult to model.

(1b) *The dissolution and growth of a single crystal, bubble, or droplet (collectively, a particle).* The many-body problem is much more complicated, but if they do not interact (e.g., they are far away from each other), each of the many particles can be treated using the theory for a single particle. Examples include olivine dissolution in a melt, xenolith digestion, contamination of magma by rocks, the growth of existing bubbles in magma, the dissolution of methane hydrate released from marine sediment, and the dissolution of injected carbon dioxide droplets in oceans. The rate may be controlled either by interface reaction or mass transfer. For example, Figure 1-12 shows the controlling mechanism for the dissolution of some minerals in pure water (with large departure from equilibrium). Mineral dissolution in silicate melts at high temperature is often controlled by mass transport when the departure from equilibrium is large. When departure from equilibrium is extremely small, interface reaction rate is very small and controls the whole reaction rate. The kinetic treatment for interface-controlled dissolution is different from that for diffusion-controlled or convection controlled dissolution. Because igneous petrologists and volcanologists often deal with such reactions, and because the kinetics of the processes is well understood, this class of problems will be developed in depth later.

(1c) *Coarsening* of crystals, also called *Ostwald ripening*.

(1d) *Reactions between gas and solid at the interface.* For example, the oxidation of metal in air is such a reaction.

(2) *Heterogeneous reactions that require nucleation.* Quantitative prediction of the rates of these reactions is not available because nucleation has not been quantified well. Examples include the following.

(2a) *Simple phase transitions in which one phase converts to another of identical composition.* For example,

diamond $\rightleftharpoons$ graphite,
quartz $\rightleftharpoons$ coesite,
calcite $\rightleftharpoons$ aragonite,
water $\rightleftharpoons$ vapor,
water $\rightleftharpoons$ ice,
melting of a mineral.

Nucleation is necessary for the new phase to form, and is often the most difficult step. Because the new phase and old phase have the same composition, mass transport is not necessary. However, for very rapid interface reaction rate, heat transport may play a role. The growth rate may be controlled either by interface reaction or heat transport. Because diffusivity of heat is much greater than chemical diffusivity, crystal growth controlled by heat transport is expected to be much more rapid than crystal growth controlled by mass transport. For vaporization of liquid (e.g., water $\rightarrow$ vapor) in air, because the gas phase is already present (air), nucleation is not necessary except for vaporization (bubbling) beginning in the interior. Similarly, for ice melting (ice $\rightarrow$ water) in nature, nucleation does not seem to be difficult.

A note is in order about differences between *dissolution* and *melting*. In this book, melting and dissolution are distinguished as follows. If the temperature is above the melting temperature (*liquidus*) of the solid, then the solid undergoes melting. If the temperature is below the melting temperature (*solidus*) of the solid, but the solid is in contact with a liquid and dissolving into it, then the solid undergoes dissolution. Between the solidus and liquidus, the process is called *partial melting* for lack of a better term. Melting is a simple phase transition, but dissolution and partial melting are complex phase transformations (see below). Melting occurs with or without the presence of another phase. Dissolution occurs only when there is a fluid phase. For example, NaCl at a temperature of 1200 K undergoes melting. NaCl in water at room temperature undergoes dissolution. Because dissolution happens only when there is an external phase, dissolution is sometimes referred to as external instability, and melting is sometimes referred to as inherent instability (Zhang and Xu, 2003). Melting rate is controlled by either interface reaction or heat transfer, and dissolution rate is controlled by either interface reaction or mass transfer. Because heat transfer is much more rapid than mass transfer, melting rate is usually much greater than dissolution rate. An example of partial melting is a plagioclase crystal (An40Ab60) heated to 1600 K; it would undergo partial melting to produce a melt with composition of An20 and a solid phase with composition of An58. When equilibrium is reached, there would be about 47% melt and 53% solid. Because partial melting requires compositional modification of the solid phase, it is often slower than both melting and dissolution.

Simple phase transitions may be classified as first-order and second-order phase transitions (not to be confused with first-order and second-order homogeneous reactions). For all phase transitions at the equilibrium temperature (or pressure), Gibbs free energy is continuous (meaning the old and new phases have identical Gibbs free energy). *First-order phase transitions* are those in which there is a discontinuity in enthalpy and first derivatives of Gibbs free energy with respect to temperature and pressure (entropy and volume). There is a major change in the structure. Examples include graphite to diamond, aragonite to calcite, water to ice, melting of a mineral, or crystallization of a mineral in its own melt. Such phase transitions always involve nucleation and interface reaction and hence can be very slow. For example, the transition from diamond to graphite requires the breakage of the three-dimensional C–C bonds. Because these bonds are very strong, the transition is very slow at room temperature and pressure.

A *second-order phase transition* is one in which the enthalpy and first derivatives are continuous, but the second derivatives are discontinuous. The C_p versus T curve is often shaped like the Greek letter λ. Hence, these transitions are also called λ-transitions (Figure 2-15b; Thompson and Perkins, 1981). The structure change is minor in second-order phase transitions, such as the rotation of bonds and order–disorder of some ions. Examples include melt to glass transition, λ-transition in fayalite, and magnetic transitions. Second-order phase transitions often do not require nucleation and are rapid. On some characteristics, these transitions may be viewed as a homogeneous reaction or many simultaneous homogeneous reactions.

(2b) *Complex phase transformations in which some components in a phase or in multiple phases combine to form a new phase or multiple new phases.* This class includes most of the heterogeneous reactions, such as

- the precipitation of calcite from an aqueous solution,
  $Ca^{2+}(aq) + CO_3^{2-}(aq) \rightarrow calcite$

- growth of a mineral, a droplet, or a bubble in water or melt

- condensation of minerals from solar nebular gas

- crystallization of olivine from a basaltic magma

- oxidation of the fayalite component in olivine:
  $3Fe_2SiO_4(olivine) + O_2(gas) \rightarrow 2Fe_3O_4(spinel) + 3SiO_2(quartz)$

- decomposition of one phase into several phases (e.g., spinodal decomposition)

- combination of several phases into one phase (e.g., melting at a eutectic point)

- reaction of multiple phases to form multiple new phases, such as
  $MgAl_2O_4(spinel) + 4MgSiO_3(opx) \rightleftharpoons Mg_2SiO_4(olivine) + Mg_3Al_2Si_3O_{12}$
  (pyrope)

- partial melting of a polymineralic rock

- most metamorphic reactions

- crystallization of natural silicate melts

- volcanic eruptions

From the above list, one can see that kinetics of complex heterogeneous reactions are intimately related to important geological processes such as igneous rock formation, volcanic eruptions, and metamorphism.

Complex phase transformation requires nucleation, interface reaction, and mass transport; the interplay of these factors controls the rate of complex phase transformations. Because nucleation, interface reaction, and mass transport are sequential steps for the formation and growth of new phases, the slowest step controls the reaction rate. Table 4-1 shows some examples of phase transformations and the sequential steps.

In this chapter, the essential aspects of kinetics of heterogeneous reactions (nucleation, interface reaction, and mass/heat transfer) are first presented. Then one class of heterogeneous reactions, the dissolution and growth of crystals, bubbles, and droplets, is elaborated in great detail. Some other heterogeneous reactions are then discussed with examples. Many complex problems in heterogeneous reactions remain to be solved.

Table 4-1. Steps for phase transformations

Simple Phase Transition Aragonite to Calcite	*Complex Phase Transformation Magma to Rock*	*Complex Phase Transformation Volcanic Eruption*
Aragonite is decompressed to calcite stability field	Magma is cooled to below the liquidus	Gas-bearing magma is decompressed to oversaturation
↓	↓	↓
Calcite embryos nucleate	Crystals nucleate	Bubbles nucleate
↓	↓	↓
Calcite crystals grow at the expense of aragonite	Crystals grow; Other minerals nucleate and grow	Bubbles grow; volume of bubbly magma expands rapidly
↓	↓	↓
No more aragonite	No more magma	A foam is formed; almost no dissolved volatiles in magma
↓	↓	↓
Calcite crystals coarsen	Solid state reactions and coarsening of crystals	Bubbles coalesce or fragment into explosive eruption

4.1 Basic Processes in Heterogeneous Reactions

4.1.1 Nucleation

For a reaction to produce a new phase, the new phase must first form (nucleate) from an existing phase or existing phases. Nucleation theory deals with how the new phase nucleates and how to predict nucleation rates. The best characterization of the present status of our understanding on nucleation is that we do not have a quantitative understanding of nucleation. The theories provide a qualitative picture, but fail in quantitative aspects. We have to rely on experiments to estimate nucleation rates, but nucleation experiments are not numerous and often not well controlled. In discussion of heterogeneous reaction kinetics and dynamics, the inability to predict nucleation rate is often the main obstacle to a quantitative understanding and prediction. The nucleation theories are nonetheless discussed here because they do provide a qualitative picture and because a book on geochemical kinetics would not be complete without a discussion of nucleation.

4.1.1.1 Homogeneous nucleation

Homogeneous nucleation refers to the nucleation of a new phase inside an existing phase. Our current understanding is that nucleation is rarely completely homogeneous because there are almost always impurities in the system, which provide interfaces for heterogeneous nucleation. The classical theory for homogeneous nucleation is based on the concept of *heterophase fluctuation*. From statistical physics, microscopically, the density and composition in a single phase are constantly changing around the mean value because of the vibrational, rotational, and translational motion of the atoms, ions, and molecules in a phase. The relative variation on such properties locally may be very small, such as 10^{-15}. Such variations are called *homophase fluctuations*. The small fluctuations are transient, that is, they continuously form and disappear.

If the transient variations are large enough, the clusters made of many molecules take on the characteristics of a new phase. For example, a cluster of Mg^{2+}, Fe^{2+}, and SiO_4^{4-} in a basaltic melt may form a structure that is similar to an olivine structure. These large fluctuations are called *heterophase fluctuations*. Nucleation theory characterizes how these heterophase fluctuations are distributed and how they grow. Even though the clusters may have greater energy and hence are energetically not favored, their presence increases the entropy of the system and hence there is a finite (and usually very small) probability for clusters with higher energy to form. If the new phase is not stable, the heterophase fluctuation is never stable and hence will disappear with time (e.g., decaying exponentially, Equation 3-63b). When the new phase has lower molar free energy, a small cluster that takes on the characteristics of the new phase may still be energetically unfavored because of interface energy. So a small cluster may disappear with time. However, a large cluster (low probability) that takes on the characteristics of the new phase may be energetically favored and hence grow. This approach yields the *classical nucleation theory*, which is summarized below.

Free Energy of a Cluster For clarity of discussion, crystal nucleation from a melt is used to derive the following relations. For nucleation of liquid droplets, the derivation is similar. For nucleation of bubbles, the formulation is slightly different and is summarized separately below. Let the Gibbs free energy difference between the crystalline and the melt state per mole of the crystalline composition be $\Delta G_{c-m} = \mu_c - \mu_m$, where μ_c and μ_m are the chemical potential (partial molar free energy) for the crystalline composition. $\Delta G_{c-m} < 0$ if the crystalline phase is more stable than the melt; it is positive if the melt is more stable. Let the interface energy per unit area between the crystal and the melt be σ (usually ranging between 0.05 and 2 J/m^2). Let the molar volume of the crystalline composition be V_c. The total energy to produce a spherical cluster of radius r consists of two terms, one due to the bulk energy of the crystalline phase and the second due to surface energy. The total may be expressed as

$$\Delta G_r = \left(\tfrac{4}{3}\right)\pi r^3 \Delta G_{c-m}/V_c + 4\pi r^2 \sigma, \tag{4-1}$$

where ΔG_r is the ΔG value to produce a spherical crystalline cluster of radius r. The first term in the above equation is related to the bulk free energy difference between the new crystalline phase and the old melt phase, which is negative when the crystalline phase is stable, but positive otherwise. The second term is related to the interface energy of the cluster, which is always positive. Because Gibbs free energy is minimized at equilibrium, the second term (that is, the interface energy term) always impedes nucleation. To estimate ΔG_{c-m}, one starts with the equilibrium condition at which $\Delta G_{c-m} = 0$, and uses the equation that $d\Delta G = -\Delta S\,dT + \Delta V\,dP$. If the equilibrium temperature (T_e) between the crystalline phase and the melt under the given pressure is known to be T_e, then ΔG_{c-m} may be estimated as

$$\Delta G_{c-m} \approx -\Delta S_{c-m}(T - T_e) = \Delta S_{m-c}(T - T_e), \tag{4-2a}$$

where $\Delta S_{m-c} = (S_m - S_c) = \Delta H_{m-c}/T_e > 0$ is the fusion entropy. If the equilibrium pressure between the crystalline phase and the melt under the given temperature is known to be P_e, then

$$\Delta G_{c-m} \approx \Delta V_{c-m}(P - P_e), \tag{4-2b}$$

where $\Delta V_{c-m} = (V_c - V_m)$. If the equilibrium concentration (i.e., saturation concentration) of the crystalline component (such as SiO_2 for quartz crystallization) in the melt is C_e, then for ideal solutions

$$\Delta G_{c-m} \approx -RT \ln(C/C_e). \tag{4-2c}$$

More accurately, activity instead of concentration should be used. When V_c and σ are known, with estimation of ΔG_{c-m}, the dependence of ΔG_r versus radius can be calculated (Figure 4-1).

Critical Cluster Size The size of the cluster at which the free energy reaches a maximum is called the *critical cluster size*. Greater than this size, adding more molecules to the cluster reduces Gibbs free energy and hence makes it more stable. Therefore, such clusters tend to grow. Below this size, adding more molecules to the cluster makes it less stable. Hence, such clusters tend to shrink. The critical size with the maximum Gibbs free energy is an example of unstable equilibrium. Adding one more molecule, the resulting cluster tends to grow. Losing one molecule, the resulting cluster tends to shrink. That is, any departure from this state would make the system more stable and the departure hence

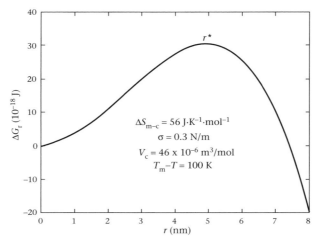

Figure 4-1 Extra Gibbs free energy of clusters as a function of cluster radius. The critical cluster size is when the extra free energy reached the maximum. $\Delta S_{m-c} = 56$ J/K/mol, $V_c = 46$ cm^3/mol, $\sigma = 0.3$ J/m^2, $T_e =$ melting temperature $= 1600$ K, and system temperature $= 1500$ K. $\Delta G_{c-m} \approx \Delta S_{m-c}(T - T_e) = -5600$ J/mol. The radius of the critical cluster is: $r^* = 2\sigma V_c/(\Delta G_{m-c}) = (2) \cdot (0.3) \cdot (46 \times 10^{-6})/5600$ m $= 4.93$ nm. The Gibbs free energy of the critical cluster relative to the melt is $\Delta G^* = (16/3)\pi\sigma^3/(\Delta G_{m-c}/V_c)^2 = 3.05 \times 10^{-17}$ J.

tends to grow. To find the critical cluster size from Equation 4-1, take the derivative of ΔG_r with respect to r and set the derivative to be zero. The critical cluster radius thus found is

$$r* = -2\sigma V_c/\Delta G_{c-m}. \tag{4-3}$$

Because $\Delta G_{c-m} < 0$ for crystallization to occur, r^* is greater than zero. The free energy of the critical cluster (relative to the melt) is

$$\Delta G^* = \left(\frac{16}{3}\right)\pi\sigma^3(V_c/\Delta G_{c-m})^2. \tag{4-4}$$

Note that the critical radius is when the free energy of the cluster is at maximum, not when $\Delta G_r = 0$. The latter occurs when the cluster radius is $1.5r^*$.

In the above derivation, the size of a cluster is expressed using the radius. The size of a cluster may also be characterized by the number of molecules i in the cluster. The total free energy to produce a cluster i is

$$\Delta G_i = (i/N_a)\Delta G_{c-m} + A_i\sigma, \tag{4-5}$$

where N_a is the Avogadro number (6.022×10^{23}) and A_i is the surface area of cluster i. If the cluster is spherical, then volume $= \frac{4}{3}\pi r^3 = (i/N_a)V_c$. That is, $r = [3iV_c/(4\pi N_a)]^{1/3}$. Hence, $A_i = 4\pi r^2 = (4\pi)^{1/3}(3iV_c/N_a)^{2/3}$. Equation 4-5 becomes

$$\Delta G_i = (i/N_a)\Delta G_{c-m} + (4\pi)^{1/3}\sigma(3iV_c/N_a)^{2/3}. \tag{4-6}$$

The critical cluster size in terms of number of molecules is

$$i* = -(\tfrac{4}{3})\pi N_a (V_c)^2 (2\sigma/\Delta G_{c-m})^3. \tag{4-7}$$

And the free energy of the critical cluster is still Equation 4-4. If the cluster is not spherical (e.g., the cluster could be a cube, or some specific crystalline shape), then the specific relations between i and cluster volume and surface area are necessary to derive the critical cluster size.

Statistical distribution of clusters The statistical distribution of clusters is described by the Boltzmann distribution:

$$N_i \approx N_1 \exp[-(\Delta G_i - \Delta G_1)/(kT)], \tag{4-8a}$$

where k is the Boltzmann constant (1.3807×10^{-23} J/K), N_1 is the number of molecules of the components of the crystalline phase in the melt per unit melt volume, N_i is the number of clusters with i molecules, and ΔG_i is the free energy required to form a cluster with i molecules. The value of ΔG_1 may be regarded to be zero. The number of critical nuclei per unit volume is

$$N^* \approx N_1 \exp[-\Delta G^*/(kT)]. \tag{4-8b}$$

Nucleation rate based on the classical nucleation theory The nucleation rate is the steady-state production of critical clusters, which equals the rate at which critical clusters are produced (actually the production rate of clusters with critical number of molecules plus 1). The growth rate of a cluster can be obtained from the transition state theory, in which the growth rate is proportional to the concentration of the activated complex that can attach to the cluster. This process requires activation energy. Using this approach, Becker and Doring (1935) obtained the following equation for the nucleation rate:

$$I = \frac{dN^*}{dt} = \frac{n^*}{i^*} \nu N_1 \left(\frac{\Delta G^*}{3\pi kT} \right)^{1/2} \exp\left(-\frac{E + N_a \Delta G^*}{RT} \right), \tag{4-9}$$

where I is the nucleation rate per unit volume ($m^{-3} s^{-1}$), E is the activation energy, ΔG^* is the free energy required to form the critical clusters and depends on the interface energy σ (Equation 4-4), R is the gas constant, ν is the fundamental frequency ($=k_B T/h$), N_1 is the number of molecules per unit volume ($=N_a/V_m$ for crystal nucleation in its own melt), i^* is the number of molecules in the critical embryo [$=N_a 4\pi r^{*3}/(3 V_m)$], and n^* is the number of molecules next to the embryo ($=N_a 4\pi r^{*2} l/V_m$, where l is the thickness of the layer next to the embryo from which molecules may enter the embryo and may be taken to be 2×10^{-10} m). When the degree of saturation is small, $N_a \Delta G^*$ is very large and dominates the $E + N_a \Delta G^*$ term. For example, $N_a \Delta G^* = 18,367$ kJ/mol for the case shown in Figure 4-1, and typical activation energy E at magmatic temperatures is about 250 kJ/mol.

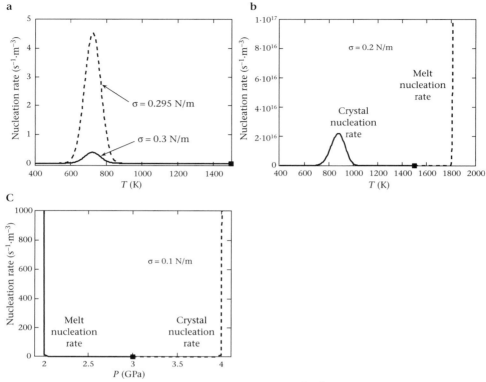

Figure 4-2 Calculated nucleation rate for $V_c = 46 \times 10^{-6}\,\text{m}^3/\text{mol}$, $E = 250\,\text{kJ/mol}$, $\Delta S_{m-c} = 50\,\text{J·K}^{-1}\text{·mol}^{-1}$, $\Delta V^{m-c} = 5 \times 10^{-6}\,\text{m}^3$, the equilibrium temperature of 1500 K for (a) and (b), and the equilibrium pressure of 3 GPa for (c). (a) The dependence of crystal nucleation rate on the interface energy. Note that for a small change in interface energy from 0.300 to 0.295 J/m², the peak nucleation rate increases by more than one order of magnitude. If the interface energy changes from 0.3 to 0.2 J/m², the peak nucleation rate would increase by 17 orders of magnitude. (b) The nucleation rate of crystal and melt as a function of temperature. (c) The nucleation rate of crystal and melt as a function of pressure.

Some calculations using Equation 4-9 are shown in Figure 4-2. The results show the following:

(1) Nucleation rate is zero right at saturation (because $\Delta G^\star = \infty$).

(2) Huge undercooling is necessary for homogeneous nucleation.

(3) The nucleation rate depends strongly on the interface energy (Figure 4-2a).

(4) On a nucleation rate versus temperature diagram (Figure 4-2b), crystal nucleation rate below the crystal–melt equilibrium temperature and melt nucleation above the temperature are not symmetric. For crystal nucleation there is a peak nucleation rate at a very large degree of undercooling, but not for melt nucleation. That is, for crystal nucleation as temperature decreases, the nucleation rate first increases with decreasing

temperature, reaches a maximum, and decreases because at low temperature, the exponential term becomes very small. On the other hand, for melt nucleation as temperature increases from crystal–melt equilibrium temperature, the nucleation rate increases with increasing temperature rapidly.

(5) On a nucleation rate versus pressure diagram (Figure 4-2c), melt nucleation rate below the crystal–melt equilibrium pressure and crystal nucleation above the pressure are roughly symmetric. In Equation 4-9, only ΔG^* would vary with pressure or concentration. Hence, both melt nucleation rate and crystal nucleation rate increase monotonically with departure from equilibrium. There is no peak nucleation rate.

The above qualitative predictions are consistent with experiments. However, in terms of absolute nucleation rate, Equation 4-9 usually predicts too low a rate by many orders of magnitude (see below).

A variation of Equation 4-9 is to approximate the many terms in it by viscosity η (Kirkpatrick, 1975). Assuming that the activation energy for viscous flow is the same as that for nucleation, then $\eta = A' \exp[E/(RT)]$ where A' is a constant. Substituting it into Equation 4-9 leads to

$$I = \frac{A}{\eta} v \, \exp\!\left(-\frac{\Delta G^*}{k_B T}\right), \tag{4-10}$$

where A is a parameter that is almost independent of temperature, and k_B is the Boltzmann constant.

Experimental Nucleation Rate Numerous experimental studies have been carried out on homogeneous nucleation rate. The experimental results are almost always many orders of magnitude greater than theoretical calculations. Figure 4-3 shows experimental data of Neilson and Weinberg (1979) on homogeneous nucleation rates in lithium disilicate melt. The shape of the experimental nucleation rate versus temperature curve follows the general shape of the theoretical curve, but in detail, the experimental data differ from theory in two aspects: (i) the magnitude of experimental nucleation rate is much larger (by 10 orders of magnitude) than the theoretical prediction, and (ii) the curve defined by the experimental data has a narrower peak, and the peak occurs at a higher temperature.

Failure of the Classical Nucleation Theory There are several suggested explanations for the failure of the classical nucleation theory to quantitatively predict the nucleation rate, including the following:

(1) Experimental *internal* nucleation rates are due not to homogeneous nucleation but to heterogeneous nucleation (which would require a different theory) instead. The argument is that it is extremely difficult to

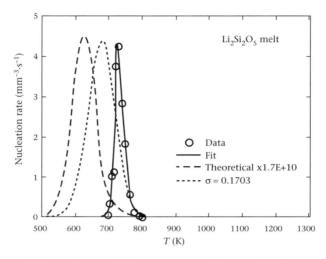

Figure 4-3 Experimental data on nucleation rate in lithium disilicate melt (points) compared to theory. The solid curve is a fit to the experimental data. The short dashed curve (middle curve) is the theoretically calculated curve multiplied by 1.7×10^{10}. Parameters used in calculating the theoretical curve using Equation 4−9: $T_e = 1306$ K, $\sigma = 0.201$ J/m^2, $V_c = 61.2$ cm^3/mol, $\Delta H_{m-c} = 61.1$ kJ/mol, $N_1 = 10^{28}$ m^{-3} (Neilson and Weinberg, 1979). The long dashed curve (left curve) is the theoretical curve by changing σ to 0.17036 J/m^2 so that the calculated maximum nucleation rate is the same as the experimental data.

get rid of all impurities. To this end, experimentalists have made great effort to show that the homogeneous nucleation experiments are indeed due to homogeneous, not heterogeneous nucleation.

(2) The theory is adequate but the interface energy σ for small clusters may differ from macroscopically measured interface energy. For example, if one assumes that the interface energy is constant and adjust the interface energy so that the calculated maximum nucleation rate matches the experimental data, the calculated curve (short-dashed curve in Figure 4-3) would agree better with the experimental data (the huge factor of discrepancy of 10^{10} is removed). By varying σ, one can always fit the maximum nucleation rate, but not the whole curve.

(3) Properties of nanoparticles (nuclei) are different from those of the bulk counterparts. Metastable phases may nucleate first because nanoparticles of these phases are more stable (e.g., due to low surface energy) than the nanoparticles of phases that are more stable in bulk (Ranade et al., 2002).

(4) The classical nucleation theory itself is inadequate. Hence, effort has been made to develop nonclassical theories of nucleation, often by

allowing the interface energy to vary with the size of the clusters and the temperature, or other semi-empirical models (e.g., Granasy and James, 1999). These theories are not yet able to predict nucleation rate.

The best characterization of our current knowledge of homogeneous nucleation is that the problem remains a challenge for future scientists. It is likely that the solution will come from studies of nanoparticles.

Transient Nucleation If a liquid is cooled continuously, the liquid structure at a given temperature may not be the equilibrium structure at the temperature. Hence, the cluster distribution may not be the steady-state distribution. Depending on the cooling rate, a liquid cooled rapidly from 2000 to 1000 K may have a liquid structure that corresponds to that at 1200 K and would only slowly relax to the structure at 1000 K. Therefore, Equation 4-9 would not be applicable and the transient effect must be taken into account. Nonetheless, in light of the fact that even the steady-state nucleation theory is still inaccurate by many orders of magnitude, transient nucleation is not discussed further.

Bubble Nucleation in a Liquid Phase The above classical nucleation theory can be easily extended to melt nucleation in another melt. It can also be extended to melt nucleation in a crystal but with one exception. Crystal grains are usually small with surfaces or grain boundaries. Melt nucleation in crystals most likely starts on the surface or grain boundaries, which is similar to heterogeneous nucleation discussed below. Homogeneous nucleation of bubbles in a melt can be treated similarly using the above procedures. Because of special property of gases, the equations are different from those for the nucleation of a condensed phase, and are hence summarized below for convenience.

Gas bubble nucleation in a melt is often cast in terms of a pressure decrease. If the actual process is due to a temperature increase or concentration increase, it is equivalent to a pressure decrease as long as the saturation pressure (that is, the equilibrium pressure) corresponding the concentration and temperature of the melt can be found. Let the equilibrium pressure be P_e, the ambient melt pressure be P, and the gas pressure in the bubble be P_g. The gas pressure in a bubble of radius r and the ambient melt pressure are related as

$$P_g - P = 2\sigma/r, \tag{4-11}$$

Assume the gas phase is ideal. The molar Gibbs free energy difference between the gas and melt phase is

$$\Delta G_{g-m} = \int (V_g - V_m) dP = RT \ln (P_g/P_e) - V_m(P - P_e), \tag{4-12a}$$

where V_g $(=RT/P_g)$ is the molar volume of the gas component in the gas phase, V_m is the partial molar volume of the gas component in the melt, and R is the gas constant. Because $V_m \ll V_g$, the above may be simplified as

$$\Delta G_{g-m} \approx \int V_g dP = RT \ln(P_g/P_e). \tag{4-12b}$$

If the absolute value of $(P - P_e)$ is small, the molar volume of the gas is roughly constant, then

$$\Delta G_{g-m} \approx V_g(P - P_e). \tag{4-12c}$$

The above approximate relation is not very accurate but is often made because it can significantly simplify relations below. The total energy to produce a cluster of gas molecules of radius r is

$$\Delta G_r = \left(\frac{4}{3}\right)\pi r^3 \Delta G_{g-m}/V_g + 4\pi r^2 \sigma = \left(\frac{4}{3}\right)\pi r^3 [RT \ln (P_g/P_e) $$
$$- V_m(P - P_e)]/V_g + 4\pi r^2 \sigma. \tag{4-13}$$

Using the approximation in Equation 4-12c, the above can be simplified as

$$\Delta G_r \approx \left(\frac{4}{3}\right)\pi r^3 (P - P_e) + 4\pi r^2 \sigma. \tag{4-14}$$

Take the derivative of ΔG_r with respect to r and set it to zero to find the critical bubble radius:

$$r\star \approx 2\sigma/(P_e - P). \tag{4-15}$$

The above is similar to the result of Toramaru (1989) by equating the chemical potential of the gas component in the gas phase to that in the melt phase:

$$\mu_g(P_g) = \mu_g(P_e) + RT \ln \frac{P + \frac{2\sigma}{r\star}}{P_e} = \mu_m(P) = \mu_m(P_e) + V_m(P - P_e). \tag{4-16}$$

Because $\mu_g(P_e) = \mu_m(P_e)$, we obtain

$$r\star = \frac{2\sigma}{P_e \exp\left[\frac{V_m(P - P_e)}{RT}\right] - P} \approx \frac{2\sigma}{P_e - P}. \tag{4-17}$$

The critical $\Delta G\star$ is

$$\Delta G\star \approx \left(\frac{16}{3}\right)\pi\sigma^3(V_g/\Delta G_{g-m})^2 = \left(\frac{16}{3}\right)\pi\sigma^3/(\Delta P)^2 = \sigma A^*/3 = (P_e - P)V^*/2, \tag{4-18}$$

where $A\star$ and $V\star$ are the surface area ($4\pi r\star$) and volume ($4\pi r\star^3/3$) of the critical nucleus. The above expression can be plugged into Equation 4-9 to obtain the classical nucleation rate.

4.1.1.2 Heterogeneous nucleation

Homogeneous nucleation is very difficult because of the large interface energy involved. If there are already interfaces in the system, an embryo may grow from

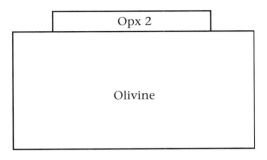

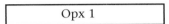

Figure 4-4 Heterogeneous nucleation versus homogeneous nucleation of orthopyroxene in a melt. (*Top and middle*) Heterogeneous nucleation of an orthopyroxene embryo (opx 2) on olivine–melt interface. The total interface energy for the formation of the embryo is $A(\sigma_{\text{opx/melt}} - \sigma_{\text{ol/melt}} + \sigma_{\text{ol/opx}}) + \Delta G_{\text{sides}} \approx \Delta G_{\text{sides}}$, where "sides" mean the four surfaces other than the upper and lower surfaces of opx 2. This energy can be small under the right conditions, such as epitaxial growth ($\sigma_{\text{ol/opx}}$ is small, leading to small $\sigma_{\text{opx/melt}} - \sigma_{\text{ol/melt}} + \sigma_{\text{ol/opx}}$) of a thin layer of opx on olivine (ΔG_{sides} is small due to small side surface area). (*Bottom*) Homogeneous nucleation of an orthopyroxene embryo (opx 1) in melt. The total interface energy for the formation of the embryo is $2A\sigma_{\text{opx/melt}} + \Delta G_{\text{sides}}$.

the interface because it may not require so much energy. Impurities, which are often present in a system, introduce interfaces, on which an embryo may nucleate heterogeneously. If impurities are present, nucleation on these impurities (i.e., at the interface between the impurities and the existing host phase) often dominates relative to homogeneous nucleation because the new embryo may orient itself in such a way so that the interface energy between the embryo and the impurity is small. For example, consider the nucleation of a parallelepiped orthopyroxene embryo in an olivine-bearing silicate melt (Figure 4-4). If it nucleated homogeneously (opx 1 in Figure 4-4), the interface energy for the embryo is $2A\sigma_{\text{opx/melt}} + \Delta G_{\text{sides}}$, where A and σ are the surface area and interface energy of the upper and lower surfaces, and ΔG_{sides} indicates the interface energy of the four side surfaces. (To minimize interface energy, interfaces with low energy would be favored.) If it nucleates on an olivine–melt interface (opx 2 in Figure 4-4), the total extra interface energy for the embryo in addition to the olivine–melt interface energy is $A(\sigma_{\text{opx/melt}} - \sigma_{\text{ol/melt}} + \sigma_{\text{ol/opx}}) + \Delta G_{\text{sides}}$, where "opx" means orthopyroxene and "ol" means olivine. In such nucleation, to minimize total interface energy, orthopyroxene nucleus would orient itself so that the olivine–

orthopyroxene interface is coherent. Hence, $\sigma_{ol/opx}$ is smaller than $\sigma_{ol/melt}$. On the other hand, $\sigma_{opx/melt}$ is roughly the same as $\sigma_{ol/melt}$. Hence, the total extra interface energy is much smaller than the case for homogeneous nucleation, meaning heterogeneous nucleation requires much less supersaturation (undercooling).

From the above, one can see that for heterogeneous nucleation, the interface energy term is different. Heterogeneous nucleation rate may be expressed by Equation 4-9 in which ΔG^* is found by Equation 4-4, but with modification of the interface energy term as (Christian, 1975):

$$\sigma' = \sigma[(2 - 3\cos\theta + \cos^3\theta)/4]^{1/3}, \tag{4-19}$$

where σ' is the interface energy for heterogeneous nucleation (i.e., in the presence of heterogeneity), σ is the interface energy for homogeneous nucleation, and θ is the *contact angle* (or interface angle) between the new embryo and the heterogeneity. The concept of contact angle stems from the contact between a liquid phase and a solid phase. If the liquid wet the surface completely (meaning a liquid droplet would spread perfectly and form a thin liquid on the solid surface), then the contact angle θ is zero, leading to $\cos\theta = 1$ and $\sigma' = 0$. If the contact angle is 90°, then $\sigma' = \sigma/2^{1/3} = 0.7937\sigma$. If the contact angle is 180°, then $\sigma' = \sigma$. Because the nucleation rate depends strongly on the interface energy, a small change in θ changes the calculated nucleation rate significantly. Mathematically, $\cos\theta$ is defined as

$$\cos\theta = (\sigma_{12} - \sigma_{23})/\sigma_{13}, \tag{4-20}$$

where σ_{ij} means interface energy between phases i and j, "1" means the original phase (melt in Figure 4-4), "2" means the impurity phase (olivine in Figure 4-4), and "3" means the new phase to nucleate (opx in Figure 4-4).

Although the effective interface energy (σ') is quantifiable, prediction of heterogeneous nucleation rate requires knowing the number and kinds of impurities in the host phase. Because these are difficult to quantify for a given application (e.g., for bubble nucleation in magma, or new mineral nucleation in metamorphic reactions), heterogeneous nucleation rates are also difficult to predict. Furthermore, because Equation 4-9 cannot even predict homogeneous nucleation rates, it is not clear whether replacing σ in the equation by σ' would be the right equation to predict heterogeneous nucleation rates.

4.1.2 Interface reaction

Interface reaction is another necessary step for crystal growth and dissolution. After formation of crystal embryos, their growth requires attachment of molecules to the interface. The attachment and detachment of molecules and ions to and from the interface are referred to as interface reaction. (During nucleation, the attachment and detachment of molecules to and from clusters are similar to interface reaction.) For an existing crystal to dissolve in an existing melt,

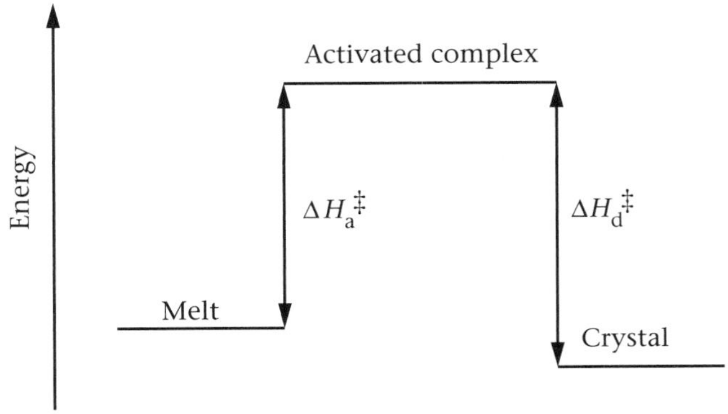

Figure 4-5 Energy diagram for reactants (melt or aqueous solution), activated complex, and products (crystal) for the case of crystal growth.

nucleation is not necessary but interface reaction is necessary. Crystal growth and dissolution also requires another step, transfer of nutrients toward or away from the interface. Because the two are sequential steps, the slowest step determines the overall rate. If crystal growth or dissolution is controlled by interface reaction, there is a simple rate law: under constant temperature, pressure, and composition, the rate is constant. The dependence of the rate on temperature or pressure or concentration can be derived as follows. Specifically, because the interface reaction consists of both forward and backward reaction (similar to reversible homogeneous reactions), the net interface reaction rate is not simply proportional to the concentration raised to some power, but can be linear to the concentrations raised to some power.

In the context of transition-state theory, for ions and molecules in the liquid to attach to the crystal, they must first become an activated complex. The same is true for detachment (Figure 4-5). For clarity of discussion, use calcite (Cc), assumed to be pure $CaCO_3$, growth from an aqueous solution as an example. The reaction is

$$Ca^{2+}(aq) + CO_3^{2-}(aq) \rightleftharpoons CaCO_3(Cc). \tag{4-21}$$

The attachment reaction requires the following steps:

$$Ca^{2+}(aq) + CO_3^{2-}(aq) \rightleftharpoons CaCO_3^{\ddagger} \rightarrow CaCO_3(Cc), \tag{4-22}$$

where $CaCO_3^{\ddagger}$ is the activated complex at the interface (the symbol ‡ denotes activated complex). The first step above is to form the activated complex. The second step is to attach the activated complex to the crystal surface. According to the transition-state theory, the activated complex is in equilibrium with the reactants. Let $K_a^{\ddagger}$ be the equilibrium constant between the activated complex and the reactants for the attachment reaction. Hence, $K_a^{\ddagger} = [CaCO_3^{\ddagger}]/\{[Ca^{2+}][CO_3^{2+}]\}$ where $[Ca^{2+}]$ and $[CO_3^{2-}]$ are the concentrations of the species near the interface, meaning $[CaCO_3^{\ddagger}] = K_a^{\ddagger}[Ca^{2+}][CO_3^{2-}]$. Let $k_a^{\ddagger}$ be the reaction rate constant from the activated complex to the crystal. According to the transition-state theory, $k_a^\ddagger = \kappa_a v$, where v is the fundamental frequency ($k_B T/h$ with k_B being the Boltzmann constant, and h being the Planck constant), and κ_a is the transmission coefficient (the fraction of vibrations which will result in attachments). Therefore, the attachment rate r_a is

$$r_a = a_a k_a^\ddagger [CaCO_3^\ddagger] = a_a k_a^\ddagger K_a^\ddagger [Ca^{2+}][CO_3^{2-}] = a_a (\kappa_a v) K_a^\ddagger [Ca^{2+}][CO_3^{2-}], \qquad (4\text{-}23)$$

where a_a is the thickness of the layer of the solution at the interface that can react to be attached to the crystal. Based on Equation 4-23, the attachment reaction is a second-order reaction. The units of $[CaCO_3^\ddagger]$, r_a, a_a, and $k_a^\ddagger$ are mol/m^3, mol m^{-2}s^{-1}, m, and s^{-1}.

The backward reaction (detachment) must also be considered to obtain the net crystal growth (or dissolution) rate. The detachment goes through the same activated complex:

$$CaCO_3(Cc) \rightleftharpoons CaCO_3^\ddagger \rightarrow Ca^{2+}(aq) + CO_3^{2-}(aq). \qquad (4\text{-}24)$$

The concentration of the activated complex $[CaCO_3^\ddagger]$ is $K_d^\ddagger$ if calcite is pure $CaCO_3$, where subscript "d" stands for detachment. Hence, for pure $CaCO_3$, the detachment rate r_d is

$$r_d = a_d k_d^\ddagger [CaCO_3^\ddagger] = a_d k_d^\ddagger K_d^\ddagger = a_d (\kappa_d v) K_d^\ddagger, \qquad (4\text{-}25)$$

where a_d is the thickness of the layer of the crystal at the interface that can react to be detached (a_d is the distance between subsequent crystalline layers), $k_d^\ddagger$ is the reaction rate constant from the activated complex to $Ca^{2+}(aq) + CO_3^{2-}$ (aq), $K_d^\ddagger$ is the "equilibrium" constant between the activated complex and $CaCO_3(Cc)$, and κ_d is the "transmission" coefficient. From Equation 4-25, the detachment reaction may be referred to as zeroth-order reaction. The net crystal growth rate $u' = r_a - r_d$ (mol per unit area per unit time) is

$$u' = a_a (\kappa_a v) K_a^\ddagger [Ca^{2+}][CO_3^{2-}] - a_d (\kappa_d v) K_d^\ddagger. \qquad (4\text{-}26)$$

The net reaction rate does not behave as a simple second-order reaction or as a zeroth-order reaction. The net rate is linear to $[Ca^{2+}][CO_3^{2-}]$, but not proportional to $[Ca^{2+}][CO_3^{2-}]$. At constant composition, temperature, and pressure, the net reaction rate is constant. The concentrations approach equilibrium and hence the net reaction rate approaches zero as reaction proceeds.

Using the condition that the net reaction rate is zero at equilibrium and noting that the solubility product $K_{sp} = [Ca^{2+}]_e[CO_3^{2-}]_e$, one may simplify the above equation further. At equilibrium, we have

$$0 = u' = a_a (\kappa_a v) K_a^\ddagger K_{sp} - a_d (\kappa_d v) K_d^\ddagger. \qquad (4\text{-}27)$$

Hence,

$$a_a(\kappa_a v)K_a^\ddagger = a_d(\kappa_d v)K_d^\ddagger/K_{sp}. \tag{4-28}$$

Assuming that the parameters $a_a(\kappa_a v)K_a^\ddagger$ and $a_d(\kappa_d v)K_d^\ddagger$ do not depend on the concentrations, and substituting the above into Equation 4-26, we obtain

$$u' = a_d(\kappa_d v)K_d^\ddagger\{[Ca^{2+}][CO_3^{2-}]/K_{sp} - 1\} = a_d(\kappa_d v)K_d^\ddagger(w - 1), \tag{4-29}$$

where $w = [Ca^{2+}][CO_3^{2-}]/K_{sp}$ is the degree of saturation for calcite. Note that the equilibrium constant for the growth reaction $Ca^{2+}(aq) + CO_3^{2-}(aq) \rightleftharpoons CaCO_3(Cc)$ is $K = 1/K_{sp}$, and the quotient $Q = 1/([Ca^{2+}][CO_3^{2-}])$. Hence, $w = [Ca^{2+}][CO_3^{2-}]/K_{sp} = K/Q$. If we define Q/K to be w', then $w' = 1/w$. The above formula is for crystal growth rate at constant temperature but different degree of oversaturation. For generality, the parameter w can be related to

$$w = \exp[-\Delta G_{cc-aq}/(RT)]. \tag{4-30}$$

With this replacement, net crystal growth rate may be written as

$$u' = a_d(\kappa_d v)K_d^\ddagger\{\exp[-\Delta G_r/(RT)] - 1\}. \tag{4-31}$$

where ΔG_r is the Gibbs free energy of the reaction (G of products minus that of reactants). Equation 4-31 is general and can account for supersaturation due to concentration change, temperature change, and/or pressure change.

For the growth rate to be greater than zero, the products must be more stable than the reactants, meaning that ΔG_r must be negative. The value of ΔG_r may be estimated as $-\Delta S_r(T - T_e)$ if crystal growth is caused by temperature change, or $\Delta V_r(P - P_e)$ if crystal growth is caused by pressure change, or $-RT\ln\{[Ca^{2+}][CO_3^{2-}]/K_{sp}\}$ if crystal growth is caused by concentration variation (Equations 4-2a,b,c).

The unit of the growth rate u' in Equation 4-31 is mole per unit area per unit time. If growth rate in m/s (referred to as *linear growth rate*) is needed, Equation 4-31 must be multiplied by the molar volume of the crystal (V_c):

$$u = V_c a_d(\kappa_d v)K_d^\ddagger\{\exp[-\Delta G_r/(RT)] - 1\}. \tag{4-32}$$

If the rate of melt consumption (corresponding to crystal growth) or melt growth (corresponding to crystal dissolution) is needed, then V_c in the above equation should be replaced by the molar volume of the melt V_m. Because $K_d^\ddagger = e^{\Delta S_d^\ddagger/R - \Delta H_d^\ddagger/(RT)}$, where $\Delta H_d^\ddagger$ is the enthalpy of formation of the activated complex for detachment (that is, activation energy for detachment, Figure 4-5), $v = k_B T/h$, V_c and a_d are constants, and κ_d is assumed constant, we have

$$u = V_c a_d(\kappa_d k_B T/h)e^{\Delta S_d^\ddagger/R}e^{-E/(RT)}(e^{-\Delta G_r/(RT)} - 1). \tag{4-33a}$$

Collecting the various constants as one single constant A, then

$$u = ATe^{-E/(RT)}(w - 1), \tag{4-33b}$$

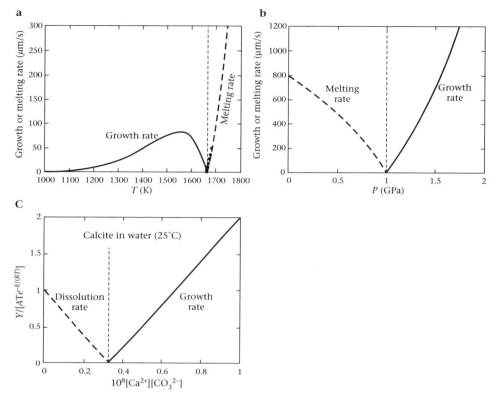

Figure 4-6 Interface reaction rate as a function of temperature, pressure, and composition. The vertical dashed line indicates the equilibrium condition (growth rate is zero). (a) Diopside growth and melting in its own melt as a function of temperature with the following parameters: $T_e = 1664\,K$ at 0.1 MPa, $\Delta S_{m-c} = 82.76\,J \cdot mol^{-1}\,K^{-1}$, $E/R = 30000\,K$, $A = 12.8\ m\,s^{-1}\,K^{-1}$, and $\Delta V_{m-c} = 12.1 \times 10^{-6}\,m^3/mol$. The dots are experimental data on diopside melting (Kuo and Kirkpatrick, 1985). (b) Diopside growth and melting in its own melt as a function of pressure at 1810 K ($T_e = 1810\,K$ at 1 GPa from the equilibrium temperature at 0.1 MPa and the Clapeyron slope for diopside). (c) Calcite growth and dissolution rate in water at 25°C as a function of Ca^{2+} and CO_3^{2-} concentrations.

where $A = V_c a_d (\kappa_d k_B / h) e^{\Delta S_d^\ddagger / R}$ and $w = e^{-\Delta G_r/(RT)}$. The interface reaction rate u is positive (meaning crystal growth) for $w > 1$ and negative (meaning crystal dissolution or melting) for $w < 1$. In literature, sometimes the interface reaction rate is given as $u = ATe^{-E/(RT)}(1 - 1/w)$, which works OK if w is not much different from 1, but does not work well as $w \to 0$, which can be achieved in experiments by, e.g., dissolving calcite into pure water, or forsterite olivine in an MgO-free melt. The dependence of net crystal growth rate on temperature, pressure, and concentrations is diagrammed in Figure 4-6. At a fixed temperature, the above equation may be written as

$$u = k(w - 1),\tag{4-34}$$

where k is a constant (growth rate at $w = 2$, or dissolution rate at $w = 0$).

The parameter A in Equation 4-33b contains $\Delta S_d^{\ddagger}$ and κ that cannot be calculated. Hence, A is usually regarded as a fitting parameter to treat experimental data. For order of magnitude estimate of A, κ may be assumed to be 1 (especially for continuous growth mechanism, see below), and ΔS_{m-c} may be employed to approximate $\Delta S_d^{\ddagger}$, which would mean that the activated complex is somewhat similar to the liquid state. For example, for diopside melting in its own melt, take $V_c = 69.74 \times 10^{-6}\,\text{m}^3/\text{mol}$, $\kappa = 1$, $\Delta S_d^{\ddagger} = \Delta S_{m-c} = 82.76\,\text{J K}^{-1}\,\text{mol}^{-1}$, $a_d = 4 \times 10^{-10}$ m, then A can be found to be 12.2 m s^{-1} K^{-1}, which is similar to the value 12.8 m s^{-1} K^{-1} obtained from fitting the experimental data of Kuo and Kirkpatrick (1985) assuming an activation energy of 250 kJ/mol (Figure 4-6a).

Another simplification that is often made is to assume that the activation energy E for interface reaction is the same as that for viscosity η. Then Equation 4-33b may be rewritten as

$$u = u_r(w - 1)/\eta, \tag{4-35}$$

where u_r is the *reduced growth rate* and has the unit of Pa·m.

Examination of Figure 1-12 provides some clue to qualitatively gauge the interface reaction rate for reactions in water. Figure 1-12 shows that, for mineral with low solubility and high bond strength (characterized by $(z_+ z_-)_{max}$, where z_+ and z_- are valences of ions to be dissociated), the overall dissolution rate is controlled by interface reaction; otherwise, it is controlled by mass transport. Because diffusivities of common cations and anions in water do not differ much (by less than a factor of 10; Table 1-3a), when the overall reaction rate is controlled by interface reaction, it means that interface reaction is slow; when the overall reaction rate is controlled by mass transport, the interface reaction rate is rapid. Therefore, from Figure 1-12, we may conclude that the interface reaction rate increases with mineral solubility and decreases with bond strength $(z_+ z_-)_{max}$ to be dissociated.

In summary, crystal growth or dissolution can be caused by a change in concentration, temperature, or pressure. For each case, the interface reaction rate may be calculated by estimating ΔG_r (Equations 4-2a,b,c) and assuming an activation energy (such as activation energy for viscosity) and the parameter A. Figure 4-6 shows such a calculation, where the parameter A is estimated from the diopside melting data (points in Figure 4-6a) of Kuo and Kirkpatrick (1985). The calculated curves share some features of the nucleation curves (Figure 4-2). At equilibrium, the growth or melting rate is zero. The crystal growth rate below T_e first increases as temperature decreases, reaches the maximum, and then decreases. The decrease is due to the $e^{-E/(RT)}$ term. The melt growth rate above T_e increases monotonically and rapidly with increasing temperature. That is, the curve for crystal growth rate below T_e and that for melt growth rate above T_e are not symmetric. However, when plotted as a function of pressure or degree of saturation (w), the curve for crystal growth and that for melt growth are roughly symmetric.

Experiments to determine the interface reaction rate must insure that the growth or melting rate is not limited by other processes such as mass or heat transfer. Because mass transfer is much slower than heat transfer, mass transfer is more often the limiting factor for crystal growth compared to heat transfer. Hence, kinetic experiments at high temperatures to investigate the interface reaction rates are often designed as crystal growth or melting in its own melt. For example, the experiments with diopside melting in its own melt (not diopside dissolving in a different melt) of Kuo and Kirkpatrick (1985) provide interface reaction rates. These authors also conducted melting experiments under various speeds of rotation and found that the melting rate does not change. Hence, the melting is controlled by interface reaction. Their data (Figure 4-6a) indicate that the interface reaction rate is very high for diopside melting. At a moderate undercooling of 10 K, the melting rate is about 20 μm/s, or 72 mm/h (Kuo and Kirkpatrick, 1985). Table 4-2 (see page 349) lists more interface reaction rate data based on the growth rate of a crystal in its own melt.

When the interface reaction rate is compared to homogeneous nucleation rate, the general trend of rate vs. temperature or rate vs. pressure is similar. However, quantitatively, there is a large difference between crystal growth rate and homogeneous nucleation rate as a function of temperature or pressure. For nucleation, the rate depends exponentially on the degree of oversaturation, and a huge degree of oversaturation (such as 700°C undercooling) is necessary for the rate to be noticeable. For crystal growth or melt growth, at small degree of oversaturation, the growth rate is linear to the degree of oversaturation. Hence, a small degree or oversaturation (such as 10°C undercooling) is enough for the rate to be noticeable. Figure 4-7 (see page 350) compares the temperature dependence of crystal growth rate (for constant κ) and homogeneous nucleation rate.

Microscopically, the interface reaction during crystal growth may be through various mechanisms. One mechanism is called the *continuous model*. Two other models are *layer-spreading models*.

In the context of the *continuous growth model*, the crystal surface is assumed to be atomically rough. Hence, molecules may attach to all sites on the surface (Figure 4-8a, see page 351). The transmission coefficient in Equation 4-23 may be regarded to be 1. It appears that solids with low fusion enthalpy, such as metals, grow in this mechanism. The growth rate equations derived above apply best to this growth mechanism. For small undercooling, the interface reaction rate (Equations 4-33a and 4-35) may be written as

$$u\eta = k_1 \Delta T, \tag{4-36}$$

where $\Delta T = T_e - T$ (undercooling) and k_1 is a constant. The parameter $u\eta$ is proportional to ΔT.

There are two *layer-spreading models*. In these models, the crystal surface is atomically flat except at screw dislocations or steps of a partially grown surface layer. If there are screw dislocations, growth would continue on the screw

Table 4-2 Measured crystal growth rates of substances in their own melt

Mineral	Liquidus T (K)	ΔT at peak rate (K)	Peak rate (μm/s)
Cristobalite	1996	50	0.02
Anorthite	1830	300	150
Diopside	1664	>100	220
$Na_2Si_2O_5$	1147	60	10
GeO_2	1389	100	0.1
SrB_4O_7	1270	100	160
PbB_4O_7	1048	120	2

Note. From Dowty (1980a).

dislocations, and the growth mechanism is referred to as the *screw dislocation mechanism* (Figure 4-8b). With this mechanism, the interface reaction rate may be expressed as

$$u\eta = k_2(\Delta T)^2, \tag{4-37}$$

where k_2 is another constant. The rate is proportional to the square of ΔT because the transmission coefficient (the proportion of surface for attachment or detachment) is not a constant but is proportional to ΔT in the screw dislocation mechanism.

In the second layer-spreading model, there are no screw dislocations. Growth occurs at the steps of a partially grown surface layer (Figure 4-8c). When the layer is fully grown, a new molecule must be added to the flat surface, which is similar to nucleation because new surfaces are produced. The beginning of a new layer is hence referred to as surface nucleation. After the formation of the nucleus on the surface, other molecules can be added to the steps until the layer is fully grown. Hence, the growth rate depends on surface nucleation rate as well as spreading rate. This mechanism is referred to as the *surface nucleation mechanism*. In the context of this mechanism, the interface reaction rate is related to undercooling as

$$u\eta = k_3 \exp[-B/(T\,\Delta T)], \tag{4-38}$$

where B is related to fusion enthalpy and nucleation energy, k_3 is another constant, and the exponential term is due to Boltzmann distribution of nucleation clusters. Plotting $\ln(u\eta)$ against $1/(T\,\Delta T)$ would yield a straight line with a negative slope (Kirkpatrick, 1975).

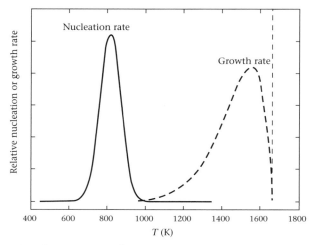

Figure 4-7 Comparison of crystal growth rate (dashed curve) and nucleation rate (solid curve) as a function of temperature. The equilibrium temperature (marked by the vertical dashed line) is 1664.15 K. The peak crystal growth rate is attained at an under-cooling of 120 K, but the peak nucleation rate is attained at an undercooling of 845 K. At a mere undercooling of 10 K, the crystal growth rate is 20% of the peak crystal growth rate. For the nucle-ation rate to be 20% of the peak nucleation rate, an undercooling of 750 K is necessary.

In the above discussion, crystal growth occurs by atom-by-atom (or molecule-by-molecule) addition to a template in various interface growth models. Another mode of crystal growth is by *aggregation and assembly of nanoparticles* (Banfield et al., 2000). That is, instead of atom-by-atom addition, nanoparticles are aggregated together either with the aid of bacteria, or by their own collision (e.g., in colloidal solutions). The result is a coarse polycrystalline material with high concentrations of point defects and dislocations, plus slabs of distinct materials. Such material may be highly reactive.

4.1.3 Role of mass and heat transfer

For crystal growth in its own melt, heat transfer is another necessary step. For crystal growth in a melt that is compositionally different from the crystal, mass transfer is another necessary step. (Although crystal growth is specifically mentioned, it also applies to crystal dissolution or melting, bubble growth and dissolution, droplet growth and dissolution, etc.) To understand this, consider the example of crystal growth in its own melt. Crystal growth releases heat. If the released heat is not transferred away, the interface crystal and melt would heat up, and eventually the temperature would be too high for crystal growth. Therefore, heat transfer is necessary for the crystal to grow.

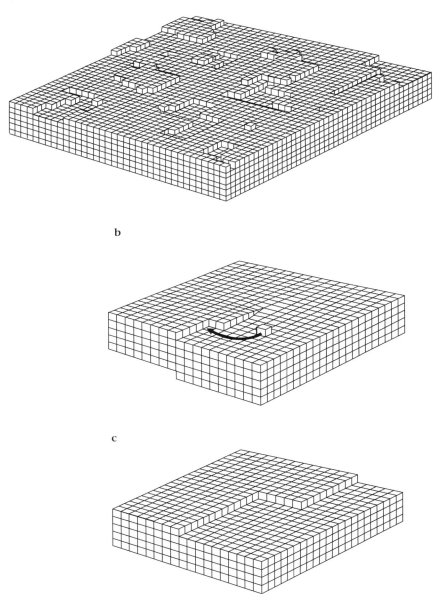

Figure 4-8 Various interface reaction mechanisms. (a) Continuous growth mechanism. (b) Screw dislocation mechanism. (c) Surface nucleation mechanism.

If the interface reaction rate is extremely small so that mass/heat transfer is rapid enough to transport nutrients to the interface, then interface reaction rate (Equation 4-33) is the overall heterogeneous reaction rate (Figure 1-11a). If the interface reaction is relatively rapid and if the crystal composition is different from the melt composition, the heterogeneous reaction rate may be limited or slowed down by the mass transfer rate because nutrients must be transported to the interface and extra junk must be transported away from the interface (Figures 1-11b and 1-11c). If the crystal composition is the same as the melt composition, then mass transfer is not necessary. When interface reaction rate and mass transfer rate are comparable, both interface reaction and mass transfer would control the overall heterogeneous reaction (Figure 1-11d).

There are a few rules of thumb for evaluating whether interface reaction or mass transport is the controlling process. (i) If the crystal is near saturation in the initial melt or water, meaning that the initial departure from equilibrium is very small, then interface reaction is slow and controls the overall reaction rate. Exactly what constitutes "very small" departure from equilibrium will have to be assessed case by case. (ii) For mineral dissolution in pure water (meaning large departure from equilibrium) at room temperatures, interface reaction controls the dissolution rate of minerals with low solubility and high bond strength, and mass transport controls the dissolution rate of minerals with high solubility and low bond strength (Figure 1-12). (iii) For mineral dissolution in magma, the dissolution rate is often controlled by mass transport. (iv) If interface reaction controls the dissolution of a mineral in a fluid of different composition (meaning interface reaction rate is slower than mass transport), then interface reaction also controls the melting of the mineral in its own melt or fluid because heat transport is much more rapid than mass transport.

Mass or heat transport may involve diffusion, convection, or both (e.g., Zhang et al. 1989; Kerr 1995; Zhang and Xu, 2003). The diffusion distance is proportional to the square root of time, and the diffusion rate is inversely proportional to the square root of time. Hence, the diffusion rate and diffusive crystal growth rate is infinity at $t=0$, which is clearly impossible because growth rate is also limited by interface reaction rate. Hence, for a very short initial period (e.g., 0.1 s), interface reaction rate is slower than diffusion rate and crystal growth rate is controlled by the slower step, interface reaction. After this initial period, for some transient time (e.g., 2 s) crystal growth would be controlled by both interface reaction and diffusion. During the initial and transient periods, the diffusion equation is complicated because the interface concentration (and hence the degree of saturation at the interface) and the growth rate are both varying in a complicated way with time. Afterward, crystal growth would be controlled by diffusion, which is easy to treat.

Zhang et al. (1989) treated the interplay between diffusion and interface reaction during the initial and transient stages of crystal dissolution in a silicate melt. Using the interface reaction rate of diopside, they found that the period for crystal dissolution to be controlled by both interface reaction and diffusion (the transient period) is very short, on the order of seconds or less, after which the process is controlled by diffusion. The rate of a diffusion-controlled process is often inversely proportional to $t^{1/2}$ if the length of the diffusion medium is much larger than $(Dt)^{1/2}$. This leads to the often-referred parabolic reaction law (e.g., Figure 1-13a).

The heterogeneous reactions most often encountered are crystal growth and dissolution in igneous petrology, and bubble growth and dissolution in volcanology. In these problems, the boundary between the two phases (crystal and melt or bubble and melt) changes with time due to interface reaction and due to density difference between the two phases. Therefore, the diffusion problems for heterogeneous reactions are moving boundary problems. Sometimes, diffusion in the crystal may also play a role. The diffusion aspect of crystal growth was considered in Section 3.4.6. The problem crystal dissolution and growth is considered in great detail in Section 4.3.1. In this section, the main steps, paths, mathematical descriptions, and results (without detailed derivation) about crystal growth and dissolution are overviewed.

Crystal growth and dissolution in silicate melts are multicomponent problems, but the treatment below is simplified as effective binary diffusion (Section 3.3). Hence, the results only apply approximately to those components that can be treated in such a simple way, which usually means the components with high concentrations in the crystal and relatively low concentration in the melt (principal equilibrium-determining components, such as MgO for the growth and dissolution of forsteritic olivine, and ZrO_2 for the growth and dissolution of zircon). There are numerous components in a natural silicate melt, and the effective binary approach does not work for many components. More rigorous treatment must consider the multicomponent effect (Liang, 1999, 2000).

Below, the melt consumption rate u is distinguished from the crystal growth rate u^{cryst} (they differ by the density ratio), and the concentration in terms of kg/m^3 (denoted as C) is also distinguished from mass fraction (the same as weight percent, denoted as w).

Crystal growth Consider the case for crystal growth along one direction (hence a one-dimensional problem). Define the initial interface to be at $x=0$ and the crystal is on the side with negative x (left-hand side) and the melt is on the positive side (Section 3.4.6). Due to crystal growth, the interface advances to the positive side. Define the interface position at time t to be at $x=x_0$, where $x_0 \geq 0$ is a function of time. Let w be the mass fraction of the main equilibrium-determining component; then the diffusion equation in the melt is

$$\frac{\partial w}{\partial t} = D\frac{\partial^2 w}{\partial x^2}, \quad t > 0, \ x > x_0, \tag{4-39a}$$

with mass balance

$$D\frac{\partial w}{\partial x}\Big|_{x=x_0} + u(w - w^{\text{cryst}})_{x=x_0} = 0, \tag{4-39b}$$

where u is the melt consumption rate and is related to crystal growth rate as

$$\rho^c u^c = \rho u, \tag{4-39c}$$

where ρ is density and u^c is crystal growth rate. (The superscript "c" means crystal. The superscript "m" for melt is ignored.) To remove the moving boundary so as to solve the equation, a new coordinate (reference frame[1]) is defined as

$$t' = t \tag{4-40a}$$

$$x' = x - x_0(t). \tag{4-40b}$$

The interface is now fixed at $x' = 0$. Equation 4-39a becomes

$$\frac{\partial w}{\partial t'} = D\frac{\partial^2 w}{\partial x'^2} + u\frac{\partial w}{\partial x'}, \quad t' > 0, \; x' > 0. \tag{4-41a}$$

The initial condition is

$$w(x', 0) = w_\infty. \tag{4-41b}$$

The boundary condition is

$$w(0, t') = f(t'). \tag{4-41c}$$

Mass balance at the boundary provides an equation relating crystal growth rate to other parameters:

$$D\frac{\partial w}{\partial x'}\Big|_{x'=0} + u(w - w^c)_{x'=0} = 0 \tag{4-41d}$$

The new reference frame is known as the interface-fixed reference frame, and the old reference frame is called the laboratory-fixed reference frame. The melt consumption rate u depends on whether the growth is controlled by interface reaction, or by diffusion, or by externally imposed conditions such as cooling.

Crystal dissolution For crystal dissolution, one may use Equation 4-41a and remember that u (the melt consumption rate) is negative. One may also use the following equation in which u is the melt growth rate instead of the consumption rate:

$$\frac{\partial w}{\partial t'} = D\frac{\partial^2 w}{\partial x'^2} - u\frac{\partial w}{\partial x'} \quad t' > 0, \; x' > 0. \tag{4-42}$$

---

[1] A reference frame is a frame in which a diffusion profile is measured. It is discussed in more detail in Section 4.2.1.

Note that because the diffusion equations are for the melt phase, the rate is also that for melt motion. Therefore, during crystal growth, instead of crystal growth rate, melt consumption rate is used in the diffusion equation. During crystal dissolution, instead of crystal dissolution rate, the melt growth rate is used. Equation 4-39c may be applied to convert the rates.

4.1.3.1 Steady state

Diffusive crystal growth at a fixed temperature would not result in a constant crystal growth rate (see below). However, under some specific conditions, such as continuous slow cooling, or in the presence of convection with diffusion across the boundary layer, time-independent growth rate may be achieved. Similarly, time-independent dissolution rate may also be achieved.

For crystal growth at constant rate, if the crystal composition can respond to interface melt composition through surface equilibrium, steady state may be reached (Smith et al., 1956). At steady state, $(\partial C/\partial t)_{x'} = 0$ by definition. Hence,

$$D\frac{\partial^2 w}{\partial x'^2} + u\frac{\partial w}{\partial x'} = 0, \tag{4-43a}$$

where u is the melt dissolution rate (differing from the crystal growth rate by the density ratio). Let $z = \partial C/\partial x'$, and solve for z and then for w for constant u; we obtain,

$$w = w_\infty + (w_0 - w_\infty)\exp(-ux'/D), \tag{4-43b}$$

or

$$\frac{w}{w_\infty} = 1 + \frac{w_0 - w_\infty}{w_\infty}e^{-ux'/D} = 1 + \frac{1-K}{K}e^{-ux'/D}, \tag{4-43c}$$

where $K = w_\infty/w_0 = w^{cryst}/w_0$ (partitioning coefficient). Hence, at the steady state, the concentration profile is an exponential function, and the concentration of the component in the crystal is the same as the initial concentration in the melt. For a fixed crystal composition (such as SiO_2 concentration during quartz growth), or for a component with very small K (such as Al_2O_3 component during olivine growth), the concentration in the crystal can never be the same as the initial melt concentration, and there would be no steady state. For components that can reach steady state during crystal growth, D may be obtained by fitting experimental concentration profiles to Equation 4-43c and by independently obtaining melt consumption rate u.

During crystal dissolution at constant rate, some authors used an equation similar to Equation 4-43b to extract D values. However, because of the opposite sign for crystal dissolution versus growth, the concentration profile would be

$$w = w_\infty + (w_0 - w_\infty)\exp(ux'/D), \tag{4-44}$$

where u is the melt growth rate (crystal dissolution rate multiplied by the density ratio). Because $u > 0$, $D > 0$, and in the melt $x' > 0$, the above concentration profile does not approach a constant concentration when x' is large. Because concentration cannot be infinite at large x', there is hence no steady-state profile except for the trivial case of $w_0 = w_\infty$ (i.e., a uniform concentration profile in the melt $w = w_\infty$). That is, contrary to the case of crystal growth, there is no steady-state solution in the melt during crystal dissolution. When there is convection, there is additional melt motion whose velocity must be included in the u term so that now u is not simply melt growth rate but also includes the convective effect. Hence, steady state is possible. However, unless convective effect can be independently quantified and added to the melt growth rate, D cannot be obtained by fitting experimental data to Equation 4-44. Hence, D values extracted from steady-state concentration profiles during crystal dissolution are not reliable unless convection rate is quantified.

4.1.3.2 Growth or dissolution controlled by diffusion or heat conduction

If crystal growth or dissolution or melting is controlled by diffusion or heat conduction, then the rate would be inversely proportional to square root of time (Stefan problem). It is necessary to solve the appropriate diffusion or heat conduction equation to obtain both the concentration profile and the crystal growth or dissolution or melting rate. Below is a summary of how to treat the problems; more details can be found in Section 4.2.

One-dimensional diffusive growth of a crystal of fixed composition For constant crystal composition (such as quartz growth), the partition effect does not need to be considered. If the crystal growth rate is controlled by diffusion and if the problem is one-dimensional, the diffusion problem is a moving-boundary problem and may be written as (Equation 4-41a)

$$\text{Diffusion equation:} \quad \frac{\partial w}{\partial t'} = D \frac{\partial^2 w}{\partial x'^2} + u \frac{\partial w}{\partial x'}, \quad t' > 0, \ x' > 0. \tag{4-45a}$$

$$\text{Initial condition:} \quad w|_{t'=0} = w_\infty, \tag{4-45b}$$

$$\text{Boundary condition:} \quad w|_{x'=0} = w_0, \tag{4-45c}$$

$$\text{Mass balance:} \quad D \frac{\partial w}{\partial x'}|_{x'=0} = u(w^c|_{x'=0} - w|_{x'=0}), \tag{4-46}$$

$$\text{Stefan condition:} \quad u = \alpha(D/t)^{1/2}, \tag{4-47}$$

where α is a dimensionless constant to be determined.

If the short initial transient period with complicated behavior is ignored, the solution for the concentration profile is (Equation 3-116)

$$w = w_\infty + (w_0 - w_\infty)\text{erfc}\left(\frac{x'}{\sqrt{4Dt}} + \alpha\right)/\text{erfc}(\alpha), \tag{4-48}$$

where the parameter α satisfies (Equation 3-117c)

$$\pi^{1/2}\alpha\exp(\alpha^2)\mathrm{erfc}(\alpha) = (w_\infty - w_0)/(w^c - w_0).$$ (4-49)

After solving α from the above equation, the growth rate is known (Equation 4-47), and the growth distance is $2\alpha(Dt)^{1/2}$. Hence, the diffusion–growth problem is fully solved. For example, for quartz crystal growth in rhyolitic melt, if $w_\infty = 78\%$, $w_0 = 76\%$, and $w^c = 100\%$ for the SiO_2 component, then $(w_\infty - w_0)/(w^c - w_0) = 0.0833$, $\alpha = 0.0497$, and the melt consumption distance $\approx 0.1(Dt)^{1/2}$. If $D_{SiO_2} = 0.2$ $\mu m^2/s$ for a dry rhyolitic melt, the melt consumption distance is $2.7\,\mu m$ in $1\,h$. The growth distance of quartz equals $(\rho/\rho^{qtz})\,(2.7) = (2.34/2.65)\,(2.7) = 2.4\,\mu m$ in 1 h.

One-dimensional diffusive crystal dissolution During crystal dissolution, the surface concentration of the crystal may be treated as constant. Define $u = \alpha(D/t)^{1/2}$ to be the melt growth rate (instead of melt consumption rate). Then the concentration profile is

$$w = w_\infty + (w_0 - w_\infty)\mathrm{erfc}\left(\frac{x}{\sqrt{4Dt}} - \alpha\right)\Big/\mathrm{erfc}(-\alpha),$$ (4-50)

where the parameter α is to be solved from

$$\pi^{1/2}\alpha\exp(\alpha^2)\mathrm{erfc}(-\alpha) = (w_0 - w_\infty)/(w^c - w_0).$$ (4-51)

The melt growth distance is $2\alpha(Dt)^{1/2}$. For example, for forsteritic olivine (Fo90) dissolution in a basaltic melt, if $w_0 = 14$ wt%, $w_\infty = 7$ wt%, and $w^c = 50$ wt% for MgO, then $(w_0 - w_\infty)/(w^c - w_0) = 0.194$ and $\alpha \approx 0.098$. Hence, the dissolution distance is $0.196(Dt)^{1/2}$. If $D = 6\,\mu m^2/s$ for a basaltic melt, then the melt growth distance is $29\,\mu m$ in an hour, and the olivine dissolution distance is $(\rho/\rho^{oliv})29 = (2.7/3.28)29 = 24\,\mu m$ in 1 h.

One-dimensional crystal growth at constant growth rate The crystal growth rate may be controlled by factors other than the diffusion process itself. In such a case, the growth rate may be constant. Assume constant D and uniform initial melt. The diffusion problem can be described by the following set of equations:

Diffusion equation : $\quad \dfrac{\partial w}{\partial t'} = D\dfrac{\partial^2 w}{\partial x'^2} + u\dfrac{\partial w}{\partial x'}, \quad t' > 0,\ x' > 0.$ (4-52a)

Initial condition : $\quad w|_{t'=0} = w_\infty,$ (4-52b)

Boundary condition : $\quad D\dfrac{\partial w}{\partial x'}\Big|_{x'=0} = u(w^c|_{x'=0} - w|_{x'=0})$

$\qquad\qquad\qquad\qquad = u(K-1)w_0,$ (4-52c)

and

$$u = \text{constant},$$ (4-52d)

where $K=(w^c/w)|_{x'=0}$ is the partition coefficient. Note that the constant u is not to be determined from the solution (unlike in the cases of diffusive crystal growth) because it is not controlled by diffusion (diffusion controlled u would be inversely proportional to $t^{1/2}$). Note also that w_0 is not necessarily constant and has to be obtained from the solution.

The solution to the above problem has been given by Carslaw and Jaeger (1959, p. 389):

$$\frac{w}{w_\infty} = 1 - \frac{1}{2}\left\{\text{erfc}\frac{x'+ut}{2\sqrt{Dt}} + \frac{K-1}{K}e^{-ux'/D}\text{erfc}\frac{x'-ut}{2\sqrt{Dt}}\right\}$$
$$+ \frac{2K-1}{2K}e^{u(K-1)[x'+Kut]/D}\text{erfc}\frac{x'+ut(2K-1)}{2\sqrt{Dt}}. \qquad (4\text{-}53)$$

The surface concentration in the crystal is hence

$$w^c|_{x'=0} = Kw|_{x'=0}$$
$$= \frac{w_\infty}{2}\left\{\text{erfc}\frac{-ut}{2\sqrt{Dt}} + (2K-1)e^{K(K-1)u^2t/D}\text{erfc}\frac{ut(2K-1)}{2\sqrt{Dt}}\right\}. \qquad (4\text{-}54)$$

Let $\theta = ut/\sqrt{4Dt}$; then

$$w^c|_{x'=0} = 0.5w_\infty\left\{\text{erfc}(-\theta) + (2K-1)e^{4K(K-1)\theta^2}\text{erfc}[(2K-1)\theta]\right\}. \qquad (4\text{-}55)$$

Because $w^c|_{x'=0}$ is a function of t, the crystal grown from melt at constant rate will be zoned. This differs from the case of diffusive crystal growth in which the crystal is not zoned (except for a short initial period).

When $t \to \infty$, $\text{erfc}(-\infty) = 2$, $\text{erfc}(\infty) = 0$, $e^{u(K-1)[x'+Kut]/D}\text{erfc}\frac{x'+ut(2K-1)}{2\sqrt{Dt}} \to 0$; hence,

$$\frac{w}{w_\infty} = 1 - \frac{K-1}{K}e^{-ux'/D}$$

and

$$w^c|_{x'=0} = w_\infty,$$

which is the same as the steady-state solution (Equation 4-43c), as it should be. Figure 4-9 shows the evolution of concentration profiles and interface crystal composition as a function of time for a hypothetical trace element. If $K > 1$, Lasaga (1982) showed that the concentration profile in the melt would roughly achieve steady state at $t > 2.75D/u^2$. The above solution implies that the concentration in the crystal composition would increase or decrease without bound to be the same as that in the initial melt. However, this applies only to trace elements. For major elements, the concentration in the crystal is limited by crystalline structure constraints. For example, Al_2O_3 concentration in olivine can never reach the level in the initial melt (of the order 15%).

More examples and mathematical details of crystal growth and dissolution under various conditions can be found in Section 4.2. Furthermore, Lasaga

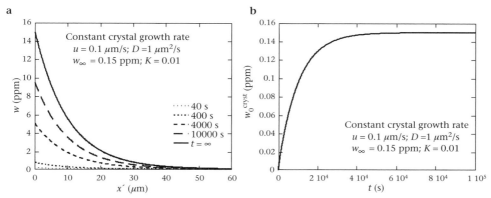

Figure 4-9 Calculated (a) concentration profiles in the melt and (b) crystal surface concentration for the case of constant crystal growth rate.

(1982) solved the problem of crystal growth with any growth rate $u(t)$, which can be consulted if such a solution is needed.

One-dimensional crystal growth controlled by heat conduction If diffusion is not necessary (such as crystal melting or growth in its own melt), then crystal growth or melting rate is often controlled by the interface reaction rate (i.e., constant growth rate for a given temperature and pressure). However, in the case of extremely rapid interface reaction rate (e.g., 0.1 mm/s), the growth or melting rate may be limited by heat transfer. In this section, only heat conduction is considered (i.e., convection is ignored). The boundary motion controlled by heat conduction also follows the parabolic law. The one-dimensional heat conduction problem during crystal growth is as follows:

$$\text{Heat conduction equation}: \quad \frac{\partial T}{\partial t'} = \kappa \frac{\partial^2 T}{\partial x'^2} + u \frac{\partial T}{\partial x'}, \quad t' > 0, \ x' > 0. \quad (4\text{-}56a)$$

$$\text{Initial condition}: \quad T|_{t'=0} = T_\infty, \quad (4\text{-}56b)$$

$$\text{Boundary condition}: \quad T|_{x'=0} = T_0, \quad (4\text{-}56c)$$

$$\text{Energy conservation}: \quad k \frac{\partial T}{\partial x'}|_{x'=0} = -u\rho \, \Delta H_f, \quad (4\text{-}56d)$$

and

$$u = \alpha(\kappa/t')^{1/2}, \quad (4\text{-}56e)$$

where k is heat conductivity (SI unit $W\,m^{-1}\,K^{-1}$), κ is heat diffusivity (SI unit m^2/s), $\kappa = k/(\rho c)$ with ρ being the density and c being the heat capacity (SI unit $J\,kg^{-1}\,K^{-1}$), ΔH_f is the latent heat of fusion (enthalpy of melting; SI unit J/kg), and α is a dimensionless constant to be determined. The solution for the temperature profile is (Equation 4-48)

$$T = T_\infty + (T_0 - T_\infty)\text{erfc}\left(\frac{x}{\sqrt{4\kappa t}} + \alpha\right) / \text{erfc}(\alpha), \tag{4-57}$$

where the parameter α is to be solved from the following:

$$\pi^{1/2}\alpha \exp(\alpha^2)\text{erfc}(\alpha) = c(T_0 - T_\infty)/(\Delta H_f). \tag{4-58}$$

The melt consumption distance is $2\alpha(\kappa t)^{1/2}$. For diopside growth, suppose overheating $(T_0 - T_\infty)$ is 10 K, because $c \approx 1.613$ kJ kg^{-1} K^{-1} and $\Delta H_f = 636$ kJ/kg, we have $c(T_0 - T_\infty)/(\Delta H_f) = 0.0254$. Hence, $\alpha = 0.0145$. Suppose $\kappa = 1$ mm^2/s; then the melt consumption distance per second is 9.2 μm. The diopside growth distance is $(2.7/3.3)(9.2) = 7.5\ \mu$m/s. Convection can enhance the crystal growth rate.

4.1.3.3 Convection

Convection refers to bulk directional (instead of random) motion of a fluid (see Chapter 3). In the presence of convection, a one-dimensional mass transport (including both diffusion and convection) equation can be obtained by adding a convective term to the diffusion equation:

$$\frac{\partial w}{\partial t} = D\frac{\partial^2 w}{\partial x^2} - v\frac{\partial w}{\partial x}, \tag{4-59}$$

where w is the concentration in the melt, v is the velocity of the directional flow along the x-direction. The above equation is in the laboratory-fixed reference frame. Convection may greatly enhance crystal growth or dissolution rates. At least two types of convection may be distinguished: free or forced. *Free convection* arises due to the dissolution process itself, which generates a boundary layer (interface melt layer) that has a different density than the bulk melt and hence may rise or sink. *Forced convection* is due to processes other than the dissolution process itself, such as a particle sinking or rising through the fluid (the relative motion of the crystal in the melt) due to buoyancy, or magma flow in the conduit relative to conduit wall. During crystal growth or dissolution, there is boundary motion (motion of the interface between the melt and crystal) in addition to fluid flow. To obtain the concentration profile relative to the interface, we again use the interface-fixed reference frame. The diffusion equation then becomes

$$\frac{\partial w}{\partial t} = D\frac{\partial^2 w}{\partial x'2} + (u - v)\frac{\partial w}{\partial x'}, \tag{4-60}$$

where u is melt consumption rate (crystal growth rate multiplied by the density ratio).

When a steady state is reached, the boundary layer thickness is independent of time and so is the crystal growth rate (or melt consumption rate u). The concentration profile at the steady state is

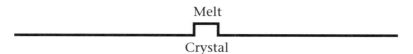

Figure 4-10 Protuberance on a crystal surface.

$$\frac{w}{w_\infty} = 1 + \left(\frac{w_0}{w_\infty} - 1\right)e^{-(u-v)x'/D},\tag{4-61}$$

where w_∞ is the initial melt concentration (of concentration at distance ∞), and w_0 is the interface melt concentration. To reach steady state, it is necessary for $(u - v)$ to be positive, which may be achieved during either crystal growth or crystal dissolution. (Remember that without convection, steady state can be reached only for crystal growth.) More detailed analysis for specific convective regimes of crystal dissolution and growth will be presented in Section 4.2.

4.1.4 Dendritic crystal growth

Dendritic growth is a special type of growth often observed in glass and in snowflakes. This type of growth is due to the interplay between interface reaction and mass or heat transfer. When mass or heat transfer is much slower compared to the interface reaction rate (hence, when mass or heat transfer controls the overall growth rate) and when there is large degree of oversaturation, there may be dendritic growth.

The key to dendritic growth is a high degree of oversaturation or undercooling, which leads to (i) high interface reaction rate, so high so that mass or heat transfer cannot keep up with growth, and (ii) high rate of nucleation on existing crystal surface. Consider a protuberance on a crystal surface as shown in Figure 4-10 and compare it with any flat part on the surface. (A corner would show the same effect.) Two factors favor the growth of the protuberance compared to the flat surface. One is that the protuberance is into the part of the melt with higher nutrient concentration. Secondly, nutrients are supplied to the protuberance by transport from the upper side, left, right, front, and back, whereas any site on the flat surface is supplied from only one direction (upper side). One factor hinders the growth of the protuberance: it has high surface energy, which leads to similar difficulty as nucleation. When the degree of oversaturation is high, nucleation is relatively easy (Figure 4-7), and the protuberance can grow more rapidly than any flat part on the surface. The growth from this protuberance leads to more protrusion into the melt, which grows further, leading to further growth of the protuberance. The cumulative result is dendritic growth (Figure 4-11). Dendritic growth may be modeled using Monte Carlo simulation (Lasaga, 1998).

Dendritic patterns do not usually occur during mineral dissolution. A crystal corner or protuberance would dissolve more rapidly because of more efficient mass or heat transfer if the process is controlled by mass or heat transfer in the melt. Higher surface energy further reduces the stability of the protuberance. Therefore, the corner would become round, and the protuberance would be eliminated. Flat or smooth interfaces between the crystal and melt will be produced (e.g., Zhang et al., 1989) and there would be no dendritic crystal dissolution for dissolution controlled by either interface reaction or mass or heat transfer in the melt. The only exception is for crystal partial melting (melt growth) controlled by diffusion in the crystal, during which dendritic crystal partial melting is possible because the argument for dendritic growth in the above paragraph can be made analogously by switching crystal and melt. Tsuchiyama and Takahashi (1983) observed dendritic pattern during partial melting of plagioclase. In summary, dendritic growth may occur if the growth of a phase is controlled by diffusion in the other phase.

Experimental results on crystal growth might be affected by dendritic growth and the effect is difficult to evaluate. Hence, experimental data on the melting rate of a pure crystal in its own melt are expected to be a more reliable indication of interface reaction rate than those on crystal growth rate.

4.1.5 Nucleation and growth of many crystals

The above discussion of the individual steps and paths (interface reaction and crystal growth) focuses on the growth or dissolution of a single crystal in a melt. For the solidification of magma, many individual crystals of several minerals grow and both the mineral and melt compositions vary with time. Hence, the complete treatment would require a set of equations, one for the growth of each single crystal if the location and time of the nuclei are known a priori. Because this is not possible, we settle for rough estimates. There are two approaches.

(1) One approach is to roughly estimate how the degree of crystallization would vary with time by making main simplifications in treating solidification, leading to the *Avrami equation*.

(2) The second approach is for the case of a single nucleation event, leading to simultaneous growth of many equal-size bubbles. Bubbles are assumed to distribute regularly similar to a crystal lattice (Figure 4-12). With the assumption, every bubble is surrounded by a melt shell whose inner surface is a spherical surface and outer surface is a polyhedral surface. This shell is further simplified to a spherical shell. With these conditions, all bubbles grow at the same rate. Hence, one only has to solve the problem of the growth of one bubble in a spherical shell. Proussevitch, Sahagian, and Anderson were the first to attack

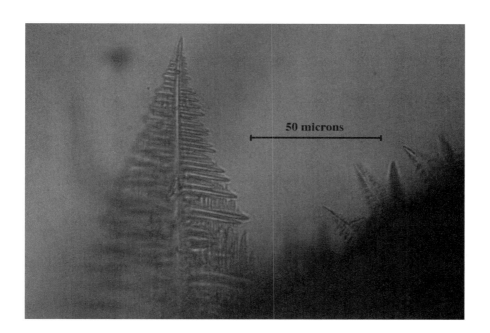

50 microns

50 microns

Figure 4-11 (a) A picture of dendritic growth in a rhyolite glass.
(b) Another picture of dendritic growth in a rhyolite glass.

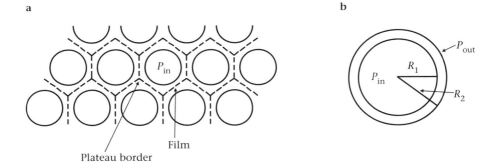

Figure 4-12 Many-body problem for bubble growth. (a) All bubbles are assumed to have nucleated at the same time, to have the same size, and to be distributed regularly. (b) The melt shell is further simplified to be a spherical shell. From Proussevitch et al. (1993) and Zhang (1999a).

the problem (Proussevitch et al., 1993; Proussevitch and Sahagian, 1998).

Both approaches are briefly summarized below.

4.1.5.1 Avrami equation

The total volume growth of crystals in a melt is the sum of the volume growth of all individual crystals. For a crystal nucleated at time τ with linear growth rate u (which may depend on time), the radius and volume of crystal i at time t can be written as

$$r_i = r_c + \int_\tau^t u_{\tau,\,t'}dt', \tag{4-62}$$

and

$$V_i = \frac{4\pi}{3}\left(r_c + \int_\tau^t u_{\tau,\,t'}dt'\right)^3, \tag{4-63}$$

where $r_{\tau,t}$ is the radius at time t for crystals nucleated at time τ, r_c is the critical radius of nucleation, and u is the crystal growth rate at time t' (between τ and t) for crystals nucleated at time τ. Because r_c is small, we ignore it in the approximate treatment below. Make the approximation that the growth rate depends only on the time at which the crystal nucleated (hence, all crystals that nucleated at the same time have the same size and growth rate). The total growth per unit volume of melt is denoted as y and can be expressed as

$$y = \int_0^t I_\tau \frac{4\pi}{3}\left(\int_\tau^t u_{\tau,\,t'}dt'\right)^3 d\tau \tag{4-64}$$

where I_τ is the nucleation rate at time τ. If the shrinkage of melt volume with time is ignored, we would have

$$V_t^{\text{solid}} = V_0^{\text{melt}} \int_0^t I_\tau \frac{4\pi}{3} r_{\tau,t'}^3 d\tau = V_0^{\text{melt}} \int_0^t I_\tau \frac{4\pi}{3} \left(\int_\tau^t u_{\tau,t'} dt' \right)^3 d\tau. \quad (4\text{-}65)$$

However, the melt volume shrinks as crystallization proceeds. Hence, one can write

$$dV^{\text{solid}} = V^{\text{melt}} dy. \quad (4\text{-}66)$$

Ignoring the density difference between the melt and the solid, we have $V_s + V_m = V_0$, where V_0 is the initial volume of the melt. Let $F = V_s/V_0$; we have

$$dF = (1 - F) dy. \quad (4\text{-}67)$$

$$\ln(1 - F) = -y. \quad (4\text{-}68)$$

Hence,

$$F = 1 - e^{-y}. \quad (4\text{-}69)$$

The parameter y (given in Equation 4-64) is zero at time zero and increases with time. In different nucleation and growth regimes, the rate of increase is different. Hence, it is assumed that y is proportional to time raised to some power n, leading to

$$F = V_t/V_\infty = 1 - \exp[-(t/t_c)^n], \quad (4\text{-}70)$$

where F is the degree of crystallization, V_∞ is the crystal volume at $t = \infty$, n is related to how the nucleation and growth rates depend on time and is often between 0.5 and 5, and t_c is a characteristic time. When $t = t_c$, $F = 1 - 1/e \approx 0.632$. Equation 4-70 is called the *Avrami equation*. For example, for isothermal solidification (crystal growth), if all crystals nucleate at a single time and all grow at the same rate, and the growth is diffusion controlled so that $u = A/t^{1/2}$, then $n = 1.5$ in the Avrami equation. If diffusive growth rate gradually slows down due to many-body interaction, n would be smaller than 1.5. On the other hand, if nucleation rate is constant, and growth rate of every crystal is constant, then $n = 4$ in the Avrami equation, which is sometimes referred to as the *JMA equation*. Figure 4-13 shows the relation between F and t for some n values.

The above discussion is for the crystallization of one single mineral (i.e., many crystals of the same mineral). If several minerals are crystallizing and they have different crystallization temperature (and hence different crystallization time), then the relation between the degree of crystallization and time would be much more complicated.

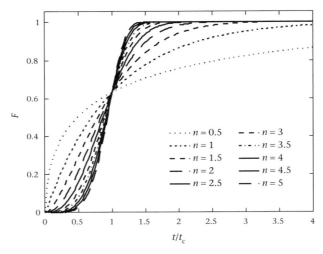

Figure 4-13 Degree of crystallization based on the Avrami equation.

4.1.5.2 Growth of many equal-size bubbles

The Avrami equation is a very rough solution to the problem of the growth of many crystals. No better treatment is available except for one class of problem: the growth of many regularly distributed and equal-sized H_2O bubbles (Figure 4-12). The numerical algorithm is due to Proussevitch et al. (1993) and Proussevitch and Sahagian (1998). The governing equations include the diffusion equation that includes a term accounting for flow of the melt due to bubble growth, the hydrodynamic equation about the pressure in the bubble, the solubility law, the mass balance condition for bubble growth rate, and other conditions. The complexities include the many bubbles and the strong dependence of viscosity and H_2O diffusivity on H_2O content. The full treatment is covered in more detail in Section 4.2.5. Figure 4-14 shows calculated bubble radius versus time, recast in terms of F versus t/t_c to compare with the Avrami equation (Equation 4-70). The corresponding n factor is about 0.55. The set of equations (Section 4.2.5) may be adapted to treat growth of many crystals nucleated at a single time.

4.1.6 Coarsening

Nucleation and growth are usually followed by *coarsening*, in which many small crystals are replaced by fewer larger crystals to minimize the interface area and total free energy. This phenomenon was first described by Wilhelm Ostwald (1853–1932), and is hence also known as *Ostwald ripening*. Coarsening begins when the concentration profiles due to growth of different crystals overlap and if the crystals (bubbles) are of different sizes. For the case discussed in Section

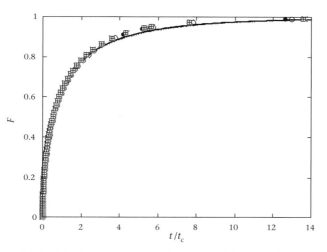

Figure 4-14 Calculated F versus t/t_c for bubble growth using the program of Proussevitch and Sahagian (1998) modified by Liu and Zhang (2000). F is the volume of the bubble versus the final equilibrium volume of the bubble. The calculated trend may be fit by the Avrami equation with an n value of 0.551.

4.1.5.2, because all bubbles are of the same size (which is hypothetical), there would be no coarsening. For the case discussed in Section 4.1.5.1, there would be coarsening, although it was not discussed there. Mostly, coarsening occurs when the volume fraction of the crystals is almost the equilibrium fraction. For example, if the equilibrium mineral fraction is 40%, coarsening might occur when the degree of crystallization is 35% (depending on many factors). If all the magma should crystallize, coarsening might occur when the degree of crystallization is 90%. During coarsening, the overall volume growth is less significant, but the average size of the crystals (or bubbles) increases with time and the total number of crystals decreases with time.

Coarsening occurs because of surface tension, which leads to greater chemical potential and hence less stability for smaller crystals (bubbles). For a spherical crystal with radius r, the interface energy is $4\pi r^2 \sigma$, and the volume is $4\pi r^3/3$ for each crystal. The chemical potential contribution from the interface energy can be found as

$$\mu_{surface} = \left(\frac{\partial G}{\partial n}\right)_{T,P,\,etc} = \left(\frac{\partial(4\pi r^2 \sigma)}{\partial(4\pi r^3/3)/V_c}\right)_{T,P,\,etc} = V_c \frac{2\sigma}{r}, \tag{4-71}$$

where σ is the interface energy and V_c is the molar volume of the crystal. Hence, the chemical potential μ of a component in a crystal of radius r is given by

$$\mu_r = \mu_\infty + V_c(2\sigma/r), \tag{4-72}$$

Table 4-3 Calculated solubility as a function of crystal size

r (μm)	0.01	0.1	1	10	100	1000
S_r (M)	0.33	0.113	0.1012	0.10012	0.100012	0.1000012
$S_r/S_\infty - 1$	2.3	0.13	0.012	0.0012	1.2×10^{-4}	1.2×10^{-5}
$S_r - S_{2r}$ (M)	0.15	0.0066	6.1×10^{-4}	6×10^{-5}	6×10^{-6}	6×10^{-7}
$S_{r/2} - S_r$ (M)	0.78	0.0144	0.00122	1.2×10^{-4}	1.2×10^{-5}	1.2×10^{-6}
$r(S_r - S_{2r})$	0.0015	0.00066	0.00061	0.00060	0.00060	0.00060

Note. $S_\infty = 0.1$ M, $\sigma = 0.5$ J/m^2, $V_c = 50 \times 10^{-6}$ m^3/mol, $T = 500$ K.

where μ_∞ is the chemical potential of a component for an infinitely large crystal. Therefore, smaller crystals have higher chemical potential and are hence less stable than large crystals. If an interstitial fluid is present, the solubility of an infinitely large crystal is S_∞ and that of a small crystal with radius r is $S_r > S_\infty$. The relation between S_r and S_∞ may be found as follows. Because

$$\mu_\infty = \mu^{\text{liq}} + RT \ln S_\infty, \tag{4-73}$$

$$\mu_r = \mu_\infty + V_c(2\sigma/r) = \mu^{\text{liq}} + RT \ln S_r, \tag{4-74}$$

we obtain

$$\ln \frac{S_r}{S_\infty} = \frac{2\sigma V_c}{rRT}, \tag{4-75}$$

or

$$S_r = S_\infty \exp\left(\frac{2\sigma V_c}{rRT}\right), \tag{4-76}$$

where R is the gas constant and T is temperature. Table 4-3 shows the calculated solubility as a function of crystal size for the case of $S_\infty = 0.1$ M, $\sigma = 0.5$ J/m^2, $V_c = 50 \times 10^{-6}$ m^3/mol, $T = 500$ K. As can be seen from Equation 4-76 and Table 4-3, smaller crystals have a higher solubility. Hence, there is a concentration gradient from a small crystal to a large crystal. This gradient causes mass transfer from the small crystal to the large crystal. As the small crystal becomes smaller, its chemical potential increases and it dissolves more rapidly. Large crystals hence grow at the expense of smaller grains. Because the crystals are already present, nucleation is not a step in the coarsening process. The rate of coarsening may be controlled by either interface reaction or mass transfer.

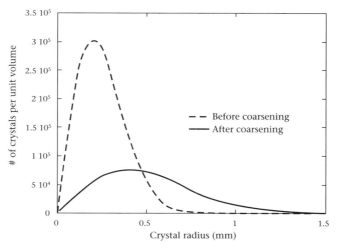

Figure 4-15 Schematic variation of crystal size distribution with time during coarsening following Chai (1974).

Quantification of coarsening is complicated. During coarsening, all crystals are not the same size. There is a crystal size distribution (Figure 4-15). The distribution function may be log-normal:

$$f(r) = \frac{1}{r\sigma\sqrt{2\pi}} e^{-(\ln r - A)^2/(2\sigma^2)}, \tag{4-77}$$

or some other function, such as

$$f(r) = A(r/r_{opt})^n \exp[-B(r/r_{opt})^2], \tag{4-78}$$

where A, B, and n are constants, r is the radius of the crystal, and r_{opt} is the optimum radius (the radius at which f is maximum). For example, Chai (1974) carried out hydrothermal coarsening experiments of calcite at 923 K and 200 MPa. The crystal size distribution was found to follow

$$\frac{f(r)}{f_{max}} = \left(\frac{32}{9\pi}\right)^2 \left(\frac{r}{\langle r \rangle}\right)^4 \exp\left[2 - \left(\frac{64}{9\pi}\right)\left(\frac{r}{\langle r \rangle}\right)^2\right], \tag{4-79}$$

where $\langle r \rangle$ is the average radius, and f_{max} is the maximum $f(r)$ value.

Some rough estimations of coarsening rate versus optimum crystal size are given below. Consider coarsening in the presence of pore fluid (either aqueous solution or hydrothermal solution or melt). Assume that during coarsening, crystal size distribution follows the same functional shape when plotted as f/f_{max} vs. r/r_{opt} as shown by the experiments of Chai (1974). Then, the coarsening rate can be simply characterized by how the optimum crystal radius r_{opt} (or average crystal radius) increases with time. Whether the coarsening is controlled by interface reaction or by mass transfer, the coarsening rate depends on crystal size.

Coarsening from 0.1- to 1-μm size is relatively rapid, whereas coarsening from 1- to 10-mm size is slow.

If coarsening is controlled by interface reaction (meaning rapid diffusion), concentration in the pore fluid is uniform (but depends on time), roughly corresponding to the saturation concentration of the optimal crystal radius. Smaller crystals are undersaturated and larger crystals are supersaturated. The dissolution rate of smaller crystals is proportional to the degree of undersaturation, and the growth rate of larger crystals is proportional to the degree of supersaturation. The degree of supersaturation for a crystal with twice the optimal radius is inversely proportional to the optimal radius (as can be seen in Table 4-3, $r(S_r - S_{2r})$ is roughly constant). Hence, the growth rate of the large crystal is also inversely proportional to the optimal radius. Similarly, the dissolution rate of a crystal with half the optimal radius is roughly proportional to the optimal radius. Therefore, the rate of increase for the optimal radius is roughly inversely proportional to the optimal radius. Hence,

$$dr_{opt}/dt = A'/r_{opt}, \tag{4-80}$$

where A' is a constant. Therefore,

$$r_{opt}^2 = 2A't. \tag{4-81}$$

That is, r_{opt} is proportional to $t^{1/2}$. As the average crystal size increases, the interface reaction rate becomes smaller because the degree of oversaturation becomes smaller. Hence, the coarsening rate dr_{opt}/dt for mean crystal size of 1 mm is 10^3 times slower than that for mean crystal size of 1 μm; coarsening timescale for a rock to grow from a mean crystal size of 1 to 2 mm is 10^6 times that for a rock to grow from a mean crystal size of 1 to 2 μm.

Now consider coarsening controlled by mass transfer. The gradient may be expressed as $\Delta w/\Delta x$, where Δw is the solubility difference and Δx is mean distance between the grains. Using Table 4-3, Δw between a small crystal of radius r and a crystal of radius $2r$ is roughly inversely proportional to r. When crystal grains are larger, the mean distance between them is also larger. If we assume that the mean distance between crystal grains is proportional to crystal size, then $\Delta w/\Delta x$ is inversely proportional to the square of mean crystal radius. Hence, we would have

$$dr_{opt}/dt = A/r_{opt}^2, \tag{4-82}$$

leading to

$$r_{opt}^3 = 3At, \tag{4-83}$$

where A is another constant. That is, r_{opt} is proportional to $t^{1/3}$, or the volume of the optimal crystal size is proportional to t. Hence, we may estimate that the coarsening timescale from 1 to 2 mm radius is 10^9 times that from 1 to 2 μm.

A more mathematical derivation of Equation 4-83 can be found in Lifshitz and Slyozov (1961).

The coarsening data of Chai (1974) using hydrothermal experiments show that $\langle r \rangle^3$ is roughly proportional to t, consistent with coarsening controlled by mass transfer. The proportionality depends on the type of solution, as well as temperature. Many other experimental data also show this relation. That is, Ostwald ripening is often controlled by mass transfer instead of interface reaction.

4.1.7 Kinetic control for the formation of new phases

In a system (aqueous solution, magma, or rock), if two or more phases are oversaturated and are more stable than the existing phase or phases, the new phase that forms first will be more stable than the existing phase or phases, but is not necessarily the most stable phase (with highest degree of oversaturation). That is, thermodynamics does not completely control the formation of new phases. Kinetics plays an important role.

It is observed that in the case of simultaneous saturation of two or more phases, the phase that forms first is often the least stable, or the most disordered, especially at room temperatures. For example, in aqueous solutions, opal (disordered) often forms but the more stable quartz rarely forms. Over a very long time, opal may "mature" to become quartz. The same is true for the formation of calcite (as compared dolomite), and analbite (as compared to albite). From the vapor phase, phosphorous vapor condenses first to yellow phosphorus (high entropy), instead of the more stable red phosphorous (low entropy)

Ostwald proposed that when two or more new phases may form from existing phase or phases, that is, when new phases are more stable than the existing phase(s), the least stable new phase would form first and then transform into more stable phases. This is called the *Ostwald rule*, the *Ostwald step rule*, or *the law of successive reactions*. An alternative statement of the Ostwald rule is as follows:

> If two or more phases may form from existing phase(s), the phase that requires the least activation energy to form would form first. If the new phase is metastable, it would transform into more stable phases. Therefore, phase transformation is a step process, with each step leading to a more stable phase, but not necessarily the most stable phase.

One example of the least activation energy is when the structure of the new phase is closest to that of the existing phase(s) or when the structure is disordered so that it does not require elaborate and precise rearrangement. The Ostwald rule is especially applicable to low-temperature phase transformations because at these low temperatures it is difficult to overcome the high activation energies required to form a new phase. At high temperatures such as igneous temperatures, the Ostwald rule is less often encountered.

One way to rationalize and remember the rule is to think that "nature is lazy" and hence would like to accomplish a process with the least effort. Because ions and molecules in a liquid are more or less randomly distributed, one may guess that the phases with simple structure and with low degree of order (meaning ions and molecules do not have to be arranged in a specific way) tend to form more easily than the phases with complicated structure and high degree of order. This often means the formation of metastable phases such as opal.

It is important to emphasize that thermodynamics is never violated in the kinetic control for the formation of new phases. If only one new phase is more stable than the existing phase, the new phase would form if the kinetics allows it (i.e., if nucleation and growth rates are high enough). However, if two or more new phases are more stable than and can form from the existing phase(s), thermodynamics dictates only that the new phase be more stable than the old phase; it does not dictate that it be the most stable phase. Kinetics determines which phase would form. Figure 4-16 compares the thermodynamic and kinetic control of the reactions. The reactant is unstable with respect to both product 1 and product 2. Product 2 is more stable (has a lower G) than product 1. Which product would form depends on which process has a lower activation energy. If activated complex 1 requires less activation energy, then product 1 would form, instead of the more stable product 2 because forming it requires a higher activation energy.

4.1.8 Some remarks

Heterogeneous reactions come in many varieties and most are complicated. Many steps and pathways may be involved in the kinetics of a heterogeneous reaction. Some of these steps or paths can be quantified well, such as diffusion and heat conduction. Convection may also be empirically quantified. For some other processes such as interface reaction rates, adequate theory is available, although more experimental data are needed, especially at high temperatures (best done by investigating melting of a crystal in its own melt). Among the most difficult tasks in heterogeneous reaction kinetics are (i) nucleation, (ii) prediction of new phases that would form first (Ostwald step rule), and (iii) growth of many interacting crystals in a magma or an existing rock. It is possible that a relatively simple nucleation theory (although it may also be complex) will be developed in the near future, especially with recent progress in nanomaterials. Understanding the energy surface to predict a priori which phase would form first will likely require quantum mechanic progress in chemistry. To quantify the growth rates of many randomly distributed crystals of different minerals in a magma or rock (the many-body problem), brute force of computation power to handle the mathematical complexity will be necessary. In the remaining part of this chapter, some specific heterogeneous reactions that have been solved are investigated in depth.

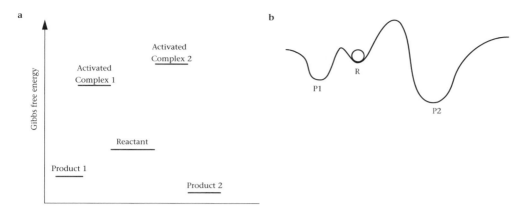

Figure 4-16 Role of kinetics in determining which of the two stable products to form, compared with stability of a ball on uneven ground. (a) Because the activation energy for forming product 1 is smaller, product 1 will form even though it is less stable than product 2. (b) Stability of a ball on uneven ground. The ball is initially in hole R. It would be gravitationally more stable if it goes to either hole P1 or P2. The most stable position would be hole P2. However, if the ball was given an initial push (similar to thermal motion of molecules), it is much more likely that it would end up in hole P1.

4.2 Dissolution, Melting, or Growth of a Single Crystal, Bubble, or Droplet Controlled by Mass or Heat Transfer

Crystal dissolution, melting, and growth in a fluid reservoir (melt or water) are an important class of problems in igneous petrology and aqueous geochemistry. The difference between dissolution and melting lies in the fact that crystal dissolution occurs when the temperature is below the melting temperature of the crystal (below the solidus for a solid solution) and melting occurs when the temperature is above the melting temperature (above the liquidus for a solid solution). For a crystal that is a solid solution, partial melting occurs when the temperature is between the solidus and liquidus. Dissolution of a crystal requires the presence of a melt or fluid that is undersaturated with the crystal. Melting occurs with or without a fluid phase.

Crystal growth is the opposite of dissolution and melting, and the treatments are similar, but there are at least two differences that make crystal growth more complicated. One is that during growth, crystal composition responds to the melt composition. Secondly, the interplay between growth and mass or heat transfer may result in dendritic growth, but crystal dissolution leads to smooth or flat interfaces. (Partial melting may lead to dendritic texture too.) Hence, crystal growth is more difficult to treat than crystal dissolution.

Crystal dissolution/melting/growth may be controlled by interface reaction rate (Figure 1-11a), meaning that mass/heat transfer rate is very high and interface reaction rate is low. Examples include dissolution of minerals with low solubility and high bond strength (z_+z_-) in water (Section 1.5.1.1). In such a case, the crystal dissolution rate is time-independent at a given temperature, pressure, and other conditions, such as pH and degree of undersaturation. Interface reaction rate may depend on the surface orientation, such as (100) or (111). Along a surface, the dissolution distance is proportional to time, resulting in a *linear reaction law*. The composition of the interface solution is the same as that of the initial solution, and the concentration profile in the solution is flat. Stirring the solution would not increase the dissolution rate. The dissolution rate can be calculated from the interface reaction rate equation (Equation 4-33) if the necessary data are available, and no additional modeling is necessary.

Crystal dissolution and growth may be controlled by mass transport (and crystal melting and growth in its own melt may be controlled by heat transfer), meaning that the mass/heat transfer rate is low and interface reaction rate is high. Examples include dissolution of minerals with high solubility and low bond strength in water (Section 1.5.1.1), as well as dissolution of many minerals in melts at high temperature. In such a case, the crystal dissolution rate depends on mass transfer rate. Stirring the solution would increase the dissolution rate. In the absence of convection, the dissolution is controlled by diffusion, referred to as *diffusive crystal dissolution* (Figure 1-11b). In the presence of convection, mass transfer is enhanced by convection, and dissolution is referred to as *convective crystal dissolution*. In both cases, the interface melt composition differs from the initial melt composition, and is near the equilibrium composition. There are concentration gradients in the melt near the dissolving crystal. Furthermore, diffusive or convective crystal dissolution rate does not vary with crystal orientation. For diffusive crystal dissolution and growth, the rate is inversely proportional to square root of time and the distance is proportional to square root of time, leading to a *parabolic reaction law*. The rate and distance can be predicted by solving the diffusion problem. Convective crystal dissolution rate and distance can be predicted from both dissolution kinetics and fluid dynamics. Steady-state convection leads to a time-independent concentration profile (Figure 1-11c), a time-independent dissolution rate, and a dissolution distance proportional to time. That is, *steady-state convective dissolution* or growth results in the linear reaction law, similar to interface-controlled dissolution. The two mechanisms can be distinguished as follows: (i) there are concentration gradients in the melt in convective dissolution and no gradient in interface-controlled dissolution; (ii) the composition of the interface melt is similar to the saturation composition in convective dissolution, but is similar to the initial composition in interface-controlled dissolution; and (iii) stirring would increase convective dissolution rate, but would not change the interface-controlled dissolution rate.

Crystal dissolution and growth may also be controlled by both mass or heat transport and interface reaction (Figure 1-11d). In this case, the interface reaction rate is comparable to mass transport rate. The interface melt composition is not the same as the initial composition, and is not near the equilibrium composition either. The interface melt composition varies with time (moving toward equilibrium composition). The dissolution distance may be between the linear law and the parabolic law. Because this problem is slightly more complex, it will be discussed after we investigate crystal dissolution controlled by mass transfer.

The rate of both diffusive and convective dissolution can be quantified. This topic is often encountered either as an independent problem, or as part of a larger problem in diffusion studies (Chapter 3), in trying to understand the kinetics and dynamics of volcanic eruptions or processes in beverages (Section 4.3), and in geospeedometry (Chapter 5). In this section, we focus on the treatment of crystal dissolution and growth controlled by mass and heat transfer, and examine the various aspects in great detail. We will not only address new questions, but also expound further upon some of the previously discussed issues for completeness and thoroughness. Readers who are not familiar with previous sections are encouraged to go through overviews in Sections 4.1.3.

To avoid confusion, in this section we explicitly distinguish concentrations expressed in kg/m^3, and dimensionless concentration such as weight fraction. The former will be denoted as C, and the latter as w (weight fraction). The relation between C and w is

$$C = \rho w, \tag{4-84}$$

where ρ is density.

4.2.1 Reference frames

There are many subtleties in adopting the *reference frame* (Brady, 1975a), some of which are discussed here so that the different forms of equations during crystal dissolution or growth and during bubble dissolution or growth can be understood.

4.2.1.1 One-dimensional crystal dissolution or growth

First, consider one-dimensional diffusion in the melt during crystal dissolution (the case of crystal growth is similar except for a negative versus positive sign) along the direction of x. Because melt and crystal densities are different, the melt growth rate is different from the crystal dissolution rate. Furthermore, because melt density may vary from one position to another, the melt motion velocity may depend on x. Use superscript "c" to represent the crystal phase, and ignore superscript "m" for melt. Let the crystal dissolution rate be u^c (in this definition, $u^c > 0$ if the crystal dissolves). Let the melt growth rate at the interface be u_0. Let that at any x be u. Based on the continuity equation, using steady-state approximation, we have

$$\rho^c u^c = \rho_0 u_0 \text{ (at the interface)}, \tag{4-85a}$$

$$\rho_0 u_0 A_0 = \rho u A \text{ (in the melt)}, \tag{4-85b}$$

where ρ^c (assumed to be constant) is density of the crystal, ρ_0 is density of melt at the interface, ρ is density of the melt at any x, and A is the cross-section area. Both u and u^c depend on time during diffusive crystal dissolution (parabolic law). In most applications, melt density variation is small ($\leq 4\%$). Hence, for simplicity, we tolerate this small error and assume constant melt density (leading to $u = u_0$ for constant cross-section area) to simplify the equations. The hurdle for improving the accuracy by considering melt density variation is high. For example, previous diffusivity data are obtained assuming constant melt density. The concentration profile and hence diffusivity would change slightly if density variation across a profile is considered. For self-consistency, if density variation across a profile is treated, the diffusivity (and other relevant parameters) should also be redetermined by considering melt density variation.

In the case of one-dimensional crystal dissolution with $u = u_0$, if the reference frame is fixed at the faraway melt ($x = \infty$), the melt does not flow even though the melt is generated at the interface at velocity u. (The interface moves as a rate of u.) Hence, the diffusion equation is Equation 3-9 without a velocity term:

$$\frac{\partial C}{\partial t} = \frac{\partial}{\partial x}\left(D\frac{\partial C}{\partial x}\right), \tag{4-86a}$$

where x is the coordinate fixed at the faraway melt.

Still in the case of one-dimensional dissolution, if the reference frame is fixed at the nondissolving part of the crystal ($x = -\infty$), the interface moves at a velocity of u^c. However, any point in the melt is moving at a velocity of $u > u^c$. That is, relative to the reference frame fixed to the nondissolving part of the crystal, the melt flows at a velocity of $(u - u^c)$. Hence, the equation to describe diffusion in the melt is the flow–diffusion equation (Equation 3-19b),

$$\frac{\partial C}{\partial t} = \frac{\partial}{\partial x_1}\left(D\frac{\partial C}{\partial x_1}\right) - (u - u^c)\frac{\partial C}{\partial x_1}, \tag{4-86b}$$

where x_1 is the coordinate fixed at the crystal. In Section 3.5.5, we simply used the term "lab-fixed" reference frame to write down Equation 4-86a but did not explain. With the elaboration above, it can be seen that the laboratory-fixed reference frame in Section 3.5.5 is actually the reference frame fixed at the faraway melt, and is different from the crystal-fixed reference frame because of the density difference between the crystal and melt.

No matter which reference frame we start with, when transformed into the same interface-fixed reference frame, the results should be the same. Hence, starting from Equation 4-86b, we should also arrive at Equation 3-114a. Let y be the coordinate fixed at the crystal–melt interface; then

$$y = x_1 + \int u^c dt. \tag{4-87}$$

Transforming Equation 4-86b into the interface-fixed reference frame, then

$$\frac{\partial C}{\partial t} = \frac{\partial}{\partial y}\left(D\frac{\partial C}{\partial y}\right) - (u - u^c)\frac{\partial C}{\partial y} - u^c\frac{\partial C}{\partial y} = \frac{\partial}{\partial y}\left(D\frac{\partial C}{\partial y}\right) - u\frac{\partial C}{\partial y}, \tag{4-88}$$

which is similar to Equation 3-114a (the difference in sign is because Equation 3-114a is for crystal growth and the above equation is for crystal dissolution).

4.2.1.2 Three-dimensional crystal dissolution or growth

Now consider the case of three-dimensional crystal dissolution. Let the radius of the crystal be a (which depends on time). In this case, the most often-used reference frame is fixed at the center of the crystal, i.e., lab-fixed reference frame (different from the case of one-dimensional crystal growth for which the reference frame is fixed at the interface) so that the problem has spherical symmetry. Ignore melt density variation. The crystal dissolution rate (u^c) and melt growth rate at the interface (u_a) are related by the continuity equation with approximation of steady state:

$$\rho^c u^c = \rho_a u_a \text{ (at the interface) ,} \tag{4-89a}$$

$$4\pi a^2 \rho_a u_a = 4\pi r^2 \rho_r u_r \text{ (in the melt),} \tag{4-89b}$$

where $r > a$ (in the melt), and u_r is melt motion velocity at the radial position of r. As viewed from the center of the crystal, the interface melt moves at a velocity of $-u^c$ (negative sign because it moves inward), of which $-u_a$ is due to melt growth and $-(u^c - u_a)$ is flow. The melt at radial distance r moves at a velocity of $-u^c(a/r)^2$, of which $-u_a(a/r)^2$ is due to melt growth and $-(u^c - u_a)(a/r)^2 = (a/r)^2 (u_a - u^c) = (a/r)^2(\rho^c/\rho_a - 1)u^c$ is flow. Therefore, the equation to describe diffusion in the melt ($r > a$) during spherical crystal dissolution using the reference frame fixed at the center of the crystal is the diffusion–flow equation (Equation 3-19a):

$$\frac{\partial C}{\partial t} = \frac{1}{r^2}\frac{\partial}{\partial r}\left(Dr^2\frac{\partial C}{\partial r}\right) - u^c\left(\frac{\rho^c}{\rho_a} - 1\right)\frac{a^2}{r^2}\frac{\partial C}{\partial r}, \tag{4-90}$$

where a is a function of t, $r > a$, $t > 0$, and u^c is positive for crystal dissolution and negative for crystal growth. If one prefers to use crystal growth rate so that u^c is positive for crystal growth, the negative sign in front of u^c would become positive.

4.2.1.3 Three-dimensional bubble dissolution or growth

Finally, consider the case of three-dimensional bubble dissolution. Let the radius of the bubble be a. Again, the most-often used reference frame is fixed at the center of the bubble so that the problem has spherical symmetry. Ignore melt density variation. The bubble dissoultion rate (u^g) and melt growth rate (u_a) are related by the continuity equation:

$$\rho^g u^g = \rho_a u_a, \tag{4-91a}$$

$$4\pi a^2 u_a \rho_a = 4\pi r^2 u_r \rho_r, \tag{4-91b}$$

where $r > a$ (in the melt), and u_r is melt motion velocity at the position of r. As viewed from the center of the bubble, the interface melt moves at a rate of $-u^g$ (negative sign because it moves inward), of which $-u_a$ is due to melt consumption and $-(u^g - u_a)$ is flow. Following the procedures above, the diffusion equation for bubble dissolution in the center-fixed reference frame is

$$\frac{\partial C}{\partial t} = \frac{1}{r^2}\frac{\partial}{\partial r}\left(Dr^2\frac{\partial C}{\partial r}\right) - u^g\left(\frac{\rho^g}{\rho_a} - 1\right)\frac{a^2}{r^2}\frac{\partial C}{\partial r} \approx \frac{1}{r^2}\frac{\partial}{\partial r}\left(Dr^2\frac{\partial C}{\partial r}\right) + u^g\frac{a^2}{r^2}\frac{\partial C}{\partial r}, \tag{4-92}$$

where a is a function of t, $r > a$, $t > 0$, and u^g is positive for bubble dissolution and negative for bubble growth. If one prefers to use bubble growth rate so that u^g is positive for bubble growth (e.g., Proussevitch and Sahagian, 1998), the sign in front of u^g would be negative. The approximation above is because ρ^g is much smaller than melt density. For example, if the density ratio ρ^g/ρ_a is 0.005, then u_a is only 0.5% of u^g. That is, melt growth rate during bubble dissolution is negligible.

In summary, flow velocity is relative and depends on the reference frame. By changing the reference frame, flow velocity changes. The reference frame to be chosen is the one that makes the problem easier to solve. For three-dimensional cases with spherical symmetry, the reference frame is almost always fixed at the center of the sphere (i.e., the frame does not move). For one-dimensional cases, the reference frame is usually a moving frame fixed at the interface.

4.2.2 Diffusive crystal dissolution in an infinite melt reservoir

Diffusive crystal dissolution means that crystal dissolution is controlled by diffusion, which requires high interface reaction rate and absence of convection. In nature, diffusive crystal dissolution is rarely encountered, because there is almost always fluid flow, or crystal falling or rising in the fluid. That is, crystal dissolution in nature is often convective dissolution, which is discussed in the next section. One possible case of diffusive crystal dissolution is for crystals on the roof or floor of a magma chamber if melt produced by dissolution does not sink or rise. For these cases, one surface of the crystal is facing the melt and hence diffusion may be treated as one dimensional, both for simplicity and for practical applications. (Convective crystal dissolution is often treated as three dimensional.)

Although diffusive crystal dissolution is seldom encountered in nature, its theoretical development is instructive for understanding convective crystal dissolution, and it is often encountered in experimental studies. Such experiments are easy to conduct, and can be applied to infer diffusion coefficients, to establish equilibrium conditions, and to investigate the rate of diffusive crystal dissolution. Furthermore, the interface–melt composition and diffusivity obtained from diffusive crystal dissolution experiments are of use to estimate convective crystal dissolution rates (Section 4.2.3).

If temperature or pressure varies during crystal dissolution, the problem becomes more complicated because both the diffusivity and the interface melt concentration vary, causing the dissolution rate to vary. Although the diffusivity dependence on time is not difficult to tackle analytically, the variation in the interface condition and the consequent change in dissolution rate cannot be treated simply. Hence, the treatment here is for constant temperature and pressure. Numerical method is necessary to handle crystal dissolution with variable temperature and pressure.

The discussion in this section focuses on the dissolution of a single crystal grain with a uniform composition. For an aggregate of the same mineral (such as quartzite cemented by quartz), its dissolution may be treated the same way as long as there is no disintegration along grain boundaries. For an aggregate of different minerals (such as a mantle xenolith made of olivine, orthopyroxene, and clinopyroxene, or a wall rock made of quartz and feldspars), dissolution rate varies from one mineral to another. Due to different dissolution rates, the interface between the rock and the melt would have bays and protrusions as dissolution goes on even if the initial interface is smooth. Hence, the method outlined here can only be applied very roughly.

Mathematically, diffusive crystal dissolution is a moving boundary problem, or specifically a Stefan problem. It was treated briefly in Section 3.5.5.1. During crystal dissolution, the melt grows. Hence, there are melt growth distance and also crystal dissolution distance. The two distances differ because the density of the melt differs from that of the crystal. For example, if crystal density is 1.2 times melt density, dissolution of 1 μm of the crystal would lead to growth of 1.2 μm of the melt. Hence, $\Delta x_c = (\rho_{melt}/\rho_{cryst}) \Delta x$, where Δx_c is the dissolution distance of the crystal and Δx is the growth distance of the melt.

In this section, the detailed analysis of the problem is first given. Next, the results are summarized with comments. Those who do not wish to go through the detailed analysis may proceed to the summary directly (Section 4.3.1.2). The summary is written so that it is basically independent for the benefit of those who do not wish to read the detailed analysis, even if this means some repetition. After the summary, examples are shown.

4.2.2.1 Detailed analysis of the problem

The complete treatment of crystal dissolution in a melt requires full consideration of multicomponent diffusion in the melt (Liang, 1999). Using Liang's approach, the interface melt composition may be estimated from the diffusivity matrix and the thermodynamic equilibrium between the dissolving crystal and the melt. Because diffusivity matrix is not available for natural silicate melts and the thermodynamic description of natural silicate melts is not accurate enough, the full consideration does not yet have much practical value. Fortunately, for the purpose of estimating dissolution rate, during the dissolution of a crystal, the diffusion of the principal equilibrium-determining component can be treated as effective binary. The principal equilibrium-determining component is the component that determines the saturation of the crystal, such as ZrO_2 during the dissolution of zircon. Treating the diffusion of other components (including trace elements) is more difficult because they cannot be treated as effective binary (Zhang et al., 1989). Zhang (1993) developed a compromise method (modified effective binary approach) for treating these other components but it has not been applied much.

The diffusion equation for three-dimensional diffusive crystal dissolution in the spherical case (Eq. 4-90) is rarely encountered and too complicated. Hence, such problems will not be treated here.

One-dimensional diffusive dissolution With the above general discussion, we now turn to the special case of one-dimensional crystal dissolution. Use the interface-fixed reference frame. Let melt be on the right-hand side ($x > 0$) in the interface-fixed reference frame. Crystal is on the left-hand side ($x < 0$) in the interface-fixed reference frame. Properties in the crystal will be indicated by superscript "c". For simplicity, the superscript "m" for melt properties will be ignored. Diffusivity in the melt is D. Diffusivity in the crystal is D^c. The concentration in the melt is C (kg/m^3) or w (mass fraction). The initial concentration in the crystal is C^c_∞ or w^c_∞, simplified as C^c or w^c if there would be no confusion from the context. It is assumed that the interface composition rapidly reaches equilibrium. In the following, diffusion in the melt is first considered, and then diffusion in the crystal.

Diffusion in the melt. Ignoring melt density variation, the diffusion equation in the melt during crystal dissolution is

$$\frac{\partial C}{\partial t} = \frac{\partial}{\partial x}\left(D\frac{\partial C}{\partial x}\right) - u\frac{\partial C}{\partial x}, \quad x > 0, \ t > 0. \tag{4-93a}$$

Initial condition: $C|_{t=0} = C_\infty$ for $x > 0$. \hfill (4-93b)

Boundary condition: $C|_{t=0} = C_0$ for $t > 0$. \hfill (4-93c)

The melt growth rate u satisfies the following:

Stefan condition (parabolic law): $u = \alpha(D/t)^{1/2}$.   (4-93d)

Mass balance at the interface: $u^c C^c - u C_0 = -D(\partial C/\partial x)_{x=0}$.   (4-93e)

That is,

$$u^c \rho^c w^c - u \rho_0 w_0 = -D(\partial C/\partial x)_{x=0},$$ (4-93f)

or,

$$\rho u(w^c - w_0) \approx -\rho D(\partial w/\partial x)_{x=0},$$ (4-93g)

or,

$$u(w^c - w_0) \approx -D(\partial w/\partial x)_{x=0}.$$ (4-93h)

Assume D is constant. Both $w(x, t)$ and u can be solved from Equations 4-93a,b,c,h. Use Boltzmann transformation by letting $\eta = \sqrt{4t}$. Then,

$$\frac{\partial w}{\partial t} = \frac{dw}{d\eta}\frac{\partial \eta}{\partial t} = -\frac{\eta}{2t}\frac{dw}{d\eta},$$ (4-94a)

$$\frac{\partial w}{\partial x} = \frac{dw}{d\eta}\frac{\partial \eta}{\partial x} = \frac{1}{2t^{1/2}}\frac{dw}{d\eta},$$ (4-94b)

$$D\frac{\partial^2 w}{\partial x^2} = \frac{D}{4t}\frac{d^2 w}{d\eta^2}.$$ (4-94c)

Note that the partial differential has been replaced by the total differential because it is assumed that w depends on only one variable η. Hence, the diffusion equation can be written as

$$-\frac{\eta}{2t}\frac{dw}{d\eta} = \frac{D}{4t}\frac{d^2 w}{d\eta^2} - \alpha\sqrt{\frac{D}{t}}\frac{1}{2\sqrt{t}}\frac{dw}{d\eta}.$$ (4-95)

Simplify; then

$$D\frac{d^2 w}{d\eta^2} + 2(\eta - \alpha\sqrt{D})\frac{dw}{d\eta} = 0.$$ (4-96)

Let

$$\eta' = \eta - \alpha D^{1/2} = x/\sqrt{4t} - \alpha D^{1/2};$$ (4-96a)

then

$$D\frac{d^2 w}{d\eta'^2} - 2\eta'\frac{dw}{d\eta'} = 0.$$ (4-97)

The above is similar to Equation 3-48. Let $\xi = \eta'/D^{1/2} = (x/\sqrt{4Dt}) - \alpha$. The solution is

$$w = A \operatorname{erfc}(\xi) + B = A \operatorname{erfc}\left(\frac{x}{\sqrt{4Dt}} - \alpha\right) + B. \tag{4-98}$$

From the initial and boundary conditions (Equations 4-94b,c), A and B satisfy the following:

$$w|_{t=0} = w_\infty = B,$$

$$w|_{x=0} = w_0 = A \operatorname{erfc}(-\alpha) = B.$$

That is,

$$A = (w_0 - w_\infty)/\operatorname{erfc}(-\alpha).$$

Hence,

$$w = w_\infty + (w_0 - w_\infty)\operatorname{erfc}\left(\frac{x}{\sqrt{4Dt}} - \alpha\right)/\operatorname{erfc}(-\alpha). \tag{4-99}$$

Next, the parameter α needs to be found using the mass balance condition. Because

$$(\partial w/\partial x)_{x=0} = (2/\pi^{1/2})[(w_0 - w_\infty/(4Dt)^{1/2}]\exp(-\alpha^2)/\operatorname{erfc}(-\alpha), \tag{4-99a}$$

combining with Equation 4-93h, we obtain

$$\pi^{1/2}\alpha \exp(\alpha^2)\operatorname{erfc}(-\alpha)(w_0 - w_\infty)/(w^c - w_0) \equiv b, \tag{4-100}$$

where w^c is the bulk crystal composition (not the interface crystal composition, and this point will be discussed further when comparing with crystal growth). The relation between the parameters α and b above is shown in Figure 4-17. As the parameter α is solved from the above equation, the melt growth rate u is solved and the concentration profile is also uniquely solved. Hence, this completes the solution of the diffusion and growth problem.

Note that Equation 4-99 means that the solution is an error function with respect to the lab-fixed reference frame ($x' = x - 2\alpha\sqrt{Dt}$). In the interface-fixed reference frame, the solution appears like an error function, and its shape is often error function shape, but the diffusion distance is not simply $(Dt)^{1/2}$, especially when the absolute value of α is large (to be discussed later using more extreme examples).

Diffusion in the crystal. If a crystal has a fixed composition, such as quartz, there is no need to consider diffusion in the crystal except for isotopic exchange. For a crystal that is a solid solution, such as olivine, the equilibrium composition at the crystal surface may be different from the initial composition. There would be diffusion in the crystal. Although this problem has not been investigated before in the literature, it is not a difficult problem and it can be solved using the same steps as diffusion in the melt. The diffusion equation is

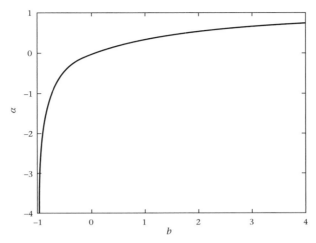

Figure 4-17 The relation between α and b. The parameters α and b satisfy $\pi^{1/2}\alpha\exp(\alpha^2)\operatorname{erfc}(-\alpha)=b$. A few simple relations of the above equation may be derived: (1) When $|b| < 0.01$, $\alpha \approx b/\pi^{1/2} \approx 0.564b$. (2) As α approaches ∞, b approaches $2\pi^{1/2}\alpha\exp(\alpha^2)$. (3) As α approaches $-\infty$, $b \to -(1-0.5/\alpha^2 + 1.5/\alpha^4 - \cdots) \to -1$.

$$\frac{\partial w^c}{\partial t} = \frac{\partial}{\partial x}\left(D^c\frac{\partial w^c}{\partial x}\right) - u^c\frac{\partial w^c}{\partial x}, \quad x<0,\ t>0. \tag{4-101a}$$

Initial condition : $\quad w^c|_{t=0} = w^c_\infty$ for $x<0$. $\hfill$ (4-101b)

Boundary condition : $\quad w^c|_{x=0} = w^c_0$ for $t>0$. $\hfill$ (4-101c)

where w^c_0 is the interface concentration in equilibrium with the melt. The growth rate u_c satisfies the following:

$$u^c = \alpha^c(D^c/t)^{1/2}. \tag{4-101d}$$

Because $u^c = (\rho_{\mathrm{melt}}/\rho_{\mathrm{cryst}})u$, $\alpha^c(D^c)^{1/2} = (\rho_m/\rho_c)\alpha D^{1/2}$. That is,

$$\alpha_c = \alpha(\rho_m/\rho_c)(D/D_c)^{1/2}. \tag{4-101e}$$

Because D (diffusivity in the melt) is often many orders of magnitude greater than D_c (diffusivity in the crystal), α_c is usually a large number, about 100 to 1000 times that of α. Let $y=-x$ so that $y>0$. The above diffusion problem thus becomes

$$\frac{\partial w^c}{\partial t} = \frac{\partial}{\partial y}\left(D^c\frac{\partial w^c}{\partial y}\right) + u^c\frac{\partial w^c}{\partial y} \quad y>0,\ t>0. \tag{4-102a}$$

$$w^c|_{t=0} = w^c_\infty \text{ for } y>0. \tag{4-102b}$$

$$w^c|_{y=0} = w^c_0 \text{ for } t>0. \tag{4-102c}$$

Using Boltzmann transformation and following similar steps as in the case of diffusion in the melt, the solution is

$$w^c = w_\infty^c + (w_0^c - w_\infty^c)\mathrm{erfc}\left(\frac{y}{\sqrt{4D_c t}} + \alpha_c\right)\bigg/\mathrm{erfc}(\alpha_c). \tag{4-103}$$

Calculated concentration profiles in the melt and in the crystal are shown in an example below. Equation 4-103 means that the solution is an error function with respect to the lab-fixed boundary ($y' = y + 2\alpha^c\sqrt{D^c t}$). In the interface-fixed reference frame, the solution appears like an error function, and its shape is often error function shape, but the diffusion distance is not simply $(D^c t)^{1/2}$. For a "normal" error function profile, the length of the concentration profile would be characterized by $(Dt)^{1/2}$ in the melt and $(D^c t)^{1/2}$ in the crystal. This is not so in the crystal. To gauge the length of the diffusion profile described by the above equation, we define the mid-concentration distance y_mid that satisfies

$$\mathrm{erfc}\left(\frac{y_\mathrm{mid}}{\sqrt{4D^c t}} + \alpha^c\right)\bigg/\mathrm{erfc}(\alpha^c) = 0.5. \tag{4-104}$$

The value $y_\mathrm{mid}/(D^c t)^{1/2}$ can be solved from the above equation. For large positive values of α^c (e.g., $\alpha^c \geq 10$), $y_\mathrm{mid}/(D^c t)^{1/2} \approx \ln(2)/\alpha^c$. For $\alpha_c \leq -2$, $y_\mathrm{mid}/(D^c t)^{1/2} \approx 2|\alpha^c|$. The following table gives the relation between $y_\mathrm{mid}/(D^c t)^{1/2}$ and α^c:

α^c	10	8	6	4	2	0	−0.5	−1	−1.5
$y_\mathrm{mid}/(D^c t)^{1/2}$	0.0689	0.0858	0.1135	0.1667	0.3041	0.95387	1.4316	2.1396	3.030

That is, if $\alpha^c = 0$ (normal error function profile), $y_\mathrm{mid} = 0.954(D^c t)^{1/2}$, and the mid-concentration distance is roughly $(D^c t)^{1/2}$. If α^c is a large positive value, the mid-concentration distance is much smaller than $(D^c t)^{1/2}$. If α^c is negative, the mid-concentration distance is much larger than $(D^c t)^{1/2}$. Therefore, we have the following relations:

(1) During crystal dissolution, α^c is typically a large positive value. Hence, the concentration profile in the crystal is much shorter than $(D^c t)^{1/2}$. The shorter profile is because the surface layer of the crystal is continuously dissolved or peeled off, thinning the profile. An example of such a profile is shown in Figure 4-18a.

(2) During crystal growth, α^c is a large negative number. Hence, the length of the profile in the crystal is much longer than $(D^c t)^{1/2}$. The long profile is due to the growth of the layer with new interface composition. The shape of the concentration profile can be seen later (Figure 4-22a).

(3) For the profile in the melt during crystal dissolution (melt growth) or crystal growth, the absolute value of α is small. Hence, the diffusion

profile in the melt differs only slightly from a normal error function profile. During crystal dissolution, the mid-distance of diffusion in the melt is *longer* than $0.954(Dt)^{1/2}$. During crystal growth, the mid-distance of diffusion in the melt is *shorter* than $0.954(Dt)^{1/2}$.

4.2.2.2 Summary

To predict crystal dissolution or melt growth distance, it is first necessary to determine the principal equilibrium-determining component controlling the saturation of the mineral. Examples include ZrO_2 during dissolution of zircon, MgO during dissolution of an Mg-rich olivine or orthopyroxene in a basaltic melt, FeO during fayalite dissolution in rhyolitic melt, SiO_2 during dissolution of quartz, and TiO_2 during dissolution of rutile. The diffusion of such a component can be characterized as effective binary. For simplicity, the melt density is assumed to be constant. The melt growth distance Δx may be calculated as

$$\Delta x = 2\alpha(Dt)^{1/2}, \tag{4-105}$$

where D is effective binary diffusivity of the principal equilibrium-determining component in the melt, t is time, and α is a dimensionless parameter to be solved from the following equation:

$$\pi^{1/2} \cdot \alpha \exp(\alpha^2) \operatorname{erfc}(-\alpha) = b \equiv (w_0 - w_\infty)/(w^c - w_0), \tag{4-106}$$

where w_∞ is the mass fraction of the major component in the initial melt, w_0 is the mass fraction in the interface melt, and w^c is the mass fraction in the initial (or bulk) crystal. The relation between α and b (defined in Equation 4-106) is graphed in Figure 4-17. The melt growth rate u is

$$u = \alpha(D/t)^{1/2}. \tag{4-107}$$

The crystal dissolution distance Δx^c and rate u^c are different from the melt growth distance and rate because of density difference. Hence, the crystal dissolution distance and rate are

$$\Delta x^c = (\rho/\rho^c)\Delta x = 2\alpha^c(D^c t)^{1/2} = 2\alpha(\rho/\rho^c)(Dt)^{1/2}, \tag{4-108}$$

and

$$u^c = (\rho/\rho^c)u = \alpha^c(D^c t)^{1/2} = \alpha(\rho/\rho^c)(D/t)^{1/2}, \tag{4-109}$$

where the parameter α^c is

$$\alpha^c = \alpha(\rho/\rho^c)(D/D^c)^{1/2}. \tag{4-110}$$

The dissolution distance is proportional to the square root of time (parabolic reaction law), and the dissolution rate is inversely proportional to the square root of time. This law does not apply at $t=0$ because dissolution rate cannot be infinity. At $t=0$, the dissolution rate is limited by the interface reaction rate and hence is finite. In other words, it takes a finite time (though very short) for the interface melt concentration to increase from C_∞ to C_0, and hence the concentration gradient at the interface is not infinity and the dissolution rate is not infinity at $t=0$.

Using the interface-fixed reference frame (i.e., $x=0$ at the interface) and defining melt to be at the right-hand side ($x>0$) and crystal to be at the left-hand side ($x<0$), the diffusion profile for the major component in the melt is

$$w = w_\infty + (w_0 - w_\infty)\mathrm{erfc}\left(\frac{x}{\sqrt{4Dt}} - \alpha\right)/\mathrm{erfc}(-\alpha). \tag{4-111}$$

The diffusion profile in the crystal is

$$w^c = w_\infty^c + (w_0^c - w_\infty^c)\mathrm{erfc}\left(\frac{y}{\sqrt{4D^c t}} + \alpha^c\right)/\mathrm{erfc}(\alpha^c). \tag{4-112}$$

where $y = -x > 0$.

The diffusion behavior of components that are not the principal equilibrium-determining component is difficult to model because of multicomponent effect. Many of them may show uphill diffusion (Zhang et al., 1989). To calculate the interface-melt composition using full thermodynamic and kinetic treatment and to treat diffusion of all components, it is necessary to use a multicomponent diffusion matrix (Liang, 1999). The effective binary treatment is useful in the empirical estimation of the dissolution distance using interface-melt composition and melt diffusivity, but cannot deal with multicomponent effect and components that show uphill diffusion.

4.2.2.3 Examples and applications

The results above have the following applications: (i) estimation of diffusive crystal dissolution distance for given crystal and melt compositions, temperature, pressure, and duration if diffusivities are known and surface concentrations can be estimated; and (ii) determination of diffusivity (EBDC) and interface-melt concentrations. Those diffusivities and interface concentrations can be applied to estimate crystal dissolution rates in nature.

Diffusive dissolution of MgO-rich olivine and diffusion profiles MgO is the principal equilibrium-determining component and its diffusion behavior is treated as effective binary. Consider the dissolution of an olivine crystal (Fo90, containing 49.5 wt% MgO) in an andesitic melt (containing 3.96 wt% MgO) at 1285°C and 550 MPa (exp#212 of Zhang et al. 1989). The density of olivine is 3198 kg/m³, and that of the initial melt is 2632 kg/m³. Hence, the density ratio is 1.215. To estimate the dissolution parameter α, it is necessary to know the interface melt

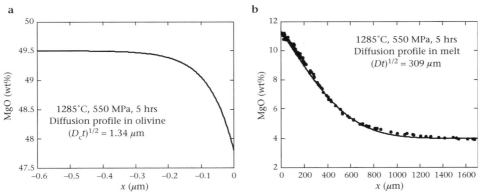

Figure 4-18 MgO profile in olivine and in melt during olivine dissolution in an andesitic melt. (a) Calculated MgO profile in olivine. Note that the length of the profile is much shorter than $(D^c t)^{1/2}$ because the surface layer is continuously peeled off (dissolved). (b) Experimental MgO diffusion profile in melt and fit (Zhang et al., 1989). The crystal dissolution distance is 48 ± 5 μm.

concentration. This concentration may be estimated from thermodynamics. Here we use the experimental value of 11.3 wt%. Hence,

$$b = (w_0 - w_\infty)/(w^c - w_0) = (11.3 - 3.96)/(49.5 - 11.3) = 0.192.$$

Then the parameter α can be solved to be

$$\alpha = 0.097.$$

Hence, the melt growth distance is $\Delta x = 0.194(Dt)^{1/2}$ and the olivine dissolution distance is $\Delta x_c = 0.194(\rho/\rho^c)(Dt)^{1/2} = 0.160(Dt)^{1/2}$. Suppose $D = 5.3$ μm²/s. For $t = 5$ h, the melt growth distance is 60 μm and the olivine dissolution distance is 49 μm. The experimentally measured dissolution distance is 48 ± 5 μm.

The calculated concentration profiles in olivine and in the melt are shown in Figure 4-18. The mean length of the concentration profile in the crystal is very short, about 0.1 μm, much shorter than the dissolution distance (48 μm) or $(D^c t)^{1/2} = 1.3$ μm. Hence, the MgO deficiency of the profile in olivine is negligible compared to the amount of olivine dissolved. Therefore, during crystal dissolution, w^c in calculating the parameter $b \equiv (w_0 - w_\infty)/(w^c - w_0)$ should be the concentration of the initial or bulk crystal, not the interface crystal concentration, which makes the calculation simple.

In experimental studies of crystal dissolution, the sample must be quenched to stop the experiment. Because the interface melt is near saturation during the experiment, interface melt is oversaturated during quench, resulting in olivine growth. For normal cooling rate in a piston–cylinder apparatus, the cooling rate is about 100 K/s, and the growth distance is of the order of 0.3 μm (depending on the experimental temperature). That is, olivine growth during quench would produce a layer of olivine thicker than the diffusion profile in olivine! Hence, if one carries out an olivine dissolution experiment and measures the MgO profile (suppose some microbeam method has the requisite spatial resolution), it would differ from the calculated profile in Figure 4-18a because of olivine growth during quench.

In calculating the diffusive or convective dissolution rate of a crystal, the most appropriate effective binary diffusivities are from the dissolution experiments of the same mineral in the same melt. Diffusivities from diffusion-couple experiments or other methods may suffer from (i) compositional effect on melt diffusivity, and (ii) the multicomponent effect due to different concentration gradients.

Determination of diffusivity and surface concentrations using crystal dissolution experiments Because the theoretical solution of the diffusion profile in the melt during crystal dissolution is known, dissolution experiments may be used to obtain diffusivity and surface concentration. The setup of a dissolution experiment is somewhat similar to a diffusion-couple experiment. Use olivine dissolution in an andesitic melt as an example. An olivine crystal cylinder (or disc) and an andesitic glass cylinder with the same diameter are first prepared. They are then placed together vertically (with the interface horizontal). Because the interface melt during olivine dissolution is denser than the initial melt, olivine should be at the bottom during the experiment. The sample is then pressurized and heated up to the desired temperature (glass should melt and the temperature should be high so that olivine dissolves rather than grows) for a desired duration to generate a long enough profile but not too long to affect the end of the melt (for simple treatment, the melt reservoir should be large compared to the diffusion distance so that the melt may be treated as an infinite reservoir). The sample is then quenched and the melt becomes glass. The charge is then sectioned perpendicular to the interface. Concentration profile in the melt and the dissolution distance of the crystal are measured.

The experimental profiles are more complicated than the calculated profiles because there is some olivine growth during quench of the experimental charge. Hence, right near the interface (e.g., within 5 μm, depending on the quench rate), the MgO concentration in the melt might decrease toward the interface. This part of the profile should not be used in the fitting (it is not shown in Figure 4-18b but is shown in Figure 3-32a). The concentration profile in olivine would be too short to be measured, and would in fact often be dominated by the layer of olivine growth during quench. Hence, the interface-melt or interface-crystal composition cannot be obtained from direct measurement close to the interface but must be obtained from fitting or extending the part of the profile unaffected by quench. Furthermore, diffusivity in the crystal cannot be obtained from crystal dissolution experiments.

After calculation of the parameters b and α, the diffusivity in the melt may be obtained by two ways, which provide cross-check on data consistency. One is to use $\Delta x = (\rho^c/\rho_m)\Delta x^c = 2\alpha(Dt)^{1/2}$. For the example given above, $\alpha = 0.097$, $\Delta x^c = 48 \pm 5\,\mu$m, $\rho^c/\rho_m \approx$ and $t = 18{,}000$ s. Hence, $D \approx (48 \times 1.2/0.194)^2/18{,}000\,\mu$m^2/s $= 4.9 \pm 1.0\,\mu$m^2/s. The second method is to fit the measured concentration profile using Equation 4-111. Figure 4-18b shows a fit to the experimental concentration profile and $D = 5.3 \pm 0.3 \; \mu m^2/s$ based on the fitting. Hence, the two methods give roughly the same D, and the value from fitting the profile is preferred. The theoretical curve does not fit the experimental data perfectly (Figure 4-18b): there are noticeable deviations at $x \approx 1000 \; \mu m$ and $x \approx 200 \; \mu m$. This is likely due to (i) the significant compositional variation across the diffusion profile, leading to variation in D, and (ii) the multicomponent effect because the concentration gradients of other components are varying along the profile. These effects are small enough to be ignored.

The diffusivity and surface concentrations obtained from crystal dissolution experiments can be applied to investigate dissolution rates in nature for the dissolution of the same mineral–melt pair, either diffusive or convective. Because diffusivity in the melt depends on the melt composition, to estimate dissolution rate in a melt, one should use diffusivity determined in the same melt. Furthermore, because effective binary diffusivities depend on concentration gradients of other components, it is necessary to use diffusivities determined from dissolution of the same mineral in the same melt in order to estimate dissolution rate of a mineral in a melt. For example, diffusivities obtained from olivine dissolution are often orders of magnitude greater than those based on quartz dissolution because of two effects: (i) the compositional effect because interface melt composition is basaltic during olivine dissolution but is rhyolitic during quartz dissolution, and (ii) multicomponent effect because the concentration gradients are very different affecting the effective binary diffusivities. The compositional effect is probably the major effect.

Another example for treating concentration profiles during mineral dissolution can be found in Figure 3-32b, which shows a Zr concentration profile during zircon dissolution. In this case, the dissolution distance is very small compared to the diffusion profile length. Hence, the diffusion profile is basically an error function.

4.2.2.4 Diffusive dissolution of many crystals

For the dissolution of many crystals when their diffusion profiles overlap, the bulk melt can no longer be treated as an infinite reservoir. An approximate treatment is to assume that the crystals are regularly distributed in the melt, and every crystal is enclosed by a spherical melt shell. The problem may then be solved using the method developed for bubble growth by Proussevitch et al. (1993) and Proussevitch and Sahagian (1998) (Section 4.2.5.2).

4.2.2.5 Complete melting of a single crystal in its own melt (an infinite liquid reservoir)

Melting of a single crystal in its own melt may be treated similarly if it is controlled by heat conduction. Assume that the melt reservoir is infinite. Because heat diffusivity κ in the melt is about 6 orders of magnitude larger than mass diffusivity, the interface reaction rate must be extremely rapid for melting to be controlled by heat conduction instead of interface reaction. Melting of many silicate minerals is likely controlled by interface reaction rather than heat transfer. On the other hand, ice melting is often assumed to be controlled by heat transfer. If so, the melting rate controlled by heat conduction may be solved using the same type of equations (Equations 4-93a,b,c) but replacing concentration by temperature. (Melting controlled by convective heat transfer requires different equations; Section 4.2.3.4.) For one-dimensional heat conduction in the interface-fixed reference frame, the equation is

$$\frac{\partial T}{\partial t} = \kappa \frac{\partial^2 T}{\partial x^2} - u \frac{\partial T}{\partial x}, \qquad x > 0, t > 0. \tag{4-113a}$$

$$T|_{t=0} = T_\infty \quad \text{for } x > 0. \tag{4-113b}$$

$$T|_{x=0} = T_m \quad \text{for } t > 0. \tag{4-113c}$$

$$u = \alpha(\kappa/t)^{1/2}. \tag{4-113d}$$

$$k \frac{\partial T}{\partial x}\Big|_{x=0} = u^c \rho^c \Delta H_f = u \rho \ \Delta H_f, \tag{4-113e}$$

where T is temperature, k is heat conductivity (SI unit $W\,m^{-1}\cdot K^{-1}$), κ is heat diffusivity (SI unit m^2/s), $\kappa = k/(\rho c)$ with ρ being the density and c being the heat capacity (SI unit $J\,kg^{-1}\,K^{-1}$), ΔH_f is the latent heat of fusion, T_∞ is the initial melt temperature, T_m is the melting temperature of the crystal and is $<T_\infty$, u is melt growth rate (the crystal melting rate is slightly smaller, by the density ratio), and α is a dimensionless constant to be determined. The solution for the concentration profile is (Equation 4-48)

$$T = T_\infty + (T_0 - T_\infty)\mathrm{erfc}\left(\frac{x}{\sqrt{4\kappa t}} - \alpha\right)/\mathrm{erfc}(-\alpha), \tag{4-114}$$

where the dimensionless parameter α is to be solved from the following:

$$\pi^{1/2} \alpha \exp(\alpha^2) \mathrm{erfc}(-\alpha) = c(T_\infty - T_0)/\Delta H_f. \tag{4-115}$$

The melt growth distance is $2\alpha(\kappa t)^{1/2}$. The crystal melting distance is (ρ/ρ^c) times $2\alpha(\kappa t)^{1/2}$.

4.2.2.6 Partial melting of a solid solution controlled by diffusion in the solid

The partial melting of a solid solution such as a plagioclase crystal may be controlled by diffusion in the crystal (Tsuchiyama and Takahashi, 1983). Figure 4-19a shows the phase diagram of plagioclase. Suppose plagioclase C has a composition of An40 (meaning 40% anorthite and 60% albite) and is heated to 1573 K. It would undergo partial melting, and the final equilibrium state would consist of melt B and plagioclase D. Two parallel paths may lead to the final equilibrium

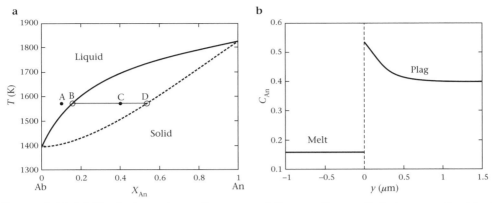

Figure 4-19 (a) Plagioclase phase diagram and the reaction path between a liquid of composition A (50% of the total mass) and a solid of composition C (50% of the total mass). (b) Diffusion profile in plagioclase during partial melting of plagioclase composition C in (a) at 1573 K controlled by diffusion in plagioclase.

state: (i) partial melting controlled by diffusion in plagioclase, and (ii) dissolution and recrystallization. Below we discuss partial melting controlled by diffusion in plagioclase.

As plagioclase C (An40) melts, because the melt contains lower An content, in plagioclase the An content at the interface builds up rapidly until composition D (An53.6) is almost reached. The An content of the interface plagioclase cannot exceed that of D; otherwise, it would be below its solidus and there would be no partial melting. For further partial melting, diffusion in plagioclase must transport extra An content to the interior of plagioclase and keep the An content in the interface plagioclase to be no more than 53.6%. Because diffusion in plagioclase is extremely slow, partial melting controlled by diffusion in plagioclase is also extremely slow, leading to uniform melt composition roughly at composition B. Hence, the diffusion problem may be written as (Equation 4-102a)

$$\frac{\partial w^{\text{plag}}}{\partial t} = \frac{\partial}{\partial y}\left(D^{\text{plag}}\frac{\partial w^{plag}}{\partial y}\right) + u^{\text{plag}}\frac{\partial w^{\text{plag}}}{\partial y}, \quad y > 0, \ t > 0 \tag{4-116a}$$

$$w^{\text{plag}}\big|_{t=0} = w_{\infty}^{\text{plag}} \quad \text{for } y > 0, \tag{4-116b}$$

$$w^{\text{plag}}\big|_{y=0} = w_{0}^{\text{plag}} \quad \text{for } t > 0, \tag{4-116c}$$

where y is a coordinate in the plagioclase crystal, u^{plag} is the rate of partial melting, D^{plag} is Ab–An interdiffusivity in plagioclase, w^{plag} is the mass fraction of $CaAl_2Si_2O_8$ in plagioclase, w_{∞}^{plag} is the An content of the initial plagioclase (0.4), and w_{0}^{plag} is the An content of the interface plagioclase (0.536). This is a Stefan problem and the rate of partial melting may be expressed as

$$u^{\text{plag}} = \alpha^{\text{plag}}(D^{\text{plag}}/t)^{1/12}. \tag{4-116d}$$

The following mass balance condition may be applied to relate u^{plag} and other parameters:

$$u^{\text{plag}}(w_0^{\text{plag}} - w^{\text{melt}}) = -D^{\text{plag}}(\partial w^{\text{plag}}/\partial y)_{y=0}, \tag{4-116e}$$

where w^{melt} is the An content of the melt (0.16). Note that although the diffusion problem is similar to that for crystal growth controlled by diffusion in the melt, the partial melting rate controlled by diffusion in the crystal is exceedingly slow. The concentration profile can be solved as (Equation 4-103)

$$w^{\text{plag}} = w_\infty^{\text{plag}} + (w_0^{\text{plag}} - w_\infty^{\text{plag}})\text{erfc}\left(\frac{y}{\sqrt{4D^{\text{plag}}t}} + \alpha^{\text{plag}}\right)/\text{erfc}(\alpha^{\text{plag}}). \tag{4-117}$$

To find α^{plag}, the mass balance relation is applied, leading to

$$\pi^{1/2}\alpha^{\text{plag}}\exp[(\alpha^{\text{plag}})^2]\text{erfc}(\alpha^{\text{plag}}) = b^{\text{plag}} \equiv (w_0^{\text{plag}} - w_\infty^{\text{plag}})/(w_0^{\text{plag}} - w^{\text{melt}}). \tag{4-118}$$

By solving for α^{plag}, the partial melting rate and partial melting distance can be calculated. For the problem of partial melting of plagioclase at 1573 K (Figure 4-19a), $w_0^{\text{plag}}=0.536$, $w_0^{\text{plag}}=0.40$, $w^{\text{melt}}=0.16$, leading to $b=0.362$, and hence α^{plag} may be solved to be

$$\alpha^{\text{plag}} = 0.270.$$

A calculated compositional profile is shown in Figure 4-19b. The partial melting distance of plagioclase is $\Delta y^{\text{plag}} = 2\alpha^{\text{plag}}(D^{\text{plag}}t)^{1/2}$. Using the Ab–An interdiffusivity of Grove et al. (1984), $D \approx \exp(-6.81 - 62{,}100/T)$ m²/s at 1373 to 1673 K. At 1573, $D \approx 8 \times 10^{-21}$ m²/s $= 8 \times 10^{-9}$ μm²/s. Hence, the partial melting distance $\Delta y^{\text{plag}} = 5 \times 10^{-5}t^{1/2}$, where t is in second and Δy is in μm. This means a partial melting distance of about 0.3 μm in one year, which is extremely slow, much slower than the observed partial melting rate of plagioclase by Tsuchiyama and Takahashi (1983). The partial melting rate can be increased by melt nucleation in the interior of the plagioclase crystal because the crystal interior is also unstable with respect to partial melting (unlike the case of crystal dissolution which occurs only at the interface). Furthermore, the partial melting process might be accomplished by a different path: crystal dissolution and reprecipitation. In this process, crystal of composition C would dissolve in the melt B, from which plagioclase with composition D would crystallize. The dissolution and reprecipitation process is more difficult to model, but is controlled by transport in the melt and interface reaction rate, which are much more rapid than diffusion in the crystal.

If the initial plagioclase is more enriched in the anorthite composition than the solidus composition D in Figure 4-19a, it is stable and would not undergo partial melting. The above solution also satisfies this, as it should, because $(w_0^{\text{plag}} - w_\infty^{\text{plag}})$ in Equation 4-118 would be negative, leading to negative b and α values, meaning negative partial melting rate (that is, it grows rather than melts).

Comparing the rate of crystal dissolution versus complete melting versus partial melting, one finds that complete melting is the most rapid (controlled by heat transfer), dissolution is slower, and partial melting controlled by diffusion in the solid phase is the slowest.

4.2.3 Convective dissolution of a falling or rising crystal in an infinite liquid reservoir

Convective crystal dissolution means that crystal dissolution is controlled by convection, which requires (i) a high interface reaction rate so that crystal dissolution is controlled by mass transport (see previous section), and (ii) that mass transport be controlled by convection. In nature, convective crystal dissolution is common. In aqueous solutions, the dissolution of a falling crystal with high solubility (Figure 1-12) is convective. In a basaltic melt, the dissolution of most minerals is likely convection-controlled.

There are different forms of convection and fluid dynamic regimes. One example is convection during the dissolution of a single spherical crystal in an infinite melt reservoir. If many crystals are in an infinite melt reservoir, as long as the flow fields and the compositional boundary layers of the crystals do not interact or overlap, each crystal may be treated as a single crystal in an infinite melt. As the crystal sinks or rises due to gravity (i.e., density difference), the crystal motion leads to the removal of the interface melt. This is called *forced convection*. If the crystal motion in the melt is negligible (either neutral buoyancy or extremely small size or extremely high viscosity), there may be another kind of convection. Because the interface-melt composition is different from the bulk melt composition, there is a density difference. Under the right conditions, the density difference would lead to gravitational instability and the interface melt would sink or rise away from the crystal. This kind convection is called *free convection*.

A second example of convective dissolution is the dissolution of a solid floor or roof. *Forced convection* means that the fluid is moving relative to the solid floor or roof such as magma convection in a magma chamber, or bottom current over ocean sediment. *Free convection* means that there is no bulk flow or convection, but the interface melt may be gravitationally unstable, leading to its rise or fall.

Some convective crystal dissolution problems can be treated by combined consideration of dissolution kinetics and hydrodynamics. Hydrodynamic consideration is necessary because it is necessary to know how rapidly the interface fluid is removed. In considering the problems, steady state is assumed. Irregular and unsteady convection is not treated. The following problems have been tackled:

(1) Convective dissolution of a falling or rising single crystal in an infinite fluid reservoir. The theory has been developed by Kerr (1995) and Zhang and Xu (2003).

(2) Convective dissolution of a solid floor or roof when the overlying or underlying fluid moves uniformly at a constant velocity.

The first of these two problems will be treated in detail, and the second will be treated briefly. The calculation is numerical in nature and simple enough to be handled by a spreadsheet program. Temperature or pressure variation during crystal dissolution may be handled in a numerical scheme. The discussion in this section focuses on the dissolution of a single crystal. For aggregates such as mantle xenoliths, comments made earlier (Section 4.2.2) apply here too.

The structure of this section is similar to that on diffusive crystal dissolution: First, the detailed analysis of the problem of convective dissolution of a single rising or falling crystal in an infinite fluid reservoir will be given. Multicomponent effect is ignored. Diffusion of the major component is treated as effective binary. Next a summary is presented. Those who do not wish to go through the detailed analysis may proceed to the summary directly. Thirdly, examples are shown. Finally, convective dissolution of many rising or sinking crystals and of a solid floor is briefly discussed.

In convective dissolution, the crystal dissolution rate is again denoted by u (or $-da/dt$) for consistency with earlier sections, and the ascent or descent velocity of the crystal is denoted by U.

4.2.3.1 Detailed analysis of the problem

Hydrodynamics of free fall or rise of a spherical crystal The following is the method to calculate the free fall or rise velocity of a spherical crystal (Clift et al., 1978). For a small particle (see below) or viscous fluid, the ascent or descent velocity U can be calculated using Stokes' law:

$$U = \frac{2ga^2\Delta\rho}{9\eta_f},$$

(4-119)

where g is acceleration due to the Earth's gravity, a is the radius of the particle, $\triangle\rho$ is the absolute value of the density difference between the crystal and the fluid, and η_f is the viscosity of the fluid. If the crystal is denser, it sinks. If the crystal is less dense, it rises. The above equation applies when the Reynolds number Re is ≤ 0.1. The Reynolds number is defined as

$$Re = \frac{2aU\rho_f}{\eta_f},$$

(4-120)

where ρ_f is the density of the fluid (magma or water). In applications, one first calculates the velocity using Equation 4-119. Then Re is calculated. If Re < 0.1,

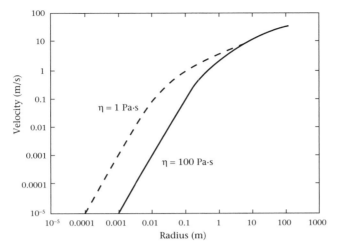

Figure 4-20 Falling velocity of a mantle xenolith (density 3200 kg/m³) in a basaltic melt (density 2700 km/m³) for viscosity of 1 Pa·s and 100 Pa·s. The calculation does not continue to greater sizes because the applicability of the formulation is limited to Re $\leq 3 \times 10^5$. At small radius, the velocity is proportional to the square of the radius (Stokes' law). For larger radius, the velocity does not increase so rapidly with radius, and roughly increases with square root of radius.

the result is accurate. Otherwise, Stokes' law is not accurate, and U (and two other unknowns) must be solved from a set of three equations: one is Equation 4-120, and the other two equations are

$$C_D = \frac{24}{Re}(1 + 0.15Re^{0.687}) + \frac{0.42}{1 + 42,500Re^{-1.16}}, \tag{4-121}$$

$$U = \sqrt{\frac{8ga\Delta\rho}{3\rho_f C_D}}, \tag{4-122}$$

where C_D is the drag coefficient. Note that the first equation among the three is the definition of the Reynolds number, the second equation (Clift et al., 1978) relates the drag coefficient and the Reynolds number (the accuracy in calculating C_D using the equation is $\pm 5\%$ when Re $\leq 3 \times 10^5$), and the third equation relates the rise/fall velocity with the drag coefficient, densities of the two phases, and the crystal size. Three unknowns (Re, C_D, and U) are to be solved from the above three equations by numerical methods. The above method can be applied only when Re $\leq 3 \times 10^5$ (including when Re < 0.1). For Re $> 3 \times 10^5$, a different expression of C_D is necessary. Figure 4-20 shows a calculated example on how the falling velocity of a mantle xenolith in a basaltic melt depends on radius and viscosity.

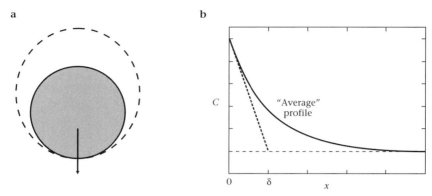

Figure 4-21 The concept of boundary layer and boundary layer thickness δ. (a) Compositional boundary layer surrounding a falling and dissolving spherical crystal. The arrow represents the direction of crystal motion. The shaded circle represents the spherical particle. The region between the solid circle and the dashed oval represents the boundary layer. For clarity, the thickness of the boundary layer is exaggerated. (b) Definition of boundary layer thickness δ. The compositional profile shown is "averaged" over all directions. From the average profile, the "effective" boundary layer thickness is obtained by drawing a tangent at $x = 0$ $(r = a)$ to the concentration curve. The δ is the distance between the interface $(x = 0)$ and the point where the tangent line intercepts the bulk concentration.

Compositional boundary layer As a falling or rising crystal dissolves in a fluid, it is assumed that at the interface there is rough equilibrium between the crystal and fluid. That is, dissolution is controlled by convection, rather than by interface reaction. The interface-melt composition differs from that of the bulk melt. During olivine dissolution, the interface melt would have greater MgO concentration. This layer with different composition is called the *compositional boundary layer*. (There may also be a *thermal boundary layer*, meaning that the temperature in the interface melt is different from that in the bulk melt. The thermal boundary layer may be encountered when dissolution is controlled by heat transfer, rather than by mass transfer.) Without convection, the boundary layer thickness would be proportional to square root of time (Section 3.4.3.1). When there is steady-state convection due to crystal falling or rising in a fluid, the compositional boundary layer thickness would start from zero thickness, grow diffusively to a fixed thickness controlled by hydrodynamics, and then be kept at this thickness when steady state is reached. For a falling crystal dissolving in or growing from a melt, the schematic shape of the boundary layer is shown in Figure 4-21a. The boundary layer thickness varies with direction: it is thin on the leading side and thick on the trailing side (Levich, 1962). For simplicity, an effective boundary layer thickness δ is defined, as explained in Figure 4-21b. Mathematically, the boundary layer thickness δ is defined by the following equation:

$$\int \mathbf{F} \, d\mathbf{S} \equiv -4\pi a^2 D(\partial C/\partial r)_{r=a} \equiv 4\pi a^2 D(C_0 - C_\infty)/\delta$$

$$= 4\pi a^2 D(\rho_0 w_0 - \rho_\infty w_\infty)/\delta. \tag{4-123}$$

where $\int \mathbf{F} \, d\mathbf{S}$ is total compositional flux toward the interface (the integration is over the whole spherical interface area), a is the radius, $(\partial C/\partial r)_{r=a}$ is the average slope at the interface melt (dashed tangent line in Figure 4-21b), w_0 and ρ_0 are the concentration and density of the interface liquid (and is also the liquid saturated by the crystal because convective dissolution means that the interface is near equilibrium), and w_∞ and ρ_∞ are the bulk concentration and density (or those of the initial melt). The first equal sign above defines the average concentration gradient, and the second equal sign defines the boundary layer thickness δ.

Convective dissolution rate of a falling or rising crystal in an infinite melt reservoir
Using the above concept of compositional boundary layer, dissolution rate of a falling or rising crystal may be written as

$$d[4\pi a^3 \rho^c (w^c - w_0)/3]dt = \int \mathbf{F} \, d\mathbf{S} \equiv 4\pi a^2 D(\rho_0 w_0 - \rho_\infty w_\infty)/\delta, \tag{4-124}$$

where $4\pi a^3/3$ is the volume of the crystal, w^c is the mass fraction of the major component in the crystal (remember that we are using the effective binary approach), melt density ρ is assumed to be constant, and $4\pi a^3 \rho^c (w^c - w_0)/3$ is extra mass that must be transported away. Take the derivative and simplify:

$$u = -da/dt = \beta D/\delta, \tag{4-125}$$

where u is the convective dissolution rate (during dissolution, u is positive), and β is a dimensionless compositional parameter defined as $\beta = (\rho_0 w_0 - \rho_\infty w_\infty)/[\rho^c(w^c - w_0)]$. If liquid density variation is small, then $\beta = b(\rho/\rho^c)$, where b is the same as the earlier defined $b = (w_0 - w_\infty)/(w^c - w_0)$.

Because D is independently determined, and β is obtainable from initial conditions and thermodynamic equilibrium, the problem of determining the convective dissolution rate now becomes the problem of estimating the boundary layer thickness. In fluid dynamics, the boundary layer thickness appears in a dimensionless number, the Sherwood number Sh:

$$Sh = 2a/\delta. \tag{4-126a}$$

That is,

$$\delta = 2a/Sh. \tag{4-126b}$$

From experimental investigations, the Sherwood number for crystal falling or rising in a fluid can be found as follows for Re $\leq 10^5$:

$$\text{Sh} = 1 + (1 + \text{Pe})^{1/3} \left(1 + \frac{0.096\text{Re}^{1/3}}{1 + 7\text{Re}^{-2}} \right), \tag{4-127}$$

where Pe is the compositional Peclet number defined as

$$\text{Pe} = 2aU/D. \tag{4-128}$$

Equation 4-127 is from Zhang and Xu (2003). Therefore, with Sh and δ calculated, the convective dissolution rate can be calculated. The calculation procedure is summarized next.

4.2.3.2 Summary

Based on the above results, the following is a summary of steps to calculate the convective dissolution rate of a single falling or rising crystal in an infinite melt reservoir:

(1) Give initial conditions, including (i) the melt composition, density, diffusivity, and viscosity, (ii) crystal composition and density, and (iii) the initial crystal radius.

(2) Use the hydrodynamics equations to calculate Re and the crystal falling or rising velocity u by solving Equations 4-120, 4-121, and 4-122.

(3) Calculate $\text{Pe} = 2aU/D$.

(4) Estimate Sh (Equation 4-127).

(5) Obtain $\delta = 2a/\text{Sh}$.

(6) Use equilibrium data to calculate w_0, and then determine $\beta = (\rho_0 w_0 - \rho_\infty w_\infty)/[\rho^c(w^c - w_0)]$.

(7) Compute the convective dissolution rate $u = -da/dt = \beta D/\delta$.

(8) If the purpose is to calculate the dissolution rate for this crystal size, then we are done. If the purpose is to find how the crystal size changes as the crystal moves in the melt, then one chooses a small time interval dt, and obtains new depth as $h + U\,dt$ and new crystal radius as $a - u\,dt$. Then go to step (2) and iterate.

For the calculation of convective dissolution rate of a falling crystal in a silicate melt, the diffusion is multicomponent but is treated as effective binary diffusion of the major component. The diffusivity of the major component obtained from diffusive dissolution experiments of the same mineral in the same silicate melt is preferred. Diffusivities obtained from diffusion-couple experiments or other types of experiments may not be applicable because of both compositional effect on diffusivity and/or the multicomponent effect (cross terms in the diffusion matrix) on the effective binary diffusivity. The interface-melt concentration of the major component can also be estimated from diffusive dissolution experiments.

Comparison of the above method with experimental data in aqueous solutions shows that the calculation is accurate to about 20% relative. Possible errors arise from the following: (i) the variation of fluid density in the boundary layer (as a function of r) is not considered, (ii) the variation of the fluid viscosity in the boundary layer is not considered, (iii) the variation of fluid diffusivity in the boundary layer is not considered, (iv) empirical relations between C_D and Re, and between Sh and Pe and Re have errors of $\pm 5\%$, and (v) uncertainties in understanding of the hydrodynamics. Based on these considerations, for dissolution in water, if input data are well known and variation of density, viscosity, and diffusivity across the boundary layer is negligible (such as CO_2 liquid and hydrate dissolution in water; Zhang, 2005b), the method is accurate to within 20% relative. In aqueous fluid, with variations in density, viscosity, and diffusivity across the boundary layer, the accuracy is about 30% relative (Zhang and Xu, 2003). For mineral dissolution in silicate melt, if the interface-melt composition is similar to bulk melt composition (such as dissolution of zircon), the above theory would be applicable with accuracy of about 20%. If the interface-melt composition is significantly different from the initial melt composition (such as olivine or quartz dissolution), the variation in melt viscosity across the boundary layer may be large (order of magnitude). Geometric average viscosity may be used but the calculation would have large error because no theory has been developed for variable viscosity. One possible improvement in the future would be to treat convective crystal dissolution when viscosity and diffusivity vary by orders of magnitude across the boundary layer.

4.2.3.3 Examples on mineral dissolution in water and in silicate melts

Convective dissolution of a falling KCl crystal in water Because KCl solubility is high, according to Figure 1-11, the dissolution in water is controlled by mass transport. A KCl crystal would fall freely in water and dissolve. The dissolution rate of a falling 0.3-mm-radius KCl crystal in pure water at 25°C may be calculated as follows. We first collect the basic information. The density at 25°C is 1984 kg/m^3 for KCl and 997 kg/m^3 for water. The viscosity of water at 25°C is 0.00089 Pa·s. The diffusivity of KCl in water at 25°C is 1.96×10^{-9} m^2/s. The solubility of KCl in water at 25°C is 35.5 g per 100 g of water. The density of KCl solution = $\rho_{water} + 670.6w$ (see Zhang and Xu, 2003). From these, we first solve Equations 4-120, 4-121, and 4-122 simultaneously to obtain

Re = 46.9 (hence, Stokes' law cannot be applied);

Descent velocity $U = 0.0698$ m/s.

Then,

$Pe = 2aU/D = 21367;$

$Sh = 38.41;$

Boundary layer thickness: $\delta = 1.56 \times 10^{-5}$ m.

$w_0 = 35.5/(100 + 35.5) = 0.262.$

The density of KCl solution is hence $997 + 670.6 \times 0.262 = 1173$ kg/m^3.

Parameter $\beta = (\rho_0 w_0 - \rho_\infty w_\infty)/[\rho^c(w^c - w_0)] = 0.210.$

Finally, the dissolution rate of the falling KCl crystal is

$$u = -da/dt = \beta D/\delta = 2.6 \times 10^{-5} \text{ m/s} = 0.026 \text{ mm/s}.$$

The calculated result is in good agreement with experimental KCl dissolution rate at this temperature (~ 0.025 mm/s, Zhang and Xu, 2003).

The steady-state convective dissolution rate calculated above applies only when the unperturbed diffusion distance $(Dt)^{1/2}$ is greater than the boundary layer thickness δ. If diffusion distance $(Dt)^{1/2}$ is smaller than the boundary layer thickness (15.6 μm), i.e., if $t < 0.12$ s, the dissolution would be controlled by diffusion. For $t > 0.12$ s, the dissolution is controlled by steady-state convection and can be calculated as above.

Convective dissolution of a rising CO$_2$ droplet in seawater To mitigate the greenhouse effect by atmospheric CO$_2$, one proposal is to collect CO$_2$ from power plants and inject it in the liquid form into oceans (requiring pressure at ≥ 400 m seawater depth). CO$_2$ droplets are less dense than ambient seawater and would rise in seawater if water depth is <2000 m, and are denser than seawater and would sink if injected to >3000 m depth. Hence, injected CO$_2$ droplets would rise or sink depending on the depth of injection, and undergo convective dissolution. One complexity is that CO$_2$ reacts with seawater to form CO$_2$ hydrate at depth > 300 m. (All these depths are approximate because they also depend on temperature.) That is, a droplet would have a hydrate shell (which protects the interior of the droplet from further reaction). The formation of hydrate shell makes the droplet behave more like a rigid sphere, which can be modeled appropriately using the method in this section.

Consider a CO$_2$ droplet of radius 3 mm injected at 600 m seawater depth with temperature of 5.2°C (Zhang, 2005b). Under these conditions, density and viscosity of seawater are 1026 kg/m^3 and 0.00161 Pa·s, and density of liquid CO$_2$ is 916 kg/m^3, or 20.82 mol/L. Because of the formation of hydrate shell, the solubility of CO$_2$ in seawater should be that of CO$_2$ hydrate, which is 1.00 mol/L (CO$_2$ liquid solubility is significantly greater), or $w_0 = 0.0429$. Because solubility of CO$_2$ is small, density of the interface water is similar to the bulk seawater. Hence, the parameter $\beta = \rho b/\rho_c = (1026/916) \times 0.0429/(1 - 0.0429) = 0.0502$. Diffusivity of CO_2 in seawater is 1.16×10^{-9} m^3/s. Then we find

Re = 455;
Ascent velocity $U = 0.119$ m/s;
Pe = $2aU/D = 6.16 \times 10^5$;
Sh = 149.5;
Boundary layer thickness: $\delta = 4.01 \times 10^{-5}$ m;
Dissolution rate: $u = -da/dt = \beta D/\delta = 1.45 \times 10^{-6}$ m/s $= 1.45$ μm/s.

The calculated result is in good agreement with CO_2 droplet dissolution rate obtained by in situ experiments (1.44 μm/s, Brewer et al., 2002).

Convective dissolution of falling MgO-rich olivine in an andesitic melt MgO is the controlling component and its diffusion behavior is treated as effective binary. Basic data can be obtained from diffusive crystal dissolution experiments of Zhang et al. (1989) (Section 4.2.2.3). For the dissolution of an olivine (Fo90) crystal in an andesitic melt at 1285°C and 550 MPa (exp #212 of Zhang et al. 1989), MgO concentration is 3.96 wt% in the initial melt, 11.3 wt% in the interface melt, and 49.5 wt% in the initial olivine. The density of olivine is 3198 kg/m^3, and that of the initial melt is 2632 kg/m^3. Because of significant compositional variation across the boundary layer, the viscosity of the melt at 1285°C also varies. Take a viscosity of about 40 Pa·s. The parameter $\beta \approx b\rho_0/\rho_c = 0.158$. The diffusivity $D_{MgO} = 5.3 \times 10^{-12}$ m^2/s. Let the initial radius of the olivine crystal be 0.002 m (2 mm). Then,

Re = 3.24×10^{-6};
Descent velocity $U = 1.23 \times 10^{-4}$ m/s;
Pe = $2aU/D = 9.30 \times 10^4$;
Sh = 46.3;
Boundary layer thickness: $\delta = 8.64 \times 10^{-5}$ m;
Dissolution rate: $u = -da/dt = \beta D/\delta = 9.7 \times 10^{-9}$ m/s.

Dissolution distance in 18,000 s would be 174 μm, greater than the diffusive dissolution distance of 48 μm obtained earlier. There are no experimental data to compare. The convective dissolution rate can be applied only when the diffusion distance $(Dt)^{1/2}$ is greater than the boundary layer thickness. If diffusion distance $(Dt)^{1/2}$ is smaller than the boundary layer thickness (86.4 μm), i.e., if $t < 1408$ s, the dissolution would be controlled by diffusion even for a falling crystal, and the method in Section 4.2.2.3 should be used.

Convective dissolution rate for quartz in an andesitic melt may be calculated similarly, but the error may be larger than the normal 20% relative because quartz dissolution increases SiO_2 content so much, leading to orders of magnitude increase in viscosity for the interface melt (viscosity is about 120 Pa·s for the initial andesitic melt and 1.7×10^4 Pa·s for the interface rhyolitic melt). Because the convective dissolution model does not account for viscosity variation in the boundary layer, the error may be large.

4.2.3.4 Convective melting of a rising or falling crystal in its own melt

One example would be ice melting or methane hydrate dissociation when rising in seawater. Convective melting rate may be obtained by analogy to convective dissolution rate. Heat diffusivity κ would play the role of mass diffusivity. The thermal Peclet number (defined as $Pe_t = 2au/\kappa$) would play the role of the compositional Peclet number. The Nusselt number (defined as $Nu = 2a/\delta_t$, where δ_t is the thermal boundary layer thickness) would play the role of Sherwood number. The thermal boundary layer (thickness δ_t) would play the role of compositional boundary layer. The melting equation may be written as

$$d[4\pi a^3 \rho^c L/3]dt = \int \mathbf{F}\, d\mathbf{S} \equiv 4\pi a^2 \rho D (T_0 - T_\infty)/\delta, \tag{4-129}$$

Hence,

$$u = -da/dt = \beta_t \kappa/\delta_t, \tag{4-130}$$

where β_t is a parameter defined as

$$\beta_t = \frac{\rho}{\rho^c} \frac{c(T_\infty - T_0)}{L}, \tag{4-131}$$

where c is heat capacity, T_0 is the interface temperature (the melting or dissociation temperature), and L is heat of fusion. Below, the calculation procedures are summarized without derivation.

(1) Give initial conditions, including the melt composition, density, heat diffusivity, viscosity, crystal density, latent heat of fusion, and the initial crystal radius.

(2) Use the hydrodynamics equations to calculate Re and the crystal falling or rising velocity U by solving Equations 4-120, 4-121, and 4-122.

(3) Calculate the thermal Peclet number $Pe_t = 2aU/\kappa$.

(4) By analogy to Equation 4-127, Nu is related to Pet and Re as follows:

$$Nu = 1 + (1 + Pe_t)^{1/3} \left(1 + \frac{0.096 Re^{1/3}}{1 + 7Re^{-2}} \right), \tag{4-132}$$

(5) Calculate $\delta_t = 2a/Nu$.

(6) Use the equilibrium condition to calculate T_0, and then calculate the parameter β_t.

(7) Calculate the convective melting rate $u = -da/dt = \beta_t \kappa/\delta_t$.

4.2.3.5 Convective dissolution of many rising or sinking crystals

In nature, it is likely to encounter convective dissolution of many crystals. In this case, if their boundary layers do not overlap and the flow velocity fields do not overlap, each crystal may be viewed as dissolving individually without interacting with other crystals. However, if their boundary layers overlap or their flow velocity fields overlap, the above treatment would not be accurate. Furthermore, when there are many crystals, the whole parcel of crystal-containing fluid may sink or rise (large-scale convection), leading to completely different fluid dynamics. Such problems remain to be solved.

4.2.3.6 Convective dissolution of a floor or roof

Convective dissolution rate of a solid floor (or roof) in the presence of forced convection due to fluid flow over (or under) it may be calculated as follows (Holman, 2002; Zhang and Xu, 2003):

$$u = 0.03 b D^{2/3} U^{4/5} (\eta_f / \rho_f)^{-7/15} L^{-1/5}, \tag{4-133}$$

where u is the convective dissolution rate, b is the dimensionless compositional parameter defined earlier, D is the diffusivity in the fluid, U is the flow rate of the overlying fluid, L is the length of the system (such as the length of the floor to be dissolved). The formulation has not been experimentally verified.

4.2.3.7 Some remarks on controlling factors of crystal dissolution rates

For convective crystal dissolution, the dissolution rate is $u = (\rho/\rho^c) b D / \delta$. For diffusive crystal dissolution, the dissolution rate is $u = \alpha (\rho/\rho^c)(D/t)^{1/2} = \alpha (\rho/\rho^c) D / (Dt)^{1/2}$. By defining the diffusive boundary layer thickness as $\delta = (Dt)^{1/2}$, the diffusive crystal dissolution rate can be written as $u = \alpha D / \delta$, where α is positively related to b through Equation 4-100. Therefore, mass-transfer-controlled crystal dissolution rates (and crystal growth rates, discussed below) are controlled by three parameters: the diffusion coefficient D, the boundary layer thickness δ, and the compositional parameter b. The variation and magnitude of these parameters are summarized below.

(1) *The diffusion coefficient D.* Effective binary diffusivities in silicate melts (10^{-10} to 10^{-15} m^2/s) vary by orders of magnitude (Table 4-4) because of melt composition variation, and because of the characteristics of the diffusing species. In a given silicate melt, the diffusivity D increases with decreasing absolute values of ionic charge, with largest diffusivity being neutral molecules. The size of the diffusing species also plays a role. On the other hand, in aqueous solutions, the diffusivity ($\sim 2 \times 10^{-9}$ m^2/s) variation for commonly encountered species is small (Table 1-3a).

Table 4-4 Effective binary diffusivities (µm²/s) in "dry" silicate melts at 1573 K

Diffusing Species or Component	D	Diffusing Species or Component	D
Molecular H_2O in rhyolite	127	Mg^{2+} in basalt to andesite	6.6
Molecular Ar in dry rhyolite	11	Al^{3+} in basalt to andesite	4.2
Zr^{4+} in rhyolite	0.016	Ti^{4+} in basalt to andesite	3.4
P^{5+} in rhyolite	0.0019	Si^{4+} in basalt to andesite	2.3
Al^{3+} in basalt	5.0	Si^{4+} in rhyolite to andesite	0.28
K^+ in rhyolite	~30	K^+ in rhyolite to andesite	~0.9
Ca^{2+} in andesite	8.6	Ca^{2+} in rhyolite to andesite	0.21

Note. Data source: molecular H_2O in rhyolite (Zhang and Behrens, 2000); molecular Ar in rhyolite (Behrens and Zhang, 2001); Zr^{4+} in rhyolite (Harrison and Watson, 1983); P^{5+} in dry rhyolite (Harrison and Watson, 1984); Al^{3+} in basalt (Kress and Ghiorso, 1995); K^+ in rhyolite and in rhyolite to andesite (estimated from Van Der Laan et al., 1994); and the rest are from Zhang et al. (1989). Basalt to andesite means the melt composition along the diffusion profile spans from basalt to andesite.

(2) *The boundary layer thickness δ.* For convective crystal dissolution, the steady-state boundary layer thickness increases slowly with increasing viscosity and decreasing density difference between the crystal and the fluid. It does not depend strongly on the crystal size. Typical boundary layer thickness is 10 to 100 μm. For diffusive crystal dissolution, the boundary layer thickness is proportional to square root of time.

(3) *The compositional parameter.* The compositional parameter $b = (w_0 - w_\infty)/(w^c - w_0)$. It may be rewritten as

$$b = (1 - w_\infty/w_0)/(w^c/w_0 - 1).$$

By the above definition, b is positive for crystal dissolution, and negative for crystal growth. During convective crystal dissolution, the dissolution rate u is directly proportional to b. During diffusive crystal dissolution, the dissolution rate is proportional to parameter α, which is positively related to b. Hence, for the dissolution of a given mineral in a melt, the size of parameter b is important. The numerator of b is proportional to the degree of undersaturation. If the initial melt is saturated, $b = 0$ and there is no crystal dissolution or growth. The denominator characterizes the concentration difference between the crystal and the saturated interface melt. By the choice of the most major equilibrium-determining component in the crystal (and the largest w_c/w_0) so that the diffusion of the component in the melt may be treated as effective binary, $(w_c/w_0 - 1)$ is always positive. Therefore, the parameter b increases as the degree of undersaturation increases, but decreases as the concentration difference between the crystal and the saturated melt increases. The degree of saturation depends on temperature and melt composition. Other conditions being equal, the larger the denominator $(w_c/w_0 - 1)$, the smaller the dissolution rate. Table 4-5 lists some values of $(w_c/w_0 - 1)$ and D that are useful for estimating relative dissolution rate. The diffusive and convective dissolution of accessory minerals is slow and that of major minerals is rapid. Furthermore, for solid solution series, the dissolution rate of intermediate members is greater than that of pure end members (Zhang et al., 1989). Because difference in Gibbs free energy is larger for pure endmembers, the results highlight that reaction rate is not proportional to Gibbs free energy difference, but is controlled by kinetics.

Table 4-5 Some typical dissolution parameters at 1300°C

Mineral	Major component	Typical w_c (wt%)	Typical w_0 (wt%)	$w_c/w_0 - 1$	D in melt ($\mu m^2/s$)
Zircon in rhyolite	Zr	50	0.45	110	0.016
Chromium spinel in basalt	Cr_2O_3	50	0.5	100	~4
Monazite in rhyolite (1 wt% H_2O)	Ce_2O_3	47	0.51	91	0.025
Apatite in rhyolite (1 wt% H_2O)	P_2O_5	42	1.6	25	0.023
Rutile in andesite	TiO_2	90	5	17	3.4
Olivine (Fo90) in andesite	MgO	50	11.7	3.3	6.6
Diopside in andesite	CaO	25	11	1.3	9.0
Quartz in andesite	SiO_2	100	75	0.33	0.28[a]
Plagioclase in basalt	Al_2O_3	30	20	0.5	4.2

[a] D_{SiO_2} is about 2.3 $\mu m^2/s$ during olivine dissolution in andesite, and 0.28 $\mu m^2/s$ during quartz dissolution in andesite. The large difference is because of the interface melt composition difference: during olivine dissolution, the interface melt is basaltic; but during quartz dissolution, the interface melt is rhyolitic.

4.2.4 Diffusive and convective crystal growth

Compared to crystal dissolution for which an introduced foreign crystal would dissolve, it is less common to encounter the growth of a single grain or aggregate because many crystals may form and grow when the melt is supersaturated. Theory of crystal growth is in many aspects similar to crystal dissolution but with additional complexities. One is that the composition of the growing crystal responds to interface melt composition. Sometimes, this complexity is important, and sometimes it is not important. Another is that at very high degree of supersaturation, there may be dendritic growth. For relatively small degree of supersaturation, one may use the theory for crystal dissolution to treat crystal growth, in which the crystal dissolution rate is negative growth rate, and melt growth rate is negative melt consumption rate. For those who prefer crystal growth rate as positive and hence an independent description of crystal growth theory, the theories are briefly summarized below.

4.2.4.1 Diffusive crystal growth in an infinite melt reservoir

Crystal growth distance and behavior of major component This problem is similar to diffusive crystal dissolution. Hence, only a summary is shown here. Consider the principal equilibrium-determining component, which can be treated as effective binary diffusion. The density of the melt is often assumed to be constant. The density difference between the crystal and melt is accounted for.

The melt consumption distance and the crystal growth distance differ because the density of the melt differs from that of the crystal, with $\Delta x_c = (\rho_{melt}/\rho_{cryst})\Delta x$, where Δx_c is the growth distance of the crystal (note that growth distance is positive here) and Δx is the consumption distance of the melt.

For *one-dimensional crystal growth in an infinite melt reservoir at constant temperature and pressure with constant melt density*, and using the interface-fixed reference frame, the diffusion equation in the melt is

$$\frac{\partial w}{\partial t} = \frac{\partial}{\partial x}\left(D\frac{\partial w}{\partial x}\right) + u\frac{\partial w}{\partial x}, \qquad x > 0,\ t > 0. \tag{4-134a}$$

Initial condition : $\qquad w|_{t=0} = w_\infty \quad$ for $x > 0$. $\tag{4-134b}$

Boundary condition : $\qquad w|_{x=0} = w_0 \quad$ for $t > 0$. $\tag{4-134c}$

Melt consumption rate u : $\qquad u = \alpha(D/t)^{1/2}. \tag{4-134d}$

Crystal growth rate : $\qquad u_c = \alpha(\rho_{melt}/\rho_{cryst})(D/t)^{1/2}. \tag{4-134e}$

Mass balance : $\qquad u(w_0 - w_0^c) = -D(\partial w/\partial x)_{x=0}, \tag{4-134f}$

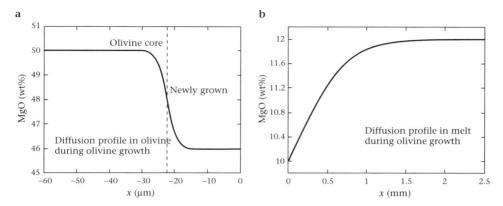

Figure 4-22 Calculated MgO profile in (a) olivine and (b) melt during olivine growth in a basaltic melt using Equations 4-138 and 4-139. Note that the unit of the x-axis is μm in (a) and mm in (b).

where w_0^c is the mass fraction in the crystal at the interface. Note that the above mass balance equation for crystal growth differs from that for crystal dissolution (Equation 4-94e) for more than a sign change: the crystal composition in the mass balance equation for crystal growth is the interface crystal composition w_0^c (or newly grown crystal composition), rather than the initial crystal composition for the case of crystal dissolution. The difference between the crystal compositions in crystal dissolution and growth used in the mass balance equation and in calculating parameter of b is due to the fact that during crystal growth the newly grown crystal composition responds to melt composition but during crystal dissolution the dissolved crystal composition is essentially independent of melt composition. This point will become clearer later (Figure 4-22a).

The melt consumption distance Δx and crystal growth distance Δx_c may be written as

$$\Delta x = 2\alpha(Dt)^{1/2}, \tag{4-135a}$$

$$\Delta x_c = (\rho/\rho_c)\Delta x = 2\alpha(Dt)^{1/2}(\rho/\rho_c) = 2\alpha_c(D_c t)^{1/2}, \tag{4-135b}$$

where D is diffusivity of the major component in the melt, D_c is diffusivity in the crystal, t is time, α is a parameter to be solved from the following equation:

$$\pi^{1/2}\alpha \exp(\alpha^2)\mathrm{erfc}(\alpha) = b \equiv (w_0 - w_\infty)/(w_0 - w_0^c), \tag{4-136}$$

and α_c is expressed as

$$\alpha_c = \alpha(\rho_m/\rho_c)(D/D_c)^{1/2}. \tag{4-137}$$

Note that u, α, and b above for crystal growth are opposite (by a negative sign) to those for crystal dissolution. The crystal growth distance is proportional to the square root of time (parabolic reaction law), and the growth rate is inversely proportional to the square root of time. This law does not apply at $t=0$ because growth rate cannot be infinity. At $t=0$, the growth rate is limited by the interface reaction rate and hence is finite.

The diffusion profile in the melt is

$$w = w_\infty + (w_0 - w_\infty)\mathrm{erfc}\left(\frac{x}{\sqrt{4Dt}} + \alpha\right)/\mathrm{erfc}(\alpha). \tag{4-138}$$

The diffusion profile in the crystal is

$$w_c = w_\infty^c + (w_0^c - w_\infty^c)\mathrm{erfc}\left(\frac{y}{\sqrt{4D_c t}} + \alpha_c\right)/\mathrm{erfc}(-\alpha_c). \tag{4-139}$$

where $y = -x$.

As an example, we consider one-dimensional diffusive growth of MgO-rich olivine in an infinite melt reservoir. Assume the following conditions: (i) the initial MgO concentrations in olivine and the melt are 50 and 12 wt%; (ii) the interface MgO concentrations in olivine and melt are 46 and 10 wt%; (iii) densities of olivine and melt are 3200 and 2750 kg/m³; and (iv) the diffusivities of MgO in olivine and melt are 10^{-16} and 5×10^{-12} m²/s. Then we have

$\rho^m/\rho^c = 0.859$,
$b = (10 - 12)/(10 - 46) = 0.0556$,
$\alpha = 0.0325$,
Melt consumption distance: $\Delta x = 2\alpha(Dt)^{1/2} = 0.065(Dt)^{1/2}$,
$\alpha_c = \alpha(\rho^m/\rho^c)(D/D^c)^{1/2} = 6.245$,
Olivine growth distance: $\Delta x_c = 0.0559(Dt)^{1/2} = 12.49(D_c t)^{1/2}$.

For example, if $t = 32{,}000$ s, then melt consumption distance $\Delta x = 26\,\mu$m, and olivine growth distance is $\Delta x^c = 22\,\mu$m. The diffusion profiles in olivine and melt are shown in Figure 4-22 (see page 407).

The MgO profile in olivine during olivine growth is peculiar: rather than a profile that is steepest at the current crystal–melt interface ($x = 0$), rapid concentration variation occurs near the original crystal–melt interface. The olivine thus contains a core of uniform composition, plus a newly grown layer (mantle or rim) that has a new fixed composition when the melt reservoir is infinite. Between the two zones, there is diffusion, giving a diffusion-couple profile. For example, the profile in Figure 4-22a matches well an error function with diffusivity of D_c if the origin of the profile is redefined at the initial crystal–melt interface. Hence, *if the diffusion distance in the crystal is significantly smaller than the growth distance* $[(D^c t)^{1/2} \ll \Delta x^c]$, *the growth-diffusion profile is a diffusion-couple profile.* This conclusion may be applied to model major element zonation in crystals in terms of estimation of the initial concentration profile. For example, if the initial time is regarded to be the beginning of the growth of a second layer, and if the composition of the second layer is roughly fixed, the whole concentration profile may be treated as a diffusion profile (meaning that none is due to growth). Furthermore, the shape of the profile is similar to an error function, and the diffusion length is roughly given by $(\int D^c dt)^{1/2}$. For an element with very small diffusivity (such as P in olivine), $(\int D^c dt)^{1/2}$ can be very small and hence the profile would be sharp.

Behavior of trace element that can be treated as effective binary diffusion The above discussion is for the behavior of the principal equilibrium-determining component. For minor and trace elements, there are at least two complexities. One is the multicomponent effect, which often results in uphill diffusion. This is because the cross-terms may dominate the diffusion behavior of such components. The second complexity is that the interface-melt concentration is not fixed by thermodynamic equilibrium. For example, for zircon growth, Zr concentration in the interface-melt is roughly the equilibrium concentration (or zircon saturation concentration). However, for Pb, the concentration would not be fixed.

If the diffusion of a minor or trace element can be treated as effective binary (not uphill diffusion profiles) with a constant effective binary diffusivity, the concentration profile may be solved as follows. The growth rate u is determined by the major component to be $\alpha(D/t)^{1/2}$, and is given, not to be solved. Use i to denote the trace element. Hence, w_i and D_i are the concentration and diffusivity of the trace element. Note that D_i for trace element i is not necessarily the same as D for the major component. The interface-melt concentration is not fixed by an equilibrium phase diagram, but is to be determined by partitioning and diffusion. Hence, the boundary condition is the mass balance condition. If the boundary condition is written as $w_i|_{x=0} = w_{i,0}$, the value of $w_{i,0}$ must be found using the mass balance condition. In the interface-fixed reference frame, the diffusion problem can be written as

$$\frac{\partial w_i}{\partial t} = D_i \frac{\partial^2 w_i}{\partial x^2} + \alpha \sqrt{\frac{D}{t}} \frac{\partial w_i}{\partial x} \qquad x > 0, \ t > 0. \tag{4-140}$$

Initial condition : $\qquad w_i|_{t=0} = w_{i\infty} \quad$ for $x > 0$. $\tag{4-141a}$

Boundary condition : $\qquad \alpha(D/t)^{1/2}(w_{i0} - w_{i0}^c) = -D_i(\partial w_i/\partial x)_{x=0}.$ $\tag{4-141b}$

The dissolution rate and α are fixed (rather than to be solved from the equation) from the solution for the major oxide component. Assume that there is simple partitioning between crystal and melt so that $w_{i0}^c/w_{i0} = K_i$. If $K_i < 1$, the trace element is incompatible in the crystal. If $K_i > 1$, the trace element is compatible. The boundary condition becomes

$$\alpha(D/t)^{1/2} w_{i0}(1 - K_i) = -D_i(\partial w_i/\partial x)_{x=0}. \tag{4-142}$$

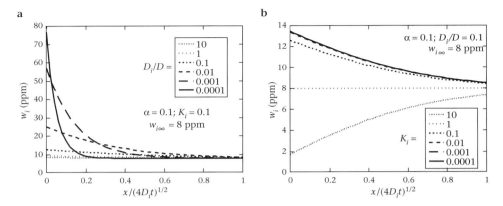

Figure 4-23 Trace element diffusion profiles during diffusive crystal growth for (a) various D_i/D ratios and (b) various K_i values.

The problem may be solved using Boltzmann transformation, and the solution for w_i is

$$w_i = w_{i\infty} + (w_{i0} - w_{i\infty})\text{erfc}\left(\frac{x}{\sqrt{4D_it}} + \gamma\right)/\text{erfc}(\gamma), \qquad (4\text{-}143)$$

where $\gamma = \alpha(D/D_i)^{1/2}$. The interface-melt concentration satisfies

$$w_{i0} = w_{i\infty} - \sqrt{\pi}\gamma e^{\gamma^2}\text{erfc}(\gamma)[w_{i0}(K_i - 1)]. \qquad (4\text{-}144)$$

Solving for w_{i0} from the above leads to

$$w_{i0} = \frac{w_{i\infty}}{\{1 + \sqrt{\pi}\gamma e^{\gamma^2}\text{erfc}(\gamma)(K_i - 1)\}}. \qquad (4\text{-}145)$$

The above solution shows that the interface-melt concentration is a constant (independent of time). Figure 4-23 shows some calculated concentration profiles.

As D_i/D decreases, there is more buildup of the trace element near the interface for $K_i < 1$. If $K_i > 1$ during crystal growth, the interface melt concentration is less than the initial concentration; otherwise, it is greater than the initial concentration. Some limiting cases are listed below:

(1) In the limiting case of $D_i \ll D$, leading to $\gamma \gg 1$ and $\sqrt{\pi}\gamma e^{\gamma^2}\text{crfc}(y) \approx 1$,

$$w_{i0} \approx w_{i\infty}/K_i, \qquad (4\text{-}146)$$

and

$$w_{i0}^c = K_i w_{i0} \approx w_{i\infty}. \qquad (4\text{-}147)$$

That is, the concentration of trace element i in the crystal is the same as that in the initial melt because an incompatible trace element would build up and a compatible trace element would be depleted at the interface until the concentration in the newly grown crystal is the same as that in the initial melt.

(2) In the other limiting case of $D_i \gg D$, the trace element profile approaches uniform concentration because the trace element diffuses much more rapidly than the equilibrium-determining component.

(3) In the limiting case of $K_i = 0$, then the interface melt concentration satisfies

$$w_{i0} = \frac{w_{i\infty}}{\{1 + \sqrt{\pi}\gamma e^{\gamma^2} \operatorname{erfc}(\gamma)\}}. \qquad (4\text{-}148)$$

(4) In the trivial case of $K_i = 1$, the trace element concentration profile is uniform (the same as the initial concentration).

4.2.4.2 Convective growth of a single falling or rising crystal in an infinite melt reservoir

For a single falling or rising crystal in an infinite melt reservoir that is uniformly oversaturated with respect to the crystal, estimation of convective growth rate can be made following the treatment on convective crystal dissolution by using the equilibrium-determining component. It is assumed that nucleation is difficult so that no new crystals form in the oversaturated melt. Below is a summary of steps to calculate the convective growth rate of a single rising or falling crystal in an infinite melt reservoir.

(1) Give initial conditions, including the melt composition, density, diffusivity, viscosity, crystal composition and density, and the initial crystal radius.

(2) Use the hydrodynamics equations to calculate Re and the crystal falling or rising velocity U by solving Equations 4-120, 4-121, and 4-122.

(3) Calculate $\text{Pe} = 2aU/D$, where D is the diffusivity of the principal equilibrium-determining component.

(4) Calculate Sh (Equation 4-127).

(5) Calculate $\delta = 2a/\text{Sh}$.

(6) Calculate the parameter $b = (w_0 - w_\infty)/(w_0 - w_{c,0})$.

(7) Calculate the convective growth rate $u = da/dt = b\rho_0 D/(\rho_c \delta)$.

(8) If the purpose is to calculate the growth rate at this crystal size, then we are done. If the purpose is to find how the crystal size changes as the crystal moves in the melt, then one chooses a small time interval dt, and integrates $\int U\,dt$ to obtain the new height of the crystal in the melt, and $\int u$ dt to obtain the new crystal radius. Then go to step (2) and iterate.

If input data are known accurately and there is not much variation in viscosity, diffusivity, and density of the boundary layer, the calculation is likely accurate to within 20% relative. If the viscosity, diffusivity, and density vary significantly across the boundary layer, then some average values of these parameters may be used, and the degree of accuracy is not known.

4.2.5 Diffusive and convective bubble growth and dissolution

Diffusive and convective bubble growth and dissolution in a liquid are explored in this section with specific problems in mind. Only growth is discussed because dissolution is similar. We will discuss growth of one bubble and the diffusive growth of many bubbles. A bubble in a liquid usually rises, meaning diffusive growth is rare. However, in a rhyolitic melt of high viscosity such as a lava dome, bubble ascent velocity may be negligible. Then bubble growth might be viewed as diffusive. For example, at 1073 K, a rhyolite melt with 1 wt% H_2O has a viscosity of 2×10^7 Pa·s. The ascent velocity of a 2-mm diameter bubble is

$$U = 2g\Delta\rho R^2/9\eta = 2.5\times10^{-10} \text{ m/s=0.9 } \mu\text{m/h},$$

which is small enough to be ignored. For bubble growth in silicate melt, because the gas in the bubbles is mostly H_2O (with some CO_2) and because H_2O diffusivity depends on H_2O content, the concentration dependence of diffusivity is often part of the modeling effort. Below we first discuss diffusive bubble growth. Then we examine convective bubble growth. For simplicity, only spherical bubbles are treated below. Given the size, whether a bubble is spherical or not may be inferred from three dimensionless numbers: Reynolds number (Re), Eotvos number (Eo), and Morton number (Mo) (Clift et al., 1978, pp. 26–27). For example, for bubbles in water, when the radius is < 1 mm, the bubble is roughly spherical. Above this critical size, the bubble has an irregular shape and wobbles as it rises.

4.2.5.1 Diffusive growth of a single spherical bubble in an infinite liquid reservoir

The following is a summary of the problem of isothermal growth of a single bubble in an infinite liquid reservoir. For clarity, the example of H_2O bubble growth in a silicic melt is shown below. In nature, CO_2 may also contribute to bubble growth, which is ignored here for simplicity.

(1) The H_2O diffusion equation in spherical coordinates is as follows (Equation 4-92):

$$\frac{\partial w}{\partial t} = \frac{1}{r^2}\frac{\partial}{\partial r}\left(Dr^2\frac{\partial w}{\partial r}\right) - u\frac{a^2}{r^2}\frac{\partial w}{\partial r}, \qquad t > 0 \text{ and } a \leq r < \infty, \tag{4-149}$$

where w is the concentration (mass fraction) of total H_2O, t is time, r is the radial coordinate, D is total H_2O diffusivity that depends on w (which makes the solution difficult), a is bubble radius and increases with time, and u is the bubble growth rate da/dt.

(2) When the viscosity is a function of H_2O content and hence varies as a function of r, the pressure in the bubble may be found as follows (Proussevitch and Sahagian, 1998):

$$P_g = P_f + \frac{2\sigma}{a} - 4ua^2 \int_{z(a)}^{z(\infty)} \eta(z)dz, \tag{4-150}$$

where P_g is the gas pressure in the bubble, P_f is the ambient pressure (pressure in the melt), σ is the surface tension, z is a function of r $[z(r) = 1/r^3]$, and η is the melt viscosity that depends on w (which makes the problem difficult). The integration above accounts for the effect of variable viscosity. For constant viscosity, the above equation reduces to

$$P_g - P_f = 2\sigma/a + 4ua^2\eta/a^3 = (2\sigma + 4\eta u)/a \to 4\eta u/a, \tag{4-151}$$

where the arrow applies to large bubbles ($a \geq 10\,\mu\text{m}$) for which surface tension can be ignored. According to the above equation, if viscosity controls bubble growth, and if $P_g - P_f$ is roughly constant, bubble growth rate u is roughly $a(P_g - P_f)/(4\eta)$ and proportional to bubble radius a. On the other hand, if diffusion controls bubble growth, bubble growth rate is inversely proportional to t. In many melts, diffusion and viscosity both play a main role in bubble growth. Hence, the relation between bubble growth rate and bubble radius is not so simple.

(3) Initial condition in the melt:

$$w|_{t=0} = w_\infty \qquad \text{for } r > a_0. \tag{4-152}$$

The initial bubble pressure must also be given, and is usually assumed to be

$$P_{g,t=0} = P_f + 2\sigma/a. \tag{4-153}$$

(4) Boundary conditions: At the boundary $r = a$, interface equilibrium between the melt and gas phases dictates that total H_2O concentration at

the interface melt is the solubility corresponding to pressure of P_g. Hence, one condition is

$$w|_{r=a} = \text{solubility at } P_g. \tag{4-154}$$

The solubility is usually a complicated function of P_g (e.g., Liu et al., 2005).

(5) Mass balance condition at the bubble–melt interface:

$$\frac{d}{dt}\left(\frac{4}{3}\pi a^3 \frac{\omega P_g}{RT}\right) = 4\pi a^2 D\rho \left(\frac{\partial w}{\partial r}\right)_{r=a}, \tag{4-155}$$

where ω is the molar mass of H_2O, R is the gas constant, and ρ is the density of the melt. This boundary condition is used to determine bubble growth rate da/dt.

The above set of equations can be solved numerically given input parameters, including initial bubble radius a_0, temperature, ambient pressure P_f, surface tension σ, solubility relation, D and η as a function of total H_2O content and temperature, and initial total H_2O content in the melt.

4.2.5.2 Diffusive growth of many equal-size spherical bubbles

For diffusive growth of many bubbles, one case has been treated in which all the bubbles are assumed to be of equal size and distributed regularly as in a lattice grid (Figure 4-12). With further simplification, each bubble is assumed to grow in a spherical shell of melt. The inner radius of the spherical shell is the radius of the bubble a, and the outer radius of the shell is denoted as S. As the bubble grows, both a and S increase. The shell thickness $(S - a)$ is not a constant, but the shell volume $(4\pi/3)(S^3 - a^3)$ is roughly constant (the volume decreases slightly because of H_2O loss, which can be accounted for if more precision is desired). The treatment of the problem is similar to the case of the growth of a single bubble in an infinite melt, but there is an outer boundary at $r = S$. Hence, the equations are summarized below without much explanation.

(1) H_2O diffusion equation in spherical coordinates is

$$\frac{\partial w}{\partial t} = \frac{1}{r^2}\frac{\partial}{\partial r}\left(Dr^2\frac{\partial w}{\partial r}\right) - u\frac{a^2}{r^2}\frac{\partial w}{\partial r}, \qquad t > 0 \quad \text{and } a \leq r \leq S. \tag{4-156a}$$

(2) The pressure in the bubble may be found as follows (Proussevitch and Sahagian, 1998):

$$P_g = P_f + \frac{2\sigma}{a} - 4ua^2 \int_{z(a)}^{z(S)} \eta(z)dz. \tag{4-156b}$$

(3) Initial conditions : $w|_{t=0}=w_\infty$ for $r > a_0$.   (4-156c)

$P_{g,\,t=0}=P_f + 2\sigma/a$.   (4-156d)

(4) Boundary conditions : $w|_{r=a}=$ solubility at P_g.   (4-156e)

$(\partial w/\partial r)_{r=S}=0$.   (4-156f)

(5) Mass balance condition at the bubble–melt interface (to determine bubble growth rate da/dt):

$$\frac{d}{dt}\left(\frac{4}{3}\pi a^3 \frac{\omega P_g}{RT}\right)=4\pi a^2 D\rho \left(\frac{\partial w}{\partial r}\right)_{r=a}.$$ (4-156g)

The above set of equations can be solved numerically given input parameters, including surface tension σ, temperature, solubility relation, D and η as a function of total H_2O content (and pressure and temperature), initial bubble radius a_0, initial outer shell radius S_0, initial total H_2O content in the melt, and ambient pressure P_f. For example, Figure 4-14 shows the calculated bubble radius versus time, recast in terms of F versus t/t_c to compare with the Avrami equation (Equation 4-70).

4.2.5.3 Convective growth of a single spherical bubble in an infinite liquid reservoir

This section follows Zhang and Xu (2008). In most liquids, a bubble rises rapidly under buoyancy, which induces forced convection. For rising bubbles, two factors cause the bubble to become larger: mass increase in the bubble and the pressure decrease as the bubble rises. The second factor is significant only when rising distance is large (e.g., >10 m). For clarity of discussion, CO_2 bubble growth in water is considered. The mass in the bubble increases as

$$dn/dt=4\pi a^2 D(C_\infty - C_{sat})/\delta,$$ (4-157)

where n is the number of moles of CO_2 in a bubble, a is the radius of the bubble, $4\pi a^2$ is the surface area of the bubble, D is the diffusivity of the dissolved gas in the liquid, C_{sat} and C_∞ are respectively the dissolved gas concentration in liquid at saturation (that is, at the interface) and faraway (that is, the initial concentration), and δ is the effective compositional boundary layer thickness. For bubble growth, $C_\infty > C_{sat}$. The critical parameter to be found is δ, which is obtained through the Sherwood number (defined as $Sh = 2a/\delta$) to be calculated from Peclet and Reynolds numbers using a relation developed by Zhang and Xu (2003). The calculation involves the following steps.

(1) At a given depth z, fluid pressure is

$$P_f = P_{atm} + \rho g z, \tag{4-158}$$

and gas pressure P_g inside the bubble is

$$P_g = P_{atm} + \rho g z + 2\sigma/a + 4\eta u/a, \tag{4-159}$$

where P_{atm} is the local atmospheric pressure, ρ is liquid density, g is acceleration due to the Earth's gravity, z is the depth of the bubble in the liquid, σ is surface tension, η is fluid viscosity, and u is bubble growth rate. Hence, the mass (number of moles) of gas inside the bubble is

$$n = (4\pi a^3/3)P_g/(RT), \tag{4-160}$$

where R is the gas constant, and T is temperature in kelvins.

(2) From the initial size of the bubble, calculate the ascent velocity U. Experience shows that in terms of rising dynamics, most bubbles may be treated as rigid spheres. Hence, the ascent velocity U may be obtained by solving three unknowns (Re, C_D, and U) from the following three equations (Equations 4-120, 4-121, and 4-122, which are written below for convenience):

$$Re = \frac{2aU\rho_f}{\eta_f},$$

$$C_D = \frac{24}{Re}(1 + 0.15\ Re^{0.687}) + \frac{0.42}{1 + 42500\ Re^{-1.16}},$$

$$U = \sqrt{\frac{8ga\ \Delta\rho}{3\rho_f C_D}}.$$

(3) Knowing the ascent velocity U from above, the Peclet number (Pe) is calculated,

$$Pe = 2Ua/D.$$

Then the Sherwood number (Sh) is calculated (Zhang and Xu, 2003),

$$Sh = 1 + (1 + Pe)^{1/3}\left(1 + \frac{0.096\ Re^{1/3}}{1 + 7\ Re^{-2}}\right).$$

Then the effective compositional boundary layer thickness δ is calculated:

$$\delta = 2a/Sh.$$

(4) Knowing δ, the rate for mass loss from or gain by the bubble can be calculated from Equation 4-157.

(5) Given Δt, new n at new time can be obtained by integrating Equation 4-157, new bubble radius can be evaluated from Equation 4-160, new depth can

be found as $z - U\Delta t$, and new pressures can be calculated using Equations 4-158 and 4-159. The process can then be iterated.

4.2.6 Other problems that can be treated similarly

Other problems that may be treated in a similar way as crystal dissolution and growth (and melting) include

(1) Convective liquid droplet dissolution and growth;

(2) Freezing of a lake or lava lake by cooling from the above;

(3) Condensation in the solar nebula;

(4) Congruent and incongruent melting of a single crystal;

(5) Partial melting at the interface of two solid phases.

4.2.7 Interplay between interface reaction and diffusion

In this section, crystal dissolution controlled by both interface reaction and diffusion is considered, generally following the treatment by Zhang et al. (1989). Let w be the degree of saturation, which is not the mass fraction of a component, but roughly the concentration of the principal equilibrium-determining component divided by the saturation concentration. Because w is roughly proportional to the concentration of the principal equilibrium-determining component, it may be viewed to behave similarly as the concentration. The diffusion-interface reaction equation in the interface-fixed reference frame may be written in the following form:

$$\frac{\partial w}{\partial t} = D\frac{\partial^2 w}{\partial x^2} - u\frac{\partial w}{\partial x},$$
(4-161a)

with initial condition

$$w|_{t=0} = w_\infty,$$
(4-161b)

and boundary condition at $x = 0$

$$D\frac{\partial w}{\partial x}\bigg|_{x=0} - u(w_0 - w^c) = 0,$$
(4-161c)

where w is the degree of saturation with respect to the crystal to be dissolved, w_∞ is that of the initial melt, w_0 is that of the interface melt, and w^c is that of the hypothetical melt with the composition of the crystal. The interface reaction (melt growth) rate u during crystal dissolution can be written as (Equation 4-34)

$$u = u_0(1 - w_0),$$
(4-161d)

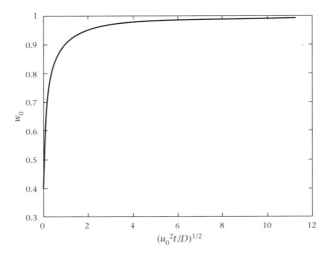

Figure 4-24 A numerical solution of the diffusion-interface reaction equation. The conditions are $w_\infty = 0.4$ and $w^c = 4$. After Zhang et al. (1989).

where u_0 is the interface reaction rate when $w_0 = 0$. The above equation differs from Equation 4-34 by a negative sign because here we are interested in crystal dissolution, whereas Equation 4-34 is for crystal growth. Figure 4-24 shows an example of calculated evolution of w_0 with a dimensionless time $u_0^2 t/D$. As this parameter is about 100 (or its square root is about 10 as shown in Figure 4-24), $w_0 = 0.99$, meaning the interface is within 1% of the saturation concentration. If u_0 and D are known, real time for the interface to be within 1% of the saturation concentration can be found. For example, using experimental data on diopside melting rate (Kuo and Kirkpatrick, 1985), u_0 is estimated to be about 0.3 mm/s. For a D of $10\,\mu m^2/s$, then the real time to reach saturation is about 0.01 s, meaning that diffusion control is essentially instantaneously established. On the other hand, if u_0 is reduced by a factor of 1000 (Table 4-2), then the real time to reach saturation is about 10,000 s = 2.8 h, meaning that there would be a finite duration during which crystal dissolution is controlled by both interface reaction and diffusion (Acosta-Vigil et al., 2002, 2006; Shaw, 2000, 2004).

4.3 Some Other Heterogeneous Reactions

4.3.1 Bubble growth kinetics and dynamics in beer and champagne

To relax a bit, the first topic to be discussed in this section is the kinetic processes in champagne and beer. This is a fun subject, but there is serious science. When a bottle of champagne or beer is opened, myriads of kinetic and dynamic processes

Figure 4-25 A schematic drawing of champagne bubbling.

take place, including nucleation of bubbles, bubble ascent and growth, and convection.[2] The beautiful trains of bubbles (Figure 4-25) and the foamy head add to the aesthetic value. Many of the kinetic processes have been revealed only recently, thanks to the untiring effort of beer and champagne lovers (e.g., Shafer and Zare, 1991; Liger-Belair, 2004; Zhang and Xu, 2008). Some aspects of bubble kinetics and dynamics in beer and champagne can be quantitatively modeled. However, the whole bubbling process cannot be predicted a priori because the nucleation part cannot be quantified.

Before it is opened, champagne usually contains about 6 bars of CO_2 gas, and beer usually contains about 2 bars of CO_2 or N_2 gas. For example, Budweiser beer contains 2.1 bars of CO_2 at 9°C, and Guinness beer contains 2 bars of N_2 and 0.014 bar of CO_2 (based on information provided by the manufacturers). Inside the bottle, the dissolved gas is in equilibrium with the gas phase (unless disturbed). Hence, no bubbles would grow. Once open, the gas phase escapes. The decompression leads to oversaturation of the gas component in the beverage. The degree of oversaturation may be calculated using C/C_e, where C is the concentration of the gas in the drink, and C_e is the equilibrium concentration. Because the degree of saturation is small, homogeneous nucleation rate is expected to be negligible, and heterogeneous nucleation dominates. If beer or champagne is poured into glass, there is usually rapid bubbling, forming a foam on the surface, followed by many trains of bubbles, each train coming from a specific site. The initial rapid bubbling is due to air bubbles trapped inside beer or champagne during pouring, which rise and grow (as CO_2 gas diffuses into the bubbles) to the top. The bubble trains are due to specific heterogeneous "nucleation" sites. Previously, it was thought that bubbles nucleate on scratches or

---

[2] The dynamic processes include bubble ascent and fluid convection. The kinetic processes include bubble nucleation and growth.

roughness on the glass. However, the careful work by Liger-Belair (2004) shows that bubbles nucleate heterogeneously on dirt particles, usually elongated, hollow, and roughly cylindrical cellulose (such as paper or cloth fibers) on the glass wall. The fibers often trap air pocket(s) inside. The presence of air pockets inside a fiber tube means that no nucleation is necessary: gas molecules simply attach to the air pocket at the tip of the fiber. When the bubble becomes large enough so the buoyant force exceeds the adhesion force to the tube, it departs the fiber and begins its ascent with further growth. A given site (given fiber) would deliver bubbles about the same size and at a given time interval. Fibers with different radii would issue different-size bubbles at different intervals. The longer the interval between two successive bubbles, the larger is the bubble because there is more time for more gas to accumulate on the "nucleation" site.

Once issued from the fiber tip, the bubble rises because of buoyancy and grows because CO_2 gas diffuses into the bubble. Expansion as a bubble rises to lower pressure also contributes to the bubble size increase, but the effect is negligible in the case of champagne and beer bubbles. To understand the ascent dynamics, it is necessary to know the viscosity of beer and champagne. Beer viscosity depends on its sugar content. A typical viscosity of beer is about 1.44 times that of pure water (Zhang and Xu, 2008). If the temperature is 9°C, the viscosity is 0.0019 Pa·s. The ascent velocity of a bubble depends on its size (the specific size limit is based on the physical property of beer) as follows:

(1) If bubble radius is <0.045 mm (45 μm), the bubble is spherical, the Reynolds number Re is ≤ 0.1, and the ascent velocity U can be obtained from the Stokes' law: $U = 2ga^2\Delta\rho/(9\eta)$, where g is acceleration due to the Earth's gravity, a is the radius of the bubble, $\Delta\rho$ is the absolute value of the density difference between the bubble and the beverage, and η is the viscosity of the fluid.

(2) If the radius of the bubble is between 0.045 to 1 mm, the bubble is still spherical, but Re is between 0.1 and 180. Stokes' law is not accurate anymore for calculating the rising velocity. One way to solve the ascent velocity is to solve three equations simultaneously (Section 4.2.5.3).

(3) If the radius of the bubble is larger than 1 mm, the bubble is not spherical anymore, and becomes wobbling and irregularly shaped. The average ascent velocity must be estimated empirically.

Beer and champagne bubbles are usually smaller than 1 mm in radius and are hence spherical. Because a beer bubble grows as it rises, the growth is controlled by convective CO_2 transport into the bubble. With theories developed in Section 4.2.5.3, both the ascent velocity U and growth rate u of a single bubble in an infinite reservoir of beer or champagne can be calculated. The two must be calculated together because bubble size affects the ascent velocity, which in turn

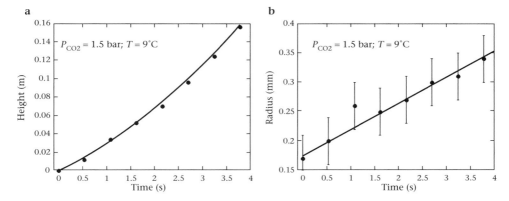

Figure 4-26 (a) The height and (b) radius of a rising bubble in a Budweiser beer as a function of time. Data points are from Shafer and Zare (1991). The viscosity of beer is measured to be 1.44 times that of pure water. The diffusivity and solubility of CO_2 in beer are assumed to be the same as those in pure water. The temperature is assumed to be 9°C. The initial CO_2 content is assumed to be 1.5 bar. From Zhang and Xu (2008).

affects the intensity of convection and hence the bubble growth rate. The effective boundary layer thickness δ for a beer bubble is usually about $16\,\mu m$ (δ would doubles if viscosity increases by a factor of 10). The dependence of δ on the bubble size and other factors is weak. Because the boundary layer thickness is roughly constant, the bubble growth rate is roughly constant. Shafer and Zare (1991) experimentally measured the height and size of rising and growing bubbles in a glass of Budweiser beer after pouring beer into the glass. From a given site, a bubble is issued every 0.54 s. As the bubble rises and grows, the height and radius of the bubble are measured. The bubble growth rate is roughly constant, about 0.04 mm/s. The rising velocity depends on bubble size and hence on time. That is, bubble radius versus time is roughly linear, but bubble height versus time is a curve. Their data are shown in Figure 4-26, together with theoretical calculations using the methods described in Section 4.2.5.3.

In the theoretical calculation (Figure 4-26), to match the bubble growth rate, the CO_2 content is 1.5 bar (0.15 MPa, or 0.0815 mol/L). Undegassed Budweiser beer contains 2.1 bars of CO_2 per volume of beer. The smaller CO_2 concentration in a glass of beer is reasonable because some CO_2 is lost during rapid bubbling when beer is poured into glass. Increasing temperature by 10°C at the same CO_2 content roughly doubles the bubble growth rate, meaning volume growth rate would increase by a factor of eight.

Because the time interval between successive bubbles is fixed but larger bubbles rise more rapidly, the distance between successive bubbles increases as one moves from the bottom to the top of the train of bubbles. For example, Figure 4-27 is a calculated relation between bubble size and height, showing the

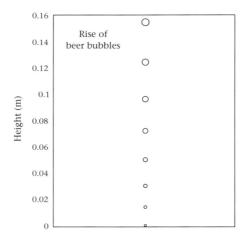

Figure 4-27 Calculated bubble rise and ascent in beer. The radius of the bubble at the top is 0.34 mm.

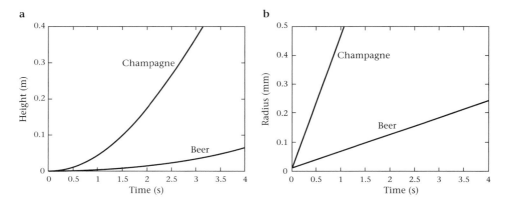

Figure 4-28 Comparison of calculated bubble rise and growth in beer (with 2 bars of CO_2) and in champagne (with 6 bars of CO_2). From Zhang and Xu (2008).

increase in distance between successive bubbles. The absolute distance between successive bubbles also depends on (i) the size of the bubble at release, and (ii) the time interval between bubble release, both of which depend on the characteristics of the "nucleation" site (dirt particle).

Because of high CO_2 concentration in champagne, bubble growth in champagne is much more rapid. Figure 4-28 compares bubble growth and ascent in a beer containing 2 bars of CO_2 and champagne containing 6 bars of CO_2. Bubble growth rate in champagne is about 6 times faster. Because of the larger bubble size in champagne, it also rises more rapidly. The rapid bubble growth leads to rapid volume expansion. Hence, champagne is able to erupt much more violently than beer.

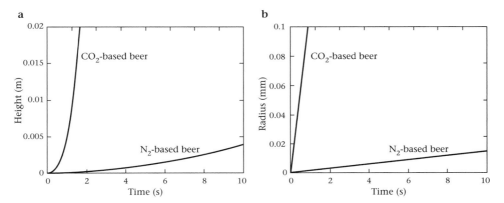

Figure 4-29 Comparison of calculated bubble rise and growth in CO_2-based and N_2-based beer. The saturation pressure is 2 bars for both cases. Due to low solubility of N_2, its concentration is small compared to that of CO_2 at the same pressure, leading to much smaller bubble growth rate and rising velocity. From Zhang and Xu (2008).

For some types of beer, the gas component is mostly N_2. Such bubbles are released from a ball that initially contains the gas. The released bubbles then rise and grow. However, because the solubility of N_2 is low, at the same pressure, the dissolved gas content is much smaller in such beer than in CO_2-based beer. With similar diffusivity, a much smaller amount of gas would be convectively transferred to each bubble in N_2-based beer, and the bubble hence grows very slowly. Figure 4-29 compares bubble growth and ascent in CO_2-based beer and N_2-based beer. Because bubbles in N_2-based beer are small, they rise very slowly (Figure 4-29). Therefore, convection in beer (upflow near the center line and downflow near glass wall) can easily entrain such bubbles and downflow can bring them down. This explains the observation that bubbles in Guinness beer can sink instead of rise.

Congratulations for mastering the "fizzics" of beer and champagne! Here is a glass to you. Cheers!

4.3.2 Dynamics of explosive volcanic eruptions

An ascending volcanic eruption column into the stratosphere (Figure 4-30) is often marveled as the most spectacular sight on the Earth. The dynamics of volcanic eruptions is exceedingly complicated. It might be surprising to learn that such might and sight arise from a relatively simple heterogeneous reaction: the exsolution of H_2O gas from a magma.

Dynamics of explosive volcanic eruptions are in many aspects similar to beer and champagne dynamics. They are driven by the exsolution of a gas from a liquid. Figure 4-31 describes the various stages of an explosive eruption. In the magma chamber, much H_2O (such as 6 wt%) and some CO_2 (such as 40 ppm) are

Figure 4-30 Eruption column of Mount St. Helens, 1980. Photograph by USGS photographer of David A. Johnston Cascades Volcano Observatory, USGS, Vancouver, Washington, USA.

dissolved in the magma under high pressure. At some shallow level, such as <200 MPa pressure, the pressure is not high enough to hold the high dissolved gas content. Hence, bubbles nucleate and grow. The nucleation is often heterogeneous, and the rate cannot be predicted. Assuming diffusion control, the growth of bubbles in rhyolitic melt may be modeled as shown in Section 4.2.5. As bubbles grow, the volume of bubble–magma system increases, pushing the bubbly magma into the exit, such as the volcanic conduit. The rise of the bubbly magma leads to a decrease in pressure, an increase in the degree of oversaturation, and hence more bubble growth and expansion. Because of the positive feedback, volume expands rapidly. Furthermore, in the case of volcanic eruptions, because of the large pressure range (from 0.1 to 200 MPa), gas volume expansion due to pressure decrease is a significant component for bubble volume increase, contrary to bubbles in beer and champagne.

When bubble growth reaches some stage, bubbly magma would fragment into a gas phase carrying magma droplets. Before magma fragmentation, the eruption is a magma flow carrying bubbles (meaning the magma is the continuous phase, and bubbles are isolated inside the magma). After magma fragmentation, the

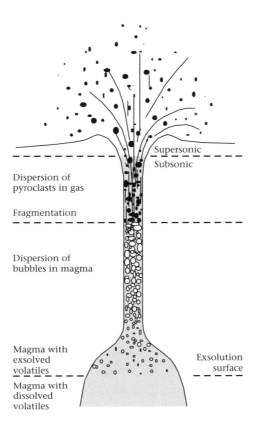

Supersonic

Subsonic

Dispersion of
pyroclasts in gas

Fragmentation

Dispersion of
bubbles in magma

Magma with
exsolved
volatiles

Exsolution
surface

Magma with
dissolved
volatiles

Figure 4-31 Schematic drawing of eruption stages. Deep inside the
magma chamber, much H_2O (plus other minor gases) is dissolved
in the magma. At the exsolution pressure, the gases become supersaturated and bubbles begin to nucleate and grow. At the fragmentation level, differential pressure between the bubble pressure
and magma pressure is high enough to cause magma fragmentation, which turns a bubbly magma flow to a gas flow carrying
magma droplets. At the exit, the eruption velocity may cross the
sound speed to become a supersonic flow. From Zhang et al. (2007).

eruption becomes a gas flow carrying magma droplets (meaning that the gas is
the continuous phase and magma droplets are isolated inside the gas). That is,
fragmentation defines the transition from a relatively quiet liquid flow to an
explosive gas flow. Spieler et al. (2004) showed that fragmentation of pumices
occurs when $\Delta P \geq S/\Phi$, where ΔP is the difference between the bubble pressure
and the ambient pressure, Φ is the vesicularity (volume fraction of the gas phase),
and S is related to the strength of magma and takes a value of 1 MPa. Although
Figure 4-31 shows that fragmentation occurs inside the eruption conduit, it is
not necessarily so. Sometimes, fragmentation of small volcanic flows may also
occur on the Earth's surface (such as Unzen eruption in 1991; Sato et al., 1992)
because surface lava flows may still contain high H_2O concentration due to

disequilibrium, leading to bubble growth, volume expansion, and fragmentation. After fragmentation, the eruption velocity increases rapidly. For a large eruption, the explosive gas flow velocity may pass the sound speed barrier when the conduit curvature is appropriate, which is often encountered at the volcanic exit. Hence, eruption velocity of major eruptions is often estimated by the sound speed of the gas–magma mixture:

$$U = \sqrt{\frac{B_{\text{melt}} B_{\text{gas}}}{\rho_{\text{mix}}[(1-\Phi)B_{\text{gas}} + \Phi B_{\text{melt}}]}} \approx \sqrt{FRT}, \tag{4-162}$$

where B_{melt} and B_{gas} are the bulk modulus of the melt and the gas ($B_{\text{melt}} \approx 11$ GPa for rhyolitic melt; $B_{\text{gas}} = P_{\text{gas}}$ for ideal gas), Φ is the vesicularity, ρ_{mix} is the density of the gas–magma mixture and equals $\Phi\rho_{\text{gas}} + (1-\Phi)\rho_{\text{melt}}$, F is the mass fraction of the gas phase, R is the gas constant expressed per kilogram instead of per mole (for H_2O gas, $R = 461.5$ J kg^{-1} K^{-1}), and T is temperature in kelvins.

Modeling the dynamics of explosive volcanic eruptions requires an understanding of magma properties and quantification of the various steps of processes. The most important magma properties for understanding explosive volcanic eruptions include the solubility and diffusivity of major volatiles in the melt, viscosity and strength of magma, and interface energy between magma and gas phase. Among these properties, interface energy is the least well understood. The steps of processes in an explosive volcanic eruption include nucleation, bubble growth, magma flow, fragmentation, gas flow, and volatile transport from the magma to the gas phase. Among these steps, nucleation is the least understood. In fact, the inability to predict bubble nucleation rate (either homogeneous or heterogeneous) precludes the quantitative prediction of the dynamics of prefragmentation gas-driven volcanic eruptions. On the other hand, bubble growth rate can be modeled realistically using the method shown in Section 4.2.5 (Proussevitch and Sahagian, 1998; Liu et al., 2000). Assuming that the exit velocity is the sound speed (Equation 4-162), it depends on the mass fraction of the gas phase (F in Equation 4-162). Because gas exsolution occurs largely before fragmentation, it is therefore necessary to accurately model the prefragmentation bubble nucleation and growth rates to evaluate F and the exit velocity.

4.3.3 Component exchange between two contacting crystalline phases

Simple component exchange between solid phases is accomplished by diffusion. If only two components (such as Fe^{2+} and Mg) are exchanging, the diffusion is binary. The boundary condition is often such that the exchange coefficient between the surfaces of two phases is constant at constant temperature and pressure. The concentrations of the components on the adjacent surfaces may be constant assuming interface equilibrium. The solution to the diffusion equation is hence straightforward. If three or more components (such as Fe^{2+}, Mn^{2+}, and Mg) are exchanging, the problem is then a multicomponent diffusion problem and the rigorous treatment is more complicated.

If one of the two phases is a fluid phase, such as $^{18}O-^{16}O$ exchange between water vapor and a magnetite crystal, component exchange can be accomplished by two parallel paths. One path is simple mass transfer through both the fluid phase and the solid phase. Mass transfer in the fluid phase is rapid, many orders of magnitude greater than that in the solid phase. Therefore, the component exchange rate in this path is determined by diffusion in the solid. The fluid phase would be roughly uniform (hence, one does not have to consider diffusion in the fluid phase) and the solid phase would show concentration profiles due to diffusion. This problem becomes a simple diffusion problem in one phase, and the fluid phase would determine the boundary condition. Many sorption and desorption experiments employed to determine diffusivity in a solid phase are of this type. The second path is by dissolution of the solid phase that does not have an equilibrium composition and reprecipitation of the same solid phase but with a composition in equilibrium with the fluid phase. The second path can occur if, in addition to the disequilibrium in terms of component exchange, the solid phase is also undersaturated in the fluid phase, but with dissolution of the solid phase, the fluid phase may be oversaturated with the solid phase of a different composition. For example, in the plagioclase phase diagram of Figure 4-19a, a liquid of composition A and plagioclase grains of composition C are brought together at 1573 K. The liquid is undersaturated with respect to the plagioclase solid. Hence, plagioclase crystals would dissolve into the melt. The melt composition gradually moves to composition B, at which point plagioclase of composition D would crystallize. The final liquid composition would be at point B, and solid composition would be at point D. Although pure diffusion in plagioclase (and liquid) may also accomplish the task of changing compositions, because albite–anorthite interdiffusion in plagioclase is extremely slow, the dissolution–reprecipitation mechanism would likely dominate the component exchange rate. Thus, in the presence of fluids, it is necessary to consider the various paths of component exchange.

For component exchange between two crystals, under isothermal and one-dimensional (meaning a planar contact interface between the two phases with diffusion along principal axes) conditions, the problem is a diffusion and partition problem. If the partition coefficient $K = w_i^B / w_i^A$ (where A and B are the two phases) is a constant, the problem was solved in Section 3.2.4.6. Here we solve a slightly more difficult problem of constant exchange coefficient K_D. One example is Fe–Mg exchange between olivine and garnet, for which the exchange coefficient $(Fe/Mg)^{ol}/(Fe/Mg)^{gt}$ (which is related to the thermodynamic equilibrium constant) is roughly constant, but the partition coefficient Fe^{ol}/Fe^{gt} is not a constant. (The assumption of constant partition coefficient may work fine for minor and trace elements.)

The two crystals in contact at $x=0$ can be considered to be two halves of a diffusion couple. The two interdiffusion species are denoted as 1 and 2, and the left and right halves (phases) are denoted as A and B. The interdiffusion coefficient in the left-hand side, or phase A ($x<0$), is D^A, and that in the other side, or phase B ($x>0$), is D^B. Use subscripts to denote components 1 and 2 and superscripts to denote phases A and B. The initial condition is

$$X_1^A|_{t=0,\,x<0}=X_{1,-\infty}^A, \tag{4-163a}$$

and

$$X_1^B|_{t=0,\,x>0}=X_{1,+\infty}^B, \tag{4-163b}$$

where $X_{1,-\infty}^A$ and $X_{1,+\infty}^B$ are the initial mole fractions of component 1 in the two phases.

If diffusion distance is small compared to the size of either phase, because each side satisfies separately the conditions for applying the Boltzmann transformation ($\eta=x/\sqrt{4Dt}$), the solution for species 1 in each phase is

$$
\begin{aligned}
X_1^A &= X_{1,-\infty}^A + b^A \operatorname{erfc} \frac{|x|}{2\sqrt{D^A t}} \\
&= X_{1,-\infty}^A + (X_{1,-0}^A - X_{1,-\infty}^A)\operatorname{erfc}\frac{|x|}{2\sqrt{D^A t}}, \quad x<0
\end{aligned}
\tag{4-164a}
$$

$$
\begin{aligned}
X_1^B &= X_{1,+\infty}^B + b^B \operatorname{erfc} \frac{x}{2\sqrt{D^B t}} \\
&= X_{1,+\infty}^B + (X_{1,+0}^B - X_{1,+\infty}^B)\operatorname{erfc}\frac{x}{2\sqrt{D^B t}}, \quad x>0
\end{aligned}
\tag{4-164b}
$$

where $X_{1,-0}^A$ and $X_{1,+0}^B$ are the mole fractions of component 1 in the two phases at the interface, with "-0" meaning x approached the interface from the left-hand side (meaning phase A), and "$+0$" meaning x approached the interface from the right-hand side (meaning phase B). The solution for component 2 can be written similarly. Next $X_{1,-0}^A$ and $X_{1,+0}^B$ in the above two equations, and $X_{2,-0}^A$ and $X_{2,+0}^B$ from similar equations for component 2, must be solved to obtain the full solution. To solve for the four unknowns, boundary conditions and mass balance conditions are used. Assuming interface equilibrium, one boundary condition is

$$K_D=\frac{(X_2^B/X_1^B)_{x=+0}}{(X_2^A/X_1^A)_{x=-0}}=\frac{(X_{2,+0}^B/X_{1,+0}^B)}{(X_{2,-0}^A/X_{1,-0}^A)}. \tag{4-165}$$

Stoichiometry in each phase means

$$X_{1,-0}^A + X_{2,-0}^A=A, \tag{4-166a}$$

$$X_{1,+0}^B + X_{2,+0}^B=B, \tag{4-166b}$$

where A and B are constants. For example, for olivine, $X^{ol}_{Fe,-0} + X^{ol}_{Mg,-0} \approx 1$. One more condition is necessary, and it may be viewed as a mass balance condition, which states that mass flux from one side must equal mass flux to the other side, i.e.,

$$D^A \rho^A \frac{\partial X^A_1}{\partial x}\Big|_{x=-0} = D^B \rho^B \frac{\partial X^B_1}{\partial x}\Big|_{x=+0}, \qquad (4\text{-}167)$$

where ρ is the molar density of cations being considered (such as all divalent cation density in mol/L). Using Equations 4-164a,b, the above equation can be converted to

$$(X^A_{1,-0} - X^A_{1,-\infty})\rho^A \sqrt{D^A} = -(X^B_{1,+0} - X^B_{1,+\infty})\rho^B \sqrt{D^B} \qquad (4\text{-}168)$$

When Equations 4-165, 4-166a,b, and 4-168 are solved for parameters $X^A_{1,-0}$, $X^B_{1,+0}$, $X^A_{2,-0}$, and $X^B_{2,-0}$, the concentration profiles can be calculated using Equations 4-164a,b.

Example 4.1. Suppose olivine and garnet are in contact and olivine is on the left-hand side $(x < 0)$. Ignore the anisotropic diffusion effect in olivine. Suppose Fe–Mg interdiffusion between the two minerals may be treated as one dimensional. Assume olivine is a binary solid solution between fayalite and forsterite, and garnet is a binary solid solution between almandine and pyrope. Hence, $C_{Fe} + C_{Mg} = 1$ for both phases, where C is mole fraction. Let initial Fe/(Fe + Mg) = 0.12 in olivine and 0.2 in garnet. Let $K_D = (Fe/Mg)_{gt}/(Fe/Mg)_{ol} = 3$, $D_{Fe-Mg,ol} = 10^{-10}$ mm^2/s, and $D_{Fe-Mg,gt} = 10^{-12}$ mm^2/s. Calculate the diffusion profile at $t = 100$ years.
Solution: The calculated concentration profiles at $t = 100$ years are plotted in Figure 4-32. The calculation of the interface concentrations is explained in the figure caption. Because diffusion in olivine is much more rapid, the profile is longer in olivine and deviation from the initial concentration is smaller.

In rocks cooled down from a high temperature, there is continuous partitioning and diffusion of the components between the minerals. This is again a diffusion problem, but for time-dependent temperature and hence time-dependent D^A, D^B, and K_D. Time-dependent D in the diffusion equation for one phase is relatively easy to handle, but time-dependent K_D leads to complicated boundary conditions. The resulting diffusion equation is often impossible to solve analytically. Given the temperature history and the temperature dependence of D^A, D^B, and K_D, diffusion profiles can be solved numerically. The problem can also be turned around: Knowing the concentration profiles in a mineral pair, it may be possible to do the inverse problem to infer cooling rates (Lasaga, 1983). This is discussed in the next chapter.

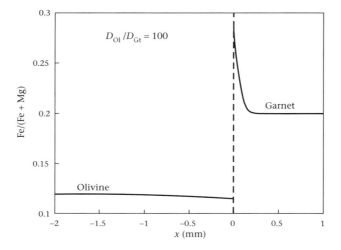

Figure 4-32 Calculated concentration profiles in olivine and garnet for the conditions given in Example 4-1. Molar volume of olivine is about 0.046 L/mol, leading to $Fe + Mg$ concentration of $(2/0.046) = 43.48$ mol/L. Molar volume of garnet is 0.117 L/mol, leading to $Fe + Mg$ concentration of $(3/0.117) = 25.64$ mol/L. The surface concentrations are solved from the following equations: $[y/(1-y)]/[x(1-x)] = K_D = 3$ and $434.8(x - 0.12) = -25.64(y - 0.2)$, where $x = X_{Fe, -0}^{ol}$ and $y = X_{Fe, +0}^{gt}$. The solution is $x = 0.1152$, $y = 0.2809$.

A much more complicated problem is for the species exchange among multiple coexisting minerals, such as ^{18}O–^{16}O exchange in a multimineral rock. The problem cannot be treated rigorously. Approximate models are available. Eiler et al. (1992, 1993, 1994) developed a model to treat oxygen isotope exchange among many coexisting minerals. In the model, it is assumed that grain boundary diffusion is very fast, leading to instantaneous equilibrium at all grain boundaries. The model is hence referred to as the FGB (fast grain boundary) diffusion model. The solution has important applications in geospeedometry and is discussed in Chapter 5.

4.3.4 Diffusive reequilibration of melt and fluid inclusions

Melt and fluid inclusions in minerals are important sources for inferring conditions of the melt and fluid when the mineral grew. One example is the pre-eruptive volatile content in a magma chamber. Because solubility of volatiles decreases as pressure decreases (that is, as magma erupts), erupted magma often does not preserve the original volatile contents in the melt. Melt inclusions may provide such information. Among common minerals, quartz in silicic rocks and olivine in mafic and rhyolitic rocks are the best to preserve volatiles in melt inclusions because they are relatively strong (not easy to crack) and there are no cleavages. Nonetheless, volatiles and other components partition between the melt inclusion and the host mineral, and may diffuse through the host mineral to reach equilibrium with the melt outside the mineral. It is hence necessary to estimate the conditions for the concentration of volatiles to be preserved in melt inclusions. Qin et al. (1992) presented a detailed analysis of the problem, which is summarized in Box 4-1. The problem is set up and some simple cases are treated below.

Box 4.1 Detailed analysis by Qin et al. (1992) on inclusion re-equilibration

The inclusion is assumed to be a sphere with radius $r < R_1$. The host is a spherical shell concentric with the inclusion with $R_1 < r < R_2$. The diffusion equation in the host mineral is

$$\frac{\partial C}{\partial t} = D\left(\frac{\partial^2 C}{\partial r^2} + \frac{2}{r}\frac{\partial C}{\partial r}\right), \quad R_1 < r < R_2.$$

The initial condition in the inclusion is uniform concentration $C_{I,0}$, and the host is in equilibrium:

$$C(r, 0) = KC_{I,0}, \quad R_1 < r < R_2.$$

The boundary conditions at the two boundaries are

$$C(R_1, t) = KC_I,$$

$$C(R_2, t) = KC_e,$$

where C_e is concentration of the component in the outside melt (equilibrium concentration) and is a constant, and C_I is concentration of the component in the melt inclusion and varies with time as

$$\frac{d}{dt}\left[4\pi R_1^3 \rho_m C_I(t)/3\right] = 4\pi R_1^2 D \frac{\partial(\rho_c C)}{\partial r}\Big|_{r=R_1}.$$

The solution is (Qin et al., 1992)

$$C(r, t) = KC_e + \frac{4\beta K R_2}{r}(C_{I,0} - C_e)$$

$$\times \sum_{n=1}^{\infty} \frac{\sin[(1 - r/R_2)q_n]\exp(-q_n^2 Dt/R_2^2)}{\{2\beta(1 - \alpha)q_n + 4\alpha q_n \sin^2[(1 - \alpha)q_n] - \beta\sin[2(1 - \alpha)q_n]\}},$$

where $\alpha = R_1/R_2$, $\beta = 3K\rho_c/\rho_m$ (where ρ_c is density of the host crystal and ρ_m is density of the melt), and q_n ($n = 1, 2, 3, \ldots$ from small to large) is the nth positive root of

$$\tan[(1 - \alpha)q] = \alpha\beta q/(\alpha^2 q^2 - \beta).$$

The concentration in the melt is evaluated as $C_I(t) = C(R_1, t)/K$:

(Continued on next page)

(Continued from previous page)

$$C_I(t) = C_e + \frac{4\beta}{\alpha}(C_{I,0} - C_e)$$

$$\times \sum_{n=1}^{\infty} \frac{\sin[(1-\alpha)q_n]\exp(-q_n^2 Dt/R_2^2)}{\{2\beta(1-\alpha)q_n + 4\alpha q_n \sin^2[(1-\alpha)q_n] - \beta\sin[2(1-\alpha)q_n]\}}.$$

First simplifying the above, then defining the extent toward equilibrium as

$$\phi = [C_I(t) - C_{I,0}]/(C_e - C_{I,0}),$$

then,

$$\phi(t) = 1 - \frac{2\beta}{\alpha}\sum_{n=1}^{\infty} \frac{\exp(-q_n^2\tau)}{\{\gamma q_n \sin[(1-\alpha)q_n] + [\alpha(1-\alpha)q_n^2 - \beta/\alpha]\cos[(1-\alpha)q_n]\}}.$$

If $\tau \geq 0.1$, only the first two terms in the series are necessary; if $\tau \geq 0.5$, only the first term is necessary.

The procedure of calculation is as follows. Given the problem and the parameters, including R_1, R_2, D, K, ρ_m, ρ_c, $C_{I,0}$, and C_e, the first step is to calculate α and β. The second step is to find the solutions of q_n from $\tan[(1-\alpha)q] = \alpha\beta q/(\alpha^2 q^2 - \beta)$. Then $\phi(t)$ may be plotted as a function of t.

For simplicity, the melt inclusion is assumed to be (i) spherical, and (ii) concentric with the spherical crystal shell (Figure 4-33). Furthermore, the host mineral is assumed to be isotropic in terms of diffusion. The concentric assumption and the isotropic assumption are rarely satisfied. Nonetheless, for order of magnitude estimate, these assumptions make the problem easy to treat.

Because diffusion in the melt inclusion is orders of magnitude faster than in the crystal, we assume that the melt composition is uniform. The inclusion radius is R_1, and the outer radius of the host is R_2. For the component under consideration, the concentration in the melt inclusion is C_I, the concentration in the outside melt (bulk melt) is C_e. The partition coefficient between the crystal and melt is K so that the concentration of the component in the mineral host is $C_1 = KC_I$ at $r = R_1$ (inclusion–host interface) and $C_2 = KC_e$ at $r = R_2$. The partition coefficient K may be larger or smaller than 1.

Two end-member cases of approximate treatments are considered. In one end-member case, the extra mass of the component in the melt inclusion is not very high compared to the mass of the component in the crystal (i.e., K is not

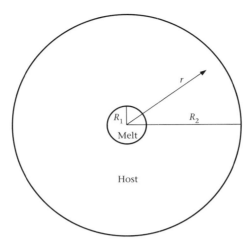

Figure 4-33 A schematic diagram showing a melt inclusion inside a host mineral. For simplicity, the inclusion and the host are assumed to be concentric.

$\ll 1$, or $R_1/R_2 \ll 1$). Hence, when the diffusion distance is about the radius of the crystal, much of the extra mass in the melt inclusion would be lost. This timescale is

$$\tau_1 = (R_2 - R_1)^2/D. \tag{4-169}$$

In the other end-member scenario, the mass of the component in the host mineral is negligible. A quasi-steady-state diffusion profile is established in the host crystal and the melt inclusion maintains the concentration on the inner surface. The steady-state concentration profile would be (Equation 3-31g)

$$C = C_1 + (C_2 - C_1)\frac{1 - R_1/r}{1 - R_1/R_2}. \tag{4-170}$$

The diffusive flux across the whole spherical surface is

$$-4\pi R_1^2 D(\partial C/\partial r)_{r=R_1} = 4\pi D R_1 R_2 (C_1 - C_2)/(R_2 - R_1). \tag{4-171}$$

Hence, the timescale for diffusive equilibration is

$$\tau_2 = \text{mass/flux} = \tfrac{4}{3}\pi R_1^3 (C_I - C_e)/[4\pi D R_1 R_2 (C_1 - C_2)/(R_2 - R_1)]. \tag{4-172}$$

That is,

$$\tau_2 = R_1^2 (1 - R_1/R_2)/(3DK). \tag{4-173}$$

In this end-member case, the timescale for reaching diffusive equilibration is proportional to $R_1^2(1 - R_1/R_2)$, and inversely proportional to D and K. Therefore, large inclusions may preserve the initial content better than smaller inclusions.

The timescale for the melt inclusion to move significantly toward equilibrium is max(τ_1, τ_2). The exact percentage toward equilibrium at $t = \max(\tau_1, \tau_2)$ is not quantified above, but is expected to be significant, e.g., between 20 and 80%.

> *Example 4.2.* Consider H_2O in a rhyolite melt inclusion in quartz. The radius of the inclusion is $100\,\mu m$. The quartz radius is $1\,mm$. The partition coefficient between quartz and melt is estimated to be 10^{-4}. Assume that the diffusivity of H_2O in quartz is $10^{-10}\,m^2/s$. Find the re-equilibration timescale.
>
> *Solution:*
>
> $$\tau_1 = (R_2 - R_1)^2 / D = (9.0 \times 10^{-4})^2 / (10^{-10}) = 8100 \text{ s} = 2.25 \text{ h.}$$
>
> $$\tau_2 = R_1^2(1 - R_1/R_2)/(3DK) = (10^{-4})^2(1 - 0.1)/(3 \times 10^{-10} \times 10^{-4}) \approx 3 \times 10^5 \text{ s} = 83 \text{ h.}$$
>
> Hence, the re-equilibration timescale is roughly 83 h.

Using the more rigorous treatment of Qin et al. (1992) in Box 4-1, when $t = 3 \times 10^5$ s, the departure from equilibrium is about 32% (i.e., the reaction has proceeded 68% toward equilibrium), roughly consistent with the definition of timescale to reach equilibrium. That is, the approximate method to estimate the reequilibration timescale works in this case.

4.3.5 Melting of two crystalline phases or reactions between them

Melting of two crystalline phases initially occurs at their contact because the melting point of a pure phase is high and the solidus of two phases is lower (Figure 4-34). After a melt is produced, the melting continues as each mineral melts (or dissolves) into the melt. Depending on the type of systems, the melting may be controlled by mass transport in the melt, interface reaction, or diffusion in the crystalline phase.

Figure 4-34 is a phase diagram for the system titanite–anorthite. Suppose a crystal of titanite is initially in contact with a crystal of anorthite. The two are heated to 1350°C. Either phase by itself would not melt. But because the temperature is higher than the eutectic point of the two phases, at the interface there is melting. As melting proceeds, a thin melt layer would form between the two crystals. The melting of the two phases continues and the rate may be controlled by different factors. The rate would depend on the controls, as outlined below.

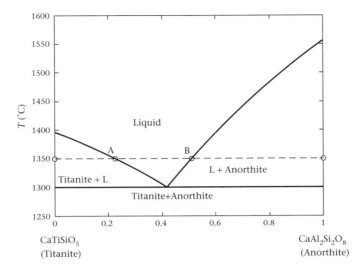

Figure 4-34 Phase diagram of the titanite–anorthite system calculated using thermodynamic data of Robie and Hemingway
(1995). The horizontal axis is mass fraction of An. At 1350°C, titanite is in equilibrium with a melt of composition A, and anorthite is in equilibrium with a melt of composition B. In other
words, melt A is saturated with respect to titanite, and melt B is
saturated with respect to anorthite.

4.3.5.1 Melting controlled by interface reactions

With only a thin melt layer between two crystals, mass transport in the melt is
rapid. Suppose the melting is controlled by interface reaction rate, meaning that
the $CaAl_2Si_2O_8$ concentration (mass fraction) in the melt is roughly uniform
(Figure 4-35a). The $CaAl_2Si_2O_8$ mass fraction in the melt is denoted as w_o, which
is greater than w_A and smaller than w_B(where A and B are points in Figure 4-34).
The $CaTiSiO_5$ concentration in the melt is hence $1 - w_o$. The melting rates of
anorthite and titanite are (Equation 4-34)

$$u^{An} = k^{An}(w^{An} - 1) = k^{An}(w_o/w_B - 1), \tag{4-174a}$$

$$u^{Tt} = k^{Tt}(w^{Tt} - 1) = k^{Tt}[(1 - w_o)/(1 - w_A) - 1]. \tag{4-174b}$$

The melt composition w_o may be related to the melting rate as

$$w_o = \rho^{An}u^{An}/(\rho^{An}u^{An} + \rho^{Tt}u^{Tt}). \tag{4-175}$$

That is,

$$w_o = (w_o/w_B - 1)/\{(w_o/w_B - 1) + \gamma[(1 - w_o)/(1 - w_A) - 1]\}, \tag{4-176}$$

where $\gamma = (\rho^{Tt}k^{Tt})/(\rho^{An}k^{An})$. Note that γ, w_A, and w_B vary with temperature. Suppose $k^{Tt} \gg k^{An}$; then $\gamma \gg 1$, leading to w_o close to w_A and far away from w_B. That is,

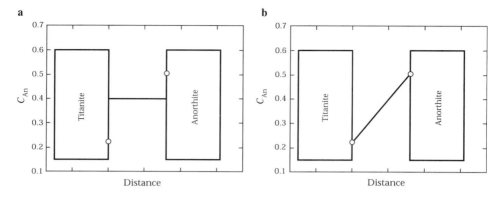

Figure 4-35 Concentration profile during melting of two crystals controlled by (a) interface reaction and (b) diffusion in the melt. The open circles indicate melt compositions in equilibrium with titanite (position A in Figure 4-34) and with anorthite (B in Figure 4-34).

w_o is not necessarily midway between w_A and w_B. If k^{An} and k^{Tt} are known, w_o may be solved from the above equation. Then both melting rates may be calculated. The ratio of the two melting rates is

$$\frac{u^{An}}{u^{Tt}} = \frac{k^{An}}{k^{Tt}} \frac{w_o/w_B - 1}{(1 - w_o)/(1 - w_A) - 1}. \tag{4-177}$$

If there are many grains of anorthite and titanite in a melt matrix, then the total mass dissolution rate of anorthite is

$$u^{An} A^{An} \rho^{An}, \tag{4-178}$$

where A^{An} is the total surface area of anorthite undergoing melting, and the melt composition satisfies

$$w_o = (w_o/w_B - 1)/\{(w_o/w_B - 1) + \lambda[(1 - w_o)/(1 - w_A) - 1]\}, \tag{4-179}$$

where $\lambda = (\rho^{Tt} A^{Tt} k^{Tt})/(\rho^{An} A^{An} k^{An})$.

4.3.5.2 Melting controlled by diffusion in the melt

If the melting is controlled by diffusion in the melt (which means extremely rapid interface reaction rate), the melt composition at the titanite–melt interface would be A, and that at the anorthite–melt interface would be B (Figure 4-35b). Treat the diffusion as binary diffusion. The diffusive flux across the melt at steady state is

$$\mathbf{J} = -\rho D \, \Delta w/\delta, \tag{4-180}$$

where δ is the thickness of the melt layer, Δw = mass fraction difference between point A and point B, and D is the binary diffusivity. The melting rate of anorthite can be expressed as

$$\rho^{An}(w^{An} - w_B)u^{An} = \rho D(w_B - w_A)/\delta, \tag{4-181}$$

where w is the mass fraction of $C_aAl_2Si_2O_8$ component, superscript An means the anorthite phase (the superscript for the melt phase is ignored), ρ is density, and u^{An} is the linear dissolution rate of anorthite crystal. Therefore, the melting rate of anorthite is

$$u^{An} = \frac{\rho}{\rho^{An}} \frac{w_B - w_A}{w^{An} - w_B} \frac{D}{\delta}. \tag{4-182a}$$

Similarly, the melting rate of titanate can be expressed as

$$u^{Tt} = \frac{\rho}{\rho^{Tt}} \frac{w'_A - w'_B}{w'_{Tt} - w'_A} \frac{D}{\delta}. \tag{4-182b}$$

where w' is the mass fraction of the C_aTiSiO_5 component. Because $(w'_A - w'_B = (w_B - w_A)$ and $w'_{Tt} - w_A = w_A$, the ratio of the dissolution rate of the two minerals is

$$\frac{u^{An}}{u^{Tt}} = \frac{\rho^{Tt}}{\rho^{An}} \frac{w_A}{w^{An} - w_B}. \tag{4-183}$$

The mean mass fraction of $CaAl_2Si_2O_8$ in the melt is $\rho^{An}u^{An}/(\rho^{An}u^{An} + \rho^{Tt}u^{Tt}) = w_A/(1 + w_A - w_B)$.

If there are many grains, then by treating δ to be the mean distance between anorthite and titanite grains, the linear melting rate can be estimated using Equations 4-182a,b. The total mass melting rate is given by Equation 4-178, and the mean mass fraction of $CaAl_2Si_2O_8$ in the melt is $\rho^{An}A^{An}u^{An}/(\rho^{An}A^{An}u^{An} + \rho^{Tt}A^{Tt}u^{Tt})$.

4.3.5.3 Melting controlled by diffusion in a crystalline phase

If one of the crystalline phases is a solid solution, the melting may be controlled by diffusion in the crystalline phase. Because partial melting controlled by diffusion in a crystalline phase is extremely slow, it may be easier to accomplish the feat by dissolution and reprecipitation, which is not considered here. The melt composition would be uniform due to the slowness of the reaction. If only one phase is a solid solution, such as partial melting of a plagioclase (An30) and diopside pair at 1200°C in the pseudo-ternary system of albite–anorthite–diopside (Figure 4-36), the rate of partial melting would be controlled by diffusion in this

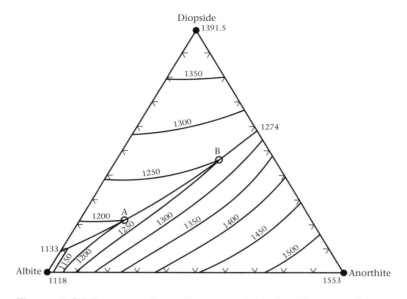

Figure 4-36 Ternary phase diagram of Ab–An–Di. From Morse (1980). Two points on the cotectic curve at 1200 and 1250°C are marked as points A and B.

phase and can be treated using the method in Section 4.2.2.6. If both phases are solid solutions, then the phase with slower diffusion would control the rate of partial melting of the two phases.

4.3.5.4 Reaction between two solid phases in the presence of a fluid phase

Consider a reaction between two solid phases in the presence of a fluid phase

$$A(\text{solid}) + B(\text{solid}) \rightleftharpoons D(\text{solid}) + C(\text{gas or solid}). \tag{4-184}$$

The following metamorphic reaction is of the above type:

$$CaCO_3(\text{calcite}) + SiO_2(\text{quartz}) \rightleftharpoons CaSiO_3(\text{wollastonite}) + CO_2(\text{gas}). \tag{4-185}$$

Suppose there are many calcite grains and quartz grains in a rock. At each surface, there is interface reaction. For metamorphic reactions that take many years, mass transport in the fluid phase may be assumed to be rapid and the reaction rate is controlled by the interface reaction rate. The fluid composition may be regarded as uniform. At the surface of phase A (such as calcite), the interface reaction rate may be written as (Equation 4-34)

$$u^A = k^A(w^A - 1), \tag{4-186}$$

where superscript A means phase A, k^A depends on temperature (see Equation 4-33), and w^A is the degree of saturation. For calcite, $w^A = [Ca^{2+}][CO_3^{2-}]/K_{sp,Cc} < 1$. The interface reaction rate at the surface of phase B can be written similarly:

$$u^B = k^B(w^B - 1), \tag{4-187}$$

Because one mole of calcite reacts with one mole of quartz, the molar reaction rates of phase A and B are identical, which equals the molar growth rate of wollastonite (D). Therefore, the linear reaction rate u^A and u^B are related as follows:

$$-u^A A^A / V^A = -u^B A^B / V^B = u^D A^D / V^D, \tag{4-188}$$

where unit of u is m/s, A^A is total surface area of phase A for the reaction, and V^A is molar volume of phase A. Therefore, we obtain

$$-A^A k^A(w^A - 1)/V^A = -A^B k^B(w^B - 1)/V^B. \tag{4-189}$$

For Reaction 4-185, the above equation becomes

$$-A^A k^A([Ca^{2+}][CO_3^{2-}]/K_{sp,Co} - 1)/V^A = -A^B k^B([SiO_2]/[SiO_2]_{eq} - 1)/V^B. \tag{4-190}$$

Suppose the grain size for phase A is very small, leading to very large total surface area for reaction. That is, $A^A \gg A^B$. Then $(w^A - 1) \ll (w^B - 1)$, meaning the fluid composition is such that it is very close to the saturation composition of phase A, whereas phase B is significantly undersaturated. After solving for w^A and w^B, the reaction rate of A and B can be calculated if k^A and k^B are known.

4.4 Remarks About Future Research Needs

Heterogeneous reactions are the most often encountered reactions in geology. Because of their complexity, the investigation of the kinetics of most heterogeneous reactions is still in its infancy. The complexity comes in various forms. One is in terms of basic understanding: for example, the homogeneous nucleation theory, although well developed, does not work. The second is in terms of the complexity of the treatment: for example, starting from a silicate melt and given a cooling rate (or even isothermal crystallization), the complete description of the kinetics of crystallization (including the location of crystal grains, different minerals, and their growth and coarsening) is next to impossible. The third complexity arises because minor amounts of fluid or other components may significantly affect the reaction rates. The fourth complexity is due to the presence of surfaces and interfaces, whose microscopic characteristics (such as the presence of kinks) influence the reaction rates. Because of these difficulties, currently scientists cannot even predict the kinetics of a simple phase transition such as diamond to graphite, or gas exsolution from magma. This lack of understanding calls for concentrated efforts by talented geochemists. Because heterogeneous reaction kinetics is essentially the goal of igneous petrology, metamorphic petrology, and volcanology, it is hoped that much more work will be focused on these areas.

In terms of theoretical advancement of fundamental principles related to the kinetics of heterogeneous reactions, the most important is to understand nucleation rates so that the rates can be quantitatively predicted given the initial conditions. For example, the understanding of explosive eruption dynamics critically depends on bubble nucleation rates. The key problem is probably homogeneous nucleation because such a theory is not necessarily mathematically complicated but it requires new insight into the nucleation process. On the other hand, heterogeneous nucleation may be understood with the help of homogeneous nucleation theory, but quantitative prediction of heterogeneous nucleation rate is likely very complicated because of the requirement of knowing the types and numbers of interfaces present. In addition to nucleation, another basic problem is to quantitatively understand the Ostwald reaction principle so that given any system, the sequences of new phases to form can be predicted a priori.

In terms of understanding crystallization and bubble growth, it is necessary to deal with the mathematically complicated many-body problems. The work by Proussevitch et al. (1993) and Proussevitch and Sahagian (1998) is one significant advancement. In the future, more complicated many-body problems will have to be dealt with, most likely through brute force computer simulations. Simplifications will be necessary, but the critical complexities, such as size distribution and different kinds of minerals, must be included so that the simulations are realistic. The modeling of such many-body problems has many applications, including the dynamics of explosive eruptions.

Because interfaces are often where the actions are during heterogeneous reactions, microscopic characterization of surface properties and their effect on reaction rates will be vital to the understanding and quantification of these effects. Development in scanning probe microscopy, such as the scanning tunneling microscope (STM) and the atomic force microscope (AFM), is making such characterization possible. Major breakthroughs in our understanding of the kinetics of heterogeneous reactions will likely come from such investigations.

Experimental investigations will continue to play a critical role in understanding heterogeneous reaction kinetics, such as mineral dissolution rates in silicate melts and in aqueous solutions, the melting rates at the interfaces of two minerals and at the intersection of three minerals (either pure phases or solid solutions), and solid phase reactions. The experimental conditions to be varied include temperature, pressure, the presence or absence of a fluid phase, as well as various compositional parameters of the melt or aqueous solution. Furthermore, it is critical to examine the conditions for a reaction path to be operative. These experimental data will ground truth sophisticated numerical models, and will allow, e.g., the inference of the controlling mechanisms of the given reaction under different conditions.

Problems

4.1 Consider homogeneous nucleation of crystals using the classic nucleation theory. Suppose the molar volume of the crystal is $V_c = 50\ \mathrm{cm^3/mol}$, the melting entropy $\Delta S_{m-c} = 50\ \mathrm{J\ K^{-1}\ mol^{-1}}$, the surface energy $\sigma = 0.2\ \mathrm{J/m^2}$, the melting temperature $T_e = 1700\ \mathrm{K}$, the melt is at a temperature of 1500 K, and the activation energy for nucleation is 250 kJ/mol.

  a. Plot ΔG (the formation of a cluster) versus the cluster radius r.

  b. Calculate the critical nucleus size of a crystal.

  c. Calculate the nucleation rate.

4.2 This problem concerns the homogeneous nucleation of H_2O bubbles in rhyolitic melt using the classic nucleation theory. Suppose the surface energy $\sigma = 0.2\ \mathrm{J/m^2}$, the temperature is 1100 K, the saturation pressure $P_e = 100\ \mathrm{MPa}$, the pressure of the melt is 80 MPa, and the activation energy for nucleation is 250 kJ/mol. Use the relation of $r^* \approx 2\sigma/(P_e - P)$. Partial molar volume of H_2O in melt may be calculated as $22.89 + 0.00964(T - 1273.15) - 0.00315(P - 0.1)\ \mathrm{cm^3/mol}$, where T is in K and P is in MPa (Ochs and Lange, 1999).

  a. Calculate the critical nucleus size of a bubble.

  b. Calculate the nucleation rate.

4.3 This question is about heterogeneous nucleation of orthopyroxene on olivine–melt interface in a basaltic melt. Suppose $\sigma_{opx/melt} = 0.35\ \mathrm{J/m^2}$, $\sigma_{oliv/melt} = 0.3\ \mathrm{J/m^2}$, and $\sigma_{oliv/opx} = 0.05\ \mathrm{J/m^2}$. Find the contact angle and the modified interface energy σ'.

4.4 The following are experimental data for diopside melting in its own melt (Kuo and Kirkpatrick, 1985). Diopside melting temperature is 1664 K. The fusion entropy is $82.76\ \mathrm{J\ mol^{-1}\ K^{-1}}$. Assume that the activation energy is 300 kJ/mol.

a. Use the data to estimate the pre-exponential factor A in Equation 4-33.

b. Using the parameter A obtained above, extrapolate to estimate the interface reaction rate during crystallization when temperature is 100 K below the melting temperature of diopside.

c. Using the parameter A obtained above, estimate and plot the interface reaction rate during crystallization as a function of temperature from 1664 to 1400 K.

$\Delta T\ (T - T_m)$ (K)	2	5	8	12	16	21
Melting rate (μm/s)	4.0	11	17	27	35	49

4.5 Suppose crystal growth rate is constant with growth rate $u = 0.1\ \mu$m/s. For a trace element, the initial concentration in the melt is 10 ppm, and the partition coefficient between the crystal and the melt is 2.5. The diffusion coefficient in the melt is $10\ \mu\text{m}^2$/s.

a. If steady state is reached, plot the concentration profile of the trace element.

b. Plot the concentration profile at $t = 1$ s, 10 s, 100 s, 500 s, 1000 s, 2000 s, and 5000 s.

4.6 Suppose effective binary diffusivity of MgO in basaltic melt is $D = 10\ \mu\text{m}^2$/s, MgO in the initial melt is 8 wt%, MgO concentration in olivine is 48 wt%, and the estimated MgO concentration at the interface melt is 11.5 wt%. Density of olivine is $3300\ \text{kg/m}^3$ and that of the melt is $2720\ \text{kg/m}^3$.

a. Find the diffusive olivine dissolution rate and melt growth rate. Find olivine dissolution distance in two hours.

b. Calculate the MgO concentration profile in olivine and in the melt at $t = 1$h. Assume that MgO concentration in the interface olivine is 45 wt%.

4.7 Suppose effective binary diffusivity of MgO in basaltic melt is $D = 10\ \mu\text{m}^2$/s, MgO in the initial melt is 8 wt%, MgO concentration in olivine is 48 wt%, and the estimated MgO concentration at the interface melt is 11.5 wt%. Density of olivine is $3300\ \text{kg/m}^3$ and that of the melt is $2720\ \text{kg/m}^3$. The viscosity of the melt is 10 Pa·s. Olivine radius is 2 mm and is freely falling in the melt. Find the convective olivine dissolution rate, and olivine dissolution distance in two hours.

4.8 If mass transfer in the melt controls olivine growth and dissolution rate, explain why there may be dendritic growth but no dendritic dissolution.

4.9 Suppose coarsening is controlled by diffusion in the melt phase. If mean crystal size grew from 0.1 to 0.11 mm in 10 years, estimate the time needed for the mean crystal size to grow from 5 to 5.1 mm.

4.10 You drop a halite crystal (common salt) in water at 20°C. It sinks in water and dissolves. Suppose the radius of halite crystal is 1 mm. Density of halite is 2165 kg/m^3. Solubility of halite in water is 0.359 kg of halite per kilogram of water, leading to 26.4 wt% NaCl in water at equilibrium. Density of NaCl solution is $\rho = \rho_{water} + 750w$, where $\rho_{water} = 998$ kg/m^3 and w is mass fraction of dissolved NaCl. Diffusivity of NaCl in water may be expressed as $D = \exp(-13.73 - 1950/T)$ m^2/s. Viscosity of water is 0.0010 Pa·s. Ignore the effect of dissolved NaCl on the diffusivity and viscosity (which causes some error in the calculation below).

a. Find the falling velocity of the halite crystal.

b. Find the convective dissolution rate of halite as it descends in water.

c. Change the radius of halite crystal to 0.1 mm. Find the convective dissolution rate.

4.11 Consider the growth of a rising CO_2 bubble in beer at 10°C. Suppose the bubble diameter is 0.2 mm. Beer density is 1010 kg/m^3. CO_2 concentration in beer is 0.1 mol/L, and the solubility at 1 atm is 0.053 mol/L. Viscosity of beer is 0.0019 Pa·s. Diffusivity of CO_2 in beer at this temperature is 1.33×10^{-9} m^2/s. Find the growth rate of the rising CO_2 bubble.

4.12 Suppose the mass fraction of the gas phase (primarily H_2O steam) is 0.04 for an explosive rhyolitic eruption at 850°C. Assume that the exit eruption velocity is the sound speed of the magma–gas mixture. Estimate the eruption velocity as the eruption exits the ground surface.

4.13 A melt inclusion and its olivine host are concentric spheres of radius R_1 and R_2. Consider H_2O in the melt inclusion and its re-equilibration with ambient melt outside olivine. Ignore anisotropy of olivine. If $R_1 = 0.1$ mm, $R_2 = 1$ mm, H_2O diffusivity in olivine is $D = 10^{-14}$ m^2/s, $K = 0.0001$, find the reequilibration timescale.

4.14 Consider Ca in a basaltic melt inclusion concentric with the host mineral olivine. The radius of the inclusion is 50 μm. The olivine radius is 1 mm. The Ca partition coefficient between olivine and melt is estimated to be 0.01. Assume that Ca diffusivity in olivine is 10^{-17} m^2/s. Find the reequilibration timescale.

4.15 Consider He in a fluid inclusion in olivine in a basaltic melt. The radius of the inclusion is 50 μm. The olivine radius is 1 mm. The He partition coefficient between olivine and melt is estimated to be 0.0001. Assume that He diffusivity in olivine is 10^{-12} m^2/s. Find the reequilibration timescale.

5 Inverse Problems: Geochronology, Thermochronology, and Geospeedometry

All geochemical tools involving time and rates are based on kinetics, including all isotopic dating methods (geochronology), thermochronology, and geospeedometry. These are also inverse problems in geochemical kinetics. They have been touched upon in respective sections in previous chapters dealing with forward problems. In this chapter, we systematically expound upon these inverse problems.

Radioactive decay and radiogenic growth are two sides of a single process. The kinetics of decay and growth provides the most powerful method for determining the age of rocks and minerals. There are many naturally occurring unstable nuclides undergoing α-decay or β-decay. The decays are first-order reactions. Because each decay constant is independent of temperature, pressure, and the chemical environment (with the exception of electron capture, Section 1.3.8.1), the decay and growth provide a record of time. Inferring the age of a mineral, a rock, a tree, or a painting (*geochronology*) through radioactive decay and radiogenic growth is the most widely applied inverse problem in geochemical kinetics. Furthermore, the initial conditions (such as the initial isotopic ratios) may also be obtained in such inversion. The initial conditions provide crucial information on the origin of the rocks and the history of the materials prior to the formation of the rocks. The applications of the initial conditions such as the initial isotopic ratio are not discussed much in this book.

Although the radioactive decay constants are independent of temperature and pressure, the retention of the radiogenic daughter in a mineral depends strongly on temperature because at high temperatures diffusivity is high, resulting in diffusive loss. Hence, the cumulative amount of radiogenic daughter depends not only on the content of the radioactive parent, time, and the decay constant, but also on the transport of the daughter. The loss or gain of the parent (radioactive nuclide), which also affects age determination, is not discussed much because the parent is often a structural component of the mineral. Therefore, if the cumulative amount of radiogenic daughter is utilized in dating, it is essential to understand the effect of diffusive transport of the daughter nuclide as a function of temperature on dating. If cooling after the formation of the mineral is rapid, loss may be negligible and the formation age can be determined. On the other hand, if cooling after the formation of the mineral is slow, significant diffusive loss of the daughter may occur, depending on the temperature, cooling rate, radiogenic species, and mineral. The temperature dependence of diffusive loss of the daughter means the radiogenic isotope in the mineral records temperature information. Because radiogenic growth also carries age information, the radiogenic growth and diffusive loss of daughter nuclides may yield information on both age and temperature, which is the basis of *thermochronology* (temperature–time history).

Chemical reaction rates (or more specifically, the rate coefficients) depend strongly on temperature. Hence, the total extent of a reaction recorded by a mineral or rock depends on the integrated thermal history. For example, the mean diffusion distance is proportional to $(\int D \, dt)^{1/2}$, where D is diffusivity that depends on temperature; the extent of a unidirectional first-order reaction is related to $\int k \, dt$, where k is the rate coefficient, also depending on temperature. Therefore, the extent of a reaction may be related to the thermal history, or more specifically, the cooling timescale or cooling rate. By investigating the extent of chemical reactions (including homogeneous reactions and heterogeneous reactions) and diffusion, it may be possible through inversion to obtain the cooling rate of a rock, which is referred to as *geospeedometry*.

Although reaction rate coefficients and diffusivities depend somewhat on pressure, the major cause of variation is temperature. Hence, inverse problems in geochemical kinetics are insensitive to pressure and cannot in general constrain pressure history. Some rate coefficients may depend strongly on H_2O content. In such cases, if the H_2O content and its variation with time are independently inferred, temperature history may be constrained by geochemical kinetics. Otherwise, at least in theory, if temperature history is independently inferred, H_2O content history may be constrained by geochemical kinetics. The variation of pressure with time, or the uplift history or erosion rate, may be indirectly inferred from thermal history, either by assuming a typical geothermal gradient, or through more advanced thermo-kinematic modeling.

Essentially all thermochronology and geospeedometry modeling starts by assuming a cooling history. The asymptotic cooling function $T = T_0/(1 + t/\tau_c)$ is often chosen, where τ_c is a characteristic time for cooling. This cooling function allows Arrhenian kinetic coefficients (including rate coefficients and diffusion coefficients) to be expressed as a simple exponential function of time such as $k = k_0 e^{-t/\tau}$, which sometimes allows analytical solution to be obtained or some simplification of the kinetic problem so that insights may be gained. If numerical methods are adopted to solve a given kinetic problem, then the thermal history could be any function, such as an exponential function $T = T_\infty + (T_0 - T_\infty) \exp(-t/\tau_c)$.

5.1 Geochronology

One of the most important tools of geochemistry (also one of the most important contributions of geochemistry to geology and other sciences) is geochronology, the science of dating ancient and recent events and processes. Many geochronology methods have been developed based on the radioactive decay of an unstable nuclide, or based on the growth of the stable daughter nuclide. Here, we first briefly review the principles of radioactive decay and radiogenic growth. The various methods are covered subsequently.

The decay (α- and β-decay) of an unstable nuclide can be described by the decay equation (first-order reaction):

$$dP/dt = -\lambda P, \tag{5-1}$$

where P is the number of atoms (or moles) of the unstable parent nuclide at time t, $-dP/dt$ is the decay rate (the unit is, e.g., number of decays per second per gram of material), which is proportional to the number of atoms (i.e., every atom has the same probability to decay) of the radioactive nuclide. The parameter λ is a constant, called the decay constant. Integration of the above equation leads to

$$P = P_0 e^{-\lambda t}, \tag{5-2}$$

where the subscript "0" means the initial number of atoms of the radioactive parent. That is, the number of atoms of the unstable nuclide decreases with time exponentially; and the decay constant λ characterizes the decay rate. Often, the half-life ($t_{1/2}$) is also used to characterize the decay timescale. The half-life is the time required to reduce the number of atoms of the parent nuclide by 50%. In other words, after one half-life, the number of atoms of the unstable nuclide is reduced to 1/2 of the initial number. After n half-lives, the number of atoms of the unstable nuclide is reduced to $(1/2)^n$ of the initial number. For example, after 10 half-lives, the number of atoms of the unstable nuclide is reduced to $1/1024 \approx 0.1\%$ of the initial number. Because this small fraction of residual atoms is often difficult to determine, 10 half-lives is often considered to be the time required for the unstable nuclide to become "extinct." The relation between half-life and decay constant is (Equation 1-58)

$$t_{1/2} = \ln(2)/\lambda. \tag{5-3}$$

In geology, often, the present-day number of parent atoms can be measured, but the initial number of the parent atoms needs to be inferred (i.e., inverse problems). Then Equation 5-2 is rewritten as

$$P_0 = Pe^{\lambda t}. \tag{5-4}$$

The decay equation can also be expressed in terms of the radioactive activity (A), i.e., the number of decays per unit time per unit mass of sample. By definition, activity is the same as the decay rate, and can be written as

$$A = -dP/dt. \tag{5-5}$$

Combining the above with Equation 5-1, the activity can be expressed as

$$A = \lambda P, \tag{5-6}$$

One can easily derive

$$dA/dt = -\lambda A, \tag{5-7}$$

and

$$A = A_0 e^{-\lambda t}. \tag{5-8}$$

One unit for A is dpm (number of disintegrations per minute). Another unit is the curie (Ci), which is defined as 3.7×10^{10} disintegrations per second.

The other side of radioactive decay is radiogenic growth. As the parent decays away, the number of atoms of the daughter increases. Let D denote the number of daughter atoms (note that D is not diffusivity here), then D grows as

$$D = D_0 + D^\star = D_0 + n(P_0 - P) = D_0 + nP_0(1 - e^{-\lambda t}), \tag{5-9}$$

where D_0 is the initial number of daughter atoms, $D^\star$ is radiogenic contribution, and n is the number of daughter nuclides produced per parent nuclide. The value of n typically equals 1, such as in the ^{87}Rb–^{87}Sr, ^{147}Sm–^{143}Nd, ^{176}Lu–^{176}Hf, and ^{187}Re–^{187}Os systems, but in special cases, it can be either smaller or greater than 1. One special case is for the decay of ^{40}K to ^{40}Ar, for which $n = 0.1048$ because only 10.48% of ^{40}K decays to ^{40}Ar, and the rest to ^{40}Ca. Another special case is for the production of ^{4}He from ^{238}U, ^{235}U, and ^{232}Th, for which $n = 8$ (^{238}U undergoes 8 α-decays to ^{206}Pb), 7, and 6, respectively. From Equation 5-9, the number of daughter atoms at any time in the future (t) can be calculated in forward calculations.

In inverse problems, the present is the key to infer the past (including the age and the initial conditions). That is, given a rock, one measures the present-day number of atoms, from which one would like to infer the age as well as the initial amount. To facilitate this, Equation 5-9 is written in the following form:

$$D = D_0 + nP(e^{\lambda t} - 1), \tag{5-10}$$

in which D and P are the present-day number of the daughter and parent atoms (which can be measured). Because there are no simple ways to replace D_0 using a known parameter, there are two unknowns in the above equation, the age t and the initial concentration D_0. Geochemists developed ways to obtain both the initial condition and the age.

For use in geochronology, the decay constant of a radioactive nuclide must be constant and must be accurately known. For α-decay and most β-decays, the decay constant does not depend on the chemical environment, temperature, or pressure. However, for one mode of β-decay, the electron capture (capture of K-shell electrons), the decay "constant" may vary slightly from compound to compound, or with temperature and pressure. This is because the K-shell (the innermost shell) electrons may be affected by the local chemical environment, leading to variation in the rate of electron capture into the nucleus. The effect is typically small. For example, for ^7Be, which has a small number of electrons and hence the K-shell is easily affected by chemical environments, Huh (1999) showed that the decay constant may vary by about 1.5% relative (Figure 1-4b). Among decay systems with geochronological applications, the branch decay constant of ^{40}K to ^{40}Ar may vary very slightly (<1% relative).

Based on the principles outlined above, several methods for age determination have been developed. If the initial number of the parent nuclides can be guessed, then Equation 5-2 may be used to obtain the age. If the initial number of the daughter nuclides can be guessed, then Equation 5-10 may be applied to obtain the age. If none can be guessed, then both the age and the initial conditions must be inferred. Geochemists developed the *isochron method* to achieve this. All three methods are discussed below. Furthermore, some nuclides that are now extinct but were present in the early solar system may also be employed to determine relative age, which will be presented as the fourth dating method.

5.1.1 Dating method 1: The initial number of parent nuclides may be guessed

The simplest method is based on Equation 5-2, in which the initial number of parent atoms is assumed to be known. Therefore, the calculation of the age is simple:

$$\text{Age} = t = \frac{1}{\lambda} \ln \frac{P_0}{P} \tag{5-11}$$

This method is used mainly for short-lived radioactive nuclides produced by cosmic ray spallation, such as ^{14}C, ^{10}Be, ^{26}Al, ^{32}Si, ^{36}Cl, and ^{39}Ar (Table 5-1). Because these nuclides have relatively short half-lives, if there was any initial amount of the nuclides at the beginning of Earth history, the initial amount would have completely decayed away. The small amount that can be found in

Table 5-1 Cosmogenic radionuclides

Nuclide	# of Protons	# of Neutrons	Daughter	Half-life (yr)	Decay Constant (yr^{-1})
^{3}H	1	2	^{3}He	12.43	0.05576
^{10}Be	4	6	^{10}B	1.51×10^6	4.59×10^{-7}
^{14}C	6	8	^{14}N	5730	1.2097×10^{-4}
^{26}Al	13	13	^{26}Mg	7.1×10^5	9.8×10^{-7}
^{32}Si	14	18	^{32}S	140	0.00495
^{36}Cl	17	19	^{36}S	3.01×10^5	2.30×10^{-6}
^{39}Ar	18	21	^{39}K	269	0.00258
^{41}Ca	20	21	^{41}K	1.04×10^5	6.66×10^{-6}
^{53}Mn	25	28	^{53}Cr	3.7×10^6	1.87×10^{-7}
^{81}Kr	36	45	^{81}Br	2.3×10^5	3.01×10^{-6}
^{129}I	53	76	^{129}Xe	1.57×10^7	4.41×10^{-8}

surface environments today is produced by the bombardment of cosmic ray particles on nuclides in the atmosphere and on the surface of the Earth. (Cosmic ray spallation also produces many other stable nuclides, but the amount is negligible compared to the amount already present except for ^{3}He.) If the production rate in the atmosphere does not vary much with time, the concentration in the atmosphere reaches a steady state between production, decay, and exchange with other reservoirs. Once the nuclides are removed from the atmosphere and surface environment, and incorporated into plant, sediment, coral, ice, etc., the nuclides decay away. Therefore, the age of a tree, sediment, or ice can be determined by measuring the present-day parent nuclide concentration, and comparing it with estimated initial concentration. These cosmogenic nuclides have found many applications in Earth sciences.

5.1.1.1 Radiocarbon dating

There are two stable isotopes of carbon: ^{12}C and ^{13}C. The unstable ^{14}C is produced in the atmosphere by cosmic ray bombardment, and the main reaction is

$$^{14}\text{N} + {}^1\text{n} \rightarrow {}^{14}\text{C} + {}^1\text{H}, \tag{5-12}$$

where ^{1}n is a neutron and ^{1}H is a proton. ^{14}C (also called radiocarbon) so produced becomes part of atmospheric carbon in the form of CO_2. ^{14}C undergoes β-decay to ^{14}N with a half-life of 5730 years and a decay constant of 0.00012097 yr^{-1}.

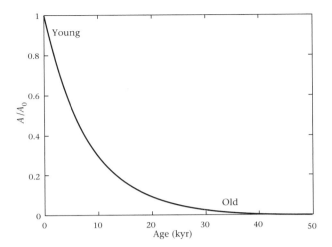

Figure 5-1 Relation between $^{14}C/^{14}C_0$ and age. A high activity or a high $^{14}C/^{12}C$ ratio means that C in the sample has not been separated from the atmosphere reservoir for long, and hence means a younger age. A low activity or a low ratio means an old age. Although this sounds trivial at this moment, it could become confusing when dating using extinct nuclides is discussed.

In the atmosphere, the concentration of ^{14}C is controlled by the production, the decay, and the exchange and cycle of ^{14}C between atmosphere and other reservoirs such as oceans and the biosphere. The steady-state concentration is roughly 6×10^{10} atoms of ^{14}C per gram of carbon in the pre-industrial atmosphere, and may vary slightly with time due to various reasons (see below). In simple calculations, the variation is ignored. In more accurate calculations, the variation must be taken into account.

When a new piece of a tree or a coral grows, it takes carbon (including ^{14}C) from the atmosphere or ocean. The concentration of ^{14}C a tree takes in is smaller than that in the atmosphere on a per-gram of carbon basis, because of carbon isotope fractionation. In age calculation, this effect has to be accounted for. Once the piece of the tree stops exchanging carbon with the atmosphere, ^{14}C simply decays away (Figure 5-1). Hence, from the measured ^{14}C concentration in the piece of the tree at present, the age (time elapsed since the piece stopped exchange with the atmosphere) can be calculated using Equation 5-11:

$$\text{Age} = t = \frac{1}{\lambda} \ln \frac{^{14}C_0}{^{14}C}, \tag{5-12a}$$

or

$$\text{Age} = t = \frac{1}{\lambda} \ln \frac{A_0}{A}, \tag{5-12b}$$

where A is activity of ^{14}C. The initial concentration of ^{14}C may be obtained by measuring ^{14}C in tree or coral of "zero age," assuming ^{14}C concentration does not vary with time. The definition of "zero age" is discussed later. Knowing the initial ^{14}C activity and the present-day ^{14}C activity, the ratio of A/A_0 or $^{14}C/^{14}C_0$ can be calculated, and the age can hence be calculated. The ^{14}C method can provide dates back to about 40,000 years before present.

^{14}C measurements may be made by counting β-particles emitted by the decay of ^{14}C (which provides the activity A), or by linear accelerator mass spectrometer (AMS) measurement of ^{14}C atoms (which provides concentrations). The development of linear accelerator mass spectrometer has significantly increased the sensitivity of ^{14}C measurements, leading to an increase of the limit of ages to be dated, but more significantly to a decrease of sample size to be measured and the measurement time.

Details of ^{14}C dating There is a strange and confusing convention in ^{14}C dating. Although it is accepted that the most accurate half-life is 5730 years (mean life of 8267 years and decay constant of $0.00012097 \text{ yr}^{-1}$), an older value of the half-life of 5568 years (and hence a mean life of 8033 years and a decay constant of $0.00012449 \text{ yr}^{-1}$, 2.9% different from the correct value) is conventionally used in calculating the ^{14}C *age*. This convention is adopted to avoid confusion between age reported in old literature and that reported in new literature.

The concentration of ^{14}C in the atmosphere has changed over time for various reasons. From 1955 to 1963, nuclear bomb tests almost doubled ^{14}C concentration in the atmosphere, which then decreased mostly due to exchange with other reservoirs, including the dilution from fossil fuel carbon combustion. Hence, the initial atmospheric concentration used in ^{14}C dating is pre-1950 ^{14}C concentrations. Prior to 1950, the burning of fossil fuel carbon, which preferentially sends ^{14}C-free carbon to the atmosphere, lowered the concentration of ^{14}C. This is referred to as the *Suess effect*. Before fossil fuel burning by humans, there were also small natural fluctuations of ^{14}C concentration in the atmosphere due to fluctuation in solar wind intensity and in carbon cycle, which must be accounted for in accurate calculation of the age. By convention, this is done by first calculating a ^{14}C age using an assumed initial ^{14}C concentration (or activity) in the year of 1950, and then correcting the ^{14}C age to the real age (called the *calibrated age*) using internationally adopted calibration curves as below.

^{14}C ages are reported relative to the reference year of 1950, or before present (BP), where 1950 is defined as the present. In the calculation of the ratio A/A_0 and hence of the ^{14}C age, the initial ^{14}C activity (or concentration) is based on measurement of an NBS standard, which is an oxalic acid with activity of A_{ox}. Repeated analyses of this standard show that the initial ^{14}C activity in 1950 may be related to the activity of this standard as

$$A_0 = 0.95 A_{ox} e^{(y-1950)/8267},$$

(5-13a)

where y is the year of measurement and the factor $e^{(y-1950)/8267}$ is to correct the activity to the year of 1950. However, because the combustion of the oxalic acid causes fractionation of C isotopes, and the sample to be measured may also be fractionated in terms of C isotopes, further correction is needed. For the standard,

$$A_{0,\,corr} = 0.95A_{ox}^m[1 - 2(19‰ + \delta^{13}C_{ox}‰)]e^{(y-1950)/8267}, \qquad (5\text{-}13b)$$

where A_{ox}^m is the measured activity in the NBS standard after combustion, $-19‰$ is the assumed $\delta^{13}C$ value of the oxalic acid, $\delta^{13}C_{ox}$ is the measured value of the CO_2 gas prepared from the oxalic acid, and the factor 2 is because fractionation in terms of $\delta^{14}C$ is twice that of $\delta^{13}C$. For the sample, the activity is corrected for isotope fractionation as

$$A_{smp,\,corr} = A_{smp}[1 - 2(25‰ + \delta^{13}C_{smp}‰)]e^{(y-1950)/8267}, \qquad (5\text{-}14)$$

where $-25‰$ is the assumed $\delta^{13}C$ value of wood. Finally, the ratio of A/A_0 is calculated as $A_{smp,corr}/A_{0,corr}$. The factor $e^{(y-1950)/8267}$ is in both the sample and standard and hence cancels out.

For AMS analyses, the $^{14}C/^{12}C$ ratio is measured and divided by the ratio in the standard. Then correction for the measurement background is made. Further correction uses a multiplier accounting for the factors in Equation 5-13b, including correction of $\delta^{13}C$ from -19 to $-25‰$, leading to $(1/0.95)[(1 - 25‰)/(1 - 19‰)]^2 = 1.0398$. After multiplying the normalized $^{14}C/^{12}C$ ratio by this factor, the ratio $(^{14}C/^{12}C)_{smp,corr}/(^{14}C/^{12}C)_{0,corr}$ is used to calculate the ^{14}C age.

The ^{14}C age is calculated using Equation 5-12 with the wrong decay constant of $1/8033$ yr^{-1}:

$$^{14}C \text{ age} = -8033\ln(\text{ratio}). \qquad (5\text{-}15)$$

where "ratio" may be either $[(^{14}C/^{12}C)_{smp}/(^{14}C/^{12}C)_0]_{corr}$ or $(A_{smp}/A_0)_{corr}$. This ^{14}C age is given as years before present. Because of the variation of ^{14}C concentration in the atmosphere with time and because the decay constant for arriving at this age is the wrong decay constant, ^{14}C age is not the real age, and it needs to be calibrated.

To obtain the real age from the ^{14}C age, tree rings and corals have been used. The principles are as follows. The age of each ring of a tree may be obtained by counting the number of growth rings. The ^{14}C age can be obtained by measuring ^{14}C concentration in the ring. Plotting the ^{14}C age versus the tree ring age (which is the accurate age) from various samples provides a calibration curve. For ages older than trees, corals may be used. The real age of a layer of coral is determined by U–Th disequilibrium dating (see Section 5.1.5), and the ^{14}C age is obtained by ^{14}C measurement of the same piece of coral. Thus, ^{14}C age can be converted to the real age (also called the calibrated age) through the calibration. The most recent calibration curve (IntCal04) adopted by the radiocarbon community is

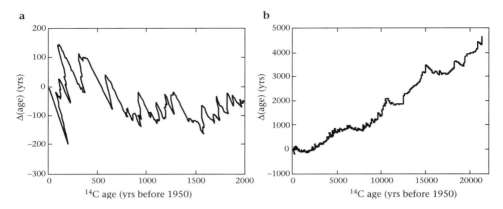

Figure 5-2 Δ(age) versus ^{14}C age, where Δ(age) = calibrated age $-$ ^{14}C age. That is, given ^{14}C age, one can first find Δ(age), and then add it to the ^{14}C age to obtain the calibrated age. Data are from the IntCal04 of Reimer et al. (2004). More detailed calibration curves can be found in Reimer et al. (2004).

shown in Figure 5-2, where the age difference (calibrated age minus ^{14}C age) Δ(age) is plotted against the ^{14}C age. Hence, after finding the ^{14}C age, one uses the calibration curve in Figure 5-2 to find Δ(age), and adds Δ(age) to the ^{14}C age to obtained the calibrated age. Because of the complicated fluctuation of initial ^{14}C concentration, there may be multiple age solutions for a given ^{14}C analysis.

If the real age and the measured present-day ^{14}C concentration are known, the initial ^{14}C concentration can be calculated using Equation 5-4. Thus, how the initial ^{14}C concentration varied with time can be obtained. Figure 5-3 shows such variations (Δ^{14}C) as a function of time: for the last 1000 years Δ^{14}C is small (a couple of percent), but for earlier times (older ages) the initial Δ^{14}C systematically increases with age to 25,000 years ago, though with some fluctuations.

> ***Example 5.1.*** ^{14}C in a sample is determined by AMS. The $(^{14}C)_{smp}/(^{14}C)_{std}$ is found to be 0.1629 ± 0.0004. The measurement background is 0.00332 ± 0.00010. The multiplier factor is 1.0398. Find the ^{14}C age and the calibrated age.
>
> *Solution*: First, the measured $(^{14}C)_{smp}/(^{14}C)_{std}$ ratio must be reduced by the background of the measurement to obtain $(0.1629 - 0.00332)/(1 - 0.00332) = 0.1601 \pm 0.0004$. (The factor $1 - 0.00332$ accounts for reduction in the measurement of the standard.) The result is multiplied by the multiplier 1.0398, resulting in $(^{14}C)_{smp}/(^{14}C)_{std}$, which is 0.1665 ± 0.0005. Hence, $\Delta^{14}C = -833.5‰$. ^{14}C age is then calculated as
>
> $$^{14}\text{C age} = 8033 \ln(1/0.1665) = 14,401 \pm 25 \text{ years BP}.$$

Using the IntCal04 table (or Figure 5-2), the age correction is +2834 years. Hence, the calibrated age is $17,235 \pm 25$ years BP (before 1950).

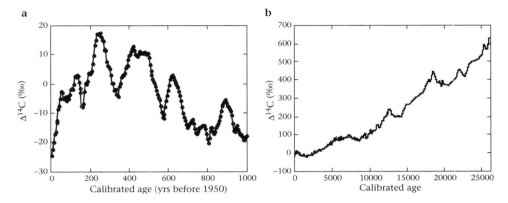

Figure 5-3 Δ^{14}C as a function of calibrated age. Δ^{14}C $= (^{14}$C$_{0,t}/^{14}$C$_{std} - 1)1000$‰, where ^{14}C$_{0,t}$ means initial ^{14}C concentration at age t. Data are from Reimer et al. (2004).

5.1.1.2 ^{10}Be dating; sedimentation rate

In addition to ^{14}C, cosmic ray spallation also produces many other radioactive nuclides. ^{10}Be is another example. Once cosmogenically produced, atoms of ^{10}Be are rapidly removed from the atmosphere by meteoric precipitation, and are absorbed onto surfaces of solid particles such as clay minerals. Hence, newly formed marine sediment contains some initial concentration of ^{10}Be. After removal from the atmosphere, the concentration of ^{10}Be in sediment decays away by β-decay to ^{10}B with a half-life of 1.51 million years (and a decay constant of 4.59×10^{-7} yr^{-1}).

There are a couple of applications of ^{10}Be. One is to determine the sedimentation rate assuming ^{10}Be production rate is constant. Given the decay law

$$^{10}\text{Be} = {}^{10}\text{Be}_0 e^{-\lambda t},$$

and assuming a constant sedimentation rate u, then age $= t = z/u$, where z is the depth of the sediment, and the above equation becomes

$$^{10}\text{Be} = {}^{10}\text{Be}_0 e^{-\lambda z/u}, \tag{5-16}$$

or

$$\ln(^{10}\text{Be}) = \ln(^{10}\text{Be}_0) - (\lambda/u)z. \tag{5-17}$$

This is another example that geochemists love to turn a relation into a linear equation. Note that ^{10}Be$_0$ for a different layer reflects the initial ^{10}Be concentration at a different time. As long as ^{10}Be$_0$ and u are roughly constant and there is no disturbance after deposition, $\ln(^{10}$Be$)$ versus the sediment depth z would be a straight line. That is, in this method, it is not necessary to know the initial ^{10}Be$_0$ concentration; but the concentration may be inferred from the intercept in such a plot. Complexities of this method include (i) variation in the initial ^{10}Be concentration in each layer of sediment with time (^{10}Be$_0$ concentration varies by about a factor 2, Frank et al., 1997), (ii) variation of the sedimentation rate with time, and (iii) the perturbation of sediments (such as subduction-zone accretionary prisms).

Another application, which is a very special application and has been instrumental in elucidating the subduction and recycling process, is to investigate ^{10}Be signal in subduction zone volcanic rocks to infer how subducted sediment came back to the surface through volcanism. It has been shown that subduction zone volcanic rocks contain ^{10}Be. Because all presently measurable ^{10}Be is cosmogenically produced in the atmosphere, any ^{10}Be measured in volcanic rocks also comes from the surface reservoirs, either by near-surface contamination, or from mixing of materials derived from subducted sediment into deep magmagenic zone. If ^{10}Be in volcanic rocks is due to surface contamination, the ^{10}Be/^{9}Be ratio would vary randomly for a suite of cogenetic volcanic rocks. However, if ^{10}Be comes from the magma chamber (meaning ^{10}Be is mixed into the magma before eruption), ^{10}Be/^{9}Be ratio would be roughly constant (isotopic fractionation effect is small) in cogenetic rocks. Measurements show that in many cases, ^{10}Be/^{9}Be ratio in a suite of volcanic rocks is constant (e.g., Ryan and Langmuir, 1988). Therefore, the studies demonstrate that in the short lifetime of ^{10}Be (half-life 1.51 million years), sedimentary ^{10}Be is able to subduct into the mantle, participate in the mantle partial melting process beneath island/continental arcs (most likely by going into dehydration fluids that rises to the hot overlying mantle to cause mantle partial melting), and be brought to the surface by volcanic eruptions. This is one more line of evidence supporting the framework of plate tectonics. Furthermore, the partition of ^{10}Be in the mantle partial melting process in subduction zones helps us to understand the mechanism of partial melting in such environments.

5.1.1.3 Dating using U decay series

Radioactive decay series may be viewed as chain reactions. Section 2.2.1 summarized the three decay series: ^{238}U series, ^{235}U series, and ^{232}Th series. In the ^{238}U decay series, three intermediate nuclides (^{234}U, ^{230}Th, and ^{226}Ra) have half-lives longer than 1000 years. In the ^{235}U decay series, there is only one intermediate nuclide (^{231}Pa) with half-life longer than 1000 years. In the ^{232}Th decay series, there is no intermediate nuclide with half-life longer than 10 years.

If the decay series is not disturbed, secular equilibrium will be reached after a duration of about 10 times the longest half-life of the all the intermediate nuclides, which means 2.4 Myr for the ^{238}U series, 0.33 Myr for the ^{235}U series, and 60 years for the ^{232}Th series. After reaching secular equilibrium, the series would not contain any information on the history of the system. However, a disturbed series before reaching secular equilibrium (i.e., a disequilibrium decay series) contains history information, and the age of the disturbance may be inferred. In addition, the disturbed decay series of ^{238}U and ^{235}U have also found applications in elucidating the dynamics of partial melting and magma transport. The method of age determination using disturbed decay series is somewhat similar to ^{14}C or ^{10}Be dating, and is hence discussed here.

Dating using intermediate nuclides in decay series requires an understanding of the evolution of the concentrations of intermediate nuclides after disturbance. The full evolution becomes increasingly more complicated for an intermediate nuclide that requires more steps from the long-lived parent. ^{231}Pa is the third nuclide (^{235}U $\rightarrow$ ^{231}Th $\rightarrow$ ^{231}Pa $\rightarrow$ ^{227}Ac) in the ^{235}U decay series, leading to relatively simple evolution equations. Hence, ^{231}Pa is discussed first to elucidate the principles of dating.

Because of differences in chemical properties of U, Th, and Pa, the elements are fractionated in many geochemical processes, such as sedimentation, mantle partial melting, and coral precipitation from water. With fractionation, the nuclide activities of ^{235}U, ^{231}Th, and ^{231}Pa do not equal one another. Define the time of disturbance to be time zero. Use A_1, A_2, and A_3 to denote the decay activity of ^{235}U, ^{231}Th, and ^{231}Pa, respectively, and λ_1, λ_2, and λ_3 to denote the decay constants of ^{235}U, ^{231}Th, and ^{231}Pa. Start from the full evolution equation for ^{231}Pa in Box 2-6,

$$
\begin{aligned}
A_3 = {} & A_3^0 e^{-\lambda_3 t} + \lambda_3 A_2^0 \left[\frac{e^{-\lambda_2 t}}{\lambda_3 - \lambda_2} + \frac{e^{-\lambda_3 t}}{\lambda_2 - \lambda_3} \right] \\
& + \lambda_2 \lambda_3 A_1^0 \left[\frac{e^{-\lambda_1 t}}{(\lambda_2 - \lambda_1)(\lambda_3 - \lambda_1)} + \frac{e^{-\lambda_2 t}}{(\lambda_1 - \lambda_2)(\lambda_3 - \lambda_2)} \right. \\
& \left. + \frac{e^{-\lambda_3 t}}{(\lambda_1 - \lambda_3)(\lambda_2 - \lambda_3)} \right]
\end{aligned}
\tag{5-18a}
$$

where the superscript 0 means at time zero (meaning at disturbance). Because $\lambda_2 \gg \lambda_3 \gg \lambda_1$ (these are true only for the ^{235}U decay series for which $\lambda_2 = 1.1 \times 10^7 \lambda_3$ and $\lambda_3 = 2.1 \times 10^4 \lambda_1$), and because $A_1^0 e^{-\lambda_1 t} = A_1$, the above general equation may be simplified to

$$
A_3 = A_3^0 e^{-\lambda_3 t} + \frac{\lambda_3}{\lambda_2} A_2^0 [e^{-\lambda_3 t} - e^{-\lambda_2 t}] + A_1^0 \left[e^{-\lambda_1 t} + \frac{\lambda_3 e^{-\lambda_2 t}}{\lambda_2} - e^{-\lambda_3 t} \right].
\tag{5-18b}
$$

$$
A_3 = A_3^0 e^{-\lambda_3 t} + A_1^0 [e^{-\lambda_1 t} - e^{-\lambda_3 t}].
\tag{5-18c}
$$

$$
A_{^{231}Pa} - A_{^{235}U} = \left(A_{^{231}Pa}^0 - A_{^{235}U}^0 \right) e^{-\lambda_3 t}.
\tag{5-18d}
$$

That is,

$$
\Delta A = \Delta A^0 e^{-\lambda_3 t},
\tag{5-18e}
$$

where $\Delta A_3 = A_{^{231}Pa} - A_{^{235}U}$, which is the excess ^{231}Pa activity over the activity of the long-lived parent (^{235}U), or the excess activity over the activity at the secular equilibrium. The "excess" activity may be positive or negative. Therefore, after disturbance, the excess activity decreases exponentially to zero. This equation is mathematically the same as the decay equation of ^{14}C (Equation 5-2) and can hence be applied to infer age similarly.

For the nuclides in the ^{238}U decay series, ^{234}U and ^{230}Th are of particular interest because of their long half-lives. ^{234}U is the fourth nuclide in the ^{238}U decay series (Table 2-2a) with a half-life of 244,000 years, and ^{230}Th is the fifth nuclide with a half-life of 75,400 years. Denote ^{238}U as nuclide 1, ^{234}Th as 2, ^{234}Pa as 3, ^{234}U as 4, and ^{230}Th as 5. For ^{234}U, because λ_1 (decay constant of ^{238}U) is the smallest, and λ_4 (decay constant of ^{234}U) is the second smallest and far smaller than λ_2 and λ_4, the following may be easily derived:

$$\Delta A_4 = \Delta A_4^0 e^{-\lambda_{234}t}, \tag{5-19a}$$

where $\Delta A_4 = A_{234_U} - A_{238_U}$. The above may also be written as

$$\frac{A_{234}}{A_{238}} - 1 = \left(\frac{A_{234}^0}{A_{238}^0} - 1 \right) e^{-\lambda_{234}t}. \tag{5-19b}$$

The initial activity ratio of $(^{234}U/^{238}U)$ in seawater is often 1.144 (Bard et al., 1990).

For the next nuclide in the decay series, ^{230}Th, because λ_5 (decay constant of ^{230}Th) is not smaller but larger than λ_4, the activity evolution equation is more complicated. Starting from the full equation (Box 2–6), with $\lambda_3 \gg \lambda_2 \gg \lambda_5 \gg \lambda_4 \gg \lambda_1$, the following may be derived:

$$1 - \frac{A_5}{A_1} = \left(1 - \frac{A_5^0}{A_1^0} \right) e^{-\lambda_5 t} - \left(\frac{A_4^0}{A_1^0} - 1 \right) \frac{\lambda_5}{\lambda_5 - \lambda_4} (e^{-\lambda_4 t} - e^{-\lambda_5 t}), \tag{5-20}$$

which may also be written as

$$1 - \frac{A_5}{A_1} = \left(1 - \frac{A_5^0}{A_1^0} \right) e^{-\lambda_5 t} - \left(\frac{A_4}{A_1} - 1 \right) \frac{\lambda_5}{\lambda_5 - \lambda_4} (1 - e^{-(\lambda_5 - \lambda_4)t}). \tag{5-20a}$$

or

$$A_5 - A_1 = (A_5^0 - A_1^0) e^{-\lambda_5 t} + (A_4^0 - A_1^0) \frac{\lambda_5}{\lambda_5 - \lambda_4} (e^{-\lambda_4 t} - e^{-\lambda_5 t}). \tag{5-20b}$$

Among intermediate nuclides in decay series, ^{230}Th dating is the most widely used because of large fractionation between Th and U in various processes. Two applications are especially notable. One is to date corals, and the second is to determine the age of ocean sediment and sedimentation rate. Because the half-life of ^{230}Th is 75,400 years, it is useful in determining ages younger than 0.5 Ma.

Age of corals and calibration of ^{14}C ***age*** One application of ^{230}Th dating is to date corals (e.g., Edwards et al., 1986/87), which has been used successfully to calibrate ^{14}C ages using corals (e.g., Reimer et al., 2004). Because of strong fractionation between Th and U, Th concentration in seawater is extremely low (Th/U ratio in seawater is 10^{-5} to 10^{-4}; Chen et al., 1986). The ratio of Th/U in corals (carbonates) grown from seawater is similar to that in seawater, 10^{-5} to 10^{-4}. Ignoring the small amount of initial ^{230}Th in corals, Equation 5-20a may be written as

$$1 - \frac{A_{230\text{Th}}}{A_{238\text{U}}} = e^{-\lambda_{230}t} - 1.4458 \left(\frac{A_{234\text{U}}}{A_{238\text{U}}} - 1 \right)(1 - e^{-(\lambda_{230}-\lambda_{234})t}). \tag{5-21}$$

Hence, the age may be solved from Equation 5-21 by simultaneous measurement of $(^{230}\text{Th}/^{238}\text{U})$ and $(^{234}\text{U}/^{238}\text{U})$ activity ratios. Note that Equation 5-21 is specifically for ^{230}Th in corals (by ignoring initial ^{230}Th), and does not apply to other cases such as ^{230}Th in ocean sediment in which initial ^{230}Th activity is very high.

> ***Example 5.2*** Measurement of a coral sample gives the following activity ratios: $(^{230}\text{Th}/^{238}\text{U}) = 0.00190 \pm 0.00005$ and $(^{234}\text{U}/^{238}\text{U}) = 1.149 \pm 0.006$ (Edwards et al., 1986/87). Find the age. *Solution*: Let activities of ^{238}U, ^{234}U, and ^{230}Th be A_1, A_4, and A_5. Assume that the initial activity of ^{230}Th is negligible compared to ^{238}U activity. Applying Equation 5-21 leads to
>
> $$0.99810 = e^{-\lambda_5 t} - 1.4458 \times 1.149 \times (1 - e^{-(\lambda_5 - \lambda_4)t}).$$
>
> Solving the above equation numerically, the age is
>
> $$t = 180 \text{ years.}$$
>
> The error may be estimated to be 5 years based on the relative error of the $(^{230}\text{Th}/^{238}\text{U})$ measurement.

Sedimentation rate In sediment, initial ^{230}Th activity (A_5) is very high because Th is rapidly removed from seawater, but U in the highly oxidized form (UO_2^{2+}) dissolves in seawater, which results in very high U/Th ratio in seawater but very low U/Th in ocean sediment. Hence, there is large excess ^{230}Th activity $(^{230}$Th activity minus ^{238}U activity is very high), opposite to the case of coral with extremely low initial ^{230}Th. As the sediment is buried, the excess activity decays away. Starting from Equation 5-20, A_5^0 is many times larger than A_1^0 (e.g., $A_5^0 = 50A_1^0$), and A_4^0 is only slightly different from A_1^0 (e.g., $A_4^0 = 1.14A_1^0$). Hence, the second term on the right-hand side of Equation 5-20 may be ignored, and the decay of excess ^{230}Th activity can be described by

$$\Delta A = \Delta A^0 e^{-\lambda_{230\text{Th}}t}, \tag{5-22}$$

where $\Delta A = A_{230\text{Th}} - A_{238\text{U}}$ and the superscript 0 means the initial state.

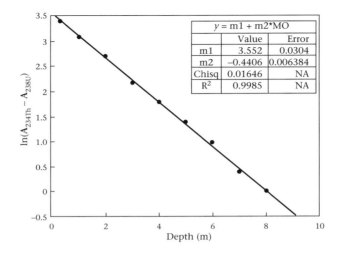

Figure 5-4 Dating sediment using U-disequilibrium series.

Assume that the sedimentation rate v is constant. Hence, the age t of the sediment is z/v, where z is the depth of sedimentary layer. Hence,

$$\Delta A = \Delta A^0 e^{-\lambda_{230\text{Th}} z/v}, \tag{5-22a}$$

or

$$\ln(A_{230\text{Th}} - A_{238\text{U}}) = \ln(A^0_{230\text{Th}} - A^0_{238\text{U}}) - (\lambda_{230\text{Th}}/v)z, \tag{5-22b}$$

By plotting $\ln(A_{230\text{Th}} - A_{230\text{U}})$ versus depth of sediment, we would obtain a straight line with the slope of $-\lambda_{230\text{Th}/v}$. From the slope, the sedimentation rate v can be calculated. From the sedimentation rate, the age of sediment at any given depth z can be calculated. If the sedimentation rate is not constant, or if $(A^0_{230\text{Th}} - A^0_{238\text{U}})$ varies significantly from one layer to another, then $\ln(A^0_{230\text{Th}} - A^0_{230\text{U}})$ versus depth of sediment would not give a straight line.

Example 5.3 Find the sedimentation rate from the following data. The excess ^{230}Th activity varies with depth of an ocean sediment column as follows:

Depth (m below seafloor)	0.3	1	2	3	4	5	6	7	8
$A_{230\text{Th}} - A_{238\text{U}}$ (dpm/g)	30	22	15	9	6	4	2.7	1.5	1.0
error in $\Delta A_{230\text{Th}}$	2	2	1.5	1.5	1	1	0.8	0.5	0.3

Solution: Plot $\ln(A_{230\text{Th}} - A_{238\text{U}})$ versus depth z; the relation is linear (Figure 5-4). Using a simple linear regression to fit the data leads to $\ln(\Delta A_{230\text{Th}}) = (3.55 \pm 0.03) - (0.4406 \pm 0.0064)z$.

Because the slope $= -\lambda 230_{Th}/v = -(0.4406 \pm 0.0064)$, the sedimentation rate is

$v = -\lambda_{230Th}/\text{slope} = 9.19 \times 10^{-6}/(0.4406 \pm 0.0064)$ m/yr $= (2.09 \pm 0.03)10^{-5}$ m/yr.

If York's program is used, and assuming an error of 0.01 m in depth, the best-fit equation is

$$\ln(\Delta A_{230Th}) = (1.24 \pm 0.06) - (0.436 \pm 0.023)z.$$

Hence, the sedimentation rate is $(2.11 \pm 0.11)10^{-5}$ m/yr.

An innovative application of the U-series disequilibrium is to investigate dynamics of mantle partial melting and magma transport (McKenzie, 1985). For example, one conclusion based on U-series disequilibrium in mid-ocean ridge basalts is that mantle partial melting is slow, or the timescale of partial melting is much longer than the half-life of ^{230}Th, and the degree of partial melting is small (a few percent). This and other applications to dynamics are not covered in this book.

5.1.2 Dating method 2: The initial number of atoms of the daughter nuclide may be guessed

If the initial number of daughter nuclides (D_0) can be estimated in Equation 5-9, $D = D_0 + nP(e^{\lambda t} - 1)$, then the age can be calculated from measurements of D and P at present day:

$$\text{Age} = t = (1/\lambda)\ln[1 + (D - D_0)/(nP)]. \tag{5-23}$$

This method has been most widely used in (i) the ^{40}K–^{40}Ar system, (ii) the U–Th–He system, (iii) the U–Pb dating of zircon, and (iv) the U–Th–Pb dating of monazite using electron microprobe measurements. The last is a developing method with large errors because of the low analytical precision and because isotopes are not measured.

5.1.2.1 The ^{40}K–^{40}Ar system

A newly formed mineral (such as biotite or hornblende) from magma incorporates K in its structure, but the initial Ar concentration is often negligible because as a noble gas, Ar does not go into any mineral in appreciable amount compared to K. The growth equation for ^{40}Ar is

$$^{40}\text{Ar} = {}^{40}\text{Ar}_0 + 0.1048 {}^{40}\text{K}(e^{\lambda t} - 1), \tag{5-24}$$

where λ is the decay constant of ^{40}K, and the factor $0.1048 = \lambda_e/\lambda_{40}$ with λ_e being the branch decay constant from ^{40}K to ^{40}Ar, and λ_{40} being the total decay constant of ^{40}K. Assume that the initial concentration of ^{40}Ar $(^{40}\text{Ar}_0)$ is zero (this assumption works better for minerals with older age). Then from the present day K and Ar concentration, the age of the mineral can be calculated as follows:

$$\text{Age} = t = \frac{1}{\lambda} \ln\left(1 + \frac{^{40}\text{Ar}^*}{0.1048\,^{40}\text{K}}\right), \tag{5-25}$$

where $^{40}\text{Ar}^*$ is radiogenic ^{40}Ar, usually treated as total ^{40}Ar, but initial ^{40}Ar may be corrected using measured ^{36}Ar and an assumption of the initial $^{40}\text{Ar}/^{36}\text{Ar}$ ratio.

In practice, dating using $^{40}\text{K}-^{40}\text{Ar}$ system often uses a special method called $^{40}\text{Ar}-^{39}\text{Ar}$ method, which is well developed and widely applied. In this method, part of ^{39}K is converted into ^{39}Ar by the following reaction (neutron irradiation) in a nuclear reactor:

$$^{39}\text{K} + {}^1\text{n} \rightarrow {}^{39}\text{Ar} + {}^1\text{H}. \tag{5-26}$$

^{39}Ar is unstable and decays to ^{39}K by β-decay with a half-life of 269 years. Because the age calculation is based on a standard going through the same procedure, the decay of ^{39}Ar is accounted for. (Furthermore, samples are typically analyzed within months after irradiation in a nuclear reactor. Hence, the correction would not be large.) The main advantage of the $^{40}\text{Ar}-^{39}\text{Ar}$ method over the $^{40}\text{K}-^{40}\text{Ar}$ method is that $^{40}\text{Ar}/^{39}\text{Ar}$ ratio can be measured in a mass spectrometer with a much smaller sample and higher precision compared to measurements of K and Ar concentrations. To obtain the equation for age calculation, we have

$$t = \frac{1}{\lambda_{40}} \ln\left[1 + \frac{^{40}\text{Ar}^*}{^{40}\text{K}}\frac{\lambda_{40}}{\lambda_e}\right] = \frac{1}{\lambda_{40}} \ln\left[1 + \frac{^{40}\text{Ar}^*}{^{39}\text{Ar}}\frac{^{39}\text{Ar}}{^{39}\text{K}}\frac{^{39}\text{K}}{^{40}\text{K}}\frac{\lambda_{40}}{\lambda_e}\right].$$

By letting $J = (^{39}\text{Ar}/^{39}\text{K})(^{39}\text{K}/^{40}\text{K})(\lambda_{40}/\lambda_e)$, which is a constant depending on the several ratios (the fraction of ^{39}K that is converted to ^{39}Ar, the isotopic ratio of $^{39}\text{K}/^{40}\text{K}$, and the branch decay ratio $\lambda_e/\lambda_{40} = 0.1048$), the equation used for age calculation is

$$t = \frac{1}{\lambda_{40}} \ln\left[1 + \frac{^{40}\text{Ar}^*}{^{39}\text{Ar}}J\right], \tag{5-27}$$

where J has to be calibrated for each set of nuclear reactor conditions.

After neutron irradiation, the sample is heated up to release Ar (including ^{40}Ar that is mostly radiogenic, ^{39}Ar that is proportional to K concentration, and ^{36}Ar and ^{38}Ar, which are two other stable and nonradiogenic Ar isotopes) from the mineral to be dated. The released Ar at a given temperature or a given heating stage is led into a mass spectrometer for the measurement of Ar isotopic ratios. Each measurement of the $^{40}\text{Ar}/^{39}\text{Ar}$ ratio (which represents a small fraction of the total Ar in the mineral) can be converted to an age using Equation 5-27. After all Ar is released, age for each fraction of Ar release is plotted against the fraction of Ar released (sometimes against the temperature during heating), which is called an age spectrum. With such data, Ar loss from the mineral in the geological

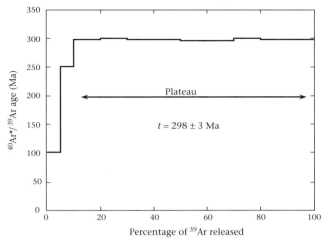

Figure 5-5 An example of Ar release age spectrum. There is a wide plateau with plateau age of 298 Ma. (1 Ma = 1 million years old.)

history after the formation of the mineral (not during heating in the mass spectrometer) can be identified. Since Ar loss or gain affects mostly the boundary of the mineral, one may expect that the initial fraction of Ar to have either too young or too old ages, which is often observed. The final fraction may also show strange features. In the middle, there may be a wide region with roughly a constant age, which is referred to as the plateau age (Figure 5-5). This plateau age is often interpreted to represent the formation age of the mineral.

Example 5.4 $^{40}Ar*/^{40}K$ ratio in a hornblende mineral is 0.00384 ± 0.00004 (2σ error). Find the age.

Solution: Using Equation 5-25, with $\lambda = 5.543 \times 10^{-10}$ yr^{-1}, the age can be found as

$$t = [\ln(1 + 0.00384/0.1048)]/(5.543 \times 10^{-10}) = 6.49 \times 10^7 \text{ years old} = 64.9 \text{ Ma.}$$

The relative error is

$$\frac{dt}{t} = \frac{d(^{40}Ar*/^{40}K)}{0.1048\lambda\left(1 + \frac{^{40}Ar*}{0.1048 - ^{40}K}\right)}\left[\frac{1}{\lambda}\ln\left(1 + \frac{^{40}Ar*}{0.1048 - ^{40}K}\right)\right].$$

Because $^{40}Ar*/^{40}K$ ratio $\ll 1$, the above may be simplified as

$$\frac{dt}{t} \approx \frac{d(^{40}Ar*/^{40}K)}{\left(1 + \frac{^{40}Ar*}{0.1048 - ^{40}K}\right)\frac{^{40}Ar*}{^{40}K}} \approx \frac{d(^{40}Ar*/^{40}K)}{^{40}Ar*/^{40}Ar}.$$

That is, the relative error in the calculated age is roughly the same as the relative error in the $^{40}Ar*/^{40}K$ ratio, leading to

error in age $= dt = 64.9 \times (0.00004/0.00384) = 0.7$ Ma.

Hence, the age is 64.9 ± 0.7 Ma.

The K–Ar method or Ar–Ar method can also be cast in terms of *isochrons*, which may reveal loss of Ar. This is discussed in the next section. A special problem with the K–Ar system is that the properties of the parent and daughter elements are very different and the daughter (a noble gas element) does not like the crystal structure. Hence, loss of the daughter Ar can be a major problem, much more so than with other dating techniques. That is, among the isotopic systems, K–Ar is one of the most easily disturbed systems by later events. Geochemists are good at turning complexity into advantage. The disturbance has been turned to a powerful tool to probe the thermal history (such as cooling rates) of a rock after the formation, and is one of the main *thermochronology* tools, which is discussed in a later section.

5.1.2.2 The ^{238}U–^{235}U–^{235}Th–^4He system

The nucleus of ^{4}He is the α-particle. That is, all α-decays generate ^{4}He. The major nuclides that contribute to ^{4}He include ^{238}U, ^{235}U, and ^{232}Th. The decay of ^{238}U produces 8 α-particles, that of ^{235}U produces 7 α-particles, and that of ^{232}Th produces 6 α-particles. There are also other nuclides undergoing α-decay (such as ^{147}Sm and ^{190}Pt), but their contribution to ^{4}He concentration is much smaller than that of U and Th. Hence, the growth equation for ^{4}He is roughly

$$^4\text{He} = {}^4\text{He}_0 + 8^{238}\text{U}(e^{\lambda_{238}t} - 1) + 7^{235}\text{U}(e^{\lambda_{235}t} - 1) + 6^{232}\text{Th}(e^{\lambda_{232}t} - 1). \qquad (5\text{-}28)$$

(Contribution by ^{147}Sm and ^{190}Pt can be easily added to the above equation if necessary.) If it is assumed that initial ^{4}He in the mineral is zero ($^4\text{He}_0 = 0$), then by measuring the amount of U, Th, and ^{4}He, an age can be calculated. Because ^{4}He is easily lost (much easier than ^{40}Ar), the meaning of this age is almost never the formation age but is an indication of cooling rate. This system is hence mostly applied in thermochronology to infer cooling rate or erosion rate.

5.1.2.3 U–Pb dating of zircon

Zircon ($ZrSiO_4$) is a common accessory mineral in igneous and metamorphic rocks. It can take a significant amount of U in its structure (in the Zr site) at the thousands of ppm level, but takes very little original Pb. Because there is an unradiogenic stable isotope of Pb (^{204}Pb), the very minor amount of initial ^{206}Pb and ^{207}Pb may be roughly corrected by assuming reasonable initial ratios of ^{206}Pb/^{204}Pb and ^{207}Pb/^{204}Pb (based on the evolution of these ratios with time in the Earth). Two U isotopes decay to two Pb isotopes: ^{238}U decays to ^{206}Pb, and ^{235}U to ^{207}Pb. The system is hence called a coupled system, or a double clock. This coupled system is more powerful than a single system because the two decay systems provide redundancy and hence a crosscheck of each other. The equations are

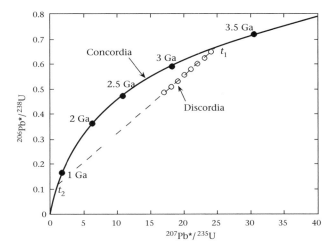

Figure 5-6 The concordia and an example of discordia for U–Pb dating. The *x*-axis is $^{207}Pb*/^{235}U$. The *y*-axis is $^{206}Pb*/^{238}U$. The origin is at age zero. The solid curve is the concordia, on which each point indicates an age. Some specific ages are marked by solid points (1, 2, 2.5, 3, and 3.5 Ga). An example of discordia is shown by analyses plotted as the open circles, which form a straight line (dashed line). One explanation of the data falling on a linear discordia is for zircons crystallized at age t_1 (3.3 Ga) and then experienced diffusive loss at age t_2 (0.7 Ga).

$$^{206}Pb = {^{206}}Pb_0 + {^{238}}U(e^{\lambda_{238}t} - 1), \tag{5-29a}$$

$$^{207}Pb = {^{207}}Pb_0 + {^{235}}U(e^{\lambda_{235}t} - 1), \tag{5-29b}$$

Note that ^{206}Pb, ^{207}Pb, ^{238}U, and ^{235}U all refer to atomic concentrations (or number of atoms). Because $^{206}Pb_0$ and $^{207}Pb_0$ may be roughly estimated from $^{204}Pb_0$, the above two equations can be rewritten as

$$^{206}Pb^*/^{238}U = (e^{\lambda_{238}t} - 1), \tag{5-30a}$$

$$^{207}Pb^*/^{235}U = (e^{\lambda_{235}t} - 1), \tag{5-30b}$$

where the superscript * means the radiogenic part of the nuclide (the present amount minus the initial amount). Either of Equations 5-30a,b may be used to determine the age. Dividing Equation 5-30b by Equation 5-30a leads to another equation for age determination (Pb/Pb age):

$$^{207}Pb^*/^{206}Pb^* = (e^{\lambda_{235}t} - 1)/[137.88(e^{\lambda_{238}t} - 1)]. \tag{5-31}$$

where 137.88 is the present-day $^{238}U/^{235}U$ ratio. Given $^{207}Pb^*/^{206}Pb^*$ ratio, age (*t*) can be solved from Equation 5-31.

Furthermore, Equations 5-30a and 5-30b define a curve when $^{206}Pb^*/^{238}U$ is plotted against $^{207}Pb^*/^{235}U$ (Figure 5-6). Each point on the curve means an age that satisfies both equations. In other words, in a plot of $^{206}Pb^*/^{238}U$ versus $^{207}Pb^*/^{235}U$, all rocks or minerals would follow a single curve as long as there was no other event after the formation of the rock or mineral. The curve is called the *concordia curve*, or the *U–Pb concordia*. If the measured datum plots on this concordia, then a single age can be obtained from the coupled U–Pb system; that is, three ages calculated from Equations 5-30a,b and 5-31 would be identical within error, and they are concordant. If the measured data plot off the curve, it means that the three ages are not the same, and they are said to be discordant. Discordance is caused by the loss or gain of either U or Pb after the formation. Sometimes, the measurements fall off the curve but form a linear trend. Such a linear trend is often interpreted by extending the linear trend to intersect the Concordia in two points, of which the older age is interpreted as the formation age of the mineral, and the younger age is interpreted as the age when there was a metamorphic (or other) event that caused the diffusive loss of Pb.

> **Example 5.5** Measurements show that for a point in a zircon sample (analysis 36-7, Wilde et al., 2001), $^{204}Pb/^{206}Pb < 0.00001$, $^{207}Pb^*/^{206}Pb^* = 0.5587 \pm 0.0028$, $^{206}Pb^*/^{238}U = 0.968 \pm 0.038$, $^{207}Pb^*/^{235}U = 74.6 \pm 3.0$, where the superscript * means the radiogenic part of the nuclide and the errors are given at 2σ level. Find the formation age for this part of the zircon sample.
>
> *Solution*: The three ages calculated from Equations 5-30a,b and 5-31 are
>
> Using Equation 5-30a: ^{238}U–$^{206}Pb^*$ age is $t_1 = [\ln(1+0.968)]/(0.155125)$ Ga $= 4.364 \pm 0.124$ Ga.
> Using Equation 5-30b: ^{235}U–$^{207}Pb^*$ age is $t_2 = [\ln(1+74.6)]/(0.98485)$ Ga $= 4.392 \pm 0.040$ Ga.
> Using Equation 5-31 with trial and error method: $^{207}Pb^* - ^{206}Pb^*$ age is $t_3 = 4.404 \pm 0.007$ Ga.
>
> The above results show that (i) all three ages are consistent within error (hence, the three ages are concordant); and (ii) the age calculated using the Pb/Pb method has the smallest error.
>
> *Comment*: The mineral age of 4.40 Ga is the oldest age ever discovered in single minerals on the Earth. The oldest age of whole-rock samples on the Earth is about 4.03 Ga (Bowring and Williams, 1999), significantly younger.

5.1.2.4 U–Th–Pb dating of monazite

Monazite ($CePO_4$ in which the Ce site also contains much La, Nd, and other REE elements) is another accessory mineral in igneous and metamorphic rocks. Monazite may take in significant amount of Th (several weight percent) and U (of the order of one weight percent), but not much Pb. Assuming that all Pb in old monazite comes from the decay of Th (which contributes to ^{208}Pb) and U (which contributes to ^{206}Pb and ^{207}Pb), then by measuring the elemental concentrations of Th, U, and Pb by electron microprobe, the age of the monazite may be estimated. Using an electron microprobe to date monazite is a developing method. The main advantages include the high spatial resolution (a few micrometers, whereas spatial resolution of secondary ion mass spectrometry measurement is tens of micrometers) and easiness of measurements (electron microprobes are widely available). The main disadvantage is inaccuracy of Pb concentration measurements especially because the PbO concentration is typically low. Hence, the method is best applied to samples with old ages (≥ 1 Ga) because the Pb concentration increases with age.

The growth equation is

$$Pb = {}^{206}Pb + {}^{207}Pb + {}^{208}Pb = {}^{238}U(e^{\lambda_{238}t} - 1) + {}^{235}U(e^{\lambda_{235}t} - 1)$$
$$+ {}^{232}Th(e^{\lambda_{232}t} - 1), \tag{5-32}$$

where Pb, Th, and U are in atomic concentrations. Hence,

$$Pb = (137.88/138.88)U(e^{\lambda_{238}t} - 1) + (1/138.88)U(e^{\lambda_{235}t} - 1)$$
$$+ Th(e^{\lambda_{232}t} - 1), \tag{5-33}$$

Using the above equation, from the measured Pb, Th, and U concentration by electron microprobe, the age t can be calculated.

Example 5.6 Analyses of a monazite sample show the following results: 5.236 ± 0.063 wt% ThO_2, 0.882 ± 0.024 wt% UO_2, and 0.1315 ± 0.0066 wt% PbO. Find the age.

Solution: First convert the wt% data into moles per kilogram of sample. Thus, 5.201 wt% ThO_2 means 0.05201 kg of ThO_2 per kilogram of monazite. By dividing the molar mass of ThO_2 (0.2640369 kg/mol), we find that Th concentration in monazite is 0.1993 mol/kg. Similarly, we find that U concentration is 0.0327 mol/kg, and Pb concentration is 0.005893 mol/kg of monazite. Therefore, Equation 5-33 becomes

$$0.005893 = (137.88/138.88)0.0327(e^{\lambda_{238}t} - 1) + (1/138.88)0.0327(e^{\lambda_{235}t} - 1)$$
$$+ 0.1993(e^{\lambda_{232}t} - 1)$$

The age is then numerically solved to be 382 Ma.

Comment: More accurate calculation requires adjustment of Pb atomic mass using the calculated Pb isotopic composition based on Th and U abundance, and then calculating the number of moles of Pb per kilogram of monazite. Iterate until the age converges. The result is still 382 Ma. Error calculation is relatively complicated and not discussed here.

5.1.3 Dating method 3: The isochron method

The above two methods require knowing the initial amount of either the radioactive parent or the radiogenic daughter. Although several powerful methods have been developed, they are not general because for many systems the initial amount of neither the parent nor the daughter is constrained. Under such circumstances, we need a method to determine all of these unknowns. This method is the *isochron method*, and it is the most powerful method in dating.

The method is based on Equation 5-10, with one simple but necessary modification. Use the ^{147}Sm–^{143}Nd system as an example. ^{147}Sm decays to ^{143}Nd with a half-life of 106 billion years. The growth equation of ^{143}Nd is (Equation 5-10)

$$^{143}\text{Nd} = {}^{143}\text{Nd}_0 + {}^{147}\text{Sm}(e^{\lambda_{147}t} - 1), \tag{5-34}$$

where λ_{147} is the decay constant of ^{147}Sm. The above equation looks like a linear equation of $y = a + bx$ if we were to plot ^{143}Nd versus ^{147}Sm for different minerals formed at the same time: the slope would be $(e^{\lambda_{147}t} - 1)$, and the intercept would be ^{143}Nd$_0$. Because the minerals are assumed to have formed at the same time t, the slope $(e^{\lambda_{147}t} - 1)$ would be a constant, from which the age t may be calculated. The problem is, for the above equation to be a straight line, the initial ^{143}Nd concentration ^{143}Nd$_0$ must be a constant, meaning that different minerals must have the same initial ^{143}Nd concentration, which almost never holds because different minerals would incorporate different concentrations of Nd. Hence, some transformation must be made to the above equation to turn it into a straight line.

The transformation turns out to be simple. Divide both sides by ^{144}Nd, which is an almost stable (half-life of 2.38×10^{15} yr) and nonradiogenic isotope of Nd, meaning ^{144}Nd = ^{144}Nd$_0$. We obtain

$$\frac{^{143}\text{Nd}}{^{144}\text{Nd}} = \left(\frac{^{143}\text{Nd}}{^{144}\text{Nd}}\right)_0 + \frac{^{147}\text{Sm}}{^{144}\text{Nd}}(e^{\lambda_{147}t} - 1). \tag{5-35}$$

A nice property of isotopes is that when minerals form from a common source such as a magma, all minerals will have identical isotopic ratios (the same as that in the magma) if all isotopes are corrected for mass-dependent isotopic fractionation. That is, if several minerals in a volcanic rock crystallized at roughly the same time (within a day, a month, a year, or 1000 years, depending on the resolution of age determination), they would all have the same ^{143}Nd/^{144}Nd ratio. Hence, in the above equation, $(^{143}$Nd/^{144}Nd$)_0$ is a constant. Furthermore, for rocks formed at the same time, $(e^{\lambda_{147}t} - 1)$ is also a constant. Because Sm and Nd have different chemical properties, their concentrations and hence the Sm/Nd ratio vary from one mineral to another. With different Sm/Nd ratios in different minerals, after some time (such as one billion years), the ^{143}Nd/^{144}Nd ratio would vary from one mineral to another. Hence, when we measure ^{143}Nd/^{144}Nd

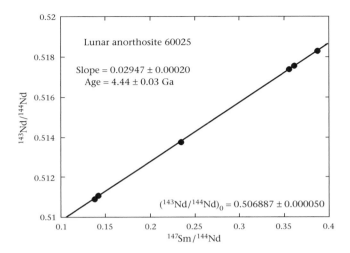

Figure 5-7 A ^{143}Sm–^{144}Nd isochron for lunar anorthosite 60025, with an age of 4.44 ± 0.03 Ga, the oldest rock on the Moon. The errors are given at the 2σ level. Data are from Carlson and Lugmair (1988). See also Example 5-7.

and ^{147}Sm/^{144}Nd ratios in billion-year old minerals at the present day, they co-vary, but $(^{143}$Nd/^{144}Nd$)_0$ and $(e^{\lambda_{147}t} - 1)$ are constants if the minerals formed at the same time and from the same source. Letting $y = {}^{143}$Nd/^{144}Nd and $x = {}^{147}$Sm/^{144}Nd, Equation 5-35 can be written as

$$y = a + bx,$$

where $a = (^{143}$Nd/^{144}Nd$)_0$, and $b = (e^{\lambda_{147}t} - 1)$. If there are two or more minerals, meaning two or more pairs of (x, y), plotting y versus x will give a straight line (Figure 5-7), and both a and b can be obtained by carrying out a linear regression. From the slope b, the age t can be calculated as $t = (1/\lambda_{147})\ln(1 + b)$. The intercept a gives the initial isotopic composition, which is useful for understanding the origin of the rock. Equation 5-35 is referred to as *the isochron equation*, and the plot of ^{143}Nd/^{144}Nd versus ^{147}Sm/^{144}Nd is called the *isochron plot*, because the minerals formed at the same (*iso-*) time (*-chron*). The linear regression for isochrons is typically carried out using the rigorous program of York (1969) (see Section 1.3.6).

In the isochron plot, the horizontal axis is the atomic ratio of ^{147}Sm/^{144}Nd, but measurements are often reported as Sm and Nd weight concentrations. The atomic ratio of two nuclides may be calculated from the weight concentrations of the two elements as follows:

$$\frac{^{147}\text{Sm}}{^{144}\text{Nd}} = \frac{F_{143\text{Sm}}C_{\text{Sm}}/W_{\text{Sm}}}{F_{144\text{Nd}}C_{\text{Nd}}/W_{\text{Nd}}} = \frac{0.150C_{\text{Sm}}/150.36}{0.2380C_{\text{Nd}}/144.24} = 0.6046\frac{C_{\text{Sm}}}{C_{\text{Nd}}}, \tag{5-36}$$

Table 5-2 Isochron systems

Parent nuclide	Half-life (yr)	Decay constant (yr^{-1})	Daughter nuclide(s)	Branch fraction	Branch decay constant (yr^{-1})	Normalizing isotope
^{40}K	1.25×10^9	5.543×10^{-10}	^{40}Ar, ^{40}Ca	0.1048	5.81×10^{-11}	
^{40}K			^{40}Ar	0.8952	4.962×10^{-10}	^{36}Ar
^{40}K			^{40}Ca			^{42}Ca
^{87}Rb	4.88×10^{10}	1.42×10^{-11}	^{87}Sr			^{86}Sr
^{138}La	1.03×10^{11}	6.75×10^{-11}	^{138}Ba, ^{138}Ce	0.6548	4.42×10^{-11}	
^{138}La			^{138}Ba	0.3452	2.33×10^{-11}	^{137}Ba
^{138}La			^{138}Ce			^{132}Ce
^{147}Sm	1.06×10^{11}	6.54×10^{-11}	^{143}Nd			^{144}Nd
^{176}Lu	3.75×10^{10}	1.848×10^{-11}	^{176}Hf			^{177}Hf
^{187}Re	4.16×10^{10}	1.666×10^{-11}	^{187}Os			^{188}Os
^{190}Pt	4.5×10^{11}	1.542×10^{-12}	^{186}Os			^{188}Os
^{232}Th	1.40×10^{10}	4.9475×10^{-11}	^{208}Pb			^{204}Pb
^{235}U	7.038×10^8	9.8485×10^{-10}	^{207}Pb			^{204}Pb
^{238}U	4.468×10^9	1.55125×10^{-10}	^{206}Pb			^{204}Pb

where $F_{^{147}\text{Sm}}$ is the isotopic fraction of ^{147}Sm in Sm, C_{Sm} is the concentration of Sm by weight (such as weight ppm), W_{Sm} is the molar weight of Sm, and similarly for Nd.

Now one can see that the simple operation of dividing Equation 5-34, $^{143}\text{Nd} = {}^{143}\text{Nd}_0 + {}^{147}\text{Sm} (e^{\lambda_{147} t} - 1)$, by ^{144}Nd completely changed the character of the equation. Without the division, ^{143}Nd versus ^{147}Sm would not form a straight line because $^{143}\text{Nd}_0$ varies from mineral to mineral. The simple normalization to ^{144}Nd nicely transforms $^{143}\text{Nd}_0$ into $(^{143}\text{Nd}/^{143}\text{Nd})_0$, which does not vary from mineral to mineral if the minerals formed from the same magma. Therefore, the equation is transformed into a straight line.

After transformation into the isochron equation, both the age and the initial isotopic ratio $(^{143}\text{Nd}/^{144}\text{Nd})_0$ can be obtained. That is, in a given mineral, the initial concentration of the daughter nuclide ^{143}Nd as well as the initial concentration of the parent nuclide in each mineral can be found. Hence, with the isochron method, we determine not only the age, but also the initial amount of the daughter and parent nuclides.

The choice of the normalizing isotope (such as ^{144}Nd for the ^{147}Sm–^{143}Nd system) is by convention, although many considerations went into the choice. The most important criterion is that the normalizing isotope is stable and non-radiogenic. Other considerations include the mass difference between the radiogenic isotope and the normalizing isotope, and the convenience to make the isotopic ratio be of the order of unity. Sm–Nd is but one of the examples of the isochron systems. Table 5-2 lists major isochron systems. A summary of each of the major isochron systems is given in Section 5.1.3.3.

Example 5.7. The oldest lunar rock known to date is a lunar highland anorthosite 60025. The data for different minerals in a single rock are shown in the table below (Carlson and Lugmair, 1988). Find the age.

Sample	Weight (mg)	Sm (ppm)	Nd (ppm)	$^{147}Sm/^{144}Nd$	Error	$^{143}Nd/^{144}Nd$	Error
Pl#1	31.45	0.07846	0.3348	0.1417	0.0006	0.511099	0.000014
Pl#2	58.3	0.07819	0.3435	0.1376	0.0003	0.510919	0.000016
Oliv	59.06	0.01048	0.02699	0.2346	0.0005	0.513773	0.000016
Maf#1	26.39	0.04024	0.06828	0.3562	0.0020	0.517391	0.000030
Maf#2	86.19	0.04218	0.07046	0.3619	0.0003	0.517561	0.000022
Pyroxene	42.95	0.07083	0.11038	0.3878	0.0006	0.518303	0.000022

Solution: Because the $^{147}Sm/^{144}Nd$ ratio has already been calculated, one simply plots $^{143}Nd/^{144}Nd$ versus $^{147}Sm/^{144}Nd$, and examines whether it is a straight line. The plot is Figure 5-7. Because the data defines an excellent linear trend, the necessary condition for an isochron is satisfied. When the data are fit using the linear equation (York's program), the intercept is found to be 0.506887 ± 0.000050 (2σ), which is the initial $^{143}Nd/^{144}Nd$ ratio. The slope is found to be 0.02947 ± 0.00020, from which the age is calculated as

$$t = [\ln(1 + slope)]/\lambda_{147} = 4.44 \pm 0.03 \text{ Ga.}$$

5.1.3.1 Mineral versus whole-rock isochrons

In isochron dating, the different samples may be from different minerals in the same rock (usually a single hand specimen). The resulting isochron would be referred to as a mineral isochron. Sometimes, several rocks may be inferred to be from the same source. For example, starting from a single magma chamber, there may be several eruptions with slightly different magma composition (and hence different whole-rock composition) due to magma fractionation. Also, with a large intrusion, there may be a chilled margin (rapidly cooling leading to fine-crystalline rock) and interior coarse rocks. The composition of the chilled margin would be similar to the initial magma composition. The whole rock composition of the interior rocks may differ from that because of crystal fractionation. These rocks have slightly different ages (such as 0.01 Myr). But for rocks of >100 Ma old, such age difference is trivial and they may be regarded to have formed at the same time and from the same source. Therefore, samples from these rocks may be employed to obtain an isochron and hence age. Such an isochron is called the whole-rock isochron. Because the elemental ratio variation from one rock to another cogenetic rock is small, the whole-rock isochron requires better analytical precision. The advantage of the whole-rock isochron age is that it may resist alteration by later events. Suppose after the formation of the volcanic rocks at 1000 Ma, there was a metamorphic event but the temperature was not extremely high. On a scale of some centimeters, there may be complete homogenization. On a scale of decimeters, there may be some alteration. But rock samples meters or tens of meters away might not have exchanged with one another. Hence, the whole rock isochron might give the old formation age, and the mineral isochron might give the younger metamorphic age.

5.1.3.2 Pseudo-isochrons

One cautionary note about isochrons is that even if data of $^{143}Nd/^{144}Nd$ versus $^{147}Sm/^{144}Nd$ follow a straight line, the line is consistent with, but is not proof that all the samples formed at the same time. That is, the linearity between $^{143}Nd/^{144}Nd$ and $^{147}Sm/^{144}Nd$ is a necessary but not sufficient condition of an isochron. It can be shown that if a suite of samples formed by mixing of different proportions of two reservoirs, data of $^{143}Nd/^{144}Nd$ versus $^{147}Sm/^{144}Nd$ would also follow a straight line. Such a straight line is called a *pseudo-isochron*. The slope of a pseudo-isochron may even be negative, which would be a clear indication that the line is not an isochron; otherwise, it would yield a negative (future) age. Therefore, before dating, it is important to understand the geologic relations, and ascertain that the samples to be dated indeed formed at the same time and from a single source (not a mixture of two sources). Even the best straight line does not eliminate the need for careful geologic work.

5.1.3.3 Summary of isochron systems

(1) $^{147}Sm–^{143}Nd$ *system*. For easy reference, this system is also summarized here even though the information was given in the earlier text. ^{147}Sm undergoes α-decay to ^{143}Nd with a half-life of 1.06×10^{11} years (decay constant 6.54×10^{-12} yr^{-1}). ^{144}Nd is used as the normalizing isotope.[1] The isochron equation is (Equation 5-35)

---

[1]The nuclide ^{144}Nd is actually slightly radioactive with a half-life of 2.38×10^{15} years, and also slightly radiogenic receiving contribution from the α-decay of ^{148}Sm with a half-life of 7×10^{15} years. These effects are negligible in deriving the isochron equation because the age of the Earth is much shorter compared to the half lives.

$$\frac{^{143}\text{Nd}}{^{144}\text{Nd}} = \left(\frac{^{143}\text{Nd}}{^{144}\text{Nd}}\right)_0 + \frac{^{147}\text{Sm}}{^{144}\text{Nd}}(e^{\lambda_{147}t} - 1).$$ (5-35)

The atomic ratio of $^{147}\text{Sm}/^{144}\text{Nd}$ may be calculated from the elemental weight ratio of Sm/Nd (Equation 5-36):

$$\left(\frac{^{147}\text{Sm}}{^{144}\text{Nd}}\right)_{\text{atom}} = 0.6046\left(\frac{\text{Sm}}{\text{Nd}}\right)_{\text{wt}}.$$ (5-36)

Because Sm and Nd are both rare Earth elements and have similar chemical properties, and because they often occupy crystalline sites that are not easily altered, this system is resistant to alteration by later events (such as a later metamorphic event). Hence, this system is often applied to determine old and formation ages, whereas other systems (such as K–Ar) may be applied to obtain metamorphic ages.

(2) ^{87}Rb–^{87}Sr *system.* ^{87}Rb undergoes β^-–decay to ^{87}Sr with a half-life of 4.9×10^{10} years (decay constant 1.42×10^{-11} yr^{-1}). ^{86}Sr is used as the normalizing isotope. The isochron equation is

$$\frac{^{87}\text{Sr}}{^{86}\text{Sr}} = \left(\frac{^{87}\text{Sr}}{^{86}\text{Sr}}\right)_0 + \frac{^{87}\text{Rb}}{^{86}\text{Sr}}(e^{\lambda_{87}t} - 1).$$ (5-37)

The atomic ratio of $^{87}\text{Rb}/^{86}\text{Sr}$ may be calculated from the elemental weight ratio of Rb/Sr:

$$\left(\frac{^{87}\text{Rb}}{^{86}\text{Sr}}\right)_{\text{atom}} = 2.895\left(\frac{\text{Rb}}{\text{Sr}}\right)_{\text{wt}}.$$ (5-38)

This system is used to determine very old ages.

(3) ^{176}Lu–^{176}Hf *system.* ^{176}Lu undergoes β^-decay to ^{176}Hf with a half-life of 3.75×10^{10} years (decay constant 1.848×10^{-11} yr^{-1}). ^{177}Hf is used as the normalizing isotope. The isochron equation is

$$\frac{^{176}\text{Hf}}{^{177}\text{Hf}} = \left(\frac{^{176}\text{Hf}}{^{177}\text{Hf}}\right)_0 + \frac{^{176}\text{Lu}}{^{177}\text{Hf}}(e^{\lambda_{176}t} - 1).$$ (5-39)

The atomic ratio of $^{176}\text{Lu}/^{177}\text{Hf}$ may be calculated from the elemental weight ratio of Lu/Hf:

$$\left(\frac{^{176}\text{Lu}}{^{177}\text{Hf}}\right)_{\text{atom}} = 0.1420\left(\frac{\text{Lu}}{\text{Hf}}\right)_{\text{wt}}.$$ (5-40)

This system is widely used and also mostly applicable to the determination of old ages.

(4) ^{187}Re–^{187}Os *system.* ^{187}Re undergoes β^--decay to ^{187}Os with a half-life of 4.16×10^{10} years (decay constant 1.666×10^{-11} yr^{-1}; Smoliar et al., 1996; there is

some uncertainty with this decay constant). ^{188}Os is used as the normalizing isotope. The isochron equation is

$$\frac{^{187}\text{Os}}{^{188}\text{Os}} = \left(\frac{^{187}\text{Os}}{^{188}\text{Os}}\right)_0 + \frac{^{187}\text{Re}}{^{188}\text{Os}}(e^{\lambda_{187}t} - 1). \tag{5-41}$$

The atomic ratio of ^{187}Re/^{188}Os may be calculated from the elemental weight ratio of Re/Os:

$$\left(\frac{^{187}\text{Re}}{^{188}\text{Os}}\right)_{\text{atom}} \approx 4.808 \left(\frac{\text{Re}}{\text{Os}}\right)_{\text{wt}}. \tag{5-42}$$

This system is widely used in dating metals (such as iron meteorites).

(5) ^{190}Pt–^{186}Os system. ^{190}Pt undergoes α-decay to ^{186}Os with a half-life of 4.5×10^{11} years (decay constant 1.542×10^{-12} yr^{-1}). ^{188}Os is used as the normalizing isotope. The isochron equation is

$$\frac{^{186}\text{Os}}{^{188}\text{Os}} = \left(\frac{^{186}\text{Os}}{^{188}\text{Os}}\right)_0 + \frac{^{190}\text{Pt}}{^{188}\text{Os}}(e^{\lambda_{190}t} - 1). \tag{5-43}$$

This system is not widely used.

(6) ^{40}K–^{40}Ar system. There are three K isotopes that are naturally present, ^{39}K, ^{40}K, and ^{41}K, of which ^{40}K is radioactive. The decay of ^{40}K is complicated, with two branches to different daughters. One branch (about 89.52%) is β^--decay to ^{40}Ca, and the other branch (about 10.48%) is by mostly electron capture to ^{40}Ar. The half-life of ^{40}K is 1.25×10^9 years, and the overall decay constant is $\lambda_{40} = 5.543 \times 10^{-10}$ yr^{-1}. The branch decay constant to ^{40}Ar is $\lambda_e = 0.1048 \times 5.543 \times 10^{-10}$ yr$^{-1} = 5.81 \times 10^{-11}$ yr^{-1}. Dating using ^{40}K–^{40}Ar system has been discussed before by assuming initial ^{40}Ar to be zero. The system may also be used as the isochron method. The normalizing isotope is ^{36}Ar. The isochron equation is

$$\frac{^{40}\text{Ar}}{^{36}\text{Ar}} = \left(\frac{^{40}\text{Ar}}{^{36}\text{Ar}}\right)_0 + \frac{\lambda_e}{\lambda_{40}}\frac{^{40}\text{K}}{^{36}\text{Ar}}(e^{\lambda_{40}t} - 1). \tag{5-44}$$

In the Ar–Ar method (by converting ^{39}K to ^{39}Ar, see Section 5.1.2.1)

$$\frac{^{40}\text{Ar}}{^{36}\text{Ar}} = \left(\frac{^{40}\text{Ar}}{^{36}\text{Ar}}\right)_0 + \frac{1}{J}\frac{^{39}\text{K}}{^{36}\text{Ar}}(e^{\lambda_{40}t} - 1), \tag{5-45}$$

where J is a conversion factor obtained from the standard. Hence, the plot of ^{40}Ar/^{36}Ar versus ^{39}Ar/^{36}Ar is a straight line with a slope of $(e^{\lambda_{40}t} - 1)/J$, from which the age can be calculated. Note that ^{39}Ar/^{36}Ar ratio is directly measured by mass spectrometry with high precision, meaning there is no need for calculation of ^{40}K/^{36}Ar ratio from K and Ar concentrations.

(7) ^{40}K–^{40}Ca system. The branch decay constant of ^{40}K to ^{40}Ca is $\lambda_\beta = 0.8952 \times 5.543 \times 10^{-10}$ yr$^{-1} = 4.962 \times 10^{-11}$ yr^{-1}. The normalizing Ca isotope is ^{42}Ca. The isochron equation is

$$\frac{^{40}\text{Ca}}{^{42}\text{Ca}} = \left(\frac{^{40}\text{Ca}}{^{42}\text{Ca}}\right)_0 + \frac{\lambda_\beta}{\lambda_{40}}\frac{^{40}\text{K}}{^{42}\text{Ca}}(e^{\lambda_{40}t} - 1). \tag{5-46}$$

The atomic ratio of $^{40}\text{K}/^{42}\text{Ca}$ may be calculated from the elemental weight ratio of K/Ca:

$$\frac{\lambda_\beta}{\lambda_{40}}\left(\frac{^{40}\text{K}}{^{40}\text{Ca}}\right)_{\text{atom}} \approx 0.01660\left(\frac{\text{K}}{\text{Ca}}\right)_{\text{wt}}. \tag{5-47}$$

This system is not widely used.

(8) ^{138}La–^{138}Ce *system.* The decay of ^{138}La is a branch decay, with one branch (β^- decay) going to ^{138}Ce, and the other branch (electron capture) going to ^{138}Ba. The half-life of ^{138}La is 1.03×10^{11} years, and the total decay constant is 6.75×10^{-12} yr^{-1}. The branch decay constant from ^{138}La to ^{138}Ce is $\lambda_\beta = 2.33 \times 10^{-12}$ yr^{-1}. The normalizing isotope is ^{142}Ce. The isochron equation is

$$\frac{^{138}\text{Ce}}{^{142}\text{Ce}} = \left(\frac{^{138}\text{Ce}}{^{142}\text{Ce}}\right)_0 + \frac{\lambda_\beta}{\lambda_{138}}\frac{^{138}\text{La}}{^{142}\text{Ce}}(e^{\lambda_{138}t} - 1). \tag{5-48}$$

The atomic ratio of $^{138}\text{La}/^{142}\text{Ce}$ may be calculated from the elemental weight ratio of La/Ce:

$$\left(\frac{^{138}\text{La}}{^{142}\text{Ce}}\right)_{\text{atom}} \approx 0.008194\left(\frac{\text{La}}{\text{Ce}}\right)_{\text{wt}}. \tag{5-49}$$

This system is not very widely used.

(9) ^{138}La–^{138}Ba *system.* The branch decay constant from ^{138}La to ^{138}Ba is $\lambda_e = 4.42 \times 10^{-12}$ yr^{-1}. The normalizing isotope is ^{137}Ba. The isochron equation is

$$\frac{^{138}\text{Ba}}{^{137}\text{Ba}} = \left(\frac{^{138}\text{Ba}}{^{137}\text{Ba}}\right)_0 + \frac{\lambda_e}{\lambda_{138}}\frac{^{138}\text{La}}{^{137}\text{Ba}}(e^{\lambda_{138}t} - 1). \tag{5-50}$$

The atomic ratio of $^{138}\text{La}/^{137}\text{Ba}$ may be calculated from the elemental weight ratio of La/Ba:

$$\left(\frac{^{138}\text{La}}{^{137}\text{Ba}}\right)_{\text{atom}} \approx 0.007923\left(\frac{\text{La}}{\text{Ba}}\right)_{\text{wt}}. \tag{5-51}$$

This system is not very widely used.

(10) ^{238}U–^{206}Pb *system.* The last three isochron systems to be discussed involve U, Th, and Pb isotopes. There are four stable isotopes of Pb (whose atomic number is a magic number 82): ^{204}Pb ($\sim$1.4%), ^{206}Pb ($\sim$24.1%), ^{207}Pb ($\sim$22.1%), and ^{208}Pb ($\sim$52.4%). Three of the four are radiogenic: ^{206}Pb from ^{238}U, ^{207}Pb from ^{235}U, and ^{208}Pb from ^{232}Th. The Pb systems are discussed last because of the complexities of the U–Th–Pb system in several aspects. One is that each of the nuclides ^{235}U, ^{238}U, and ^{232}Th undergoes a long chain of decay to the final stable nuclide: ^{238}U decays

to ^{206}Pb through 8 α-decays and 6 β-decays; ^{235}U decays to ^{207}Pb through 7 α-decays and 4 β-decays; ^{232}Th decays to ^{208}Pb through 6 α-decays and 4 β-decays. All three decay chains involve complicated branching. The second complexity is the presence of three individual isochron systems: ^{238}U–^{206}Pb, ^{235}U–^{207}Pb, and ^{232}Th–^{208}Pb. The third complexity comes from the fact that two U isotopes decay to two Pb isotopes. This complexity provides a powerful special dating method called the Pb–Pb method (discussed in the next subsection).

First the ^{238}U–^{206}Pb system is discussed. ^{238}U undergoes 8 α-decays and 6 β-decays to finally reach the stable nuclide ^{206}Pb (see Table 2-2a). The half-life of ^{238}U is 4.468×10^9 years, and the decay constant is 1.55125×10^{-10} yr^{-1}. Assuming the changes in the concentrations of the intermediate nuclides (such as ^{234}U, ^{230}Th, etc.) are negligible, the growth of ^{206}Pb can be written as

$$^{206}\text{Pb} = {}^{206}\text{Pb}_0 + {}^{238}\text{U}(e^{\lambda_{238}t} - 1). \tag{5-52}$$

Using ^{204}Pb as the normalizing isotope of Pb, the isochron equation may be written as

$$\frac{^{206}\text{Pb}}{^{204}\text{Pb}} = \left(\frac{^{206}\text{Pb}}{^{204}\text{Pb}}\right)_0 + \frac{^{238}\text{U}}{^{204}\text{Pb}}(e^{\lambda_{238}t} - 1). \tag{5-53}$$

The atomic ratio of ^{238}U/^{204}Pb may be calculated from the elemental weight ratio of U/Pb:

$$\left(\frac{^{238}\text{U}}{^{204}\text{Pb}}\right)_{\text{atom}} \approx 61.73 \left(\frac{\text{U}}{\text{Pb}}\right)_{\text{wt}}. \tag{5-54}$$

Equation 5-53 assumes that every decayed ^{238}U nuclide becomes ^{206}Pb. This is approximate because of intermediate nuclides in the decay chain. Based on the discussion in Section 2.2.1.1, if (i) a relative precision of 1% in age is required, (ii) both ^{238}U and ^{234}U are incorporated with equal activity, and (iii) the activities of intermediate nuclides lie between 0 and 2 times the activity of ^{238}U, then for ages greater than 11 Ma it would not be necessary to consider the intermediate species. Otherwise, there will be a need to account for these species, especially if the intermediate species concentration could be high initially. For example, if initially ^{230}Th activity is 50 times ^{238}U activity, the age calculated from the ^{238}U–^{206}Pb geochronometer would be older than the real age by as much as 5.3 Ma.

(11) ^{235}U–^{207}Pb system. ^{235}U undergoes 7 α-decays and 4 β-decays to finally reach the stable nuclide ^{207}Pb (see Table 2-2b). The half-life of ^{235}U is 7.038×10^8 years, and the decay constant is 9.8485×10^{-10} yr^{-1}. Assuming the changes in the concentrations of the intermediate nuclides (such as ^{231}Pa) are negligible, the growth of ^{207}Pb can be written as

$$^{207}\text{Pb} = {}^{207}\text{Pb}_0 + {}^{235}\text{U}(e^{\lambda_{235}t} - 1). \tag{5-55}$$

Using ^{204}Pb as the normalizing isotope of Pb, the isochron equation may be written as

$$\frac{^{207}\text{Pb}}{^{204}\text{Pb}} = \left(\frac{^{207}\text{Pb}}{^{204}\text{Pb}}\right)_0 + \frac{^{235}\text{U}}{^{204}\text{Pb}}(e^{\lambda_{235}t} - 1). \tag{5-56}$$

The atomic ratio of $^{235}\text{U}/^{204}\text{Pb}$ may be calculated from the elemental weight ratio of U/Pb:

$$\left(\frac{^{235}\text{U}}{^{204}\text{Pb}}\right)_{\text{atom}} \approx 0.4477\left(\frac{\text{U}}{\text{Pb}}\right)_{\text{wt}}. \tag{5-57}$$

Equation 5-55 assumes that every decayed ^{235}U nuclide actually becomes ^{207}Pb. This is approximate because of intermediate nuclides in the decay chain. Based on the discussion in Section 2.2.1.1, if a relative precision of 1% in age is required, the application of the $^{235}\text{U}-^{207}\text{Pb}$ geochronometer to ages younger than 4.8 Ma (if instrumental analytical accuracy allows such determination) would require a careful account of the intermediate species. If initially ^{231}Pa activity is 50 times ^{235}U activity, age obtained from the $^{235}\text{U}-^{207}\text{Pb}$ geochronometer would be older than the real age by as much as 2.3 Ma.

(12) $^{232}Th-^{208}Pb$ *system.* ^{232}Th undergoes 6 α-decays and 4 β-decays to finally reach the stable nuclide ^{208}Pb (see Table 2-2c). The half-life of ^{232}Th is 1.40×10^{10} years, and the decay constant is 4.9475×10^{-11} yr^{-1}. Assuming the changes in the concentrations of the intermediate nuclides are negligible, the growth of ^{208}Pb can be written as

$$^{208}\text{Pb} = {}^{208}\text{Pb}_0 + {}^{232}\text{Th}(e^{\lambda_{232}t} - 1). \tag{5-58}$$

Using ^{204}Pb as the normalizing isotope of Pb, the isochron equation may be written as

$$\frac{^{208}\text{Pb}}{^{204}\text{Pb}} = \left(\frac{^{208}\text{Pb}}{^{204}\text{Pb}}\right)_0 + \frac{^{232}\text{Th}}{^{204}\text{Pb}}(e^{\lambda_{232}t} - 1). \tag{5-59}$$

The atomic ratio of $^{232}\text{Th}/^{204}\text{Pb}$ may be calculated from the elemental weight ratio of Th/Pb:

$$\left(\frac{^{232}\text{Th}}{^{204}\text{Pb}}\right)_{\text{atom}} \approx 63.78\left(\frac{\text{Th}}{\text{Pb}}\right)_{\text{wt}}. \tag{5-60}$$

5.1.3.4 Coupled systems

Because two U isotopes, ^{235}U (0.7200% of U; half-life 0.704 billion years) and ^{238}U (99.2745% of U; half-life 4.468 billion years) decay to isotopes of Pb, the U–Pb system is often referred to as a *coupled system*. Two isochrons can be generated with the U–Pb dating system (Equations 5-53 and 5-56):

$$\frac{^{206}Pb}{^{204}Pb} = \left(\frac{^{206}Pb}{^{204}Pb}\right)_0 + \frac{^{238}U}{^{204}Pb}(e^{\lambda_{238}t} - 1),$$

$$\frac{^{207}Pb}{^{204}Pb} = \left(\frac{^{207}Pb}{^{204}Pb}\right)_0 + \frac{^{235}U}{^{204}Pb}(e^{\lambda_{235}t} - 1).$$

Each isochron will give an age. If the two ages agree within error, then this is great and we will trust the age. Otherwise, the differences have to be resolved. Hence, the two systems provide independent cross-check of the reliability of age determination. Because the two equations involve two U isotopes decaying into two Pb isotopes, if disturbed, they would both be disturbed. Therefore, the coupled system is especially powerful for checking the consistency of the ages using the concordia (Section 5.1.2.2). Furthermore, the two U–Pb decay systems (i.e., the above two equations) can be combined to obtain

$$\frac{^{207}Pb}{^{204}Pb} - \left(\frac{^{207}Pb}{^{204}Pb}\right)_0 = \frac{1}{137.88}\frac{(e^{\lambda_{235}t} - 1)}{(e^{\lambda_{238}t} - 1)}\left[\frac{^{206}Pb}{^{204}Pb} - \left(\frac{^{206}Pb}{^{204}Pb}\right)_0\right], \tag{5-61}$$

where 137.88 is the isotopic ratio of $^{238}U/^{235}U$ at the present day. This is a linear equation when $^{207}Pb/^{204}Pb$ is plotted against $^{206}Pb/^{204}Pb$, and the slope is a function of the age t:

$$\text{Slope} = \frac{1}{137.88}\frac{(e^{\lambda_{235}t} - 1)}{(e^{\lambda_{238}t} - 1)}. \tag{5-62}$$

To solve the age from the slope requires numerical method (such as trial and error, iteration, etc.).

Equation 5-61 involves only Pb isotopic ratios and is called the Pb–Pb isochron. There are a few advantages of this Pb–Pb isochron compared to the individual U–Pb isochrons. One is that Pb isotopic ratios can be measured more accurately than U/Pb ratios. Hence, the above equation has the advantage of Ar–Ar method compared to the K–Ar method. Secondly, if there was recent (such as yesterday or in the last million years) U loss or gain or Pb loss, the loss or gain would not affect the Pb isotopic composition and hence would not affect Pb–Pb dating (but it would affect U–Pb isochrons). Pb gain would affect the Pb–Pb isochron.

The Pb–Pb isochron was made famous by the determination of the age of the Earth: Patterson (1956) grouped meteorite samples with a sediment sample that is supposed to represent the bulk silicate Earth in terms of Pb isotopes (Figure 5-8). The assumption is that the Earth formed at roughly the same time as the meteorites. The colinearity of the data in Figure 5-8 is viewed as verification of the assumption. The age given by Patterson (1956) is 4.55 Ga.

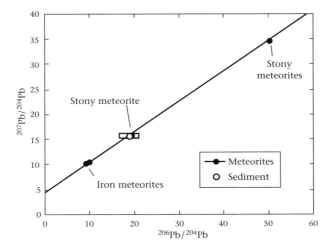

Figure 5-8 A Pb–Pb isochron that determined the age of the Earth to be about 4.55 Ga. Stony and iron meteorites as well as a sediment of the Earth are plotted on a Pb–Pb isochron. The sediment, as a "bulk" sample of the silicate Earth in terms of Pb isotopes, plots on the same line as the meteorites, suggesting that the Earth and meteorites formed at the same time and are the same age. From Patterson (1956). Later studies reveal a more detailed evolution history of the Earth, including core formation (about 4.53 Ga), atmospheric formation (about 4.45 Ga), and crustal evolution.

Example 5.8 The data used by Patterson (1956) to determine the age of the Earth are in the following table. Find the age.

Sample	$^{206}Pb/^{204}Pb$	$^{207}Pb/^{204}Pb$	$^{208}Pb/^{204}Pb$
Stony meteorite 1	50.28	34.86	67.97
Stony meteorite 2	19.27	15.95	39.05
Stony meteorite 3	19.48	15.76	38.21
Iron meteorite 1	9.55	10.38	29.54
Iron meteorite 2	9.46	10.34	29.44
Terrestrial sediment	19.0	15.8	

Solution: Plot $^{207}Pb/^{204}Pb$ versus $^{206}Pb/^{204}Pb$ as shown in Figure 5-8. Because the error bars are not given, a simple linear regression method is used. If all the data are used, the slope is 0.6027 ± 0.0158 (2σ). (If only meteorite data are used, the slope is 0.6026 ± 0.0180.)

The age numerically solved from the slope (Equation 5-62) is

4.51 ± 0.04 Ga.

This age differs from 4.55 Ga obtained by Patterson (1956) because he used an old set of decay constants.

Some comments are in order about Figure 5-8 and Example 5-8. After the work of Patterson (1956), subsequent investigations showed that different meteorites have slightly different ages. A combination of new isotopic systems (such as extinct nuclides, to be discussed next) and better analytical accuracies allow small age differences to be resolved. These studies indicate that (i) the oldest solar system materials are the calcium–aluminum-rich inclusions in chondrites, about 4.566 ± 0.002 Ga (Chen and Wasserburg, 1981); (ii) the age of meteorites span a narrow range (4.54 to 4.566 Ga); and (iii) the age of the Earth is younger than that of a reference meteorite (age 4.56 Ga) and depends on the event one is interested in. For example, the mean age of core formation (which may be regarded to be the most important event in the Earth's evolution) is about 4.53 Ga (Kleine et al., 2002; Yin et al., 2002). The mean age of atmospheric formation (or the closure age) is about 4.45 Ga (Staudacher and Allegre, 1982; Zhang, 1998c, 2002). Ignoring such small differences allowed Patterson (1956) to determine the age of the Earth.

Another coupled system is the Sm–Nd system, with two Sm isotopes (^{147}Sm and ^{146}Sm) undergoing α-decay to become two Nd isotopes (^{143}Nd and ^{142}Nd). The half-life of ^{147}Sm is 106 billion years and that of ^{146}Sm is 103 million years. In principle, the concepts for the U–Pb system (such as concordia and discordia, Nd–Nd isochron) can also be applied to the Sm–Nd system. However, the Sm–Nd coupled system has not found many applications. One reason is that the half-life of ^{146}Sm is so short that it is an extinct nuclide. Secondly, the half-lives of ^{147}Sm and ^{146}Sm are very different, by a factor of 1000 (in contrast, the half-lives of ^{238}U and ^{235}U differ only by a factor of 6.3). Hence, the coupled system has found only limited applications to very old rocks, such as meteorites and very old terrestrial rocks.

5.1.4 Dating method 4: Extinct nuclides for relative ages

The half-lives of many nuclides are much shorter than the age of the Earth. For example, the half-life of ^{26}Al is 0.71 million years, ^{107}Pd 6.5 million years, ^{129}I 15.7 million years, ^{146}Sm 103 million years, ^{182}Hf 9 million years, and ^{244}Pu 80 million years. If they were present initially when the solar system formed (there are reasons to believe so), by now they would have completely decayed away. However, for those nuclides with long enough half-lives (millions of years), geochemists have developed methods to investigate the decay product to infer their initial presence. Furthermore, by assuming that the initial isotopic abundance of an extinct nuclide was uniform in the solar system and that it depends only on time (i.e., decays exponentially with time), comparison of the initial isotopic ratio of the extinct nuclide in different meteorites may reveal relative age differences (such as which one formed earlier and by how much). The discovery of many extinct nuclides is one of the many great achievements by geochemists. Below, we first derive the extinct nuclide isochron equation, and then discuss how to determine relative ages.

Start from the equation for the growth of the daughter: $D = D_0 + P_0(1 - e^{-\lambda t})$. If the parent is an extinct nuclide, it would have completely decayed to the daughter. That is, for extinct nuclide,

$$D = D_0 + P_0. \tag{5-63}$$

Below we use the $^{26}\text{Al}-^{26}\text{Mg}$ system as an example for clarity in derivation. ^{26}Al undergoes β^+-decay to ^{26}Mg. With a half-life of 0.71 million years, if there was any initial ^{26}Al, it would have completely decayed away to ^{26}Mg. In terms of isotopic ratios, it is assumed that the $^{26}\text{Al}/^{27}\text{Al}$ ratio in the early solar system was uniform spatially and decreased exponentially with time to zero. The amount of ^{26}Mg at the present day may be written as

$$^{26}\text{Mg} = {^{26}\text{Mg}_0} + {^{26}\text{Al}_0}. \tag{5-64}$$

Converting the above equation into a linear equation similar to the isochron equation requires a couple of steps. First, we normalize the above equation to a stable and nonradiogenic Mg isotope (^{24}Mg), yielding

$$\frac{^{26}\text{Mg}}{^{24}\text{Mg}} = \left(\frac{^{26}\text{Mg}}{^{24}\text{Mg}}\right)_0 + \frac{^{26}\text{Al}_0}{^{24}\text{Mg}}. \tag{5-65}$$

Second, we rewrite the last term by dividing and multiplying ^{27}Al (a stable and nonradiogenic isotope of Al), yielding

$$\frac{^{26}\text{Mg}}{^{24}\text{Mg}} = \left(\frac{^{26}\text{Mg}}{^{24}\text{Mg}}\right)_0 + \left(\frac{^{26}\text{Al}}{^{27}\text{Al}}\right)_0 \frac{^{27}\text{Al}}{^{24}\text{Mg}}. \tag{5-66}$$

The above equation is now a linear equation if we let $y = {^{26}\text{Mg}}/{^{24}\text{Mg}}$ and $x = {^{27}\text{Al}}/{^{24}\text{Mg}}$. If several (x, y) pairs are measured and if these minerals formed at the same time and from the same reservoir (hence, identical initial $^{26}\text{Al}/^{27}\text{Al}$ and $^{26}\text{Mg}/^{24}\text{Mg}$ ratios in different minerals after correction of mass-dependent fractionation), plotting y versus x gives a straight line (Figure 5-9), which is an isochron. The intercept of the isochron gives the initial isotopic composition of $^{26}\text{Mg}/^{24}\text{Mg}$, and the slope gives the initial $^{26}\text{Al}/^{27}\text{Al}$ ratio. A nonzero slope means the presence of initial ^{26}Al, an extinct nuclide. Not only is its presence demonstrated by a diagram such as Figure 5-9, but the initial isotopic ratio involving the extinct nuclide is also obtained, which is an indication of age.

Examining Equation 5-66 for its physical meaning shows that if there is high concentration of Al/Mg in a mineral, for a given initial $^{26}\text{Al}/^{27}\text{Al}$ ratio, it would mean more initial ^{26}Al relative to ^{24}Mg in the mineral. Because all ^{26}Al turned

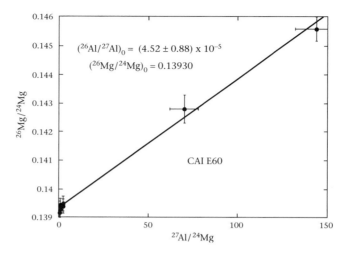

Figure 5-9 ^{26}Al–^{26}Mg isochron for a calcium–aluminum-rich inclusion (CAI) E60. The initial isotopic ratio of $(^{26}Al/^{27}Al)_0 = 4.52 \times 10^{-5}$. Adapted from Amelin et al. (2002).

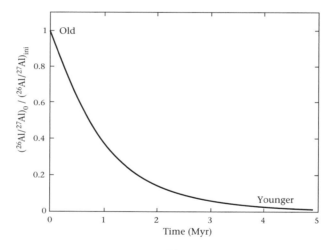

Figure 5-10 Relation between $(^{26}Al/^{27}Al)_0$ and age. $(^{26}Al/^{27}Al)_0$ means an initial ratio determined from a meteorite, and $(^{26}Al/^{27}Al)_{ini}$ means the earliest ratio of the solar nebula. A high $(^{26}Al/^{27}Al)_0$ ratio means an older age, and a low $(^{26}Al/^{27}Al)_0$ ratio means a younger age. This is in contrast with dating using cosmogenic nuclide such as ^{14}C (Figure 5-1).

into ^{26}Mg, the present-day $^{26}Mg/^{24}Mg$ ratio in the mineral would be high. This positive correlation between the present-day $^{26}Mg/^{24}Mg$ ratio and Al/Mg ratio is the basis of the *extinct nuclide isochron*.

To obtain the age from the slope in Figure 5-9, it is assumed that in the early solar nebula, the $^{26}Al/^{27}Al$ ratio was uniform but decreased with increasing time (or decreasing age). The evolution of the ^{26}Al/^{27}Al ratio follows Figure 5-10. That is, if two meteorites have different initial ^{26}Al/^{27}Al, it means that there is an age difference between the two meteorites. The age difference between them can be calculated using method 1. An example is given below.

Example 5.9 In meteorite 1, the initial ^{26}Al/^{27}Al ratio is found to be 5×10^{-5}. In meteorite 2, the initial ^{26}Al/^{27}Al ratio is found to be 2.5×10^{-5}. Which meteorite formed earlier (i.e., has an older age), and by how much?

Solution: Because the initial ^{26}Al/^{27}Al ratio in meteorite 1 is greater, meteorite 1 formed earlier. That is, meteorite 1 has an older age. Because the initial ratio in meteorite 2 is half of that in meteorite 1, the age difference is the half-life, which is 0.71 million years. That is, meteorite 1 is older than meteorite 2 by 0.71 Ma. If the age of meteorite 1 is 4562 Ma, then the age of meteorite 2 is 4561.3 Ma. This example shows that using extinct nuclides can distinguish small age differences in the early evolution history of the solar system.

In extinct nuclide dating, the greater the isotopic ratio of the extinct nuclide to its stable isotope, the older the age is, because the ratio always decreases with time after the initial production. This is in contrast to cosmogenic nuclide dating (method 1). In cosmogenic nuclide dating, the smaller the measured concentration of the cosmogenic nuclide, the older the age. The difference is because cosmogenic nuclides are continually produced before incorporation into a sample to be dated and hence the ratio is highest in the most recent sample, but extinct nuclides were present initially and decayed continually without reproduction. Many extinct nuclides have been identified to be present in the early solar system. Table 5-3 lists some extinct nuclides. Readers may notice that many extinct nuclides are also cosmogenic nuclides (Table 5-1). It is necessary to distinguish whether the ^{26}Al one measures is cosmogenic (present-day surface samples on the Earth) or extinct (old meteorite samples).

5.1.5 Requirements for accurate dating

Some requirements for accurate dating are obvious. For example, the decay constant must be accurately known and the present-day elemental concentrations and isotopic ratios must be measured accurately. For the isochron methods, other requirements include the following:

(1) The event to be dated must fractionate the parent and daughter elements, and homogenize the daughter isotopic ratios so that the ratio was the same in every mineral at the beginning of the system. Only an event that results in identical isotopic ratios and the differentiation of parent and daughter nuclides can be dated.

Table 5-3 Extinct nuclides

Nuclide	# of protons	# of neutrons	Daughter	Half-life (yr)	Decay constant (yr^{-1})	Positively identified?
^{10}Be	4	6	^{10}B	1.51×10^6	4.59×10^{-7}	
^{26}Al	13	13	^{26}Mg	7.1×10^5	9.8×10^{-7}	Yes
^{36}Cl	17	19	^{36}S	3.01×10^5	2.30×10^{-6}	
^{41}Ca	20	21	^{41}K	1.04×10^5	6.66×10^{-6}	Yes
^{53}Mn	25	28	^{53}Cr	3.7×10^6	1.87×10^{-7}	Yes
^{60}Fe	26	34	^{60}Ni	1.5×10^6	4.62×10^{-7}	Yes
^{81}Kr	36	45	^{81}Br	2.3×10^5	3.01×10^{-6}	
^{92}Nb	41	51	^{92}Zr	3.5×10^7	1.98×10^{-8}	Yes
^{93}Zr	40	53	^{93}Nb	1.5×10^6	4.62×10^{-7}	
^{97}Tc	43	54	^{97}Mo	2.6×10^6	2.67×10^{-7}	
^{98}Tc	43	55	^{98}Ru	4.2×10^6	1.65×10^{-7}	
^{99}Tc	43	56	^{99}Ru	2.13×10^5	3.25×10^{-6}	
^{107}Pd	46	61	^{107}Ag	6.5×10^6	1.07×10^{-7}	Yes
^{129}I	53	76	^{129}Xe	1.57×10^7	4.41×10^{-8}	Yes
^{135}Cs	55	80	^{135}Ba	2.3×10^6	3.01×10^{-7}	
^{146}Sm	62	84	^{142}Nd	1.03×10^8	6.73×10^{-9}	Yes
^{182}Hf	72	110	^{182}W	9×10^6	7.7×10^{-8}	Yes
^{205}Pb	82	123	^{205}Tl	1.5×10^7	4.62×10^{-8}	
^{244}Pu	94	150	^{232}Th	8×10^7	8.66×10^{-9}	Yes

(2) There must be several members of the system (such as several minerals in a single rock, or several rocks in a suite of rocks), which formed at the same time but with different parent/daughter ratio.

(3) The formation event must be well defined and rapid. For example, the formation event of a volcanic rock (eruption of magma to the surface and cooling down of the magma) is well defined and rapid. But the formation of a plutonic rock is not well defined and is slow.

(4) After formation, the parent and daughter should have not been lost from or gained by the system to be dated. That is, the system must be a closed system after the event. In some ideal cases, for a system disturbed by later events, there can be two different isochrons, a mineral isochron, from which the age of the later disturbance (e.g., metamorphic event) can be calculated, and a whole rock isochron, from which the age of the original igneous event can be calculated. That is, the ages of both events may be inferred.

Generally speaking, volcanic rocks (magma eruption and rapid cooling on the surface of the Earth) are the easiest to date, but plutonic rocks are more difficult due to slow cooling, which means that daughter nuclides may be lost gradually. For sedimentary rocks, the formation event is well defined but there may not be isotopic reequilibration and elemental fractionation. Hence, dating is not straightforward. Metamorphic rocks are the most difficult to date because there is no single formation event. When an age is determined for a metamorphic rock, it is critical to know what the age means (that is, what the timing means: was it the time when the rock was at the peak temperature in its metamorphic history, was it a time related to some premetamorphic event, or was it a time during retrograde metamorphism?).

By studying the meanings of the measured ages, the difficulties in determining the formation ages have been turned by geochemists into advantages. Thus, for plutonic and metamorphic rocks, by using several minerals or several dating systems, not only the age, but also the full thermal history may be revealed. This is the subject of thermochronology discussed in the next section.

5.2 Thermochronology

In the previous section, the radioactive parent and radiogenic daughter systems are applied to infer a single age and the initial conditions. For some rocks, most notably volcanic rocks that cooled rapidly on the surface (in air or water) and that have not experienced events such as later heating, a single age is all that it is necessary to know and different dating systems would yield the same age. However, for most other rocks that have a more complicated thermal history, the meaning of the age determined must be clarified. For such rocks, it is possible that different dating systems would result in different ages. These ages are called apparent ages, and also closure ages. An apparent age means the closure of the parent and daughter nuclides in the mineral that was dated, which is related to temperature–time history. The determination of temperature–time history is referred to as thermochronology. Even though the thermal history of rocks may be very complicated (e.g., Figure 1-18), the temperature history that can be inferred is usually the part at and post the maximum temperature because most previous information was likely wiped out at the maximum temperature. That is, thermochronology is mostly applied to infer the cooling part of the thermal history. For intrusive rocks, the cooling history is what interests us. For metamorphic rocks, this means the retrograde temperature history. The prograde thermal history may be inferred in some cases, often through the mineral assemblage and zoning profiles preserved (e.g., Figure 3-7). In this section on thermochronology, the focus is on the inference of cooling history of the rock using isotopic systems. Cooling from 500 K to surface temperature is often associated with exhumation of rocks (one exception is cooling of volcanic rocks). Hence, it may be possible to use thermochronology to understand exhumation, including orogenic erosion (Reiners and Brandon, 2006).

The central concept in thermochronology is the closure temperature (Figure 1-20), which is related to the closure (or apparent) age. Any age obtained by geochronology (previous section) is an apparent age. For rapid cooling, the apparent age is the formation age or peak temperature age, and the closure temperature is the formation or peak temperature. For slow cooling, diffusive exchange (i.e., open behavior) must be considered. At high temperature, diffusion is rapid. There may hence be diffusive loss or diffusive exchange of the daughter element and isotopes between the mineral to be dated and the surroundings. For the parent, the element is usually part of the structure (such as K in biotite) or is already in equilibrium with the surroundings. Thus, the concentration of the parent nuclide often does not change much with time (except for decay) during cooling. Furthermore, the isotopic ratio of the parent nuclide (such as $^{40}K/^{39}K$ ratio) is essentially constant in different phases (although it does depend on time because of decay), and hence diffusive exchange would not affect the isotopic ratio of the parent nuclide. Therefore, when we consider the effect of diffusive loss or exchange, only the daughter nuclide is considered. For some systems, such as $^{40}K-^{40}Ar$ system, the loss of the daughter affects the age determination. For other systems, such as $^{147}Sm-^{143}Nd$, the exchange of isotopic ratios with the surroundings (i.e., rehomogenization of isotopic ratios) affects the age determination. The closure temperature is defined to be the temperature of the rock at the time given by the apparent age (Figure 1-20).

The most commonly used radioactive–radiogenic system in thermochronology is the $^{40}K-^{40}Ar$ system, often refined as the $^{40}Ar-^{39}Ar$ method. Recently, the U–Th–He method has been developed with applications to understand erosion. Below, the closure temperature relation is first derived using a simple method. Then, diffusive loss and radiogenic growth of ^{40}Ar are examined in more detail.

5.2.1 Closure temperature and closure age

Dodson (1973) developed the concept of closure temperature and closure age. The concepts of the closure temperature and closure age are discussed in Sections

1.7.3 and 3.5.2, and are explained here in more detail. The closure age is the age of the rock if we use normal dating method to calculate the age, assuming the system was closed from the beginning regardless of thermal history and nuclide loss. That is, Equation 5-25 is used for ^{40}K–^{40}Ar dating and Equation 5-27 is used for ^{40}Ar–^{39}Ar dating, even if we are not sure whether the mineral has retained all of the radiogenic ^{40}Ar. Therefore, by definition, the age from all geochronology methods is the closure age (or apparent age). The closure age would be the formation age of an igneous rock only if the rock cooled very rapidly (such as volcanic rocks); it would be the peak temperature age of a metamorphic rock only if the rock after reaching the peak temperature cooled down rapidly. In many cases, cooling from high temperature is slow. Because of high diffusion rate at high temperature, there would be significant loss of the radiogenic daughter nuclide at high temperature. As the system cools down, the diffusivity decreases and diffusive loss gradually becomes insignificant. Hence, the closure age is often younger than either the formation age or the peak temperature age. The difference between the formation age (or peak temperature age) and the closure age depends on the cooling rate and the diffusion property of the daughter nuclide in the given mineral. That is, the closure age for the case of slow cooling is an indication of the cooling rate, which can be quantified if the diffusion property is known. By using isotopic systems with slow diffusion at high temperatures (such as Pb in zircon), the closure age may approach the formation age.

The closure temperature T_c is defined to be the temperature of the rock when the rock was at the time of the closure age (Figure 1-20). T_c is a calculated property; it has never been verified experimentally. The calculation of the closure temperature is based on modeling of the diffusive loss (or exchange) and radiogenic growth in a given mineral for a given cooling rate. The T_c concept is most often applied to whole mineral grains, but it can also be applied to the center of a mineral, or any point in a mineral. Below we first discuss the closure of whole mineral grains.

The mean diffusion distance over the cooling period subsequent to T_c is proportional to $(\int D \, dt)^{1/2}$, where the integration is from the time of T_c to the time of room temperature. The concept of T_c dictates that mid-diffusion distance $(\int D \, dt)^{1/2}$ subsequent to T_c is significantly smaller than the half-thickness of a thin slab (one-dimensional diffusion), or the radius of a sphere or long cylinder. Denote the half-thickness and radius as a. Therefore,

$$a^2 \gg \int_0^\infty D \, dt. \qquad (5\text{-}67a)$$

To quantify, we define that T_c satisfies

$$a^2 = G \int_0^\infty D \, dt, \qquad (5\text{-}67b)$$

where G is a parameter (>1) to be determined and depends on the shape (thin slab, long cylinder, or spherical grains). Hence, it is called the shape factor. Assume that cooling subsequent to T_c follows the asymptotic cooling law:

$$T = T_c/(1 + t/\tau_c), \tag{5-68}$$

where τ_c is the cooling timescale, and q at T_c can be found as

$$q = -(\mathrm{d}T/\mathrm{d}t)_{t=0} = T_c/\tau_c. \tag{5-69}$$

The diffusivity can then be expressed as

$$D = Ae^{-E/(RT)} = D_c e^{-t/\tau}, \tag{5-70}$$

where E and A are the activation energy and pre-exponential factor for diffusion, R is the universal gas constant, and

$$D_c = Ae^{-E/(RT_c)}, \tag{5-71}$$

and

$$\tau = \tau_c RT_c/E = RT_c^2/(qE). \tag{5-72}$$

The parameter τ characterizes how rapidly D decreases with time. Applying Equation 5-70, we obtain

$$\int_0^\infty D\,\mathrm{d}t = D_c\tau. \tag{5-73}$$

Replacing the above into Equation 5-67b leads to

$$a^2 = GD_c\tau. \tag{5-74}$$

Express the above in terms of T_c explicitly

$$a^2 = GA\ e^{-E/(RT_c)}\tau_c RT_c/E. \tag{5-74a}$$

or

$$a^2 = GA\ e^{-E/(RT_c)}RT_c^2/(qE). \tag{5-74b}$$

By rearranging the above, we obtain (Dodson, 1973)

$$\frac{E}{RT_c} = \ln\left[\frac{GAT_c^2}{a^2 qE/R}\right], \tag{5-75a}$$

or

$$\frac{E}{RT_c} = \ln\left(\frac{GA\tau}{a^2}\right), \tag{5-75b}$$

or

$$T_c = \frac{E/R}{\ln\left[\dfrac{GAT_c^2}{a^2qE/R}\right]}, \tag{5-76a}$$

or

$$T_c = \frac{E/R}{\ln\left(\dfrac{GA\tau_c T_c}{a^2E/R}\right)}, \tag{5-76b}$$

or

$$q = \frac{GT_c^2 D_{T_c}}{a^2E/R}, \tag{5-77a}$$

or

$$q = \frac{GT_c^2}{\tau_d E/R}, \tag{5-77b}$$

where E and A are the activation energy and pre-exponential factor for diffusion, R is the gas constant, τ_c is the timescale for T to decrease from T_0 to $T_0/2$, τ is the time for D to decrease by a factor of e, q is the cooling rate when the temperature was at T_c, a is the grain size, $\tau_d = a^2/D_{T_c}$, and G is the shape factor. Equations 5-75a to 5-77b are all equivalent expressions relating T_c to diffusion and cooling parameters. They may be derived from one another, but they are nonetheless written down explicitly for easy reference. They are based on the work of M. H. Dodson (Dodson, 1973) and may be referred to as *Dodson's equations*. Based on Dodson's solutions to the isotropic volume diffusion and radiogenic growth equation following asymptotic cooling, and assuming slow decay of the parent nuclide (i.e., τ is much less than the half-life of the parent) as well as high initial temperature T_0 (significantly higher than T_c), the shape factor G equals 55 for a sphere (a is the radius), 27 for an infinitely long cylinder (a is the radius), and 8.65 for a plane sheet (a is the half-thickness). The initial temperature (T_0) does not appear in the above expressions, meaning that T_0 does not affect T_c, which would be the case if T_0 is high enough so that there is essentially complete loss of the daughter nuclide or complete isotopic equilibrium with the surroundings. If this condition is not satisfied, then modification of the above formulations is necessary (Ganguly and Tirone, 1999, 2001), which is discussed in Section 5.2.3.2.

The next section discusses diffusive loss and radiogenic growth in more detail. Because the full problem of diffusive loss and radiogenic growth is complicated, to build our understanding of the problem, we start from simple cases and move to more realistic cases. Readers who do not wish to go through the detailed mathematical analyses may jump to Section 5.2.3.

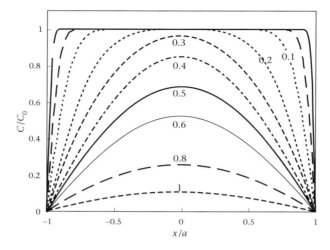

Figure 5-11 The evolution of the concentration profile as a function of α $\{=[\int D(t\prime)\mathrm{d}t\prime]^{1/2}/a\}$ with the integration from 0 to time t. Each curve corresponds to an α value indicated.

5.2.2 Mathematical analyses of diffusive loss and radiogenic growth

The full mathematical diffusion problem of thermochronology includes both radiogenic growth and cooling. It turns out that a diffusion problem with cooling is not difficult to solve, and a diffusion problem with a constant radiogenic production rate is not difficult to solve, but the combination of the two makes the problem difficult to treat mathematically. Even though analytical solutions in the form of summation of infinite terms of integrals are available, the solution is not particularly instructive and numerical method is required to calculate specific values. Below we go though problems of diffusive loss during cooling and problems of diffusion with constant radiogenic growth rate to help readers to gain some intuitive understanding of the processes.

5.2.2.1 Diffusive loss during cooling without radiogenic growth

Without considering radiogenic growth, effectively we are considering a non-radiogenic isotope such as ^{36}Ar. Two effective shapes, plane sheet and solid sphere, are considered here. The effective shape is not necessarily the physical shape; diffusive anisotropy must also be considered in determining the effective shape (Section 3.2.11 and Figure 3-13).

Plane sheet bounded by two parallel plane surfaces Suppose the mineral grains can be treated as thin wafers so that ^{36}Ar loss is through two parallel surfaces. Define the two surfaces to be $x = \pm a$. For the initial condition of uniform initial concentration C_0 and the boundary condition of zero surface concentration, Ar diffusion profile in a thin wafer of half-thickness a with isotropic diffusivity D is as follows (Appendix A3.2.4f):

$$C = \frac{4C_0}{\pi} \sum_{n=0}^{\infty} \frac{(-1)^n}{(2n+1)} \cos \frac{(2n+1)\pi x}{2a} e^{-(2n+1)^2 \pi^2 \alpha^2 / 4}, \tag{5-78}$$

where $\alpha = [\int_0^t D(t')dt']^{1/2}/a$ is a dimensionless parameter. For small $\alpha \leq 0.3$, the above converges slowly, and the following expression may be used:

$$C = C_0 \left\{ 1 - \sum_{n=0}^{\infty} (-1)^n \left[\mathrm{erfc} \frac{(2n+1)a - x}{\sqrt{4 \int D \, dt}} + \mathrm{erfc} \frac{(2n+1)a + x}{\sqrt{4 \int D \, dt}} \right] \right\}$$

$$\approx C_0 \left(\mathrm{erf} \frac{a - x}{\sqrt{4 \int D \, dt}} - \mathrm{erfc} \frac{a + x}{\sqrt{4 \int D \, dt}} \right), \tag{5-79}$$

where the approximation applies if $\alpha \leq 0.3$. The evolution of the concentration profile with time is shown in Figure 5-11.

The fraction of ^{36}Ar loss can be calculated from the above equation as

$$F = 1 - \frac{8}{\pi^2} \sum_{n=0}^{\infty} \frac{1}{(2n+1)^2} e^{-(2n+1)^2 \pi^2 \alpha^2 / 4}. \tag{5-80}$$

For $\alpha \leq 0.3$, the above expression does not converge rapidly, and the following may be used (by letting $L = 2a$ and $Dt = \int D(t')dt'$ in Equation 3-126):

$$F = \frac{2\alpha}{\sqrt{\pi}} \left[1 + 2\sqrt{\pi} \sum_{n=1}^{\infty} (-1)^n \mathrm{ierfc} \frac{n}{\alpha} \right] \approx \frac{2\alpha}{\sqrt{\pi}}, \tag{5-81}$$

where the approximate relation is valid for $F < 0.5$. The fractional loss in this case (plane sheet or slab) is plotted as the solid curve in Figure 5-12. According to the above, if $\alpha^2 = 0.1967$, then $F = 0.5$ (50% ^{36}Ar loss); if $\alpha^2 = 0.8481$, then $F = 0.9$ (90% ^{36}Ar loss). From the solid curve in Figure 5-12 and from Equation 5-81, F is almost perfectly proportional to α when $\alpha \leq 0.5$. This is because for small times, the diffusion has not much changed the center concentration of the slab, meaning diffusion from each surface may be viewed as diffusion into a semi-infinite medium, for which F is perfectly proportional to α. As the diffusion profile reaches the center of the slab, the diffusion medium can no longer be viewed as a semi-infinite medium, and F is no longer proportional to α.

In addition to the fraction of mass loss as a function of $\alpha = (\int D \, dt)^{1/2}/a$, it is of interest to examine how the fraction of mass loss change with time and temperature during cooling to understand the concept of closure. Figure 5-13 shows how the remaining fraction in the phase $(1 - F)$ depends on time and temperature for a specific cooling history. In this example, the whole history of the mineral is 100 Myr. There was mass loss in the first 5 Myr, or at $T > 850$ K. As the system is cooled below 850 K, no more mass loss occurred. Hence, the system became closed at the temperature of about 850 K.

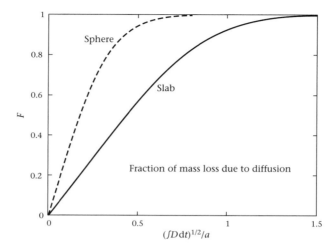

Figure 5-12 Fraction of mass loss due to diffusion in a thin slab with half-thickness a (solid curve) and a solid sphere with radius a (dashed curve) for uniform initial concentration and zero surface concentration with no growth. Some specific values are as follows (F_1 for slab; F_2 for sphere):

α	F_1	F_2
0	0	0
0.1	0.11284	0.30851
0.2	0.22568	0.55703
0.3	0.33861	0.74554
0.4	0.45124	0.87440
0.5	0.56223	0.94844
0.6	0.66653	0.98259
0.8	0.83290	0.99890
1.0	0.93126	0.999969

Solid sphere For isotropic diffusion in a spherical mineral of radius a with uniform initial concentration C_0 and zero surface concentration, ^{36}Ar diffusion profile is as follows (Equation 3-68g):

$$C = \frac{2aC_0}{\pi r} \sum_{n=1}^{\infty} \frac{(-1)^{n+1}}{n} \sin\frac{n\pi r}{a} e^{-n^2\pi^2\alpha^2}, \tag{5-82}$$

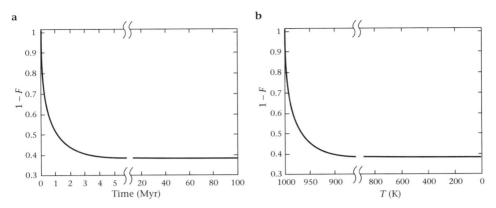

Figure 5-13 The fraction of mass still kept in the phase as a function of (a) time and (b) temperature. The temperature history is $T = T_0/(1 + t/\tau_c)$ with $T_0 = 1000$ K and $\tau_c = 30$ Myr, leading to $D = D_{0e}^{-t/\tau}$, where $\tau = 1$ Myr, and $D_0\tau/a^2 = 0.3$. The total history of the mineral (i.e., the age) is assumed to be 100 Ma.

where $\alpha = [\int_0^t D(t')dt']^{1/2}/a$. The fraction of ^{36}Ar loss (F) from the grain may be estimated from (Equation 3-68c):

$$F = 1 - \frac{6}{\pi^2} \sum_{n=1}^{\infty} \frac{1}{n^2} e^{-n^2\pi^2\alpha^2}. \tag{5-83}$$

For $\alpha < 0.22$ or $F \leq 0.6$, it is easier to calculate the fractional loss from Equation 3-68f:

$$F \approx \frac{6}{\sqrt{\pi}}\alpha - 3\alpha^2. \tag{5-84}$$

The fractional loss is plotted as the dashed curve in Figure 5-12. According to the above, if $\alpha^2 = 0.0305$, then $F = 0.5$ (50% ^{36}Ar loss); if $\alpha^2 = 0.183$, then $F = 0.9$ (90% ^{36}Ar loss).

Compared with one-dimensional diffusion in a plane sheet, for the same α value (such as $\alpha = 0.5$), the fraction of mass loss is much larger for the solid sphere ($F = 0.9484$) than for the planar slab ($F = 0.5622$). To reach the same degree of mass loss (such as 50%), a smaller α^2 value is needed for the sphere ($\alpha^2 = 0.0305$) than for the thin slab ($\alpha^2 = 0.1963$). The difference in terms of α^2 is a factor of 6.4. Because the definition of α^2 is $\int D(t')dt'/a^2$, by comparing with Equation 5-67b, we see that α^2 is proportional to $1/G$. The difference in the value of α^2 for the same degree of mass loss for spherical and plane sheet shapes roughly explains the difference between G values ($55/8.65 = 6.4$) for spherical and plane sheet shapes.

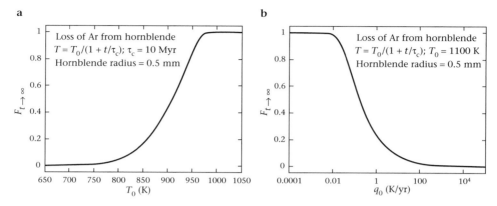

Figure 5-14 Diffusive loss of Ar that was initially in hornblende during cooling after complete cooling down ($t = \infty$) for asymptotic cooling history with (a) a fixed cooling timescale but varying the initial temperature and (b) a fixed initial temperature but varying the cooling rate.

For a practical example, consider ^{36}Ar diffusion in hornblende. Harrison (1981) determined Ar diffusivity assuming hornblende grains are isotropic spheres:

$$D = \exp\left(-12.94 - 32,257/T\right) \text{ m}^2/\text{s}. \tag{5-85}$$

At 1100 K, D in hornblende is 4.4×10^{-19} m^2/s. For hornblende grains of 1 mm in diameter, $a = 0.5$ mm. Loss of 90% of ^{36}Ar would occur in 3300 years. Hence, if the sample cooled from 1110 to 1090 K in 3300 years (6100 K/Myr, a very high cooling rate for intrusive rocks), the average D would be about 4.4×10^{-19} m^2/s, and 90% of the initial ^{36}Ar would be lost. That is, even at relatively high cooling rate, ^{36}Ar would be lost at 1100 K almost completely.

On the other hand, at a lower temperature of 500 K, D in hornblende is 2.3×10^{-34} m^2/s (this value requires large extrapolation and, hence, is not expected to be accurate, but our interest is only an order of magnitude estimation). Because this diffusivity is a finite number, diffusive loss in infinite time would be complete. However, the age of the Earth is not infinite, and hornblende age cannot exceed that of the Earth. Even if hornblende has an age of 4.4×10^9 years (the age of the oldest zircon on the Earth), $(Dt)^{1/2}$ is only 5.7 nm. If $a = 0.5$ mm, then $Dt/a^2 = 1.3 \times 10^{-10}$, and ^{36}Ar loss would be only about 0.004%. That is, ^{36}Ar would be quantitatively retained in the mineral.

Now consider ^{36}Ar loss from hornblende grains of radius 0.5 mm, but for the case of continuous cooling. For asymptotic cooling history of $T = T_0/(1 + t/\tau_c)$, $D = Ae^{-E/(RT)} = D_0 e^{-t/\tau}$, where $\tau = \tau_c RT_0/E = \tau_c T_0/32,257$ for hornblende. After cooling down, the parameter $\alpha^2 = \int D(t')dt'/a^2 = D_0\tau/a^2$. Suppose $\tau_c = 10$ Myr. If the initial temperature $T_0 = 1100$ K, then $D_0 = 4.4 \times 10^{-19}$ m^2/s, $\tau = 0.341$ Myr, $D_0\tau/a^2 = 19.0$, and essentially all initial ^{36}Ar is lost (note that radiogenic growth is not considered here). If $T_0 = 900$ K, then $D_0 = 6.5 \times 10^{-22}$ m^2/s, $\tau = 0.279$ Myr, $D_0\tau/a^2 = 0.0230$, and 44.4% of initial ^{36}Ar would be lost. If $T_0 = 700$ K, only 0.27% of ^{36}Ar would be lost. The fraction of ^{36}Ar loss as a function of initial temperature is shown in Figure 5-14a. For this cooling timescale, the transition from essentially complete ^{36}Ar loss to essentially complete ^{36}Ar closure would occur at T_0 between 700 and 950 K.

Continue the above consideration of ^{36}Ar loss from hornblende of radius 0.5 mm under continuous cooling by varying the cooling rate at a fixed initial temperature T_0 of 1100 K. Note that the initial cooling rate $q_0 = -(dT/dt)_{t=0} = T_0/\tau_c$. The fraction of ^{36}Ar loss as a function of cooling rate is shown in Figure 5-14b. A typical volcanic rock cools down in a few days to years; that is, τ_c is of order 0.01 to 1 yr, and q_0 of order 10^3 to 10^5 K/yr. From Figure 5-14b, there is essentially no ^{36}Ar loss, and the closure age is the eruption age. As q_0 decreases, the fraction of ^{36}Ar loss increases.

The above solutions are for diffusive loss only, without considering radiogenic growth. For the case of continuous radiogenic growth of ^{40}Ar, the fraction of Ar loss is significantly smaller because recently produced Ar would not have had much time to diffuse away.

5.2.2.2 Isothermal diffusive loss and constant radiogenic growth rate

Because ^{40}Ar is radiogenic, that is, it is continuously produced by the decay of ^{40}K, the situation is more complicated than the treatment above. Below we consider another simple case: ^{40}Ar generation and isothermal diffusion. Suppose the production rate of ^{40}Ar is time-independent, corresponding to the slow decay assumption of Dodson (1973). This assumption is valid if the cooling timescale τ_c is much shorter than the half-life of ^{40}K (1250 Myr), such as a τ_c of less than 25 Myr.

Plane sheet bounded by two parallel surfaces For the case of one-dimensional diffusion in a thin slab with half-thickness of a, the diffusion equation is

$$\frac{\partial C}{\partial t} = D\frac{\partial^2 C}{\partial x^2} + p, \qquad t > 0, \ -a < x < a, \tag{5-86a}$$

where C is the concentration of ^{40}Ar, and $p = \lambda_e{}^{40}$K and assumed to be time-independent. The initial condition is

$$C|_{t=0} = C_0. \tag{5-86b}$$

The boundary conditions are

$$C|_{x=\pm a} = 0. \tag{5-86c}$$

Let $w = C - pt$. Then the above set of equations becomes

$$\frac{\partial w}{\partial t} = D\frac{\partial^2 w}{\partial x^2}, \quad t > 0, \; -a < x < a, \tag{5-87a}$$

$$w|_{t=0} = C_0. \tag{5-87b}$$

$$w|_{x=\pm a} = -pt. \tag{5-87c}$$

If D is time-dependent, although one may also divide both sides of Equation 5-86a by D to remove the time dependence of the $D\partial^2 C/\partial x^2$ term, the resulting boundary condition depends on time. Hence, the equation is still complicated to solve. That is, even though time-dependent D may be simply treated, and a constant production rate may be simply treated, the combination of the two cannot be simply treated.

The above diffusion problem may be separated into two problems. Define w_1 and w_2 so that each satisfies Equation 5-87a. Let w_1 satisfy the initial condition of $w_1|_{t=0} = C_0$, and the boundary condition of $w_1|_{x=\pm a} = 0$. Let w_2 satisfy the initial condition of $w_1|_{t=0} = 0$, and the boundary condition of $w_2|_{x=\pm a} = -pt$. Hence, the summation $w = w_1 + w_2$ satisfies Equations 5-87b,c. Because of the principle of superposition, one can first find the solution of w_1 and w_2, and then obtain the solution of w by $w = w_1 + w_2$. Then C can be found as $C = w + pt = w_1 + w_2 + pt$.

The solution of w_1 is given by Equation 5-78

$$w_1 = \frac{4C_0}{\pi}\sum_{n=0}^{\infty}\frac{(-1)^n}{(2n+1)}\cos\frac{(2n+1)\pi x}{2a}e^{-(2n+1)^2\pi^2 Dt/(4a^2)}. \tag{5-88a}$$

The solution of w_2 can be found in Appendix A3.2.4g as

$$w_2 = -pt + \frac{p(a^2 - x^2)}{2D} - \frac{16pa^2}{D\pi^3}\sum_{n=0}^{\infty}\frac{(-1)^n}{(2n+1)^3}$$
$$\cos\frac{(2n+1)\pi x}{2a}e^{-(2n+1)^2\pi^2 Dt/(4a^2)}. \tag{5-88b}$$

Hence, C can be found as

$$C = \frac{4C_0}{\pi}\sum_{n=0}^{\infty}\frac{(-1)^n}{(2n+1)}\cos\frac{(2n+1)\pi x}{2a}e^{-(2n+1)^2\pi^2 Dt/(4a^2)} + \frac{p(a^2 - x^2)}{2D}$$
$$- \frac{16pa^2}{D\pi^3}\sum_{n=0}^{\infty}\frac{(-1)^n}{(2n+1)^3}\cos\frac{(2n+1)\pi x}{2a}e^{-(2n+1)^2\pi^2 Dt/(4a^2)} \tag{5-88c}$$

The average concentration in the slab at any given time t is

$$\overline{C} = \frac{1}{2a}\int_{-a}^{a}C\,dx, \tag{5-89a}$$

leading to

$$\overline{C} = \frac{pa^2}{3D}\left[1 - \frac{96}{\pi^4}\sum_{n=0}^{\infty}\frac{e^{-(2n+1)^2\pi^2 Dt/(4a^2)}}{(2n+1)^4}\right] + \frac{8C_0}{\pi^2}\sum_{n=0}^{\infty}\frac{e^{-(2n+1)^2\pi^2 Dt/(4a^2)}}{(2n+1)^2}. \tag{5-89b}$$

The special case of $p=0$ has already been discussed (Equations 5-78 and 5-79). For the special case of $C_0 = 0$ (no initial radiogenic daughter), fractional ^{40}Ar loss is

$$F = 1 - \frac{\overline{C}}{pt} = 1 - \frac{a^2}{3Dt}\left[1 - \frac{96}{\pi^4}\sum_{n=0}^{\infty}\frac{1}{(2n+1)^4}e^{-(2n+1)^2\pi^2 Dt/(4a^2)}\right]. \tag{5-90}$$

Interestingly, the fractional mass loss depends on only one parameter Dt/a^2, and is independent of p. Nonetheless, the fractional mass loss given by the above equation differs from the case of uniform initial distribution and no growth (Equation 5-80), which is not surprising. The fraction of mass loss is plotted as the dashed curve in Figure 5-15, and compared with simple diffusive loss without growth (solid curve in the figure). At a given time, the fraction of mass loss when there is continuous growth is smaller than simple loss from the initial uniform distribution. This is expected because much of the recently produced ^{40}Ar has not had much time to escape.

Now the behavior of the general solution of Equation 5-89b is examined. As $t \to \infty$, the volume-averaged concentration reaches a constant value (steady state):

$$\overline{C}_{t \to \infty} = \frac{pa^2}{3D}. \tag{5-91}$$

The time required for achieving the steady-state concentration depends on the value of C_0 and the precision required. Take the required precision to be 1% relative. If $C_0 \leq pa^2/D$, then the condition of $t \to \infty$ is roughly satisfied when $Dt/a^2 \geq 2$. If $C_0 > pa^2/(3D)$, then steady state is reached when Dt/a^2 is greater than $(4/\pi^2)\ln[100|8Q/\pi^2 - 96/\pi^4|]$, where $Q = C_0/[pa^2/(3D)]$. The steady-state volume-average concentration corresponds to a steady-state apparent age of

$$\bar{t}_{ss} = \overline{C}/p = a^2/(3D). \tag{5-92}$$

That is, in the case of isothermal diffusive loss and continuous growth with constant growth rate, the apparent age of the bulk mineral asymptotically approaches a constant determined by the size of the mineral grain and the diffusivity alone, and independent of the initial concentration or the growth rate (or concentration of the parent nuclides). For example, if $a = 0.5$ mm and $D = 10^{-20}$ m^2/s, the steady-state apparent age is: $\bar{t}_{ss} = 0.264$ Myr. Because D decreases with decreasing temperature, the steady-state apparent age increases with decreasing temperature for isothermal diffusion and constant growth rate. Figure 5-16 shows how the volume-averaged apparent age of the mineral grains approaches the steady-state value for the case of $C_0 = 0$ and of $C_0 = pa^2/D$.

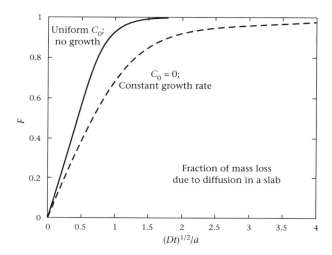

Figure 5-15 Fraction of mass loss due to diffusion in a thin slab: comparison between (i) uniform initial concentration and no growth (solid curve), and (ii) zero initial concentration and constant growth rate (dashed curve). Some specific values are as follows (F_1 for solid curve; F_2 for dashed curve):

α	F_1	F_2
0.1	0.11284	0.07523
0.2	0.22568	0.15045
0.3	0.33861	0.22568
0.4	0.45124	0.30089
0.5	0.56223	0.37584
0.6	0.66653	0.44947
0.8	0.83290	0.58498
1.0	0.93126	0.69452
1.5	0.99685	0.85242

In addition to volume-averaged concentration in the mineral, the concentration distribution inside the mineral may also be investigated. From Equation 5-88c, the steady-state concentration profile is

$$C = p(a^2 - x^2)/(2D). \tag{5-93}$$

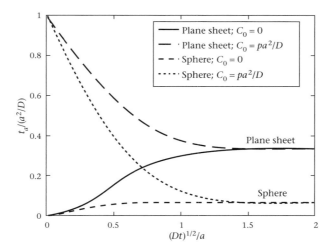

Figure 5-16 The dependence of the volume-average apparent age on the dimensionless parameter $(Dt)^{1/2}/a$. The unit of the apparent age is a^2/D. The volume-average apparent age is defined for the whole mineral grains (both plane sheets and spheres). If $C_0 = 0$, then the volume-average apparent age increases gradually to the steady-state value. If $C_0 = pa^2/D$, then the volume-average apparent age decreases gradually to the steady-state value. The steady-state volume-average apparent age is reached when $(Dt)^{1/2}/a > 1.6$.

Hence, every point in the concentration profile corresponds to a steady-state apparent age of

$$t_{ss} = (a^2 - x^2/(2D).$$ (5-94)

At the center, the steady-state apparent age is $t_{ss} = a^2/(2D)$, 1.5 times the volume-averaged apparent age of the whole grain.

Solid sphere Next we consider the solution to the diffusion equation in a solid sphere of radius a with constant D, uniform initial concentration, zero surface concentration, and a constant production rate of p. The diffusion equation is

$$\frac{\partial C}{\partial t} = \frac{D}{r^2} \frac{\partial}{\partial r} \left(r^2 \frac{\partial C}{\partial r} \right) + p, \qquad t > 0, 0 < r < a,$$ (5-95a)

where C is the concentration of ^{40}Ar, and $p = \lambda_e{}^{40}$K and assumed to time-independent. The initial condition is

$$C|_{t=0} = C_0.$$ (5-95b)

The boundary conditions are

$$C|_{r=a} = 0.$$ (5-95c)

The solution for the zero initial and boundary conditions may be found from Carslaw and Jaeger (1959, p. 242). The nonzero initial condition may be treated the same way as for the case of plane sheets. The procedure for simplifying the problem is as follows. First, let $w = rC$; then Equation 5-95a becomes

$$\frac{\partial w}{\partial t} = D\frac{\partial^2 w}{\partial r^2} + pr, \qquad t > 0, 0 < r < a. \tag{5-96}$$

Then let $u = w - pr^3/6 = rC - pr^3/6$, and the above equation becomes

$$\frac{\partial u}{\partial t} = D\frac{\partial^2 u}{\partial r^2}, \qquad t > 0, 0 < r < a. \tag{5-97a}$$

The initial and boundary conditions become

$$u|_{t=0} = rC_0 - pr^3/6. \tag{5-97b}$$

$$u|_{r=0} = 0. \tag{5-97c}$$

$$u|_{r=a} = aC_0 - pa^3/6. \tag{5-97d}$$

The function u can hence be solved. The final solution for $C = u/r + pr^2/6$ is

$$C = \frac{p}{6D}(a^2 - r^2) + \frac{2pa^3}{D\pi^3}\sum_{n=1}^{\infty}(-1)^n\frac{1}{n^3}e^{-n^2\pi^2 Dt/a^2}\sin\frac{n\pi r}{a}$$
$$+ \frac{2aC_0}{\pi r}\sum_{n=1}^{\infty}\frac{(-1)^{n+1}}{n}\sin\frac{n\pi r}{a}e^{-n^2\pi^2 Dt/a^2}. \tag{5-98}$$

For small values of Dt/a^2, the following equation converges more rapidly

$$C = pt - \frac{4apt}{r}\sum_{n=0}^{\infty}\left\{i^2\text{erfc}\frac{(2n+1)a - r}{\sqrt{4Dt}} - i^2\text{erfc}\frac{(2n+1)a + r}{\sqrt{4Dt}}\right\}$$
$$+ C_0 - \frac{aC_0}{r}\sum_{n=0}^{\infty}\left\{\text{erfc}\frac{(2n+1)a - r}{\sqrt{4Dt}} - \text{erfc}\frac{(2n+1)a + r}{\sqrt{4Dt}}\right\}. \tag{5-99}$$

The volume-averaged concentration in the solid sphere may be found as (Wolf et al., 1998)

$$\overline{C} = \frac{pa^2}{15D} - \frac{6pa^2}{D\pi^4}\sum_{n=1}^{\infty}\frac{1}{n^4}e^{-n^2\pi^2 Dt/a^2} + \frac{6C_0}{\pi^2}\sum_{n=1}^{\infty}\frac{1}{n^2}e^{-n^2\pi^2 Dt/a^2} \tag{5-100}$$

As $t \to \infty$, the steady-state concentration profile is

$$C = p(a^2 - r^2)/(6D). \tag{5-101}$$

and the volume-averaged steady-state concentration is $pa^2/(15D)$. At the steady state, the apparent age at every point in the sphere is

$$t_{ss} = (a^2 - r^2)/(6D). \tag{5-102}$$

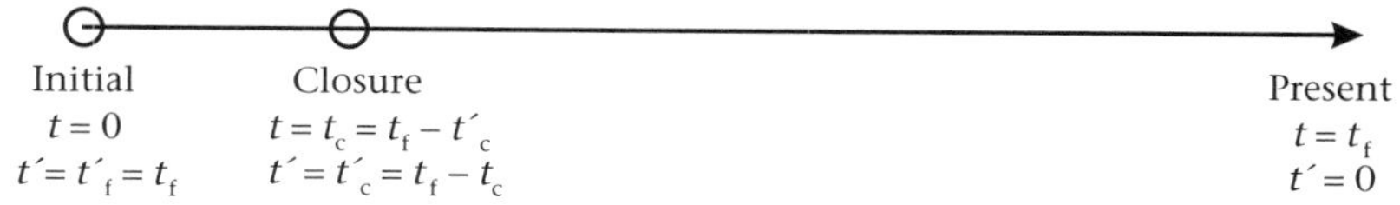

Initial Closure Present

$t = 0$ $t = t_c = t_f - t'_c$ $t = t_f$

$t' = t'_f = t_f$ $t' = t'_c = t_f - t_c$ $t' = 0$

Figure 5-17 The relation between time and age.

The volume-averaged apparent age is $a^2/(15D)$, and the apparent age at the center of the sphere is $a^2/(6D)$, 2.5 times the volume average.

When a sphere of radius a is compared to plane sheets with half-thickness a, the volume-averaged apparent age for the case of sphere is 1/5 of that for the case of plane sheet, and the apparent age at the center for the case of sphere is 1/3 of that for the case of plane sheet. Figure 5-16 shows how the volume-averaged apparent age of the mineral grains approaches the steady-state value for the case of (a) $C_0 = 0$ and (b) $C_0 = pa^2/D$.

5.2.2.3 Diffusive loss upon cooling and radiogenic growth

In this section, we treat the full problem of time-dependent D and time-dependent radiogenic growth. To avoid confusion, we use symbol t to denote time and t' to denote age. The relation between them is shown in Figure 5-17, where t_f is time since formation (also the formation age), t_c is the time of closure, and t'_c is the closure age.

The growth rate for ^{40}Ar is $\lambda_e P_0 e^{-\lambda t}$, where P_0 is the initial concentration of ^{40}K. The solution to this problem follows Dodson (1973). Assume asymptotic cooling is

$$T = T_0/(1 + t/\tau_c).$$

where τ_c is the cooling timescale. The diffusivity is then expressed as (Equation 3-55a)

$$D = Ae^{-E/(RT)} = D_0 e^{-t/\tau},$$

where A is the pre-exponential factor, E is the activation energy for diffusion, R is the gas constant, $D_0 = A\,e^{-E/(RT_0)}$ is the initial diffusivity (not the pre-exponential factor), and τ is a timescale for D to decrease by a factor of e and equals

$$\tau = \tau_c RT_0/E.$$

For the general case of one-dimensional diffusion of a radiogenic component in a slab of half-thickness a under asymptotic cooling, the diffusion equation is, hence,

$$\frac{\partial C}{\partial t} = D_0 e^{-t/\tau}\frac{\partial^2 C}{\partial x^2} + \lambda_e P_0 e^{-\lambda t}, \qquad t > 0, \ -a < x < a, \tag{5-103a}$$

where C is the concentration of ^{40}Ar, P_0 is the initial concentration of ^{40}K, λ is decay constant of ^{40}K (5.543×10^{-10} yr^{-1}), and λ_e is the branch decay constant of ^{40}K to ^{40}Ar (5.81×10^{-11} yr^{-1}). The initial condition is

$$C|_{t=0} = 0. \tag{5-103b}$$

The boundary conditions are

$$C|_{x=\pm a} = 0. \tag{5-103c}$$

Define dimensionless concentration $u = \lambda C/(\lambda_e P_0)$, dimensionless time $\theta = t/\tau$, and dimensionless distance $\xi = x/a$; then

$$\frac{\partial u}{\partial \theta} = \frac{D_0 \tau}{a^2} e^{-\theta} \frac{\partial^2 u}{\partial \xi^2} + \lambda \tau e^{-\lambda \tau \theta}, \qquad \theta > 0, \ -1 < \xi < 1. \tag{5-104}$$

Let $Q = 1 - e^{-\lambda t} - u = 1 - e^{-\lambda \tau \theta} - u$, in which $1 - e^{-\lambda \tau \theta}$ is the radiogenic production term, u is the remaining fraction, and Q is a measure of diffusive mass loss. The above equation becomes

$$\frac{\partial Q}{\partial \theta} = \frac{D_0 \tau}{a^2} e^{-\theta} \frac{\partial^2 Q}{\partial \xi^2}, \qquad \theta > 0, \ -1 < \xi < 1. \tag{5-105}$$

Let $\alpha^2 = \int (D_0 \tau/a^2) e^{-\theta} d\theta = (D_0 \tau/a^2)(1 - e^{-\theta})$. Then

$$\frac{\partial Q}{\partial \alpha} = \frac{\partial^2 Q}{\partial \xi^2}, \qquad \theta > 0, \ -1 < \xi < 1. \tag{5-106a}$$

The initial condition is

$$Q|_{\alpha=0} = 0. \tag{5-106b}$$

The boundary conditions are

$$Q|_{\xi=\pm 1} = 1 - e^{-\lambda \tau \theta} = 1 - [1 - \alpha^2 a^2/(D_0 \tau)]^{\lambda \tau}. \tag{5-106c}$$

This diffusion problem is a standard problem and the analytical solution is given as a summation of integral terms (Carslaw and Jaeger, 1959, p. 104). The solution for $\theta \to \infty$ so that $e^{-\theta} = 0$ (t is finite but is $\gg \tau$; $\alpha \to D_0 \tau/a^2$) is (Dodson, 1973)

$$Q = 2 \sum_{n=1}^{\infty} (-1)^{n+1} \frac{\cos[(n - \frac{1}{2})\pi x/a]}{(n - \frac{1}{2})\pi} \left\{ 1 - \frac{\Gamma(1 + \lambda \tau)}{[(n - \frac{1}{2})^2 \pi^2 M]^{\lambda \tau}} \right\}, \tag{5-107}$$

where $\Gamma(z)$ is the gamma function and M is a dimensionless parameter defined as α^2 at $t \to \infty$:

$$M = D_0 \tau/a^2. \tag{5-108}$$

The volume average is

$$\overline{Q} = 1 - \frac{2\Gamma(1+\lambda\tau)}{M^{\lambda\tau}\pi^{2(1+\lambda\tau)}} \sum_{n=1}^{\infty} \frac{1}{(n-\frac{1}{2})^{2(1+\lambda\tau)}}. \tag{5-109}$$

Using similar procedures, the solution for infinitely long cylinders as $t \to \infty$ is

$$Q = 2 \sum_{n=1}^{\infty} \frac{J_0(\mu_n r/a)}{\mu_n J_1(\mu_n)} \left\{ 1 - \frac{\Gamma(1+\lambda\tau)}{(\mu_n^2 M)^{\lambda\tau}} \right\}, \tag{5-110}$$

where $J_0(z)$ and $J_1(z)$ are Bessel functions of zeroth and first order, and μ_n is the nth root of $J_0(z)$. The volume average is

$$\overline{Q} = 1 - \frac{4\Gamma(1+\lambda\tau)}{M^{\lambda\tau}} \sum_{n=1}^{\infty} \frac{1}{\mu_n^{2(1+\lambda\tau)}}. \tag{5-111}$$

For spheres, the solution for $t \to \infty$ is

$$Q = 2 \sum_{n=1}^{\infty} (-1)^{n+1} \frac{\sin(n\pi r/a)}{n\pi r/a} \left[1 - \frac{\Gamma(1+\lambda\tau)}{[n^2\pi^2 M]^{\lambda\tau}} \right]. \tag{5-112}$$

The volume average is

$$\overline{Q} = 1 - \frac{6\Gamma(1+\lambda\tau)}{M^{\lambda\tau}\pi^{2(1+\lambda\tau)}} \sum_{n=1}^{\infty} \frac{1}{n^{2(1+\lambda\tau)}}. \tag{5-113}$$

In all three geometries, $\overline{Q}$ can be written as

$$\overline{Q} = 1 - \frac{\Gamma(1+\lambda\tau)}{M^{\lambda\tau}} \sum_{n=1}^{\infty} \frac{B}{\omega_n^2(1+\lambda\tau)}, \tag{5-114}$$

where $\omega_n = n - \frac{1}{2}$ for plane sheets, μ_n for long cylinders, and n for spheres, and $B = 2/\pi^{2(1+\lambda\tau)}$ for plane sheets, 4 for cylinders, and $6/\pi^{2(1+\lambda\tau)}$ for spheres. The concentration C can be calculated as

$$C = (1 - e^{-\lambda t} - Q)\lambda_e P_0/\lambda. \tag{5-115}$$

Dodson (1973, 1986) adopted that at closure time t_c, the accumulation of Ar is zero ($C=0$), leading to

$$Q_c = 1 - e^{-\lambda t_c}. \tag{5-116}$$

From $T = T_0/(1 + t/t_c)$, therefore,

$$\frac{E}{RT_c} = \frac{E}{RT_0}\left(1 + \frac{t_c}{\tau_c}\right) = \frac{E}{RT_0} + \frac{t_c}{\tau}. \tag{5-117}$$

Use Equation 5-116 to obtain t_c and replace it in Equation 5-117:

$$\frac{E}{RT_c} = \frac{E}{RT_0} - \frac{1}{\lambda\tau}\ln(1 - Q_c). \tag{5-118}$$

Because

$$\ln M = \ln(D_0\tau/a^2) = \ln[(A\tau/a^2)e^{-E/(RT_0)}] = \ln(A\tau/a^2) - E/(RT_0), \qquad (5\text{-}119)$$

hence,

$$\frac{E}{RT_c} = \ln\frac{A\tau}{a^2} - \ln M - \frac{1}{\lambda\tau}\ln(1 - Q_c) = \ln\frac{A\tau}{a^2} - \ln[M(1 - Q_c)^{1/(\lambda\tau)}]. \qquad (5\text{-}120)$$

The above equation is in the form of Equation 5-75b if we recognize that

$$G = \frac{1}{M(1 - Q)^{1/(\lambda\tau)}}. \qquad (5\text{-}121)$$

From Equation 5-120 all other closure temperature equations may be obtained. The evaluation of G values takes some effort because the series in Equations 5-107 to 5-113 converge slowly. For the limiting case of $\lambda\tau \to 0$, Dodson (1973) obtained the values of G (shape factor) to be 8.65 for plane sheets with infinite area, 27 for infinitely long cylinders, and 55 for spheres.

5.2.2.4 ^{40}Ar concentration profile and age

In the ^{40}Ar–^{39}Ar method of dating (which is the most often used method for ^{40}K–^{40}Ar system), ^{40}Ar/^{39}Ar age spectrum is obtained from the release of ^{40}Ar and ^{39}Ar. Because ^{39}Ar is from conversion of ^{39}K, the concentration of ^{39}Ar is proportional to that of ^{39}K. Because ^{39}K is part of the mineral structure, its concentration may be treated as uniform throughout the mineral. Hence, ^{39}Ar concentration is also uniform in the mineral. On the other hand, ^{40}Ar concentration reflects the growth and diffusive loss and would not be uniform.

A specific case of the evolution of ^{40}Ar concentration profile is shown in Figure 5-18. The assumed parameters are $T = T_0/(1 + t/\tau_c)$, where $T_0 = 1000$ K and $\tau_c = 30$ Myr, and $D = Ae^{-E/(RT)} = D_0 e^{-t/\tau}$, where $A = 1.07 \times 10^{-6}$ m^2/s $= 3.37 \times 10^7$ mm^2/yr, $E/R = 30{,}000$ K, $D_0 = 10^{-19}$ m^2/s $= 3.16 \times 10^{-6}$ mm^2/yr, and $\tau = 1$ Myr. ^{40}Ar is gradually produced, and diffuses away from the surfaces, producing a smooth profile. The present day is at $t = 100$ Myr.

Figure 5-19 compares the ^{40}Ar concentration profile with the case of ^{40}Ar concentration without diffusive loss. It can be seen that at high temperatures (at 0.791 Myr into cooling, or at a temperature of 974 K), most (about 90%) of radiogenic ^{40}Ar is lost; if the initial diffusivity D_0 is greater, there would be even more complete ^{40}Ar loss. After cooling down, there is essentially complete retention of ^{40}Ar. The average ^{40}Ar concentration in the mineral can be found by numerical integration of the whole profile at 100 Myr. Not only can the whole mineral closure temperature and closure age be calculated, but so can the closure age and closure temperature of every point along the profile. Figure 5-20 shows such results.

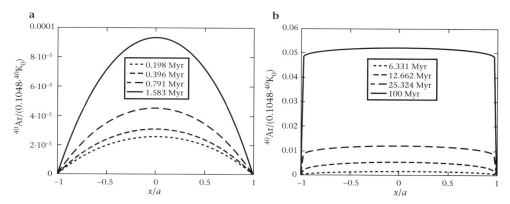

Figure 5-18 Numerically calculated evolution of ^{40}Ar concentration profile at various times. The input parameters are: $D = D_0 e^{-t/\tau}$, where $D_0 = 10^{-19}$ m^2/s and $\tau = 1$ Myr, half-thickness $a = 0.5$ mm, and the total time span (true age) is 100 Myr. The corresponding temperature history is $T/K = 1000/(1 + t/\tau_c)$, where $\tau_c = 30$ Myr.

5.2.3 More developments on the closure temperature concept

In the initial development by Dodson (1973), only the whole mineral properties (closure temperature and closure age) are considered because at that time single-point age determination was not available. Nonetheless, the solutions to the diffusion and radiogenic growth equation indicate that closure age and closure temperature may be defined for every point along the profile (if the profile can be measured). Dodson (1986) developed a simple way to obtain closure temperature and age of every point in the interior of a mineral. With laser heating, it is now possible to obtain concentration profiles of K and 40Ar in a mineral (such as phlogopite or hornblende), and hence the theory can be applied. Another new development concerns an assumption in the theory by Dodson (1973, 1986) that the initial temperature is high enough so that the closure temperature does not depend on the initial temperature. Ganguly and Tirone (1999, 2001) treated the cases where this assumption is not satisfied. The consideration of these additional effects takes the form of a correction factor, which can be incorporated in the shape factor G in Equations 5-75a to 5-77b to allow the calculation of closure temperature and age. Another assumption is that the timescale for D to decrease is much smaller than the half-life of the radioactive parent ($\lambda\tau \ll 1$). Usually this is not a problem because the half-lives of the radioactive nuclides are of the order of the age of the Earth.

5.2.3.1 Closure temperature profile in a single grain

As shown in Figure 5-20, the closure age and closure temperature can be calculated at each point along the ^{40}Ar concentration profile using forward modeling.

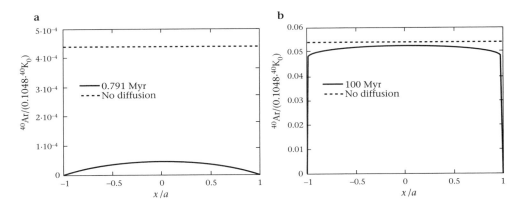

Figure 5-19 Numerically calculated ^{40}Ar concentration profile (solid curves) compared with pure growth profile (dashed lines) at two times during a single cooling history. Same input parameters as Figure 5-18.

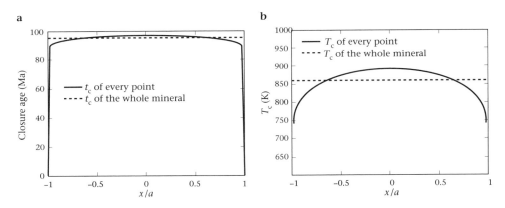

Figure 5-20 Closure age and closure temperature at every point along the profile (solid curves), and the bulk mineral closure age and closure temperature (dashed lines). Same input parameters as Figure 5-18.

To calculate the closure temperature of any point in a mineral without carrying out the full forward numerical calculations, Dodson (1986) analyzed the problems for different effective shapes systematically and modified his closure temperature equation for the whole minerals slightly to apply to individual points. His formulation is by adding a correction term to Equation 5-75b. This correction term will be referred to as g_1 in this book and another correction term by Ganguly and Tirone (1999, 2001) will be referred to as g_2. The formulation of Dodson (1986) for the calculation of closure temperature at every point of a profile is

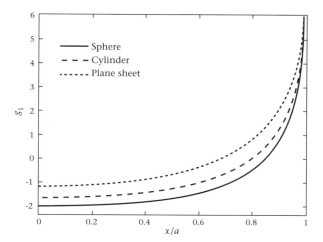

Figure 5-21 Plot of g_1 value versus position in the crystal, x/a, where a is the radius for sphere and long cylinder and half-thickness for plane sheet, and x is the radial coordinate for sphere and long cylinder and abscissa for plane sheet.

$$\frac{E}{RT_c} = \ln\left(\frac{GA\tau}{a^2}\right) + g_1, \tag{5-122}$$

where g_1 is a correction function depending on the normalized distance x/a, where a is the half-thickness for plane sheets and the radius for long cylinders and spheres. The numerical values of the g_1 function are given in Table 5-4 and plotted in Figure 5-21. Increasing g_1 value means a decrease of the closure temperature. A negative g_1 value (e.g., at the center of the grain) means the closure temperature at the position is higher than that of the whole grain, and a positive g_1 value means T_c at this location is lower than that of the whole grain. The closure temperature of the whole mineral grain is roughly the same as T_c at $x/a = 0.858$ if the effective shape of the mineral grain is a sphere, at $x/a = 0.803$ if the effective shape is a long cylinder, and at $x/a = 0.683$ if the effective shape is a plane sheet. In practical calculations, one may incorporate the correction function g_1 into G to obtain a new parameter $G' = G\,e^{g_1}$, and use this new G' to replace G in Equations 5-75a to 5-77. See Example 5-10 for calculation procedures.

Example 5.10 Treat hornblende grains as isotropic spheres of radius 0.5 mm. Diffusivity is given by $D = \exp(-12.94 - 32{,}257/T)$ m²/s. Calculate T_c of the whole mineral grains and T_c at the center of the grains for cooling rate of 30 K/Myr.

Solution: Use Equation 5-76a. From the conditions given, $E/R = 32{,}257$ K; $A = \exp(-12.94)\,\text{m}^2/\text{s} = 2.4 \times 10^{-6}\,\text{m}^2/\text{s}$; $a = 5 \times 10^{-4}\,\text{m}$; $q = 30\,\text{K/Myr} = 9.51 \times 10^{-13}\,\text{K/s}$.

Table 5-4 Values of the correction function g_1

x/a	Sphere	Cylinder	Plane sheet
0.00 (center)	−2.04309	−1.69345	−1.16905
0.05	−2.03958	−1.67953	−1.16446
0.10	−2.02899	−1.67794	−1.15063
0.15	−2.01112	−1.65840	−1.12732
0.20	−1.98567	−1.63056	−1.09414
0.25	−1.95215	−1.59391	−1.05052
0.30	−1.90992	−1.54773	−0.99564
0.35	−1.85807	−1.49108	−0.92845
0.40	−1.79546	−1.42269	−0.84752
0.45	−1.72051	−1.34089	−0.75100
0.50	−1.63114	−1.24345	−0.63641
0.55	−1.52454	−1.12733	−0.50042
0.60	−1.39676	−0.98835	−0.33845
0.65	−1.24217	−0.82052	−0.14400
0.70	−1.05234	−0.61493	0.09253
0.75	−0.81395	−0.35758	0.38611
0.80	−0.50427	−0.02467	0.76192
0.85	−0.07988	0.42890	1.26727
0.90	0.55858	1.10566	2.00852
0.95	1.73517	2.33687	3.32537
0.98	3.40464	4.05617	5.11635
0.99	4.73190	5.39699	6.48878
0.995	6.06723	6.75723	7.86814

Note.Obtained from Dodson (1986) by subtracting the volume average value from the position-dependent $G(x)$ value.

(1) For the whole grains, $G = 55$. Use Equation 5-76a to iterate, $T_c = 869.35$ K.

(2) For the center of the grain, $g_1 = -2.04309$. Hence, $G' = G\,e^{g_1} = 7.13$. Use Equation 5-76a to iterate, $T_c = 917.20$ K, which is higher than the whole mineral T_c by 48 K.

Near the center of a crystal, the correction function g_1 does not change much with position. For example, for $x/a \leq 0.3$, the variation in g_1 is ≤ 0.17, which would cause an error of only a couple of degrees (depending on the activation energy) in T_c. Hence, it is particularly robust to use the center part of a grain to extract T_c. Near the rim, g_1 varies rapidly with position and it is not robust to obtain T_c. The method of using T_c at the center of a grain and along an $^{40}Ar-^{39}Ar$ profile is advantageous to that of bulk mineral grains because the full profile contains more information than the average grains. Furthermore, the uncertainty on whether the whole grains are broken or have cracks can be avoided by the choice of the most pristine, crack-free, and inclusion-free grain for concentration profile measurement.

5.2.3.2 Dependence on the initial temperature

In the treatment of Dodson (1973, 1979, 1986), the initial temperature is assumed to be high enough for the given cooling rate so that there is complete loss or reequilibrium (i.e., no Ar accumulation) at the initial temperature even at the center of the mineral grain. Retention or closure occurs at a lower temperature. Because of complete loss at the initial temperature, the value of the initial temperature does not matter, meaning that the closure temperature depends only on the cooling rate and diffusion properties, but independent of the initial temperature.

The assumption holds if the initial temperature is much greater than the closure temperature, e.g., at least 100 K above the closure temperature (depending on the activation energy). Hence, it usually holds for volcanic and igneous rocks. However, it may not work for metamorphic rocks with low peak temperature. Ganguly and Tirone (1999) considered the case for small degree of initial loss of the daughter isotope, and derived the correction factor g (referred to as g_2 below), in addition to the shape factor G. The following equation is obtained to calculate the closure temperature (T_c):

$$\frac{E}{RT_c} = \ln\left(\frac{GA\tau}{a^2}\right) + g_2, \tag{5-123}$$

where $g_2 \geq 0$ is a correction factor for finite degree of initial loss of the daughter nuclide, and the rest of the equation is Equation 5-75b. The above equation reduces to the simple closure temperature equation (Equation 5-75b) if $g_2 = 0$. As g_2 increases, T_c decreases. The closure temperature for the case of low T_0 is less than that for the case of high T_0 (i.e., Dodson's T_c).

Table 5-5 Values of the correction function g_2

M	Sphere			Cylinder			Plane sheet		
	$x/a = 0.25$	$x/a = 0.5$	$x/a = 1.0$	$x/a = 0.25$	$x/a = 0.5$	$x/a = 1.0$	$x/a = 0.25$	$x/a = 0.5$	$x/a = 1.0$
0.001	4.891	4.709	3.105	5.331	5.189	3.914	5.879	5.571	4.815
0.005	3.281	3.100	1.755	3.715	3.565	7.433	4.269	4.141	3.297
0.010	2.588	2.407	1.246	3.019	2.864	1.847	3.576	3.448	2.671
0.020	1.895	1.715	0.804	2.321	2.164	1.311	2.883	2.755	2.072
0.040	1.206	1.047	0.447	1.623	1.474	0.841	2.191	2.067	1.511
0.060	0.823	0.697	0.286	1.221	1.091	0.606	1.788	1.674	1.207
0.080	0.579	0.484	0.195	0.950	0.841	0.461	1.508	1.405	1.005
0.100	0.416	0.345	0.137	0.754	0.664	0.361	1.296	1.204	0.858
0.120	0.303	0.251	0.099	0.607	0.533	0.288	1.129	1.046	0.743
0.140	0.224	0.185	0.073	0.495	0.433	0.234	0.993	0.919	0.652
0.160	0.167	0.138	0.054	0.407	0.356	0.192	0.880	0.814	0.577
0.180	0.126	0.104	0.041	0.337	0.294	0.158	0.785	0.726	0.514
0.200	0.095	0.079	0.031	0.281	0.245	0.132	0.703	0.650	0.460
0.300	0.026	0.021	0.008	0.121	0.105	0.057	0.430	0.397	0.281
0.400	0.008	0.006	0.003	0.056	0.049	0.026	0.278	0.257	0.182
0.600	0.001	0.001	0.000	0.013	0.012	0.006	0.128	0.118	0.083

Note. From Ganguly and Tirone (2001).

The value of g_2 depends on $M = D_0 \tau / a^2$ (Equation 5-108) and the position x/a in the mineral. Because $D_0 = Ae^{-E/(RT_0)}$, M depends on the initial temperature T_0. As $M \to \infty$, $g_2 \to 0$. The relation between g_2 and M is shown in Table 5-5 for various positions inside a crystal. Equation 5-109 reduces to Equation 5-75b (that is, $g_2 \leq 0.01$) when $M > 0.3$ for a sphere, 0.6 for a cylinder, and 1.2 for a plane sheet. Example 5-11 shows how to evaluate the effect of low initial temperature on the closure temperature.

Example 5.11 Treat hornblende grains as isotropic spheres of radius 0.5 mm. Diffusivity is given by $D = \exp(-12.94 - 32{,}257/T)$ m^2/s. Calculate T_c of for a point in hornblende with $x/a = 0.25$. Assume a cooling rate of 30 K/Myr and initial temperature of 900 K.

Solution: From the conditions given, $E/R = 32,257$ K; $A = \exp(-12.94)$ $m^2/s = 2.4 \times 10^{-6}$ m^2/s; $a = 5 \times 10^{-4}$ m; $q = 30$ K/Myr $= 9.51 \times 10^{-13}$ K/s.

(1) For the case of high initial temperature, $G' = 55e^{g_1} = 7.81$ (where $g_1 = -1.952$ is found in Table 5-4). Use Equation 5-76a to iterate and find that Dodson $T_c = 915.96$ K. This temperature is higher than the assumed initial temperature of 900 K, and cannot be correct. (T_c must be $\leq T_0$.)

(2) For $T_0 = 900$ K, T_c can be found as follows. First find $D_0\tau/a^2$. From the conditions given,
$D_0 = \exp(-12.94 - 32,257/900) = 6.5 \times 10^{-22}$ m^2/s;
$\tau_c = T_0/q = 9.5 \times 10^{14}$ s;
$\tau = \tau_c T_0/(E/R) = 2.64 \times 10^{13}$ s.

Hence, $M = D_0\tau/a^2 = 0.06895$.

Use Table 5-5 and interpolation to obtain $g_2 \approx 0.71$. Hence, $G' = 55$, $e^{g_1}e^{g_2} \approx 15.9$. Using Equation 5-76a to iterate, we obtain $T_c = 897.82$ K. This T_c is very close to the initial temperature, meaning not much Ar loss occurred by cooling down from such low T_0.

5.2.3.3 Closure in other systems

Although the above discussion specifically concerns ^{40}Ar, the results also apply to other radiogenic isotopes. For example, the application to ^4He, which satisfies zero surface concentration, is straightforward. The closure temperature concept also applies to exchange equilibrium between a mineral and its homogeneous surroundings. Considering Sm–Nd isotopic system, there would be isotopic exchange between the mineral of interest and the surroundings. At high temperatures, Nd isotopes would exchange with the surroundings rapidly, leading to homogenization of $^{143}Nd/^{144}Nd$ ratio in all minerals with different Sm/Nd ratios. That is, there would be no differential radiogenic increase of $^{143}Nd/^{144}Nd$ ratio in minerals. Therefore, the isochron clock would begin to tick only when the temperature is low enough, leading to differential $^{143}Nd/^{144}Nd$ growth from different Sm/Nd ratio.

The closure temperature concept is also applicable to systems other than radiogenic isotopes. For example, for oxygen isotope fractionation (or Fe–Mg fractionation among ferromagnesian minerals), at high temperatures, oxygen isotope diffusion is rapid and all phases are roughly at isotopic equilibrium (slightly different $^{18}O/^{16}O$ ratios due to stable isotope fractionation). At lower temperatures, diffusion is slow or insignificant, $^{18}O/^{16}O$ ratio in a given mineral would not change with temperature even though the fractionation factor depends on temperature. Hence, the closure temperature may be calculated for oxygen isotope exchange in a given mineral, and the closure time may also be defined. One difference with the radiogenic system is that the system cannot be used to determine the closure age. Another difference is that for a given mineral the other minerals do not behave as a uniform infinite reservoir. Therefore, the boundary conditions are different, leading to complexities. Some of these complexities are explored in Section 5.3.4.2.

5.2.4 Applications

Although the mathematics is complicated in deriving and quantifying the closure temperature concept, the final results for thermochronology applications (Equations 5-75a to 5-77) are easy to use if the conditions are satisfied. For a given set of conditions, solving for T_c using Equation 5-76a requires numerical method, such as bisection, or iteration, which can be easily achieved using a spreadsheet program. With the closure temperature concept, by obtaining apparent age (or closure age) from thermochronology and estimating closure temperature using Equation 5-76a, one obtains one point (t_c, T_c) in temperature–time history. With the same isotopic system but different minerals (hence, with different T_c) or different isotopic system (also with different T_c), several (t_c, T_c) points can be obtained. With enough data points, by plotting these data on a temperature versus time diagram, the cooling history is obtained. This is the most widely used method for inferring thermal history if cooling rate is slow, and it has been widely applied for understanding tectonic events, because it provides both absolute time and temperature (methods in the next section provide only cooling rates, but not absolute timing). One example of cooling history based on closure age concept is shown in Figure 1-21.

To calculate the closure temperature using Equation 5-76a, it is necessary to know the diffusion parameters (activation energy E and pre-exponential factor A) of the mineral, the grain size and shape, and the cooling rate. Below are some considerations about these parameters.

(1) For minerals with diffusion anisotropy, the shape of the mineral is the effective shape after coordinate transformation (Section 3.2.11). For Ar and He diffusion in minerals, often the diffusivities along different crystallographic directions are difficult to obtain. In experimental studies, an effective shape of the crystal is assumed to obtain diffusivity. For example, Harrison (1981) inferred Ar diffusivity from experiments by assuming that hornblende grains are isotropic spheres. When one applies the diffusivity of Harrison (1981), it is necessary to maintain consistency and assume that hornblende grains are isotropic spheres (unless new diffusion data accounting for anisotropy are available) even if the isotropic assumption is likely wrong (e.g., oxygen isotope diffusion in hornblende is anisotropic). Table 1-3c lists some diffusion data and assumed shapes of minerals.

(2) Based on Equation 5-76a, knowing the diffusion properties (E and A) and grain size, we must know q to calculate T_c. However, the cooling rate is not known a priori and must be guessed. Nonetheless, because of the weak

dependence of T_c on q, T_c can often be estimated from a very rough estimate of q (such as 1 K/Myr). For example, calculation using a specific set of diffusion properties shows that if q varies by a factor of 10, T_c would vary by only 46 K. With enough T_c versus t_c data, q at a given temperature may be estimated from the slope of the T_c versus t_c diagram (i.e., q does not have to be guessed). Then T_c for each system can be recalculated with confidence.

(3) The mineral grains selected for closure age determination must be whole grains for Equation 5-76a to be applicable. The grain size must be estimated. For interior point analysis, it is necessary to determine the relative position of the point in the grain (x/a), and the appropriate correction factor (g_1 value) must be applied.

(4) The most critical parameters in thermochronology applications are diffusion parameters (activation energy E and pre-exponential factor A). Although many users of thermochronology simply treat the expression of diffusivity as known, several issues may affect the accuracy of thermochronology results.

(4a) The accuracy of the experimental diffusivity expression is important. Because of the typically low concentrations of Ar and He in minerals, the diffusivity of species (such as Ar and He) important for thermochronology is usually obtained using bulk extraction experiments. Mineral powders are heated up to extract Ar or He from the powders. Then the diffusivity is calculated (see Section 3.6.1.2) by assuming an effective shape of the powders (see Section 3.2.11). As discussed in Section 3.6.1.2, the bulk method may overestimate the diffusivity by a huge factor. It might be argued that natural mineral grains always have imperfections. With this argument, as long as the imperfections in experimental mineral grains are the same as those in natural samples, the diffusivities would be applicable.

(4b) The D values are determined at relatively high temperatures where the diffusivity is large enough to produce significant diffusion for measurement. In geologic applications, it is often necessary to extrapolate the expressions obtained at high temperatures to low temperatures. Sometimes huge extrapolation is employed. There may be two kinds of uncertainties in extrapolation down temperature. One is that at lower temperatures, the diffusion mechanism may be different with a lower activation energy. Thus, extrapolation may be systematically off by a large factor. This systematic error is difficult to estimate. The second is the errors in extrapolation because the activation energy itself has uncertainties. It is best to experimentally determine diffusivity at temperatures near T_c to minimize the extrapolation.

(4c) Most minerals are anisotropic in terms of diffusion. When oxygen isotope diffusivity is determined using the profiling method, the diffusivity along different directions often differs by an order of magnitude. For Ar and He diffusion, the diffusive anisotropy is not quantified, but an effective shape is assumed to treat diffusion in a given mineral. The assumed effective shape may be incorrect, which would cause uncertainty in obtaining accurate thermal history.

(4d) For ^{4}He diffusion, because ^{4}He is the product of α-decay, ^{4}He is ejected from the original site of the parent by 10 to 30 μm, depending on the decay energy and the type of the mineral. Farley et al. (1996) considered the effect of α-particle ejection. For a homogeneous distribution of parent nuclides, when the characteristic length of the mineral is much larger than α-particle stopping distance, the fraction of α-particle loss (F) from a given phase is about

$$F \approx \left(\tfrac{1}{4}\right) x_\alpha (S/V), \tag{5-124}$$

where x_α is the α-stopping distance (about 20 μm), and S/V is the ratio of the total surface area over total volume. For example, for spherical particles (this shape is the real physical shape, not the effective shape for diffusion) of radius a, the surface area to volume ratio is $4\pi a^2/(4\pi a^3/3) = 3/a$. Hence, $F \approx 3x_\alpha/(4a)$. If $a = 100$ μm, then $F \approx 15\%$. This loss needs to be corrected in thermochronology calculations. More advanced treatment will need to consider the interaction of the loss with diffusion.

(4e) Radiation damage of the mineral is expected to increase the diffusivity, especially for He and Pb diffusion (because of high U and Th concentrations). For young samples, this effect would not be significant. For old samples, ignoring the effect may produce some uncertainty.

Despite some difficulties, the closure temperature concept and apparent age measurements are the pillars to the inference of thermal history for slow cooling. For rapid cooling with cooling timescale of less than a year (such as cooling of volcanic rocks, which is important for understanding the degassing, welding, crystallinity, and other textures of volcanic rocks), the isotopic systems do not have enough resolution and hence other methods (geospeedometry methods in Section 5.3) are necessary. However, these other methods can provide only the cooling rate, and not temperature–time history. The various thermochronology systems are briefly outlined below.

^{40}K–^{40}Ar system (and ^{40}Ar–^{39}Ar method) This system is the mostly widely employed in thermochronology. The minerals of interest are K-bearing minerals of hornblende (with T_c about 770 K, depending on cooling rate), phlogopite (with T_c about 650 K), biotite (with T_c about 550 K), and orthoclase (with T_c about 530 K) (Table 1-3c). The T_c range corresponds to depths of 10 to 20 km. Hence, the system has found extensive applications in inferring cooling history of plutonic and metamorphic rocks.

U–Th–^{4}He system Recently there has been much work on this system in zircon and apatite. Each ^{238}U produces eight ^{4}He particles, ^{235}U seven ^{4}He particles, and ^{232}Th six ^{4}He particles. In addition, each of the ^{147}Sm and ^{190}Pt atoms produce one ^{4}He particle. He is a small atom (molecule) and diffuses rapidly. Therefore, the closure temperatures are low (330 to 500 K), corresponding to depths of a few to 10 km. The low closure temperatures are especially useful to investigate the cooling rate near the surface. By assuming that the temperature is related to depth through the geothermal gradient or through sophisticated thermokinematic modeling, the cooling rate may be related to the erosion rate.

U–Th–Pb system The closure temperature of this system is higher than that of the ^{40}K–^{40}Ar system. Hence, the system is applicable to cooling at greater depth. For the U–Th–Pb system in zircon, the closure temperature is very high. Peak metamorphic age and or zircon crystallization age may be inferred. Zircon is so resistant to re-equilibration that we may encounter the opposite problem: instead of a young closure age, it may record multiple growth ages if zircons are xenocrysts from older rocks. Zircon crystals in igneous rocks may record ages older than the magmatic age. Hence, the zircon U–Th–Pb method is very powerful in mapping out detailed thermal history even prior to the recent cooling. This system is usually coupled with other systems with lower closure temperature, so as to infer both the high-temperature and cooling history.

Fission track method Fission tracks are radiation damage due to spontaneous fission of ^{238}U and induced fission of ^{235}U. The number of fission tracks is related to U concentration, the age, and whether tracks have healed. Usually the annealing or healing temperature of fission tracks is relatively low, about 500 K for zircon, and 400 K for apatite, similar to the closure temperature range of the U–Th–^{4}He system. Unless the minerals formed at such low temperatures or cooled extremely rapidly (such as volcanic rocks), fission track dating is most useful in determining the closure temperature and closure age at relatively low temperatures. This method has been around for a long time but has been plagued by uncertainties. The recently developed U–Th–^{4}He system is applicable to similar closure temperature ranges and seems to be more accurate.

Inference of thermal history is most easily done for simple monotonic cooling. Nonmonotonic temperature–time history is difficult to resolve. Recently, there have been developments in inferring disturbances such as that by wildfire using fission track and U–Th–He techniques because these have low closure temperatures and can be easily disturbed. That is, the complexity may be turned into a tool to probe wildfire frequency in the geologic past.

Other isotopic systems, such as ^{147}Sm–^{143}Nd, ^{87}Rb–^{87}Sr, ^{176}Lu–^{176}Hf systems, can in principle all be applied to thermochronology. However, because of (i) long half-lives of the parent nuclide and (ii) either the small fractionation between the parent and daughter elements or the high abundance of the daughter nuclide, they are useful only for very old ages, where some tens of million years cooling history is difficult if not impossible to resolve. Hence, in practice, they are rarely applied to thermochronology.

To determine the entire thermal history from high temperature to low temperature, it is best to use as many systems as possible to obtain many points in the closure temperature versus closure age curve, as shown in Figure 1-21. For specific purposes, such as trying to understand the low-temperature history, appropriate systems can be used.

The theory of thermochronology is still rapidly developing. New developments include three-dimensional modeling of heat conduction–convection coupled with crustal exhumation history, using the full age spectrum to model continuous thermal history, etc. Volume 58 of *Reviews in Mineralogy and Geochemistry*, edited by Reiners and Ehlers (2005), provides an excellent coverage of various thermochronology methods.

After obtaining the thermal history, it is necessary to interpret it, which is often the more important goal. For plutonic rocks, the initial part of the cooling history from a relatively high temperature such as 1100 to 600 K (depending on the depth of the plutonic rock) is likely due to cooling by heat loss from the newly crystallized igneous body to the ambient rock. The gradual uplift of the plutonic body might play a role, but it is usually secondary in this initial stage. When the temperature of the plutonic rock is similar to that of the ambient rocks, cooling of the igneous body relative to the ambient rocks becomes insignificant, and slow cooling during this stage is more likely attributable to uplift or erosion of the whole region. For example, in Figure 1-21, the initial rapid cooling of the granitoid may be attributed to cooling of magma body in country rock by heat transfer. Because the country rock temperature is likely no less than 200°C (about 8 km depth), the slow cooling at temperatures below 200°C may be attributed to the uplift of the granitoid due to erosion. Volume 58 of *Reviews of Mineralogy and Geochemistry*, edited by Reiners and Ehlers (2005), and a review article by Reiners and Brandon (2006) discussed how to use thermochronology to understand erosion.

> **Example 5.12** Use Figure 1-21 to estimate the average cooling rate from 200 to 100°C. Assuming that the cooling is due to slow uplift following a normal geothermal gradient of 25°C/km, estimate the uplift rate.
>
> *Solution*: The closure age corresponding to the temperature of 200°C is about 100 Ma, and that to 100°C is about 80 Ma. Hence, the cooling rate is 100°C over 20 Myr, or 5°C/Myr. Along a normal geothermal gradient of 25°C/km, the uplift rate is about 0.2 km/Myr, or 0.2 mm/yr. According to Ehlers (2005), this erosion rate is not rapid enough to produce significant departure from normal geotherm. Hence, the estimate of erosion or uplift rate using the normal geotherm is OK.

5.3 Geospeedometry

As is often the case, the most powerful and useful methods are often the simplest and most elegant. Among inverse problems in geochemical kinetics, geochronology is the most elegant in terms of mathematical treatment and the most useful in geology. Thermochronology is a powerful tool in inferring the full thermal history. The methods covered in this section, geospeedometry methods to infer cooling rates, are in general less powerful. The geospeedometry methods, coupled with thermobarometry, may provide complementary thermal information of the rocks, such as the formation temperature of minerals or mineral assemblages and subsequent cooling rate.

Thermochronology may be said to be the best geospeedometer by providing both temperature and time from isotopic measurements. Nonetheless, there are many situations for which thermochronology does not work. In these situations, the geospeedometry methods are especially useful. For example, for rapidly cooled rocks (such as volcanic rocks), thermochronology does not have the resolution to infer the cooling rate, but geospeedometry comes handy for such a situation. Furthermore, for prograde metamorphic history, thermochronology may not provide much help either, but geospeedometry coupled with thermobarometry methods are often able to offer some constraints.

There are many geospeedometry methods. Essentially, all temperature-dependent reaction rates are bases for developing cooling rate indicators. Geospeedometers may be based on homogeneous reaction kinetics, diffusion kinetics, or heterogeneous reaction kinetics. Homogeneous reactions here refer to chemical homogeneous reactions, not including nuclear decays. The theory for homogeneous reaction geospeedometry is relatively simple and well developed, but only a few reactions have been investigated in enough detail to be of practical use because of measurement difficulties on homogeneous reactions. This type of geospeedometry is especially useful at inferring rapid cooling rates. The theory for diffusion-based geospeedometry is also well developed and many applications have been found. Because of the complexity of heterogeneous reaction kinetics, even though the heterogeneous reactions record more information about the thermal history, it is more difficult to quantify such information.

5.3.1 Quantitative geospeedometry based on homogeneous reactions

For a given homogeneous reaction that is thermally activated (i.e., the reaction rate coefficient depends on temperature), the extent of the reaction depends on the thermal history (especially the cooling rate) of the rock. If the cooling rate is high, the extent of the reaction reflects an apparent equilibrium at high temperature. If the cooling rate is low, the extent of the reaction reflects an apparent equilibrium at low temperature (i.e., more reaction as the system cooled down). The dependence of the extent of the reaction on the cooling rate is the basis for geospeedometers using homogeneous reactions. The apparent equilibrium temperature (Section 1.7.3; Figure 1-22) plays the same role in geospeedometry as the closure temperature in thermochronology.

Although the theory of geospeedometry is well developed for simple homogeneous reactions, the applications are limited because only a few homogeneous reactions have been investigated to enough detail for this application. These geospeedometers apply well to rapid cooling, but do not apply well to slow cooling because calibration is done on the timescale of less than 10 years and extrapolation to a timescale of millions of years would result in large uncertainty. Therefore, for slow cooling, it is best to apply thermochronology to infer cooling rate and the temperature–time history; for rapid cooling for which thermochronology may be able to determine the age but may not be able to resolve the thermal history, cooling rate may be inferred from homogeneous reaction geospeedometers.

In Chapter 2, the concentration evolution as a function of time for reversible reactions during cooling was investigated. The task of geospeedometry is opposite to that of forward modeling. In forward modeling, the thermal history is known and the final species concentrations are calculated. In geospeedometry, the final species concentrations are known by measuring the composition of minerals, and we want to find the cooling history. For both forward and inverse modeling, it is necessary to know the equilibrium constant and the reaction rate coefficient as a function of temperature. Unless other information is available, usually only the cooling rate at a single temperature (the apparent equilibrium temperature, or T_{ae}) in a continuous cooling function may be inferred from the final species distribution. Theoretically, different homogeneous reactions can provide cooling rates at different T_{ae}. However, in practice, this is not possible because for a given geologic problem, finding one homogeneous reaction to be applicable is lucky enough.

In the literature, various methods to infer cooling rate from the measured species concentrations of a homogeneous reaction have been developed. These methods are summarized and assessed below. Then two specific geospeedometers based on two homogeneous reactions are presented.

5.3.1.1 Various methods of geospeedometry

Four methods are available in the literature for inferring cooling rate or cooling timescale from measured species concentrations in homogeneous reaction geospeedometers: temperature–time transformation, Ganguly's method, Zhang's equation, and the empirical method. They are outlined below.

Temperature–time transformation The temperature–time transformation, or T-t-T method (e.g., Seifert and Virgo, 1975), is the oldest method in geospeedometry. In this method, a reasonably high initial temperature is given, and equilibrium species concentrations are calculated. This speciation is assumed to be the initial speciation. The final species concentrations after cooling down (i.e., at present day) are measured and hence known. To reach the present-day species concentration from the assumed initial concentration requires time, and this time would depend on the assumed temperature at which reaction happens. At a given temperature, the time to reach the observed species concentration (such as concentration of Fe in M1 site in orthopyroxene) in a rock from the given initial species concentration through isothermal reaction is calculated. Then the temperature is varied, and a new time is calculated. The results (T versus t required to reach the observed species concentrations) are plotted on a T versus log(t) diagram. A set of cooling history curves, such as $T(t) = T_0/(1 + t/\tau_c)$ with different cooling timescales τ_c, are also plotted on the same diagram. The cooling history curve tangential to the curve of the observed speciation is assumed to give the cooling history of the sample. This method is approximate because it assumes that the time to reach the observed speciation during cooling to a given temperature is the same as that at the given constant temperature (Ganguly, 1982). In practice, it often recovers τ to within a factor of two of the accurate τ (Zhang, 1994). However, because this is an approximate method and because it involves fairly complicated calculations, the method is not used anymore in geospeedometry calculations based on homogeneous reaction kinetics.

Ganguly's method The second method was developed by Ganguly (1982). In this method, a cooling history, such as the asymptotic cooling function $T(t) = T_0/(1 + t/\tau_c)$ with a specific τ_c, is given. The final species concentrations corresponding to the cooling history are solved numerically as follows. The cooling history is divided into many small time divisions. In each small time interval, the temperatures k_f and k_b are assumed to be constant and the reaction progress in that time interval is calculated by solving the reaction rate law equation with constant k_f and k_b (Section 2.1.4.2). The method can be made to reach a given precision if sufficiently small time steps are chosen. After numerically solving the equation, if the final species concentrations do not match the observed concentrations, then the cooling timescale τ_c is changed, and the calculation is redone. The process is iterated until the calculated final species concentrations are in agreement with the observed concentrations. This method can reach the required precision by decreasing the Δt (time step) of the calculation.

Zhang's equation The third method was developed by Zhang (1994). This method is based on a theoretically derived relation between T_{ae}, cooling rate (q), and kinetic parameters for a special case, which is generalized by examining and synthesizing results of many numerical simulations. The resulting equation (Equation 1-117) is analogous to the closure temperature equation and is written below for easy reference:

$$q \approx \frac{2RT_{ae}^2}{\tau_r E},$$

(5-125)

where q is the cooling rate when the temperature was T_{ae}, E is the greater of the forward and backward reaction activation energies, and τ_r is the mean reaction time at T_{ae}. The equations for calculating the parameter τ_r are given in Table 2-1. In Chapter 1, the equation was given but not derived. In this section, the equation is derived mathematically for a special case, first-order reversible reactions with $E_f = 2E_b$ and with an asymptotic cooling history, because for this special case there is a simple analytical solution. The derivation is in Box 5-1. Zhang (1994) carried out many numerical simulations using various types of reactions and cooling history and shows that Equation 5-125 recovers cooling rate to within a factor of 1.25. More recent simulations covering more extreme conditions (Zhang, unpublished work) show that Equation 5-125 recovers cooling rate within a factor of 1.6 (i.e., the error is significantly smaller than the T-t-T method).

The similarity between the apparent equilibrium temperature equation and the closure temperature equation can be seen by comparing Equation 5-125 and Equation 5-77b: by letting T_{ae} and T_c be equivalent, τ_r (reaction timescale) and τ_d (diffusion timescale) be equivalent, and $G = 2$, the two equations become the same.

Given measured species concentrations for a homogeneous reaction in a rock, cooling rate at T_{ae} can be found as follows if the equilibrium constant K and the forward reaction rate coefficient k_f as a function of temperature are known. First, the apparent equilibrium temperature is calculated from the species concentrations. Then k_f and k_b at T_{ae} are calculated. Then the mean reaction time τ_r at T_{ae} is calculated using expressions in Table 2-1. From τ_r, the cooling rate q at T_{ae} can be obtained using Equation 5-125. Two examples are given below.

Example 5.13 This example tests the applicability and accuracy of Equation 5-125 using forward calculation results of Example 2-1. We want to use the resulting T_{ae} from the forward calculation to infer the cooling rate. From Example 2-1, for a reversible first-order reaction, $k_f = 40 \exp(-15{,}000/T)$ and $k_b = 0.02 \exp(-7500/T)$, where k_f and k_b are in yr^{-1}. T_{ae} has been found to be 857.93 K in Example 2-1. Find the cooling rate at T_{ae} using Equation 5-125 and compare it with the assumed cooling function $T = 1500/(1 + t/10^6)$, where t is in years.

Solution: From Table 2-1, the reaction timescale for reversible first-order reaction is $\tau_r = 1/(k_f + k_b)$. Hence, knowing $T_{ae} = 857.93$ K, we find the mean reaction time τ_r at T_{ae} as follows:

$$\tau_r = \frac{1}{k_f + k_b} = \frac{1}{40e^{-15{,}000/T_{ae}} + 0.02e^{-7500/T_{ae}}} = 2.37 \times 10^5 \text{ yr.}$$

The greater of the forward and backward $E/R = 15{,}000$ K. Hence, q at T_{ae} can be found as follows (Equation 5-125):

$$q \approx \frac{2RT_{ae}^2}{\tau_r E} = \frac{2 \times 857.93^2}{2.37 \times 10^5 \times 15{,}000} = 4.14 \times 10^{-4} \text{ K/yr.}$$

Box 5.1 Derivation of Equation 5-125 for the special case of first-order reversible reactions with $E_f = 2E_b$, and an asymptotic cooling history with $T_\infty = 0$ K.

The mean reaction time for a first-order reversible reaction is given by Equation 2-7, $\tau_r = 1/(k_f + k_b)$. That is,

$$\tau_r|_{T_{ae}} = \frac{1}{A_f e^{-E_f/(RT_{ae})} + A_b e^{-E_b/(RT_{ae})}}.$$

Because $E_f = 2E_b$, $E_b = \Delta H$, where ΔH is the standard state enthalpy change of the reaction, we have,

$$e^{-E_b/(RT_{ae})} = e^{-\Delta H/(RT_{ae})} = \frac{1}{A_K} \frac{[B]_\infty}{[A]_\infty}.$$

$$e^{-E_f/(RT_{ae})} = e^{-2\Delta H/(RT_{ae})} = \left(\frac{1}{A_K} \frac{[B]_\infty}{[A]_\infty}\right)^2.$$

Therefore,

$$\tau_r|_{T_{ae}} = \frac{1}{A_f\left(\dfrac{1}{A_K}\dfrac{[B]_\infty}{[A]_\infty}\right)^2 + A_b\dfrac{1}{A_K}\dfrac{[B]_\infty}{[A]_\infty}} = \frac{A_K}{A_b}\frac{[A]_\infty}{[B]_\infty}\frac{[A]_\infty}{[A]_\infty + [B]_\infty}.$$

From Equation 2-44, if $\eta_\infty \gg 1$, then

$$[A]_\infty \approx ([A]_0 + [B]_0)\sqrt{\pi}\eta_\infty \exp(\eta_\infty^2)\mathrm{erfc}(\eta_\infty) \approx ([A]_0 + [B]_0)(1 - 0.5\eta_\infty^{-2}).$$

Hence,

$$[A]_\infty/([A]_\infty + [B]_\infty) = [A]_\infty/([A]_0 + [B]_0) \approx (1 - 0.5\eta_\infty^{-2}),$$
$$[B]_\infty/([A]_0 + [B]_0) \approx 0.5\eta_\infty^{-2},$$
$$[A]_\infty/[B]_\infty \approx (1 - 0.5\eta_\infty^{-2})/(0.5\eta_\infty^{-2}) = 2\eta_\infty^2 - 1.$$

Therefore,

$$\tau_r|_{T_{ae}} = \frac{A_K}{A_b}\frac{[A]_\infty}{[B]_\infty}\frac{[A]_\infty}{[A]_\infty + [B]_\infty} \approx \frac{A_K}{A_b}(2\eta_\infty^2 - 1)\left(1 - \frac{1}{2\eta_\infty^2}\right) \approx 2\frac{A_K}{A_b}\eta_\infty^2.$$

Because $\eta_\infty^2 = k_{f0}\tau_f(k_{b0}/k_{f0})^2$, we have

$$\tau_r|_{T_{ae}} \approx 2\frac{A_K}{A_b}k_{f0}\tau_f\left(\frac{k_{b0}}{k_{f0}}\right)^2 = 2\frac{A_f}{A_b^2}\tau_f\frac{k_{b0}^2}{k_{f0}} = 2\tau_f = \tau_b = \frac{RT_{ae}^2}{E_b q}.$$

That is,

$$q \approx \frac{RT_{ae}^2}{E_b\tau_r} = \frac{2RT_{ae}^2}{E_f\tau_r}.$$

Because E_f is the greater of the forward and backward reaction activation energies, the above equation is identical to Equation 5-125. Hence, the equation is proven for this special case.

From the assumed cooling history of $T = 1500/(1 + t/10^6)$ with t in years, the input cooling rate q at T_{ae} is:

$$q|_{T_{ae}} = -\frac{dT}{dt}\bigg|_{T_{ae}} = \frac{T_0}{\tau_c(1 + t/\tau_c)^2}\bigg|_{T_{ae}} = \frac{T_{ae}^2}{\tau_c T_0} = \frac{857.93^2}{10^6 \times 1500} = 4.91 \times 10^{-4} \text{ K /yr.}$$

The cooling rate at T_{ae} obtained from Equation 5-125 is 4.14×10^{-4} K/yr and deviates from the input cooling rate of 4.91×10^{-4} K/yr by 16% relative.

Example 5.14 For a hypothetical first-order reaction $A \rightleftharpoons B$ in a mineral, suppose

$$k_f = \exp(-10 - 22{,}000/T) \text{ and } k_b = \exp(-10 - 24{,}000/T)$$

where k is in s^{-1} and T is in K. After the mineral is cooled, T_{ae} is found to be 900 K. Find the cooling rate at T_{ae}.

Solution: Based on the information given, $E/R = \max(22{,}000,\ 24{,}000) = 24{,}000$ K. To find τ_r, the reaction timescale at T_{ae}, we use the equation for Reaction 1 in Table 2-1: $\tau_r = 1/(k_f + k_b)$. At T_{ae}, $k_f = 1.10 \times 10^{-15}$ s^{-1} and $k_b = 1.19 \times 10^{-16}$ s^{-1}. Hence, $\tau_r = 1/(k_f + k_b) = 8.2 \times 10^{14}$ $s = 26$ Myr. Therefore,

$$q \approx \frac{2RT_{ae}^2}{\tau_r E} = 2.6 \text{ K/Myr.}$$

Empirical method Zhang et al. (1997b, 2000) used this method to calibrate the hydrous species geospeedometer. In the above discussion, it is assumed that the reaction rate law is known and the dependence of the equilibrium constant and rate coefficients on temperature and composition are also known. These constants and coefficients are determined from isothermal experiments. That is, the isothermal experiments provide theoretical and indirect calibration to the geospeedometer. For some homogeneous reactions in silicate melts and minerals, however, the reaction rate law may not be perfectly known. Then it would be difficult to understand kinetic results from isothermal experiments, let alone to infer cooling history. For these reactions, calibration of the geospeedometer may be carried out empirically as follows. Controlled cooling rate experiments are conducted. From such experimental data, a direct but empirical relation between cooling rate and speciation can be obtained (the relation may depend on the composition). From this relation, the cooling rate can be obtained from measured speciation of natural samples (Zhang et al., 1997b, 2000).

Unlike the theoretical method, which can be extrapolated to some degree (as long as the temperature dependences of K and k_f apply to lower temperatures), it is best not to extrapolate a geospeedometer calibrated using the empirical method. Hence, it is necessary that experiments cover the cooling rates to be encountered in nature. Because experiments can last at most a few years, the mean cooling time of natural processes cannot be much longer than a few years for the empirical method to be applicable. Hence, such empirical calibration of geospeedometers can be applied only to systems cooled very rapidly, such as volcanic rocks.

Comparison of various methods For the first three methods, it is necessary to know how the equilibrium constant of the reaction depends on temperature (and often on the composition of the phase), the reaction rate law, and how the rate coefficients depend on temperature (and the composition). The empirical method directly relates cooling rate with cooled species concentrations. The first three methods have better extrapolation capabilities, whereas the empirical method does not have much extrapolation ability. The empirical method, hence, only works on a cooling timescale of several years or less.

Among the first three methods, the *T-t-T* method is approximate and requires fairly complicated calculations. Hence, it is no longer used and is not recommended for future use. Ganguly (1982) numerically solved the kinetic equation under cooling. Zhang (1994) extracted a relation (Equation 5-125) from the numerical simulations. It is easy to use Equation 5-125 to extract q, with a maximum uncertainty of ≤ 0.47 in $\ln q$ (or 0.20 in terms of $\log q$), better than the current precision in experimental calibration. Hence, the use of Equation 5-125 is recommended unless a program to numerically solve the appropriate differential equation is already available. Note that 0.20 log units are the maximum error in using Equation 5-125 to approximate the real solution. There are additional errors in analytical data precision and in the accuracy of the equilibrium and kinetic constants.

With the above general background, we are now ready to apply some widely used geospeedometers based on homogeneous reactions. The equilibrium and kinetics of these reactions have already been discussed earlier. The geospeedometry application is the focus in the following two subsections.

5.3.1.2 Geospeedometry based on the Fe–Mg order–disorder reaction in orthopyroxene

This section focuses on how the Fe–Mg order–disorder reaction (Section 2.1.4) is applied as a geospeedometer. The equilibrium and kinetics of the reaction are discussed in Section 2.1.4 and only a brief review is provided here. Although there is some complexity in the kinetics of this reaction (e.g., Figure 2-5), it is minor, and is hence usually ignored so that the forward and backward reactions are treated as elementary reactions. The rate coefficient for the forward reaction of this reaction (Reaction 2-55)

$$Fe^{(M2)}Mg^{(M1)}Si_2O_6 \rightleftharpoons Mg^{(M2)}Fe^{(M1)}Si_2O_6 \tag{5-126}$$

has been assessed to follow the expression (Equation 2-60; Kroll et al., 1997)

$$\ln k_f = 23.33 - (32,241 - 6016X_{Fs}^2)/T, \tag{5-127}$$

where X_{Fs} is the mole fraction of the ferrosilite component and k_f is in s^{-1}. The "exchange equilibrium constant" depends on the method of measurement. For Mössbauer measurements (Equation 2-57; Wang et al., 2005),

$$K_D = \exp(0.391 - 2205/T). \tag{5-128}$$

For X-ray diffraction measurement (Equations 2-58a,b; Stimpfl et al., 1999),

$$K_D = \exp(0.547 - 2557/T), \quad 0.19 < X_{Fs} < 0.75; \tag{5-129a}$$

$$K_D = \exp(0.603 - 2854/T), \quad 0.11 < X_{Fs} < 0.17. \tag{5-129b}$$

The backward reaction rate coefficient can be calculated from the forward reaction rate coefficient and the exchange equilibrium constant.

The calculation of cooling rate is straightforward using Equation 5-125. From measured Fe and Mg concentrations in M1 and M2 sites in orthopyroxene by either Mössbauer or XRD method, the apparent equilibrium constant $K_{ae} = (Fe/Mg)^{(M1)}/(Fe/Mg)^{(M2)}$ may be calculated. Then T_{ae} may be calculated because the dependence of K on T is known (one of Equations 5-128 to 5-129b). Then k_f and k_b at T_{ae} are calculated. Next the mean reaction time at T_{ae} is calculated using the appropriate expression in Table 2-1 (Reaction 2, third cell) and letting both the instantaneous and the equilibrium concentrations be the measured species concentrations. Then the cooling rate at T_{ae} may be calculated using Equation 5-125. The procedures are easy to include in a spreadsheet program and are shown in Example 5-15. The example also verifies that experimental cooling rate may be retrieved from the geospeedometer to within a factor of 2 from the equilibrium and kinetic constants given above.

Example 5.15 Schlenz et al. (2001) investigated the reaction $Fe_{M2}^{2+} + Mg_{M1}^{2+} \rightleftharpoons Fe_{M1}^{2+} + Mg_{M2}^{2+}$ with continuous cooling $T = T_0/(1 + t/\tau_c)$, where $T_0 = 1023$ K and $\tau_c = 48.68$ d. After cooling down, the species concentrations based on XRD measurements are $X_{Fe}^{M1} = 0.0397$, $X_{Mg}^{M1} = 0.9568$, $X_{Fe}^{M2} = 0.3450$, and $X_{Mg}^{M2} = 0.6518$.

(1) Find T_{ae}.

(2) Find q at T_{ae}.

(3) Compare q calculated from the final species concentrations with experimental q at T_{ae}.

Solution:

(1) First, $K_{ae} = (Fe/Mg)_{M1}/(Fe/Mg)_{M2} = 0.0784$, and $X_{Fs} = (0.0397 + 0.3450)/2 = 0.192$. Hence, Equation 5-129a should be used, leading to
$$T_{ae} = 2557/(0.547 - \ln K_{ae}) = 826.7 \text{ K. This is the answer to question (1).}$$

(2) Next, from $X_{Fs} = 0.192$ and Equations 5-127 and 5-129a, we obtain
$E_f/R = 32{,}241 - 6016 \cdot 0.192^2 = 32{,}018$ K,

$E_b/R = 32{,}018 - 2557 = 29{,}461$ K,

$k_f = \exp(23.33 - 32{,}019/T)$, and

$k_b = k_f/K = \exp(22.783 - 29{,}462/T)$.

Then $k_f|_{T_{ae}} = 2.049 \times 10^{-7} \text{s}^{-1}$, and $k_b|_{T_{ae}} = k_f/K_{ae} = 2.614 \times 10^{-6} \text{s}^{-1}$. From Table 2-1,

$$\tau_r = \frac{1}{k_f(X_{Fe}^{M2} + X_{Mg_\infty}^{M1}) + k_b(X_{Fe}^{M1} + X_{Mg_\infty}^{M2})}.$$

where the concentrations are the observed concentrations. Hence,

$$\tau_r \approx \frac{1}{2.049 \times 10^{-7}(0.3450 + 0.9568) + 2.614 \times 10^{-6}(0.0397 + 0.6518)}$$
$$= 482{,}100 \text{ s.}$$

Therefore, q at T_{ae} can be found from Equation 5-125:

$$q \approx \frac{2T_{ae}^2}{\tau_r(E_{max}/R)} = \frac{2 \times 826.7^2}{4.821 \times 10^5 \times 32{,}018} = 8.85 \times 10^{-5} \text{ K/s} = 7.7 \text{ K/d.}$$

(3) The experimental cooling rate when $T = T_{ae}$ is

$$q|_{T_{ae}} = -\left.\frac{dT}{dt}\right|_{T_{ae}} = \left.\frac{T_0}{\tau_c(1 + t/\tau_c)^2}\right|_{T_{ae}} = \frac{T_{ae}^2}{\tau_c T_0} = \frac{826.7^2}{48.68 \times 1023} = 13.7 \text{ K/d.}$$

The inferred cooling rate (7.7 K/d) is within a factor of two of the experimental cooling rate (13.7 K/d). The difference of a factor 1.8 is due to (i) the inaccuracy of Equation 5-125, which is likely minor, (ii) uncertainty in the calculation of T_{ae} from species concentrations (Equation 5-129a), and (iii) errors in the dependence of the kinetic coefficient on temperature (Equation 5-127). This difference of a factor of 1.8 is considered small, taking into consideration of the various uncertainties. (Usually, when cooling rate can be estimated to within a factor of 2, it is considered excellent agreement.)

The following factors should be considered in using this geospeedometer:

(1) The experimental timescale is no more than a few years, but the cooling timescale of interest is often of the order of millions of years. That is, experimental kinetic data often must be extrapolated by 6 orders of magnitude in timescale (about 300 K in temperature). Because the formulation of this geospeedometer has a theoretical basis, some extrapolation is OK. However, the reliability of huge extrapolations by six orders of magnitude cannot be evaluated.

(2) Because of the long cooling timescale of most rocks, natural orthopyroxenes are often highly ordered (i.e., have a very low T_{ae}). Hence, the Fe concentration in M1 site may be extremely low, often barely detectable. One extreme case can be found in Ganguly et al. (1994) who reported Fe(M1) content to be between 0.0019 to 0.0025 in orthopyroxene from a meteorite (Bondoc, a mesosiderite). Thus, measurement accuracy in the measured Fe concentration in M1 site may be poor, leading to huge uncertainties as well as unreliability in inferred cooling rates. Examples can be found below. The conclusion is that with such low Fe(M1) concentration, the geospeedometer should not be applied.

(3) Nonmonotonic thermal history may mess up the calculation.

In summary, the orthopyroxene geospeedometer is best applied to rocks that cooled rapidly so that Fe(M1) can be measured to enough precision and extrapolation of experimental results is small.

Example 5.16. The following data show some XRD data on Fe–Mg distribution in orthopyroxene. The first two samples are from Skaergaard Intrusion (Ganguly and Domeneghetti,1996), and the last two rows are two repeated analyses on orthopyroxene in the meteorite Bondoc (Ganguly et al., 1994):

Sample	Fe(M1)	Mg(M1)	Fe(M2)	Mg(M2)	X_{Fs}
IVN-14	0.050 ± 0.004	0.929	0.646	0.274	0.348
CG-379	0.060 ± 0.005	0.917	0.597	0.318	0.328
Bondoc 1	0.0019	0.9791	0.3295	0.6495	0.165
Bondoc 2	0.0025	0.9755	0.3365	0.6415	0.164

Errors are given at the 2σ level. Find the cooling rate of the four samples.

Solution: Using the K_D expression of Stimpfl et al. (1999) and the k_f expression of Kroll et al. (1997), cooling rates and other parameters of the samples are found:

Sample	K_{ae}	T_{ae} (K)	$k_f(T_{ae})$, s^{-1}	$k_b(T_{ae})$, s^{-1}	τ_r (yr)	q (K/kyr)
IVN-14	0.0228	591	9.4×10^{-14}	4.2×10^{-12}	2.1×10^4	$1.04 \overset{\times}{\div} 2.7$
CG-379	0.0348	655	1.54×10^{-11}	4.4×10^{-10}	167	$163 \overset{\times}{\div} 2.7$
Bondoc 1	0.0038	463	1.05×10^{-20}	2.7×10^{18}	1.8×10^{10}	7.6×10^{-7}
Bondoc 2	0.0049	482	1.67×10^{-19}	3.4×10^{-17}	1.4×10^9	1.0×10^{-5}

Comments. The results of the first two samples agree with those of Ganguly and Domeneghetti (1996) (1 and 273 K/kyr) to within 40%. The small difference is due to (i) different formulation of k_f and K_D, and (ii) approximations introduced in the simple Equation 5-125. Based on error propagation, 2σ relative error on calculated q is a factor of about 2.7. These are very high cooling rates and hence the precision is acceptable. Although there are uncertainties associated with extrapolation, the extrapolation is not huge and the results are expected to be OK.

On the other hand, the two orthopyroxene samples (repeated analyses) of Bondoc meteorite give cooling rates from 7.6×10^{-4} to 0.01 K/Myr, which differ by a factor of 13, highlighting the large uncertainty when Fe(M1) concentration is low. Furthermore, both cooling rates are extremely slow, too slow to be real (even at the higher cooling rate of 0.01 K/Myr, total cooling from the beginning of the solar system to the present day would be only 45.7 K), indicating that the results are not accurate, due to (i) extrapolation and (ii) error in analyzing Fe(M1).

5.3.1.3 Geospeedometry based on the hydrous species reaction in rhyolitic melt

The hydrous species reaction (Reaction 2-79),

$$H_2O_m(melt) + O(melt) \rightleftharpoons 2OH(melt), \tag{5-130}$$

is discussed in Section 2.1.5. Because the reaction kinetics is complicated and no complete understanding of the reaction kinetics is available, the geospeedometer was calibrated empirically (Zhang et al., 1997b, 2000). The method of empirical calibration of geospeedometers is described in Section 5.3.1.1. To avoid extra inaccuracy introduced by converting band intensities to species concentrations, a parameter Q' is defined as follows:

$$Q' = \frac{(\overline{A}_{452})^2}{\overline{A}_{523}}, \tag{5-131}$$

where $\overline{A}_{452}$ and $\overline{A}_{523}$ are infrared peak intensities per unit thickness of a cooled sample. Because $\overline{A}_{452}$ is roughly proportional to OH content and $\overline{A}_{523}$ to

H_2O_m content, Q' is roughly proportional to K_{ae} in Equation 2-80 because usually [O] is not much different from 1. From experimental data based on controlled cooling rates, Q' is found to depend on both the cooling rate q and H_2O_t content. Again because of the desire to avoid uncertainty in IR calibration, $\overline{A}_{452} + \overline{A}_{523}$ is used to roughly represent H_2O_t. Let $x = \ln(\overline{A}_{452} + \overline{A}_{523})$; $y = \ln q$, where q is in K/s, and $z = \ln Q'$, where unit of $\overline{A}_{452}$ and $\overline{A}_{523}$ is mm^{-1}. Data from controlled cooling rate experiments were fit to obtain the following (Zhang et al., 2000):

$$z = m_0 + m_1 x + m_2 y + m_3 xy + m_4 \exp(m_5 x + m_6 y) + m_7 \exp(m_8 x), \tag{5-132}$$

with $m_0 = -5.4276$, $m_1 = -1.196$, $m_2 = -0.044536$, $m_3 = -0.023054$, $m_4 = 3.7339$, $m_5 = 0.21361$, $m_6 = 0.030617$, $m_7 = -0.37119$, and $m_8 = 1.6299$. From $\overline{A}_{452}$ and $\overline{A}_{523}$ based on an IR spectrum (which must use curved baseline to be consistent with the calibration, see Zhang, 1999b), one first calculate $x = \ln(\overline{A}_{452} + \overline{A}_{523})$ and $z = \ln Q'$. Then $y = \ln q$ may be solved from the above equation using numerical method (such as iteration). Figure 5-22 shows the relation between $\overline{A}_{452}$, $\overline{A}_{523}$, and q. Zhang et al. (2000) also presented another method of calculation for easiness of incorporation in a spreadsheet program. In this algorithm, from the $\ln Q'(x,y)$ value at x and y defined above, the $\ln Q'$ value at a fixed x value of -1.7 is first estimated as

$$\ln Q'|_{x=-1.7} = \ln Q'(x,y) + z(-1.7, y) - z(x, y), \tag{5-133}$$

where $z(-1.7, y)$ and $z(x,y)$ are calculated using Equation 5-132. Let $\xi = \ln Q'|_{x=-1.7}$. Then $\ln q$ is calculated from

$$\ln q = 8.7905 + 7.8096\xi - 3.4937\xi^2. \tag{5-134}$$

Even though calculation of $\ln q$ using Equation 5-132 and that using Equations 5-133 and 5-134 reproduced experimental data of Zhang et al. (2000) equally well, new data by Zhang and Xu (2007) show that the second algorithm is more accurate when extrapolating to lower cooling rates (down to 10^{-6} K/s). Hence, the algorithm using Equations 5-133 and 5-134 is recommended. The 2σ uncertainty in predicting $\ln q$ is about 0.5 for cooling rate range of 10^{-6} to 100 K/s.

Figure 5-22 can be used to estimate q. For more accurate calculation, the equations can be used (Example 5-17). Because this is an empirical calibration, extrapolation outside the cooling rate range of 10^{-6} to 100 K/s is not advised. Most volcanic glasses cooled rapidly and over this cooling rate range; hence, the method can be readily applied. Some volcanic glasses might have experienced nonmonotonic thermal history, such as first cooling slowly in the volcanic conduit and then being picked up by the next eruption with transient heating and rapid cooling. The nonmonotonic thermal history cannot be inferred using the hydrous species geospeedometer.

In summary, the cooling rate range for rapidly cooled volcanic glasses is large, about 10 orders of magnitude (10^{-8} to 100 K/s), and contains information on the cooling environment and welding process. Because the cooling rate variation of slowly cooled plutonic rocks is about 6 orders of magnitude, from 0.1 K/Myr to 0.1 K/yr, the relative variation in cooling rate or cooling timescale is even greater for the case of volcanic rocks than for plutonic rocks. Thermochronology based on radiogenic growth and diffusion can only resolve timescales of many thousand years, but cannot resolve rapid cooling with a timescale of seconds to a hundred years. Currently the techniques to infer such rapid cooling rates include the hydrous species geospeedometer, the heat capacity geospeedometer (Wilding et al., 1995; see below), and the oxidation geospeedometer (Tait et al., 1998; see below).

> **Example 5.17** An IR spectrum for a 0.500-mm-thick sample was measured and the absorbances of the two NIR peaks are $A_{523} = 0.100$, and $A_{452} = 0.100$. Find cooling rate q.
>
> *Solution:* First find $\overline{A}_{452} = 0.200$ and $\overline{A}_{523} = 0.200$. From Figure 5-22, $q \approx 0.01$ K/s. For more accurate calculation, use Equation 5-132 or Equations 5-133 and 5-134. Because $\ln(\overline{A}_{452} + \overline{A}_{523}) = -0.916$ and $\ln Q' = -1.609$, $\ln q$ can be solved by trial and error to be -4.07 using Equation 5-132. If Equations 5-133 and 5-134 are used, then $\ln q = -4.08$. Hence, cooling rate $q = 0.017$ K/s $= 61$ K/h.

5.3.2 Cooling history of anhydrous glasses based on heat capacity measurements

The hydrous species geospeedometer discussed above applies only to hydrous glasses, and is calibrated for hydrous rhyolitic glass only. It cannot be applied to anhydrous glasses. In Section 2.4.3.3, the C_p ($C_p = \partial H / \partial T$, where H is enthalpy) versus T curve upon heating (specifically, in the glass transition region) is shown to depend on the prior cooling rate of the glass (or, more generally, the thermal history). Figure 5-23 shows how T_{ae} and C_p versus temperature curves upon heating depend on prior cooling rate. In particular, the maximum C_p value during heating increases as the prior cooling rate decreases, which may be used in inverse applications to infer cooling rate. Because there is a whole curve, it is in principle possible to obtain the full thermal history, more than just a single cooling rate (Wilding et al., 1995, 1996a,b). This method is best applied to anhydrous glass because heat capacity measurements of hydrous glass must account for heat absorbed by dehydration. The implementation of the method using empirical and experimental approach is straightforward. For a given glass composition and a fixed heating rate (h_0), calibration to obtain the relation between the maximum C_p value and the prior cooling rate is required for empirical approach. That is, first cool down the given melt at a designated cooling rate (q). Then heat up at heating rate

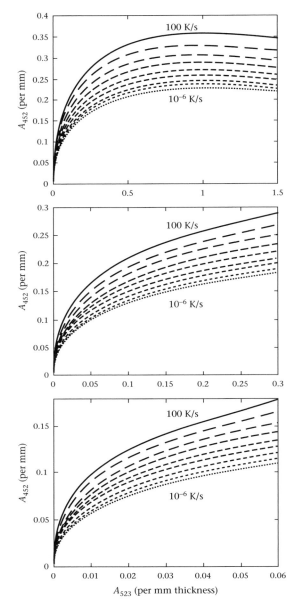

Figure 5-22 The relation between IR band intensities (peak heights per millimeter sample thickness) of the two NIR bands and quench rate q. The cooling rates between adjacent curves differ by a factor of 10. Three figures are shown so that there is enough resolution at both high and low H_2O_t contents. The figures can be used to estimate cooling rates from IR band intensities (absorbance peak height) of the 452- and 523-mm^{-1} bands per millimeter of rhyolitic glass. For example, if $A_{523} = 0.05$, $A_{452} = 0.15$, and the sample thickness is 1.25 mm, then per millimeter absorbances are (0.04, 0.12). Using the lower figure, the cooling rate is found to be close to 0.01 K/s. Using Equation 5-132, the cooling rate is found to be 0.016 K/s, or about 1°C/min. From Zhang et al. (2000) and Zhang and Xu (2007).

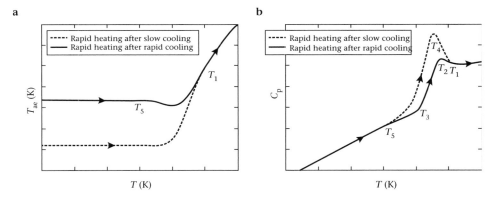

Figure 5-23 Schematic diagram showing how the apparent equilibrium temperature and heat capacity vary with temperature during heating for two samples with different prior thermal history. From Zhang (unpublished).

h_0 to obtain the C_p versus temperature curve and the maximum C_p value. Repeat this procedure for many different cooling rates (such as $q = 0.0001$ to 100 K/s). After the maximum C_p value as a function of prior cooling rate is quantified as the calibration curve, measurement of the C_p versus temperature curve upon heating of any natural glass sample with the same composition as the calibration may be used to obtain the cooling rate of the natural glass.

The quantitative geospeedometer developed by Wilding et al. (1995, 1996a,b) is based on more advanced modeling of the measured C_p curve as a function of temperature using a structural relaxation model for glass (Narayanaswamy, 1971). The model involves several fit parameters of timescales, powers, enthalpies, and fractions. Cooling rate may be obtained from the fitting results.

Using the same principles, the cooling rate of a natural glass or even a mineral may be constrained by studying the heating behavior of a homogeneous reaction in the phase. The geospeedometry methods in Section 5.3.1 use only the quenched species concentrations to infer cooling rate. More information might be inferred by (i) heating up the sample at a given rate and monitoring how T_{ae} of the reaction depends on temperature, and (ii) heating up the sample rapidly to a fixed temperature and hold the sample at the temperature, and monitoring how T_{ae} of the reaction depends on time. For example, Zhang et al. (1995) showed that there is indeed extra information stored in volcanic glass by Reaction 5-130 but quantitative reading of the information is not straightforward unless the kinetics of the reaction is fully understood.

5.3.3 Geospeedometry based on diffusion and zonation in a single phase

For a given cooling history, the diffusivity depends on time, and the mean diffusion distance may be estimated by $(\int D \, dt)^{1/2}$. Thus, if we know the dependence

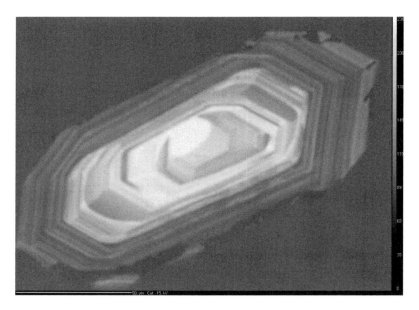

Figure 5-24 A BSE image of zircon showing a core and many growth zones. The height of the image is 240 μm. Courtesy of Charles W. Carrigan.

of D on T, the effect of diffusion may be estimated using forward modeling. In inverse problems, it is possible to use diffusion profiles, or lack of diffusion profiles, to indicate cooling rate. Because compositional zoning of a crystal may be due to crystal growth (Figure 4–22) or to diffusion after crystal growth, care must be taken to determine what is responsible for a given zoning profile by understanding the geologic circumstances before geospeedometry is applied.

Diffusion-couple profiles

Many minerals, such as garnet and zircon, show a core and a rim, or core/mantle/ rim (or even more layers, Figure 5-24), with each layer roughly uniform in composition, but a compositional jump from one layer to the other. If it can be demonstrated that the concentration distribution from one layer to the next was initially a step function so that the profile is a postgrowth diffusion profile (not zoning generated by growth), then the profile can be employed to estimate cooling rate. One way to show that the initial concentration distribution was a step function is to examine a large suite of elements across the boundary. If the concentration distribution is a step function for elements with very small diffusivity (usually high-valence cations, such as P in garnet), but a smooth gradual profile for elements with larger diffusivities (such as univalent and divalent cations), then the concentration profiles for elements with larger diffusivities are almost certainly due to diffusion. If the lengths of profiles for elements with very different diffusivities are similar, it may be assumed that all the profiles are due to

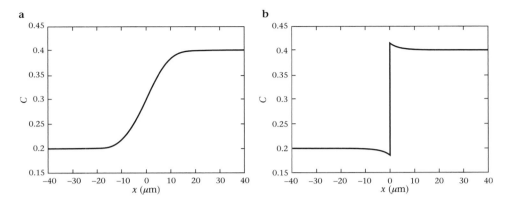

Figure 5-25 (a) Diffusion profile across a diffusion couple for a given cooling history. This profile is an error function even if temperature is variable as long as D is not composition dependent. (b) Diffusion profile across a miscibility gap for a given cooling history. Because the interface concentration changes with time, each half of the profile is not necessarily an error function.

growth, instead of due to diffusion. If the lengths of the profiles of different elements are roughly proportional to the square root of their respective diffusivities, then all the profiles may be interpreted to be due to diffusion. If it cannot be ruled out that part of the gradient is due to growth, then the approach assuming that the entire profile is postgrowth diffusion would set a lower limit on the cooling rate or upper limit on the cooling timescale.

Knowing that the profile is due to diffusion, and if the diffusion distance is small compared to the size of the crystal, we can view the diffusion as a one-dimensional diffusion-couple problem. The solution for the diffusion-couple problem is (Equation 3-38)

$$C = \frac{C_1 + C_2}{2} + \frac{C_2 - C_1}{2} \operatorname{erf} \frac{x}{2\sqrt{\int D \, dt}}, \tag{5-135}$$

where C_1 is the concentration in the first (inner) layer, and C_2 is the concentration in the second (outer) layer, $x = 0$ is at the layer boundary (either defined by the midconcentration point, or from concentration profile of a slowly diffusing component), and $\int D \, dt$ is integrated from $t = 0$ (i.e., the time the second layer grew) to the present. That is, even though temperature was variable, the diffusion profile across a diffusion couple is still an error function (Figure 5-25a). By fitting the profile one obtains $\int D \, dt$. Define $l^2 = \int D \, dt$. That is, l is roughly the mean diffusion distance. Knowing l^2, if the initial temperature at which the second layer grew is known, and, hence, D at that temperature can be calculated, one may use l^2 to infer either one of the following:

(1) For phenocrysts in volcanic rocks, it may be assumed that the mineral spent much time at the initial temperature as and after the second layer

grew, and then cooled down rapidly. Hence, $l^2 = Dt$, where D is the diffusivity at the magmatic temperature, and t is the residence time of the mineral at the magmatic temperature. In this situation, the residence time may be inferred.

(2) For plutonic and metamorphic rocks, the mineral cooled down gradually after the growth of the second layer. Assuming that the cooling is asymptotic, $T = T_0/(1 + t/\tau_c)$ with cooling rate $q|_{t=0} = -(dT/dt)_{t=0} = T_0/\tau_c$, the diffusivity would depend on time as $D = Ae^{-E/(RT)} = D_0 e^{-t/\tau}$, where E is the activation energy, $D_0 = Ae^{-E/(RT_0)}$ (diffusivity at T_0), and $\tau = RT_0\tau_c/E$. Hence,

$$l^2 = \int D \, dt = D_0 \tau. \tag{5-136}$$

From the measured diffusion profile, we obtain $D_0\tau$. By independently estimating T_0 and hence D_0, τ can be found to be l^2/D_0. Then, τ_c can be found to be $El^2/(RT_0 D_0)$, and cooling rate at $T = T_0$ can be found to be

$$q = RT_0^2 D_0/(l^2 E) = RT_0^2/(\tau E). \tag{5-137}$$

This equation may be compared with the equation to calculate cooling rate using homogeneous reaction kinetics and T_{ae} (Equation 5-125) and using thermochronology (Equations 5-77b). The similarity is obvious, except for the difference of a constant factor.

Some minerals display *miscibility gaps*. If the boundary between two compositional zones corresponds to a miscibility gap, then there will be both partitioning and diffusion across the boundary. Because the miscibility gap widens as temperature decreases, the concentrations on the two sides are further separated rather than smoothed out. Figure 5-25b shows a hypothetical example. Wang et al. (2000) used this property to make the first direct observation of immiscibility in garnet.

Often it is necessary to treat diffusion between different layers as three dimensional diffusion. For isotropic minerals such as garnet and spinel (including magnetite), diffusion across different layers may be considered as between spherical shells, here referred to as "spherical diffusion couple." Oxygen diffusion in zircon may also be treated as isotropic because diffusivity $\|\mathbf{c}$ and that $\perp\mathbf{c}$ are roughly the same (Watson and Cherniak, 1997). If each shell can be treated as a semi-infinite diffusion medium, the problem can be solved (Zhang and Chen, 2007) as follows:

$$C(r, t) = C_2 + \frac{C_1 - C_2}{2} \left\{ \frac{2\sqrt{Dt}}{r\sqrt{\pi}} \left[e^{-(r+a)^2/(4Dt)} - e^{-(r-a)^2/(4Dt)} \right] \right.$$

$$\left. + \text{erf} \frac{r+a}{\sqrt{4Dt}} - \text{erf} \frac{r-a}{\sqrt{4Dt}} \right\}, \tag{5-138}$$

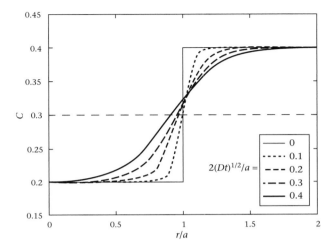

Figure 5-26 The concentration evolution for a "spherical diffusion couple." The radius of the initial core is a. The initial concentration is $C_1 = 0.2$ in the core and $C_2 = 0.4$ in the mantle. Note that the position for the midconcentration between the two halves moves toward smaller radius, which is due to the much larger volume per unit thickness in the outer shell. From Zhang and Chen (2007).

where r is the radial coordinate, a is the initial interface position (radius) between layer 1 (which may be the core) and layer 2, C_1 and C_2 are the initial concentrations in layer 1 and layer 2, and D is diffusivity. For time-dependent D, Dt should be replaced by $\int D\, dt$. That is, the measured concentration profile may be fit by the above equation to obtain $\int D\, dt$ defined to be l^2. Then Equations 5-136 and 5-137 may be applied. Equation 5-138 may be expressed by dimensionless parameters $y = r/a$ and $z = 2(\int D\, dt)^{1/2}/a$:

$$
\begin{aligned}
C(r, t) = C_2 + \frac{C_1 - C_2}{2} \Big\{ &\frac{z}{y\sqrt{\pi}} [e^{-(y+1)^2/z^2} - e^{-(y-1)^2/z^2}] \\
&+ \operatorname{erf}\frac{y+1}{z} - \operatorname{erf}\frac{y-1}{z} \Big\}.
\end{aligned}
\tag{5-139}
$$

The above solution is shown in Figure 5-26.

Although the shape of the profile of a "spherical diffusion couple" is similar to that of a one-dimensional diffusion couple, one difference is that, whereas the midconcentration position stays mathematically at the initial interface for the normal diffusion couple, the midconcentration position moves with time in the "spherical diffusion couple." Initially, the concentration at the initial interface $(r = a)$ is the mid-concentration $C_{mid} = (C_1 + C_2)/2$. However, as diffusion progresses, the concentration at $r = a$ is no longer the mid-concentration. Rather, the location of the mid-concentration moves to a smaller r. Define the mid-concentration location as r_0. Then $r_0 \approx a(1 - z^2/2)$ for small times. If layer 1 is the solid core (meaning r extends to 0), the concentration at the center begins to be affected by 1%, defined to be $C|_{r=0} = C_1 + 0.01 \cdot (C_{mid} - C_1)$, when $z = 0.3947$. At this time, $r_0 = 0.92a$. For $z > 0.3947$, the above solution becomes increasingly less accurate because the diffusion medium can no longer be treated as semi-infinite.

The above results may be applied to infer the critical cooling rate for the concentration of the core to be affected by diffusion. It is necessary to define precisely what is meant when we say "the center is affected by diffusion." If we use center concentration of $C_1 + 0.01 \cdot (C_{mid} - C_1)$ as the criterion for center concentration to be affected by diffusion, then it would occur at $z = 2(\int D\,dt)^{1/2}/a = 0.3947$. For an asymptotic cooling history, this means that $2(D_0\tau)^{1/2}/a = 0.3947$, or $D_0\tau/a^2 = 0.0389$. Combining with Equation 5-137 that $q = RT_0^2/(\tau E)$, we obtain the critical q:

$$q = AT_0^2 e^{-E/(RT_0)} R/(0.0389a^2 E). \tag{5-140}$$

Therefore,

$$\ln q = -E/(RT_0) - 2\ln a + \ln[ART_0^2/(0.0389E)]. \tag{5-141}$$

If the expression of oxygen diffusivity in zircon under wet conditions from Watson and Cherniak (1997) is applied, the following expression would be obtained:

$$\ln q = -27,095/T_0 - 2\ln a + 41.55, \tag{5-142}$$

where a is in μm and q is in K/Myr.

Watson and Cherniak (1997) used numerical solutions to investigate the critical cooling rate for the center concentration of zircon core to be affected. Equation 5-142 is similar to their result and may be viewed as the analytical proof of it. The general equation (Equation 5-141) may be applied to diffusion of other species in zircon, as well as other minerals.

In Chapter 3, an Fe concentration profile in garnet was fit by an error function (Figure 3-7). Because diffusion between the core and the neighboring shell of garnet is better modeled as isotropic diffusion in a sphere, Figure 5-27 is a fit to the same data using Equation 5-138. Good fit is achieved by $2(D_0\tau)^{1/2}/a = 0.1$, meaning $D_0\tau = 0.19$, similar to the one-dimensional fit with $D_0\tau = 0.20$ (Example 3-4). The similarity is expected because the diffusion profile is short compared to the radius of garnet.

Sometimes, the profile is so short that it cannot be resolved by the measurement technique. Such information may also be applied to constrain cooling rate. For example, if the spatial resolution of the measurement is l, the absence of a profile (i.e., a step-function profile) means that $\int D\,dt < l^2$. For an asymptotic cooling history, then $D_0\tau < l^2$, leading to

$$\tau < l^2/D_0, \tag{5-143}$$

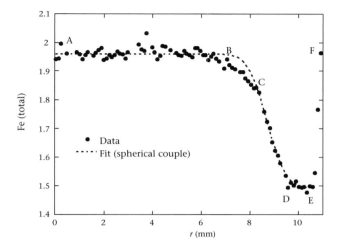

Figure 5-27 Measured Fe concentration (moles of Fe in the garnet formula) profile in a large garnet grain (Figure 3-7). The position $r=0$ roughly corresponds to the center of garnet. The profile from the center to point E ($r=10.5\,\mathrm{mm}$) is interpreted to be due to prograde garnet growth, with relatively low temperature at the beginning of garnet growth (such as 500°C) at $r=0$, and peak temperature at point E. The part of the profile from E to F corresponds to retrograde garnet growth. The fit is by Equation 5-138 with $2(D_0\tau)^{1/2}/a=0.1$ with $a=8.75\,\mathrm{mm}$ (midconcentration point). The part of the profile between points B and C is not well fit (spherical diffusion-couple fit does not differ much from one-dimensional diffusion-couple fit), which might be related to the growth part of the profile. From Zhang and Chen (2007).

or

$$RT_0\tau_c/E < l^2 D_0. \tag{5-144}$$

An application of the above can be found in Example 5-18.

Example 5.18. Watson and Cherniak (1997) reported ^{18}O diffusivity in zircon under wet conditions as $D=\exp(-25.93-25{,}280/T)$ m²/s. In a natural zircon crystal, oxygen isotope ratio shows a step function across the core and mantle. Suppose the spatial resolution is $5\,\mu m$. The initial temperature is constrained independently (e.g., from the mineral assemblage) to be 1100 K and the cooling may be assumed to be asymptotic. Constrain the cooling timescale.

Solution: From the conditions given, $l=5\,\mu m=5\times10^{-6}$ m, $E/(RT_0)=25{,}280/1100=22.98$, and the diffusivity at the initial temperature is $D_0=5.73\times10^{-22}$ m²/s. Hence,

$$\tau < l^2/D_0 = 4.36\times10^{10}\ \mathrm{s} = 1383\ \text{years};$$

or

$$\tau_c = \tau[E/(RT_0)] < 0.0318 \text{ Myr}.$$

If it is known that the cooling timescale is much longer, then either the wet diffusivity does not apply (Peck et al., 2003; Page et al., 2006), or the initial temperature estimate is inaccurate.

Example 5.19. This example shows how diffusivity may be inferred from natural samples. Suppose an Mg–Fe interdiffusion profile is measured in a mineral, and it can be modeled as a diffusion couple with $l = 60\,\mu m$. The temperature history is asymptotic with $T_0 = 1100\,K$, and the cooling time-scale $\tau_c = 10$ Myr. Estimate the diffusivity at 1100 K.

Solution: Use $D_0\tau = l^2$. If τ can be estimated, then D_0 can be obtained. Use $\tau = RT_0\tau_c/E$. Although E/R is not known, activation energy for Mg–Fe inter-diffusion in minerals is of the order 200 to 400 kJ. Thus, RT_0/E is between 0.0457 and 0.0229. Hence, τ is between 0.457 and 0.229 Myr. Therefore, $D_0 = l^2/\tau$ is between 2.5×10^{-22} and 5.0×10^{-22} m^2/s.

5.3.3.2 Homogenization time and residence time of a zoned crystal in the magma chamber

At magmatic temperatures, diffusion is rapid and can erase a diffusion profile in a short time. Because a volcanic rock is cooled rapidly on the Earth's surface (in a matter of days), diffusion during cooling usually can be ignored. Hence, if the crystal is zoned, the zoning provides a constraint on the residence time in a magma chamber. The following is an example. Suppose an olivine phenocryst in a basalt with a radius $a = 1$ mm is monotonically zoned from core to rim (e.g., X_{Fo} is 0.9 in the core and 0.7 in the rim, where X_{Fo} is the mole fraction of forsterite component). The matrix of the basalt is glassy, implying very rapid cooling after eruption to the surface. Therefore, growth of olivine during eruption is negligible. The zonation can be regarded as due to growth in the magma chamber. If the growth rate is very slow, or if the phenocryst resided in the magma chamber for a long time after the growth, diffusion may erase the growth profile.

To treat this more quantitatively, it is necessary to know the behavior of Fe–Mg interdiffusion in olivine. Olivine crystals are usually equidimensional. The diffusion is anisotropic with $Dc \sim 4D_a \sim 5D_b$ (Buening and Buseck, 1973), where the subscripts refer to crystallographic directions. Although the difference in diffusivity is not very large, for simplicity, diffusion in olivine is often treated to occur only along the **c**-axis. The Fe–Mg interdiffusivity in olivine along the **c**-axis at 1253–1573 K may be expressed as (Equation 3-147)

$$D_{//c} = (10^7 f_{O_2})^{1/4.25} \exp(-19.96 - 27,181/T + 6.56X_{Fa}), \tag{5-145}$$

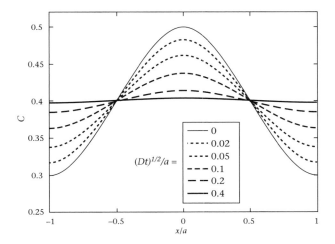

Figure 5-28 Homogenization of a symmetric profile.

where D is in m^2/s, f_{O_2} is in Pa, T is in K, and X_{Fa} is the mole fraction of the fayalite component.

For a given olivine crystal with radius a, treat it as a plane sheet along the **c**-axis with half-thickness a. Assume that the initial zonation is symmetric with respect to the center. Approximate the initial profile as $C = C_0 + C_1\cos(\pi x/a)$, where C_1 is the amplitude of the variation. Assume no flux boundary condition. The solution to the diffusion problem can be found by separation of variables as

$$C = C_0 + C_1 \cos\frac{\pi x}{a} e^{-\pi^2 Dt/a^2}. \tag{5-146a}$$

The concentration profiles stays as a symmetric cosine function, but the amplitude decreases as $C_1\exp(-\pi^2 Dt/a^2)$. When $Dt/a^2 = 0.4666$, the amplitude decreases 1% of the initial amplitude. The concentration evolution is shown in Figure 5-28.

If $T = 1550$ K, $f_{O_2} = 0.001$, Pa (QFM $- 0.6$), and $X_{Fa} = 0.14$, we obtain $D_{//c} = 1.1 \times 10^{-15}$ m^2/s. Define the critical time for homogenization as the time to reduce the initial heterogeneity amplitude by a factor of 100, meaning

$$Dt/a^2 = 0.4666. \tag{5-146b}$$

Hence, homogenization of an olivine crystal with radius of 1 mm would occur in $t = 0.4666a^2/D = 4.2 \times 10^8$ s $\approx$ 13 years. Therefore, a zoned olivine phenocryst of this size must have resided in the magma chamber for less than 13 years since its formation, meaning that the zoned olivine phenocryst must have grown rapidly and been brought up by eruption shortly after its formation. Based on the above analysis, a very thin rim (of the order 10 μm) with high fayalite component would not survive long in a magma chamber or conduit, and hence likely formed during the eruption with almost no time to homogenize.

For a mineral with oscillatory zoning (such as plagioclase), the homogenization time (reduction of initial amplitude by a factor of 100) may be estimated from

$$t_h \sim 0.11665x^2/D, \tag{5-147}$$

where x is the length of each zoning period (total thickness divided by the number of periods), $0.11665 = 0.4666/4$ (because x is equivalent to $2a$ in Figure 5-28), and t_h is the homogenization time. Interdiffusivity between albite and anorthite components in plagioclase involves Al^{3+} and Si^{4+} diffusion (coupled CaAl and NaSi diffusion) and is much smaller than Fe–Mg interdiffusivity in olivine. Hence, zonation in plagioclase may be preserved for a long duration. For example, Grove et al. (1984) estimated $D \approx \exp(-6.81 - 62{,}100/T)$ m^2/s at 1373 to 1673 K. At 1550 K, $D = 4.4 \times 10^{-21}$ m^2/s. If width of each oscillatory layer is 0.1 mm, homogenization would take 8000 years. Longer zoning profiles would take more time to homogenize. A zoned plagioclase could not have resided in the magma chamber for more than the homogenization time.

Example 5.20 In a mid-ocean ridge basalt, one may sometimes find zonation of MgO in glass next to an olivine phenocryst. The distance of mid-concentration of the profile is about 20 μm away from the interface. How would one interpret the formation of the profile?

Solution: Because the profile is very short, and because diffusion of MgO in basalt is rapid (typical diffusivity is 10 μm^2/s), the time required to produce such a profile is about $x^2/D < 40$ s. Because even eruption time in conduit is longer than 40 s, the profile most likely formed due to growth of olivine during quench of the MORB sample in seawater.

5.3.3.3 Dehydration profile in a mineral such as garnet

Wang et al. (1996) measured OH concentrations in mantle-derived garnet crystals and modeled one of the profiles to infer the cooling history. The garnet megacrysts are thought to be brought up by diatreme eruptions. The crystals are roughly spherical with a radius a. The OH concentration in garnet is uniform in the center region and decreases to zero at the edge. From the observation, we may assume that the garnet crystal initially had a uniform OH concentration (accounting for the uniform OH content in the center region) when the garnet crystal was in the mantle, but some OH was lost when the garnet crystal was brought up during a diatreme eruption (accounting for the zonation near the rim). By modeling the loss of OH upon ascent, it is possible to constrain the temperature and volatile history experienced by the garnet crystal. Wang et al. (1996) showed that the diffusivity of the hydrous component in garnet is proportional to the concentration of OH (C), which may be written as

$$D = D_0 C/C_0, \tag{5-148}$$

where C is OH concentration, C_0 is the initial uniform OH concentration, D is OH diffusivity in garnet, and D_0 is the diffusivity when the OH concentration is C_0. The boundary condition (OH concentration on the surface of the spherical garnet) must be assumed. For simplicity, assume the surface concentration to be zero. Hence, this problem is diffusion in a sphere with uniform initial concentration and zero surface concentration. The solution may be expressed as a function of $\int D_0 \, dt$. Because D is proportional to concentration, no analytical solution exists for this spherical diffusion problem. Based on numerical solution for one-dimensional diffusion,

$$x_{\mathrm{mid}} = 0.392 \left(\int D_0 \, dt \right)^{1/2} = 0.665 \left(\int D_{\mathrm{out}} \, dt \right)^{1/2}, \tag{5-149}$$

where D_{out} is the diffusion-out diffusivity and equals $0.347 D_0$ for the case of $D = D_0 \cdot C/C_0$ (Equation 3-88b and Figure 3-33a). The above equation may be compared to Equation 3-42b: the difference is on the coefficient 0.665 (versus 0.954). If the temperature in an erupting diatreme does not change much as magma rises, and quench is rapid upon reaching the surface, then $\int D_0 \, dt = D_0 t$, where t is the duration of the megacrysts spent in the eruption conduit.

For a given garnet crystal, Wang et al. (1996) found that the mid-concentration distance in the rim is about $150 \, \mu m$ (garnet radius is 2.1 mm). Suppose the magma temperature was 1173 K, at which D_{out} is about $8 \, \mu m^2/s$. Hence, the time required for garnet to ascend from mantle to the surface is

$$t = x_{\mathrm{mid}}^2 / (0.665^2 D_{\mathrm{out}}) = 1.8 \mathrm{h}. \tag{5-150}$$

This would imply extremely rapid eruption. Because the surface OH concentration may gradually decrease to zero rather than suddenly drop to zero, the above estimate of duration may be increased.

5.3.4 Geospeedometry based on diffusion between two or more phases

5.3.4.1 Quantitative geospeedometry based on component exchange between two phases

Often components in minerals may exchange through diffusion with adjacent minerals. For example, Fe and Mg in garnet may exchange with adjacent ferromagnesian minerals. Oxygen isotopes in quartz may exchange with those in magnetite. The equilibrium constants are exchange coefficients, such as $K_D = (\mathrm{Fe/Mg})^{\mathrm{gt}}/(\mathrm{Fe/Mg})^{\mathrm{ol}}$ or $\alpha = ({}^{18}\mathrm{O}/{}^{16}\mathrm{O})^{\mathrm{qtz}}/({}^{18}\mathrm{O}/{}^{16}\mathrm{O})^{\mathrm{mgt}}$. In this kind of modeling, it is assumed that there is surface equilibrium, meaning in the case of garnet–olivine equilibrium that $(\mathrm{Fe/Mg})^{\mathrm{gt}}_{x=0+}/(\mathrm{Fe/Mg})^{\mathrm{gt}}_{x=0-}$ is indeed K_D, where $x = 0$ is

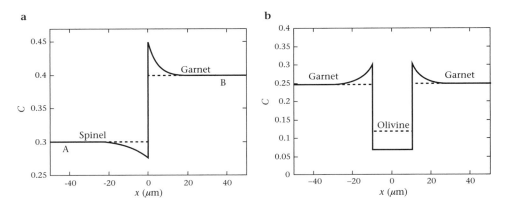

Figure 5-29 Schematic exchange diffusion profiles between (a) two large minerals and (b) a small olivine inclusion and its garnet host. The dashed lines are the assumed initial concentration distribution, and the solid curves are the resulting concentration distribution at present.

the interface between the two phases, olivine is defined to be on the left-hand side, $x = 0-$ means the surface at the left-hand side (meaning olivine surface), and $x = 0+$ means surface at the right-hand side (meaning garnet surface). Under isothermal conditions, the exchange and diffusion problem is relatively easy to handle, and the solution can be found in Section 4.3.3.

For natural samples, we need to consider continuous cooling. By measuring the concentration profiles in both phases, it may be possible to obtain both the temperature at which the two phases were at equilibrium with uniform composition in each phase (this temperature may be the formation temperature or peak temperature) and the cooling rate. This is advantageous because inference of cooling rates requires knowledge of the initial temperature T_0. We first consider two cases for which the initial equilibrium temperature may be obtained from measured profiles in the two phases. It is assumed that initially both phases were uniform in composition. Applying thermometry to infer the initial equilibrium temperature requires removing the effect of diffusion during cooling.

In the first case, both minerals are large. The concentration in each mineral is uniform except when the interface is approached, such as Fe–Mg exchange between spinel and garnet (Figure 5-29a). The uniform compositions in the center parts of the two minerals may be taken to be the compositions at the initial equilibrium temperature. Therefore, by measuring spinel composition at point A and garnet composition at point B, the initial equilibrium temperature may be inferred if the exchange coefficient as a function of temperature and composition is known. (K_D also depends on pressure. Hence, pressure also needs to be known or inferred from other reactions.) One requirement for this approach is that the mass gained by one phase should be identical to the mass lost by the other phase.

In the above case, the initial composition of each phase is preserved in the center of the mineral. Sometimes, the initial composition is not preserved but may be inferred, allowing thermometry calculations. In garnet, there are often small olivine or other mineral inclusions. Fe–Mg diffusivity is much larger in olivine than in garnet. Because of rapid diffusion in olivine, the concentration profile in olivine is almost uniform, and this uniform concentration is not the initial concentration. However, due to slow diffusion in garnet, Fe–Mg concentration gradient may exist in garnet. Without measuring concentration profiles in garnet, one would not be able to infer the initial olivine composition. However, with the measured profile in garnet, it is possible to estimate the extra FeO in garnet: $M = 4\pi \int r^2(C - C_i)dr$, where $(C - C_i)$ is the excess FeO in kg/m^3 or mol/m^3, and the integration is from a to ∞, with a being the radius of olivine. The total extra FeO in garnet is then distributed in olivine as $M/(4\pi a^3/3)$, and again the unit of FeO is kg/m^3 or mol/m^3. The new FeO concentration in olivine after making this correction is the inferred initial FeO concentration in olivine (dashed line in Figure 5-29b). The inferred olivine composition can be combined with garnet composition far away from olivine to infer the initial temperature (Wang et al., 1999).

One may also try to obtain an "equilibrium" temperature using the rim compositions, which might be interpreted as the last equilibrium temperature. However, this temperature does not have much meaning because the surface concentration changes continuously and the rim concentration depends on the resolution of the measurement.

Next we turn to the inference of cooling history. The length of the concentration profile in each phase is a rough indication of $(\int D\,dt)^{1/2} = (D_0\tau)^{1/2}$, where D_0 is calculated using T_0 estimated from the thermometry calculation. If $(D_0\tau)^{1/2}$ can be estimated, then τ, τ_c and cooling rate q may be estimated. However, because the interface concentration varies with time (due to the dependence of the equilibrium constants between the two phases, K_D and α, on temperature), the concentration profile in each phase is not a simple error function, and often may not have an analytical solution. Suppose the surface concentration is a linear function of time, the diffusion profile would be an integrated error function i^{2}erfc$[x/(4\int D\,dt)^{1/2}]$ (Appendix A3.2.3b). Then the mid-concentration distance would occur at

$$x_{mid} = 0.286544 \left(4 \int D\,dt\right)^{1/2} = 0.573 \left(\int D\,dt\right)^{1/2} = 0.573(D_0\tau)^{1/2}. \quad (5\text{-}151)$$

A comparison of the above equation to Equation 3-42b shows that the different coefficients are due to the difference in the boundary condition. For a given mid-diffusion distance, the e-folding timescale for diffusivity may be found as

$$\tau = 3.05x_{mid}^2/D_0, \quad (5\text{-}152)$$

from which the timescale for cooling (τ_c) may be found as $\tau E/(RT_0)$, and the thermal history is approximated as $T = T_0/(1 + t/\tau_c)$. For the case illustrated in Figure 5-29a for the exchange of components between two phases, there are two profiles (one in each phase), and hence two τ_c values may be found. The two values should be the same and hence provide cross-check to each other. When the two values differ, the geometric average of the two cooling rates may be taken to approximate the true cooling rate. For the problem shown in Figure 5-29b, only one cooling rate (or τ_c) can be inferred.

More advanced modeling has been developed on using component exchange during cooling as a geospeedometer (Lasaga, 1983, 1998; Jiang and Lasaga, 1990; Ganguly et al., 2000). Mathematically, this problem cannot be simplified by defining $\int D \, dt$ to be another variable to eliminate the time dependence as in Section 3.2.8.1 because of the dependence of exchange coefficient on temperature and composition. Hence, either some simplification must be made to obtain an analytical solution (Lasaga, 1983), or the diffusion problem is solved numerically (Ganguly et al., 2000). By matching the numerical solution to the measured concentration profile, the cooling rate may be constrained. The cooling rate would provide a constraint to the exhumation history. Although efforts have been made by various authors, so far the applicability has been limited because of (i) the complexity of the problem, (ii) the high spatial resolution required in determining the diffusion profile, and (iii) lack of accurate thermodynamic partitioning and diffusion data as a function of temperature.

5.3.4.2 Quantitative geospeedometry based on bulk exchange between minerals

Oxygen isotope fractionation has been applied as a thermometer and geospeedometer. Oxygen isotopic ratios vary slightly from one phase to another. The small variations are conventionally expressed by δ-notation defined as

$$\delta^{18}O = \left[\frac{(^{18}O/^{16}O)_{\text{sample}}}{(^{18}O/^{16}O)_{\text{standard}}} - 1 \right] \cdot 1000\text{‰}. \tag{5-153}$$

That is, the $^{18}O/^{16}O$ ratio in a sample is compared with a standard, and the difference is expressed as per mil (per thousand). The standard for oxygen isotopes is standard mean ocean water (SMOW). The exchange reaction of ^{18}O and ^{16}O between two minerals is a heterogeneous reaction and may be written as, e.g.,

$$^{18}O(\text{mgt}) + {}^{16}O(\text{qtz}) \rightleftharpoons {}^{16}O(\text{mgt}) + {}^{18}O(\text{qtz}), \tag{5-154}$$

where "mgt" stands for magnetite and "qtz" stands for quartz. The equilibrium constant is called the isotopic fractionation factor and denoted as α:

$$\alpha_{^{18}O/^{16}O}^{\text{qtz}/\text{mgt}} = \frac{(^{18}O/^{16}O)_{\text{qtz}}}{(^{18}O/^{16}O)_{\text{mgt}}}. \tag{5-155}$$

The value of α is usually very close to 1 because isotopic fractionation is usually small. Using the δ-notation, the fractionation factor may be expressed as

$$\alpha_{^{18}O/^{16}O}^{qtz/mgt} = \frac{1000\%_0 + \delta^{18}O_{qtz}}{1000\%_0 + \delta^{18}O_{mgt}} \tag{5-156}$$

and

$$\left(\ln\alpha_{^{18}O/^{16}O}^{qtz/mgt}\right)(1000\%_0) \approx \delta^{18}O_{qtz} - \delta^{18}O_{mgt} \equiv \Delta^{18}O_{qtz-mgt}. \tag{5-157}$$

The dependence of α on temperature is often expressed as

$$\ln\alpha = B/T^2, \tag{5-158}$$

where B is a constant. For example (Chiba et al., 1989),

$$\ln\alpha_{^{18}O/^{16}O}^{qtz/mgt} \approx \frac{6290}{T^2}. \tag{5-159}$$

Among common rock-forming minerals, oxygen isotope fractionation between a quartz and magnetite pair is the largest, with quartz being the most enriched and magnetite being the most depleted in ^{18}O. The magnitude of isotopic fractionation may be roughly estimated from the difference in bond strength, with the heavier isotope enriched in the stronger bond. In quartz, oxygen ions are bonded to Si^{4+} and the bond (mostly covalent) is strong. In magnetite, oxygen ions are bonded to Fe^{2+} and Fe^{3+} and the bonds (mostly ionic) are relatively weak. Hence, the fractionation is large and ^{18}O is enriched in quartz. In feldspar, oxygen ions are largely bonded to Si^{4+} and Al^{3+} and the bonds are strong (but slightly weaker than in quartz). Hence, the fractionation between quartz and feldspar is small, with ^{18}O slightly enriched in quartz.

Most oxygen isotope data in literature are for whole mineral grains. (Only recently, is it possible to measure oxygen isotopic ratio profiles.) Based on the bulk isotopic measurements of two minerals in a rock, the apparent equilibrium constant α may be calculated, from which the apparent equilibrium temperature (T_{ae}) may be calculated. This temperature would be the real formation temperature if equilibrium was reached at the formation and subsequent cooling was rapid, such as volcanic rocks. For rocks that cooled slowly, the meaning of T_{ae} needs to be clarified. On the other hand, for each mineral, if the diffusion property is known, the bulk-mineral closure temperature (T_c) can be calculated using any one of Equations 5-75a to 5-77b. For a rock containing N minerals, theoretically, N values of T_c and $N(N-1)/2$ values of T_{ae} may be defined and obtained. (If there was perfect equilibrium between the minerals, only $N-1$ of the T_{ae} equations would be independent and they would all give the same T_{ae}.) T_c's are calculated properties related to cooling rate and T_{ae}'s are obtained from measured isotopic ratios. For simplicity, let's call T_{ae} a measured property and T_c a calculated property. If T_{ae} can be related to T_c, it may be possible to use measured T_{ae} to constrain T_c and to further estimate cooling rate.

A full treatment of this problem is unavailable. The simplest case would be a bimineralic rock. The mineral grains are randomly distributed with variable grain sizes. Even if both minerals are diffusionally isotropic, all grains of the same mineral are equal in size and shape, and the mineral grains are regularly spaced, the problem still has not been solved. One complication is that grain boundary diffusion is much more rapid than volume diffusion.

An early attempt by Giletti (1986) considered a multimineral rock as a closed system and assumed that when T_c of one mineral is considered, all other minerals with lower T_c behave as an infinite reservoir with rapid mass transport (so that Dodson's theory can be applied to calculate T_c). With this simple model, it was found that T_{ae} between two minerals corresponds to neither T_c nor the formation/peak temperatures, but for a bimineralic rock, or for two minerals with the lowest closure temperatures in a rock, the two minerals close at the same temperature (the higher of the two T_c values), which would be T_{ae}.

Afterward, Eiler et al. (1992) pointed out that the treatment of Giletti (1986) does not satisfy the mass balance constraint because of the assumption that other minerals with lower T_c are an infinite reservoir (the infinite reservoir assumption is especially problematic when the volume of other minerals is smaller than that of the mineral under consideration). To further understand the isotopic behavior during cooling of a rock, Eiler et al. (1992, 1993, 1994) made use of the fact that grain boundary diffusivity is often many orders of magnitude faster than volume diffusivity, and developed a fast grain boundary (FGB) diffusion model. In this model, oxygen isotopes at the boundary of mineral grains are assumed to remain in equilibrium. With mass balance constraint, the coupled diffusion equations for multiple mineral phases are solved to obtain the bulk oxygen isotopic ratios in all minerals. For a bimineralic rock, T_{ae} is shown to vary from the T_c value of one mineral (when the proportion of this mineral approaches zero) to that of the other mineral depending on the modal abundance of the two minerals, more complicated than the result of Giletti (1986). For a rock containing three or more minerals, the relation between T_{ae} and composition is more complicated: T_{ae} between two minerals does not seem to have much significance. Simulations show that isotopic fractionation between some mineral pairs may be opposite to that determined by experiments, meaning that T_{ae} is not defined.

Nonetheless, the conclusion about bimineralic rocks is useful: Because T_{ae} lies between the T_c values of the two minerals, if the two minerals happen to have similar T_c values (meaning their diffusivities are similar), then measured T_{ae} would be a good approximation for both T_c values, from which two cooling rates may be estimated (Equation 5-77a). Because the rock should have one single cooling history, the two cooling rates provide cross-check to each other. If the two cooling rates are different, geometric average of the two values may be taken to approximate the cooling rate of the rock, and the difference from the geometric average may be taken as an uncertainty estimate.

Using the same FGB model of Eiler et al. (1992, 1993, 1994), Ni and Zhang (2003) examined multimineralic rocks for simple relations out of the complex system. They found that in a rock with three or more minerals, T_{ae} values between most mineral pairs do not mean much, but for the mineral pair with the largest isotopic fractionation (or largest isotopic fractionation pair, LIFP, such as quartz and magnetite), T_{ae} value is always between the T_c values of the two minerals. Hence, if the two minerals also have similar diffusion properties, which happens to be the case for quartz and magnetite, measured T_{ae} allows estimation of two T_c values and, hence, the inference of cooling rate. That is, for each mineral in the LIFP, q can be calculated from Equation 5-77a by letting $T_c = T_{ae}$:

$$q = \frac{GT_{ae}^2 A}{a^2 E/R} \exp\left(-\frac{E/R}{T_{ae}}\right) \tag{5-160}$$

where A and E/R are diffusion parameters ($D = A\,e^{-E/(RT)}$), G is the shape factor (Section 5.2.1), and a is the grain size.

The accuracy of the FGB model has not been verified. Even though grain boundary diffusion is indeed rapid, it is not necessary that all grain surfaces of the same mineral would be at the same concentration, allowing diffusion to be treated simply as from the fastest diffusion direction. For example, Usuki (2002) showed that Mg concentration on the surface of biotite depends on the direction of contact with garnet, and that Fe–Mg diffusion profiles in biotite also depends on the direction of contact. Along the cleavage plane ($\perp\mathbf{c}$), the profile in biotite is roughly flat due to high diffusivity; parallel to $\mathbf{c}$, there is steep profile in biotite. If grain boundary diffusion were rapid enough, the surface Fe–Mg concentration in biotite would be the same whether or not it is in contact with garnet, and diffusion along the cleavage plane would have homogenized biotite regardless of the contact surface between biotite and garnet. It is hence possible that the FGB model applies only under certain conditions (e.g., in the presence of fluid or small enough grains so that surface to volume ratio is large), but not under other conditions.

With the improvement of analytical techniques, it is now possible to measure oxygen isotopic profiles in minerals. By analogy to the concepts of bulk mineral closure temperature versus closure temperature at every point in the interior of a mineral for the case of thermochronology, future theoretical development will need to clarify the concepts and meanings of T_c and T_{ae} based on interior isotopic or elemental compositions of two contacting minerals. Furthermore, the principles of the above discussion may also be applied to elemental exchange reactions, such as Fe–Mg exchange between minerals.

5.3.5 Cooling history based on other heterogeneous reactions

Heterogeneous reactions are the most common type of reactions in petrology and volcanology. Component exchange between the two phases discussed above is a heterogeneous reaction, but the mathematical treatment is essentially diffusion (with a complicated boundary condition). Other heterogeneous reactions are often more complicated to treat. Because complete equilibrium is not easy for heterogeneous reactions, they provide rich information on temperature and pressure history. For igneous rocks, the initial state is a melt, which is an equilibrated liquid and does not store information on prior temperature and pressure history except for some crystals that survive the melting. Therefore, only the crystallization history may be inferred from the heterogeneous reactions (crystallization). The melt composition may preserve information about the partial melting conditions and sources. The isotopic composition may preserve some information on the history of the source region. For metamorphic rocks, it is very difficult for solid-state reactions to completely reach equilibrium and some minerals may be well preserved (such as mineral inclusions in host minerals). Hence, metamorphic mineral assemblages may store both prograde temperature–pressure history and retrograde history. Nonetheless, "reading" (quantitative interpretation of) the information is difficult because many aspects of heterogeneous reactions are not quantified yet. Some of the heterogeneous reactions have been investigated to provide constraints on cooling rate or heating time.

5.3.5.1 Oxidation geospeedometer

One geospeedometer is based on the color of pumice, which is an indicator of the degree of oxidation. During an explosive eruption, vesicular magma droplets are carried by the eruption flow into an eruption column. They are cooled to form pumice by air mixed into the eruption column (if they are not cooled enough in the eruption column, they arrive on the ground hot enough to weld). If cooling in the eruption column is rapid, the magma droplet would not oxidize much. If cooling is slow, the vesicular magma droplet would stay at high temperature for long enough time to be oxidized. The degree of oxidation is roughly indicated by the color. In some explosive eruptions, pumice may display different colors, such as white pumice and pink pumice in the 3600 BC eruption of Santorini. If the relation between the degree of oxidation (or the color), temperature, duration at the temperature, and the permeability of the pumice can be quantified, the color index may be applied to infer cooling rate or cooling timescale. The permeability comes into play because oxygen must enter the interior of pumice to oxidize Fe^{2+} into pink or red color.

Tait et al. (1998) investigated the oxidation process of Santorini pumice. They heated pumice pieces of 5 to 10 mm diameter in air. They also conducted some controlling experiments using other gas mixtures. The results are summarized in Figure 5-30. According to them, white pumice would be oxidized to pink pumice as long as the heating time at the given temperature satisfies

$$\text{Time}\,(s) \geq \exp\,(64.10 - 0.0548T), \tag{5-161}$$

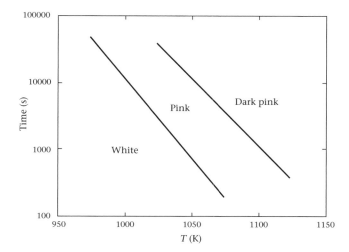

Figure 5-30 The dependence of color variation from white pumice to pink or dark pink pumice on temperature and heating time. The equation for the line separating white and pink pumice is Time $(s) = \exp(64.10 - 0.0548T)$, and the equation for the line separating pink and dark pink pumice is Time $(s) = \exp(57.94 - 0.0463T)$, where T is in K. The uncertainty is unknown. Although one would commonly use $1/T$ instead of T in the above forms, the original authors used T. Data read from Tait et al. (1998).

and would be oxidized further to dark pink pumice when the heating time satisfies

$$\text{Time} (s) \geq \exp(57.94 - 0.0463T), \tag{5-162}$$

where T is in K. Therefore, given a piece of untreated pumice of 5 to 10 mm in diameter, if it is pink in color but not dark pink (one needs to learn from the original authors how to tell the difference), and if the temperature in the eruption column is 1073 K, the heating time would be between 200 and 3900 s. However, because cooling in the eruption column was likely continuous, a better model would be to assume asymptotic cooling and estimate the cooling timescale in the eruption column. The principles of the above method may be applied to pumices of other explosive eruptions. Whether the exact equations for color changes are also applicable depends how much the chemical composition and permeability affect the rate of oxidation.

5.3.5.2 Exsolution lamellae width (spinodal decomposition)

Many minerals show exsolution textures. For these minerals, there is complete solid solution at sufficiently high temperatures, but a miscibility gap at lower temperatures. The alkali feldspar, $(Na, K)AlSi_3O_8$, is an example. Suppose a roughly homogenous mineral of intermediate composition cools down and into the miscibility gap. The mineral would undergo spontaneous decomposition into two phases. This process is called exsolution. The exsolution of an alkali feldspar may lead to a perthite (sodium feldspar lamellae in potassium feldspar) or anti-perthite (potassium feldspar lamellae in sodium feldspar). The width of the lamellae depends on the diffusion property, the cooling rate, and other factors that may affect diffusion rate (such as water content). If the relation is calibrated, the lamellae width may be used to estimate cooling rate. Because a number of common minerals (alkali feldspar, pyroxene) exhibit such exsolution textures, potentially the method can be very useful. If the diffusion property does not depend on conditions such as water content, it would be a more reliable geospeedometer because one would not have to estimate such conditions.

Extensive work has been carried out on the growth of kamacite lamellae (also referred to as spindles) in taenite host in iron meteorite as a function of cooling rate. Iron meteorites commonly display the Widmanstatten texture, which is due to the exsolution of kamacite lamellae from taenite. At high temperature, Fe–Ni metal forms complete solid solution with the face-centered cubic structure of taenite. As the metal is cooled, Ni-poor kamacite crystals (body-centered cubic structure) nucleate and grow from taenite. The growth is controlled by Ni diffusion in taenite (Ni must diffuse away from the kamacite–taenite interface). The width of kamacite and Ni concentration profile in taenite depend on cooling rate and composition. Earlier work in the 1960s led to very slow cooling (such as 1–4 K/Myr for group I iron meteorites), which did not agree with other inferences. Later it was found that some factors (such as P content) affected Ni diffusion and kamacite growth but were not considered by earlier work. When these factors were taken into account, the cooling rates were revised upward. In the first iteration of the revisions using the width of kamacite lamellae (Narayan and Goldstein, 1985), the cooling rate was revised upward by a factor of 100 to 1000. Later it was found there are other effects. A further revision by the same group (Saikumar and Goldstein, 1988) using the Ni profile method arrived at cooling rates between the high and low values, about 10 K/Myr for Tazewell (group IIICD), 25 K/Myr for Toluca meteorite (group IA), and 150–300 K/Myr for Bristol meteorite (group IVA). The history of this geospeedometer highlights the caveats of this kind of geospeedometers and the need for careful considerations of many factors that may affect the heterogeneous reaction kinetics.

5.3.5.3 Qualitative geospeedometry based on crystallinity

Volcanic rocks (rapid cooling) are less well crystallized compared to plutonic rocks (slow cooling). The latter are coarse-grained and the former may be glassy (very rapid cooling) or porphyritic with phenocrysts in a fine matrix. For a basalt magma, a very high cooling rate (≥ 100 K/s) leads to glassy basalt (MORB), a cooling rate in the order of ~ 100 K/hr leads to microcrystalline to fine-grained basalts with phenocrysts, and a much slower cooling rate ($\sim$1000 K/Ma) leads to coarse-grained gabbro. For a rhyolite melt, a cooling rate of $\geq$100 K/day leads to glassy rhyolite (obsidian) and a cooling rate of $\sim$100 K/Myr leads to coarse-grained granite. Hence, crystallinity, or more generally, the texture (including crystal shape and crystal size distribution) of an igneous rock is an indication of cooling rate.

The application of crystallinity as a geospeedometer is usually relative and qualitative. For a given magma composition (including the same H_2O content), inference of a relative cooling rate is reliable. For example, for a given mid-ocean ridge basalt, a relative cooling rate may be inferred based on crystallinity: glassy margin quenched very rapidly but the partially crystallizing interior cooled at a slower rate.

Although the principle is simple, such a geospeedometer has not been quantified because melt composition (especially SiO_2 and H_2O contents) plays a major role in addition to cooling rate in determining the crystallinity and texture of a rock. For example, for a cooling rate of 100 K/day, a rhyolite melt would form glass but a basalt melt would form a porphyritic rock. For a cooling rate of 100 K/Myr, volatile-rich melt may crystallize into pegmatite with huge crystals (decimeter to meter size) but a rhyolite melt with smaller concentration of volatiles would crystallize as a regular granite with typically centimeter-size crystals. It is hoped that future work will quantify glass formation as well as crystallinity and texture of igneous rocks as a function of thermal history and composition.

5.3.5.4 Geospeedometry based on crystal size distribution (CSD)

Marsh (1988), Cashman and Marsh (1988), and Cashman and Ferry (1988) investigated the application of crystal size distribution (CSD) theory (Randolph and Larson, 1971) to extract crystal growth rate and nucleation density. The following summary is based on the work of Marsh (1988). In the CSD method, the crystal population density, $n(L)$, is defined as the number of crystals of a given size L per unit volume of rock. The cumulative distribution function $N(L)$ is defined as

$$N(L) = \int_0^L n(L)\,dL. \tag{5-163}$$

In other words,

$$n(L) = \frac{dN(L)}{dL}. \tag{5-164}$$

That is, $N(L)$ is the total numbers of crystals with size smaller than L per unit volume of the rock.

To investigate the controlling factors of CSD, crystal population balance is considered. In a magma chamber of volume V, suppose n crystals of size L grow at a linear rate u (m/s) at time t. In addition to crystal growth, magma may flow in and out. The influx (volume per unit time) is denoted as Q_{in}, and the outflux is denoted as Q_{out}. The general popolation balance equation is complicated and difficult to apply. To simplify, assume (i) a steady state in terms of magma volume, meaning $Q_{in} = Q_{out}$, and (ii) that incoming magma does not contain crystals. Define the residence time τ to be $V/Q_{in} = V/Q_{out}$. Then population balance leads to

$$\frac{\partial n}{\partial t} + \frac{\partial(un)}{\partial L} + \frac{n}{\tau} = 0. \tag{5-165}$$

With two further assumptions, (iii) the crystal growth rate is independent of crystal size L, and (iv) CSD reached steady state so that $\partial n/\partial t = 0$, the above equation becomes

$$u\frac{\partial n}{\partial L} + \frac{n}{\tau} = 0. \tag{5-166}$$

The solution to the above equation is

$$n = n^0 e^{-L/(u\tau)}, \tag{5-167a}$$

or

$$\ln(n) = \ln(n^0) - L/(u\tau), \tag{5-167b}$$

where n^0 is the number density of crystals with $L=0$, meaning the nucleation density. Equation 5-167b predicts that a plot of $\ln(n)$ versus L is a straight line, with the intercept of $\ln(n^0)$, and a negative slope of $-1/(u\tau)$. From the intercept, the nucleation density can be obtained. From the slope, if the growth rate u is known, then the residence time τ can be inferred. If the residence time τ is known, then the growth rate u can be inferred. If the crystal size distribution follows Equation 5-167, the distribution is said to be log-linear.

Although the above derivation is for crystals, the theory is also applicable to bubble size distribution. In addition to the above four assumptions, the other conditions for its application include (v) no Ostwald ripening, which would modify CSD, and (vi) no coalescence of bubbles.

In the application of the theory, it is critical to accurately estimate the three-dimensional crystal size distribution. For pumice clasts, Bindeman (2003) used HF and HBF_4 acids to dissolve the glass material, and leaving behind acid-resistant phenocrysts such as zircon and quartz. He then measured the length, width, and shape of these grains using camera and computer assistance. For each kind of phenocrysts, the crystal size distribution can hence be obtained. Another method to obtain 3-dimensional crystal (or bubble) size distribution is to use 3-D imaging similar to CAT scan (e.g., Sahagian et al., 2002; Gualda and Rivers, 2006). Without direct determination of the three-dimensional crystal size distribution, one may use two-dimensional observations (thin-section observations) to infer three-dimensional size distribution. This inference is not straightforward because cutting a cross section from uniform-sized spheres in 3-D would still generate a distribution of radius in 2-D. Some authors have investigated approximate methods to infer 3-D distribution from 2-D measurements (e.g, Sahagian and Proussevitch, 1998; Higgins, 2000).

If 3-D distribution can be obtained accurately, uncertainty may still arise because some of the six assumptions are not necessarily satisfied. Sometimes the CSD does not follow a log-linear relation. Then the above theory cannot be applied. Even if the CSD follows a log-linear distribution, whether the crystal growth rate can be reliably inferred has not been verified experimentally. Higgins (2002) and Pan (2002) pointed out other possible uncertainties in applying the CSD theory.

5.3.6 Comments on various geospeedometers

Based on the above discussion on various geospeedometers, a rock contains many clues from which its thermal history may be read. Some of these processes, such as homogeneous reactions and diffusion, are simpler and better understood, and hence can be more easily quantified as geospeedometers. Other processes are more complicated, and information stored by those remains to be deciphered. Often the more complicated processes may store more information on the thermal history.

The geospeedometer based on the kinetics of Fe–Mg order–disorder reaction in orthopyroxene is well developed. The inconsistency in analytical data (Mössbauer versus X-ray diffraction) is one problem, but the main difficulty in applying this geospeedometer is that experimental data are obtained at relatively high temperatures, but the apparent equilibrium temperature in most natural rocks is much lower. The problem cannot be avoided because the experimental timescale cannot exceed years, but the natural cooling timescale is often many thousands to several million years. The difference in terms of experimental and natural timescales causes at least two problems. One is the required large extrapolation of equilibrium and kinetic constants obtained at high temperature to such low temperature, which may not be reliable (and the uncertainty cannot even be evaluated). Secondly, at such low T_{ae}, the Fe concentration in M1 site is extremely low because the reaction approaches complete order, causing large uncertainties in the measured Fe concentration in M1 site. The best application of this geospeedometer is to infer cooling rates of rapidly cooled rocks, such as basalt.

The geospeedometer based on the kinetics of interconversion of hydrous species in rhyolitic melt is also well developed although the reaction mechanism and rate law are not known. The empirical calibration covers a cooling rate range of 50 K/yr to 100 K/s, about eight orders of magnitude. Theoretically, this homogeneous reaction geospeedometer has the same limitation as the order–disorder reaction geospeedometer. In practice, because the speedometer can be applied only to hydrous rhyolitic glasses, and because such samples cooled rapidly, not much extrapolation is necessary. Therefore, as long as the melt composition is similar to that of the calibrated samples, the geospeedometer is applicable. Similar geospeedometers may be developed for other natural hydrous glasses.

The diffusion-based geospeedometers require accurate knowledge of the temperature dependence of the diffusivity. One might think that large down-temperature extrapolation, which plagues homogeneous reaction geospeedometers, would also be a major problem in diffusion-based geospeedometers. This is sometimes the case but not necessarily so if diffusion profiles of very short length can be measured. The extent of diffusion, which is characterized by diffusion distance, may be resolved to very small distances in some profiling techniques, such as 10 nm with Rutherford backscattering spectrometry and ion microprobe depth profiling. Using such techniques, diffusivity at relatively low temperature may be obtained. For example, if a diffusion profile of 10 nm long is produced in an experiment lasting for a month, a 10-μm-long profile would require about 0.1 Myr. Hence, not much down-temperature extrapolation would be necessary if natural profile lengths are of 10 μm and experimental profile lengths are of 10 nm. This ability to obtain low-temperature diffusivity is in contrast with the inability to obtain low-temperature reaction rate coefficients of homogeneous reactions. (If bulk extraction technique is applied to obtain diffusivity, then large down-temperature extrapolation is usually necessary.)

A natural concentration profile may be utilized to infer cooling rate if it is due to diffusion (not due to growth), *the initial temperature is independently and accurately known*, other initial conditions (such as initial composition) are known, the boundary condition is simple, and the diffusivity as a function of temperature is known. The effect of initial temperature uncertainty on the cooling rate inferred from diffusion profiles can be very large. Because only $D_0\tau$ (where D_0 is the diffusivity at the initial temperature, and τ is the timescale for D to reduce by a factor of e) is constrained from the length of the diffusion profile, the relative error in inferred τ is the same as that in D_0 (the diffusivity at the initial temperature). If the initial temperature estimate is off by 50 K, the resulting relative error in D_0 may be about a factor of 5 (depending on the temperature and activation energy), which would lead to a relative error of a factor of 5 in q. Therefore, independent and accurate determination of the initial temperature is critical in estimating the cooling rate, in addition to the availability of experimental diffusion data as well as other aspects in diffusion modeling. This dependence on the initial temperature is in contrast to homogeneous-reaction geospeedometers, which provide a cooling rate independent of the initial temperature (as long as the initial temperature is high enough).

Complicated heterogeneous reactions may contain the most information on the thermal history of a rock. Currently, few such reactions have been quantified as geospeedometers because rates of such reactions are more difficult to evaluate. Nonetheless, future development of kinetic theory may demonstrate the rich resources in the kinetics of such reactions and the possibility to infer complex thermal history from such reactions.

Problems

5.1. The cosmogenic nuclide ^{14}C is unstable and decays into ^{14}N with a half-life of 5730 years ($\lambda = 0.00012097$ yr^{-1}). The initial activity of ^{14}C in a newly formed plant tissue is 13.56 dpm per gram of carbon (dpm = decays per minute), and in the year of 2000 you measured ^{14}C activity for a piece of tree tissue from the center of a tree and found that it is 12.8 dpm.

  a. Calculate the age of the tree using the correct decay constant.

  b. Calculate the age of the tree using the conventional decay constant ($\lambda = 0.00012449$ yr^{-1}).

  c. Convert the ^{14}C age in (b) to calibrated age.

  d. What does the age mean? Did the tree form at this time? Did it die at this time?

5.2 The following table gives measured ^{238}U, ^{234}U, and ^{230}Th activities as a function of depth in the sediment. The decay constants of ^{238}U, ^{234}U, and ^{230}Th are 1.55125×10^{-10}, 2.835×10^{-6} and 9.195×10^{-6} yr^{-1}. Find the sedimentation rate.

Depth (m)	^{238}U (dpm/g)	^{234}U (dpm/g)	^{230}Th (dpm/g)
0.03	1.4	1.5	68.2
0.20	1.4	1.5	35.1
0.40	1.4	1.5	18.0
0.60	1.4	1.5	10.0
0.80	1.4	1.5	5.6
1.00	1.4	1.5	3.5

5.3 Measurement of a coral sample gives the following activity ratios: $(^{230}Th/^{238}U) = 0.1100 \pm 0.0005$, and $(^{234}U/^{238}U) = 1.110 \pm 0.0006$.

  a. Find the age.

  b. Find the initial $(^{234}U/^{238}U)$ activity ratio.

5.4. Manganese nodules are small spherical concretions of iron and manganese oxide in ocean sediment typically about a few tens of millimeters in radius. A nodule grows from a small nucleus layer by layer to form concentric layers (sometimes there may be two growth centers). As a nodule grows, it also takes in cosmogenic ^{10}Be in seawater. Once inside the nodule, ^{10}Be decays to ^{10}B with a half-life of 1.51 Myr. The following table gives measured ^{10}Be activity as a function of depth into a nodule. Estimate the growth rate of this manganese nodule.

Depth (mm)	^{10}Be (dpm/kg)
1	88
5	54
10	28
15	15

5.5 In a mineral ^{40}Ar*/^{40}K (where "*" signifies radiogenic) ratio is 0.0250. The decay constant of ^{40}K is 5.543×10^{-10} yr^{-1}. Calculate the age of the mineral. (Do not forget that only 10.48% of ^{40}K decays to ^{40}Ar) What does the age mean? Under what conditions is this age the formation age of the mineral?

5.6 The extinct nuclide ^{26}Al decays to ^{26}Mg with a half-life of 7.3×10^5 yr.

a. Assume that the ^{26}Al/^{27}Al ratio was 0.3 when the element was synthesized in a star. Calculate the time interval between the cessation of the synthesis of ^{26}Al and the formation of Allende meteorite if the meteorite has an initial ^{26}Al/^{27}Al ratio of 5×10^{-5}.

b. Another meteorite has an initial ^{26}Al/^{27}Al ratio of 2.5×10^{-5}. Compared with the Allende meteorite, which meteorite formed earlier and by how much?

5.7 This ^{147}Sm–^{143}Nd isochron problem is best done with a computer using a graphing or spreadsheet program. The decay constant of ^{147}Sm is 6.54×10^{-12} yr^{-1}. Sm and Nd concentrations and ^{143}/^{144}Nd ratio in Komatiite samples from Cape Smith are as follows (adapted from Zindler, 1982):

Sample No.	Sm (ppm)	Nd (ppm)	$^{143/144}$Nd
1. (rock 1)	0.6531	1.776	0.513147
2. (cpx in rock 1)	0.6911	1.525	0.513727
3. (rock 2)	3.668	11.31	0.512792
5. (rock 3)	2.440	7.784	0.512689

Sample No.	Sm (ppm)	Nd (ppm)	$^{143/144}$Nd
4. (cpx in rock 2)	0.8770	1.903	0.513826
6. (rock 4)	1.436	4.176	0.512918
7. (rock 5)	2.036	6.073	0.512907
8. (rock 6)	6.119	18.16	0.512894

a. Calculate the ^{147}Sm/^{144}Nd atomic ratio for all the samples by multiplying Sm/Nd weight ratio by the factor 0.6046. Plot the isochron and find the age of the rocks (using only the whole rock data) and the initial ^{143}/^{144}Nd ratio.

b. Now make a new plot with the cpx data and the isochron you obtained above. Use the plot to address whether there was a metamorphic event that homogenized the isotopic ratio in 0.1-m scale (hand specimen scale) but not in 10-m scale (scale of different rocks) after the formation of the rocks.

5.8 Zircon grains in an igneous rock contain 2000 ppm U, 200 ppm Th, and 5 ppm Sm. If ^{4}He concentration in zircon is 2×10^{-7} mol/g, find the He-closure age of zircon.

5.9 U–Pb analytical data of igneous zircon samples from a silicic dike intruded into the Dufek layered intrusion are as follows (Minor and Mukasa, 1997), ^{204}Pb/^{206}Pb $= 0.001602$, ^{207}Pb*/^{206}Pb* $= 0.04976 \pm 0.00002$, ^{206}Pb*/^{238}U $= 0.02873 \pm 0.00013$, ^{207}Pb*/^{235}U $= 0.1971 \pm 0.0012$, where the superscript "*" means the radiogenic part of the nuclide and the errors are given at 2σ level.

a. Plot the ^{206}Pb*/^{238}U and ^{207}Pb*/^{235}U ratios in the concordia diagram. Are the data concordant?

b. Find the ^{238}U–^{206}Pb* age of the igneous zircon.

c. Find the ^{235}U–^{207}Pb* age of the igneous zircon.

d. Find the ^{207}Pb*–^{206}Pb* age of the igneous zircon.

e. (Optional) Estimate the error of each of the above age calculations. Which age has the smallest error?

5.10 The following data are from electron microprobe analyses of a monazite inclusion in garnet: 7.52 wt% ThO_2, 0.21 wt% UO_2, and 0.372 ± 0.002 wt% PbO. Estimate the age of the monazite crystal.

5.11 Pb in zircon is largely from the decay of U (^{238}U and ^{235}U). On the other hand, Pb in monazite is largely from the decay of ^{232}Th. In which mineral is the atomic mass of Pb greater? The atomic mass of the common lead is 207.2. Should the atomic mass of lead in zircon be greater or less than that of common lead? How about monazite?

5.12 Ar diffusivity in biotite perpendicular to the **c**-direction (that is, in the hexagonal plane) may be expressed as $D = \exp(-11.77 - 23{,}694/T)$ m^2/s, where T is in K. Treat the effective shape of biotite as an infinite cylinder of radius 2 mm. Calculate:

a. T_c of Ar in bulk biotite mineral and at the center of biotite for a cooling rate of 1 K/day,

b. T_c of Ar in bulk biotite mineral and at the center of biotite for a cooling rate of 1000 K/Myr,

c. T_c of Ar in bulk biotite mineral and at the center of biotite for a cooling rate of 2 K/Myr.

5.13 He diffusivity in apatite below 560 K may be expressed as $D/a^2 = \exp(17.7 - 18{,}220/T)$ s^{-1}, where T is in K (Wolf et al., 1996) and a is the radius of apatite grains. Treat the effective shape of apatite as spheres. Calculate:

a. T_c of He in bulk apatite and at the center of apatite for a cooling rate of 1000 K/Myr,

b. T_c of He in bulk apatite and at the center of apatite for a cooling rate of 10 K/Myr.

5.14 He diffusivity in zircon may be expressed as $D = (4.6 \overset{\times}{\div} 3) \times 10^{-5} \exp(-20{,}330/T)$ m^2/s, where T is in K (Reiners et al., 2004) and zircon is treated as isotropic and spherical in terms of He diffusion. Treating He diffusion in zircon as isotropic diffusion is probably wrong (Reich et al., 2007) but we will do so in this homework problem. Zircon grains of 60-μm diameter are held at 500 K for millions of years. Find the steady-state age for the bulk zircon grains and at the center of zircon grains. (The production rate may be regarded as roughly constant because the half-lives of the parents of ^{4}He, ^{238}U, ^{235}U, ^{232}Th, and ^{147}Sm are all much longer than millions of years.)

5.15 Consider the Fe–Mg order–disorder reaction between M1 and M2 sites of orthopyroxene $Fe(M2) + Mg(M1) \rightleftharpoons Fe(M1) + Mg(M2)$. The forward reaction rate coefficient $\ln k_f = 23.33 - (32{,}241 - 6016 X_{Fs}^2)/T$ (Kroll et al., 1997), and for Mössbauer measurements $\ln K_D = 0.391 - 2205/T$ (Wang et al., 2005).

a. Wang et al. (2005) reported the following composition for a natural orthopyroxene: Fe(M1) = 0.00199 ± 0.00009; Mg(M1) = 0.9797; Fe(M2) = 0.0194; Mg(M2) = 0.9780. Only the error on Fe(M1) is given because errors on the other parameters are less significant.

$X_{Fs} = [Fe(M1) + Fe(M2)]/2$. Calculate the apparent equilibrium temperature and the cooling rate.

b. Wang et al. (2005) reported the following composition for another natural orthopyroxene: $Fe(M1) = 0.0161 \pm 0.0012$; $Mg(M1) = 0.9724$; $Fe(M2) = 0.3033$; $Mg(M2) = 0.6793$. Only the error on $Fe(M1)$ is given because errors on the other parameters are less significant. Calculate the apparent equilibrium and the cooling rate.

5.16 Fe^{2+} and Mg in orthopyroxene can partition between M1 and M2 sites through the reaction $Fe(M2) + Mg(M1) \rightleftharpoons Fe(M1) + Mg(M2)$. Ganguly et al. (1994) expressed K_D for the intracrystalline exchange reaction as $\exp(0.888 - 3062/T)$ and the forward reaction rate coefficient as $k_f = \exp[(26.2 + 6.0X_{Fs}) - 31,589/T]$, where k_f is in min^{-1} and T is in K. Note that $k_f = C_0 k_1$, where C_0 is the total concentration of M1 and M2 sites (the definition of C_0 by Ganguly is 2 times the definition of C_0 in this book) and k_1 is the rate coefficient defined by

$$d\xi/dt = k_1[Fe(M2)][Mg(M1)] - k_2[Fe(M1)][Mg(M2)].$$

An orthopyroxene crystal has the following bulk composition:

$$(Mg_{1.6221}Fe_{0.3309}Ca_{0.026}Cr_{0.021})(Al_{0.021}Si_{1.979})O_6.$$

Ganguly et al. (1994) found that an opx crystal from a meteorite has the above overall composition with $Fe(M1) = 0.0060$, and $Mg(M1) = 0.9730$. Find T_{ae}. Then find the cooling rate (in °C per million years) at T_{ae} and compare it with the result of Ganguly et al. (1994). There may be a difference. If you need other information, consult the original work of Ganguly et al. (1994) or Zhang (1994).

5.17 Infrared measurement of a hydrous obsidian glass gives the band intensities as $0.040/mm$ for the $523\text{-}mm^{-1}$ band and $0.12/mm$ for the $452\text{-}cm^{-1}$ band. Find the cooling rate of the obsidian glass when it was at the temperature of T_{ae}.

5.18 The following table gives measured Fe concentrations in garnet as a function of distance from the center. Treat the diffusion profile as a spherical diffusion couple. Fit the data to find $\int D\,dt$.

r (mm)	$X(Fe)$	r (mm)	$X(Fe)$	r (mm)	$X(Fe)$
0	0.800	5.95	0.766	7.28	0.668
0.70	0.799	6.30	0.743	7.42	0.659
1.40	0.799	6.44	0.733	7.56	0.647
2.10	0.802	6.58	0.723	7.70	0.641

r (mm)	X(Fe)	r (mm)	X(Fe)	r (mm)	X(Fe)
2.80	0.799	6.72	0.710	8.05	0.625
3.50	0.801	6.86	0.700	8.40	0.611
4.20	0.799	6.93	0.696	8.75	0.607
4.90	0.796	7.00	0.686	9.10	0.601
5.25	0.792	7.07	0.686	9.80	0.598
5.60	0.780	7.14	0.679	10.5	0.600

5.19 In an andesitic rock, a large plagioclase phenocryst shows oscillatory zoning in albite and anorthite concentrations. The period of each oscillation is about $10\,\mu m$. Suppose the andesitic magma temperature in the magma chamber was 1400 K, and the albite–anorthite interdiffusivity may be expressed as $D = 0.0011 \exp(-62{,}100/T)$, where T is in K and D is in m^2/s (Grove et al., 1984). What is the maximum amount of time the plagioclase phenocryst resided in the magma chamber after its growth?

5.20 For a zoned zircon crystal with a core and a mantle, consider how oxygen isotope exchange would affect the center ^{18}O isotope composition of the core under dry conditions. The diffusivity of ^{18}O diffusivity in zircon under dry conditions is given as (Watson and Cherniak, 1997)

$$D = \exp(-8.93 - 53{,}920/T),$$

where D is in m^2/s and T is in K. Use Equation 5-141 to calculate the relation between cooling rate q (in K/Myr), the initial temperature T_0 (in K) ranging from 800 to 1200 K, and the core radius a (in μm) for the center ^{18}O concentration to be affected through diffusion by 1% relative. Then, carry out a linear regression to express $\ln q$ as a linear function of $1/T_0$ and $\ln a$.

Appendix 1 Entropy Production and Diffusion Matrix

In irreversible thermodynamics, the second law of thermodynamics dictates that entropy of an isolated system can only increase. From the second law of thermodynamics, entropy production in a system must be positive. When this is applied to diffusion, it means that binary diffusivities as well as eigenvalues of diffusion matrix are real and positive if the phase is stable. This section shows the derivation (De Groot and Mazur, 1962).

Unlike mass, which is conserved, leading to $\partial\rho/\partial t = -\mathbf{\nabla}\cdot\mathbf{J}$, entropy is not conserved. That is, in addition to entropy flux that would lead to variation in entropy in a given volume, entropy is also produced from nowhere. Hence, entropy balance is written as

$$\partial S/\partial t = -\mathbf{\nabla}\cdot\mathbf{J}_s + \sigma, \tag{A1-1}$$

where $\mathbf{J}_s$ is entropy flux, and σ is entropy production. According to the second law of thermodynamics, $\sigma \geq 0$. From the combination of the first and second laws of thermodynamics, and considering a fixed volume, we have

$$du = T\,ds + \sum \mu_i d\rho_i, \tag{A1-2}$$

where u is the internal energy per unit volume, T is absolute temperature, s is entropy per unit volume, and μ_i and ρ_i are the chemical potential and molar density of component i. Therefore,

$$T\,ds = du - \sum \mu_i d\rho_i. \tag{A1-3}$$

Differentiating with respect to time leads to

$$T\frac{\partial s}{\partial t} = \frac{\partial u}{\partial t} - \sum_{k=1}^{N} \mu_k \frac{\partial \rho_k}{\partial t}.$$ (A1-4)

Replacing $\partial s/\partial t$ by the entropy balance equation, $\partial u/\partial t$ by the energy conservation equation, and $\partial \rho_k/\partial t$ by the mass balance equation, we have

$$T(-\mathbf{\nabla} \cdot \mathbf{J}_s + \sigma) = -\mathbf{\nabla} \cdot \mathbf{J}_u + \sum \mu_k(\mathbf{\nabla} \cdot \mathbf{J}_k - \sum v_{kj}J_j),$$ (A1-5)

where J_j is reaction progress of reaction j per unit time. By analagy to $ds = du/T - \sum \mu_i d\rho_i/T$, we have

$$T\mathbf{\nabla}\mathbf{J}_s = T[\mathbf{\nabla}(\mathbf{J}_u/T) - \sum \mathbf{\nabla}(\mu_k\mathbf{J}_k/T)] = \mathbf{\nabla}\mathbf{J}_u + T\mathbf{J}_u\mathbf{\nabla}(1/T)$$
$$- \sum \mu_k\mathbf{\nabla}\mathbf{J}_k - T\sum \mathbf{J}_k\mathbf{\nabla}(\mu_k/T),$$ (A1-6)

and

$$\sigma = \frac{1}{T}\left[T\mathbf{\nabla}\mathbf{J}_s - \mathbf{\nabla}\mathbf{J}_u + \sum_k \mu_k(\mathbf{\nabla}\mathbf{J}_k - \sum_i v_{ki}J_i)\right]$$
$$= \mathbf{J}_u\mathbf{\nabla}(1/T) - \sum_k \mathbf{J}_k\mathbf{\nabla}\frac{\mu_k}{T} - \sum_k \sum_j \frac{v_{kj}\mu_k J_j}{T}.$$ (A1-7)

Since $\Delta G_j =$ the Gibbs free energy change for the reaction $j = \sum v_{kj}\cdot\mu_k$, we finally have

$$\sigma = \mathbf{J}_u\mathbf{\nabla}(1/T) - \sum_k \mathbf{J}_k\mathbf{\nabla}\frac{\mu_k}{T} - \sum_j \frac{J_j\Delta G_j}{T},$$ (A1-8)

or

$$\sigma = -\frac{1}{T^2}\mathbf{J}_u\mathbf{\nabla}T - \sum_k \mathbf{J}_k\mathbf{\nabla}\frac{\mu_k}{T} - \sum_j \frac{J_j\Delta G_j}{T}.$$ (A1-9)

That is, the entropy production in the volume consists of three terms, each of which is due to an irreversible process. The first term is the heat conduction term, the second is the mass diffusion term, and the third is the chemical reaction term. The above equation is known as the *entropy production equation*.

If we only consider diffusion under an isothermal condition, then

$$\sigma = -\sum \mathbf{J}_i \cdot \mathbf{\nabla}(\mu_i/T).$$ (A1-10)

To find entropy production rate, we need to relate flux to chemical potential gradients. To the first-order approximation, flux of each component is linearly related to gradients:

$$\mathbf{J}_i = -\sum L_{ij}\mathbf{\nabla}(\mu_j/T),$$ (A1 − 11)

where L_{ij}'s are Onsager's phenomenological coefficients for diffusion.

In a binary system, $\mathbf{J}_1$ and $\mathbf{J}_2$ are, in general, linearly dependent due to constraints such as constant density and no-void space. Choosing a reference frame we would have

$$\alpha_1 \mathbf{J}_1 + \alpha_2 \mathbf{J}_2 = \mathbf{0}, \tag{A1-12}$$

where α_1 and α_2 are constants. In appropriate reference frames, $\alpha_1 = \alpha_2 = 1$. For constant T,

$$\sigma = -\mathbf{J}_2 \nabla(\mu_2 - \alpha_2 \mu_1/\alpha_1)/T. \tag{A1-13}$$

Similar to calling $\nabla(\mu_2/T)$ a thermodynamic "force," we can call $\nabla(\mu_2 - \alpha_2\mu_1/\alpha_1)/T$ an independent force. By defining the independent flux $\mathbf{J}_2$ through this independent force, we have

$$\mathbf{J}_2 = -L\nabla(\mu_2 - \alpha_2\mu_1/\alpha_1)/T, \tag{A1-14}$$

where L is a phenomenological coefficient. Using the Gibbs-Duhem relation for a binary system $X_1\nabla\mu_1 + X_2\nabla\mu_2 = 0$, where X is mole fraction, Equation A1-14 may be expressed as

$$\mathbf{J}_2 = -L\left(1 + \frac{\alpha_2 X_2}{\alpha_1 X_1}\right)\nabla\frac{\mu_2}{T}. \tag{A1-15}$$

When Equation A1-14 is substituted into A1-13, the entropy production rate σ is

$$\sigma = L\left(\nabla\frac{\mu_2 - \alpha_2\mu_1/\alpha_1}{T}\right)^2 = L\left(1 + \frac{\alpha_2 X_2}{\alpha_1 X_1}\right)^2\left(\nabla\frac{\mu_2}{T}\right)^2. \tag{A1-16}$$

The second law of thermodynamics dictates that σ is positive when $\nabla\mu_2 \neq \mathbf{0}$; therefore, $L > 0$. Hence, based on Equation A1-15, diffusion in binary solutions is *always* down the chemical potential (or activity) gradient. Comparing $\mathbf{J}_2$ in Equation A1-14 with Fick's law and assuming constant molar density ρ, we have

$$D = \frac{LR}{a_2\rho}\left(1 + \frac{\alpha_2 X_2}{\alpha_1 X_1}\right)\frac{\nabla a_2}{\nabla x_2} = \frac{LR}{a_2\rho}\left(1 + \frac{\alpha_2 X_2}{\alpha_1 X_1}\right)\frac{da_2}{dx_2}. \tag{A1-17}$$

where R is the gas constant, and a_2 is the activity of component 2. Because σ must be positive, D is negative whenever da_2/dX_2 is negative. For regular solutions, when $W/(RT) > 2$, then there is a region where da_2/dX_2 is negative and there will be spinodal decomposition (a single phase decomposes into two phases). In the region where D is negative, the diffusive flux for a component is up against the concentration gradient (but not the activity gradient or the chemical potential gradient) of the component. This phenomenon is known as *uphill diffusion*.

Instead of Equation A1-14, we can also define $\mathbf{J}_2 = -L_2\nabla\mu_2/T$ (note that $L_2 \neq L$); then D and L_2 are related through

$$D = \frac{L_2 R d a_2}{a_2 \rho d x_2}. \tag{A1-18}$$

In multicomponent systems, the single diffusivity is replaced by a multicomponent diffusion matrix. By going through similar steps, it can be shown that the $[D]$ matrix must have positive eigenvalues if the phase is stable. In a multicomponent system, the diffusive flux of a component can be up against its chemical potential gradient except for eigencomponents.

Appendix 2 The Error Function and Related Functions

The solutions of a diffusion equation under the transient case (non-steady state) are often some special functions. The values of these functions, much like the exponential function or the trigonometric functions, cannot be calculated simply with a piece of paper and a pencil, not even with a calculator, but have to be calculated with a simple computer program (such as a spreadsheet program, but see later comments for practical help). Nevertheless, the values of these functions have been tabulated, and are now easily available with a spreadsheet program. The properties of these functions have been studied in great detail, again much like the exponential function and the trigonometric functions. One such function encountered often in one-dimensional diffusion problems is the error function, erf(z). The error function erf(z) is defined by

$$\text{erf}(z) = \frac{2}{\sqrt{\pi}} \int_0^z e^{-\xi^2} d\xi. \tag{A2-1}$$

A related function is the complimentary error function erfc(z), defined by

$$\text{erfc}(z) = \frac{2}{\sqrt{\pi}} \int_z^\infty e^{-\xi^2} d\xi. \tag{A2-2}$$

Some values of the error function are

$$\text{erf}(0) = 0; \quad \text{erf}(\infty) = 1; \quad \text{erf}(-\infty) = -1;$$

$$\text{erf}(0.5) = 0.520500; \quad \text{erf}(1) = 0.842701; \quad \text{erf}(2) = 0.995322; \quad \text{erf}(3) = 0.999978$$

Some properties of the error and complimentary error functions are

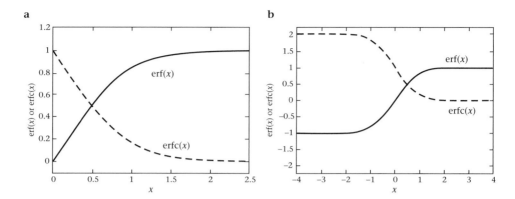

Figure A2-1 erf(x) and erfc(x) (a) at $x > 0$ and (b) in $-\infty < x < \infty$.

$$\mathrm{erf}(z) + \mathrm{erfc}(z) = 1, \qquad \text{for all } z. \tag{A2-3}$$

$$\mathrm{erf}(-z) = -\mathrm{erf}(z), \qquad \text{for all } z. \tag{A2-4}$$

From the above relations,

$$\mathrm{erfc}(-z) = 1 - \mathrm{erf}(-z) = 1 + \mathrm{erf}(z) \tag{A2-5}$$

$$\frac{d\mathrm{erf}(z)}{dz} = \frac{2}{\sqrt{\pi}} e^{-z^2}; \tag{A2-6}$$

$$\frac{d\mathrm{erfc}(z)}{dz} = -\frac{2}{\sqrt{\pi}} e^{-z^2}. \tag{A2-7}$$

To calculate the error functions one uses

$$\mathrm{erf}(z) = \frac{2z}{\sqrt{\pi}} \left(1 - \frac{z^2}{1! \cdot 3} + \frac{z^4}{2! \cdot 5} - \frac{z^6}{3! \cdot 7} + \cdots \right), \quad \text{for small } |z|, \tag{A2-8a}$$

$$\mathrm{erf}(z) = \frac{2z}{\sqrt{\pi}} e^{-z^2} \left(1 + \frac{(2z^2)}{1 \times 3} + \frac{(2z^2)^2}{1 \times 3 \times 5} + \cdots \right), \quad \text{for intermediate } |z|, \tag{A2-8b}$$

$$\mathrm{erfc}(z) = \frac{1}{z\sqrt{\pi}} e^{-z^2} \left(1 - \frac{1}{2z^2} + \frac{3!!}{2^2 z^4} - \frac{5!!}{2^3 z^6} + \cdots \right), \quad \text{for large } z > 0. \tag{A2-8c}$$

Equations A2-8a and A2-8b always converge but for large absolute values of z (e.g., >5) the convergence is slow and truncation errors may dominate. Hence, in practice, Equation A2-8a is often applied for $|z| \leq 1$, and Equation A2-8b is often applied for $1 < |z| \leq 4.5$. Equation A2-8c is an asymptotic expression and must be truncated at or before the absolute value of the term in the series reaches a minimum. For large z ($z > 10$), $ze^{z^2}\mathrm{erfc}(z)$ approaches $1/\sqrt{\pi}$.

A diagram of erf(z) vs. z is shown in Figure A2-1. Values of error functions are given in Table A2-1.

Table A2-1 Values of error function and related functions

x	$\text{erf}(x)$	$\text{erfc}(x)$	$\sqrt{\pi}\,\text{ierfc}(x)$	$4\,\text{i}^2\text{erfc}(x)$
0	0	1	1	1
0.05	0.0563720	0.9436280	0.9138763	0.8920681
0.1	0.1124629	0.8875371	0.8327380	0.7935727
0.15	0.1679960	0.8320040	0.7565479	0.7039531
0.2	0.2227026	0.7772974	0.6852447	0.6226542
0.25	0.2763264	0.7236736	0.6187435	0.5491293
0.3	0.3286268	0.6713732	0.5569378	0.4828421
0.35	0.3793821	0.6206179	0.4997001	0.4232700
0.4	0.4283924	0.5716076	0.4468846	0.3699055
0.45	0.4754817	0.5245183	0.3983285	0.3222588
0.5	0.5204999	0.4795001	0.3538549	0.2798589
0.55	0.5633234	0.4366766	0.3132745	0.2422558
0.6	0.6038561	0.3961439	0.2763883	0.2090215
0.65	0.6420293	0.3579707	0.2429900	0.1797506
0.7	0.6778012	0.3221988	0.2128686	0.1540612
0.75	0.7111556	0.2888444	0.1858103	0.1315961
0.8	0.7421010	0.2578990	0.1616012	0.1120211
0.85	0.7706681	0.2293319	0.1400287	0.0950272
0.9	0.7969082	0.2030918	0.1208843	0.0803288
0.95	0.8208908	0.1791092	0.1039649	0.0676630
1	0.8427008	0.1572992	0.0890739	0.0567901
1.1	0.8802051	0.1197949	0.0646333	0.0395710
1.2	0.9103140	0.0896860	0.0461706	0.0271685
1.3	0.9340079	0.0659921	0.0324613	0.0183748
1.4	0.9522851	0.0477149	0.0224570	0.0122388
1.5	0.9661051	0.0338949	0.0152836	0.0080262

(*continued*)

Table A2-1 (*continued*)

x	erf(x)	erfc(x)	$\sqrt{\pi}\mathrm{ierfc}(x)$	$4\,\mathrm{i}^2\mathrm{erfc}(x)$
1.6	0.9763484	0.0236516	0.0102305	0.0051814
1.7	0.9837905	0.0162095	0.0067341	0.0032919
1.8	0.9890905	0.0109095	0.0043580	0.0020580
1.9	0.9927904	0.0072096	0.0027724	0.0012657
2	0.9953223	0.0046777	0.0017335	0.0007657
2.2	0.9981372	0.0018629	0.0006431	0.0002665
2.4	0.9993115	0.0006885	2.22E-04	8.66E-05
2.6	0.9997640	0.0002360	7.15E-05	2.62E-05
2.8	0.9999250	7.50E-05	2.14E-05	7.44E-06
3	0.9999779	2.21E-05	5.94E-06	1.97E-06
3.2	0.9999940	6.03E-06	1.54E-06	4.82E-07
3.4	0.9999985	1.52E-06	3.68E-07	1.10E-07
3.6	0.9999996	3.56E-07	8.19E-08	2.32E-08
3.8	0.9999999	7.70E-08	1.69E-08	4.59E-09
4	1	1.54E-08	3.23E-09	8.32E-10
4.2	1	2.86E-09	5.72E-10	1.43E-10
4.4	1	4.89E-10	9.40E-11	2.25E-11
4.6	1	7.75E-11	1.43E-11	3.29E-12
4.8	1	1.14E-11	2.02E-12	4.40E-13
5	1	1.54E-12	2.62E-13	5.82E-14
5.5	1	7.36E-15	1.15E-15	2.25E-16
6	1	2.15E-17	3.10E-18	5.60E-19
7	1	4.18E-23	5.19E-24	8.13E-25
8	1	1.12E-29	1.22E-30	1.69E-31
10	1	2.09E-45	1.83E-46	2.04E-47

Integrated error functions

Integrated error functions are repeated integrations of the complementary error function. Define

$$i^n \mathrm{erfc}(z) = \int_z^\infty i^{n-1} \mathrm{erfc}(\xi) d\xi. \tag{A2-9}$$

Hence, $\mathrm{ierfc}(z) = \int_z^\infty \mathrm{erfc}(\xi) d\xi$, $i^2 \mathrm{erfc}(z) = \int_z^\infty \mathrm{ierfc}(\xi) d\xi$, etc.

Integrated error functions can be expressed in terms of error functions. For example, integrating by part, we can find

$$\mathrm{ierfc}(z) = \frac{1}{\sqrt{\pi}} e^{-z^2} - z\,\mathrm{erfc}(z). \tag{A2-10}$$

$$i^2 \mathrm{erfc}(z) = \tfrac{1}{4}[\mathrm{erfc}(z) - 2z \cdot \mathrm{ierfc}(z)] \tag{A2-11}$$

$$2n \cdot i^n \mathrm{erfc}(z) = i^{n-2} \mathrm{erfc}(z) - 2z \cdot i^{n-1} \mathrm{erfc}(z) \tag{A2-12}$$

Some values of the integrated error functions are

$$\mathrm{ierfc}(0) = 1/\sqrt{\pi}; \quad i^2 \mathrm{erfc}(0) = \tfrac{1}{4}; \quad i^n \mathrm{erfc}(0) = i^{n-2} \mathrm{erfc}(0)/(2n).$$

Values of $\mathrm{ierfc}(z)$ and $i^2 \mathrm{erfc}(z)$ are also listed in Table A2-1.

Although some spreadsheet programs provide values of error function and related functions, there may be limitations on the value of the independent variables. For example, in a spreadsheet program, $\mathrm{erf}(x)$ values are provided only for $0 \leq x \leq 10$. Then one may use the following to obtain $\mathrm{erf}(x)$ for any real x:

$$\mathrm{erf}(x) = \mathrm{sign}(x)\, \mathrm{erf}\,[\mathrm{abs}(x)]; \tag{A2-13a}$$

If $\mathrm{abs}(x) > 10$, then $\mathrm{erf}(x) = \mathrm{sign}(x)$. \hfill (A2-13b)

To obtain values of $\mathrm{erfc}(x)$ for $x \geq 4.5$, Equation A2-8c, instead of $\mathrm{erfc}(x) = 1 - \mathrm{erf}(x)$, should be applied.

Appendix 3 Some Solutions to Diffusion Problems

The solutions given in this Appendix may be found in Carslaw and Jaeger (1959) or Crank (1975) unless otherwise indicated.

A3.1 Instantaneous plane, line, or point source

The following solutions for instantaneous sources are useful in conjunction with the principle of superposition to derive solutions to other diffusion problems.

A3.1.1 Plane source for one-dimensional infinite medium $(-\infty < x < \infty)$

Diffusion equation: $\frac{\partial C}{\partial t} = D\frac{\partial^2 C}{\partial x^2}$
 Initial condition: Plane source at $x = 0$ with total mass M; i.e., $C|_{t=0} = M\delta(x)$,
 where

$$\delta(x) = \begin{cases} \infty & x = 0 \\ 0 & x \neq 0 \end{cases}; \qquad \int_a^b \delta(x)\mathrm{d}x = \begin{cases} 1 & \text{if} \quad 0 \in (a, b) \\ 0 & \text{if} \quad 0 \notin [a, b] \end{cases}$$

Solution: $C = \dfrac{M}{(4\pi Dt)^{1/2}}\mathrm{e}^{-x^2/4Dt}$.

A3.1.2. Plane source for one-dimensional semi-infinite medium ($0 \leq x > \infty$)

Diffusion equation:

$$\frac{\partial C}{\partial t} = D \frac{\partial^2 C}{\partial x^2}$$

Initial condition: Plane source at $x = 0$ with total mass M.

Solution: $C = \dfrac{M}{(\pi Dt)^{1/2}} \mathrm{e}^{-x^2/4Dt}$.

A3.1.3. Line source for two-dimensional infinite medium ($-\infty < x < \infty$, $-\infty < y < \infty$)

Diffusion equation: $\dfrac{\partial C}{\partial t} = D \left\{ \dfrac{\partial^2 C}{\partial x^2} + \dfrac{\partial^2 C}{\partial y^2} \right\}$.

Initial condition: Line source at $x = 0$ and $y = 0$ with total mass M.

Solution: $C = \dfrac{M}{4\pi Dt} \mathrm{e}^{-(x^2+y^2)/4Dt} = \dfrac{M}{4\pi Dt} \mathrm{e}^{-r^2/4Dt}$,

where $r^2 = x^2 + y^2$.

A3.1.4. Point source for three-dimensional infinite medium ($-\infty < x, y, z < \infty$)

Diffusion equation: $\dfrac{\partial C}{\partial t} = D \left\{ \dfrac{\partial^2 C}{\partial x^2} + \dfrac{\partial^2 C}{\partial y^2} + \dfrac{\partial^2 C}{\partial z^2} \right\}$.

Initial condition: Point source at $x = 0$, $y = 0$, and $z = 0$ with total mass M.

Solution: $C = \dfrac{M}{(4\pi Dt)^{3/2}} \mathrm{e}^{-(x^2+y^2+z^2)/4Dt} = \dfrac{M}{(4\pi Dt)^{3/2}} \mathrm{e}^{-r^2/4Dt}$,

where $r^2 = x^2 + y^2 + z^2$.

Problems with extended sources (sources over extended space) or continuous sources (sources over continuous time) can be solved by integrating the above solutions.

A3.2 One-dimensional diffusion

Diffusion equation: $\dfrac{\partial C}{\partial t} = D \dfrac{\partial^2 C}{\partial x^2}$

A3.2.1. Infinite medium, D is constant, boundary conditions at $x = -\infty$ and $x = \infty$ conform to the initial conditions

(a) General initial condition: $C|_{t=0} = f(x)$.

Solution: $C(x, t) = (4\pi Dt)^{-1/2} \int_{-\infty}^{\infty} f(\xi) \mathrm{e}^{-(x-\xi)^2/4Dt} \mathrm{d}\xi$.

(b) Extended source problem.

Initial condition: $C|_{t=0, -\delta<x<\delta} = C_0$; $C|_{t=0, x<-\delta} = C|_{t=0, x>\delta} = 0$.

Solution: $C(x, t) = \dfrac{C_0}{2} \left\{ \mathrm{erf} \dfrac{x+\delta}{2\sqrt{Dt}} - \mathrm{erf} \dfrac{x-\delta}{2\sqrt{Dt}} \right\}$.

(c) Initial condition: $C|_{t=0} = \begin{cases} C^0_- & x < 0 \\ C^0_+ & x > 0. \end{cases}$

Solution: $C = \left(\dfrac{C^0_+ + C^0_-}{2} + \dfrac{C^0_+ - C^0_-}{2} \right) \mathrm{erf}\left(\dfrac{x}{2\sqrt{Dt}} \right)$.

A3.2.2 Infinite medium, $D = D_1$ at $x < 0$, and $D = D_2$ at $x > 0$, boundary conditions at $x = -\infty$ and $x = \infty$ conform to the initial condition

(a) Initial condition: $C|_{t=0,\ x<0} = C^0_1$, and $C|_{t=0,\ x>0} = C^0_2$.

Boundary conditions: $(C_2/C_1)|_{x=0} = K$, $D_1 \dfrac{\partial C_1}{\partial x}\bigg|_{x=-0} = D_2 \dfrac{\partial C_2}{\partial x}\bigg|_{x=+0}$.

Let C_1 be the solution for $x < 0$ and C_2 be the solution for $x > 0$; then

$$C_1 = C^0_1 + \frac{C^0_2 - KC^0_1}{K + \sqrt{D_1/D_2}} \mathrm{erfc}\, \frac{|x|}{2\sqrt{D_1 t}},$$

$$C_2 = C^0_2 + \frac{KC^0_1 - C^0_2}{1 + K\sqrt{D_2/D_1}} \mathrm{erfc}\, \frac{x}{2\sqrt{D_2 t}}.$$

(b) Special case of the above when $K = 1$ (continuous at the boundary):

$$C_1 = C^0_1 + \frac{C^0_2 - C^0_1}{1 + \sqrt{D_1/D_2}} \mathrm{erfc}\, \frac{|x|}{2\sqrt{D_1 t}},$$

$$C_2 = C^0_2 + \frac{C^0_1 - C^0_2}{1 + \sqrt{D_2/D_1}} \mathrm{erfc}\, \frac{x}{2\sqrt{D_2 t}}.$$

A3.2.3 Semi-infinite medium ($x \geq 0$), constant D, boundary condition at $x = \infty$ conforms to the initial condition (i.e., $C|_{x=\infty} = C|_{t=0}$)

(a) Initial condition: $C|_{t=0} = C_\infty$.
  Boundary condition: $C|_{x=0} = C_0$.
  Solution: $C - C_\infty = (C_0 - C_\infty)\mathrm{erfc}\left(\dfrac{x}{2\sqrt{Dt}} \right)$.
(b) Initial condition: $C|_{t=0} = C_\infty$.
  Boundary condition: $C|_{x=0} = C_\infty + kt^{n/2}$, where n is a positive integer.

  Solution: $C = C_\infty + k\Gamma\left(\dfrac{n}{2} + 1 \right)(4t)^{n/2} i^n \mathrm{erfc}\, \dfrac{x}{2\sqrt{Dt}}$.

Total mass entering the medium per unit area is

$$M(t) = \frac{k}{2^n}\sqrt{Dt}(4t)^{n/2} \frac{\Gamma(n/2+1)}{\Gamma(n/2+3/2)}$$

where $\Gamma\left(\dfrac{n}{2} + 1 \right) = \begin{cases} (n/2)! & \text{for even } n, \\ n!!\sqrt{\pi}/2^{(n+1)/2} & \text{for odd } n. \end{cases}$

Note: If the surface concentration is a smooth function of t, it can be expanded as a polynomial of $t^{1/2}$; hence, the solution is a linear combination of the above solution of different n.

(c) Initial condition: $C|_{t=0} = C_\infty$.

Boundary condition: $(\partial C/\partial x)|_{x=0} = \alpha$.

Let $u = \partial C/\partial x$, where u satisfies the one-dimensional diffusion equation, and initial and boundary conditions are $u|_{t=0} = 0$, $u|_{x=\infty} = 0$, and $u|_{x=0} = \alpha$. Hence, from 3.2.3a,

$$u = \alpha \, \text{erfc}\left(\frac{x}{2\sqrt{Dt}}\right).$$

Therefore, the solution is $C = C_\infty + 2\alpha\sqrt{Dt} \cdot \text{ierfc}\left(\frac{x}{2\sqrt{Dt}}\right)$.

(d) Initial condition is general: $C_{t=0} = f(x)$.

Boundary condition: $C|_{x=0} = 0$.

Solution: $C(x,t) = \dfrac{1}{2\sqrt{\pi Dt}} \int_0^\infty f(\xi)[e^{-(x-\xi)^2/(4Dt)} - e^{-(x+\xi)^2/(4Dt)}]d\xi$.

(e) Initial condition is general: $C|_{t=0} = f(x)$.

Boundary condition: $(\partial C/\partial x)|_{x=0} = 0$.

Solution: $C(x,t) = \dfrac{1}{2\sqrt{\pi Dt}} \int_0^\infty f(\xi)[e^{-(x-\xi)^2/(4Dt)} + e^{-(x+\xi)^2/(4Dt)}]d\xi$.

(f) Initial condition is general: $C|_{t=0} = f(x)$.

Boundary condition is general: $C|_{x=0} = g(x)$.

Solution: $C(x,t) = \dfrac{1}{2\sqrt{\pi Dt}} \int_0^\infty f(\xi)[e^{-(x-\xi)^2/(4Dt)} - e^{-(x+\xi)^2/(4Dt)}]d\xi$

$+ \dfrac{x}{2\sqrt{\pi Dt}} \int_0^\infty \dfrac{g(\tau)}{(t-\tau)^{3/2}} e^{-x^2/[4D(t-\tau)]}d\tau.$

(g) Extended initial source problem.

Initial condition: $C|_{t=0, \, 0 < x < \delta} = C_0$; $C|_{t=0, \, x > \delta} = 0$.

Boundary condition: $(\partial C/\partial x)|_{x=0} = 0$.

Solution: $C(x,t) = \dfrac{C_0}{2}\left(\text{erf}\,\dfrac{x+\delta}{\sqrt{4Dt}} - \text{erf}\,\dfrac{x-\delta}{\sqrt{4Dt}}\right).$

A3.2.4 Finite medium, $0 \leq x \leq L$ or $-L \leq x \leq L$, constant D

(a) Initial condition: $C|_{t=0} = C_0$.

Boundary conditions: $C|_{x=0} = C|_{x=L} = C_1$.

Solution: $\dfrac{C - C_1}{C_0 - C_1} = \sum_{n=1}^\infty \dfrac{4}{(2n+1)\pi} \sin\dfrac{(2n+1)\pi x}{L} e^{-n^2\pi^2 Dt/L^2}$.

Hence, $M_t = \int C dx - C_0 L = (C_1 - C_0)L + \dfrac{8(C_0 - C_1)L}{\pi^2}\sum_{n=0}^\infty \dfrac{1}{(2n+1)^2} e^{-(2n+1)^2\pi^2 Dt/L^2}$.

Because $M_\infty = (C_1 - C_0)L$, then

$$\frac{M_t}{M_\infty} = \frac{M_t}{(C_1 - C_0)L} = 1 - \frac{8}{\pi^2} \sum_{n=0}^{\infty} \frac{e^{-(2n+1)^2 \pi^2 Dt/L^2}}{(2n+1)^2}.$$

The above two equations converge rapidly for large Dt/L^2 but slowly when $Dt/L^2 \ll 1$. For small Dt/L^2, one can use

$$\frac{C - C_0}{C_1 - C_0} = \sum_{n=0}^{\infty} (-1)^n \left[\operatorname{erfc} \frac{nL + x}{2\sqrt{Dt}} + \operatorname{erfc} \frac{(n+1)L - x}{2\sqrt{Dt}} \right]$$

$$= \operatorname{erfc} \frac{x}{2\sqrt{Dt}} + \operatorname{erfc} \frac{L - x}{2\sqrt{Dt}} - \operatorname{erfc} \frac{L + x}{2\sqrt{Dt}} - \operatorname{erfc} \frac{2L - x}{2\sqrt{Dt}} + \cdots,$$

$$\frac{M_t}{M_\infty} = \frac{M_t}{(C_1 - C_0)L} = \frac{4\sqrt{Dt}}{L} \left[\frac{1}{\sqrt{\pi}} + 2 \sum_{n=1}^{\infty} (-1)^n \operatorname{ierfc} \frac{nL}{2\sqrt{Dt}} \right],$$

which converges rapidly for small t but slowly for large t. For small x and small t, the above solution reduces to the solution for semi-infinite medium (3.2.3a).

(b) Initial condition: $C|_{t=0} = C_0$.
  Boundary conditions: $C|_{x=0} = C_1$, $C|_{x=L} = C_2$.
  Let $u = C - C_1 - (C_2 - C_1)x/L$. Then u satisfies the diffusion equation and the following initial and boundary conditions:

$$u|_{t=0} = C_0 - C_1 - (C_2 - C_1)x/L; \quad u|_{x=0} = 0; \quad u|_{x=L} = 0.$$

The solution for u can be written as

$$u = \sum_{n=1}^{\infty} A_n \sin \frac{n\pi x}{L} e^{-n^2 \pi^2 Dt/L^2},$$

where A_n is to be determined by the initial condition:

$$(C_0 - C_1) - (C_2 - C_1)\frac{x}{L} = \sum_{n=1}^{\infty} A_n \sin \frac{n\pi x}{L}, \quad \text{for } 0 < x < l.$$

The final solution for C is

$$C = C_1 + (C_2 - C_1)\frac{x}{L} + \frac{2}{\pi} \sum_{n=1}^{\infty} \frac{C_2 \cos(n\pi) - C_1}{n} \sin \frac{n\pi x}{L} e^{-Dn^2\pi^2 t/L^2}$$

$$+ \frac{4C_0}{\pi} \sum_{n=0}^{\infty} \frac{1}{2n+1} \sin \frac{(2n+1)\pi x}{L} e^{-D(2n+1)^2\pi^2 t/L^2}.$$

(c) Initial condition is general: $C|_{t=0} = f(x)$.
  Boundary conditions: $C|_{x=0} = C_1$, $C|_{x=L} = C_2$. Solution:

$$C = C_1 + (C_2 - C_1)\frac{x}{L} + 2 \sum_{n=1}^{\infty} \sin \frac{n\pi x}{L} e^{-Dn^2\pi^2 t/L^2}$$

$$\times \left[\frac{(-1)^n C_2 - C_1}{n\pi} + \frac{1}{L} \int_0^L f(\xi) \sin \frac{n\pi \xi}{L} d\xi. \right]$$

(d) Initial condition: $C|_{t=0} = C_0$.

Boundary conditions: $C|_{x=0} = C_1$, $(\partial C/\partial x)|_{x=L} = 0$.

First, let $u = C - C_1$, then u satisfies the following initial and boundary conditions:

$$u|_{t=0} = C_0 - C_1; \quad u|_{x=0} = 0; \quad (\partial u/\partial x)|_{x=L} = 0.$$

The solution may be written in the form of a sine and cosine series (Equation 3-51). To satisfy $u|_{x=0} = 0$, all the B_n values must be zero. To satisfy $(\partial u/\partial x)|_{x=L} \sim \cos(\lambda_n L) = 0$, $\lambda_n L$ must be $(n + \frac{1}{2})\pi$, where $n = 0, 1, 2, \ldots$, i.e., $\lambda_n = (n + \frac{1}{2})\pi/L$. Therefore,

$$C - C_1 = \sum_{n=0}^{\infty} A_n \sin \frac{(n + \frac{1}{2})\pi x}{L} e^{-(n+\frac{1}{2})^2 \pi^2 Dt/L^2}.$$

A_n's can be determined from the initial condition, which completes the solution.

(e) Initial condition: $C|_{t=0} = f(x)$.

Boundary conditions: $(\partial C/\partial x)|_{x=0} = 0$, $(\partial C/\partial x)|_{x=L} = 0$. That is, there is no flux at the boundaries, which is applicable to treat internal homogenization of a zoned crystal. The solution can be written as

$$C = \frac{B_0}{2} + \sum_{n=1}^{\infty} B_n \cos \frac{n\pi x}{L} e^{-n^2 \pi^2 Dt/L^2},$$

because the derivative of the above is a sine function series and satisfies the boundary condition. The coefficients B_n can be obtained by writing from the initial condition $f(x)$ as a cosine series:

$$f(x) = C|_{t=0} = \frac{B_0}{2} + \sum_{n=1}^{\infty} B_n \cos \frac{n\pi x}{L},$$

where B_n can be found as follows:

$$B_n = \frac{2}{L} \int_0^L f(x) \cos \frac{n\pi x}{L} dx, \quad (n = 0, 1, 2, \ldots).$$

(e1) One example is for $C|_{t=0} = f(x) = 0.5 + 0.25 \cos(\pi x/L)$. Because this is already a cosine series, we have $B_0/2 = 0.5$, $B_1 = 0.25$, and $B_n = 0$ for $n \geq 2$. Therefore,

$$C = 0.5 + 0.25 e^{-\pi^2 Dt/L^2} \cos \frac{\pi x}{L}.$$

The concentration profile evolves as a cosine function but the amplitude is $0.25 e^{-\pi^2 dt/L^2}$, decreasing with time exponentially. When $Dt/L^2 = 0.4666$, $e^{-\pi^2 Dt/L^2} = 0.01$, the sample may be considered homogenized.

(e2) Another example is for $C|_{t=0} = f(x) = 0.5 + 0.25 \cos(2m\pi x/L)$, where m is a large integer for oscillatory zoning. There are a total of m periods of oscillations from 0 to L, and the width of each period is L/m. Compared to the general solution, $B_0/2 = 0.5$, $B_{2m} = 0.25$, and $B_n = 0$ for $n \neq 2m$. The solution is

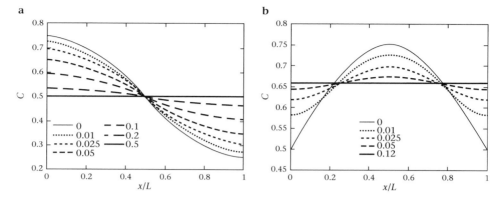

Figure A3-2-4 Concentration evolution corresponding to solution 3.2.4e1 and 3.2.4e3. The value for each curve is Dt/L^2. At $Dt/L^2 = 0.4666$, the amplitude is reduced by two orders of magnitude.

$$C = 0.5 + 0.25e^{-4m^2\pi^2Dt/L^2}\cos\frac{2m\pi x}{L}.$$

The concentration profile evolves as a cosine function but the amplitude is $0.25e^{-4m^2\pi^2Dt/L^2}$, decreasing with time exponentially. When $Dt/(L/m)^2 = 0.11665$, $e^{-4m^2\pi^2Dt/L^2} = 0.01$, the sample may be considered homogenized.

(e3) A third example is for $C|_{t=0} = f(x) = 0.5 + 0.25\sin(\pi x/L)$; then

$$C = \frac{1}{2\pi}(\pi + 1) + \frac{1}{2\pi}\left(\frac{1}{3} - 1\right)\cos\frac{2\pi x}{L}e^{-2^2\pi^2Dt/L^2}$$

$$+ \frac{1}{2\pi}\left(\frac{1}{5} - \frac{1}{3}\right)\cos\frac{4\pi x}{L}e^{-4^2\pi^2Dt/L^2} + \cdots$$

Or,

$$C = \left(0.5 + \frac{1}{2\pi}\right) - \frac{1}{\pi}\sum_{m=1}^{\infty}\left(\frac{1}{(2m-1)(2m+1)}\right)\cos\frac{2m\pi x}{L}e^{-4m^2\pi^2Dt/L^2}.$$

Figures A3.2.4a and A3.2.4b show the concentration profile of C versus x/L at several values of Dt/L^2 for (e1) and (e3). The solutions can be used to estimate the timescale to homogenize a crystal.

(f) Initial condition: $C|_{t=0} = C_0$.

Boundary conditions are symmetric: $C|_{x=\pm L} = 0$. Note that the boundary here is from $-L$ to L, not from 0 to L (when the two boundaries are symmetric, it is often easier to treat by defining the boundaries to be $\pm L$).

Solution:

$$C = \frac{4C_0}{\pi} \sum_{n=0}^{\infty} \frac{(-1)^n}{2n+1} \cos \frac{(2n+1)\pi x}{2L} e^{-(2n+1)^2 \pi^2 Dt/(4L^2)}$$

For small Dt/L^2, the following converges more rapidly:

$$C = C_0 \left\{ 1 - \sum_{n=0}^{\infty} (-1)^n \left[\mathrm{erfc} \frac{(2n+1)L-x}{\sqrt{4Dt}} + \mathrm{erfc} \frac{(2n+1)L+x}{\sqrt{4Dt}} \right] \right\}$$

$$= C_0 \left\{ 1 - \mathrm{erfc} \frac{L-x}{\sqrt{4Dt}} - \mathrm{erfc} \frac{L+x}{\sqrt{4Dt}} + \mathrm{erfc} \frac{3L-x}{\sqrt{4Dt}} + \mathrm{erfc} \frac{3L+x}{\sqrt{4Dt}} - \cdots \right\}.$$

(g) Initial condition: $C|_{t=0} = 0$.
   Boundary conditions are symmetric: $C|_{x=\pm L} = kt$ (when the two boundaries are symmetric, it is often easier to treat by defining the boundaries to be $\pm L$).
   Solution:

$$C = kt - \frac{k(L^2 - x^2)}{2D} + \frac{16kL^2}{D\pi^3} \sum_{n=0}^{\infty} \frac{(-1)^n}{(2n+1)^3} \cos \frac{(2n+1)\pi x}{2L} e^{-(2n+1)^2 \pi^2 Dt/(4L^2)}.$$

A3.3 Three-dimensional diffusion using spherical coordinates with constant D

A3.3.1 Three-dimensional diffusion in a solid sphere of radius a with spherical geometry (meaning concentration depends only on r ($0 \le r \le a$)

Diffusion equation: $\dfrac{\partial C}{\partial t} = \dfrac{D}{r^2} \dfrac{\partial}{\partial r} \left(r^2 \dfrac{\partial C}{\partial r} \right)$ or $\dfrac{\partial (rC)}{\partial t} = D \dfrac{\partial^2 (rC)}{\partial r^2}$.

   For initial condition of $C|_{t=0} = f(r)$ and boundary condition of $C|_{r=a} = C_a$, the solution for C is

$$C = C_a + \frac{2aC_a}{\pi r} \sum_{n=1}^{\infty} \frac{(-1)^n}{n} \sin \frac{n\pi r}{a} e^{-Dn^2\pi^2 t/a^2}$$

$$+ \frac{2}{ar} \sum_{n=1}^{\infty} \sin \frac{n\pi r}{a} e^{-Dn^2\pi^2 t/a^2} \int_0^a y f(y) \sin \frac{n\pi y}{a} dy.$$

If $C|_{t=0} = f(r) = C_0$, the above general solution can be integrated to obtain

$$C = C_a + (C_a - C_0) \frac{2a}{\pi r} \sum_{n=1}^{\infty} \frac{(-)^n}{n} \sin \frac{n\pi r}{a} e^{-Dn^2\pi^2 t/a^2}.$$

$$C_{r=0} = C_a + 2(C_a - C_0) \sum_{n=1}^{\infty} (-1)^n e^{-Dn^2\pi^2 t/a^2}.$$

When the above is integrated, the total amount of diffusing substance entering or leaving the sphere is

$$\frac{M_t}{M_\infty} = 1 - \frac{6}{\pi^2} \sum_{n=1}^{\infty} \frac{1}{n^2} e^{-Dn^2\pi^2 t/a^2},$$

where M_∞ is the final mass gain or loss as t approaches ∞, and equals $4\pi a^3$ $(C_a - C_0)/3$. The above two equations converge rapidly for large Dt/a^2, but slowly for small Dt/a^2. To improve the convergence rate at small t, the following two equations may be used:

$$C = C_0 + (C_a - C_0)\frac{a}{r} \sum_{n=0}^{\infty} \left\{ \text{erfc}\frac{(2n+1)a-r}{\sqrt{4Dt}} - \text{erfc}\frac{(2n+1)a+r}{\sqrt{4Dt}} \right\},$$

$$C_{r=0} = C_0 + (C_a - C_0)\frac{2a}{\sqrt{\pi Dt}} \sum_{n=0}^{\infty} e^{-(2n+1)^2 a^2/(4Dt)}$$

$$\frac{M_t}{M_\infty} = 6\frac{\sqrt{Dt}}{a} \left\{ \frac{1}{\sqrt{\pi}} + 2 \sum_{n=1}^{\infty} \text{ierfc}\frac{na}{\sqrt{Dt}} \right\} - 3\frac{Dt}{a^2}.$$

A3.3.2 Diffusion in a spherical shell ($a \leq r \leq b$)

Suppose the initial condition is $C|_{t=0} = f(r)$ and the boundary conditions are $C|_{r=a} = C_1$, $C|_{r=b} = C_2$. Let $u = rC$. Then u satisfies the following diffusion equation:

$$\frac{\partial u}{\partial t} = D\frac{\partial^2 u}{\partial r^2},$$

and the following initial and boundary conditions:

$$u|_{t=0} = rf(r); \quad u|_{r=a} = aC_1; \quad C|_{r=b} = bC_2.$$

The solution to this one-dimensional diffusion problem can be found by procedures similar to those in Example 3-5.

A3.3.3 Diffusion in an infinite sphere with a spherical hole within ($r \geq a$).

Initial condition: $C|_{t=0} = C_\infty$.
  Boundary condition: $C|_{r=a} = C_a$.
  Solution: $C = C_\infty + \dfrac{a}{r}(C_a - C_\infty)\text{erfc}\dfrac{r-a}{2\sqrt{Dt}}$.

A3.3.4 "Spherical diffusion couple"

Initial condition: $C|_{t=0} = \begin{cases} C_1, & r < a, \\ C_2, & r > a. \end{cases}$
  There is no boundary. Solution:

$$C(r,t) = C_2 + \frac{C_1 - C_2}{2} \left\{ \frac{2\sqrt{Dt}}{r\sqrt{\pi}} [e^{-(r+a)^2/(4Dt)} - e^{-(r-a)^2/(4Dt)}] \right.$$

$$\left. + \text{erf}\frac{r+a}{\sqrt{4Dt}} + \text{erf}\frac{a-r}{\sqrt{4Dt}} \right\}.$$

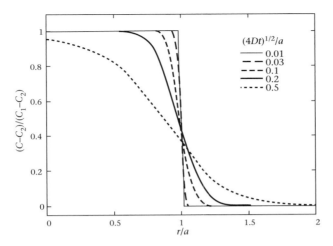

Figure A3-3-4 Diffusion profile evolution in a "spherical diffusion couple."

Let $x = r - a$ (so that $x = 0$ at the initial interface), then,

$$C(x, t) = C_2 + \frac{C_1 - C_2}{2} \left\{ \frac{2\sqrt{Dt}}{(a + x)\sqrt{\pi}} [e^{-(x + 2a)^2/(4Dt)} - e^{-x^2/(4Dt)}] \right.$$
$$\left. + \text{erfc}\frac{x}{\sqrt{4Dt}} - \text{erfc}\frac{x + 2a}{\sqrt{4Dt}} \right\}.$$

The concentration at the center of the sphere is

$$C(r = 0, t) = C_2 + (C_1 - C_2) \left\{ \text{erf}\frac{a}{\sqrt{4Dt}} - \frac{a}{\sqrt{\pi Dt}} e^{-a^2/(4Dt)} \right\}.$$

The above solutions apply to diffusion between the core and its neighboring shell, as well as any two adjacent concentric shells. Note that the above solution applies only when the concentration at the center of the core has not been altered much.

Recall that for one-dimensional diffusion in infinite medium, the concentration profile is an error function and the mid-concentration point is the interface. The above profile is also roughly an error function (e.g., fitting the profile by an error function would give D accurate to within 0.1% if $(4Dt)^{1/2}/a \leq 0.5$), but the mid-concentration point is not fixed at $r_0 = a$; rather it moves toward the center as $r_0 = a - 2Dt/a$. The evolution of concentration profile is shown in Figure A3.3.4.

Appendix 4 Diffusion Coefficients

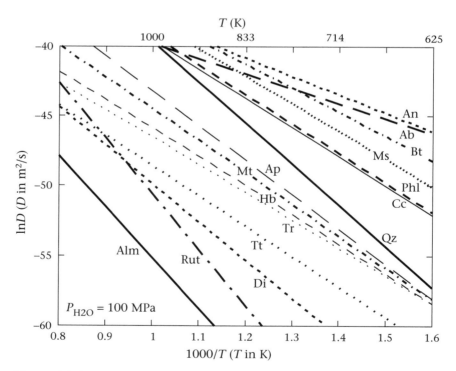

Figure A4-1 Comparison of oxygen diffusivity in various minerals under hydrothermal conditions- Mineral names (from high to low diffusivity): An, anorthite; Ab, albite; Bt, biotite; Ms, muscovite; Phl, phlogopite; Cc, calcite; Qz, quartz; Ap, apatite; Mt, magnetite; Hb, hornblende; Tr, tremolitel; Tt, titanite; Di, diopside; Rut, rutile; Alm, almandine.

Table A4-1 Diffusion coefficients in aqueous solutions

Gas molecules	D (m^2/s)	T (K) range	D (m^2/s) at 298 K
CO$_2$ (ref. 1)	$\exp(-12.462 - 2258.5/T)$	288–368	1.98×10^{-9}
H$_2$S (ref. 1)	$\exp(-14.396 - 1687.9/T)$	288–368	1.94×10^{-9}
N$_2$O (ref. 1)	$\exp(-12.646 - 2214.8/T)$	288–368	1.91×10^{-9}
CH$_4$ (ref. 2)	$\exp(-15.066 - 1545/T)$	278–308	1.60×10^{-9}
N$_2$ (ref. 3)	$\exp(-12.588 - 2225.5/T)$	283–328	1.95×10^{-9}
O$_2$ (ref. 3)	$\exp(-12.36 - 2249.4/T)$	283–328	2.26×10^{-9}
He (ref. 4)	$\exp(-13.71 - 1493/T)$	278–308	7.41×10^{-9}
Ne (ref. 4)	$\exp(-13.34 - 1784/T)$	278–308	4.04×10^{-9}
Kr (ref. 4)	$\exp(-11.96 - 2430/T)$	278–308	1.84×10^{-9}
Xe (ref. 4)	$\exp(-11.62 - 2599/T)$	278–308	1.46×10^{-9}
Ionic compounds			
NaCl (ref. 5)	$\exp(-13.725 - 1950/T)$	298–353	1.58×10^{-9}
KCl (ref. 5)	$\exp(-13.906 - 1831.5/T)$	313–363	1.96×10^{-9}
Self-diffusivity			
H$_2$O (ref. 6)	$\exp(-14.917 - 845.4/T$ $- 191{,}088/T^2)$	277–498	2.26×10^{-9}

Note. References: 1, Tamimi et al. (1994); 2, Maharajh and Walkley (1973); 3, Verhallen et al. (1984); 4, Jahne et al. (1987); 5, Fell (1971); 6, self-diffusivity at 10 MPa, Weingartner (1982).

Table A4-2 Diffusion coefficients in silicate melts

Component	Melt composition	D (m^2/s)	T and P range
EBD of CO$_2$ (ref. 1)	Basalt to rhyolite	$\exp[-13.99 - 17367/T - 1.9448P/T + (855.2 + 0.2712P)w/T]$	753–1773 K; 0.1–1500 MPa; $w \le 5$ wt%
EBD of H$_2$O (ref. 1)	Rhyolite	$w \exp(-17.14 - 10661/T - 1.772P/T)$	673–1473 K; 0.1–810 MPa; $w \le 2$ wt%
EBD of Ar (ref. 1)	Rhyolite	$\exp[-13.99 - 17367/T - 1.9448P/T + (855.2 + 0.2712P)w/T]$	753–1773 K; 0.1–1500 MPa; $w \le 5$ wt%
Self D of Ca (ref. 2)	56% SiO$_2$; 10% Al$_2$O$_3$; 5% CaO; 29% Na$_2$O	$\exp(-15.20 - 11628/T + 0.3672P - 634.72P/T]$	1373–1673 K; 0.1–3000 MPa
Interdiffusivity of Na–K (ref. 3)	Alkali feldspar melt	$\exp(-10.64 - 17536/T)$	1.0 GPa, 1473–1673 K; K/(Na + K) = 0.5
EBD of P (ref. 4)	Rhyolite	$\exp(-12.65 - 72270/T)$	0.8 GPa, 1473–1773 K
Tracer dif of K (ref. 5)	Rhyolite	$\exp(-6.46 - 12785/T)$	0.1 MPa; air; 623–1123 K

Note. EBDC, effective binary diffusivity; T, temperature (K); P, pressure (MPa); and w, total H$_2$O content (wt%) (for 2 wt% total H$_2$O, $w = 2$). References: 1, Zhang et al. (2007); 2, Watson (1979a); 3, Freda and Baker (1998); 4, Harrison and Watson (1984); 5, Jambon (1982).

Table A4-3 Selected diffusivities of radioactive and radiogenic species in minerals

Mineral	Species	Orientation	D (m^2/s)	Conditions	Ref.
Zircon	U	$\perp$**c**	$\exp(-1.31-84035/T)$	1673–1923 K	1
Zircon	Th	$\perp$**c**	$\exp(4.08-94688/T)$	1673–1923 K	1
Zircon	Hf	$\perp$**c**	$\exp(7.64-98094/T)$	1673–1923 K	1
Zircon	Pb	$\sim$Isotropic	$\exp(-2.21-66149/T)$	1273–1773 K	2
Monazite	Pb	$\sim$Isotropic	$\exp(-0.06-71200/T)$	1373–1623 K	3
Rutile	Pb	$\sim$Isotropic	$\exp(-21.66-30068/T)$	973–1373 K	4
Apatite	Pb	$\perp$**c**	$\exp(-18.18-27476/T)$	873–1473 K	5
Calcite	Pb	$\sim$Isotropic	$\exp(-32.78-14072/T)$	713–923 K	6
Calcite	Sr	$\sim$Isotropic	$\exp(-29.19-15876/T)$	713–1073 K	6
Apatite	Sr	$\sim$Isotropic	$\exp(-15.12-32709/T)$	973–1323 K	7
Plag(An23)	Sr	$\perp$ (001)	$\exp(-14.05-32721/T)$	997–1349 K	8
Plag(An43)	Sr	~Isotropic	$\exp(-15.62-31705/T)$	997–1347 K	8
Plag(An67)	Sr	$\perp$(001)	$\exp(-16.34-31922/T)$	1000–1348 K	8
Albite(Or1)	Sr	$\perp$(001)	$\exp(-19.66-26941/T)$	948–1298 K	9
Sanidine(Or61)	Sr	$\perp$(001)	$\exp(2.13-54122/T)$	998–1348 K	9
Diopside	Sr	$\perp$**c**	$\exp(-2.12-61393/T)$	1373–1573 K	10
Apatite	Nd	$\perp$**c**	$\exp(-12.94-41855/T)$	1073–1523 K	11
Calcite	Nd	$\perp\{10\bar{1}4\}$	$\exp(-31.36-18041/T)$	873–1123 K	12

Note. Type of diffusion is tracer or effective binary diffusion. The pressure is 0.1 MPa or less. For Ar diffusivity, see Table 1-3c. References: 1, Cherniak et al. 1997; 2, Cherniak and Watson (2000); 3, Cherniak et al. (2004); 4, Cherniak (2000a); 5, Cherniak et al. (1991); 6, Cherniak (1997); 7, Cherniak and Ryerson (1993); 8, Cherniak and Watson (1994); 9, Cherniak, (1996); 10, Sneeringer et al. (1984); 11, Cherniak, (2000b); 12, Cherniak (1998).

Table A4-4 Selected interdiffusion data in minerals

Mineral	Interdiffusion	Orientation	D (m^2/s)	Conditions	Ref.
(Mg, Fe)O	Fe–Mg	Isotropic	$f_{O2}^{0.19}X^{0.73} \exp[-12.75 - (25{,}137 + 11{,}546X)/T]$, where $X = $ Fe/(Fe + Mg) and f_{O_2} is in Pa	1593–1673 K; 0.1 MPa; $\log f_{O_2}$: -4.3 to -1; $X < 0.27$	1
Olivine (Mg,Fe)$_2$SiO$_4$	Fe–Mg	//**c**	$f_{O_2}^{1/4.25} \cdot \exp(-16.17 - 27{,}181/T + 6.56X_{Fa})$	1253–1573K; 0.1 MPa	2
Opx (Mg,Fe)SiO$_3$	Fe–Mg	$\perp$**a**	$\exp(-21.97 + 5.99X_{Fe} - 28{,}851/T)$	773–1073 K; 0.1 MPa; $f_{O_2} \approx$ IW	3
Perovskite (Mg,Fe)SiO$_3$	Fe–Mg	Isotropic	$\exp(-19.34 - 49{,}793/T)$	2023–2773 K; 24 GPa; IW-3	4
Garnet	Ca–(Fe,Mg)	Isotropic	$\exp(-13.62 - 32{,}521/T)$	1173–1373 K; 3 GPa	5
Plagioclase	NaSi–CaAl	Random	$\exp(-6.81 - 62{,}100/T)$	1373–1673 K	6

Note. References: 1, Mackwell et al. (2005); 2, Chakraborty (1997) and Petry et al. (2004); 3, Ganguly and Tazzoli (1994); 4, Holzapfel et al. (2005); 5, Freer and Edwards (1999); 6, Grove et al. (1984).

Table A4-5 Selected data on oxygen isotopic diffusion in minerals by exchange with a fluid phase (Figure A4-1)

Mineral	Fluid condition	Orientation	D (m^2/s)	T (K) range
Quartz (ref. 1)	$P_{H_2O} = 100$ MPa	//**c**	$\exp(-10.45-29{,}226/T)$	723–863
Quartz (ref. 2)	$P_{CO_2} = 10$ MPa	//**c**	$\exp(-26.89-19{,}123/T)$	1018–1173
Quartz (ref. 3)	$P_{O_2} = 0.1$ MPa	//**c**	$\exp(-24.25-26{,}580/T)$	1133–1273
Calcite (ref. 4)	$P_{H_2O} = 100$ MPa	~Isotropic	$\exp(-18.78-20{,}807/T)$	673–1073
Calcite (ref. 5)	$P_{CO_2} = 100$ MPa	~Isotropic	$\exp(-14.10-29{,}106/T)$	873–1073
Albite (ref. 6)	$P_{H_2O} = 100$ MPa	~Isotropic	$\exp(-29.10-10{,}719/T)$	623–1078
Anorthite (ref. 6)	$P_{H_2O} = 100$ MPa	~Isotropic	$\exp(-25.00-13{,}184/T)$	623–1078
Anorthite (ref. 7)	$P_{O_2} = 0.1$ MPa	~Isotropic	$\exp(-20.72-28{,}384/T)$	1124–1571
Anorthite (ref. 8)	$P_{CO} - P_{CO_2} = 0.1$ MPa	~Isotropic	$\exp(-27.81-19{,}484/T)$	1281–1568
Muscovite (ref. 9)	$P_{H_2O} = 100$ MPa	$\perp$**c**	$\exp(-18.68-19{,}626/T)$	785–973
Phlogopite (ref. 9)	$P_{H_2O} = 100$ MPa	$\perp$**c**	$\exp(-18.08-21{,}135/T)$	873–1173
Biotite (ref. 9)	$P_{H_2O} = 100$ MPa	$\perp$**c**	$\exp(-20.82-17{,}110/T)$	773–1073
Apatite (ref. 10)	$P_{H_2O} = 100$ MPa	//**c**	$\exp(-18.53-24{,}658/T)$	823–1473
Hornblende (ref. 11)	$P_{H_2O} = 100$ MPa	//**c**	$\exp(-25.33-20{,}632/T)$	923–1073
Tremolite (ref. 11)	$P_{H_2O} = 100$ MPa	//**c**	$\exp(-26.94-19{,}626/T)$	923–1073
Almandine (ref. 12)	$P_{H_2O} = 100$ MPa	Isotropic	$\exp(-18.93-36{,}202/T)$	1073–1273
Diopside (ref. 13)	$P_{H_2O} = 100$ MPa	//**c**	$\exp(-22.62-27{,}182/T)$	973–1473
Diopside (ref. 8)	$P_{CO} - P_{CO_2} = 0.1$ MPa	//**c**	$\exp(-7.75-54{,}965/T)$	1377–1524
Titanite (ref. 14)	$P_{H_2O} = 100$ MPa	~Isotropic	$\exp(-26.91-21{,}649/T)$	973–1173
Forsterite (ref. 15)	$P_{O_2} = 0.02$ MPa	~Isotropic	$\exp(-16.65-45{,}439/T)$	1273–1773
Rutile (ref. 16)	$P_{H_2O} = 100$ MPa	$\perp$**c**	$\exp(-10.82-39{,}690/T)$	1053–1273
Magnetite (ref. 17)	$P_{H_2O} = 100$ MPa	~Isotropic	$\exp(-21.77-22{,}645/T)$	773–1073

Note. References: 1, Farver and Yund (1991); 2, Dennis (1984); 3, Sharp et al. (1991); 4, Farver (1994); 5, Labotka et al. (2000); 6, Giletti et al. (1978); 7, Elphick et al. (1988); 8, Ryerson and McKeegan (1994); 9, Fortier and Giletti (1991); 10, Farver and Giletti (1989); 11, Farver and Giletti (1985); 12, Coghlan (1990); 13, Farver (1989); 14, Zhang et al. (2006); 15, Hallwig et al. (1982); 16, Moore et al. (1998b); 17, Giletti and Hess (1988).

Answers to Selected Problems

Chapter 1

1.1 Heterogeneous reactions: i, m, o.
Homogeneous and overall reactions; a, b, f.

1.1c Homogeneous and elementary. $d\xi/dt = k$. Molecularity $= 2$; order $= 0$.

1.3a $K = \alpha_{D/H,\text{liq/vapor}}$; $\Delta G_r^\circ = -189\,\text{J}$.

1.3b $K = 1/\alpha_{18_0}/\alpha_{16_0,\text{liq/vapor}}$; $\Delta G_r^\circ = 22.7\,\text{J}$

1.3c $\Delta G_r^\circ = 31.5\,\text{kJ}$.

1.5a $d\xi/dt = k[A][B]$. Overall reaction.

1.5b $\xi = B_0\left(1 - \dfrac{1}{1 + 2kB_0 t}\right)$.

1.5d $t_{1/2} = 1/(2kB_0) = 10\,\text{s}$.

1.7 $E = 83.1\,\text{kJ}$.

1.9 The reaction rate law and the rate coefficient: $d\xi/dt = k[^7\text{Be}]$, where $k = 0.0127825\ \text{day}^{-1}$. The order of the reaction is 1. The reaction is an elementary reaction.

1.10a 320 km. Because this is much smaller than the radius of the Earth, heat conduction is not an efficient way for the whole Earth to lose heat. (Convection is more efficient.)

1.10b 142 km; 263 km.
1.10c 1.2 mm.

1.11b For a chicken egg radius of 22 mm, time for the center to reach 98°C is about 302 s.

1.13a Diffusion control.
1.13b Interface control or convection control, which can be distinguished by examining the effect of stirring or by measuring the concentration profile next to the dissolving crystal.

1.14a Controlled by mass transfer.
1.14b and 1.14c Controlled by interface reaction.

1.16a $T_{ae} = 688$ K.
1.16b $q = 0.027$ K/yr.
1.16c This is the cooling rate when the rock temperature was at 688 K.

1.17a $T_c = 831$ K.
1.17b $T_c = 800$ K.
1.17b $T_c = 809$ K.

Chapter 2

2.1a $t_{1/2} = \ln(2)/k$.
2.1b $t_{1/2} = \ln(2)/(k_f + k_b)$.

2.2a The rate law is $d\xi/dt = k_f[CO_2] - k_b[H_2CO_3]$.
2.2b The backward rate constant is $k_b = 15\,s^{-1}$.

2.3a The assumption is reasonable because the equilibrium constant is very close to 1.
2.3c $t_{1/2} = \ln(2)/\{k([^{56}Fe^{2+}] + [^{55}Fe^{3+}] + [^{56}Fe^{3+}] + [^{55}Fe^{2+}])\} = 2490$ s.

2.4a and 2.4b $\tau_r = 0.066$ s. The relaxation timescale of first-order reactions is independent of the initial species concentrations.

2.5a $\tau_r = 6.67 \times 10^{-5}$ s.
2.5b $\tau_r = 6.67 \times 10^{-9}$ s.
2.5c $\tau_r = 6.59 \times 10^{-6}$ s.

The relaxation timescale of second-order reactions depends strongly on the initial species concentrations.

2.7a $K = 0.119$.
2.7b $k_f = 2.4 \times 10^{-7} \, s^{-1}$.
2.7c   The unit is time^{-1}, and differs from second-order reactions in aqueous solutions.

2.8a $K = 0.114$.

2.9a $^{234}U/^{238}U = 2.45 \times 10^5/(4.468 \times 10^9) = 5.48 \times 10^{-5}$, same as the observed.
2.9b   At 1.0 Ma, $^{234}U/^{238}U = 5.48 \times 10^{-5}$.
2.9c   At 4.0 Ga, $^{234}U/^{238}U = 5.48 \times 10^{-5}$.
2.9d   At 4.0 Ga, $^{238}U/^{235}U = 4.990$.
2.9e   At 4.0 Ga, 83.3% ^{238}U; 16.7% ^{235}U; 0.0047% ^{234}U.

2.10a   The production rate is 8.52×10^{18} atoms of ^{14}C.
2.10b $^{14}C/C = 1.18 \times 10^{-12}$.

2.12 $[^3He] = 0.40 \, mol/L$. PP I chain: 45%; PP II chain: 55%.
2.14 $4 \times 10^{-8} \, s$.

Chapter 3

3.1a $F = 0.00712$ (or 0.712% mass loss).
3.1b $F = 1.5 \times 10^{-10}$. Not noticeable.
3.1c $F = 2.7 \times 10^{-11}$.

3.2   Activation energy $= 224 \, kJ$. Preexponential factor $= 7.8 \times 10^{-11} \, m^2/s$.

3.3c 1.00734×10^{-23}.

3.5   Define the thickness of the layer that is affected by diffusion to be $(4Dt)^{1/2}$.
3.5a $6.8 \times 10^{-5} \, \mu m$ (negligible).
3.5b $0.135 \, \mu m$ (negligible).
3.5c $25 \, \mu m$ (significant).

3.6a At 900 K, 0.1 MPa, and $X = 0.08$, $D_{H_2O_m} = 1.81 \, \mu m^2/s$.
3.6b At 900 K, 0.1 MPa, and $X = 0.08$, $K = 0.206$.
3.6c At 900 K, 0.1 MPa, and $X = 0.08$, $dX_m/dX = 0.668$.
3.6d At 900 K, 0.1 MPa, and $X = 0.08$, $D_{H_2O_t} = 1.21 \, \mu m^2/s$.

3.7 In an experiment of 2 days, equilibrium is not reached. In the magma for 1000 years, equilibrium is reached.

3.8 Assuming that the average radius of a chicken egg is 24 mm, $t = 215$ s.

3.9 $C = C_0 \exp[-x^2/(4Dt)]$, where C_0 is proportional to $(t)^{-1/2}$.

3.10 $C = C_0 \operatorname{erfc}[x/(4Dt)^{1/2}]$.

3.11 $C = C_\infty + (C_0 - C_\infty) \dfrac{\operatorname{erfc}\left(\frac{x}{\sqrt{4Dt}} - a\right)}{\operatorname{erfc}(-a)}$, where $a = A/D^{1/2}$.

3.16 $D_{out} = 0.60\ \mu m^2/s$. $D_{out} = D_{in}$ if D is does not depend on concentration.

Chapter 4

4.1b $r^* = 2$ nm.
4.1c $I = 5 \times 10^{-38}\ m^{-3}\ s^{-1}$.

4.2a $r^* = 20$ nm.
4.2b $I = 10^{-9552}\ m^{-3}\ s^{-1}$.

4.3 $\sigma' = 0.133$ N/m.

4.4a Pre-exponential factor $= 5.2 \times 10^8\ \mu m\ s^{-1}\ K^{-1}$.
4.4b Growth rate $= 67\ \mu m/s$.

4.6a Olivine dissolution distance in 2 h is 23 μm.

4.7a Olivine dissolution distance in 2 h is 85 μm.

4.9 231 years.

4.10a 0.23 m/s.
4.10b 17 $\mu m/s$.
4.10c 23 $\mu m/s$.

4.11 90 $\mu m/s$.

4.12 144 m/s.

4.13 95 years.

4.15 0.25 years.

Chapter 5

5.1a   477 years old.

5.1b   463 years old.

5.1c   499 years BP.

5.1d   This is the age when the piece of tree was isolated from the atmosphere.

5.2 $2.6\,\mu m/yr$.

5.3a   Age $= 11,331 \pm 53$ years.

5.3b   The initial $(^{234}U/^{238}U)$ activity ratio is 1.146.

5.4 $3.6\,nm/yr$.

5.5   386 Ma. The age is the closure age. It becomes the formation age if cooling is very rapid.

5.6a $\Delta t = 9.16\,Myr$.

5.6b   Allende is older by 0.73 Myr.

5.8   Closure age $= 18.1\,Ma$.

5.9b $182.6 \pm 0.8\,Ma$.

5.9c $182.7 \pm 1.0\,Ma$

5.9d $183.8 \pm 0.9\,Ma$

5.11   Atomic mass of Pb in monazite $>$ that of common lead $>$ that of Pb in zircon.

5.12a T_c is 1218 K for the whole mineral grain, and is 1322 K for the center.

5.12b T_c is 757 K for the whole mineral grain, and is 797 K for the center.

5.12c T_c is 637 K for the whole mineral grain, and is 666 K for the center.

5.13a T_c is 380 K for the whole mineral grain, and is 396 K for the center.

5.13b T_c is 348 K for the whole mineral grain, and is 361 K for the center.

5.14   Steady-state age is 0.018 Ma for the bulk mineral, and 0.044 Ma for the center.

5.15a $T_{ae} = 826$ K. $q = 2128$ K/yr $= 5.8$ K/day.

5.15b $T_{ae} = 598$ K. $q = 0.000976$ K/yr $= 974$ K/Myr.

5.17 $q = 0.0157$ K/s.

5.20 $\ln q = 57.82 - 2 \ln a - 55{,}787/T_0$.

References

Acosta-Vigil A., London D., Dewers T. A., and Morgan G.B. VI (2002) Dissolution of corundum and andalusite in H_2O-saturated haplogranitic melts at 800°C and 200 MPa: constraints on diffusivities and the generation of peraluminous melts. *J. Petrol.* **43**, 1885–1908.

Acosta-Vigil A., London D., Morgan G.B. VI, and Dewars T.A. (2006) Dissolution of quartz, albite, and orthoclase in H_2O-saturated haplogranitic melt at 800°C and 200 MPa: diffusive transport properties of granitic melts at crustal anatectic conditions. *J. Petrol.* **47**, 231–254.

Albarede F. and Bottinga Y. (1972) Kinetic disequilibrium in trace element partitioning between phenocrysts and host lava. *Geochim. Cosmochim. Acta* **36**, 141–156.

Allegre C. J., Provost A., and Jaupart C. (1981) Oscillatory zoning: a pathological case of crystal growth. *Nature* **294**, 223–228.

Amelin Y., Krot A.N., Hutcheon I.D., and Ulyanov A.A. (2002) Lead isotopic ages of chondrules and calcium–aluminum-rich inclusions. *Science* **297**, 1678–1683.

Anbar A.D., Roe J.E., Barling J., and Nealson K.H. (2000) Nonbiological fractioantion of iron isotopes. *Science* **288**, 126–128.

Andreaozzi G.B. and Princivalle F. (2002) Kinetics of cation ordering in synthetic $MgAl_2O_4$ spinel. *Am. Mineral.* **87**, 838–844.

Anovitz L.M., Essene E.J., and Dunham W.R. (1988) Order–disorder experiments on orthopyroxenes: implications for the orthopyroxene geospeedometer. *Am. Mineral.* **73**, 1060–1073.

Arredondo E.H. and Rossman G.R. (2002) Feasibility of determining the quantitative OH content of garnet with Raman spectroscopy. *Am. Mineral.* **87**, 307–311.

Atkins P. W. (1982) *Physical Chemistry*. Oxford, UK: Freeman.

Atkinson R., Baulch D.L., Cox R.A., Hampson Jr. R.F., Kerr J.A., Rossi M.J., and Troe J. (1997) Evaluated kinetic, photochemical and heterogeneous data for atmospheric chemistry. Supplement V, IUPAC Subcommittee on Gas Kinetic Data Evaluation for Atmospheric Chemistry. *J. Phys. Chem. Ref. Data* **26**, 521–1011.

Avrami M. (1939) Kinetics of phase change, I: general theory. *J. Chem. Phys.* **7**, 1103–1112.

———. (1940) Kinetics of phase change, II: transformation–time relations for random distribution of nuclei. *J. Chem. Phys.* **8**, 212–224.

———. (1941) Kinetics of phase change, III: granulation, phase change, and microstructure. *J. Chem. Phys.* **9**, 177.

Bahcall J.N. (1989) *Neutrino Astrophysics*. Cambridge, UK: Cambridge University Press.

Bai T.B. and van Groos A.F.K. (1994) Diffusion of chlorine in granitic melts. *Geochim. Cosmochim. Acta* **58**, 113–123.

Baker D.R. (1989) Tracer versus trace element diffusion: diffusional decoupling of Sr concentration from Sr isotope composition. *Geochim. Cosmochim. Acta* **53**, 3015–3023.

———. (1991) Interdiffusion of hydrous dacitic and rhyolitic melts and the efficacy of rhyolite contamination of dacitic enclaves. *Contrib. Mineral. Petrol.* **106**, 462–473.

———. (1992) Tracer diffusion of network formers and multicomponent diffusion in dacitic and rhyolitic melts. *Geochim. Cosmochim. Acta* **56**, 617–632.

Baker D.R., Conte A.M., Freda C., and Ottolini L. (2002) The effect of halogens on Zr diffusion and zircon dissolution in hydrous metaluminous granitic melts. *Contrib. Mineral. Petrol.* **142**, 666–678.

Bamford C. H. and Tipper C.F.H. (1972) *Comprehensive Chemical Kinetics*, Vol. 6, p. 517. New York: Elsevier.

Banfield J.F., Welch S.A., Zhang H., Ebert T.T., and Penn R.L. (2000) Aggregation-based crystal growth and microstructure development in natural iron oxyhydroxie biomineralization products. *Science* **289**, 751–754.

Bard E., Hamelin B., Fairbanks R.G., and Zindler A. (1990) Calibration of the ^{14}C timescale over the past 30,000 years using mass spectrometric U–Th ages from Barbados corals. *Nature* **345**, 405–410.

Barrer R.M., Bartholomew R.F., and Rees L.V.C. (1963) Ion exchange in porous crystals, II: the relationship between self- and exchange-diffusion coefficients. *J. Phys. Chem. Solids* **24**, 309–317.

Baulch D.L., Duxbury J., Grant S.J., and Montague D.C. (1981) Evaluated high temperature kinetic data. *J. Phys. Chem. Ref. Data* **10 (suppl. 1)**.

Becker R. (1938) Nucleation during the precipitation of metallic mixed crystals. *Ann. Phys.* **32**, 128–140.

Becker R. and Doring W. (1935) Kinetische behandburg der Keim building in ubersattigten dampfen. *Ann. Phys. ser. 5* **24**, 719–752.

Behrens H. and Zhang Y. (2001) Ar diffusion in hydrous silicic melts: implications for volatile diffusion mechanisms and fractionation. *Earth Planet. Sci. Lett.* **192**, 363–376.

Behrens H., Romano C., Nowak M., Holtz F., and Dingwell D.B. (1996) Near-infrared spectroscopic determination of water species in glasses of the system $MAlSi_3O_8$ (M = Li, Na, K): an interlaboratory study. *Chem. Geol.* **128**, 41–63.

Behrens H., Zhang Y., and Xu Z. (2004) H_2O diffusion in dacitic and andesitic melts. *Geochim. Cosmochim. Acta* **68**, 5139–5150.

Behrens H., Zhang Y., Leschik M., Miedenbeck M., Heide G., and Frischat G.H. (2007) Molecular H_2O as carrier for oxygen diffusion in hydrous silicate melts. *Earth Planet. Sci. Lett.* **254**, 69–76.

Bejina F. and Jaoul O. (1997) Silicon diffusion in silicate minerals. *Earth Planet. Sci. Lett.* **153**, 229–238.

Benson S.W. and Axworthy A.E. (1957) Mechanism of the gas phase, thermal decomposition of ozone. *J. Chem. Phys.* **26**, 1718–1726.

Berman R.G. (1988) Internally-consistent thermodynamic data for minerals in the system $Na_2O-K_2O-CaO-MgO-FeO-Fe_2O_3-Al_2O_3-SiO_2-TiO_2-H_2O-CO_2$. *J. Petrol.* **29**, 445–522.

Berner R.A. (1978) Rate control of mineral dissolution under Earth surface conditions. *Am. J. Sci.* **278**, 1235–1252.

———. (1980) *Early Diagenesis: A Theoretical Approach*. Princeton, NJ: Princeton University Press.

Berner R.A., Lasaga A.C., and Garrels R.M. (1983) The carbonate–silicate geochemical cycle and its effect on atmospheric carbon dioxide over the past 100 million years. *Am. J. Sci.* **283**, 641–683.

Besancon J.R. (1981) Rate of cation disordering in orthopyroxenes. *Am. Mineral.* **66**, 965–973.

Bindeman I.N. (2003) Crystal sizes in evolving silicic magma chambers. *Geology* **31**, 367–370.

Blum J.D., Wasserburg G.J., Hutcheon I.D., Beckett J.R., and Stolper E.M. (1989) Diffusion, phase equilibria and partitioning experiments in the Ni–Fe–Ru system. *Geochim. Cosmochim. Acta* **53**, 483–489.

Boeker E. and van Grondelle R. (1995) *Environmental Physics*. Chichester, UK: Wiley.

Bolton E.W., Lasaga A.C., and Rye D.M. (1996) A model for the kinetic control of quartz dissolution and precipitation in porous media flow with spatially variable permeability: formulation and examples of thermal convection. *J. Geophys. Res.* **101**, 22157–22187.

Borders R.A. and Birks J.W. (1982) High-precision measurements of activation energies over small temperature intervals: curvature in the Arrhenius plot for the reaction $NO + O_3 = NO_2 + O_2$. *J. Phys. Chem.* **86**, 3295–3302.

Boudreau B.P. (1997) *Diagenetic Models and Their Implementation*. Berlin: Springer-Verlag.

Bowring S.A. and Williams I.S. (1999) Priscoan (4.00–4.03 Ga) orthogneisses from northwestern Canada. *Contrib. Mineral. Petrol.* **134**, 3–16.

Boyce J.W., Hodges K.V., Olszewski W.J., and Jercinovic M.J. (2005) He diffusion in monazite: implications for (U–Th)/He thermochronometry. *Geochem. Geophys. Geosyst.* **6**, Q12004, doi:10.1029/2005GC001058.

Brady J.B. (1975a) Reference frames and diffusion coefficients. *Am. J. Sci.* **275**, 954–983.

———. (1975b) Chemical components and diffusion. *Am. J. Sci.* **275**, 1073–1088.

———. (1995) Diffusion data for silicate minerals, glasses, and liquids. In *Mineral Physics and Crystallography: A Handbook of Physical Constants*, Vol. Reference Shelf 2 (ed. T. J. Ahrens), pp. 269–290. AGU.

Brady J.B. and McCallister R.H. (1983) Diffusion data for clinopyroxenes from homogenization and self-diffusion experiments. *Am. Mineral.* **68**, 95–105.

Brady J.B. and Yund R.A. (1983) Interdiffusion of K and Na in alkali feldspars: homogenization experiments. *Am. Mineral.* **68**, 106–111.

Brenan J.M. (1994) Kinetics of fluorine, chlorine and hydroxyl exchange in fluorapatite. *Chem. Geol.* **110**, 195–210.

Brenan J.M., Cherniak D.J., and Rose L.A. (2000) Diffusion of osmium in pyrrhotite and pyrite: implications for closure of the Re–Os isotopic system. *Earth Planet. Sci. Lett.* **180**, 399–413.

Brewer P.G., Peltzer E.T., Friedrich G., and Rehder G. (2002) Experimental determination of the fate of rising CO_2 droplets in seawater. *Environ. Sci. Technol.* **36**, 5441–5446.

Brizi E., Molin G., Zanazzi P.F., and Merli M. (2001) Ordering kinetics of Mg–Fe^{2+} exchange in a $Wo_{43}En_{46}Fs_{11}$ augite. *Am. Mineral.* **86**, 271–278.

Broecker W.S. and Peng T. (1982) *Tracers in the Sea*. Palisades, NY: Lamont-Doherty Geological Observatory.

Buening D.K. and Buseck P.R. (1973) Fe–Mg lattice diffusion in olivine. *J. Geophys. Res.* **78**, 6852–6862.

Burn I. and Roberts J.P. (1970) Influence of hydroxyl content on the diffusion of water in silica glass. *Phys. Chem. Glasses* **11**, 106–114.

Burnham C.W. (1975) Water and magmas; a mixing model. *Geochim. Cosmochim. Acta* **39**, 1077–1084.

Cable M. and Frade J.R. (1987a) The diffusion-controlled dissolution of spheres. *J. Mater. Sci.* **22**, 1894–1900.

———. (1987b) Diffusion-controlled growth of multi-component gas bubbles. *J. Mater. Sci.* **22**, 919–924.

———. (1987c) Diffusion-controlled mass transfer to or from spheres with concentration-dependent diffusivity. *Chem. Eng. Sci.* **42**, 2525–2530.

———. (1987d) Numerical solutions for diffusion-controlled growth of spheres from finite initial size. *J. Mater. Sci.* **22**, 149–154.

———. (1988) The influence of surface tension on the diffusion-controlled growth or dissolution of spherical gas bubbles. *Proc. R. Soc. Lond. Ser. A* **420**, 247–265.

Cahn J.W. (1966) The later stages of spinodal decomposition and the beginning of particle coarsening. *Acta Metall.* **14**, 1685–1692.

———. (1968) Spinodal decomposition. *Trans. Metall. Soc. AIME* **242**, 166–180.

Carlson R.W. and Lugmair G.W. (1988) The age of ferroan anorthosite 60025: oldest crust on a young Moon? *Earth Planet. Sci. Lett.* **90**, 119–130.

Carlson W.D. (1983) Aragonite–calcite nucleation kinetics: an application and extension of Avrami transformation theory. *J. Geol.* **91**, 57–71.

———. (2006) Rates of Fe, Mg, Mn, and Ca diffusion in garnet. *Am. Mineral.* **91**, 1–11.

Carman P.C. (1968) Intrinsic mobilities and independent fluxes in multicomponent isothermal diffusion, I: simple Darken systems; II: complex Darken systems. *J. Phys. Chem.* **76**, 1707–1721.

Carpenter M.A. and Putnis A. (1985) Cation order and disorder during crystal growth: some implications for natural mineral assemblages. *Adv. Phys. Geochem.* **4**, 1–26.

Carroll M.R. (1991) Diffusion of Ar in rhyolite, orthoclase and albite composition glasses. *Earth Planet. Sci. Lett.* **103**, 156–168.

Carroll M.R. and Stolper E.M. (1991) Argon solubility and diffusion in silica glass: implications for the solution behavior of molecular gases. *Geochim. Cosmochim. Acta* **55**, 211–225.

Carroll M.R., Sutton S.R., Rivers M.L., and Woolum D.S. (1993) An experimental study of krypton diffusion and solubility in silicic glasses. *Chem. Geol.* **109**, 9–28.

Carslaw H.S. and Jaeger J.C. (1959) *Conduction of Heat in Solids*. Oxford, England: Clarendon Press.

Cashman K.V. (1991) Textural constraints on the kinetics of crystallization of igneous rocks. *Rev. Mineral.* **24**, 259–314.

———. (1993) Relationship between plagioclase crystallization and cooling rate in basaltic melts. *Contrib. Mineral. Petrol.* **113**, 126–142.

Cashman K.V. and Ferry J.M. (1988) Crystal size distribution (CSD) in rocks and the kinetics and dynamics of crystallization, III: metamorphic crystallization. *Contrib. Mineral. Petrol.* **99**, 401–415.

Cashman K.V. and Marsh B.D. (1988) Crystal size distribution (CSD) in rocks and the kinetics and dynamics of crystallization, II: Makaopuhi lava lake. *Contrib. Mineral. Petrol.* **99**, 292–305.

Chai B.H.T. (1974) Mass transfer of calcite during hydrothermal recrystallization. In *Geochemical Transport and Kinetics*, Vol. 634 (ed. A.W. Hofmann, B.J. Giletti, H.S. Yoder, and R.A. Yund), pp. 205–218. Carnegie Institution of Washington Publ.

Chakraborty S. (1995) Diffusion in silicate melts. *Rev. Mineral.* **32**, 411–503.

———. (1997) Rates and mechanisms of Fe–Mg interdiffusion in olivine at 980°–1300°C. *J. Geophys. Res.* **102**, 12317–12331.

Chakraborty S. and Rubie D.C. (1996) Mg tracer diffusion in aluminosilicate garnets at 750–850°C, 1 atm and 1300°C, 8.5 GPa. *Contrib. Mineral. Petrol.* **122**, 406–414.

Chakraborty S., Farver J.R., Yund R.A., and Rubie D.C. (1994) Mg tracer diffusion in synthetic forsterite and San Carlos olivine as a function of P, T and fO$_2$. *Phys. Chem. Miner.* **21**, 489–500.

Chakraborty S., Knoche R., Schulze H., Rubie D.C., Dobson D., Ross N.L., and Angel R.J. (1999) Enhancement of cation diffusion rates across the 410-kilometer discontinuity in Earth's mantle. *Science* **283**, 362–365.

Chekhmir A.S. and Epelbaum M.B. (1991) Diffusion in magmatic melts: new study. In *Physical Chemistry of Magmas*, Vol. 9 (ed. L. L. Perchuk and I. Kushiro), pp. 99–119. Springer-Verlag.

Chen J.H. and Wasserburg G.J. (1981) The isotopic composition of uranium and lead in Allende inclusions and meteoritic phosphates. *Earth Planet. Sci. Lett.* **52**, 1–15.

Chen J.H., Edwards R.L., and Wasserburg G.J. (1986) ^{238}U, ^{234}U and ^{232}Th in seawater. *Earth Planet. Sci. Lett.* **80**, 241–251.

Cherniak D.J. (1996) Strontium diffusion in sanidine and albite, and general comments on strontium diffusion in alkali feldspars. *Geochim. Cosmochim. Acta* **60**, 5037–5043.

———. (1997) An experimental study of strontium and lead diffusion in calcite, and implications for carbonate diagenesis and metamorphism. *Geochim. Cosmochim. Acta* **61**, 4173–4179.

———. (1998) REE diffusion in calcite. *Earth Planet. Sci. Lett.* **160**, 273–287.

———. (2000a) Pb diffusion in rutile. *Contrib. Mineral. Petrol.* **139**, 198–207.

———. (2000b) Rare Earth element diffusion in apatite. *Geochim. Cosmochim. Acta* **64**, 3871–3885.

———. (2001) Pb diffusion in Cr diopside, augite, and enstatite, and consideration of the dependence of cation diffusion in pyroxene on oxygen fugacity. *Chem. Geol.* **177**, 381–397.

———. (2002) Ba diffusion in feldspar. *Geochim. Cosmochim. Acta* **66**, 1641–1650.

Cherniak D.J. and Ryerson F.J. (1993) A study of strontium diffusion in apatite using Rutherford backscattering spectroscopy and ion implantation. *Geochim. Cosmochim. Acta* **57**, 4653–4662.

Cherniak D.J. and Watson E.B. (1994) A study of strontium diffusion in plagioclase using Rutherford backscattering spectroscopy. *Geochim. Cosmochim. Acta* **58**, 5179–5190.

———. (2000) Pb diffusion in zircon. *Chem. Geol.* **172**, 5–24.

Cherniak D.J., Lanford W.A., and Ryerson F.J. (1991) Lead diffusion in apatite and zircon using ion implantation and Rutherford backscattering techniques. *Geochim. Cosmochim. Acta* **55**, 1663–1673.

Cherniak D.J., Hanchar J.M., and Watson E.B. (1997) Diffusion of tetravalent cations in zircon. *Contrib. Mineral. Petrol.* **127**, 383–390.

Cherniak D.J., Zhang X.Y., Wayne N.K., and Watson E.B. (2001) Sr, Y, and REE diffusion in fluorite. *Chem. Geol.* **181**, 99–111.

Cherniak D.J., Watson E.B., Grove M., and Harrison T.M. (2004) Pb diffusion in monazite: a combined RBS/SIMS study. *Geochim. Cosmochim. Acta* **68**, 829–840.

Chiba H., Chacko T., Clayton R.N., and Goldsmith J.R. (1989) Oxygen isotope fractionations involving diopside, forsterite, magnetite, and calcite: application to geothermometry. *Geochim. Cosmochim. Acta* **53**, 2985–2995.

Christian J.W. (1975) *The Theory of Transformations in Metals and Alloys*. Oxford, England: Pergamon Press.

Chuang P.Y., Charlson R. J., and Seinfeld J. H. (1997) Kinetic limitations on droplet formation in clouds. *Nature* **390**, 594–596.

Clift R., Grace J.R., and Weber M.E. (1978) *Bubbles, Drops, and Particles*. New York, NY: Academic Press.

Coghlan R.A.N. (1990) Studies in diffusional transport: grain boundary transport of oxygen in feldspar, diffusion of oxygen, strontium and the REE's in garnet, and thermal histories of granitic intrusions in south-central Maine using oxygen isotopes. Thesis, Providence, RI: Brown University.

Connolly C. and Muehlenbachs K. (1988) Contrasting oxygen diffusion in nepheline, diopside and other silicates and their relevance to isotopic systematics in meteorites. *Geochim. Cosmochim. Acta* **52**, 1585–1591.

Cook G.B. and Cooper R.F. (1990) Chemical diffusion and crystalline nucleation during oxidation of ferrous iron-bearing magnesium aluminosilicate glass. *J. Non-Cryst. Solids* **120**, 207–222.

Cooper A.R. (1965) Model for multi-component diffusion. *Phys. Chem. Glasses* **6**, 55–61.

———. (1966) Diffusive mixing in continuous laminar flow systems. *Chem. Eng. Sci.* **21**, 1095–1106.

———. (1968) The use and limitations of the concept of an effective binary diffusion coefficient for multi-component diffusion. In *Mass Transport in Oxides*, Vol. 296 (ed. J.B. Wachman and A.D. Franklin), pp. 79–84. Nat. Bur. Stand. Spec. Publ.

Cooper A.R. and Kingery W.D. (1963) Dissolution in ceramic systems, I: molecular diffusion, natural convection, and forced convection studies of sapphire dissolution in calcium aluminum silicate. *J. Am. Ceram. Soc.* **47**, 37–43.

Cooper A.R. and Schut R.J. (1980) Analysis of transient dissolution in $CaO–Al_2O_3–SiO_2$. *Metall. Trans. B* **11**, 373–376.

Cooper A.R. and Varshneya A.K. (1968) Diffusion in the system $K_2O–SrO–SiO_2$, I: effective binary diffusion coefficients. *J. Am. Ceram. Soc.* **51**, 103–106.

Cooper K.M. and Reid M.R. (2003) Reexamination of crystal ages in recent Mount St. Helens lavas: implications for magma reservoir processes. *Earth Planet. Sci. Lett.* **213**, 149–167.

Cooper R.F., Fanselow J.B., and Poker D.B. (1996a) The mechanism of oxidation of a basaltic glass: chemical diffusion of network-modifying cations. *Geochim. Cosmochim. Acta* **60**, 3253–3265.

Cooper R.F., Fanselow J.B., Weber J.K.R., Merkley D.R., and Poker D.B. (1996b) Dynamics of oxidation of a Fe^{2+}-bearing aluminosilicate (basaltic) melt. *Science* **274**, 1173–1176.

Crank J. (1975) *The Mathematics of Diffusion*. Oxford, UK: Clarendon Press.

———. (1984) *Free and Moving Boundary Problems*. Oxford, UK: Clarendon Press.

Culling W.E.H. (1960) Analytical theory of erosion. *J. Geol.* **68**, 336–344.

Cussler E.L. (1976) *Multicomponent Diffusion*. Amsterdam, Netherlands: Elsevier.

———. (1997) *Diffusion: Mass Transfer in Fluid Systems*. Cambridge, UK: Cambridge University Press.

Cygan R.T. and Lasaga A.C. (1985) Self-diffusion of magnesium in garnet at 750–900 °C. *Am. J. Sci.* **285**, 328–350.

Darken L.S. (1948) Diffusion mobility and their interrelation through free energy in binary metalic systems. *Trans. AIME* **175**, 184–201.

Davis M.J., Ihinger P.D., and Lasaga A.C. (1997) Influence of water on nucleation kinetics in silicate melt. *J. Non-Cryst. Solids* **219**, 62–69.

De Groot S.R. and Mazur P. (1962) *Non-Equilibrium Thermodynamics*. New York, NY: Interscience.

Delaney J.R. and Karsten J.L. (1981) Ion microprobe studies of water in silicate melts: concentration-dependent water diffusion in obsidian. *Earth Planet. Sci. Lett.* **52**, 191–202.

Dennis P.F. (1984) Oxygen self-diffusion in quartz under hydrothermal conditions. *J. Geophys. Res.* **89**, 4047–4057.

DeWolf C.P., Belshaw N., and O'Nions R.K. (1993) A metamorphic history from micron-scale [207]Pb/[206]Pb chronometry of Archean monazite. *Earth Planet. Sci. Lett.* **120**, 207–220.

Dimanov A., Jaoul O., and Sautter V. (1996) Calcium self-diffusion in natural diopside single crystals. *Geochim. Cosmochim. Acta* **60**, 4095–4106.

Dingwell D.B. and Webb S.L. (1989) Structural relaxation in silicate melts and non-Newtonian melt rheology in geologic processes. *Phys. Chem. Miner.* **16**, 508–516.

———. (1990) Relaxation in silicate melts. *Eur. J. Mineral.* **2**, 427–449.

Dodson M.H. (1973) Closure temperature in cooling geochronological and petrological systems. *Contrib. Mineral. Petrol.* **40**, 259–274.

———. (1979) Theory of cooling ages. In *Lectures in Isotope Geology* (ed. E. Jager and J. C. Hunziker), pp. 194–202. Berlin: Springer-Verlag.

———. (1986) Closure profiles in cooling systems. *Mater. Sci. Forum* **7**, 145–154.

Dohmen R., Chakraborty S., and Becker H.W. (2002) Si and O diffusion in olivine and implications for characterizing plastic flow in the mantle. *Geophys. Res. Lett.* **29**, (26-1)–(26-4).

Donaldson C.H. (1985) The rates of dissolution of olivine, plagioclase, and quartz in a basaltic melt. *Mineral. Mag.* **49**, 683–693.

Donaldson C.H., Usselman T.M., Williams R.J., and Lofgren G.E. (1975) Experimental modeling of the cooling history of Apollo 12 olivine basalts. *Proc. Lunar Sci. Conf. 6th*, 843–869.

Doremus R.H. (1969) The diffusion of water in fused silica. In *Reactivity of Solids* (ed. J.W. Mitchell, R.C. Devries, R.W. Roberts, and P. Cannon), pp. 667–673. Wiley.

———. (1973) *Glass Science.* New York, NY: Wiley.

———. (1975) Interdiffusion of hydrogen and alkali ions in a glass surface. *J. Non-Cryst. Solids* **19**, 137–144.

———. (1982) Interdiffusion of alkali and hydronium ions in glass: partial ionization. *J. Non-Cryst. Solids* **48**, 431–436.

———. (1983) Diffusion-controlled reaction of water with glass. *J. Non-Cryst. Solids* **55**, 143–147.

———. (2002) *Diffusion of Reactive Molecules in Solids and Melts.* New York, NY: Wiley.

Dowty E. (1976a) Crystal structure and crystal growth, I: the influence of internal structure on morphology. *Am. Mineral.* **61**, 448–459.

———. (1976b) Crystal structure and crystal growth, II: sector zoning in minerals. *Am. Mineral.* **61**, 460–469.

———. (1977) The importance of adsorption in igneous partitioning of trace elements. *Geochim. Cosmochim. Acta* **41**, 1643–1646.

———. (1980a) Crystal growth and nucleation theory and the numerical simulation of igneous crystallization. In *Physics of Magmatic Processes* (ed. R. B. Hargraves), pp. 419–486. Princeton, NJ: Princeton University Press.

———. (1980b) Crystal-chemical factors affecting the mobility of ions in minerals. *Am. Mineral.* **65**, 174–182.

Drever J.I. (1997) *The Geochemistry of Natural Waters.* Upper Saddle River, NJ: Prentice Hall.

Dreybrodt W., Lauckner J., Liu Z., Svensson U., and Buhmann D. (1996) The kinetics of the reaction $CO_2 + H_2O = H^+ + HCO_3^-$ as one of the rate limiting steps for the dissolution of calcite in the system $H_2O-CO_2-CaCO_3$. *Geochim. Cosmochim. Acta* **60**, 3375–3381.

Drury T. and Roberts J.P. (1963) Diffusion in silica glass following reaction with tritiated water vapor. *Phys. Chem. Glasses* **4**, 79–90.

Dunn T. (1982) Oxygen diffusion in three silicate melts along the join diopside–anorthite. *Geochim. Cosmochim. Acta* **46**, 2293–2299.

———. (1983) Oxygen chemical diffusion in three basaltic liquids at elevated temperatures and pressures. *Geochim. Cosmochim. Acta* **47**, 1923–1930.

Dunn T. and Ratliffe W.A. (1990) Chemical diffusion of ferrous iron in a peraluminous sodium aluminosilicate melt: 0.1 MPa to 2.0 GPa. *J. Geophys. Res.* **95**, 15665–15673.

Edwards R.L., Chen J.H., and Wasserburg G.J. (1986/87) $^{238}U-^{234}U-^{230}Th-^{232}Th$ systematics and the precise measurement of time over the past 500,000 years. *Earth Planet. Sci. Lett.* **81**, 175–192.

Ehlers T.A. (2005) Crustal thermal processes and thermochronometer interpretation. *Rev. Mineral. Geochem.* **58**, 315–350.

Eiler J.M., Baumgarter L.P., and Valley J.W. (1992) Intercrystalline stable isotope diffusion: a fast grain boundary model. *Contrib. Mineral. Petrol.* **112**, 543–557.

Eiler J.M., Valley J.W., and Baumgarter L.P. (1993) A new look at stable isotope thermometry. *Geochim. Cosmochim. Acta* **57**, 2571–2583.

Eiler J.M., Baumgartner L.P., and Valley J.W. (1994) Fast grain boundary: a Fortran-77 program for calculating the effects of retrograde interdiffusion of stable isotopes. *Comput. Geosci.* **20**, 1415–1434.

Einstein A. (1905) The motion of small particles suspended in static liquids required by the molecular kinetic theory of heat. *Ann. Phys.* **17**, 549–560.

Elphick S. C., Dennis P.F., and Graham C.M. (1986) An experimental study of the diffusion of oxygen in quartz and albite using an overgrowth technique. *Contrib. Mineral. Petrol.* **92**, 322–330.

Elphick S.C., Ganguly J., and Loomis T.P. (1985) Experimental determination of cation diffusivities in aluminosilicate garnets, I: experimental methods and interdiffusion data. *Contrib. Mineral. Petrol.* **90**, 36–44.

Elphick S.C., Graham C.M., and Dennis P.F. (1988) An ion microprobe study of anhydrous oxygen diffusion in anorthite: a comparison with hydrothermal data and some geological implications. *Contrib. Mineral. Petrol.* **100**, 490–495.

Epelbaum M.B., Chekhmir A.S., and Lyutov V.S. (1978) Component diffusion in water–albite melt during mineral assimilation. *Geokhimiya* **2**, 217–227.

Ernsberger F.M. (1980) The role of molecular water in the diffusive transport of protons in glasses. *Phys. Chem. Glasses* **21**, 146–149.

Essene E.J. and Fisher D.C. (1986) Lightning strike fusion: extreme reduction and metal-silicate liquid immiscibility. *Science* **234**, 189–193.

Ewing R.C., Meldrum A., Wang L., and Wang S. (2000) Radiation-induced amorphization. *Rev. Mineral. Geochem.* **39**, 319–361.

Eyring H. (1936) Viscosity, plasticity, and diffusion as examples of absolute reaction rates. *J. Chem. Phys.* **4**, 283–291.

Farley K.A., Wolf R.A., and Silver L.T. (1996) The effects of long alpha-stopping distances on (U–Th)/He ages. *Geochim. Cosmochim. Acta* **60**, 4223–4229.

Farver J.R. (1989) Oxygen self-diffusion in diopside with application to cooling rate determinations. *Earth Planet. Sci. Lett.* **92**, 386–396.

———. (1994) Oxygen self-diffusion in calcite: dependence on temperature and water fugacity. *Earth Planet. Sci. Lett.* **121**, 575–587.

Farver J.R. and Giletti B.J. (1985) Oxygen diffusion in amphiboles. *Geochim. Cosmochim. Acta* **49**, 1403–1411.

Farver J.R. and Giletti B.J. (1989) Oxygen and strontium diffusion in apatite and potential applications to thermal history determinations. *Geochim. Cosmochim. Acta* **53**, 1621–1631.

Farver J.R. and Yund R.A. (1990) The effect of hydrogen, oxygen, and water fugacity on oxygen diffusion in alkali feldspar. *Geochim. Cosmochim. Acta* **54**, 2953–2964.

———. (1991) Oxygen diffusion in quartz: dependence on temperature and water fugacity. *Chem. Geol.* **90**, 55–70.

———. (2000) Silicon diffusion in forsterite aggregates: implications for diffusion accommodated creep. *Geophys. Res. Lett.* **27**, 2337–2340.

Felipe M.A., Xiao Y., and Kubicki J.D. (2001) Molecular orbital modeling and transition state theory in geochemistry. *Rev. Mineral. Geochem.* **42**, 485–531.

Fell C.J.D. and Hutchison H. P. (1971) Diffusion coefficients for sodium and potassium chlorides in water at elevated temperatures. *J. Chem. Eng. Data* **16**, 427–429.

Feng X. and Savin S.M. (1993) Oxygen isotope studies of zeolites: stiblite, analcime, heulandite and clinoptilolite, II: kinetics and mechanism of isotopic exchange between zeolites and water vapor. *Geochim. Cosmochim. Acta* **57**, 4219–4238.

Feng X., Faiia A.M., Gabriel G.W., Aronson J.L., Poage M.A., and Chamberlain C.P. (1999) Oxygen isotope studies of illite/smectite and clinoptilolite from Yucca Mountain: implications for paleohydrologic conditions. *Earth Planet. Sci. Lett.* **171**, 95–106.

Firestone R.B. and Shirley V.S. (1996) *Table of Isotopes.* New York, NY: Wiley.

Fisher G.W. and Lasaga A.C. (1981) Irreversible thermodynamics in petrology. *Rev. Mineral.* **8**, 171–209.

Fortier S.M. and Giletti B.J. (1989) An empirical model for predicting diffusion coefficients in silicate minerals. *Science* **245**, 1481–1484.

———. (1991) Volume self-diffusion of oxygen in biotite, muscovite, and phlogopite micas. *Geochim. Cosmochim. Acta* **55**, 1319–1330.

Fowler W.A., Caughlan G.R., and Zimmerman B.A. (1967) Thermonuclear reaction rates. *Annu. Rev. Astron. Astrophys.* **5**, 525–570.

———. (1975) Thermonuclear reaction rates, II: *Annu. Rev. Astron. Astrophys.* **13**, 69–112.

Frank M., Schwaz B., Baumann S., Kubik P.W., Suter M., and Mangini A. (1997) A 200 kyr record of cosmogenic radionuclide production rate and geomagnetic field intensity from 10Be in globally stacked deep-sea sediments. *Earth Planet. Sci. Lett.* **149**, 121–129.

Freda C. and Baker D.R. (1998) Na–K interdiffusion in alkali feldspar melts. *Geochim. Cosmochim. Acta* **62**, 2997–3007.

Freda C., Baker D.R., Romano C., and Scarlato P. (2003) Water diffusion in natural potassic melts. *Geol. Soc. Spec. Publ.* **213**, 53–62.

Freda C., Baker D.R., and Scarlato P. (2005) Sulfur diffusion in basaltic melts. *Geochim. Cosmochim. Acta* **69**, 5061–5069.

Freer R. and Dennis P.F. (1982) Oxygen diffusion studies, I: a preliminary ion microprobe investigation of oxygen diffusion in some rock-forming minerals. *Mineral. Mag.* **45**, 179–192.

Freer R. and Edwards A. (1999) An experimental study of Ca–(Fe,Mg) interdiffusion in silicate garnets. *Contrib. Mineral. Petrol.* **134**, 370–379.

Fuss T., Ray C.S., Lesher C.E., and Day D.E. (2006) In situ crystallization of lithium disilicate glass: effort of pressure on crystal growth rate. *J. Non-Cryst. Solids* **352**, 2073–2081.

Gabitov R.I., Price J.D., and Watson E.B. (2005) Diffusion of Ca and F in haplogranitic melt from dissolving fluorite crystals at 900°–1000°C and 100 MPa. *Geochem. Geophys. Geosyst.* **6**, doi:10.1029/2004GC000832.

Gaetani G.A. and Watson E.B. (2000) Open system behavior of olivine-hosted inclusions. *Earth Planet. Sci. Lett.* **183**, 27–41.

Ganguly J. (1982) Mg–Fe order–disorder in ferromagnesian silicates, II: thermodynamics, kinetics and geological applications. In *Advances in Physical Geochemistry*, Vol. 2 (ed. S. K. Saxena), pp. 58–99. Springer-Verlag.

———. (1986) Disordering energy versus disorder in minerals: a phenomenological relation and application to orthopyroxene. *J. Phys. Chem. Solids* **47**, 417–420.

———. (ed.) (1991) *Diffusion, Atomic Ordering, and Mass Transport. Advances in Physical Geochemistry*, Vol. 8. New York: Springer Verlag.

Ganguly J. and Domeneghetti M.C. (1996) Cation ordering of orthopyroxenes from the Skaergaard intrusion: implications for the subsolidus cooling rates and permeabilities. *Contrib. Mineral. Petrol.* **122**, 359–367.

Ganguly J. and Stimpfl M. (2000) Cation ordering in orthopyroxenes from two stony-iron meteorites: implications for cooling rates and metal-silicate mixing. *Geochim. Cosmochim. Acta* **64**, 1291–1297.

Ganguly J. and Tazzoli V. (1994) Fe^{2+}–Mg interdiffusion in orthopyroxene: retrieval from data on intracrystalline exchange reaction. *Am. Mineral.* **79**, 930–937.

Ganguly J. and Tirone M. (1999) Diffusion closure temperature and age of a mineral with arbitrary extent of diffusion: theoretical formulation and applications. *Earth Planet. Sci. Lett.* **170**, 131–140.

———. (2001) Relationship between cooling rate and cooling age of a mineral: theory and applications to meteorites. *Meteorit. Planet. Sci.* **36**, 167–175.

Ganguly J., Bhattacharya R.N., and Chakraborty S. (1988) Convolution effect in the determination of compositional profiles and diffusion coefficients by microprobe step scans. *Am. Mineral.* **73**, 901–909.

Ganguly J., Yang H., and Ghose S. (1994) Thermal history of mesosiderites: quantitative constraints from compositional zoning and Fe–Mg ordering in orthopyroxenes. *Geochim. Cosmochim. Acta* **58**, 2711–2723.

Ganguly J., Chakraborty S., Sharp T.G., and Rumble D. (1996) Constraint on the time scale of biotite-grade metamorphism during Acadian orogeny from a natural garnet–garnet diffusion couple. *Am. Mineral.* **81**, 1208–1216.

Ganguly J., Cheng W., and Chakraborty S. (1998a) Cation diffusion in aluminosilicate garnets: experimental determination in pyrope–almandine diffusion couples. *Contrib. Mineral. Petrol.* **131**, 171–180.

Ganguly J., Tirone M., and Hervig R.L. (1998b) Diffusion kinetics of samarium and neodymium in garnet, and a method for determining cooling rates of rocks. *Science* **281**, 805–807.

Ganguly J., Dsasgupta S., Cheng W., and Neogi S. (2000) Exhumation history of a section of the Sikkim Himalayas, India: records in the metamorphic mineral equilibria and compositional zoning of garnet. *Earth Planet. Sci. Lett.* **183**, 471–486.

Gast P.W. (1968) Trace element fractionation and the origin of tholeiitic and alkaline magma types. *Geochim. Cosmochim. Acta* **32**, 1057–1086.

Gerard O. and Jaoul O. (1989) Oxygen diffusion in San Carlos olivine. *J. Geophys. Res.* **94**(B4), 4119–4128.

Gessmann C.K., Spiering B., and Raith M. (1997) Experimental study of the Fe–Mg exchange between garnet and biotite: constraints on the mixing behavior and analysis of the cation-exchange mechanisms. *Am. Mineral.* **82**, 1225–1240.

Ghiorso M.S. (1987a) Chemical mass transfer in magmatic processes, III: crystal growth, chemical diffusion and thermal diffusion in multicomponent silicate melts. *Contrib. Mineral. Petrol.* **96**, 291–313.

Ghiorso M.S., Evans B. W., Hirschmann M. M., and Yang H. (1995) Thermodynamics of the amphiboles: Fe–Mg cummingtonite solid solutions. *Am. Mineral.* **80**, 502–519.

Giletti B.J. (1974) Studies in diffusion, I: Ar in phlogopite mica. In *Geochemical Transport and Kinetics* (ed. A. W. Hoffman, B.J. Giletti, H.S. Yoder, and R.A. Yund), pp. 107–115. Carnegie Institution of Washington Publ.

———. (1985) The nature of oxygen transport within minerals in the presence of hydrothermal water and the role of diffusion. *Chem. Geol.* **53**, 197–206.

———. (1986) Diffusion effects on oxygen isotope temperatures of slowly cooled igneous and metamorphic rocks. *Earth Planet. Sci. Lett.* **77**, 218–228.

———. (1991) Rb and Sr diffusion in alkali feldspars, with implications for cooling histories of rocks. *Geochim. Cosmochim. Acta* **55**, 1331–1343.

Giletti B.J. and Anderson T.F. (1975) Studies in diffusion, II: oxygen in phlogopite mica. *Earth Planet. Sci. Lett.* **28**, 225–233.

Giletti B. J. and Casserly J.E.D. (1994) Strontium diffusion kinetics in plagioclase feldspars. *Geochim. Cosmochim. Acta* **58**, 3785–3793.

Giletti B.J. and Hess K.C. (1988) Oxygen diffusion in magnetite. *Earth Planet. Sci. Lett.* **89**, 115–122.

Giletti B.J. and Yund R.A. (1984) Oxygen diffusion in quartz. *J. Geophys. Res.* **89**, 4039–4046.

Giletti B.J., Semet M.P., and Yund R.A. (1978) Studies in diffusion, III: oxygen in feldspars: an ion microprobe determination. *Geochim. Cosmochim. Acta* **42**, 45–57.

Glasstone S., Laider K.J., and Eyring H. (1941) *The Theory of Rate Processes.* New York, NY: McGraw-Hill.

Glikson A. and Allen C. (2004) Iridium anomalies and fractionated siderophile element patterns in impact ejecta, Brockman Iron Formation, Hamersley Basin, Western Australia: evidence for a major asteroid impact in simatic crustal regions of the early proterozoic Earth. *Earth Planet. Sci. Lett.* **220**, 247–264.

Goldsmith J.R. (1987) Al/Si interdiffusion in albite: effect of pressure and the role of hydrogen. *Contrib. Mineral. Petrol.* **95**, 311–321.

———. (1988) Enhanced Al/Si diffusion in $KAlSi_3O_8$ at high pressures: the effect of hydrogen. *J. Geol.* **96**, 109–124.

———. (1991) Pressure-enhanced Al/Si diffusion and oxygen isotope exchange. In *Diffusion, Atomic Ordering, and Mass Transport: Selected Topics in Geochemistry* (ed. J. Ganguly), pp. 221–247. Berlin, Germany: Springer-Verlag.

Graham C.M. (1981) Experimental hydrogen isotope studies, III: diffusion of hydrogen in hydrous minerals, and stable isotope exchange in metamorphic rocks. *Contrib. Mineral. Petrol.* **76**, 216–228.

Granasy L. and James P.F. (1999) Non-classical theory of crystal nucleation: application to oxide glasses: review. *J. Non-Cryst. Solids* **253**, 210–230.

Gregg M.C., Sanford T.B., and Winkel D.P. (2003) Reduced mixing from the breaking of internal waves in equatorial waters. *Nature* **422**, 513–515.

Grove M. and Harrison T.M. (1996) [40]Ar diffusion in Fe-rich biotite. *Am. Mineral.* **81**, 940–951.

Grove T.L., Baker M.B., and Kinzler R.J. (1984) Coupled CaAl–NaSi diffusion in plagioclase feldspar: experiments and applications to cooling rate speedometry. *Geochim. Cosmochim. Acta* **48**, 2113–2121.

Gualda G.A.R. and Rivers M. (2006) Quantitative 3D petrography using x-ray tomography: application to Bishop Tuff pumice clasts. *J. Volcanol. Geotherm. Res.* **154**, 48–62.

Gupta P.K. and Cooper A.R. (1971) The [D] matrix for multicomponent diffusion. *Physica* **54**, 39–59.

Haller W. (1963) Concentration-dependent diffusion coefficient of water in glass. *Phys. Chem. Glasses* **4**, 217–220.

Hallwig D., Schachtner R., and Sockel H.G. (1982) Diffusion of magnesium, silicon, and oxygen in Mg_2SiO_4 and formation of the compound in the solid state. In *Reactivity in Solids. Proc. Int'l Sympos. (9th)* (ed. K. Dyrek, J. Habor, and J. Nowotry), pp. 166–169.

Ham F.S. (1958) Theory of diffusion-limited precipitation. *J. Phys. Chem. Solids* **6**, 335–351.

Hammouda T. and Cherniak D.J. (2000) Diffusion of Sr in fluorphlogopite determined by Rutherford backscattering spectroscopy. *Earth Planet. Sci. Lett.* **178**, 339–349.

Hammouda T. and Pichavant M. (1999) Kinetics of melting of fluorphlogopite–quartz pairs. *Eur. J. Mineral.* **11**, 637–653.

Hargraves R.B. (1980) *Physics of Magmatic Processes*, pp. 585. Princeton, NJ: Princeton University Press.

Harris M.J., Fowler W.A., Caughlan G.R., and Zimmerman B.A. (1983) Thermonuclear reaction rates, III. *Annu. Rev. Astron. Astrophys.* **21**, 165–176.

Harrison R.J. and Putnis A. (1999) Determination of the mechanism of cation ordering in magnesioferrite ($MgFe_2O_4$) from the time- and temperature-dependence of magnetic susceptibility. *Phys. Chem. Miner.* **26**, 322–332.

Harrison T.M. (1981) Diffusion of ^{40}Ar in hornblende. *Contrib. Mineral. Petrol.* **78**, 324–331.

Harrison T.M. and McDougall I. (1980) Investigations of an intrusive contact, northwest Nelson, New Zealand, I: thermal, chronological and isotopic constraints. *Geochim. Cosmochim. Acta* **44**, 1985–2003.

Harrison T.M. and Watson E.B. (1983) Kinetics of zircon dissolution and zirconium diffusion in granitic melts of variable water content. *Contrib. Mineral. Petrol.* **84**, 66–72.

———. (1984) The behavior of apatite during crustal anatexis: equilibrium and kinetic considerations. *Geochim. Cosmochim. Acta* **48**, 1467–1477.

Harrison T.M., Duncan I., and McDougall I. (1985) Diffusion of ^{40}Ar in biotite: temperature, pressure and compositional effects. *Geochim. Cosmochim. Acta* **49**, 2461–2468.

Harrison T.M., Lovera O.M., and Heizler M.T. (1991) $^{40}Ar/^{39}Ar$ results for alkali feldspars containing diffusion domains with differing activation energy. *Geochim. Cosmochim. Acta* **55**, 1435–1448.

Hart S.R. (1981) Diffusion compensation in natural silicates. *Geochim. Cosmochim. Acta* **45**, 279–291.

Haul R., Hubner K., and Kircher O. (1976) Oxygen diffusion in strontium titanate studied by solid/gas exchange. In *Reactivity of Solids* (ed. J. Wood and O. Lindquist), pp. 101–106. Plenum.

Hayashi T. and Muehlenbachs K. (1986) Rapid oxygen diffusion in melilite and its relevance to meteorites. *Geochim. Cosmochim. Acta* **50**, 585–591.

Heinemann R., Staack V., Fischer A., Kroll H., Vad T., and Kirfel A. (1999) Temperature dependence of Fe, Mg partitioning in Acapulco olivine. *Am. Mineral.* **84**, 1400–1405.

Helfferich F. and Plesset M.S. (1958) Ion exchange kinetics. a nonlinear diffusion problem. *J. Chem. Phys.* **28**, 418–425.

Henderson C.M.B., Knight K. S., Redfern S.A.T., and Wood B. J. (1996) High-temperature study of octahedral cation exchange in olivine by neutron powder diffraction. *Science* **271**, 1713–1715.

Henderson J., Yang L., and Derge G. (1961) Self-diffusion of aluminum in CaO-SiO$_2$–Al$_2$O$_3$ melts. *Trans. Met. Soc. AIME* **221**, 56–60.

Higgins M. D. (2000) Measurement of crystal size distributions. *Am. Mineral.* **85**, 1105–1116.

———. (2002) Closure in crystal size distributions (CSD), verification of CSD calculations, and the significance of CSD fans. *Am. Mineral.* **87**, 171–175.

Higgins S.R., Jordan G., and Eggleton C.M. (1998) Dissolution kinetics of the barium sulfate (001) surface by hydrothermal atomic force microscopy. *Langmuir* **14**, 4967–4971.

Higgins S.R., Boram L.H., Eggleston C.M., Coles B.A., Compton R.G., and Knauss K.G. (2002a) Dissolution kinetics, step and surface morphology of magnesite (104) surfaces in acidic aqueous solution at 60°C by atomic force microscopy under defined hydrodynamic conditions. *J. Phys. Chem. B* **106**, 6696–6705.

Higgins S.R., Jordan G., and Eggleston C.M. (2002b) Dissolution kinetics of magnesite in acidic aqueous solution: a hydrothermal atomic force microscopy study assessing step kinetics and dissolution flux. *Geochim. Cosmochim. Acta* **66**, 3201–3210.

Hobbs B.E. (1984) Point defect chemistry of minerals under hydrothermal environment. *J. Geophys. Res.* **89**, 4026–4038.

Hobbs P.V. (1974) *Ice Physics*. Oxford, UK: Clarendon Press.

Hofmann A.W. (1980) Diffusion in natural silicate melts: a critical review. In *Physics of Magmatic Processes* (ed. R.B. Hargraves), pp. 385–417. Princeton, NJ: Princeton University Press.

Hofmann A.W. and Magaritz M. (1977) Diffusion of Ca, Sr, Ba, and Co in a basaltic melt: implications for the geochemistry of the mantle. *J. Geophys. Res.* **82**, 5432–5440.

Hofmann A.W., Giletti B.J., Yoder H.S., and Yund R.A. (1974) *Geochemical Transport and Kinetics*, Vol. 634. Washington, DC: Carnegie Institution of Washington Publ.

Holman J.P. (2002) *Heat Transfer*. New York, NY: McGraw-Hill.

Holzapel C., Rubie D.C., Frost D.J., and Langenhorst F. (2005) Fe–Mg interdiffusion in (Mg,Fe)SiO$_3$ perovskite and lower mantle reequilibration. *Science* **309**, 1707–1710.

Houser C.A., Herman J.S., Tsong I.S.T., White W.B., and Lanford W.A. (1980) Sodium–hydrogen interdiffusion in sodium silicate glasses. *J. Non-Cryst. Solids* **41**, 89–98.

Huh C. (1999) Dependence of the decay rate of [7]Be on chemical forms. *Earth Planet. Sci. Lett.* **171**, 325–328.

Hui H. and Zhang Y. (2007) Toward a general viscosity equation for natural anhydrous and hydrous silicate melts. *Geochim. Cosmochim. Acta* **71**, 403–416.

Hurwitz S. and Navon O. (1994) Bubble nucleation in rhyolitic melts: experiments at high pressure, temperature, and water content. *Earth Planet. Sci. Lett.* **122**, 267–280.

Ihinger P.D., Zhang Y., and Stolper E.M. (1999) The speciation of dissolved water in rhyolitic melt. *Geochim. Cosmochim. Acta* **63**, 3567–3578.

Ingrin J., Hercule S., and Charton T. (1995) Diffusion of hydrogen in diopside: results of dehydration experiments. *J. Geophys. Res.* **100**, 15489–15499.

Ingrin J., Pacaud L., and Jaoul O. (2001) Anisotropy of oxygen diffusion in diopside. *Earth Planet. Sci. Lett.* **192**, 347–361.

Jahne B., Heinz G., and Dietrich W.E. (1987) Measurement of the diffusion coefficients of sparingly soluble gases in water. *J. Geophys. Res.* **92**, 10767–10776.

Jambon A. (1982) Tracer diffusion in granitic melts: experimental results for Na, Rb, Cs, Ca, Sr, Ba, Ce, Eu to 1300°C and a model of calculation. *J. Geophys. Res.* **87**, 10797–10810.

Jambon A. and Semet M.P. (1978) Lithium diffusion in silicate glasses of albite, orthoclase, and obsidian compositions: an ion-microprobe determination. *Earth Planet. Sci. Lett.* **37**, 445–450.

Jambon A. and Shelby J.E. (1980) Helium diffusion and solubility in obsidians and basaltic glass in the range 200–300°C. *Earth Planet. Sci. Lett.* **51**, 206–214.

Jambon A., Zhang Y., and Stolper E.M. (1992) Experimental dehydration of natural obsidian and estimation of D_{H2O} at low water contents. *Geochim. Cosmochim. Acta* **56**, 2931–2935.

Jaoul O., Bertran-Alvarez Y., Liebermann R.C., and Price G.D. (1995) Fe–Mg interdiffusion in olivine up to 9 GPa at T = 600–900°C; experimental data and comparison with defect calculations. *Phys. Earth Planet. In.* **89**, 199–218.

Jiang J. and Lasaga A.C. (1990) The effect of post-growth thermal events on growth-zoned garnet: implications for metamorphic P-T history calculations. *Contrib. Mineral. Petrol.* **105**, 454–459.

Johari G.P. (2000) An equilibrium supercooled liquid's entropy and enthalpy in the Kauzmann and the third law extrapolations, and a proposed experimental resolution. *J. Chem. Phys.* **113**, 751–761.

Jones P., Haggett M.L., and Longridge J.L. (1964) The hydration of carbon dioxide. *J. Chem. Educ.* **41**, 610–612.

Karsten J.L., Holloway J.R., and Delaney J.R. (1982) Ion microprobe studies of water in silicate melts: temperature-dependent water diffusion in obsidian. *Earth Planet. Sci. Lett.* **59**, 420–428.

Kauzmann W. (1948) The nature of glassy state and the behavior of liquids at low temperatures. *Chem. Rev.* **43**, 219–256.

Kerr J.A. and Drew R.M. (1987) *CRC Handbook of Bimolecular and Termolecular Gas Reactions*, Vol. 3 (part 2), pp. 243.

Kerr R.C. (1994a) Melting driven by vigorous compositional convection. *J. Fluid Mech.* **280**, 255–285.

———. (1994b) Dissolving driven by vigorous compositional convection. *J. Fluid Mech.* **280**, 287–302.

———. (1995) Convective crystal dissolution. *Contrib. Mineral. Petrol.* **121**, 237–246.

Kerr R.C. and Tait S.R. (1986) Crystallization and compositional convection in a porous medium with application to layered igneous intrusions. *J. Geophys. Res.* **91**, 3591–3608.

Kim H. (1969) Combined use of various experimental techniques for the determination of nine diffusion coefficients in four-component systems. *J. Phys. Chem.* **73**, 1716–1722.

King D.B. and Saltzman E.S. (1995) Measurement of the diffusion coefficient of sulfur hexafluoride in water. *J. Geophys. Res.* **100**, 7083–7088.

Kirkaldy J.S. and Purdy G.R. (1969) Diffusion in multicomponent metallic systems, X: diffusion at and near ternary critical states. *Can. J. Phys.* **47**, 865–871.

Kirkaldy J.S. and Young D.J. (1987) *Diffusion in the Condensed State*. London, England: The Institute of Metals.

Kirkaldy J.S., Weichert D., and Haq Z.U. (1963) Diffusion in multicomponent metallic systems, VI: some thermodynamic properties of the D matrix and the corresponding solutions of the diffusion equations. *Can. J. Phys.* **41**, 2166–2173.

Kirkpatrick R.J. (1974) Kinetics of crystal growth in the system $CaMgSi_2O_6$–$CaAl_2SiO_6$. *Am. Mineral.* **274**, 215–242.

Kirkpatrick R.J. (1975) Crystal growth from the melt: a review. *Am. Mineral.* **60**, 798–814.

———. (1981) Kinetics of crystallization of igneous rocks. *Rev. Mineral.* **8**, 321–389.

———. (1985) Kinetics of crystallization of igneous rocks. *Rev. Mineral.* **8**, 321–398.

Kirkpatrick R.J., Robinson G.R., and Hays J.F. (1976) Kinetics of crystal growth from silicate melts. *J. Geophys. Res.* **81**, 5715–5720.

Kirkpatrick R.J., Reck B.H., Pelly I.Z., and Kuo L.-C. (1983) Programmed cooling experiments in the system $MgO–SiO_2$: kinetics of a peritectic reaction. *Am. Mineral.* **68**, 1095–1101.

Kivelson D., Kivelson S.A., Zhao X., Nussinov Z., and Tarjus G. (1995) A thermodynamic theory of supercooled liquids. *Physica A* **219**, 27–38.

Kleine T., Munker C., Mezger K., and Palme H. (2002) Rapid accretion and early core formation on asteroids and the terrestrial planets from Hf–W chronometry. *Nature* **418**, 952–955.

Kress V.C. and Ghiorso M.S. (1993) Multicomponent diffusion in $MgO–Al_2O_3–SiO_2$ and $CaO–MgO–Al_2O_3–SiO_2$ melts. *Geochim. Cosmochim. Acta* **57**, 4453–4466.

———. (1995) Multicomponent diffusion in basaltic melts. *Geochim. Cosmochim. Acta* **59**, 313–324.

Kroll H., Lueder T., Schlenz H., Kirfel A., and Vad T. (1997) The Fe^{2+}, Mg distribution in orthopyroxene: a critical assessment of its potential as a geospeedometer. *Eur. J. Mineral.* **9**, 705–733.

Kronenberg A.K., Kirby S.H., Aines R.D., and Rossman G.R. (1986) Solubility and diffusional uptake of hydrogen in quartz at high water pressures: implications for hydrolytic weakening. *J. Geophys. Res.* **91**(B12), 12723–12744.

Kubicki J.D. and Lasaga A.C. (1988) Molecular dynamics simulations of SiO_2 melt and glass: ionic and covalent models. *Am. Mineral.* **73**, 941–955.

———. (1993) Molecular dynamics simulations of interdiffusion in $MgSiO_3$-Mg_2SiO_4 melts. *Phys. Chem. Minerals* **20**, 255–262.

Kuo L.-C. and Kirkpatrick R.J. (1985) Kinetics of crystal dissolution in the system diopside–forsterite–silica. *Am. J. Sci.* **285**, 51–90.

Kurz M.D. and Jenkins W.J. (1981) The distribution of helium in oceanic basalt glasses. *Earth Planet. Sci. Lett.* **53**, 41–54.

Labotka T.C., Cole D.R., and Riciputi L.R. (2000) Diffusion of C and O in calcite at 100 MPa. *Am. Mineral.* **85**, 488–494.

Laidler K.J. (1987) *Chemical Kinetics.* New York, NY: Harper & Row.

Lanford W.A., Davis K., Lamarche P., Laursen T., Groleau R., and Doremus R.H. (1979) Hydration of soda-lime glass. *J. Non-Cryst. Solids* **33**, 249–265.

Lange R.A. (1994) The effect of H_2O, CO_2 and F on the density and viscosity of silicate melts. *Rev. Mineral.* **30**, 331–369.

Lange R.A. and Carmichael I.S.E. (1987) Densities of $Na_2O–K_2O–CaO–MgO–FeO–Fe_2O_3$–$Al_2O_3–TiO_2–SiO_2$ liquids: new measurements and derived partial molar properties. *Geochim. Comochim. Acta* **51**, 2931–2946.

Langer J.S. (1973) Statistical methods in the theory of spinodal decomposition. *Acta Metall.* **21**, 1649–1659.

Lasaga A.C. (1979) Multicomponent exchange and diffusion in silicates. *Geochim. Cosmochim. Acta* **43**, 455–469.

———. (1981a) Implication of a concentration dependent growth rate on the boundary layer crystal–melt model. *Earth Planet. Sci. Lett.* **56**, 429–434.

———. (1981b) Rate laws and chemical reactions. *Rev. Mineral.* **8**, 1–68.

———. (1982) Toward a master equation in crystal growth. *Am. J. Sci.* **282**, 1264–1288.

———. (1983) Geospeedometry: an extension of geothermometry. In *Kinetics and Equilibrium in Mineral Reactions* (ed. S.K. Saxena). Springer-Verlag.

———. (1998) *Kinetic Theory in the Earth Sciences.* Princeton, NJ: Princeton University Press.

Lasaga A.C. and Jiang J. X. (1995) Thermal history of rocks: P-T-t paths from geospeedometry, petrologic data, and inverse theory techniques. *Am. J. Sci.* **295**, 697–741.

Lasaga A.C. and Kirkpatrick R.J. (1981) Kinetics of Geochemical Processes. (ed. P. H. Ribbe), pp. 398. Washington, DC: Mineralogical Society of America.

Lasaga A.C. and Luttge A. (2001) Variation of crystal dissolution rate based on a dissolution stepwave model. *Science* **291**, 2400–2404.

Lasaga A.C. and Rye D.M. (1993) Fluid flow and chemical reaction kinetics in metamorphic systems. *Am. J. Sci.* **293**, 361–404.

Lasaga A.C., Soler J.M., Ganor J., Burch T.E., and Nagy K.L. (1994) Chemical weathering rate laws and global geochemical cycles. *Geochim. Cosmochim. Acta* **58**, 2361–2386.

Laughlin D.E. and Cahn J.W. (1975) Spinodal decomposition in age hardening copper–titanium alloys. *Acta Metall.* **23**, 329–339.

Lesher C.E. (1986) Effects of silicate liquid composition on mineral-liquid element partitioning from Soret diffusion studies. *J. Geophys. Res.* **91**, 6123–6141.

———. (1990) Decoupling of chemical and isotopic exchange during magma mixing. *Nature* **344**, 235–237.

———. (1994) Kinetics of Sr and Nd exchange in silicate liquids: theory, experiments, and applications to uphill diffusion, isotopic equilibrium and irreversible mixing of magmas. *J. Geophys. Res.* **99**, 9585–9604.

Lesher C.E. and Walker D. (1986) Solution properties of silicate liquids from thermal diffusion experiments. *Geochim. Cosmochim. Acta* **50**, 1397–1411.

———. (1988) Cumulate maturation and melt migration in a temperature gradient. *J. Geophys. Res.* **93**, 10295–10311.

———. (1991) Thermal diffusion in petrology. In *Diffusion, Atomic Ordering, and Mass Transport*, Vol. 8 (ed. J. Ganguly), pp. 396–451. New York, NY: Springer-Verlag.

Lesher C.E., Hervig R.L., and Tinker D. (1996) Self diffusion of network formers (silicon and oxygen) in naturally occurring basaltic liquid. *Geochim. Cosmochim. Acta* **60**, 405–413.

Levich V.G. (1962) *Physicochemical Hydrodynamics*. Englewood Cliff, NJ: Prentice-Hall.

Lewis M. and Glaser R. (2003) Synergism of catalysis and reaction center rehybridization: a novel mode of catalysis in the hydrolysis of carbon dioxide. *J. Phys. Chem. A* **107**, 6814–1818.

Liang Y. (1994) Axisymmetric double-diffusive convection in a cylindrical container: linear stability analysis with applications to molten $CaO–Al_2O_3–SiO_2$. In *Double-Diffusive Convection*, pp. 115–124. AGU.

———. (1999) Diffusive dissolution in ternary systems: analysis with applications to quartz and quartzite dissolution in molten silicates. *Geochim. Cosmochim. Acta* **63**, 3983–3995.

———. (2000) Dissolution in molten silicates: effects of solid solution. *Geochim. Cosmochim. Acta* **64**, 1617–1627.

Liang Y. and Davis A.M. (2002) Energetics of multicomponent diffusion in molten $CaO–Al_2O_3–SiO_2$. *Geochim. Cosmochim. Acta* **66**, 635–646.

Liang Y., Richter F.M., Davis A.M., and Watson E.B. (1996a) Diffusion in silicate melts, I: self diffusion in $CaO–Al_2O_3–SiO_2$ at $1500°C$ and 1 GPa. *Geochim. Cosmochim. Acta* **60**, 4353–4367.

Liang Y., Richter F.M., and Watson E.B. (1996b) Diffusion in silicate melts, II: multicomponent diffusion in $CaO–Al_2O_3–SiO_2$ at $1500°C$ and 1 GPa. *Geochim. Cosmochim. Acta* **60**, 5021–5035.

Liang Y., Richter F.M., and Chamberlin L. (1997) Diffusion in silicate melts, III: empirical models for multicomponent diffusion. *Geochim. Cosmochim. Acta* **61**, 5295–5312.

Liermann H.P. and Ganguly J. (2002) Diffusion kinetics of Fe^{2+} and Mg in aluminous spinel: experimental determination and applications. *Geochim. Cosmochim. Acta* **66**, 2903–2913.

Lifshitz I.M. and Slyozoc V.V. (1961) The kinetics of precipitation from supersaturated solid solutions. *J. Phys. Chem. Solids* **19**, 35–50.

Liger-Belair G. (2004) *Uncorked: the Science of Champagne*. Princeton, NJ: Princeton University Press.

Liu Y. and Zhang Y. (2000) Bubble growth in rhyolitic melt. *Earth Planet. Sci. Lett.* **181**, 251–264.

Liu Y., Behrens H., and Zhang Y. (2004a) The speciation of dissolved H_2O in dacitic melt. *Am. Mineral.* **89**, 277–284.

Liu Y., Zhang Y., and Behrens H. (2004b) H_2O diffusion in dacitic melt. *Chem. Geol.* **209**, 327–340.

———. (2005) Solubility of H_2O in rhyolitic melts at low pressures and a new empirical model for mixed $H_2O–CO_2$ solubility in rhyolitic melts. *J. Volcanol. Geotherm. Res.* **143**, 219–235.

Livingston F.E., Whipple G.C., and George S.M. (1997) Diffusion of HDO into single-crystal $H2^{16}O$ ice multilayers: comparison with $H_2^{18}O$. *J. Phys. Chem. B* **101**, 6127–6131.

Lockheed Martin (2002) *Chart of the Nuclides*. Lockheed Martin.

Lodders K. and Fegley B. Jr. (1998) *The Planetary Scientist's Companion*. New York: Oxford University Press.

Loomis T.P. (1983) Compositional zoning of crystals: a record of growth and reaction history. In *Kinetics and Equilibrium in Mineral Reactions* (ed. S.K. Saxena), pp. 1–59. New York: Springer-Verlag.

Loomis T.P., Ganguly J., and Elphick S.C. (1985) Experimental determination of cation diffusivities in aluminosilicate garnets, II: multicomponent simulation and tracer diffusion coefficients. *Contrib. Mineral. Petrol.* **90**, 45–51.

Lovera O.M., Richter F.M., and Harrison T.M. (1989) The $^{40}Ar/^{39}Ar$ thermochronometry for slowly cooled samples having a distribution of diffusion domain sizes. *J. Geophys. Res.* **94**, 17917–17935.

———. (1991) Diffusion domains determined by ^{39}Ar released during step heating. *J. Geophys. Res.* **96**, 2057–2069.

Mackwell S.J. (1992) Oxidation kinetics of fayalite (Fe_2SiO_4). *Phys. Chem. Miner.* **19**, 220–228.

Mackwell S.J. and Kohlstedt D.J. (1990) Diffusion of hydrogen in olivine: implications for water in the mantle. *J. Geophys. Res.* **95**, 5079–5088.

Mackwell S.J. and Paterson M.S. (1985) Water related diffusion and deformation effects in quartz at pressures of 1500 and 300 MPa. In *Point Defects in Minerals*, Vol. 31 (ed. R.N. Schock), pp. 141–150. AGU, Geophysics monographs.

Mackwell S., Bystricky M., and Sproni C. (2005) Fe–Mg interdiffusion in (Mg,Fe)O. *Phys. Chem. Miner.* **32**, 418–425.

Magaritz M. and Hofmann A.W. (1978a) Diffusion of Eu and Gd in basalt and obsidian. *Geochim. Cosmochim. Acta* **42**, 847–858.

———. (1978b) Diffusion of Sr, Ba and Na in obsidian. *Geochim. Cosmochim. Acta* **42**, 595–605.

Maharajh D.M. and Walkley J. (1973) The temperature dependence of the diffusion coefficients of Ar, CO_2, CH_4, CH_3Cl, CH_3Br, and $CHCl_2F$ in water. *Can. J. Chem.* **51**, 944–952.

Majewski E. and Walker D. (1998) S diffusivity in Fe–Ni–S–P melts. *Earth Planet. Sci. Lett.* **160**, 823–830.

Marsh B.D. (1988) Crystal size distribution (CSD) in rocks and the kinetics and dynamics of crystallization, I: theory. *Contrib. Mineral. Petrol.* **99**, 277–291.

Martens R.M., Rosenhauer M., Buttner H., and von Gehlen K. (1987) Heat capacity and kinetic parameters in the glass transformation interval of diopside, anorthite and albite glass. *Chem. Geol.* **62**, 49–70.

Matano C. (1933) On the relation between the diffusion coefficient and concentrations of solid metals. *Japan J. Phys.* **8**, 109–113.

McConnell J.D.C. (1995) The role of water in oxygen isotope exchange in quartz. *Earth Planet. Sci. Lett.* **136**, 97–107.

McKenzie D. (1985) ^{230}Th–^{238}U disequilibrium and the melting processes beneath ridge axes. *Earth Planet. Sci. Lett.* **72**, 149–157.

Meek R.L. (1973) Diffusion coefficient for oxygen in vitreous SiO_2. *J. Am. Ceram. Soc.* **56**, 341–343.

Merli M., Oberti R., Caucia F., and Ungaretti L. (2001) Determination of site population in olivine: warnings on X-ray data treatment and refinement. *Am. Mineral.* **86**, 55–65.

Milke R., Wiedenbeck M., and Heinrich W. (2001) Grain boundary diffusion of Si, Mg, and O in enstatite reaction rims: a SIMS study using isotopically doped reactants. *Contrib. Mineral. Petrol.* **142**, 15–26.

Miller D.G., Ting A.W., Rard J.A., and Epstein L.B. (1986) Ternary diffusion coefficients of the brine systems NaCl (0.5 M)–Na_2SO_4 (0.5 M)–H_2O and NaCl (0.489 M)–$MgCl_2$ (0.051 M)–H_2O (seawater composition) at 25°C. *Geochim. Cosmochim. Acta* **50**, 2397–2403.

Minor D.R. and Mukasa S.B. (1997) Zircon U–Pb and hornblende ^{40}Ar–^{39}Ar ages for the Dufek layered mafic intrusion, Antarctica: implications for the age of the Ferrar large igneous province. *Geochim. Cosmochim. Acta* **61**, 2497–2504.

Molin G.M., Saxena S.K., and Brizi E. (1991) Iron–magnesium order–disorder in an orthopyroxene crystal from the Johnstown meteorite. *Earth Planet. Sci. Lett.* **105**, 260–265.

Moore D.K., Cherniak D.J., and Watson E.B. (1998a) Oxygen diffusion in rutile from 750 to 1000°C and 0.1 to 1000 MPa. *Am. Mineral.* **83**, 700–711.

Moore G., Vennemann T., and Carmichael I.S.E. (1998b) An empirical model for the solubility of H_2O in magmas to 3 kilobars. *Am. Mineral.* **83**, 36–42.

Morgan Z., Liang Y., and Hess P.C. (2006) An experimental study of anorthosite dissolution in lunar picritic magmas: implications for crustal assimilation processes. *Geochim. Cosmochim. Acta* **70**, 3477–3491.

Morioka M. and Nagasawa H. (1991) Diffusion in single crystals of melilite, II: cations. *Geochim. Cosmochim. Acta* **55**, 751–759.

Morioka M. and Nagasawa H. (1991) Ionic diffusion in olivine. In *Diffusion, Atomic Ordering, and Mass Transport* (ed. J. Ganguly), pp. 176–197. New York: Springer-Verlag.

Moriya Y. and Nogami M. (1980) Hydration of silicate glass in steam atmosphere. *J. Non-Cryst. Solids* **38 & 39**, 667–672.

Morse S.A. (1980) *Basalts and Phase Diagrams.* New York: Springer-Verlag.

Mosenfelder J.L. and Bohlen S.R. (1997) Kinetics of the coesite to quartz transformation. *Earth Planet. Sci. Lett.* **153**, 133–147.

Mueller R.F. (1969) Kinetics and thermodynamics of intracrystalline distributions. *Mineral. Soc. Am. Spec. Pap.* **2**, 83–93.

Muncill G.E. and Chamberlain C.P. (1988) Crustal cooling rates inferred from homogenization of metamorphic garnets. *Earth Planet. Sci. Lett.* **87**, 390–396.

Muncill G.E. and Lasaga A.C. (1987) Crystal-growth kinetics of plagioclase in igneous systems: one-atmosphere experiments and application of a simplified growth model. *Am. Mineral.* **72**, 299–311.

Mungall J.E., Romano C., and Dingwell D.B. (1998) Multicomponent diffusion in the molten system K_2O–Na_2O–Al_2O_3–SiO_2–H_2O. *Am. Mineral.* **83**, 685–699.

Nagy K.L. and Giletti B.J. (1986) Grain boundary diffusion of oxygen in a macroperthitic feldspar. *Geochim. Cosmochim. Acta* **50**, 1151–1158.

Narayan C. and Goldstein J.I. (1985) A major revision of iron meteorite cooling rates—an experimental study of the growth of the Widmanstatten pattern. *Geochim. Cosmochim. Acta* **49**, 397–410.

Narayanaswamy O.S. (1971) A model of structural relaxation in glass. *J. Am. Ceram. Soc.* **54**, 491–498.

———. (1988) Thermorheological simplicity in the glass transition. *J. Am. Ceram. Soc.* **71**, 900–904.

Neilson G.F. and Weinberg M.C. (1979) A test of classical nucleation theory: crystal nucleation of lithium disilicate glass. *J. Non-Cryst. Solids* **34**, 137–147.

Newman S., Stolper E.M., and Epstein S. (1986) Measurement of water in rhyolitic glasses: calibration of an infrared spectroscopic technique. *Am. Mineral.* **71**, 1527–1541.

Ni H. and Zhang Y. (2003) Oxygen isotope thermometry and speedometry. *Eos* **84**, F1530.

Nogami M. and Tomozawa M. (1984a) Diffusion of water in high silica glasses at low temperature. *Phys. Chem. Glasses* **25**, 82–85.

———. (1984b) Effect of stress on water diffusion in silica glass. *J. Am. Ceram. Soc.* **67**, 151–154.

Nowak M. and Behrens H. (1997) An experimental investigation on diffusion of water in haplogranitic melts. *Contrib. Mineral. Petrol.* **126**, 365–376.

Nowak M., Schreen D., and Spickenbom K. (2004) Argon and CO_2 on the race track in silicate melts: a tool for the development of a CO_2 speciation and diffusion model. *Geochim. Cosmochim. Acta* **68**, 5127–5138.

Nye J.F. (1985) *Physical Properties of Crystals*. Oxford, UK: Clarendon Press.

Ochs F.A. and Lange R.A. (1999) The density of hydrous magmatic liquids. *Science* **283**, 1314–1317.

Oelkers E.H. (2001) General kinetic description of multioxide silicate mineral and glass dissolution. *Geochim. Cosmochim. Acta* **65**, 3703–3719.

O'Neill H.S.C. (1994) Temperature dependence of the cation distribution in $CoAl_2O_4$ spinel. *Eur. J. Mineral.* **6**, 603–609.

Ottonello G., Princivalle F., and Giusta A.D. (1990) Temperature, composition, and fO_2 effects on intersite distribution of Mg and Fe^{2+} in olivines. *Phys. Chem. Mineral.* **17**, 301–312.

Ozawa K. (1984) Olivine-spinel geospeedometry: analysis of diffusion-controlled Mg–Fe^{2+} exchange. *Geochim. Cosmochim. Acta* **48**, 2597–2611.

Page F.Z., Deangelis M., Fu B., Kita N., Lancaster P., and Valley J.W. (2006) Slow oxygen diffusion in zircon. *Geochim. Cosmochim. Acta* **70**, A467.

Pan Y. (2002) Commens on: Higgins: "Closure in crystal size distribution (CSD), verification of CSD calculations, and the significance of CSD fans." *Am. Mineral.* **87**, 1242–1243.

Parsons B. and Sclater J.G. (1977) An analysis of the variation of ocean floor bathymetry with age. *J. Geophys. Res.* **82**, 803–827.

Patterson C. (1956) Age of meteorites and the Earth. *Geochim. Cosmochim. Acta* **10**, 230–237.

Peate D.W. and Hawkesworth C.J. (2005) U series disequilibria: insights into mantle melting and the timescales of magma differentiation. *Rev. Geophys.* **43**, RG1003.

Peck W.H., Valley J.W., and Graham C.M. (2003) Slow oxygen diffusion rates in igneous zircons from metamorphic rocks. *Am. Mineral.* **88**, 1003–1014.

Perkins W.G. and Begeal D.R. (1971) Diffusion and permeation of He, Ne, Ar, Kr, and D_2 through silicon oxide thin films. *J. Chem. Phys.* **54**, 1683–1694.

Petersen E.U., Anovitz L.M., and Essene E.J. (1985) Donpeacorite, $(Mn,Mg)MgSi_2O_6$, a new orthopyroxene and proposed phase relations in the system $MnSiO_3$–$MgSiO_3$–$FeSiO_3$. *Am. Mineral.* **69**, 472–480.

Petry C., Chakraborty S., and Palme H. (2004) Experimental determination of Ni diffusion coefficients in olivine and their dependence on temperature, composition, oxygen fugacity, and crystallographic orientation. *Geochim. Cosmochim. Acta* **68**, 4179–4188.

Phillpot S.R., Yip S., and Wolf D. (1989) How do crystals melt? *Comput. Phys.*, 20–31.

Pilling M.J. and Seakins P.W. (1995) *Reaction Kinetics*. New York: Oxford University Press.

Poe B.T., McMillan P.F., Rubie D.C., Chakraborty S., Yarger J., and Diefenbacher J. (1997) Silicon and oxygen self-diffusivities in silicate liquids measured to 15 GPa and 2800 K. *Science* **276**, 1245–1248.

Press W.H., Flannery B.P., Teukolsky S.A., and Vetterling W.T. (1992) *Numerical Recipes*. Cambridge, UK: Cambridge University Press.

Proussevitch A.A. and Sahagian D.L. (1996) Dynamics of coupled diffusive and decompressive bubble growth in magmatic systems. *J. Geophys. Res.* **101**, 17447–17455.

———. (1998) Dynamics and energetics of bubble growth in magmas: analytical formulation and numerical modeling. *J. Geophys. Res.* **103**, 18223–18251.

Proussevitch A.A., Sahagian D.L., and Anderson A.T. (1993) Dynamics of diffusive bubble growth in magmas: isothermal case. *J. Geophys. Res.* **98**, 22283–22307.

Qin Z., Lu F., and Anderson A.T. (1992) Diffusive reequilibration of melt and fluid inclusions. *Am. Mineral.* **77**, 565–576.

Randolph A.D. and Larson M.A. (1971) *Theory of Particulate Processes*. San Diego, CA: Academic Press.

Ranade M.R., Navrotsky A., Zhang H.Z., Banfield J.F., Elder S.E., Zaban A., Borse P.H., Kulkarni S.K., Doran G.S., and Whitfield H.J. (2002) Energetics of nanocrystalline TiO_2. *Proc. Natl. Acad. Sci. USA* **99**, 6476–6481.

Rapp R.P. and Watson E.B. (1986) Monazite solubility and dissolution kinetics: implications for the thorium and light rare Earth chemistry of felsic magmas. *Contrib. Mineral. Petrol.* **94**, 304–316.

Redfern S.A.T., Henderson C.M.B., Wood B. J., Harrison R.J., and Knight K.S. (1996) Determination of olivine cooling rates from metal-cation ordering. *Nature* **381**, 407–409.

Reich M., Ewing R.C., Ehlers T.A., and Becker U. (2007) Low-temperature anisotropic diffusion of helium in zircon: implication for zircon (U–Th)/He thermochronometry. *Geochim. Cosmochim. Acta* **71**, 3119–3130.

Reid J.E., Peo B.T., Rubie D.C., Zotov N., and Wiedenbeck M. (2001) The self-diffusion of silicon and oxygen in diopside ($CaMgSi_2O_6$) liquid up to 15 GPa. *Chem. Geol.* **174**, 77–86.

Reimer P., Baillie M.G.L., Bard E., Bayliss A., Beck J.W., Bertrand C.J.H., Blackwell P.G., Buck C.E., Burr G.S., Cutler K., Damon P.E., Edwards R.L., Fairbanks R.G., Friedrich M., Guilderson T.P., Hogg A.G., Hughen K.A., Kromer B., McCormac G.M., Manning S., Ramsey C.B., Reimer R.W., Remmele S., Southon J.R., Stuiver M., Talamo S., Taylor F.W., Van Der Plicht J., and Weyhenmeyer C.E. (2004) IntCal04 Terrestrial radiocarbon age calibration, 0-26 cal kyr BP. *Radiocarbon* **46**, 1029–1058.

Reiners P.W. and Brandon M.T. (2006) Using thermochronology to understand orogenic erosion. *Annu. Rev. Earth Planet. Sci.* **34**, 419–466.

Reiners P.W. and Ehlers T.A. (2005) *Low-temperature Thermochronology, Techniques, Interpretations, and Applications*. p. 622. Washington, DC: Mineralogical Society of America.

Reiners P.W., Spell T.L., Nicolescu S., and Zanetti K.A. (2004) Zircon (U–Th)/He thermochronometry: He diffusion and comparisons with $^{40}Ar/^{39}Ar$ dating. *Geochim. Cosmochim. Acta* **68**, 1857–1887.

Richter F., Liang Y., and Minarik W.G. (1998) Multicomponent diffusion and convection in molten $MgO–Al_2O_3–SiO_2$. *Geochim. Cosmochim. Acta* **62**, 1985–1991.

Richter F.M. (1993) A model for determining activity-composition relations using chemical diffusion in silicate melts. *Geochim. Cosmochim. Acta* **57**, 2019–2032.

Richter F.M., Liang Y., and Davis A.M. (1999) Isotope fractionation by diffusion in molten oxides. *Geochim. Cosmochim. Acta* **63**, 2853–2861.

Richter F.M., Davis A.M., DePaolo D.J., and Watson E.B. (2003) Isotope fractionation by chemical diffusion between molten basalt and rhyolite. *Geochim. Cosmochim. Acta* **67**, 3905–3923.

Roberts G.J. and Roberts J.P. (1964) Influence of thermal history on the solubility and diffusion of 'water' in silica glass. *Phys. Chem. Glasses* **5**, 26–32.

———. (1966) An oxygen tracer investigation of the diffusion of water in silica glass. *Phys. Chem. Glasses* **7**, 82–89.

Robie R.A. and Hemingway B.S. (1995) Thermodynamic properties of minerals and related substances at 298.15 K and 1 bar (10^5 Pascals) pressure and at high temperatures. U.S. Geological Survey Bulletin, Report: B 2131, 461 pp.

Roering J.J., Kirchner J.W., and Dietrich W.E. (1999) Evidence for nonlinear, diffusive sediment transport on hillslopes and implications for landscape morphology. *Water Resources Res.* **35**, 853–870.

Rubie D.C. and Brearley A.J. (1994) Phase transitions between b and g (Mg, Fe)$_2$SiO$_4$ in the Earth's mantle: mechanisms and rheological implications. *Science* **264**, 1445–1448.

Rubie D.C. and Thompson A.B. (1985) Kinetics of metamorphic reactions at elevated temperatures and pressures: an appraisal of available experimental data. *Adv. Phys. Geochem.* **4**, 27–79.

Rutherford M.J. and Hill P.M. (1993) Magma ascent rates from amphibole breakdown: experiments and the 1980–1986 Mount St. Helens eruptions. *J. Geophys. Res.* **98**, 19667–19685.

Ryan J.G. and Langmuir C.H. (1988) Beryllium systematics in young volcanic rocks: implications for ^{10}Be. *Geochim. Cosmochim. Acta* **52**, 237–244.

Ryerson F.J. and McKeegan K.D. (1994) Determination of oxygen self-diffusion in åkermanite, anorthite, diopside, and spinel: implications for oxygen isotopic anomalies and the thermal histories of Ca–Al-rich inclusions. *Geochim. Cosmochim. Acta* **58**, 3713–3734.

Ryerson F.J., Durham W.B., Cherniak D.J., and Lanford W.A. (1989) Oxygen diffusion in olivine: effect of oxygen fugacity and implications for creep. *J. Geophys. Res.* **94**, 4105–4118.

Sahagian D.L. and Proussevitch A.A. (1998) 3D particle size distributions from 2D observations: stereology for natural applications. *J. Volcanol. Geotherm. Res.* **84**, 173–196.

Sahagian D.L., Proussevitch A.A., and Carlson W.D. (2002) Analysis of vesicular basalts and lava emplacement processes for application as a paleobarometer/paleoaltimeter. *J. Geol.* **110**, 671–685.

Saikumar V. and Goldstein J.L. (1988) An evaluation of the methods to determine the cooling rates of iron meteorites. *Geochim. Cosmochim. Acta* **52**, 715–725.

Sato H. (1975) Diffusion coronas around quartz xenocrysts in andesite and basalt from Tertiary volcanic region in northeastern Shikoku, Japan. *Contrib. Mineral. Petrol.* **50**, 49–64.

Sato H., Fujii T., and Nakada S. (1992) Crumbling of dacite dome lava and generation of pyroclastic flows at Unzen volcano. *Nature* **360**, 664–666.

Sauer V.F. and Freise V. (1962) Diffusion in binaren Gemischen mit Volumenanderung. *Z. Elektrochem. Angew. Phys. Chem.* **66**, 353–363.

Saxena S.K. (ed.) (1982) *Advances in Physical Geochemistry*, Vol. 2. New York: Springer-Verlag.

————. (ed.) (1983a) *Kinetics and Equilibrium in Mineral Reactions. Advances in Physical Geochemistry*, Vol. 3. New York: Springer-Verlag.

————. (1983b) Exsolution and Fe^{2+}–Mg order–disorder in pyroxenes. In *Kinetics and Equilibrium in Mineral Reactions* (ed. S. K. Saxena), pp. 61–80. New York: Springer-Verlag.

Saxena S.K. and Ghose S. (1971) Mg–Fe order–disorder and the thermodynamics of the orthopyroxene crystalline solution. *Am. Mineral.* **56**, 532–559.

Saxena S.K. and Negro A.D. (1983) Petrologic application of Mg–Fe^{2+} order–disorder in orthopyroxene to cooling history of rocks. *Bull. Mineral.* **106**, 443–449.

Scherer G., Vergano P.J., and Uhlmann D.R. (1970) A study of quartz melting. *Phys. Chem. Glass.* **11**, 53–58.

Scherer G.W. (1986) *Relaxation in Glass and Composites*. New York: Wiley.

Schlenz H., Kroll H., and Phillips M.W. (2001) Isothermal annealing and continuous cooling experiments on synthetic orthopyroxenes: temperature and time evolution of the Fe, Mg distribution. *Eur. J. Mineral.* **13**, 715–726.

Schwandt C.S., Cygan R.T., and Westrich H.R. (1995) Mg self-diffusion in pyrope garnet. *Am. Mineral.* **80**, 483–490.

Seifert F. and Virgo D. (1975) Kinetics of Fe^{2+}–Mg order–disorder reaction in anthophyllites: quantitative cooling rates. *Science* **188**, 1107–1109.

Shafer N.E. and Zare R.N. (1991) Through a beer glass darkly. *Phys. Today* **44**(10), 48–52.

Shannon R.D. (1976) Revised effective ionic radii and systematic studies of interatomic distances in halides and chalcogenides. *Acta Crystallogr. A* **32**, 751–767.

Sharp Z.D. (1991) Determination of oxygen diffusion rates in magnetite from natural isotopic variations. *Geology* **19**, 653–656.

Sharp Z.D., Giletti B. J., and Yoder H. S. (1991) Oxygen diffusion rates in quartz exchanged with CO_2. *Earth Planet. Sci. Lett.* **107**, 339–348.

Shaw C.S. J. (2000) The effect of experiment geometry on the mechanism and rate of dissolution of quartz in basanite at 0.5 GPa and 1350°C. *Contrib. Mineral. Petrol.* **139**, 509–525.

————. (2004) Mechanisms and rates of quartz dissolution in melts in the CMAS (CaO–MgO–Al_2O_3–SiO_2) system. *Contrib. Mineral. Petrol.* **148**, 180–200.

Shaw D. M. (1970) Trace element fractionation during anatexis. *Geochim. Cosmochim. Acta* **34**, 237–243.

Shaw H.R. (1974) Diffusion of H_2O in granitic liquids, I: experimental data; II: mass transfer in magma chambers. In *Geochemical Transport and Kinetics*, Vol. 634 (ed. A.W. Hofmann, B.J. Giletti, H.S. Yoder, and R.A. Yund), pp. 139–170. Washington, DC: Carnegie Institution of Washington Publ.

Shelby J.E. (1972a) Helium migration in natural and synthetic vitreous silica. *J. Am. Ceram. Soc.* **55**, 61–64.

————. (1972b) Helium migration in TiO_2–SiO_2 glasses. *J. Am. Ceram. Soc.* **55**, 195–197.

————. (1972c) Neon migration in vitreous silica. *Phys. Chem. Glasses* **13**, 167–170.

————. (1973) Neon migration in TiO_2–SiO_2 glasses. *J. Am. Ceram. Soc.* **56**, 340–341.

————. (1977) Molecular diffusion and solubility of hydrogen isotopes in vitreous silica. *J. Appl. Phys.* **48**, 3387–3394.

————. (1979) Molecular solubility and diffusion. *Treatise Mater. Sci. Tech.* **17**, 1–40.

Shelby J.E. and Eagan R.J. (1976) Helium migration in sodium aluminosilicate glasses. *J. Am. Ceram. Soc.* **59**, 420–425.

Shewmon P.G. (1963) *Diffusion in Solids*. New York: McGraw-Hill.

Shimizu N. and Kushiro I. (1984) Diffusivity of oxygen in jadeite and diopside melts at high pressures. *Geochim. Cosmochim. Acta* **48**, 1295–1303.

————. (1991) The mobility of Mg, Ca, and Si in diopside–jadeite liquids at high pressures. In *Physical Chemistry of Magmas* (ed. L. L. Perchuk and I. Kushiro), pp. 192–212. New York: Springer-Verlag.

Sierralta M., Nowak M., and Keppler H. (2002) The influence of bulk composition on the diffusivity of carbon dioxide in Na aluminosilicate melts. *Am. Mineral.* **87**, 1710–1716.

Sipp A. and Richet P. (2002) Equivalence of volume, enthalpy and viscosity relaxation kinetics in glass-forming silicate liquids. *J. Non-Cryst. Solids* **298**, 202–212.

Smith V.G., Tiller W.A., and Rutter J.W. (1956) A mathematical analysis of solute redistribution during solidification. *Can. J. Phys.* **33**, 723–745.

Smoliar M.I., Walker R.J., and Morgan J.W. (1996) Re–Os ages of group IIA, IIIA, IVA, and IVB iron meteorites. *Science* **271**, 1099–1102.

Sneeringer M., Hart S.R., and Shimizu N. (1984) Strontium and samarium diffusion in diopside. *Geochim. Cosmochim. Acta* **48**, 1589–1608.

Spera F.J. and Trial A.F. (1993) Verification of Onsager's reciprocal relations in a molten silicate solution. *Science* **259**, 204–206.

Spieler O., Kennedy B., Kueppers U., Dingwell D.B., Scheu B., and Taddeucci J. (2004) The fragmentation threshold of pyroclastic rocks. *Earth Planet. Sci. Lett.* **226**, 139–148.

Staudacher T. and Allegre C. J. (1982) Terrestrial xenology. *Earth Planet. Sci. Lett.* **60**, 389–406.

Stebbins J. F., Carmichael I.S.E., and Weill D. E. (1983) The high-temperature liquid and glass heat contents and heats of fusion of diopside, albite, sanidine and nepheline. *Am. Mineral.* **68**, 717–730.

Steefel C.I. and Van Cappellen P. (1990) A new kinetic approach to modeling water–rock interaction: the role of nucleation, precursors, and Ostwald ripening. *Geochim. Cosmochim. Acta* **54**, 2657–2677.

Steefel C.I. and Lasaga A.C. (1994) A coupled model for transport of multiple chemical species and kinetic precipitation/dissolution reactions with application to reactive flow in single phase hydrothermal systems. *Am. J. Sci.* **294**, 529–592.

Stern K. H. (1954) The Liesegang phenomenon. *Chem. Rev.* **54**, 79–99.

Stillinger F. H., Debenedetti P.G., and Truskett T.M. (2001) The Kauzmann paradox revisited. *J. Phys. Chem.* **105**, 11809–11816.

Stimpfl M. (2005) The Mn, Mg-intracrystalline exchange reaction in donpeacorite ($Mn_{0.54}Ca_{0.03}Mg_{1.43}Si_2O_6$) and its relation to the fractionation behavior of Mn in Fe, Mg-orthopyroxene. *Am. Mineral.* **90**, 155–161.

Stimpfl M., Ganguly J., and Molin G. (1999) Fe^{2+}–Mg order–disorder in orthopyroxene: equilibrium fractionation between the octahedral sites and thermodynamic analysis. *Contrib. Mineral. Petrol.* **136**, 297–309.

————. (2005) Kinetics of Fe^{2+}–Mg order-disorder in orthopyroxene: experimental studies and applications to cooling rates of rocks. *Contrib. Mineral. Petrol.* **150**, 319–334.

Stoffregen R.E., Rye R.O., and Wasserman M.D. (1994a) Experimental studies of alunite, I: ^{18}O–^{16}O and D–H fractionation factors between alunite and water at 250–450°C. *Geochim. Cosmochim. Acta* **58**, 903–916.

————. (1994b) Experimental studies of alunite, II: rates of alunite-water alkali and isotope exchange. *Geochim. Cosmochim. Acta* **58**, 917–929.

Stolper E.M. (1982a) Water in silicate glasses: an infrared spectroscopic study. *Contrib. Mineral. Petrol.* **81**, 1–17.

————. (1982b) The speciation of water in silicate melts. *Geochim. Cosmochim. Acta* **46**, 2609–2620.

Stolper E.M. and Epstein S. (1991) An experimental study of oxygen isotope partitioning between silica glass and CO2 vapor. In *Stable Isotope Geochemistry: A Tribute to Samuel Epstein*, Vol. 3 (ed. H.P. Taylor, J.R. O'Neil, and I.R. Kaplan), pp. 35–51. Geochem. Soc.

Sugawara H., Nagata K., and Goto K.S. (1977) Interdiffusivities matrix of CaO–Al$_2$O$_3$–SiO$_2$ melt at 1723 K to 1823 K. *Metall. Trans. B* **8**, 605–612.

Sutherland W. (1905) A dynamical theory of diffusion for non-electrolytes and the molecular mass of albumin. *Philos. Mag.* **9**, 781–785.

Sykes-Nord J.A. and Molin G.M. (1993) Mg–Fe order–disorder reaction in Fe-rich orthopyroxene: structural variations and kinetics. *Am. Mineral.* **78**, 921–931.

Tait S., Thomas R., Gardner J., and Jaupart C. (1998) Constraints on cooling rates and permeabilities of pumices in an explosive eruption jet from color and magnetic mineralogy. *J. Volcanol. Geotherm. Res.* **86**, 79–91.

Tamimi K., Rinker E.B., and Sandall O.C. (1994) Diffusion coefficients for hydrogen sulfide, carbon dioxide, and nitrous oxide in water over the temperature range of 293–368 K. *J. Chem. Eng. Data* **39**, 330–332.

Taylor L.A., Onorato P.I., and Uhlmann D.R. (1977) Cooling rate estimations based on kinetic modeling of Fe–Mg diffusion in olivine. *Proc. Lunar Sci. Conf. 8th*, 1581–1592.

Taylor S.R. (1967) Composition of meteorite impact glass across the Henbury strewnfield. *Geochim. Cosmochim. Acta* **31**, 961–968.

Thomas R. (2000) Determination of water contents of granite melt inclusions by confocal laser Raman microprobe spectroscopy. *Am. Mineral.* **85**, 868–872.

Thompson A.B. and Perkins E.H. (1981) Lambda transition in minerals. In *Thermodynamics of Minerals and Melts* (ed. R.C. Newton, A. Navrotsky, and B.J. Wood), pp. 35–62. New York: Springer-Verlag.

Thompson A.B. and Rubie D.C. (eds) (1985) *Metamorphic Reactions: Kinetics, Textures, and Deformation. Advances in Physical Geochemistry*, Vol. 4. New York: Springer Verlag.

Tingle T.N., Green H.W., and Finnterty A.A. (1988) Experiments and observations bearing on the solubility and diffusivity of carbon in olivine. *J. Geophys. Res.* **93**, 15289–15304.

Tinker D. and Lesher C.E. (2001) Self diffusion of Si and O in dacitic liquid at high pressures. *Am. Mineral.* **86**, 1–13.

Tinker D., Lesher C.E., and Hucheon I.D. (2003) Self-diffusion of Si and O in diopside–anorthite melt at high pressures. *Geochim. Cosmochim. Acta* **67**, 133–142.

Tinker D., Lesher C.E., Baxter G.M., Uchida T., and Wang Y. (2004) High-pressure viscometry of polymerized silicate melts and limitations of the Eyring equation. *Am. Mineral.* **89**, 1701–1708.

Tomozawa M. (1985) Concentration dependence of the diffusion coefficient of water in SiO2 glass. *Am. Ceram. Soc.* **C**, 251–252.

Toplis M.J., Gottsmann J., Knoche R., and Dingwell D.B. (2001) Heat capacities of haplogranitic glasses and liquids. *Geochim. Cosmochim. Acta* **65**, 1985–1994.

Toramaru A. (1989) Vesiculation process and bubble size distributions in ascending magmas with constant velocities. *J. Geophys. Res.* **94**, 17523–17542.

Tossell J.A. (2002) Does the calculated decay constant for 7Be vary significantly with chemical form and/or applied pressure? *Earth Planet. Sci. Lett.* **195**, 131–139.

Towers H. and Chipman J. (1957) Diffusion of calcium and silicon in a lime–alumina–silica slag. *Trans. AIME, J. Metals*, 769–773.

Trial A.F. and Spera F.J. (1988) Natural convection boundary layer flows in isothermal ternary systems: role of diffusive coupling. *Int. J. Heat Transfer* **31**, 941–955.

———. (1994) Measuring the multicomponent diffusion matrix: experimental design and data analysis for silicate melts. *Geochim. Cosmochim. Acta* **58**, 3769–3783.

Tsuchiyama A. (1985a) Dissolution kinetics of plagioclase in the melt system of diopside–albite–anorthite and origin of dusty plagioclase in andesites. *Contrib. Mineral. Petrol.* **89**, 1–16.

———. (1985b) Partial melting kinetics of plagioclase–diopside pairs. *Contrib. Mineral. Petrol.* **91**, 12–23.

———. (1986) Melting and dissolution kinetics: application to partial melting and dissolution of xenoliths. *J. Geophys. Res.* **91**, 9395–9406.

Tsuchiyama A. and Takahashi E. (1983) Melting kinetics of a plagioclase feldspar. *Contrib. Mineral. Petrol.* **84**, 345–354.

Turcotte D.L. and Schubert G. (1982) *Geodynamics: Applications of Continuum Physics to Geological Problems*. New York: Wiley.

Usuki T. (2002) Anisotropic Fe–Mg diffusion in biotite. *Am. Mineral.* **87**, 1014–1017.

Van Der Laan S.R., Zhang Y., Kennedy A., and Wylie P.J. (1994) Comparison of element and isotope diffusion of K and Ca in multicomponent silicate melts. *Earth Planet. Sci. Lett.* **123**, 155–166.

Varshneya A.K. and Cooper A.R. (1972a) Diffusion in the system $K_2O–SrO–SiO_2$, II: cation self-diffusion coefficients. *J. Am. Ceram. Soc.* **55**, 220–223.

———. (1972b) Diffusion in the system $K_2O–SrO–SiO_2$, III: interdiffusion coefficients. *J. Am. Ceram. Soc.* **55**, 312–317.

———. (1972c) Diffusion in the system $K_2O–SrO–SiO_2$, IV: mobility model, electrostatic effects, and multicomponent diffusion. *J. Am. Ceram. Soc.* **55**, 418–421.

Verhallen P.T.H.M., Oomen L.J.P., v.d.Elsen A.J.J.M., Kruger A.J., and Fortuin J.M.H. (1984) The diffusion coefficients of helium, hydrogen, oxygen and nitrogen in water determined from the permeability of a stagnant liquid layer in the quasi-steady state. *Chem. Eng. Sci.* **39**, 1535–1541.

Virgo D. and Hafner S. (1969) Fe^{2+}, Mg order–disorder in heated orthopyroxenes. *MSA Special Paper* **2**, 67–81.

Wakabayashi H. and Tomozawa M. (1989) Diffusion of water into silica glass at low temperature. *Am. Ceram. Soc.* **72**, 1850–1855.

Walker D. (1983) New developments in magmatic processes. *Rev. Geophys. Space Phys.* **21**, 1372–1384.

———. (2000) Core participation in mantle geochemistry: Geochemical Society Ingerson Lecture, GSA Denver, October 1999. *Geochim. Cosmochim. Acta* **64**, 2897–2911.

Walker D. and DeLong S.E. (1984) A small Soret effect in spreading center gabbros. *Contrib. Mineral. Petrol.* **85**, 203–208.

Walker D. and Kiefer W.S. (1985) Xenolith digestion in large magma bodies. *J. Geophys. Res.* **90**, C585–C590.

Walker D., Longhi J., Lasaga A.C., Stolper E.M., Grove T.L., and Hays J.F. (1977) Slowly cooled microgabbros 15555 and 15065. *Proc. Lunar Sci. Conf. 8th*, 1521–1547.

Walker J.C.G. (1977) *Evolution of the Atmosphere*. New York: Macmillan.

Wallace P.J., Dufek J., Anderson A.T., and Zhang Y. (2003) Cooling rates of Plinian-fall and pyroclastic-flow deposits in the Bishop Tuff: inferences from water speciation in quartz-hosted glass inclusions. *Bull. Volcanol.* **65**, 105–123.

Wang L., Zhang Y., and Essene E.J. (1996) Diffusion of the hydrous component in pyrope. *Am. Mineral.* **81**, 706–718.

Wang L., Essene E.J., and Zhang Y. (1999) Mineral inclusions in pyrope crystals from Garnet Ridge, Arizona, USA: implications for processes in the upper mantle. *Contrib. Mineral. Petrol.* **135**, 164–178.

———. (2000) Direct observation of immiscibility in pyrope–almandine–grossular garnet. *Am. Mineral.* **85**, 41–46.

Wang L., Moon N., Zhang Y., Dunham W.R., and Essene E.J. (2005) Fe–Mg order–disorder in orthopyroxenes. *Geochim. Cosmochim. Acta* **69**, 5777–5788.

Wasserburg G.J. (1988) Diffusion of water in silicate melts. *J. Geol.* **96**, 363–367.

Watson E.B. (1976) Two-liquid partitioning coefficients: experimental data and geochemical implications. *Contrib. Mineral. Petrol.* **56**, 119–134.

———. (1979a) Calcium diffusion in a simple silicate melt to 30 kbar. *Geochim. Cosmochim. Acta* **43**, 313–322.

———. (1979b) Diffusion of cesium ions in H_2O-saturated granitic melt. *Science* **205**, 1259–1260.

———. (1980) Apatite and phosphorus in mantle source regions: an experimental study of apatite/melt equilibria at pressures to 25 kbar. *Earth Planet. Sci. Lett.* **51**, 322–335.

———. (1981) Diffusion in magmas at depth in the earth: the effects of pressure and dissolved H_2O. *Earth Planet. Sci. Lett.* **52**, 291–301.

———. (1982a) Basalt contamination by continental crust: some experiments and models. *Contrib. Mineral. Petrol.* **80**, 73–87.

———. (1982b) Melt infiltration and magma evolution. *Geology* **10**, 236–240.

———. (1985) Henry's law behavior in simple systems and in magmas: criteria for discerning concentration-dependent partition coefficients in nature. *Geochim. Cosmochim. Acta* **49**, 917–923.

———. (1991a) Diffusion in fluid-bearing and slightly-melted rocks: experimental and numerical approaches illustrated by iron transport in dunite. *Contrib. Mineral. Petrol.* **107**, 417–434.

———. (1991b) Diffusion of dissolved CO_2 and Cl in hydrous silicic to intermediate magmas. *Geochim. Cosmochim. Acta* **55**, 1897–1902.

———. (1994) Diffusion in volatile-bearing magmas. *Rev. Mineral.* **30**, 371–411.

———. (1996a) Dissolution, growth and survival of zircons during crustal fusion: kinetic principles, geological models, and implications for isotopic inheritance. *Geol. Soc. Am. Spec. Paper* **315**, 43–56.

———. (1996b) Surface enrichment and trace-element uptake during crystal growth. *Geochim. Cosmochim. Acta* **60**, 5013–5020.

Watson E.B. and Baker D.R. (1991) Chemical diffusion in magmas: an overview of experimental results and geochemical implications. In *Advances in Physical Geochemistry, 9* (ed. L. Perchuk and I. Kushiro), pp. 120–151.

Watson E.B. and Cherniak D.J. (1997) Oxygen diffusion in zircon. *Earth Planet. Sci. Lett.* **148**, 527–544.

———. (2003) Lattice diffusion of Ar in quartz, with constraints on Ar solubility and evidence of nanopores. *Geochim. Cosmochim. Acta* **67**, 2043–2062.

———. (1983) Zircon saturation revisited: temperature and composition effects in a variety of crustal magma types. *Earth Planet. Sci. Lett.* **64**, 295–304.

———. (1984) Accessory minerals and the geochemical evolution of crustal magmatic systems: a summary and prospectus of experimental approaches. *Phys. Earth Planet. In.* **35**, 19–30.

———. (2005) Zircon thermometer reveals minimum melting conditions on earliest Earth. *Science* **308**, 841–844.

Watson E.B. and Jurewicz S.R. (1984) Behavior of alkalies during diffusive interaction of granitic xenoliths with basaltic magma. *J. Geol.* **92**, 121–131.

Watson E.B. and Liang Y. (1995) A simple model for sector zoning in slowly grown crystals: implications for growth rate and lattice diffusion, with emphasis on accessory minerals in crustal rocks. *Am. Mineral.* **80**, 1179–1187.

Watson E.B. and Wark D.A. (1997) Diffusion of dissolved SiO_2 in H_2O at 1 GPa, with implications for mass transport in the crust and upper mantle. *Contrib. Mineral. Petrol.* **130**, 66–80.

Watson E.B., Sneeringer M.A., and Ross A. (1982) Diffusion of dissolved carbonate in magmas: experimental results and applications. *Earth Planet. Sci. Lett.* **61**, 346–358.

Wilde S.A., Valley J.W., Peck W.H., and Graham C.M. (2001) Evidence from detrital zircons for the existence of continental crust and oceans on the Earth 4.4 Gyr ago. *Nature* **409**, 175–178.

Wilding M., Webb S., and Dingwell D.B. (1996a) Tektitie cooling rates: calorimetric relaxation geospeedometry applied to a natural glass. *Geochim. Cosmochim. Acta* **60**, 1099–1103.

Wilding M., Webb S., Dingwell D.B., Ablay G., and Marti J. (1996b) Cooling rate variation in natural volcanic glasses from Tenerife, Canary Islands. *Contrib. Mineral. Petrol.* **125**, 151–160.

Wilding M.C., Webb S.L., and Dingwell D.B. (1995) Evaluation of a relaxation geospeedometer for volcanic glasses. *Chem. Geol.* **125**, 137–148.

Winchell P. (1969) The compensation law for diffusion in silicates. *High Temp. Sci.* **1**, 200–215.

Withers A.C., Zhang Y., and Behrens H. (1999) Reconciliation of experimental results on H_2O speciation in rhyolitic glass using in situ and quenching techniques. *Earth Planet. Sci. Lett.* **173**, 343–349.

Wolf R.A., Farley K.A., and Silver L.T. (1996) Helium diffusion and low-temperature thermochronometry of apatite. *Geochim. Cosmochim. Acta* **60**, 4131–4240.

Wolf R.A., Farley K.A., and Kass D.M. (1998) Modeling of the temperature sensitivity of the apatite (U–Th)/He thermochronometer. *Chem. Geol.* **148**, 105–114.

Xu Z. and Zhang Y. (2002) Quench rates in water, air and liquid nitrogen, and inference of temperature in volcanic eruption columns. *Earth Planet. Sci. Lett.* **200**, 315–330.

Yang H. and Ghose S. (1994) In-situ Fe–Mg order–disorder studies and thermodynamic properties of orthopyroxene (Mg, Fe)$_2Si_2O_6$. *Am. Mineral.* **79**, 633–643.

Yin Q., Jacobsen S.B., Yamashita K., Blichert-Toft J., Telouk P., and Albarede F. (2002) A short time scale for terrestrial planet formation from Hf-W chronometry of meteorites. *Nature* **418**, 949–952.

Yinnon H. and Cooper A.R. (1980) Oxygen diffusion in multicomponent glass forming silicates. *Phys. Chem. Glasses* **21**, 204–211.

York D. (1969) Least-squares fitting of a straight line with correlated errors. *Earth. Planet. Sci. Lett.* **5**, 320–324.

Young T.E., Green H.W, Hofmeister A.M., and Walker D. (1993) Infrared spectroscopic investigation of hydroxyl in beta-(Mg, Fe)$_2SiO_4$ and coexisting olivine: implications for mantle evolution and dynamics. *Phys. Chem. Minerals* **19**, 409–422.

Yund R.A. and Anderson T.F. (1978) The effect of fluid pressure on oxygen isotope exchange between feldspar and water. *Geochim. Cosmochim. Acta* **42**, 235–239.

Yurimoto H., Morioka M., and Nagasawa H. (1989) Diffusion in single crystals of melilite, I: oxygen. *Geochim. Cosmochim. Acta* **53**, 2387–2394.

Zeilik M., Gregory S. A., and Smith E.V.P. (1992) *Introductory Astronomy and Astrophysics.* Fort Worth, TX: Saunders College.

Zema M., Domeneghetti M.C., and Molin G.M. (1996) Thermal history of Acapulco and ALHA81261 acapulcoites constrained by Fe–Mg ordering in orthopyroxene. *Earth Planet. Sci. Lett.* **144**, 359–367.

Zema M., Domeneghetti M.C., Molin G.M., and Tazzoli V. (1997) Cooling rates of diogenites: A study of Fe^{2+}–Mg ordering in orthopyroxene by single-crystal x-ray diffraction. *Meteor. Planet. Sci.* **32**, 855–862.

Zema M., Domeneghetti M.C., and Tazzoli V. (1999) Order–disorder kinetics in orthopyroxene with exsolution products. *Am. Mineral.* **84**, 1895–1901.

Zema M., Tarantino S.C., Domeneghetti M.C., and Tazzoli V. (2003) Ca in orthopyroxene: structural variations and kinetics of the disordering process. *Eur. J. Mineral.* **15**, 373–380.

Zhang X.Y., Cherniak D.J., and Watson E.B. (2006) Oxygen diffusion in titanite: lattice diffusion and fast-path diffusion in single crystals. *Chem. Geol.* **235**, 105–123.

Zhang Y. (1988) *Kinetics of Crystal Dissolution and Rock Melting: a Theoretical and Experimental Study.* Thesis, Columbia University, New York.

———. (1993) A modified effective binary diffusion model. *J. Geophys. Res.* **98**, 11901–11920.

———. (1994) Reaction kinetics, geospeedometry, and relaxation theory. *Earth. Planet. Sci. Lett.* **122**, 373–391.

———. (1996) Dynamics of CO_2-driven lake eruptions. *Nature* **379**, 57–59.

———. (1998a) Experimental simulations of gas-driven eruptions: kinetics of bubble growth and effect of geometry. *Bull. Volcanol.* **59**, 281–290.

———. (1998b) Mechanical and phase equilibria in inclusion-host systems. *Earth Planet. Sci. Lett.* **157**, 209–222.

———. (1998c) The young age of Earth. *Geochim. Cosmochim. Acta* **62**, 3185–3189.

———. (1999a) A criterion for the fragmentation of bubbly magma based on brittle failure theory. *Nature* **402**, 648–650.

———. (1999b) H_2O in rhyolitic glasses and melts: measurement, speciation, solubility, and diffusion. *Rev. Geophys.* **37**, 493–516.

———. (1999c) Exsolution enthalpy of water from silicate liquids. *J. Volcanol. Geotherm. Res.* **88**, 201–207.

———. (1999d) Crystal growth. In *Encyclopedia of Geochemistry* (ed. C. P. Marshall and R. W. Fairbridge), pp. 120–123. Kluwer.

———. (2002) The age and accretion of the Earth. *Earth-Sci. Rev.* **59**, 235–263.

———. (2003) Methane escape from gas hydrate systems in marine environment, and methane-driven oceanic eruptions. *Geophys. Res. Lett.* **30(7)**, (51-1)–(51-4), doi 10.1029/2002GL016658.

———. (2004) Dynamics of explosive volcanic and lake eruptions. In *Environment, Natural Hazards and Global Tectonics of the Earth* (in Chinese), pp. 39–95. Higher Education Press.

———. (2005a) Global tectonic and climatic control of mean elevation of continents, and Phanerozoic sea level change. *Earth Planet. Sci. Lett.* **237**, 524–531.

———. (2005b) Fate of rising CO_2 droplets in seawater. *Environ. Sci. Technol.* **39**, 7719–7724.

Zhang Y. and Behrens H. (2000) H_2O diffusion in rhyolitic melts and glasses. *Chem. Geol. (Wasserburg volume)* **169**, 243–262.

Zhang Y. and Chen N.S. (2007) Analytical solution for a spherical diffusion couple, with applications to closure conditions and geospeedometry. *Geochim. Cosmachim. Acta* **submitted**.

Zhang Y. and Finch J.A. (2001) A note on single bubble motion in surfactant solutions. *J. Fluid Mech.* **429**, 63–66.

Zhang Y. and Stolper E.M. (1991) Water diffusion in basaltic melts. *Nature* **351**, 306–309.

Zhang Y. and Xu Z. (1995) Atomic radii of noble gas elements in condensed phases. *Am. Mineral.* **80**, 670–675.

————. (2003) Kinetics of convective crystal dissolution and melting, with applications to methane hydrate dissolution and dissociation in seawater. *Earth Planet. Sci. Lett.* **213**, 133–148.

————. (2007) A long-duration experiment on hydrous species geospeedometer and hydrous melt viscosity. *Geochim. Cosmochim. Acta* **71**, 5226–5232.

Zhang Y. and Xu Z. (2008) "Fizzics" of beer and champagne bubble growth. *Elements*, **4**, 47–49, doi: 10.2113/GSELEMENTS.4.1.47.

Zhang Y., Walker D., and Lesher C.E. (1989) Diffusive crystal dissolution. *Contrib. Mineral. Petrol.* **102**, 492–513.

Zhang Y., Stolper E.M., and Wasserburg G.J. (1991a) Diffusion of water in rhyolitic glasses. *Geochim. Cosmochim. Acta* **55**, 441–456.

————. (1991b) Diffusion of a multi-species component and its role in the diffusion of water and oxygen in silicates. *Earth Planet. Sci. Lett.* **103**, 228–240.

Zhang Y., Stolper E.M., and Ihinger P.D. (1995) Kinetics of reaction $H_2O + O = 2OH$ in rhyolitic glasses: preliminary results. *Am. Mineral.* **80**, 593–612.

Zhang Y., Belcher R., Ihinger P.D., Wang L., Xu Z., and Newman S. (1997a) New calibration of infrared measurement of water in rhyolitic glasses. *Geochim. Cosmochim. Acta* **61**, 3089–3100.

Zhang Y., Jenkins J., and Xu Z. (1997b) Kinetics of the reaction $H_2O + O = 2OH$ in rhyolitic glasses upon cooling: geospeedometry and comparison with glass transition. *Geochim. Cosmochim. Acta* **61**, 2167–2173.

Zhang Y., Sturtevant B., and Stolper E.M. (1997c) Dynamics of gas-driven eruptions: experimental simulations using CO_2–H_2O–polymer system. *J. Geophys. Res.* **102**, 3077–3096.

Zhang Y., Xu Z., and Behrens H. (2000) Hydrous species geospeedometer in rhyolite: improved calibration and application. *Geochim. Cosmochim. Acta* **64**, 3347–3355.

Zhang Y., Xu Z., and Liu Y. (2003) Viscosity of hydrous rhyolitic melts inferred from kinetic experiments, and a new viscosity model. *Am. Mineral.* **88**, 1741–1752.

Zhang Y., Xu Z., Zhu M., and Wang H. (2007) Silicate melt properties and volcanic eruptions. *Rev. Geophys.* **45**, RG 4004, doi: 10.1029/2006RG000216.

Zheng Y., Fu B., Li Y., Xiao Y., and Li S. (1998) Oxygen and hydrogen isotope geochemistry of ultrahigh-pressure eclogites from the Dabie Mountains and the Sulu terrane. *Earth Planet. Sci. Lett.* **155**, 113–129.

Ziebold T.O. and Ogilvie R.E. (1967) Ternary diffusion in copper–silver–gold alloys. *Trans. Met. Soc. AIME* **239**, 942–953.

Zindler A. (1982) Nd and Sr isotopic studies of komatiites and related rocks. In *Komatiites* (ed. N. T. Arndt and E. G. Nisbet), pp. 399–420. Allen & Unwin.

Zou H.B. (2007) *Quantitative Geochemistry.* London: Imperial College Press.

Index